Periodic Table of the Elements

1 (1A)	2 (2A) Alkaline earth metals	3	4	5	6	7	8	9	10	11	12	13 (3A)	14 (4A)	15 (5A)	16 (6A)	17 (7A) Halogens	18 (8A) Noble gases
1 H																	2 He
3 Li	4 Be											5 B	6 C	7 N	8 O	9 F	10 Ne
11 Na	12 Mg											13 Al	14 Si	15 P	16 S	17 Cl	18 Ar
19 K	20 Ca	21 Sc	22 Ti	23 V	24 Cr	25 Mn	26 Fe	27 Co	28 Ni	29 Cu	30 Zn	31 Ga	32 Ge	33 As	34 Se	35 Br	36 Kr
37 Rb	38 Sr	39 Y	40 Zr	41 Nb	42 Mo	43 Tc	44 Ru	45 Rh	46 Pd	47 Ag	48 Cd	49 In	50 Sn	51 Sb	52 Te	53 I	54 Xe
55 Cs	56 Ba	57 La*	72 Hf	73 Ta	74 W	75 Re	76 Os	77 Ir	78 Pt	79 Au	80 Hg	81 Tl	82 Pb	83 Bi	84 Po	85 At	86 Rn
87 Fr	88 Ra	89 Ac†	104 Rf	105 Db	106 Sg	107 Bh	108 Hs	109 Mt	110 Ds	111 Rg	112 Uub	113 Uut	114 Uuq	115 Uup			118 Uuo

Alkali metals (group 1, Li–Fr); Transition metals (groups 3–12)

*Lanthanides	58 Ce	59 Pr	60 Nd	61 Pm	62 Sm	63 Eu	64 Gd	65 Tb	66 Dy	67 Ho	68 Er	69 Tm	70 Yb	71 Lu
†Actinides	90 Th	91 Pa	92 U	93 Np	94 Pu	95 Am	96 Cm	97 Bk	98 Cf	99 Es	100 Fm	101 Md	102 No	103 Lr

Group numbers 1–18 represent the system recommended by the International Union of Pure and Applied Chemistry.

Table of Atomic Masses*

Element	Symbol	Atomic Number	Atomic Mass	Element	Symbol	Atomic Number	Atomic Mass	Element	Symbol	Atomic Number	Atomic Mass
Actinium	Ac	89	(227)†	Gold	Au	79	197.0	Praseodymium	Pr	59	140.9
Aluminum	Al	13	26.98	Hafnium	Hf	72	178.5	Promethium	Pm	61	(145)
Americium	Am	95	(243)	Hassium	Hs	108	(265)	Protactinium	Pa	91	(231)
Antimony	Sb	51	121.8	Helium	He	2	4.003	Radium	Ra	88	226
Argon	Ar	18	39.95	Holmium	Ho	67	164.9	Radon	Rn	86	(222)
Arsenic	As	33	74.92	Hydrogen	H	1	1.008	Rhenium	Re	75	186.2
Astatine	At	85	(210)	Indium	In	49	114.8	Rhodium	Rh	45	102.9
Barium	Ba	56	137.3	Iodine	I	53	126.9	Rubidium	Rb	37	85.47
Berkelium	Bk	97	(247)	Iridium	Ir	77	192.2	Ruthenium	Ru	44	101.1
Beryllium	Be	4	9.012	Iron	Fe	26	55.85	Rutherfordium	Rf	104	(261)
Bismuth	Bi	83	209.0	Krypton	Kr	36	83.80	Samarium	Sm	62	150.4
Bohrium	Bh	107	(264)	Lanthanum	La	57	138.9	Scandium	Sc	21	44.96
Boron	B	5	10.81	Lawrencium	Lr	103	(260)	Seaborgium	Sg	106	(263)
Bromine	Br	35	79.90	Lead	Pb	82	207.2	Selenium	Se	34	78.96
Cadmium	Cd	48	112.4	Lithium	Li	3	6.941	Silicon	Si	14	28.09
Calcium	Ca	20	40.08	Lutetium	Lu	71	175.0	Silver	Ag	47	107.9
Californium	Cf	98	(251)	Magnesium	Mg	12	24.31	Sodium	Na	11	22.99
Carbon	C	6	12.01	Manganese	Mn	25	54.94	Strontium	Sr	38	87.62
Cerium	Ce	58	140.1	Meitnerium	Mt	109	(268)	Sulfur	S	16	32.07
Cesium	Cs	55	132.9	Mendelevium	Md	101	(258)	Tantalum	Ta	73	180.9
Chlorine	Cl	17	35.45	Mercury	Hg	80	200.6	Technetium	Tc	43	(98)
Chromium	Cr	24	52.00	Molybdenum	Mo	42	95.94	Tellurium	Te	52	127.6
Cobalt	Co	27	58.93	Neodymium	Nd	60	144.2	Terbium	Tb	65	158.9
Copper	Cu	29	63.55	Neon	Ne	10	20.18	Thallium	Tl	81	204.4
Curium	Cm	96	(247)	Neptunium	Np	93	(237)	Thorium	Th	90	232.0
Darmstadtium	Ds	110	(281)	Nickel	Ni	28	58.69	Thulium	Tm	69	168.9
Dubnium	Db	105	(262)	Niobium	Nb	41	92.91	Tin	Sn	50	118.7
Dysprosium	Dy	66	162.5	Nitrogen	N	7	14.01	Titanium	Ti	22	47.88
Einsteinium	Es	99	(252)	Nobelium	No	102	(259)	Tungsten	W	74	183.9
Erbium	Er	68	167.3	Osmium	Os	76	190.2	Uranium	U	92	238.0
Europium	Eu	63	152.0	Oxygen	O	8	16.00	Vanadium	V	23	50.94
Fermium	Fm	100	(257)	Palladium	Pd	46	106.4	Xenon	Xe	54	131.3
Fluorine	F	9	19.00	Phosphorus	P	15	30.97	Ytterbium	Yb	70	173.0
Francium	Fr	87	(223)	Platinum	Pt	78	195.1	Yttrium	Y	39	88.91
Gadolinium	Gd	64	157.3	Plutonium	Pu	94	(244)	Zinc	Zn	30	65.38
Gallium	Ga	31	69.72	Polonium	Po	84	(209)	Zirconium	Zr	40	91.22
Germanium	Ge	32	72.59	Potassium	K	19	39.10				

General Chemistry 152

Custom Edition for University of Washington

Steven S. Zumdahl

CENGAGE
Learning™

Australia • Brazil • Japan • Korea • Mexico • Singapore • Spain • United Kingdom • United States

General Chemistry 152: Custom Edition for University of Washington

Steven S. Zumdahl

Executive Editor:
Maureen Staudt
Michael Stranz

Senior Project Development Manager:
Linda de Stefano

Marketing Specialist:
Sara Mercurio

Senior Production/Manufacturing Manager:
Donna M. Brown

PreMedia Supervisor:
Joel Brennecke

Rights & Permissions Specialist:
Kalina Hintz
Todd Osborne

Cover Image:

Chemical Principles, Sixth Edition

Steven S. Zumdahl

Library of Congress Catalog Card Number: 2007926986

Photo Credits Appear on page A75

ISBN-13: 978-1-111-00498-9
ISBN-10: 1-111-00498-6

Cengage Learning
5191 Natorp Boulevard
Mason, Ohio 45040
USA

Cengage Learning is a leading provider of customized learning solutions with office locations around the globe, including Singapore, the United Kingdom, Australia, Mexico, Brazil, and Japan. Locate your local office at: **international.cengage.com/region**

Cengage Learning products are represented in Canada by Nelson Education, Ltd.

For your lifelong learning solutions, visit **www.cengage.com/custom**

Visit our corporate website at **www.cengage.com**

Printed in the United States of America

Table of Contents

*(*Note – this is a "custom" textbook that has been designed specifically for this course in a joint effort between your instructor and the publisher. Please note that some chapters have been removed intentionally and some pages may be black & white as dictated by the changes.)*

From "Chemical Principles", Zumdahl:

From Student Solutions Manual to Accompany Chemical Principles, Zumdahl:

9 Energy, Enthalpy, and Thermochemistry

Rafflesia arnoldii *in Sumatra.*

Energy is the essence of our very existence as individuals and as a society. The food that we eat furnishes the energy to live, work, and play, just as the coal and oil consumed by manufacturing and transportation systems power our modern industrialized civilization.

Huge quantities of carbon-based fossil fuels have been available for the taking. This abundance of fuels has led to a world society with a voracious appetite for energy, consuming millions of barrels of petroleum every day. We are now dangerously dependent on the dwindling supplies of oil, and this dependence is an important source of tension among nations in today's world. In an incredibly short time we have moved from a period of ample and cheap supplies of petroleum to one of high prices and uncertain supplies. If our present standard of living is to be maintained, we must find alternatives to petroleum. To do so, we need to know the relationship between chemistry and energy, which we explore in this chapter.

There are additional problems with fossil fuels. The waste products from burning fossil fuels significantly affect our environment. For example, when a carbon-based fuel is burned, the carbon reacts with oxygen to form carbon dioxide, which is released into the atmosphere. Although much of this carbon dioxide is consumed in various natural processes such as photosynthesis and the formation of carbonate minerals, the amount of carbon dioxide in the atmosphere is steadily increasing. This increase is significant because atmospheric carbon dioxide absorbs heat radiated from the earth's surface and radiates some of it back toward the earth. Since this is an important mechanism for controlling the earth's temperature, many scientists fear that an increase in the concentration of carbon dioxide will warm the earth, causing significant changes in climate. In addition, impurities in the fuels react with components of air to produce air pollution.

In this chapter we will cover the fundamental concepts of energy and take a brief look at the practical aspects of the energy supply and pollution. Additional theoretical aspects of energy will be presented in Chapter 10.

9.1 The Nature of Energy

Although the concept of energy is quite familiar, energy is rather difficult to define precisely. We will define **energy** as the *capacity to do work or to produce heat.* In this chapter we will concentrate on the transfer of energy via heat flow that accompanies chemical processes.

The total energy content of the universe is constant.

One of the most important characteristics of energy is that it is conserved. The **law of conservation of energy** states that *energy can be converted from one form to another but can be neither created nor destroyed.* That is, the energy of the universe is constant. Energy can be classified as either potential energy or kinetic energy. **Potential energy** is energy due to position or composition. For example, water behind a dam has potential energy that can be converted to work when the water flows down through turbines, thereby creating electricity. Attractive and repulsive forces also lead to potential energy. The energy released when gasoline is burned results from differences in the attractive forces between nuclei and electrons in the reactants and products. The **kinetic energy** of an object is due to the motion of the object and depends on the mass of the object (m) and its velocity (v): $KE = \frac{1}{2}mv^2$.

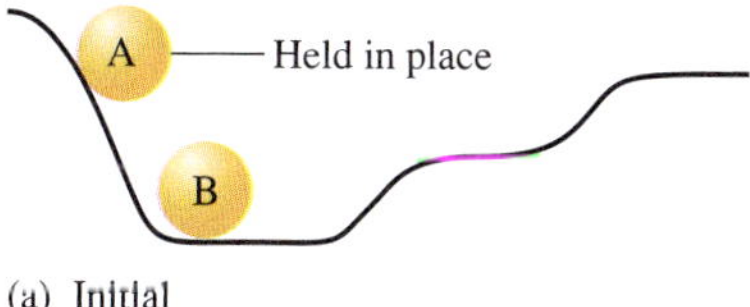

(a) Initial

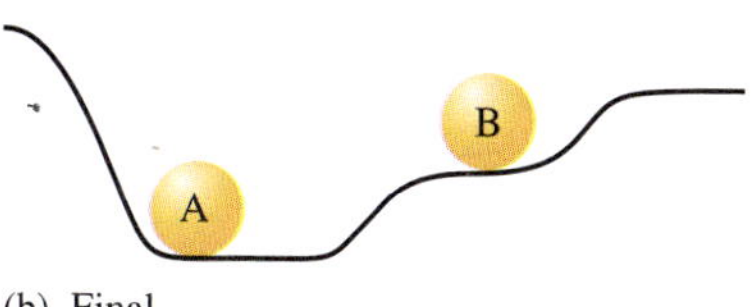

(b) Final

FIGURE 9.1

(a) In the initial positions ball A has a higher potential energy than ball B. (b) After A has rolled down the hill, the potential energy lost by A has been converted to random motions of the components of the hill (frictional heating) and to the increase in the potential energy of B.

Energy can be converted from one form to another easily. For example, consider the two balls in Fig. 9.1(a). Ball A, because of its higher position, initially has more potential energy than ball B. When A is released, it moves down the hill and strikes B. Eventually the arrangement shown in Fig. 9.1(b) is achieved. What has happened in going from the initial to the final arrangement? The potential energy of A has decreased, but since energy is conserved, all the energy lost by A must be accounted for. How is this energy distributed?

Initially, the potential energy of A is changed to kinetic energy as the ball rolls down the hill. Part of this kinetic energy has been transferred to B, causing it to be raised to a higher final position. Thus the potential energy of B has increased. However, since the final position of B is lower than the original position of A, some of the energy is still unaccounted for. Both balls are at rest in their final positions, so the missing energy cannot be due to their motions. What has happened to the remaining energy?

The answer lies in the interaction between the hill's surface and the ball. As A rolls down the hill, some of its kinetic energy is transferred to the surface of the hill as heat. This transfer of energy is called *frictional heating*. The temperature of the hill increases very slightly as the ball rolls down.

Heat involves a *transfer* of energy.

At this point it is important to recognize that heat and temperature are decidedly different. Recall that temperature is a property that reflects the random motions of the particles in a particular substance. **Heat,** on the other hand, involves the *transfer* of energy between two objects due to a temperature difference. Heat is not a substance contained in an object, although we often talk of heat as if this were true.

Note that in going from the initial to the final arrangements in Fig. 9.1, ball B gains potential energy because ball A has done work on B. **Work** is defined as *a force acting over a distance*. Work is required to raise B from its original position to a higher one. Part of the original energy stored as potential energy in A has been transferred through work to B, thereby increasing B's potential energy. Thus there are two ways to transfer energy: through work and through heat.

In rolling to the bottom of the hill as shown in Fig. 9.1, ball A always loses the same amount of potential energy. However, the way that this energy transfer is divided between work and heat depends on the specific conditions—the **pathway.** For example, the surface of the hill might be so rough that the energy of A is expended completely through frictional heating; A is moving so slowly when it hits B that it cannot move B to the next level. In this case no work is done. Regardless of the condition of the hill's surface, the *total energy* transferred will be constant. However, the amounts of heat and work will differ. Energy change is independent of the pathway; however, work and heat are both dependent on the pathway.

This brings us to a very important concept: the **state function** or *state property*. A state function refers to a property of the system that depends only on its *present state*. A state function (property) does not depend in any way on the system's past (or future). In other words, the value of a state function does not depend on how the system arrived at the present state; it depends only on the characteristics of the present state.

Stated more precisely, one very important characteristic of a state function is that a change in this function (property) in going from one state to another state is independent of the particular pathway taken between the two states.

Energy is a state function; work and heat are not.

Of the functions considered in our present example, energy is a state function, but work and heat are not state functions.

Chemical Energy

The ideas we have just illustrated using mechanical examples also apply to chemical systems. The combustion of methane, for example,

$$CH_4(g) + 2O_2(g) \longrightarrow CO_2(g) + 2H_2O(g) + \text{energy (heat)}$$

is used to heat many homes in the United States. To discuss this reaction, we divide the universe into two parts: the system and the surroundings. The

system is the part of the universe on which we wish to focus attention; the **surroundings** include everything else in the universe. In this case we define the system as the reactants and products of the reaction. The surroundings consist of the reaction container, the room, and everything else other than the reactants and products.

When a reaction results in the evolution of heat, it is said to be **exothermic** (*exo-* is a prefix meaning "out of"); that is, energy flows *out of the system.* For example, in the combustion of methane, energy flows out of the system as heat. Reactions that absorb energy from the surroundings are said to be **endothermic.** That is, when the heat flow is *into a system,* the process is endothermic. For example, the formation of nitric oxide from nitrogen and oxygen is endothermic:

$$N_2(g) + O_2(g) + \text{energy (heat)} \longrightarrow 2NO(g)$$

A familiar endothermic physical process is the vaporization of water:

$$H_2O(l) + \text{energy} \longrightarrow H_2O(g)$$

Where does the energy, released as heat, come from in an exothermic reaction? The answer lies in the difference in potential energy between the products and the reactants. In an exothermic reaction, which has lower potential energy, the reactants or the products? We know that total energy is conserved and that energy flows from the system into the surroundings in an exothermic reaction. This means that *the energy gained by the surroundings must be equal to the energy lost by the system.* For methane combustion, the energy content of the system decreases, which means that 1 mole of CO_2 and 2 moles of H_2O molecules (the products) possess less potential energy than do 1 mole of CH_4 and 2 moles of O_2 molecules (the reactants). The heat flow into the surroundings results from a lowering of the potential energy of the reaction system. This always holds true. *In any exothermic reaction, the potential energy stored in the chemical bonds is being converted to thermal energy (random kinetic energy) via heat.*

The energy diagram for the combustion of methane is shown in Fig. 9.2, where Δ(PE) represents the *change* in potential energy stored in the bonds of the products as compared with the bonds of the reactants. In other words,

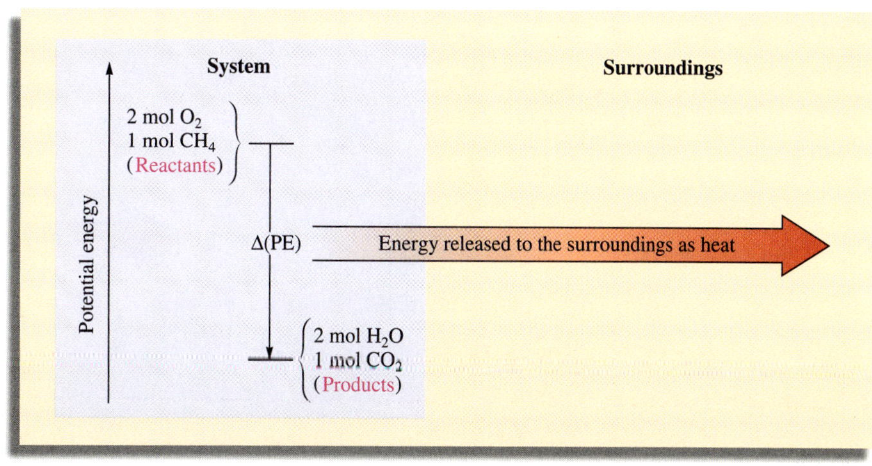

FIGURE 9.2

The combustion of methane releases the quantity of energy Δ(PE) to the surroundings via heat flow. This is an exothermic process.

CHEMICAL INSIGHTS

Bees Are Hot

A recent study at Yunnan Agricultural University in China has shown how certain species of Asian honeybees protect their colonies. Large wasps (*Vespa velutina*) with wingspans as large as 5 cm specialize in attacking social-insect nests, and they carry off the larvae as food for young wasps back at their home colony. The wasps can defeat colonies of honeybees with thousands of residents. The invader wasp sits at the entrance to the colony and kills the guard honeybees as they emerge to defend their nest. When the guard bees are all dead, the wasp then mines the honeybee nest for larvae. However, Asian honeybees (*Aspis carama*) have developed a special defense against invading wasps. They engulf the wasp in a living ball of defenders and cook the predator to death. Studies show that within about 5 minutes the center of the ball of defenders reaches a temperature of about 45°C. Separate studies of the temperature tolerances of the bees showed that the wasps die at 45.7°C, but the Asian honeybees can survive to 50.7°C. Thus the defenders come within about 5°C of a temperature that would be fatal to all of them.

this quantity represents the difference between the energy required to break the bonds in the reactants and the energy released when the bonds in the products are formed. In an exothermic process the bonds in the products are stronger (on average) than those of the reactants. That is, more energy is released in forming the new bonds in the products than is consumed in breaking the bonds in the reactants. The net result is that the quantity of energy Δ(PE) is transferred to the surroundings through heat.

For an endothermic reaction, the situation is reversed, as shown in Fig. 9.3. Energy that flows into the system as heat is used to increase the potential energy of the system. In this case the products have higher potential energy (weaker bonds on average) than the reactants.

The study of energy and its interconversions is called **thermodynamics.** The law of conservation of energy is often called the **first law of thermodynamics** and is stated as follows: *The energy of the universe is constant.*

The **internal energy** (E) of a system can be defined most precisely as the sum of the kinetic and potential energies of all of the "particles" in the system.

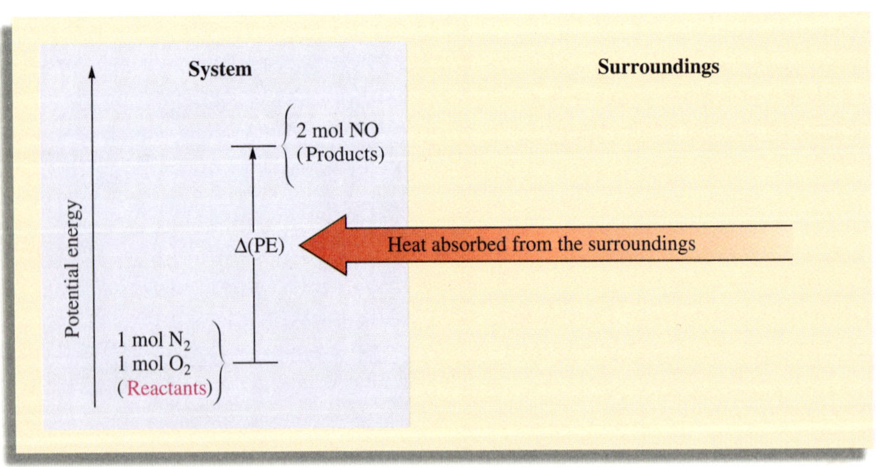

FIGURE 9.3

The energy diagram for the reaction of nitrogen and oxygen to form nitric oxide. This is an endothermic process.

The internal energy of a system can be changed by a flow of work, heat, or both. That is,

$$\Delta E = q + w$$

where ΔE represents the change in the system's internal energy, q represents heat, and w represents work.

Thermodynamic quantities always consist of two parts: a *number,* giving the magnitude of the change; and a *sign,* indicating the direction of the flow. *The sign reflects the system's point of view.* For example, if a quantity of energy flows *into* the system via heat (an endothermic process), q is equal to $+x$, where the *positive* sign indicates that the *system's energy is increasing.* On the other hand, when energy flows *out of* the system via heat (an exothermic process), q is equal to $-x$, where the *negative* sign indicates that the *system's energy is decreasing.*

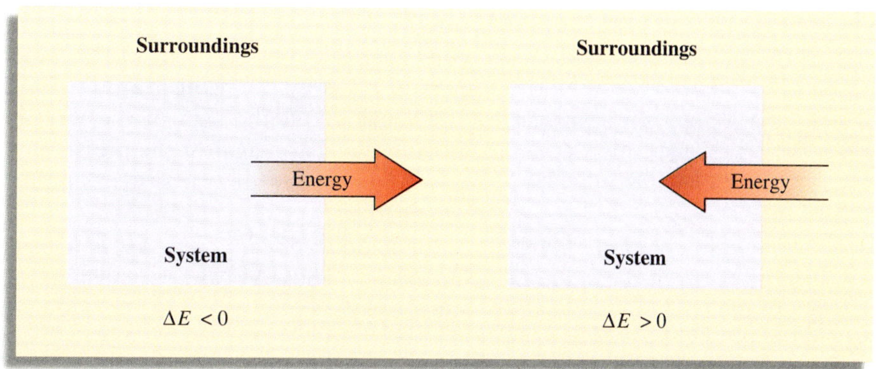

The convention in this text is to take the system's point of view: $q = -x$ denotes an exothermic process, and $q = +x$ denotes an endothermic one.

In this text the same conventions also apply to the flow of work. If the system does work on the surroundings (energy flows out of the system), w is negative. If the surroundings do work on the system (energy flows into the system), w is positive. We define work from the system's point of view to be consistent for all thermodynamic quantities. That is, in this convention the signs of both q and w reflect what happens to the system; thus we use $\Delta E = q + w$.

In this text we *always* take the system's point of view. This convention is not followed in every area of science. For example, engineers are in the business of designing machines to do work—that is, to make the system (the machine) transfer energy to its surroundings through work. Consequently, engineers define work from the surroundings' point of view. In their convention, work that flows out of the system is treated as positive because the energy of the surroundings has increased. The first law of thermodynamics is then written $\Delta E = q - w'$, where w' signifies work from the surroundings' point of view.

A common type of work associated with chemical processes is work done by a gas (through *expansion*) or work done to a gas (through *compression*). For example, in an automobile engine the heat from the combustion of the gasoline expands the gases in the cylinder, pushing back the piston. This motion is then translated into the motion of the car.

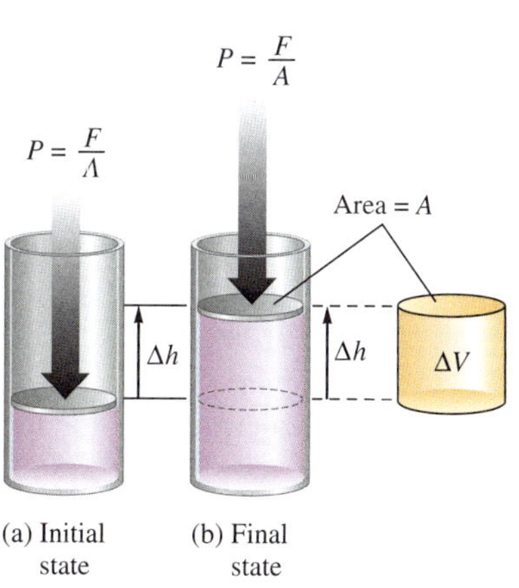

FIGURE 9.4

(a) The piston, moving a distance Δh against a pressure P, does work on the surroundings. (b) Since the volume of a cylinder is the area of the base times its height, the change in volume of the gas is given by $\Delta h \times A = \Delta V$.

Suppose we have a gas confined to a cylindrical container with a movable piston, as shown in Fig. 9.4, where F is the force acting on a piston of area A. Since pressure is defined as force per unit area, the pressure of the gas is

$$P = \frac{F}{A}$$

Work is defined as a force applied over a given distance, so if the piston moves a distance Δh, as shown in Fig. 9.4, then the magnitude of the work is

$$|\text{Work}| = |\text{force} \times \text{distance}| = |F \times \Delta h|$$

Since $P = F/A$, or $F = P \times A$, then

$$|\text{Work}| = |F \times \Delta h| = |P \times A \times \Delta h|$$

Since the volume of the cylinder equals the area of the piston times the height of the cylinder (Fig. 9.4), the change in volume ΔV resulting from the piston moving a distance Δh is

$$\Delta V = \text{final volume} - \text{initial volume} = A \times \Delta h$$

Substituting $\Delta V = A \times \Delta h$ into the expression for the magnitude of the work gives

$$|\text{Work}| = |P \times A \times \Delta h| = |P\Delta V|$$

Note that this expression gives the *magnitude* (size) of the work required to expand a gas by ΔV against a pressure P.

w and $P\Delta V$ have opposite signs since when the gas expands (ΔV is positive), work flows into the surroundings (w is negative).

What about the sign of the work? The gas (the system) is expanding, moving the piston against the pressure. Thus the system is doing work on the surroundings, so from the system's point of view, the sign of the work should be negative.

For an *expanding gas* ΔV is a positive quantity because the volume is increasing. Thus ΔV and w must have opposite signs, which leads to the equation

$$w = -P\Delta V$$

Note that for a gas expanding against an external pressure P, w is a negative quantity as required, since work flows out of the system. When a gas is *compressed*, ΔV is a negative quantity (the volume decreases), which makes w a positive quantity (work flows into the system).

The work accompanying a change in volume of a gas is often called "PV work."

In dealing with "PV work," keep in mind that the P in $P\Delta V$ always refers to the external pressure—the pressure that causes a compression or that resists an expansion.

Example 9.1

A balloon is inflated to its full extent by heating the air inside it. In the final stages of this process the volume of the balloon changes from 4.00×10^6 L to 4.50×10^6 L by addition of 1.3×10^8 J of energy as heat. Assuming the balloon expands against a constant pressure of 1.0 atm, calculate ΔE for the process.

Solution To calculate ΔE we use the equation

$$\Delta E = q + w$$

Since the problem states that 1.3×10^8 J of energy is *added* as heat,

$$q = +1.3 \times 10^8 \text{ J}$$

The joule (J) is the fundamental SI unit for energy:

$$\text{J} = \frac{\text{kg m}^2}{\text{s}^2}$$

The work done can be calculated from the expression

$$w = -P\Delta V$$

In this case $P = 1.0$ atm (the external pressure) and

$$\begin{aligned}\Delta V &= V_{\text{final}} - V_{\text{initial}} \\ &= 4.50 \times 10^6 \text{ L} - 4.00 \times 10^6 \text{ L} = 0.50 \times 10^6 \text{ L} \\ &= 5.0 \times 10^5 \text{ L}\end{aligned}$$

Thus

$$w = -1.0 \text{ atm} \times 5.0 \times 10^5 \text{ L} = -5.0 \times 10^5 \text{ L atm}$$

The conversion factor between L atm and J can be obtained from the values of R:

$$0.08206 \frac{\text{L atm}}{\text{K mol}}$$

and $$8.3145 \frac{\text{J}}{\text{K mol}}$$

Note that the negative sign for w makes sense, since the gas is expanding and thus doing work on the surroundings.

To calculate ΔE, we must add q and w. However, since q is given in units of J and w is given in units of L atm, we must change the work to units of joules:

$$w = -5.0 \times 10^5 \text{ L atm} \times \frac{101.3 \text{ J}}{\text{L atm}} = -5.1 \times 10^7 \text{ J}$$

Then

$$\Delta E = q + w = (+1.3 \times 10^8 \text{ J}) + (-5.1 \times 10^7 \text{ J}) = 8 \times 10^7 \text{ J}$$

Since more energy is added through heating than the gas expends doing work, there is a net increase in the energy of the gas in the balloon. Hence ΔE is positive.

9.2 Enthalpy

So far we have discussed the internal energy of a system. A less familiar property of a system is its **enthalpy** (H), which is defined as

$$H = E + PV$$

where E is the internal energy of the system, P is the pressure of the system, and V is the volume of the system.

Enthalpy is a state function. A change in enthalpy does not depend on the pathway between two states.

Since internal energy, pressure, and volume are all state functions, *enthalpy is also a state function.* But what exactly is enthalpy? To help answer this question, consider a process carried out at constant pressure, where the only work allowed is pressure-volume work ($w = -P\Delta V$). Under these conditions the expression

$$\Delta E = q_{\text{P}} + w$$

becomes $$\Delta E = q_{\text{P}} - P\Delta V$$

or $$q_{\text{P}} = \Delta E + P\Delta V$$

where q_{P} is the heat at constant pressure.

We will now relate q_{P} to the change in enthalpy. The definition of enthalpy is $H = E + PV$. Therefore,

$$(\text{Change in } H) = (\text{change in } E) + (\text{change in } PV)$$

or $$\Delta H = \Delta E + \Delta(PV)$$

Since P is constant, the change in PV is caused only by a change in volume. Thus

$$\Delta(PV) = P\Delta V$$

and

$$\Delta H = \Delta E + P\Delta V$$

This expression is identical to the one we obtained for q_P:

$$q_P = \Delta E + P\Delta V$$

$\Delta H = q$ at constant pressure, where only "PV work" is allowed.

Thus, for a process carried out at constant pressure, where the only work allowed is that from a volume change,

$$\Delta H = q_P$$

The change in enthalpy of a system has no easily interpreted meaning except at constant pressure, where ΔH = heat.

At constant pressure (where only PV work is allowed) the change in enthalpy (ΔH) of the system is equal to the energy flow as heat. This means that for a reaction studied at constant pressure, the flow of heat is a measure of the change in enthalpy for the system. For this reason, the terms *heat of reaction* and *change in enthalpy* are used interchangeably for reactions studied at constant pressure.

For a chemical reaction the enthalpy change is given by the equation

$$\Delta H = H_{\text{products}} - H_{\text{reactants}}$$

At constant pressure exothermic means ΔH is negative; endothermic means ΔH is positive.

In a case in which the products of a reaction have greater enthalpy than the reactants, ΔH will be positive. Thus heat is absorbed by the system, and the reaction is endothermic. On the other hand, if the enthalpy of the products is less than that of the reactants, ΔH is negative. In this case the overall decrease in enthalpy is achieved by the generation of heat, and the reaction is exothermic.

9.3 Thermodynamics of Ideal Gases

In developing the concepts of thermodynamics, we often find it useful to refer to the properties of matter in the simplest possible context. For this reason we often start with the thermodynamic characteristics of the ideal gas—the hypothetical condition approached by real gases at high temperatures and low pressures such that they obey the relationship $PV = nRT$.

To this point, we have assumed that the ideal gas particles have no internal structure—they are monatomic.

Recall from Chapter 5 that for an ideal gas

$$(\text{KE})_{\text{avg}} = \tfrac{3}{2}RT$$

where $(\text{KE})_{\text{avg}}$ represents the average, random, translational energy for 1 mole of gas at a given temperature T (in kelvins). The only way to change the kinetic energy of an ideal gas is to change its temperature. The energy ("heat") required to change the energy of 1 mole of an ideal gas by ΔT is

$$\text{Energy ("heat") required} = \tfrac{3}{2}R\Delta T$$

Note that for a temperature change of 1 K ($\Delta T = 1$), the energy required is $\frac{3}{2}R$.

The **molar heat capacity** of a substance is defined as the energy required to raise the temperature of 1 mole of that substance by 1 K. Thus we might conclude that the molar heat capacity of an ideal gas is $\frac{3}{2}R$. However, we will have to qualify this conclusion when we consider the implications of the PV work that can occur when a gas is heated.

Heating an Ideal Gas at Constant Volume

For an ideal gas, work occurs only when its volume changes. Thus, if a gas is heated at constant volume, the pressure increases but no work occurs.

When an ideal gas is heated in a rigid container in which no change in volume occurs, there can be no *PV* work ($\Delta V = 0$). Under these conditions all the energy that flows into the gas is used to increase the translational energies of the gas molecules. Thus C_v, the molar heat capacity of an ideal gas *at constant volume,* is $\frac{3}{2}R$, the result anticipated in the preceding discussion:

$$C_v = \tfrac{3}{2}R = \begin{array}{l}\text{“heat” required to change the temperature} \\ \text{of 1 mol of gas by 1 K at constant volume}\end{array}$$

Heating an Ideal Gas at Constant Pressure

When an ideal gas is heated at constant pressure, its volume increases and *PV* work occurs. Thus, when a gas is heated at constant pressure, energy must be supplied both to change the translational energy of the gas and to provide the work the gas does as it expands:

$$\text{Energy required} = \text{“heat”} = \begin{array}{c}\text{energy needed} \\ \text{to change the} \\ \text{translational energy}\end{array} + \begin{array}{c}\text{energy needed to} \\ \text{do the } PV \text{ work}\end{array}$$

The heat needed to increase the translational energy is $\frac{3}{2}R$, as we concluded above.

The *quantity* of work done as the gas expands by ΔV is $P\Delta V$. Using the ideal gas law, we see that

$$P\Delta V = nR\Delta T = R\Delta T \qquad \text{(per mole)}$$

Thus for a 1 K change in temperature ($\Delta T = 1$ K) the work is R, so

$$\begin{array}{c}\text{Heat required to increase the temperature} \\ \text{of 1 mol of gas by 1 K (constant } P)\end{array} = \tfrac{3}{2}R + R = \tfrac{5}{2}R$$

$$= C_v + R = C_p$$

$C_v = \frac{3}{2}R$
$C_p = \frac{5}{2}R = C_v + R$

Therefore, we have shown that C_p, the molar heat capacity of an ideal gas at constant pressure, is $\frac{5}{2}R$ or $C_v + R$.

Heating a Polyatomic Gas

We have established that for an ideal gas the molar heat capacity at constant volume is $\frac{3}{2}R$. This value of C_v assumes that an ideal gas consists of “particles” that have no structure. That is, we assume that the particles are monatomic (consisting of a single atom). Monatomic real gases, such as helium, have measured values of C_v very close to $\frac{3}{2}R$. However, gases such as SO_2 and $CHCl_3$ that contain polyatomic molecules have observed values for C_v that are significantly greater than $\frac{3}{2}R$. For example, the value of C_v for SO_2 is almost $4R$ at 25°C. This larger value for C_v results because polyatomic molecules absorb energy to excite rotational and vibrational motions in addition to translational motions. That is, at 25°C the molecules in such a gas are rotating, and the atoms in the molecule are vibrating, as if the bonds were springs.

As a polyatomic gas is heated, the gas molecules absorb energy to increase their rotational and vibrational motions as well as to move through space (translate) at higher speeds. Recall from our previous discussions that

C_v is greater than $\frac{3}{2}R$ for a gas in which the individual particles are molecules because some of the energy added via heat flow is "stored" in motions that do not directly raise the temperature of the gas. This effect is not related to whether or not the gas is behaving ideally.

the temperature of a monatomic ideal gas is an index of the average random *translational* energy of the gas. Thus, when a gas is heated, the temperature only increases to the extent that the translational energies of the molecules increase. Any energy that is absorbed to increase the vibrational and rotational energies does not contribute directly to the translational kinetic energy; so, for a gas that consists of diatomic or polyatomic molecules, much of the heat absorbed is used in processes that do not directly increase the temperature. Thus its heat capacity (the energy required to change its *temperature* by 1 K) is greater than $\frac{3}{2}R$.

Note that the elevated value of C_v for a gas whose particles are molecules is not caused by nonideal behavior. That is, it does not depend on whether the gas obeys the ideal gas law. Rather, it is simply that the internal structure of the molecules enables them to absorb energy for processes other than translational motions.

Recall that C_p is greater than C_v because of the work done by the heated gas as it expands at constant pressure. Thus, if we assume that a given polyatomic gas obeys the ideal gas law, the expression

$$C_p = C_v + R$$

can be used to calculate C_p if the value of C_v for the gas is known.

The observed heat capacities of several gases are shown in Table 9.1. Notice that the monatomic gases have values of C_v equal to $\frac{3}{2}R$ (12.47 J K^{-1} mol^{-1}). Note also that as the molecules become more complex (more atoms), C_v increases. This result is expected because the presence of more atoms means that more nontranslational motions are available to absorb energy. Finally, notice that in all cases $C_p - C_v = R$, as expected for gases that closely obey the ideal gas law.

Heating a Gas: Energy and Enthalpy

Recall that the average translational energy of an ideal gas E is given by the expression

$$E = \frac{3}{2}RT \qquad \text{(per mole)}$$

for a monatomic ideal gas. The energy of an ideal gas can be changed only by changing the temperature:

$$\Delta E = \frac{3}{2}R\Delta T \qquad \text{(per mole)}$$

TABLE 9.1

Molar Heat Capacities of Various Gases at 298 K

Gas	$C_v\left(\frac{J}{K\ mol}\right)$	$C_p\left(\frac{J}{K\ mol}\right)$	$C_p - C_v$
He, Ne, Ar	12.47	20.80	8.33
H_2	20.54	28.86	8.32
N_2	20.71	29.03	8.32
N_2O	30.38	38.70	8.32
CO_2	28.95	37.27	8.32
C_2H_6	44.60	52.92	8.32

Note that this expression corresponds to

$$\Delta E = C_v \Delta T \qquad \text{(per mole)}$$

or

$$\Delta E = nC_v \Delta T \qquad (n \text{ moles})$$

The constant-volume heat capacity appears in this expression because when a gas is heated at constant volume, all the input energy (heat) goes toward increasing E (no heat is needed to do work).

On the other hand, when a gas is heated at constant pressure, the volume changes and work occurs. In this case (for n moles of gas),

$$\begin{aligned} \text{“Heat” required} &= q_p = nC_p\Delta T \\ &= n(C_v + R)\Delta T \\ &= \underbrace{nC_v\Delta T}_{\Delta E} + \underbrace{nR\Delta T}_{P\Delta V = \text{work required}} \end{aligned}$$

Notice that although this process is carried out at constant pressure, ΔE is still given by $nC_v\Delta T$. This result seems contradictory at first glance, but actually it makes good sense. Because E for an ideal gas depends only on T (it does not depend on pressure or volume, for example), $\Delta E = nC_v\Delta T$ when an ideal gas is heated whether the process occurs at constant volume or constant pressure.

Next, consider the change in enthalpy when a gas is heated. Recall that by definition

$$H = E + PV$$

Thus, in general, a change in enthalpy is given by the expression

$$\Delta H = \Delta E + \Delta(PV)$$

which (using the ideal gas law) becomes

$$\Delta H = \Delta E + \Delta(nRT) = \Delta E + nR\Delta T$$

for a sample of ideal gas containing n moles. Substituting $\Delta E = nC_v\Delta T$, we have

$$\begin{aligned} \Delta H &= nC_v\Delta T + nR\Delta T \\ &= n(C_v + R)\Delta T = nC_p\Delta T \end{aligned}$$

Note that we have shown that

$$\Delta H = nC_p\Delta T$$

The only way to change H and E for an ideal gas is to change the temperature of the gas. Thus, for any process involving *an ideal gas at constant temperature,* $\Delta H = 0$ and $\Delta E = 0$.

even though we have not assumed constant pressure (or volume). Thus we have shown that for an ideal gas we can always use the expression $nC_p\Delta T$ to calculate the change in enthalpy when n moles of an ideal gas are heated, regardless of any conditions on pressure or volume.

Again, it may seem contradictory that C_p appears in this expression for ΔH even though the pressure may or may not be constant in the process. However, note that the enthalpy ($H = E + PV$) of an ideal gas depends on E and the product PV. We have seen that E depends directly on T, and from the ideal gas law we can easily show that PV depends directly on T ($PV = nRT$) for a given sample of ideal gas (containing n moles). Thus both the energy E

TABLE 9.2

Thermodynamic Properties of an Ideal Gas

Expression	Application
$C_v = \frac{3}{2}R$	Monatomic ideal gas
$C_v > \frac{3}{2}R$	Polyatomic ideal gas (value must be measured experimentally)
$C_p = C_v + R$	All ideal gases
$C_p = \frac{5}{2}R = \frac{3}{2}R + R$	Monatomic ideal gas
$C_p > \frac{5}{2}R$	Polyatomic ideal gas (specific value depends on the value of C_v)
$\Delta E = nC_v\Delta T$	All ideal gases
$\Delta H = nC_p\Delta T$	All ideal gases

and the enthalpy H of an ideal gas depend only on T, not on P or V (individually):

$$E \propto T \qquad \text{and} \qquad H \propto T$$

For energy (E), the proportionality constant is C_v (per mole), and for enthalpy (H), the proportionality constant is C_p (per mole).

In considering these ideas, we must distinguish among q, ΔH, and ΔE. In the calculation of the heat flow for an ideal gas the equation

$$q = nC\Delta T$$

applies, where C_v or C_p is used depending on the conditions. In contrast, $\Delta H = nC_p\Delta T$ and $\Delta E = nC_v\Delta T$ for a temperature change of an ideal gas regardless of whether pressure or volume (or neither) is constant. Also, note that the heat flow equals ΔE at constant volume ($\Delta E = q_v$) but the heat flow equals ΔH at constant pressure ($\Delta H = q_p$). These results are summarized in Table 9.2.

We will illustrate these concepts in Example 9.2.

Example 9.2

Consider 2.00 mol of a monatomic ideal gas that is taken from state A ($P_A = 2.00$ atm, $V_A = 10.0$ L) to state B ($P_B = 1.00$ atm, $V_B = 30.0$ L) by two different pathways:

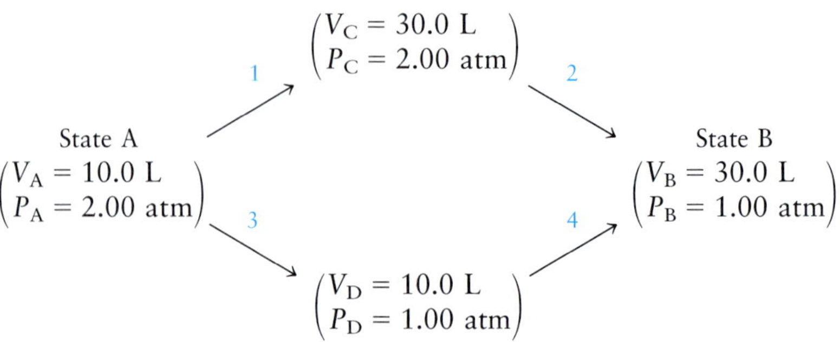

Calculate q, w, ΔE, and ΔH for both pathways.

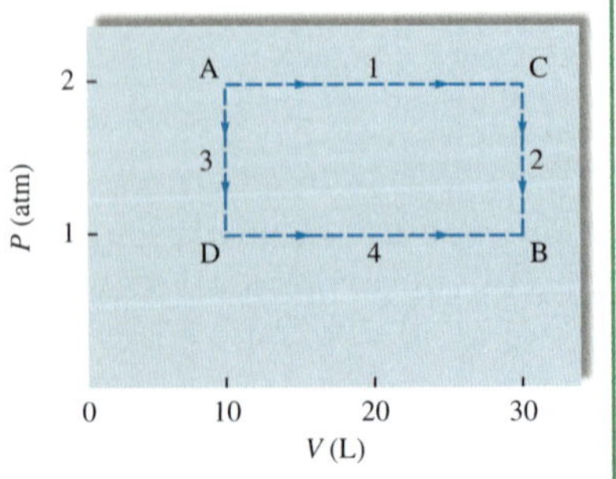

FIGURE 9.5

Summary of the two pathways discussed in Example 9.2.

Solution Before we do any calculations, it is useful to summarize the processes described above by using the "PV diagram" shown in Fig. 9.5.

Step 1

Notice from Fig. 9.5 that this step corresponds to an expansion from 10.0 to 30.0 L at a constant pressure of 2.00 atm. This process must occur by heating the gas to produce some temperature change ΔT (not specified in the given data). From the ideal gas law we know that

$$P\Delta V = nR\Delta T$$

In this case $\Delta V = 30.0\text{ L} - 10.0\text{ L}$, so

$$P\Delta V = (2.00\text{ atm})(20.0\text{ L}) = 4.00 \times 10^1\text{ L atm}$$

or if we convert to joules,

$$P\Delta V = 4.00 \times 10^1\text{ L atm} \times \frac{101.3\text{ J}}{\text{L atm}} = 4.05 \times 10^3\text{ J}$$

It follows that

$$\Delta T = \frac{P\Delta V}{nR} = \frac{4.05 \times 10^3\text{ J}}{nR}$$

We know that $w = -P\Delta V$, and because in this case the gas expands against a constant external pressure of 2.00 atm, we have

$$\begin{aligned} w_1 &= -(2.00\text{ atm})(30.0\text{ L} - 10.0\text{ L}) = -4.00 \times 10^1\text{ L atm} \\ &= -4.05 \times 10^3\text{ J} \end{aligned}$$

Also, in this case (constant P)

$$\begin{aligned} q_1 &= q_\text{p} = nC_\text{p}\Delta T \\ &= n\left(\frac{5}{2}R\right)\left(\frac{4.05 \times 10^3\text{ J}}{nR}\right) = 1.01 \times 10^4\text{ J} \end{aligned}$$

(Note that the signs of w and q are as expected: The gas expands, so work flows out of the system; the gas is heated, so heat flows into the system.)

We can calculate ΔE_1 and ΔH_1 as follows:

$$\Delta E_1 = nC_\text{v}\Delta T = n\left(\frac{3}{2}R\right)\left(\frac{4.05 \times 10^3\text{ J}}{nR}\right) = 6.08 \times 10^3\text{ J}$$

$$\Delta H_1 = nC_\text{p}\Delta T = n\left(\frac{5}{2}R\right)\left(\frac{4.05 \times 10^3\text{ J}}{nR}\right) = 1.01 \times 10^4\text{ J}$$

Note that in this case $q_1(q_\text{p})$ equals ΔH_1, as expected, because this process is carried out at constant pressure.

Step 2

In this step the gas pressure decreases from 2.00 atm to 1.00 atm at constant volume. This step must correspond to the cooling of the gas by a quantity ΔT, which we can find from the ideal gas law:

$$\Delta PV = nR\Delta T$$

$$\begin{aligned} \Delta T &= \frac{\Delta PV}{nR} = \frac{(1.00\text{ atm} - 2.00\text{ atm})(30.0\text{ L})}{nR} \\ &= \frac{-30.0\text{ L atm}}{nR} = \frac{-3.04 \times 10^3\text{ J}}{nR} \end{aligned}$$

Note that ΔT is negative, as expected for a cooling process.

Because in this step $\Delta V = 0$, thus $w_2 = 0$. In this case

$$q_2 = q_v = nC_v\Delta T = n\left(\frac{3}{2}R\right)\left(\frac{-3.04 \times 10^3 \text{ J}}{nR}\right)$$
$$= -4.56 \times 10^3 \text{ J}$$

Also,
$$\Delta E_2 = nC_v\Delta T = n\left(\frac{3}{2}R\right)\left(\frac{-3.04 \times 10^3 \text{ J}}{nR}\right)$$
$$= -4.56 \times 10^3 \text{ J} = q_v$$

and
$$\Delta H_2 = nC_p\Delta T = n\left(\frac{5}{2}R\right)\left(\frac{-3.04 \times 10^3 \text{ J}}{nR}\right)$$
$$= -7.60 \times 10^3 \text{ J}$$

Notice that in this case $q_2 = q_v = \Delta E$, as expected for a constant-volume process.

Using similar reasoning, we can compute the required quantities for steps 3 and 4.

Step 3

$$\Delta T = \frac{\Delta PV}{nR} = \frac{(-1.00 \text{ atm})(10.0 \text{ L})}{nR} = \frac{-10.0 \text{ L atm}}{nR}$$
$$= \frac{-1.01 \times 10^3 \text{ J}}{nR}$$

$$w_3 = 0 \qquad (\Delta V = 0)$$

$$q_3 = q_v = nC_v\Delta T = n\left(\frac{3}{2}R\right)\left(\frac{-1.01 \times 10^3 \text{ J}}{nR}\right)$$
$$= -1.52 \times 10^3 \text{ J}$$

$$\Delta E_3 = q_v = -1.52 \times 10^3 \text{ J}$$

$$\Delta H_3 = nC_p\Delta T = n\left(\frac{5}{2}R\right)\left(\frac{-1.01 \times 10^3 \text{ J}}{nR}\right) = -2.53 \times 10^3 \text{ J}$$

Step 4

$$\Delta T = \frac{P\Delta V}{nR} = \frac{(1.00 \text{ atm})(20.0 \text{ L})}{nR} = \frac{20.0 \text{ L atm}}{nR}$$
$$= \frac{2.03 \times 10^3 \text{ J}}{nR}$$

$$w_4 = -P\Delta V = -(1.00 \text{ atm})(20.0 \text{ L}) = -20.0 \text{ L atm}$$
$$= -2.03 \times 10^3 \text{ J}$$

$$q_4 = q_p = nC_p\Delta T = n\left(\frac{5}{2}R\right)\left(\frac{2.03 \times 10^3 \text{ J}}{nR}\right)$$
$$= 5.08 \times 10^3 \text{ J}$$

$$\Delta E_4 = nC_v\Delta T = n\left(\frac{3}{2}R\right)\left(\frac{2.03 \times 10^3 \text{ J}}{nR}\right)$$
$$= 3.05 \times 10^3 \text{ J}$$

$$\Delta H_4 = nC_p\Delta T = n\left(\frac{5}{2}R\right)\left(\frac{2.03 \times 10^3 \text{ J}}{nR}\right)$$
$$= 5.08 \times 10^3 \text{ J} = q_p$$

Summary

- Pathway one (steps 1 and 2):

$$q_{one} = q_1 + q_2 = 1.01 \times 10^4 \text{ J} - 4.56 \times 10^3 \text{ J}$$
$$= 5.5 \times 10^3 \text{ J}$$
$$w_{one} = w_1 + w_2 = -4.05 \times 10^3 \text{ J}$$
$$q_{one} + w_{one} = 1.5 \times 10^3 \text{ J} = \Delta E_{one}$$
$$\Delta H_{one} = \Delta H_1 + \Delta H_2$$
$$= 1.01 \times 10^4 \text{ J} - 7.60 \times 10^3 \text{ J}$$
$$= 2.5 \times 10^3 \text{ J}$$

- Pathway two (steps 3 and 4):

$$q_{two} = q_3 + q_4 = -1.52 \times 10^3 \text{ J} + 5.08 \times 10^3 \text{ J}$$
$$= 3.56 \times 10^3 \text{ J}$$
$$w_{two} = w_3 + w_4 = -2.03 \times 10^3 \text{ J}$$
$$q_{two} + w_{two} = 3.55 \times 10^3 \text{ J} - 2.03 \times 10^3 \text{ J}$$
$$= 1.52 \times 10^3 \text{ J} = \Delta E_{two}$$
$$\Delta H_{two} = \Delta H_3 + \Delta H_4$$
$$= -2.53 \times 10^3 \text{ J} + 5.08 \times 10^3 \text{ J}$$
$$= 2.55 \times 10^3 \text{ J}$$

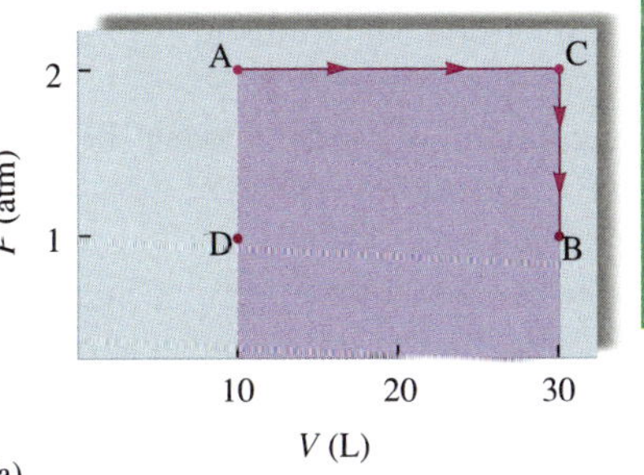

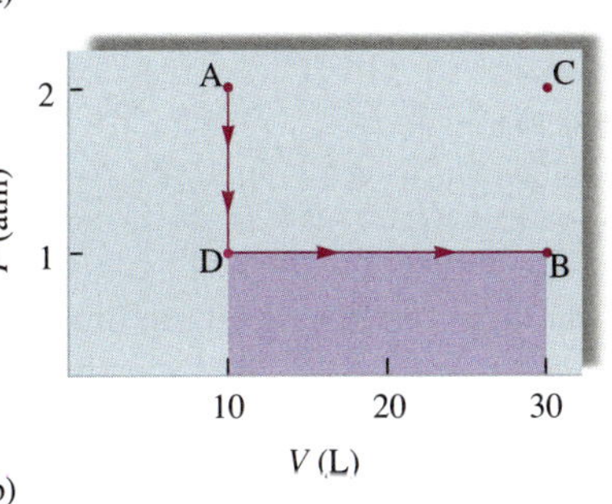

FIGURE 9.6

The magnitude of the work for pathway one (a) and pathway two (b) is shown by the colored areas: $|w| = |P\Delta V|$.

Notice from the results of Example 9.2 that the work and heat are different for the two pathways between states A and B. For example, in pathway one the expansion (by 20.0 L) is carried out at a pressure of 2.00 atm, and in pathway two the expansion (by 20.0 L) is carried out at 1.00 atm. Thus, since $|w| = |P\Delta V|$, twice as much work is obtained by way of pathway one as via pathway two. This result is shown graphically in Fig. 9.6, where the darker areas represent $|P\Delta V|$. These results again emphasize that heat and work are both pathway-dependent. On the other hand, note that the sum of q and w is the same for both pathways (within round-off differences). This is expected. Recall that

$$\Delta E = q + w$$

and that E is a state function. Note also that the overall ΔH value for pathway one equals that for pathway two (within rounding errors), as expected, because enthalpy is a state function as well.

9.4 Calorimetry

TABLE 9.3

The Specific Heat Capacities of Some Common Substances

Substance	Specific Heat Capacity ($J\ °C^{-1}\ g^{-1}$)
$H_2O(l)$	4.18
$H_2O(s)$	2.03
$Al(s)$	0.89
$Fe(s)$	0.45
$Hg(l)$	0.14
$C(s)$ (graphite)	0.71

Specific heat capacity: The energy required to raise the temperature of 1 g of a substance by 1°C.

Molar heat capacity: The energy required to raise the temperature of 1 mol of a substance by 1°C.

We can determine the heat associated with a chemical reaction experimentally by using a device called a **calorimeter. Calorimetry,** the science of measuring heat, is based on observing the temperature change when a body absorbs or discharges energy as heat. Substances respond differently to being heated. We have already discussed the response of ideal gases to heating. Now we expand that discussion to include other substances. In general terms, the **heat capacity** (C) of a substance is defined as

$$C = \frac{\text{heat absorbed}}{\text{increase in temperature}}$$

When an element or a compound is heated, the energy required to reach a certain temperature will depend on the amount of the substance present (for example, it takes twice as much energy to raise the temperature of 2 g of water by 1°C as it takes to raise the temperature of 1 g of water by 1°C). Thus, in defining the heat capacity of a substance, the amount of substance must be specified. If the heat capacity is given *per gram* of substance, it is called the **specific heat capacity** with units of $J\ K^{-1}\ g^{-1}$ or $J\ °C^{-1}\ g^{-1}$. If the heat capacity is given *per mole* of the substance, it is called the **molar heat capacity,** which has the units $J\ K^{-1}\ mol^{-1}$ or $J\ °C^{-1}\ mol^{-1}$. The specific heat capacities of some common substances are given in Table 9.3.

Although the calorimeters used for highly accurate work are precision instruments, a very simple calorimeter can be used to examine the fundamentals of calorimetry. All we need are two nested Styrofoam cups with a Styrofoam cover through which a stirrer and thermometer can be inserted, as shown in Fig. 9.7. This device is called a *coffee cup calorimeter.* The outer cup is used to provide extra insulation. The inner cup holds the solution in which the reaction occurs.

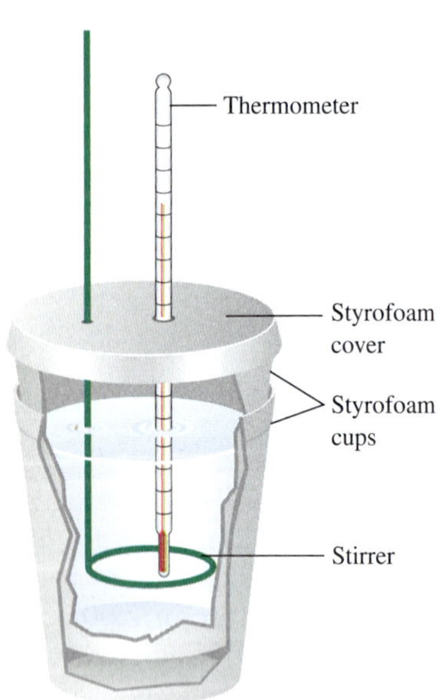

FIGURE 9.7

A coffee cup calorimeter made of two Styrofoam cups.

The measurement of heat using a simple calorimeter such as that shown in Fig. 9.7 is an example of **constant-pressure calorimetry,** since the pressure (atmospheric pressure) remains constant during the process. Constant-pressure calorimetry is used in determining the changes in enthalpy occurring in solution. Recall that under these conditions the change in enthalpy equals the heat.

For example, suppose we mix 50.0 mL of 1.0 *M* HCl at 25.0°C with 50.0 mL of 1.0 *M* NaOH also at 25°C in a calorimeter. After the reactants are mixed, the temperature is observed to increase to 31.9°C. As we saw in Chapter 4, the net ionic equation for this reaction is

$$H^+(aq) + OH^-(aq) \longrightarrow H_2O(l)$$

If two reactants at the same temperature are mixed and the resulting solution gets warmer, it means the reaction taking place is exothermic. An endothermic reaction cools the solution.

When these reactants (both originally at the same temperature) are mixed, the temperature of the mixed solution is observed to increase. Thus the chemical reaction must be releasing energy as heat. This increases the random motions of the solution components, which in turn increases the temperature. The quantity of energy released can be determined from the temperature increase, the mass of the solution, and the specific heat capacity of the solution. For an approximate result we will assume that the calorimeter does not absorb or leak any heat and that the solution can be treated as if it were pure water with a density of 1.0 g/mL.

We also need to know the heat required to raise the temperature of a given amount of water by 1°C. Table 9.3 lists the specific heat capacity of water as 4.18 J $°C^{-1}$ g^{-1}. This means that 4.18 J of energy is required to raise the temperature of 1 g of water by 1°C.

From these assumptions and definitions we can calculate the heat (change in enthalpy) for the neutralization reaction:

Note that the typical units for molar heat capacity ($\Delta E/\Delta T$ per mole) are

$$\frac{\text{J}}{\text{K mol}}$$

Because 1 K equals 1°C, the units

$$\frac{\text{J}}{°\text{C mol}}$$

are used as well.

$$\begin{aligned}\text{Energy released by the reaction} &= \text{energy absorbed by the solution}\\ &= \text{specific heat capacity} \times \text{mass of solution}\\ &\quad \times \text{increase in temperature}\end{aligned}$$

where the increase in temperature = 31.9°C − 25.0°C = 6.9°C, and where

$$\text{Mass of solution} = 100.0\ \text{mL} \times 1.0\ \text{g/mL} = 1.0 \times 10^2\ \text{g}$$

Thus

$$\text{Energy released} = \left(4.18\ \frac{\text{J}}{°\text{C g}}\right)(1.0 \times 10^2\ \text{g})(6.9°\text{C}) = 2.9 \times 10^3\ \text{J}$$

Enthalpies of reaction are often expressed in terms of moles of reacting substances. The number of moles of H^+ ions consumed in the preceding experiment is

$$50.0\ \text{mL} \times \frac{1\ \text{L}}{1000\ \text{mL}} \times \frac{1.0\ \text{mol}}{\text{L}}\ \text{H}^+ = 5.0 \times 10^{-2}\ \text{mol H}^+$$

Thus 2.9×10^3 J of heat was released when 5.0×10^{-2} mol of H^+ ions reacted. Thus

$$\frac{2.9 \times 10^3\ \text{J}}{5.0 \times 10^{-2}\ \text{mol H}^+} = 5.8 \times 10^4\ \text{J}$$

is the heat released per 1.0 mol of H^+ ions neutralized. The *magnitude* of the enthalpy of reaction per mole for

$$\text{H}^+(aq) + \text{OH}^-(aq) \longrightarrow \text{H}_2\text{O}(l)$$

at constant pressure is 58 kJ/mol. Since heat is *evolved*, $\Delta H = -58$ kJ/mol.

Notice that in this example we mentally keep track of the direction of the energy flow and assign the correct sign at the end of the calculation.

Example 9.3

When 1.00 L of 1.00 *M* $Ba(NO_3)_2$ at 25.0°C is mixed with 1.00 L of 1.00 *M* Na_2SO_4 at 25°C in a calorimeter, the white solid $BaSO_4$ forms, and the temperature of the mixture increases to 28.1°C. Assuming that the calorimeter absorbs only a negligible quantity of heat, that the specific heat capacity of the solution is 4.18 J $°C^{-1}$ g^{-1}, and that the density of the final solution is 1.0 g/mL, calculate the enthalpy change per mole of $BaSO_4$ formed.

Solution The ions present before any reaction occurs are Ba^{2+}, NO_3^-, Na^+, and SO_4^{2-}. The Na^+ and NO_3^- ions are spectator ions, since $NaNO_3$ is very soluble in water and will not precipitate under these conditions. The net ionic equation of the reaction is therefore

$$\text{Ba}^{2+}(aq) + \text{SO}_4^{2-}(aq) \longrightarrow \text{BaSO}_4(s)$$

Solutions of $Ba(NO_3)_2$ and Na_2SO_4 both initially at 25.0°C.

When barium nitrate and sodium sulfate are mixed in an insulated container, they form a white solid $BaSO_4$ (barium sulfate), and the temperature changes to 28.1°C.

Since the temperature increases, the formation of solid $BaSO_4$ must be exothermic; ΔH will be negative.

$$\begin{aligned}\text{Heat evolved by reaction} &= \text{heat absorbed by solution}\\ &= \text{specific heat capacity} \times \text{mass of solution}\\ &\quad \times \text{increase in temperature}\end{aligned}$$

Since 1.00 L of each solution is used, the total solution volume is 2.00 L, and

$$\begin{aligned}\text{Mass of solution} &= 2.00 \text{ L} \times \frac{1000 \text{ mL}}{1 \text{ L}} \times \frac{1.0 \text{ g}}{\text{mL}}\\ &= 2.0 \times 10^3 \text{ g}\\ \text{Temperature increase} &= 28.1°\text{C} - 25.0°\text{C} = 3.1°\text{C}\\ \text{Heat evolved} &= (4.18 \text{ J °C}^{-1} \text{ g}^{-1})(2.0 \times 10^3 \text{ g})(3.1°\text{C})\\ &= 2.6 \times 10^4 \text{ J}\end{aligned}$$

Thus $$q = q_P = \Delta H = -2.6 \times 10^4 \text{ J}$$

Since 1.00 L of 1.00 *M* $Ba(NO_3)_2$ contains 1 mol of Ba^{2+} ions, and 1.00 L of 1.00 *M* Na_2SO_4 contains 1.00 mol of $SO_4{}^{2-}$ ions, 1.00 mol of solid $BaSO_4$ is formed in this experiment. Thus the enthalpy change per mole of $BaSO_4$ formed is

$$\Delta H = -2.6 \times 10^4 \text{ J/mol} = -26 \text{ kJ/mol}$$

Calculation of ΔH and ΔE for Cases in Which *PV* Work Occurs

In the examples of constant-pressure calorimetry we have considered so far, the reactions have occurred in solution, where no appreciable volume changes occur (that is, the total volume of the reactant solution is the sum of the volumes of the solutions that are mixed and remains constant as the reaction

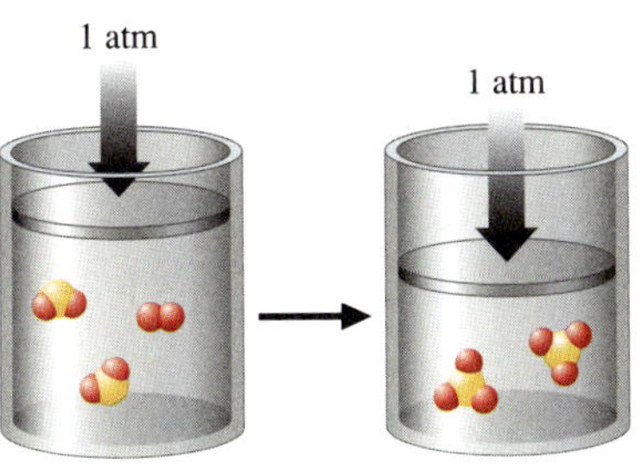

FIGURE 9.8

A schematic to show the change in volume for the reaction

$$2SO_2(g) + O_2(g) \longrightarrow 2SO_3(g)$$

proceeds). Under these conditions no work occurs (since $\Delta V = 0$, $P\Delta V = 0$, and $w = 0$). Thus, since $\Delta H = q_P$ (constant pressure) and $w = 0$,

$$\Delta E = q_P + w = \Delta H + 0$$

At constant pressure, where $\Delta V = 0$, no work is done and $\Delta E = \Delta H = q_P$.

However, when a reaction involving gases is studied at constant pressure, ΔE may not equal ΔH. The reaction

$$2SO_2(g) + O_2(g) \longrightarrow 2SO_3(g)$$

provides an example of this case. Picture this reaction being carried out at constant temperature and pressure, as shown in Fig. 9.8.

In going from reactants to products, the volume of the system decreases because the number of moles of gas decreases. Since ΔV ($= V_{\text{final}} - V_{\text{initial}}$) is negative, w is positive:

$$w = -\underbrace{P\Delta V}_{\substack{\uparrow \\ \text{Negative in this case}}}$$

Thus work flows into the system. In addition, because

$$\Delta E = q + w$$

and at constant pressure

$$\Delta H = q_P$$

then

$$\Delta E = q_P + w = \Delta H + w$$

In this case $w \neq 0$, so ΔE and ΔH are different. This case is illustrated in Example 9.4.

Example 9.4

When 2.00 mol of $SO_2(g)$ reacts completely with 1.00 mol of $O_2(g)$ to form 2.00 mol of $SO_3(g)$ at 25°C and a constant pressure of 1.00 atm, 198 kJ of energy is released as heat. Calculate ΔH and ΔE for this process.

Solution Because the pressure is constant for this process, $\Delta H = q_P$. The description of the experiment states that 198 kJ of heat is *released*. Thus $\Delta H = q_P = -198$ kJ, where the negative sign indicates that energy flows *out of* the system.

The value of ΔE can be calculated from the relationship

$$\Delta E = q + w$$

Since q is known (-198 kJ), we only need the value for w. We know that

$$w = -P\Delta V$$

Solving the ideal gas law for ΔV gives

$$\Delta V = \Delta n\left(\frac{RT}{P}\right)$$

where only n changes (T and P are constant) as the reaction occurs.

In this case

$$\Delta n = n_{\text{final}} - n_{\text{initial}}$$

$$n_{\text{final}} = \underset{\substack{\uparrow \\ \text{Moles of } SO_3}}{2 \text{ mol}}$$

$$n_{\text{initial}} = \underset{\substack{\uparrow \\ \text{Moles} \\ \text{of } SO_2}}{2 \text{ mol}} + \underset{\substack{\uparrow \\ \text{Moles} \\ \text{of } O_2}}{1 \text{ mol}}$$

So

$$\Delta n = 2 \text{ mol} - 3 \text{ mol} = -1 \text{ mol}$$

Now we can calculate w:

$$w = -P\Delta V = -P\underbrace{\left(\Delta n \times \frac{RT}{P}\right)}_{\Delta V} = -\Delta nRT$$

where

$$\Delta n = -1 \text{ mol}$$

$$R = 8.3145 \text{ J K}^{-1} \text{ mol}^{-1}$$

$$T = 25°\text{C} + 273 = 298 \text{ K}$$

Thus

$$w = -(-1 \text{ mol})\left(8.3145 \frac{\text{J}}{\text{K mol}}\right)(298 \text{ K}) = 2.48 \text{ kJ}$$

Using the values of q and w, we can calculate ΔE:

$$\Delta E = q + w = \Delta H + w = -198 \text{ kJ} + 2.48 \text{ kJ} = -196 \text{ kJ}$$

Note that ΔE and ΔH are different for this case because the volume changes (and therefore work occurs).

Calorimetry experiments can also be performed at **constant volume.** For example, when a photographic flashbulb flashes, the bulb becomes very hot, since the reaction of the zirconium or magnesium wire with the oxygen inside the bulb is exothermic. The reaction occurs inside the flashbulb, which is rigid (does not change volume). Under these conditions no work is done, since the volume must change for pressure-volume work to be performed. To study the energy changes in reactions under conditions of constant volume, a **bomb calorimeter** (Fig. 9.9) is often used. Weighed reactants are placed inside a rigid steel container (the "bomb") and ignited. The energy change is determined by measuring the increase in the temperature of the water and other calorimeter parts. For a constant-volume process the change in volume (ΔV) is equal to zero, so the work is also equal to zero. Therefore,

$$\Delta E = q + w = q = q_v \qquad \text{(constant volume)}$$

Suppose we wish to measure the energy of combustion of octane (C_8H_{18}), a component of gasoline. A 0.5269-g sample of octane is placed in a bomb calorimeter known to have a heat capacity of 11.3 kJ/°C. This means that 11.3 kJ of energy is required to raise the temperature of the water and other parts of the calorimeter by 1°C. The octane is ignited in the presence of

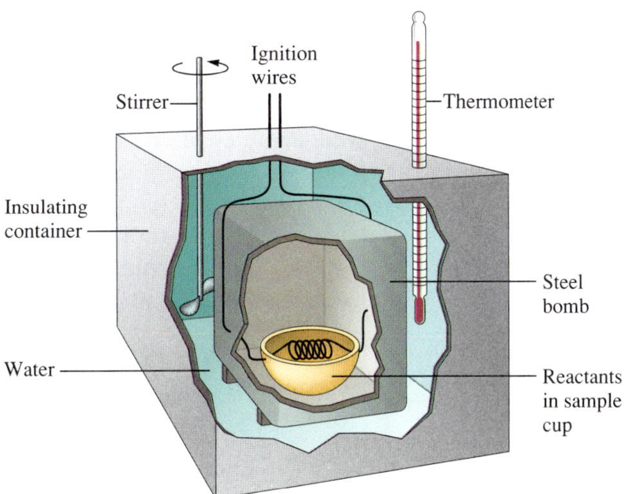

FIGURE 9.9

(left) A commercial "bomb." (right) Schematic of a bomb calorimeter. The reaction is carried out inside a rigid steel "bomb," and the heat evolved is absorbed by the surrounding water and the other calorimeter parts. The quantity of energy produced by the reaction can be calculated from the temperature increase.

excess oxygen, causing the temperature of the calorimeter to increase by 2.25°C. The amount of energy released by the reaction is calculated as follows:

$$\begin{aligned}\text{Energy released by the reaction} &= \text{temperature increase} \times \text{energy required to change the temperature by } 1°\text{C} \\ &= \Delta T \times \text{heat capacity of calorimeter} \\ &= 2.25°\text{C} \times 11.3 \text{ kJ/°C} = 25.4 \text{ kJ}\end{aligned}$$

This means that 25.4 kJ of energy was released by the combustion of 0.5269 g of octane. The number of moles of octane is

$$0.5269 \text{ g octane} \times \frac{1 \text{ mol octane}}{114.2 \text{ g octane}} = 4.614 \times 10^{-3} \text{ mol octane}$$

Since 25.4 kJ of energy was released for 4.614×10^{-3} mol of octane, the energy released per mole is

$$\frac{25.4 \text{ kJ}}{4.614 \times 10^{-3} \text{ mol}} = 5.50 \times 10^{3} \text{ kJ/mol}$$

Since the reaction is exothermic, ΔE is negative:

$$\Delta E_{\text{combustion}} = -5.50 \times 10^{3} \text{ kJ/mol}$$

Note that since no work is done in this case, ΔE is equal to the heat:

$$\begin{aligned}\Delta E &= q + w = q \\ &= -5.50 \times 10^{3} \text{ kJ/mol}\end{aligned}$$

Hydrogen's potential as a fuel is discussed in Section 9.8.

Example 9.5

It has been suggested that hydrogen gas obtained from the decomposition of water might be a substitute for natural gas (principally methane). To compare the energies of combustion of these fuels, the following experiment was carried out using a bomb calorimeter with a heat capacity of 11.3 kJ/°C. When a 1.50-g sample of methane gas was burned with excess oxygen in the calorimeter, the temperature increased by 7.3°C. When a 1.15-g sample of hydrogen gas was burned with excess oxygen, the temperature increase was 14.3°C. Calculate the energy of combustion (per gram) for hydrogen and methane.

Solution We calculate the energy of combustion for methane using the heat capacity of the calorimeter (11.3 kJ/°C) and the observed temperature increase of 7.3°C:

$$\text{Energy } \textit{released} \text{ in the combustion of 1.50 g of } CH_4 = (11.3\ \text{kJ/°C})(7.3\text{°C}) = 83\ \text{kJ}$$

$$\text{Energy released in the combustion of 1 g of } CH_4 = \frac{83\ \text{kJ}}{1.50\ \text{g}} = 55\ \text{kJ/g}$$

The direction of the energy flow is indicated by words in this example. Using signs, we have

$$\Delta E_{\text{combustion}} = -55\ \text{kJ/g}$$

for methane and

$$\Delta E_{\text{combustion}} = -141\ \text{kJ/g}$$

for hydrogen.

Similarly, for hydrogen,

$$\text{Energy } \textit{released} \text{ in the combustion of 1.15 g of } H_2 = (11.3\ \text{kJ/°C})(14.3\text{°C}) = 162\ \text{kJ}$$

$$\text{Energy released in the combustion of 1 g of } H_2 = \frac{162\ \text{kJ}}{1.15\ \text{g}} = 141\ \text{kJ/g}$$

The energy released by the combustion of 1 g of hydrogen is approximately 2.5 times that for 1 g of methane, indicating that hydrogen gas is a potentially useful fuel.

9.5 Hess's Law

ΔH is not dependent on the reaction pathway.

Since enthalpy is a state function, the change in enthalpy in going from some initial state to some final state is independent of the pathway. This means that *in going from a particular set of reactants to a particular set of products, the change in enthalpy is the same whether the reaction takes place in one step or in a series of steps.* This principle is known as **Hess's law** and can be illustrated by examining the oxidation of nitrogen to produce nitrogen dioxide. The overall reaction can be written in one step, where the enthalpy change is represented by ΔH_1:

$$N_2(g) + 2O_2(g) \longrightarrow 2NO_2(g) \qquad \Delta H_1 = 68\ \text{kJ}$$

This reaction can also be carried out in two distinct steps, with enthalpy changes designated by ΔH_2 and ΔH_3:

$$N_2(g) + O_2(g) \longrightarrow 2NO(g) \qquad \Delta H_2 = 180\ \text{kJ}$$

$$2NO(g) + O_2(g) \longrightarrow 2NO_2(g) \qquad \Delta H_3 = -112\ \text{kJ}$$

$$\text{Net reaction:} \quad N_2(g) + 2O_2(g) \longrightarrow 2NO_2(g) \qquad \Delta H_2 + \Delta H_3 = 68\ \text{kJ}$$

FIGURE 9.10

The principle of Hess's law. The same change in enthalpy occurs when nitrogen and oxygen react to form nitrogen dioxide, regardless of whether the reaction occurs in one (red) or two (blue) steps.

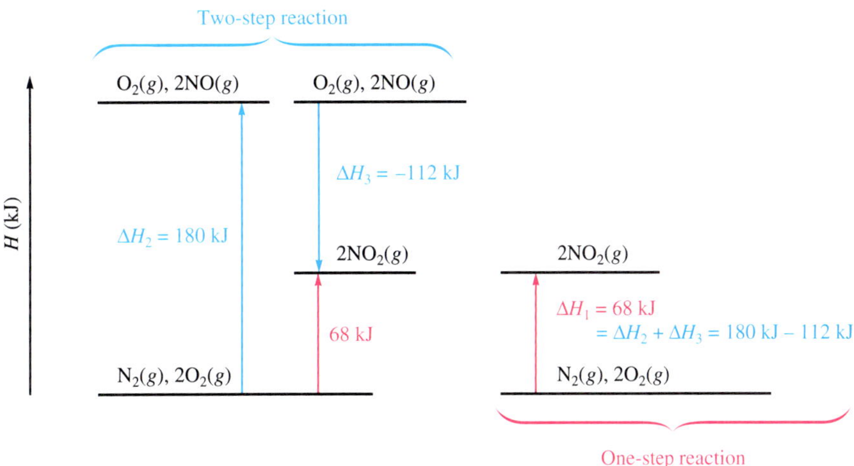

Note that the sum of these two steps gives the net, or overall, reaction and that

$$\Delta H_1 = \Delta H_2 + \Delta H_3 = 68 \text{ kJ}$$

The principle of Hess's law is shown schematically in Fig. 9.10.

Characteristics of Enthalpy Changes

To use Hess's law to compute enthalpy changes for reactions, we must understand two characteristics of ΔH for a reaction:

Reversing a reaction changes the sign of ΔH.

1. If a reaction is reversed, the sign of ΔH is also reversed.
2. The magnitude of ΔH is directly proportional to the quantities of reactants and products in a reaction. If the coefficients in a balanced reaction are multiplied by an integer, the value of ΔH is multiplied by the same integer.

Both of these rules follow in a straightforward way from the properties of enthalpy changes. The first rule can be explained by recalling that the *sign* of ΔH indicates the *direction* of the heat flow at constant pressure. If the direction of the reaction is reversed, the direction of the heat flow is also reversed. To see this, consider the preparation of xenon tetrafluoride, which was the first reported binary compound made from a noble gas:

$$Xe(g) + 2F_2(g) \longrightarrow XeF_4(s) \qquad \Delta H = -251 \text{ kJ}$$

This reaction is exothermic, and 251 kJ of energy flows into the surroundings as heat. On the other hand, if the colorless XeF_4 crystals are decomposed into the elements according to the equation

$$XeF_4(s) \longrightarrow Xe(g) + 2F_2(g)$$

the opposite energy flow occurs because 251 kJ of energy has to be added to the system in this case. Thus, for this reaction, $\Delta H = +251$ kJ.

The second rule comes from the fact that ΔH is an extensive property depending on the amount of substances reacting. For example, 251 kJ of energy is evolved for the reaction

$$Xe(g) + 2F_2(g) \longrightarrow XeF_4(s)$$

Crystals of xenon tetrafluoride, the first reported binary compound containing a noble gas element.

Thus, for a preparation involving twice the quantities of reactants and products,

$$2Xe(g) + 4F_2(g) \longrightarrow 2XeF_4(s)$$

twice as much heat would be evolved:

$$\Delta H = 2(-251\text{ kJ}) = -502\text{ kJ}$$

Hints for Using Hess's Law

Calculations involving Hess's law typically require that several reactions be manipulated and combined to finally give the reaction of interest. In doing this procedure, you should work *backward* from the required reaction, using the reactants and products to decide how to manipulate the other reactions at your disposal. Reverse any reactions as needed to give the required reactants and products, and then multiply the reactions to give the correct numbers of reactants and products. This process involves some trial and error but can be very systematic if you always allow the final reaction to guide you.

Example 9.6

Diborane (B_2H_6) is a highly reactive boron hydride that was once considered as a possible rocket fuel for the U.S. space program. Calculate ΔH for the synthesis of diborane from its elements, according to the equation

$$2B(s) + 3H_2(g) \longrightarrow B_2H_6(g)$$

using the following data:

Reaction	ΔH
(a) $2B(s) + \frac{3}{2}O_2(g) \rightarrow B_2O_3(s)$	−1273 kJ
(b) $B_2H_6(g) + 3O_2(g) \rightarrow B_2O_3(s) + 3H_2O(g)$	−2035 kJ
(c) $H_2(g) + \frac{1}{2}O_2(g) \rightarrow H_2O(l)$	−286 kJ
(d) $H_2O(l) \rightarrow H_2O(g)$	44 kJ

Solution To obtain ΔH for the required reaction, we must somehow combine equations (a), (b), (c), and (d) to produce that reaction and add the corresponding ΔH values. This procedure can best be done by focusing on the reactants and products of the required reaction. The reactants are $B(s)$ and $H_2(g)$, and the product is $B_2H_6(g)$. How can we obtain the correct equation? Reaction (a) has $B(s)$ as a reactant, as needed in the required equation. Thus reaction (a) will be used as it is. Reaction (b) has $B_2H_6(g)$ as a reactant, but this substance is needed as a product. Thus reaction (b) must be reversed, and the sign of ΔH changed accordingly. Up to this point we have

(a)	$2B(s) + \frac{3}{2}O_2(g) \longrightarrow B_2O_3(s)$	$\Delta H = -1273\text{ kJ}$
−(b)	$B_2O_3(s) + 3H_2O(g) \longrightarrow B_2H_6(g) + 3O_2(g)$	$\Delta H = -(-2035\text{ kJ})$
Sum:	$B_2O_3(s) + 2B(s) + \frac{3}{2}O_2(g) + 3H_2O(g) \longrightarrow B_2O_3(s) + B_2H_6(g) + 3O_2(g)$	$\Delta H = 762\text{ kJ}$

Deleting the species that occur on both sides gives

$$2B(s) + 3H_2O(g) \longrightarrow B_2H_6(g) + \frac{3}{2}O_2(g) \qquad \Delta H = 762\text{ kJ}$$

We are closer to the required reaction, but we still need to remove $H_2O(g)$ and $O_2(g)$ and introduce $H_2(g)$ as a reactant. We can do so by using reactions

Firewalking: Magic or Science?

For millennia people have been amazed at the ability of Eastern mystics to walk across beds of glowing coals without any apparent discomfort. Even in the United States thousands of people have performed feats of firewalking as part of motivational seminars. How is this feat possible? Do firewalkers have supernatural powers?

Actually, there are sound scientific explanations of why firewalking is possible. The first important factor concerns the heat capacity of feet. Because human tissue is mainly composed of water, it has a relatively large specific heat capacity. This means that a large amount of energy must be transferred from the coals to significantly change the temperature of the feet. During the brief contact between feet and coals, there is relatively little time for energy flow, so the feet do not reach a high enough temperature to cause damage.

A group of firewalkers in Japan.

Second, although the surface of the coals has a very high temperature, the red-hot layer is very thin. Therefore, the quantity of energy available to heat the feet is smaller than might be expected. This factor points out the difference between temperature and heat. Temperature reflects the *intensity* of the random kinetic energy in a given sample of matter. The amount of energy available for heat flow, on the other hand, depends on the quantity of matter at a given temperature—10 g of matter at a given temperature contains ten times as much thermal energy as 1 g of the same matter. For example, the tiny spark from a sparkler does not hurt when it hits your hand. The spark has a very high temperature but has so little mass that no significant energy transfer occurs to your hand. This same argument applies to the very thin hot layer on the coals.

A third factor that aids firewalkers is the presence of moisture from the perspiration on the feet of the presumably tense firewalker. In addition, because firewalking is often done at night, with moist grass surrounding the bed of coals, the firewalker's feet are probably damp before the walk. Vaporization of this moisture consumes some of the energy from the hot coals.

Thus, although firewalking is an impressive feat, there are several sound scientific reasons why anyone should be able to do it with the proper training and a properly prepared bed of coals.

(c) and (d). If we multiply reaction (c) and its ΔH value by 3 and add the result to the preceding equation, we have

$$2B(s) + 3H_2O(g) \longrightarrow B_2H_6(g) + \tfrac{3}{2}O_2(g) \qquad \Delta H = 762 \text{ kJ}$$

$$3 \times (c) \qquad 3[H_2(g) + \tfrac{1}{2}O_2(g) \longrightarrow H_2O(l)] \qquad \Delta H = 3(-286 \text{ kJ})$$

$$\text{Sum:} \quad 2B(s) + 3H_2(g) + \tfrac{3}{2}O_2(g) + 3H_2O(g) \longrightarrow B_2H_6(g) + \tfrac{3}{2}O_2(g) + 3H_2O(l) \qquad \Delta H = -96 \text{ kJ}$$

We can cancel the $\frac{3}{2}O_2(g)$ on both sides, but we cannot cancel the H_2O because it is gaseous on one side and liquid on the other. This problem can be solved by adding reaction (d), multiplied by 3:

$$2B(s) + 3H_2(g) + 3H_2O(g) \longrightarrow B_2H_6(g) + 3H_2O(l) \qquad \Delta H = -96 \text{ kJ}$$

$$3 \times (d) \qquad 3[H_2O(l) \longrightarrow H_2O(g)] \qquad \Delta H = 3(44 \text{ kJ})$$

$$2B(s) + 3H_2(g) + 3H_2O(g) + 3H_2O(l) \longrightarrow B_2H_6(g) + 3H_2O(l) + 3H_2O(g) \qquad \Delta H = +36 \text{ kJ}$$

This step gives the reaction required by the problem:

$$2B(s) + 3H_2(g) \longrightarrow B_2H_6(g) \qquad \Delta H = +36 \text{ kJ}$$

Thus ΔH for the synthesis of 1 mol of diborane is +36 kJ.

9.6 Standard Enthalpies of Formation

For a reaction studied under conditions of constant pressure, we can obtain the enthalpy change by using a calorimeter. However, this process can be very difficult. In fact, in some cases it is impossible, since certain reactions do not lend themselves to such study. An example is the conversion of solid carbon from its graphite form to its diamond form:

$$C_{\text{graphite}}(s) \longrightarrow C_{\text{diamond}}(s)$$

The value of ΔH for this process cannot be obtained readily by measurement in a calorimeter. We will show next how to *calculate* ΔH for chemical reactions and physical changes by using standard enthalpies of formation.

The **standard enthalpy of formation** (ΔH_f°) of a compound is defined as the *change in enthalpy that accompanies the formation of 1 mole of a compound from its elements with all substances in their standard states.*

The *superscript zero* on a thermodynamic function (for example, ΔH°) indicates that the corresponding process has been carried out under standard conditions. The **standard state** for a substance is a precisely defined reference state. Because thermodynamic functions often depend on the concentrations (or pressures) of the substances involved, we must use a common reference state to properly compare the thermodynamic properties of two substances. This is especially important because for most thermodynamic properties we can measure only *changes* in that property. For example, we have no method for determining absolute values of enthalpy. We can measure only enthalpy changes (ΔH values) by performing heat flow experiments.

Standard states are defined as follows:

Definitions of Standard States

1. For a gas the standard state is a pressure of exactly 1 atm.
2. For a substance present in a solution, the standard state is a concentration of exactly 1 *M* at an applied pressure of 1 atm.
3. For a pure substance in a condensed state (liquid or solid), the standard state is the pure liquid or solid.
4. For an element the standard state is the form in which the element exists (is most stable) under conditions of 1 atm and the temperature of interest (usually 25°C).

Standard state is *not* the same as standard temperature and pressure (STP) for a gas.

The standard state for oxygen is $O_2(g)$ at a pressure of 1 atm; the standard state for sodium is Na(*s*); the standard state for mercury is Hg(*l*); and so on.

The International Union of Pure and Applied Chemistry (IUPAC) has adopted 1 bar (100,000. Pa) as the standard pressure instead of 1 atm (101,325 Pa). Both standards are now widely used.

Several important characteristics of the definition of enthalpy of formation will become clearer if we again consider the formation of nitrogen dioxide from the elements in their standard states:

$$\tfrac{1}{2}N_2(g) + O_2(g) \longrightarrow NO_2(g) \qquad \Delta H_f^\circ = 34 \text{ kJ/mol}$$

Note that the reaction is written so that both elements are in their standard states, and 1 mole of product is formed. Enthalpies of formation are *always* given per mole of product with the product in its standard state.

The formation reaction for methanol is written as

$$C(s) + 2H_2(g) + \tfrac{1}{2}O_2(g) \longrightarrow CH_3OH(l) \qquad \Delta H_f^\circ = -239 \text{ kJ/mol}$$

The standard state of carbon is graphite, the standard states for oxygen and hydrogen are the diatomic gases, and the standard state for methanol is the liquid.

TABLE 9.4

Standard Enthalpies of Formation for Several Compounds at 25°C

Compound	ΔH_f° (kJ/mol)
$NH_3(g)$	−46
$NO_2(g)$	34
$H_2O(l)$	−286
$Al_2O_3(s)$	−1676
$Fe_2O_3(s)$	−826
$CO_2(g)$	−394
$CH_3OH(l)$	−239
$C_8H_{18}(l)$	−269

The H_f° values for some common substances are shown in Table 9.4. More values are found in Appendix 4. The importance of tabulated ΔH_f° values is that enthalpies for many reactions can be calculated using these numbers. To see how this is done, we will calculate the standard enthalpy change for the combustion of methane:

$$CH_4(g) + 2O_2(g) \longrightarrow CO_2(g) + 2H_2O(l)$$

Enthalpy is a state function, so we can invoke Hess's law and choose *any* convenient pathway from reactants to products and then sum the enthalpy changes. A convenient pathway, shown in Fig. 9.11, involves taking the reactants apart into the respective elements in their standard states in reactions (a) and (b) and then forming the products from these elements in reactions (c) and (d). This general pathway will work for any reaction, since atoms are conserved in a chemical reaction.

Note from Fig. 9.11 that reaction (a), where methane is taken apart into its elements,

$$CH_4(g) \longrightarrow C(s) + 2H_2(g)$$

is just the reverse of the formation reaction for methane:

$$C(s) + 2H_2(g) \longrightarrow CH_4(g) \qquad \Delta H_f^\circ = -75 \text{ kJ/mol}$$

Note that although the tabulated values of ΔH_f° usually correspond to a temperature of 25°C, values of ΔH_f° can be obtained at any temperature.

Since reversing a reaction means changing the sign of ΔH but keeping the same magnitude, ΔH for reaction (a) is $-\Delta H_f^\circ$, or 75 kJ. Thus $\Delta H^\circ_{(a)} = 75$ kJ.

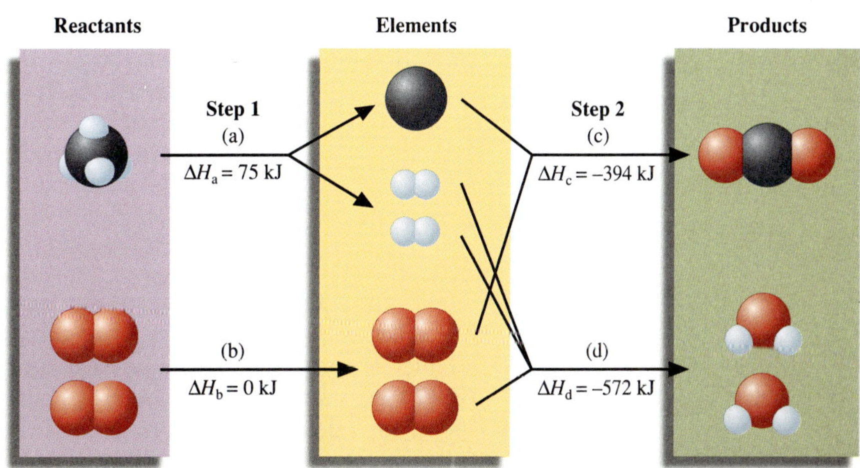

FIGURE 9.11

A schematic diagram of the energy changes for the reaction $CH_4(g) + 2O_2(g) \rightarrow CO_2(g) + 2H_2O(l)$.

Next, we consider reaction (b). Here oxygen is already an element in its standard state, so no change is necessary. Thus $\Delta H°_{(b)} = 0$.

The next steps, reactions (c) and (d), use the elements formed in reactions (a) and (b) to form the products. Note that reaction (c) is simply the formation reaction for carbon dioxide:

$$C(s) + O_2(g) \longrightarrow CO_2(g) \qquad \Delta H_f^\circ = -394 \text{ kJ/mol}$$

and

$$\Delta H°_{(c)} = \Delta H_f^\circ \text{ [for } CO_2(g)] = -394 \text{ kJ}$$

Reaction (d) is the formation reaction for water:

$$H_2(g) + \tfrac{1}{2}O_2(g) \longrightarrow H_2O(l) \qquad \Delta H_f^\circ = -286 \text{ kJ/mol}$$

However, since 2 moles of water are required in the balanced equation, we must form 2 moles of water from the elements:

$$2H_2(g) + O_2(g) \longrightarrow 2H_2O(l)$$

Thus

$$\Delta H°_{(d)} = 2 \times \Delta H_f^\circ \text{ [for } H_2O(l)] = 2(-286 \text{ kJ}) = -572 \text{ kJ}$$

We have now completed the pathway from the reactants to the products. The change in enthalpy for the overall reaction is the sum of the ΔH values (including their signs) for the steps:

$$\begin{aligned}\Delta H°_{\text{reaction}} &= \Delta H°_{(a)} + \Delta H°_{(b)} + \Delta H°_{(c)} + \Delta H°_{(d)} \\ &= -\Delta H_f^\circ \text{ [for } CH_4(g)] + 0 + \Delta H_f^\circ \text{ [for } CO_2(g)] \\ &\quad + 2 \times \Delta H_f^\circ \text{ [for } H_2O(l)] \\ &= -(-75 \text{ kJ}) + 0 + (-394 \text{ kJ}) + (-572 \text{ kJ}) \\ &= -891 \text{ kJ}\end{aligned}$$

Let's examine carefully the pathway we used in this example. First, the reactants were broken down into the elements in their standard states. This step involved reversing the formation reactions and thus switching the signs of the respective enthalpies of formation. The products were then constructed from these elements. This step involved formation reactions and thus enthalpies of formation. We can summarize the entire process as follows: *The enthalpy change for a given reaction can be calculated by subtracting the enthalpies of formation of the reactants from the enthalpies of formation of the products.* Remember to multiply the enthalpies of formation by integers as required by the balanced equation. This procedure can be represented symbolically as follows:

$$\Delta H°_{\text{reaction}} = \Sigma \Delta H_f^\circ \text{ (products)} - \Sigma \Delta H_f^\circ \text{ (reactants)} \qquad (9.1)$$

where the symbol Σ (sigma) means "to take the sum of the terms."

Elements are not included in the calculation since elements require no change in form. We have in effect *defined* the enthalpy of formation of an element in its standard state as zero, since we have chosen this as our reference point for calculating enthalpy changes in reactions.

Key Concepts for Doing Enthalpy Calculations

1. When a reaction is reversed, the magnitude of ΔH remains the same, but the sign changes.
2. When the balanced equation for a reaction is multiplied by an integer, the value of ΔH for that reaction must be multiplied by the same integer.
3. The change in enthalpy for a given reaction can be calculated from the enthalpies of formation of the reactants and products:

$$\Delta H^\circ_{\text{reaction}} = \Sigma \Delta H_f^\circ \text{ (products)} - \Sigma \Delta H_f^\circ \text{ (reactants)}$$

4. Elements in their standard states are not included in the $\Delta H_{\text{reaction}}$ calculations. That is, ΔH_f° for an element in its standard state is zero.

Example 9.7

Using the standard enthalpies of formation listed in Table 9.4, calculate the standard enthalpy change for the overall reaction that occurs when ammonia is burned in air to form nitrogen dioxide and water. This is the first step in the manufacture of nitric acid.

$$4NH_3(g) + 7O_2(g) \longrightarrow 4NO_2(g) + 6H_2O(l)$$

Solution We will use the pathway in which the reactants are broken down into elements in their standard states and then are used to form the products (see Fig. 9.12).

Step 1

Decomposition of $NH_3(g)$ into elements [reaction (a) in Fig. 9.12]. The first step involves decomposing 4 mol of NH_3 into N_2 and H_2:

$$4NH_3(g) \longrightarrow 2N_2(g) + 6H_2(g)$$

This reaction is 4 times the *reverse* of the formation reaction for NH_3:

$$\tfrac{1}{2}N_2(g) + \tfrac{3}{2}H_2(g) \longrightarrow NH_3(g) \qquad \Delta H_f^\circ = -46 \text{ kJ/mol}$$

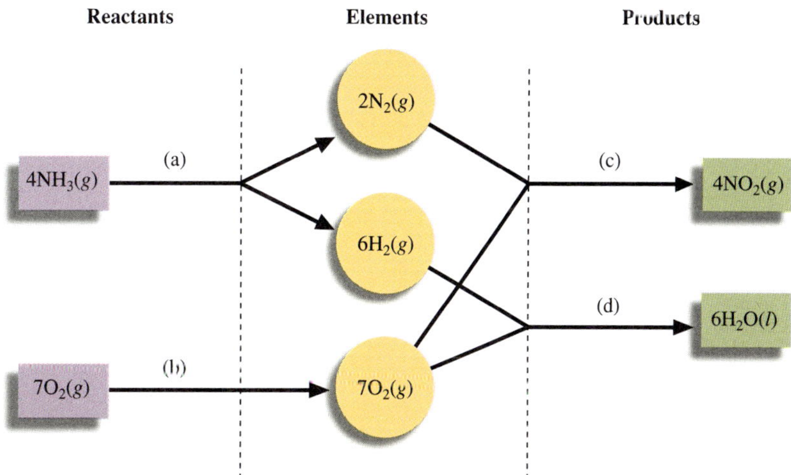

FIGURE 9.12
A pathway for the combustion of ammonia.

Thus

$$\Delta H^\circ_{(a)} = 4 \text{ mol}[-(-46 \text{ kJ/mol})] = 184 \text{ kJ}$$

Step 2

Elemental oxygen [reaction (b) in Fig. 9.12]. Since $O_2(g)$ is an element in its standard state, $\Delta H^\circ_{(b)} = 0$.

We now have the elements $N_2(g)$, $H_2(g)$, and $O_2(g)$, which can be combined to form the products of the overall reaction.

Step 3

Synthesis of $NO_2(g)$ from the elements [reaction (c) in Fig. 9.12]. The overall reaction equation has 4 mol of NO_2. Thus the required reaction is 4 times the formation reaction for NO_2:

$$4 \times [\tfrac{1}{2}N_2(g) + O_2(g) \longrightarrow NO_2(g)]$$

and

$$\Delta H^\circ_{(c)} = 4 \times \Delta H^\circ_f \text{ [for } NO_2(g)]$$

From Table 9.4, ΔH°_f [for $NO_2(g)$] = 34 kJ/mol, so

$$\Delta H^\circ_{(c)} = 4 \text{ mol} \times 34 \text{ kJ/mol} = 136 \text{ kJ}$$

Step 4

Synthesis of $H_2O(l)$ from the elements [reaction (d) in Fig. 9.12]. Since the overall reaction equation has 6 mol of $H_2O(l)$, the required reaction is 6 times the formation reaction of $H_2O(l)$:

$$6 \times [H_2(g) + \tfrac{1}{2}O_2(g) \longrightarrow H_2O(l)]$$

and

$$\Delta H^\circ_{(d)} = 6 \times \Delta H^\circ_f \text{ [for } H_2O(l)]$$

From Table 9.4, ΔH°_f [for $H_2O(l)$] = −286 kJ/mol, so

$$\Delta H^\circ_{(d)} = 6 \text{ mol}(-286 \text{ kJ/mol}) = -1716 \text{ kJ}$$

To summarize, we have done the following:

$$\begin{matrix} 4NH_3(g) \xrightarrow{\Delta H^\circ_{(a)}} \\ 7O_2(g) \xrightarrow{\Delta H^\circ_{(b)} = 0} \end{matrix} \left\{ \begin{matrix} 2N_2(g) + 6H_2(g) \\ 7O_2(g) \end{matrix} \right\} \begin{matrix} \xrightarrow{\Delta H^\circ_{(c)}} 4NO_2(g) \\ \xrightarrow{\Delta H^\circ_{(d)}} 6H_2O(l) \end{matrix}$$

Elements in their standard states

We now add the ΔH° values for the steps to obtain ΔH° for the overall reaction:

$$\begin{aligned} \Delta H^\circ_{\text{reaction}} &= \Delta H^\circ_{(a)} + \Delta H^\circ_{(b)} + \Delta H^\circ_{(c)} + \Delta H^\circ_{(d)} \\ &= 4 \times -\Delta H^\circ_f \text{ [for } NH_3(g)] + 0 + 4 \times \Delta H^\circ_f \text{ [for } NO_2(g)] \\ &\quad + 6 \times \Delta H^\circ_f \text{ [for } H_2O(l)] \\ &= 4 \times \Delta H^\circ_f \text{ [for } NO_2(g)] + 6 \times \Delta H^\circ_f \text{ [for } H_2O(l)] \\ &\quad - 4 \times \Delta H^\circ_f \text{ [for } NH_3(g)] \\ &= \Delta H^\circ_f \text{ (products)} - \Delta H^\circ_f \text{ (reactants)} \end{aligned}$$

Remember that elemental reactants and products do not need to be included, since ΔH_f° for an element in its standard state is zero. Note that we have again obtained Equation (9.1). The final solution is

$$\begin{aligned}\Delta H^\circ_{\text{reaction}} &= 6 \times (-286\ \text{kJ}) + 4 \times (34\ \text{kJ}) - 4 \times (-46\ \text{kJ}) \\ &= -1396\ \text{kJ}\end{aligned}$$

Now that we have shown the basis for Equation (9.1), we will make direct use of it to calculate ΔH for reactions in succeeding examples.

Example 9.8

Methanol (CH_3OH) is sometimes used as a fuel in high-performance engines. Using the data in Table 9.4, compare the standard enthalpy of combustion per gram of methanol with that of gasoline. Gasoline is actually a mixture of compounds, but assume for this problem that gasoline is pure liquid octane (C_8H_{18}).

Solution The combustion reaction for methanol is

$$2CH_3OH(l) + 3O_2(g) \longrightarrow 2CO_2(g) + 4H_2O(l)$$

Using the standard enthalpies of formation from Table 9.4 and Equation (9.1), we have

$$\begin{aligned}\Delta H^\circ_{\text{reaction}} &= 2 \times \Delta H_f^\circ\ [\text{for } CO_2(g)] + 4 \times \Delta H_f^\circ\ [\text{for } H_2O(l)] \\ &\quad - 2 \times \Delta H_f^\circ\ [\text{for } CH_3OH(l)] \\ &= 2 \times (-394\ \text{kJ}) + 4 \times (-286\ \text{kJ}) - 2 \times (-239\ \text{kJ}) \\ &= -1454\ \text{kJ}\end{aligned}$$

Thus 1454 kJ of heat is evolved when 2 mol of methanol burns. The molar mass of methanol is 32.0 g/mol. This means that 1454 kJ of energy is produced when 64.0 g of methanol burns. The enthalpy of combustion per gram of methanol is

$$\frac{-1454\ \text{kJ}}{64.0\ \text{g}} = -22.7\ \text{kJ/g}$$

The combustion reaction for octane is

$$2C_8H_{18}(l) + 25O_2(g) \longrightarrow 16CO_2(g) + 18H_2O(l)$$

Using the standard enthalpies of formation from Table 9.4 and Equation (9.1), we have

$$\begin{aligned}\Delta H^\circ_{\text{reaction}} &= 16 \times \Delta H_f^\circ\ [\text{for } CO_2(g)] + 18 \times \Delta H_f^\circ\ [\text{for } H_2O(l)] \\ &\quad - 2 \times \Delta H_f^\circ\ [\text{for } C_8H_{18}(l)] \\ &= 16 \times (-394\ \text{kJ}) + 18 \times (-286\ \text{kJ}) - 2 \times (-269\ \text{kJ}) \\ &= -1.09 \times 10^4\ \text{kJ}\end{aligned}$$

This value is the amount of heat evolved when 2 mol of octane burns. Since the molar mass of octane is 114.2 g/mol, the enthalpy of combustion per gram of octane is

$$\frac{-1.09 \times 10^4\ \text{kJ}}{2(114.2\ \text{g})} = -47.7\ \text{kJ/g}$$

The enthalpy of combustion per gram of octane is about twice that per gram of methanol. On this basis, gasoline appears to be superior to methanol for use in a racing car, where weight considerations are usually very important. Why, then, is methanol used in racing cars? The answer is that methanol burns much more smoothly than gasoline in high-performance engines, and this advantage more than compensates for its weight disadvantage.

9.7 Present Sources of Energy

Woody plants, coal, petroleum, and natural gas provide a vast resource of energy that originally came from the sun. By the process of photosynthesis, plants store energy that can be claimed by burning the plants themselves or the decay products that have been converted to **fossil fuels.** Although the United States currently depends heavily on petroleum for energy, this dependency is a relatively recent phenomenon, as shown in Fig. 9.13. In this section we discuss some sources of energy and their effects on the environment.

TABLE 9.5

Formulas and Names for Some Common Hydrocarbons

Formula	Name
CH_4	Methane
C_2H_6	Ethane
C_3H_8	Propane
C_4H_{10}	Butane
C_5H_{12}	Pentane
C_6H_{14}	Hexane
C_7H_{16}	Heptane
C_8H_{18}	Octane

Petroleum and Natural Gas

Although how they were produced is not completely understood, petroleum and natural gas were most likely formed from the remains of marine organisms that lived about 500 million years ago. **Petroleum** is a thick, dark liquid composed mostly of compounds called *hydrocarbons* that contain carbon and hydrogen. Table 9.5 lists the formulas and names of several common hydrocarbons. **Natural gas,** usually associated with petroleum deposits, consists mostly of methane but also contains significant amounts of ethane, propane, and butane.

The composition of petroleum varies somewhat, but it consists mostly of hydrocarbons having chains that contain from 5 to more than 25 carbons. To be used efficiently, the petroleum must be separated into fractions by boiling. The lighter molecules (having the lowest boiling points) can be boiled off, leaving the heavier ones behind. The uses of various petroleum fractions are shown in Table 9.6.

TABLE 9.6

Uses of the Various Petroleum Fractions

Petroleum Fraction in Terms of Numbers of Carbon Atoms	Major Uses
C_5–C_{10}	Gasoline
C_{10}–C_{18}	Kerosene
	Jet fuel
C_{15}–C_{25}	Diesel fuel
	Heating oil
	Lubricating oil
$>C_{25}$	Asphalt

The petroleum era began when the demand for lamp oil during the Industrial Revolution outstripped the traditional sources, animal fats and whale oil. In response to this increased demand, Edwin Drake drilled the first oil well in 1859 at Titusville, Pennsylvania. The petroleum from this well was refined to produce *kerosene* (fraction C_{10}–C_{18}), which served as an excellent lamp oil. *Gasoline* (fraction C_5–C_{10}) had limited use and was often discarded.

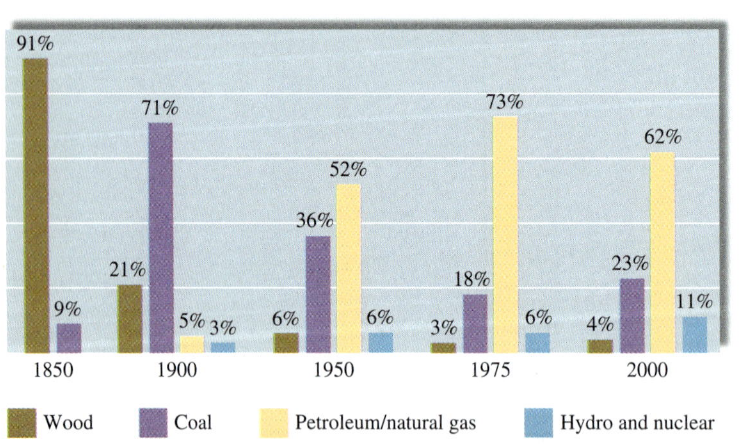

FIGURE 9.13

Energy sources used in the United States.

However, the development of the electric light decreased the need for kerosene, and the advent of the "horseless carriage" with its gasoline-powered engine signaled the birth of the gasoline age.

As gasoline became more important, new ways were sought to increase the yield of gasoline obtained from each barrel of petroleum. William Burton invented a process at Standard Oil of Indiana called *pyrolytic (high-temperature) cracking*. In this process the heavier molecules of the kerosene fraction are heated to about 700°C, causing them to break (crack) into the smaller molecules of hydrocarbons in the gasoline fraction. As cars became larger, more efficient internal combustion engines were designed. Because of the uneven burning of the gasoline then available, these engines "knocked," producing unwanted noise and even engine damage. Intensive research to find additives that would promote smoother burning produced tetraethyl lead [$(C_2H_5)_4Pb$], a very effective "antiknock" agent.

The addition of tetraethyl lead to gasoline became a common practice, and by 1960 gasoline contained as much as 3 g of lead per gallon. As we have discovered so often in recent years, technological advances can produce environmental problems. To prevent air pollution from automobile exhaust, manufacturers have added catalytic converters to car exhaust systems. The effectiveness of these converters, however, is destroyed by lead. The use of leaded gasoline has also greatly increased the amount of lead in the environment, where it can be ingested by animals and humans. For these reasons, the use of lead in gasoline has been phased out, which has required extensive (and expensive) modifications of engines and of the gasoline-refining process.

Coal

Coal has variable composition depending on both its age and where it was formed.

Coal was formed from the remains of plants that were buried and subjected to pressure and heat over long periods of time. Plant materials have a high content of cellulose, a complex molecule whose empirical formula is CH_2O but whose molar mass is around 500,000. After the plants and trees that flourished on the earth at various times and places died and were buried, chemical changes gradually lowered the oxygen and hydrogen content of the cellulose molecules. Coal "matures" through four stages: lignite, subbituminous, bituminous, and anthracite. Each stage has higher carbon-to-oxygen and carbon-to-hydrogen ratios; that is, the relative carbon content gradually increases. Typical elemental compositions of the various coals are given in Table 9.7. The energy available from the combustion of a given mass of coal increases as the carbon content increases. Anthracite is the most valuable coal, and lignite the least valuable.

Use of coal can pose pollution problems.

TABLE 9.7

Elemental Composition of Various Types of Coal

	Mass Percent of Each Element				
Type of Coal	C	H	O	N	S
Lignite	71	4	23	1	1
Subbituminous	77	5	16	1	1
Bituminous	80	6	8	1	5
Anthracite	92	3	3	1	1

Chemical Insights

Hiding Carbon Dioxide

Global warming now seems to be a reality. At the heart of this issue is the carbon dioxide produced by society's widespread use of fossil fuels. For example, in the United States CO_2 makes up 81% of greenhouse gas emissions. Thirty percent of this CO_2 comes from coal-fired power plants used to produce electricity. One way to solve this problem would be to phase out coal-fired power plants. However, this outcome is not likely because the United States possesses so much coal (at least a 250-year supply) and coal is so cheap (about a $0.01 per pound). Recognizing this fact, the U.S. government has instituted a research program to see if the CO_2 produced at power plants can be captured and sequestered (stored) underground in deep geological formations. The factors that need to be explored to determine whether sequestration is feasible are the capacities of underground storage sites and the chances that the sites will leak.

The injection of CO_2 into the earth's crust is already being undertaken by various oil companies. Since 1996, the Norwegian oil company Statoil has separated more than 1 million tons of CO_2 annually from natural gas and pumped it into a saltwater aquifer beneath the floor of the North Sea. In western Canada a group of oil companies has injected CO_2 from a North Dakota synthetic fuels plant into oil fields in an effort to increase oil recovery. The oil companies expect to store 22 million tons of CO_2 there and to produce 130 million barrels of oil over the next 20 years.

Sequestration of CO_2 has great potential as one method for decreasing the rate of global warming. Only time will tell whether it will work.

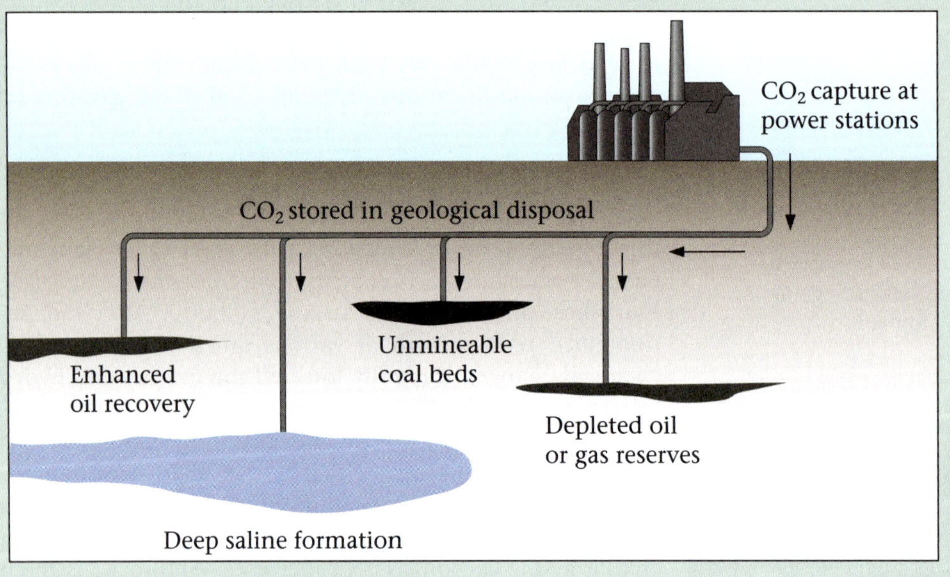

Coal is an important and plentiful fuel in the United States, currently furnishing about 20% of our energy. As the supply of petroleum dwindles, the share of the energy supply from coal is expected to increase to about 30%. However, coal is expensive and dangerous to mine underground, and the strip mining of fertile farmland in the Midwest or of scenic land in the West causes obvious problems. In addition, the burning of coal, especially high-sulfur coal, yields air pollutants such as sulfur dioxide, which in turn can lead to acid rain, as we learned in Chapter 5. However, even if coal were pure carbon, the carbon dioxide produced when it was burned would still have significant effects on the earth's climate.

The electromagnetic spectrum including visible and infrared radiation is discussed in Chapter 12.

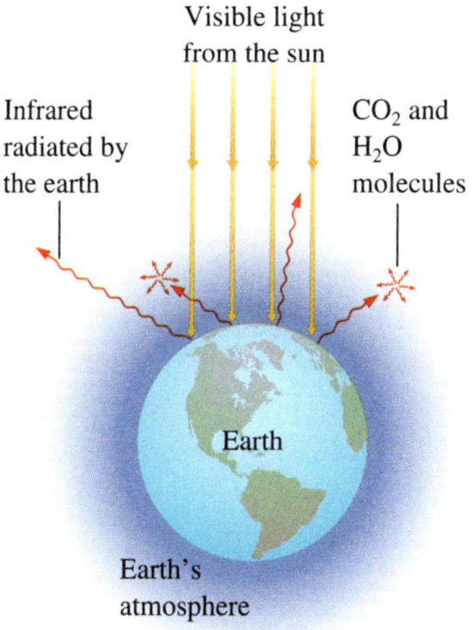

FIGURE 9.14

The earth's atmosphere is transparent to visible light from the sun. This visible light strikes the earth, and part of it is changed to infrared radiation. The infrared radiation from the earth's surface is strongly absorbed by CO_2, H_2O, and other molecules present in smaller amounts (for example, CH_4 and N_2O) in the atmosphere. In effect, the atmosphere traps some of the energy, acting like the glass in a greenhouse and keeping the earth warmer than it would otherwise be.

The average temperature of the earth's surface is 288 K. It would be ≈255 K without the "greenhouse gases."

Effects of Carbon Dioxide on Climate

The earth receives a tremendous quantity of radiant energy from the sun, about 30% of which is reflected into space by the earth's atmosphere. The remaining energy passes through the atmosphere to the earth's surface. Some of this energy is absorbed by plants to drive photosynthesis and some by the oceans to evaporate water, but most of it is absorbed by soil, rock, and water, resulting in an increase in the temperature of the earth's surface. This energy is in turn radiated from the heated surface mainly as *infrared radiation,* often called heat radiation.

The atmosphere, like window glass, is transparent to visible light but does not allow all the infrared radiation to pass through. Molecules in the atmosphere, principally H_2O and CO_2, strongly absorb infrared radiation, trapping it in the earth's atmosphere, as shown in Fig. 9.14. A net amount of thermal energy is thus retained by the earth's atmosphere, which causes the earth to be much warmer than it would be without its atmosphere. In a way the atmosphere acts like the glass of a greenhouse, which is transparent to visible light but absorbs infrared radiation, thus raising the temperature inside the building. This **greenhouse effect** is seen even more spectacularly on Venus, where the dense atmosphere is mainly responsible for the high surface temperature of that planet.

Thus the temperature of the earth's surface is controlled to a significant extent by the carbon dioxide and water content of the atmosphere. The effect of atmospheric moisture (humidity) is apparent in the Midwest. In summer when the humidity is high, the heat of the sun is retained well into the night, giving very high nighttime temperatures. On the other hand, in winter the coldest temperatures always occur on clear nights, when the low humidity allows efficient radiation of energy back into space.

The atmosphere's water content is controlled by the water cycle (evaporation and precipitation), and the average content remains constant over the years. However, as fossil fuels have come into more extensive use, the carbon dioxide concentration has increased significantly. This increase, which was 16% from 1880 to 1980, has continued in the past two decades. Some projections indicate that the carbon dioxide content of the atmosphere may be double in the twenty-first century what it was in 1880. As a result, the earth's average temperature could increase by as much as 3°C, causing dramatic changes in climate and greatly affecting the growth of food crops.

Most atmospheric scientists believe the warming trend has already started. For example, in 2005 the global average temperature of 14.6°C (58.3°F) was the warmest since records have been kept starting in the late 1800s. Global average temperatures have risen 0.6°C in the past 30 years and 0.8°C in the past century.

How well can we predict long-term effects? Because weather has been studied for a period of time that is miniscule compared with the age of the earth, the factors that control the earth's climate in the long range are not clearly understood. For example, we do not understand what causes the earth's periodic ice ages. So, indeed, it is difficult to estimate the impact of the increasing carbon dioxide levels.

In fact, the variation in the earth's average temperature over the past century is somewhat confusing. In the northern latitudes during the past century, the average temperature rose by 0.8°C over a period of 60 years, then cooled by 0.5°C during the next 25 years, and finally warmed by 0.2°C in the past

CHEMICAL INSIGHTS

Super Greenhouse Gas

A growing body of evidence indicates that human activities are causing the earth to warm up. This so-called greenhouse effect is due principally to the release of gases such as methane and carbon dioxide into the atmosphere that act to hamper the radiation of heat into space. Now scientists have found a gas in the atmosphere that is the new champion of greenhouse gases. This gas is trifluoromethyl sulfur pentafluoride (SF_5CF_3), which has the structure

```
      F
  F\  |  /F
     \C/
  F\  |  /F
     \S/
    / | \
  F   F   F
```

Notice that SF_5CF_3 can be viewed as a derivative of SF_6:

```
      F
  F\  |  /F
     \S/
    / | \
  F   F   F
```

SF_6 is also a strong absorber of infrared radiation.

On a mass basis, SF_5CF_3 absorbs 18,000 times as much infrared radiation as does carbon dioxide! The good news is that, at present, SF_5CF_3 exists in the atmosphere only at minute levels (approximately 0.1 part per trillion), although the levels are increasing about 6% per year.

Samples of air trapped in Antarctic ice indicate that SF_5CF_3 began to appear in the atmosphere in the 1960s. The source of the atmospheric SF_5CF_3 is currently unknown. The fact that the increases in SF_5CF_3 parallel the increases in SF_6 in the atmosphere, however, suggests a link between the two. The current hypothesis is that SF_6, which is used as an insulating gas in certain switches, transformers, and other high-voltage equipment, breaks down and reacts with the fluorinated carbon compounds also commonly found in these devices to produce SF_5CF_3.

The newly discovered SF_5CF_3 is merely one of a growing list of compounds, now found in trace amounts in the atmosphere, that strongly absorb infrared radiation and threaten to accelerate the greenhouse effect. Many of these gases are very stable. For example, the lifetime of SF_5CF_3 has been estimated to be several hundred years. Thus, although these gases are found at such low levels that they pose no current threat, they have the potential for long-range harm. The industrialized nations of the world need to be vigilant to stem the release of these gases as early as possible to prevent serious warming of the earth in the future.

20 years. Such fluctuations do not match the steady increase in carbon dioxide. However, in southern latitudes and in areas near the equator, the average temperature showed a steady increase totaling 0.4°C over the past century. This figure is in reasonable agreement with the predicted effect of the increasing carbon dioxide concentration over that period.

Another significant fact is that the last 10 years of the twentieth century appear to be the warmest decade on record. Although the exact relationship between the carbon dioxide concentration in the atmosphere and the earth's temperature is not known at present, one thing is clear: The increase in the atmospheric concentration of carbon dioxide is quite dramatic. We must consider the implications of this increase as we consider our future energy needs.

9.8 New Energy Sources

As we search for the energy sources of the future, we need to consider economic, climatic, and supply factors. There are several potential energy sources: the sun (solar), nuclear processes (fission and fusion), biomass (plants), and synthetic fuels. Direct use of the sun's radiant energy to heat our homes and run our factories and transportation systems seems a sensible long-term goal. But what do we do now? Conservation of fossil fuels is one obvious step, but substitutes for fossil fuels must be found eventually. We will discuss some alternative sources of energy here. Nuclear power will be considered in Chapter 20.

Coal Conversion

One alternative energy source involves using a traditional fuel—coal—in new ways. Since transportation costs for solid coal are high, more energy-efficient fuels are being developed from coal. One possibility is to produce a gaseous fuel. Substances like coal that contain large molecules have high boiling points and tend to be solids or thick liquids. To convert coal from a solid to a gas therefore requires reducing the size of the molecules; the coal structure must be broken down in a process called *coal gasification.* This process is carried out by treating the coal with oxygen and steam at high temperatures to break many of the carbon–carbon bonds. These bonds are replaced by carbon–hydrogen and carbon–oxygen bonds as the coal fragments react with the water and oxygen. This process is represented in Fig. 9.15. The desired fuel consists of a mixture of carbon monoxide and hydrogen called *synthetic gas,* or **syngas,** and methane (CH_4) gas. Since all the components of this product can react with oxygen to release heat in a combustion reaction, this gas is a useful fuel.

An industrial process must be energy-efficient.

One of the most important considerations in designing an industrial process is the efficient use of energy. In coal gasification some of the reactants are exothermic:

$$C(s) + 2H_2(g) \longrightarrow CH_4(g) \qquad \Delta H^\circ = -75\text{ kJ}$$

$$C(s) + \tfrac{1}{2}O_2(g) \longrightarrow CO(g) \qquad \Delta H^\circ = -111\text{ kJ}$$

$$C(s) + O_2(g) \longrightarrow CO_2(g) \qquad \Delta H^\circ = -394\text{ kJ}$$

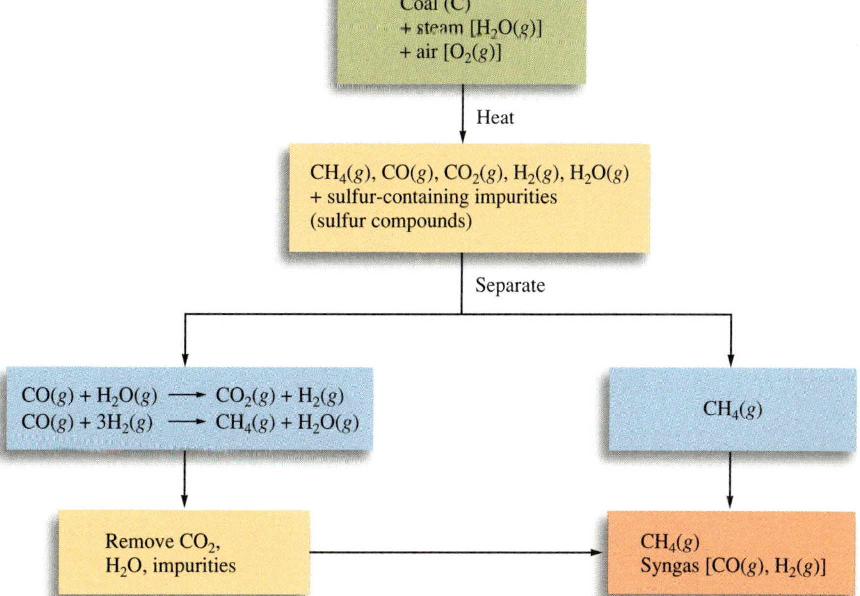

FIGURE 9.15

Coal gasification Reaction of coal with a mixture of steam and air breaks down the large hydrocarbon molecules in the coal to smaller gaseous molecules, which can be used as fuels.

Farming the Wind

In the Midwest the wind blows across fields of corn, soybeans, wheat, and wind turbines—wind turbines? It turns out that the wind that seems to blow almost continuously across the plains is now becoming the latest cash crop. One of these new-breed wind farmers is Daniel Juhl, who recently erected 17 wind turbines on six acres of land near Woodstock, Minnesota. These turbines can generate as much as 10 megawatts (MW) of electricity, which Juhl sells to the local electrical utility.

The largest wind farm in the world, located on the Oregon–Washington border, generates 300 MW. A controversial wind farm called Cape Wind, in the ocean five miles off the coast of Cape Cod, is planned to produce more than 400 MW of power.

There is plenty of untapped wind power in the United States. Wind mappers rate regions on a scale of 1 to 6 (with 6 being the best) to indicate the quality of the wind resource. Wind farms are now being developed in areas rated from 4 to 6. The farmers who own the land welcome the increased income derived from the wind blowing across their land. Economists estimate that each acre devoted to wind turbines can pay royalties to the farmers of as much as $8000 per year, or many times the revenue from growing corn on that same land. Daniel Juhl claims that farmers who construct the turbines themselves can realize as much as $20,000 per year per turbine. Globally, wind generation of electricity has nearly quadrupled in the last five years and is expected to increase by about 60% per year in the United States. The economic feasibility of wind-generated electricity has greatly improved in the last 30 years as wind turbines have become more efficient. Today's turbines can produce electricity that costs about the same as that from other sources. The most impressive thing about wind power is the magnitude of the supply. According to the American Wind Energy Association in Washington, D.C., the wind-power potential in the United States is comparable to or larger than the energy resources under the sands of Saudi Arabia.

Stateline Wind Generating Project in Walla Walla, Washington.

The biggest hurdle that must be overcome before wind power can become a significant electricity producer in the United States is construction of the transmission infrastructure—the power lines needed to move the electricity from the rural areas to the cities where most of the power is used. For example, the hundreds of turbines planned in southwest Minnesota in a development called Buffalo Ridge could supply enough electricity to power 1 million homes if transmission problems can be solved.

Another possible scenario for wind farms is to use the electrical power generated to decompose water to produce hydrogen gas that could be carried to cities by pipelines and used as a fuel. One real benefit of hydrogen is that it produces water as its only combustion product. Thus it is essentially pollution-free.

Within a few years wind power could be a major source of electricity. There could be a fresh wind blowing across the energy landscape of the United States in the near future.

Other gasification reactions are endothermic. For example,

$$C(s) + H_2O(g) \longrightarrow H_2(g) + CO(g) \qquad \Delta H^\circ = 131 \text{ kJ}$$

If the rate at which the coal, air, and steam are combined is carefully controlled, the correct temperature can be maintained in the process without using any external energy source. That is, an energy balance is attained.

As we stated earlier, syngas can be used directly as a fuel, but it is also important as a raw material for producing other fuels. For example, syngas can be directly converted to methanol:

$$CO(g) + 2H_2(g) \longrightarrow CH_3OH(l)$$

Methanol is used in the production of synthetic fibers and plastics and can also be used as a fuel. In addition, it can be converted directly to gasoline. About half of South Africa's gasoline supply comes from methanol produced from syngas.

In addition to coal gasification, the formation of *coal slurries* is another new use of coal. A slurry is a suspension of fine particles in a liquid. Coal must be pulverized and mixed with water to form a slurry. The resulting slurry can be handled, stored, and burned in ways similar to those used for *residual oil,* a heavy fuel oil from petroleum accounting for 13% of U.S. petroleum imports. One hope is that coal slurries might replace solid coal and residual oil as fuels for electricity-generating power plants. However, the water needed for slurries might place an unacceptable burden on water resources, especially in the western states.

Hydrogen as a Fuel

If you have ever seen a lecture demonstration where hydrogen–oxygen mixtures were ignited, you have witnessed a demonstration of hydrogen's potential as a fuel. The combustion reaction is

$$H_2(g) + \tfrac{1}{2}O_2(g) \longrightarrow H_2O(l) \qquad \Delta H^\circ = -286 \text{ kJ}$$

As we saw in Example 9.5, the heat of combustion of $H_2(g)$ per gram is about 2.5 times that of natural gas. In addition, hydrogen has a real advantage over fossil fuels in that the only product of hydrogen combustion is water; fossil fuels also produce carbon dioxide. But even though it appears that hydrogen is a very logical choice for a major future fuel, three main problems are associated with its use: the costs of production, storage, and transport.

First, let's look at the production problem. Although hydrogen is very abundant on earth, virtually none of it exists as the free gas. Currently, the main source of hydrogen gas is from the treatment of natural gas with steam:

$$CH_4(g) + H_2O(g) \longrightarrow 3H_2(g) + CO(g)$$

We can calculate ΔH for this reaction by using Equation (9.1):

$$\begin{aligned}\Delta H^\circ &= \Sigma\Delta H_f^\circ \text{ (products)} - \Sigma\Delta H_f^\circ \text{ (reactants)} \\ &= \Delta H_f^\circ \text{ [for } CO(g)] - \Delta H_f^\circ \text{ [for } CH_4(g)] - \Delta H_f^\circ \text{ [for } H_2O(g)] \\ &= -111 \text{ kJ} - (-75 \text{ kJ}) - (-242 \text{ kJ}) = 206 \text{ kJ}\end{aligned}$$

Note that this reaction is highly endothermic; treating methane with steam is not an efficient way to obtain hydrogen for fuel. It would be much more economical to burn the methane directly.

A virtually inexhaustible supply of hydrogen exists in the waters of the world's oceans. However, the reaction

$$H_2O(l) \longrightarrow H_2(g) + \tfrac{1}{2}O_2(g)$$

requires 286 kJ of energy per mole of liquid water, and under current circumstances large-scale production of hydrogen from water is not economically feasible. However, several methods for such production are currently being studied: electrolysis of water, thermal decomposition of water, and biological decomposition of water.

Electrolysis will be discussed in Chapter 11.

Electrolysis of water involves passing an electric current through it. The present cost of electricity makes the hydrogen produced by electrolysis too expensive to be competitive as a fuel. However, if in the future we develop more efficient sources of electricity, this situation could change.

Thermal decomposition is another method for producing hydrogen from water. This method involves heating the water to several thousand degrees, where it spontaneously decomposes into hydrogen and oxygen. However, attaining temperatures in this range would be very expensive even if a practical heat source and a suitable reaction container were available.

In the thermochemical decomposition of water, chemical reactions, as well as heat, are used to "split" water into its components. One such system involves the following reactions (the temperature required for each is given in parentheses):

$$\begin{array}{rll} 2HI \longrightarrow & I_2 + H_2 & (425°C) \\ 2H_2O + SO_2 + I_2 \longrightarrow & H_2SO_4 + 2HI & (90°C) \\ H_2SO_4 \longrightarrow & SO_2 + H_2O + \tfrac{1}{2}O_2 & (825°C) \\ \hline \text{Net reaction:} \quad H_2O \longrightarrow & H_2 + \tfrac{1}{2}O_2 & \end{array}$$

Note that the HI is not consumed in this reaction. Note also that the maximum temperature required is 825°C, a temperature that is feasible if a nuclear reactor is used as a heat source. A current research goal is to find a system for which the required temperatures are low enough that sunlight can be used as the energy source.

But what about the systems on earth that biologically decompose water without the aid of electricity or high temperatures? In the process of photosynthesis, green plants absorb carbon dioxide and water and use them along with energy from the sun to produce the substances needed for growth. Scientists have studied photosynthesis for years, hoping to get answers to humanity's food and energy shortages. At present much of this research involves attempts to modify the photosynthetic process so that plants will release hydrogen gas from water instead of using the hydrogen to produce complex compounds. Small-scale experiments have shown that under certain conditions plants do produce hydrogen gas, but the yields are far from being commercially useful. Thus the economical production of hydrogen gas remains unrealized.

The storage and transportation of hydrogen present two problems. First, on metal surfaces the H_2 molecule decomposes to atoms. Since the atoms are so small, they can migrate into the metal, causing structural changes that make it brittle. This might lead to a pipeline failure if hydrogen were pumped under high pressure.

A second problem is the relatively small amount of energy that is available *per unit volume* of hydrogen. Although the energy available per gram of hydrogen is significantly greater than that per gram of methane, the energy available per given volume of hydrogen is about one-third that available from the same volume of methane. Could hydrogen be considered as a potential fuel for automobiles? This is an intriguing question. The internal combustion engines in automobiles can be easily adapted to burn hydrogen. However, the primary difficulty is the storage of enough hydrogen to give an automobile a reasonable range. This is illustrated in Example 9.9.

Example 9.9

Assuming that the combustion of hydrogen gas provides three times as much energy per gram as gasoline, calculate the volume of liquid H_2 (density = 0.0710 g/mL) required to furnish the energy contained in 80.0 L (about 20 gal) of gasoline (density = 0.740 g/mL). Calculate also the volume that this hydrogen would occupy as a gas at 1.00 atm and 25°C.

Solution The mass of 80.0 L of gasoline is

$$80.0 \text{ L} \times \frac{1000 \text{ mL}}{1 \text{ L}} \times \frac{0.740 \text{ g}}{\text{mL}} = 59{,}200 \text{ g}$$

Since H_2 furnishes three times as much energy per gram as gasoline, only one-third as much liquid hydrogen is needed to furnish the same energy:

$$\text{Mass of } H_2(l) \text{ needed} = \frac{59{,}200 \text{ g}}{3} = 19{,}700 \text{ g}$$

Since density = mass/volume, volume = mass/density, and the volume of $H_2(l)$ needed is

$$V = \frac{19{,}700 \text{ g}}{0.0710 \text{ g/mL}} = 2.77 \times 10^5 \text{ mL} = 277 \text{ L}$$

Thus 277 L of liquid H_2 is needed to furnish the same energy of combustion as 80.0 L of gasoline.

To calculate the volume that this hydrogen would occupy as a gas at 1.00 atm and 25°C, we use the ideal gas law:

$$PV = nRT$$

In this case $P = 1.00$ atm, $T = 273 + 25°C = 298$ K, and $R = 0.08206$ L atm K^{-1} mol^{-1}. Also,

$$n = 19{,}700 \text{ g } H_2 \times \frac{1 \text{ mol } H_2}{2.02 \text{ g } H_2} = 9.75 \times 10^3 \text{ mol } H_2$$

Thus

$$V = \frac{nRT}{P} = \frac{(9.75 \times 10^3 \text{ mol})(0.08206 \text{ L atm K}^{-1} \text{ mol}^{-1})(298 \text{ K})}{1.00 \text{ atm}}$$

$$= 2.38 \times 10^5 \text{ L} = 238{,}000 \text{ L}$$

At 1 atm and 25°C, the hydrogen gas needed to replace 20 gal of gasoline occupies a volume of 238,000 L.

You can see from Example 9.9 that an automobile would need a huge tank to hold enough hydrogen gas to have a typical mileage range. Clearly, hydrogen must be stored in some other way, possibly as a liquid. Is this

CHEMICAL INSIGHTS

Heat Packs

A skier is trapped by a sudden snowstorm. After building a snow cave for protection, she realizes her hands and feet are freezing; she is in danger of frostbite. Then she remembers the four small packs in her pocket. She removes the plastic cover from each one to reveal a small paper packet. She places one packet in each boot and one in each mitten. Soon her hands and feet are toasty warm.

These "magic" packets of energy contain a mixture of powdered iron, activated carbon, sodium chloride, cellulose (sawdust), and zeolite, all moistened by a little water. The paper cover is permeable to air.

The exothermic reaction that produces the heat is a very common one—the rusting of iron. The overall reaction can be represented as

$$4Fe(s) + 3O_2(g) \longrightarrow 2Fe_2O_3(s)$$
$$\Delta H^\circ = -1652 \text{ kJ}$$

although in reality it is somewhat more complicated. When the plastic envelope is removed, O_2 molecules penetrate the paper, causing the reaction to begin.

The oxidation of iron by oxygen occurs naturally. Any steel surface exposed to the atmosphere inevitably rusts. But this oxidation process is quite slow—much too slow to be useful in hot packs. However, if the iron is ground into a fine powder, the resulting increase in surface area causes the reaction with oxygen to be fast enough to warm hands and feet. The packet can produce heat for up to six hours.

feasible? Because of hydrogen's very low boiling point (20 K), storage of liquid hydrogen requires a superinsulated container that can withstand high pressures. Storage in this manner would be both expensive and hazardous because of the potential for explosion. Thus storage of hydrogen in the individual automobile as a liquid does not seem practical.

Metal hydrides are discussed in Chapter 18.

A much better alternative seems to be the use of metals that absorb hydrogen to form solid metal hydrides:

$$H_2(g) + M(s) \longrightarrow MH_2(s)$$

In this method of storage, hydrogen gas would be pumped into a tank containing the solid metal, where it would be absorbed to form a hydride, whose volume would be little more than that of the metal. This hydrogen would then be available for combustion in the engine by release of $H_2(g)$ from the hydride as needed:

$$MH_2(s) \longrightarrow M(s) + H_2(g)$$

Several types of solids that absorb hydrogen to form hydrides are being studied for use in hydrogen-powered vehicles.

Sunflower oil can be used as motor fuel.

Other Energy Alternatives

Many other energy sources are being considered for future use. The western states, especially Colorado, contain huge deposits of *oil shale,* which consists of a complex carbon-based material called kerogen contained in porous rock formations. These deposits have the potential of being a larger energy source than the vast petroleum deposits of the Middle East. However, the main problem with oil shale is that the trapped fuel is not fluid and cannot be pumped. For recovery of the fuel, the rock must be heated to a temperature of 250°C or higher to decompose the kerogen to smaller molecules that produce gaseous and liquid products. This process is expensive and yields large quantities of waste rock, which have a negative environmental impact.

Ethanol (C_2H_5OH) is another fuel with the potential to supplement, if not replace, gasoline. The most common method of producing ethanol is fermentation, a process in which sugar is changed to alcohol by the action of yeast. The sugar can come from virtually any source, including fruits and grains, although fuel-grade ethanol would probably come mostly from corn. Car engines can burn pure alcohol or *gasohol*, an alcohol–gasoline mixture (10% ethanol in gasoline), with little modification. Gasohol is now widely available in the United States. The use of pure alcohol as a motor fuel is not feasible in most of the United States because it does not vaporize easily when temperatures are low. As a result, a fuel called E85, which is 85% ethanol and 15% gasoline, is now being used in so-called flex-fuel cars, which can run on either E85 or gasoline. Pure ethanol could be a very practical fuel in warm climates. For example, in Brazil large quantities of ethanol fuel are produced for cars.

Another potential source of liquid fuel is oil squeezed from seeds (*seed oil*). For example, some farmers in North Dakota, South Africa, and Australia are now using sunflower oil to replace diesel fuel. Oil seeds, found in a wide variety of plants, can be processed to produce a "biodiesel" oil composed mainly of carbon and hydrogen, which of course reacts with oxygen to produce carbon dioxide, water, and heat. The main advantage of seed oil as a fuel is that it is renewable. It is hoped that oil seed plants can be developed that will thrive under soil and climatic conditions unsuitable for corn and wheat. Ideally, fuel would be grown just like food crops. One major advantage of biodiesel is that its production is quite energy-efficient. When the energy required to produce biodiesel is compared to the energy gained from its use as a fuel, there is a 93% net energy gain. The net energy gain for ethanol is only about 25% because its production, which includes distillation, is relatively energy-intensive.

Discussion Questions

These questions are designed to be considered by groups of students in class. Often these questions work well for introducing a particular topic in class.

1. Objects placed together eventually reach the same temperature. When you go into a room and touch a piece of metal in that room, it feels colder than a piece of plastic. Explain.
2. What is meant by the term *lower in energy*? Which is lower in energy, a mixture of hydrogen and oxygen gases or liquid water? How do you know? Which of the two is more stable? How do you know?
3. A fire is started in a fireplace by striking a match and lighting crumpled paper under some logs. Explain the energy transfers in this scenario using the terms *exothermic, endothermic, system, surroundings, potential energy,* and *kinetic energy.*
4. Liquid water turns to ice. Is this process endothermic or exothermic? Explain what is occurring using the terms *system, surroundings, heat, potential energy,* and *kinetic energy.*
5. Consider the following statements: "Heat is a form of energy, and energy is conserved. The heat lost by the system must be equal to the amount of heat gained by the surroundings. Therefore, heat is conserved." Indicate everything you think is correct in these statements and everything you think is incorrect in these statements. Correct the incorrect statements and explain.
6. Photosynthetic plants use the following reaction to produce glucose, cellulose, and more:

   $$6CO_2(g) + 6H_2O(l) \longrightarrow C_6H_{12}O_6(s) + 6O_2(g)$$

   How might extensive destruction of forests exacerbate the greenhouse effect?
7. Explain why oceanfront areas generally have smaller temperature fluctuations than inland areas.
8. Predict the signs of q and w for the process of boiling water.
9. Hess's law is really just another statement of the first law of thermodynamics. Explain.
10. In the equation $w = -P\Delta V$, why is there a negative sign?
11. Why is C_p larger than C_v? Provide a conceptual rationale.

12. You have an ideal gas with an initial volume of 1.0 L and initial pressure of 1.0 atm. You decide to change the conditions such that P_{final} = 2.0 atm and V_{final} = 2.0 L. To make things more interesting, you and a friend each carry out this change in two steps. You first change the volume so that at one point P = 1.0 atm and V = 2.0 atm. Your friend first changes the pressure so that at one point P = 2.0 atm and V = 1.0 atm. Both of you, of course, end up with the same final conditions after the second step.
    a. How should your ΔE's, ΔH's, q's, and w's compare with those of your friend? Why?
    b. Calculate ΔE, ΔH, q, and w for each case. Do they make sense? Explain.
13. When is $\Delta H = \frac{5}{2}RT$? When is $\Delta E = \frac{5}{2}RT$? When is $\Delta H = \frac{3}{2}RT$? When is $\Delta E = \frac{3}{2}RT$? When is $\Delta H = \Delta E$? What does this say, if anything, about ΔE and ΔH as state functions?
14. For a liquid, which would you expect to be larger: $\Delta H_{vaporization}$ or ΔH_{fusion}? Explain.

Exercises

A blue exercise number indicates that the answer to that exercise appears at the back of this book and a solution appears in the *Solutions Guide.*

The Nature of Energy

15. Consider the accompanying diagram. Ball A is allowed to fall and strike ball B. Assume that all of ball A's energy is transferred to ball B at point I and that there is no loss of energy to other sources. Calculate the kinetic energy and the potential energy of ball B at point II. For a falling object, the potential energy is given by PE = mgz, where m is the mass in kilograms, g is the gravitational constant (9.81 m/s^2), and z is the distance in meters.

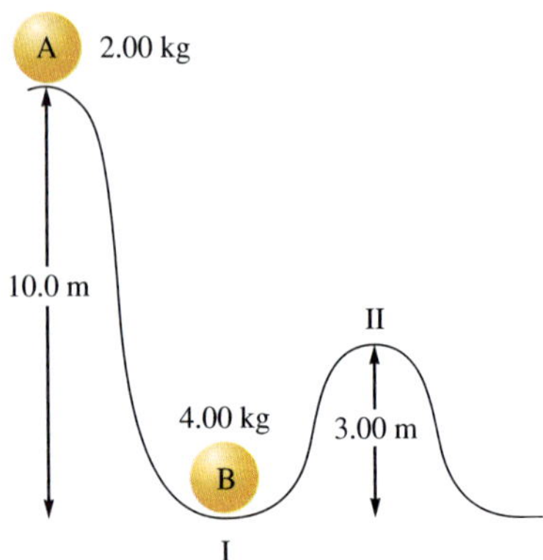

16. Consider the following potential energy diagrams for two different reactions.

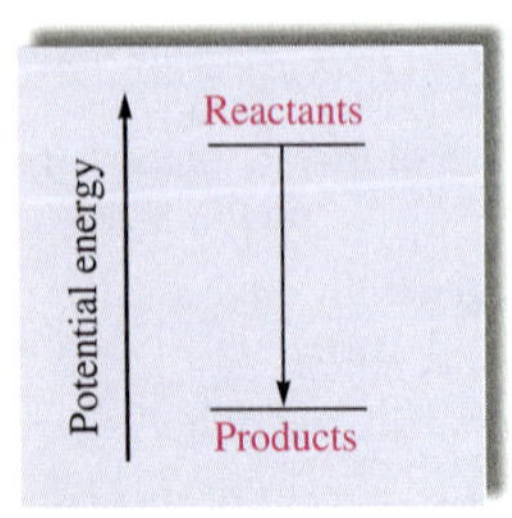

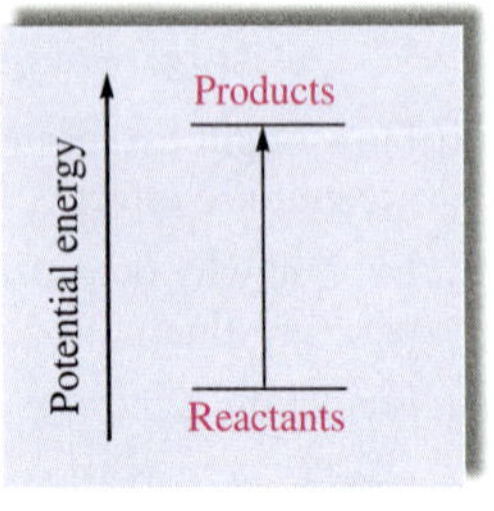

Which plot represents an exothermic reaction? In plot a, do the reactants on average have stronger or weaker bonds than the products? In plot b, reactants must gain potential energy to convert to products. How does this occur?

17. Consider an airplane trip from Chicago, Illinois, to Denver, Colorado. List some path-dependent functions and some state functions for the plane trip.
18. Consider 2.00 mol of an ideal gas that is taken from state A(P_A = 2.00 atm, V_A = 10.0 L) to state B (P_B = 1.00 atm, V_B = 30.0 L) by two different pathways.

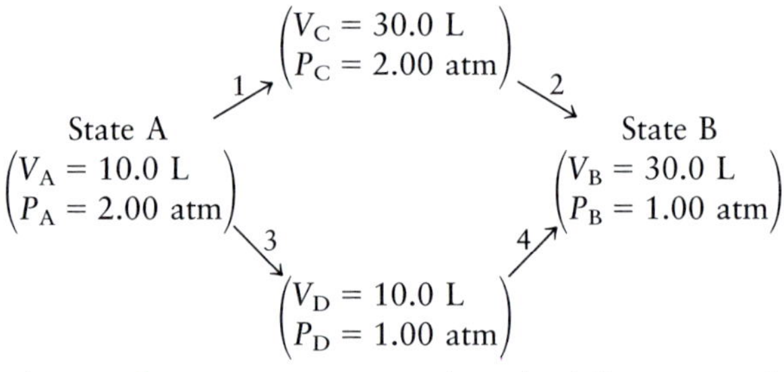

These pathways are summarized on the following graph of P versus V:

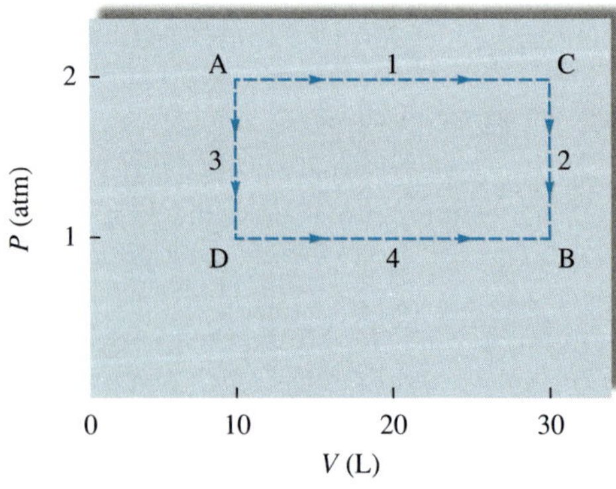

Calculate the work (in units of J) associated with the two pathways. Is work a state function? Explain.

19. A system undergoes a process consisting of the following two steps:

    Step 1: The system absorbs 72 J of heat while 35 J of work is done on it.

    Step 2: The system absorbs 35 J of heat while performing 72 J of work.

    Calculate ΔE for the overall process.

20. Calculate the internal energy change for each of the following.
    a. One hundred (100.) joules of work are required to compress a gas. At the same time, the gas releases 23 J of heat.
    b. A piston is compressed from a volume of 8.30 L to 2.80 L against a constant pressure of 1.90 atm. In the process, there is a heat gain by the system of 350. J.
    c. A piston expands against 1.00 atm of pressure from 11.2 L to 29.1 L. In the process, 1037 J of heat is absorbed.

21. A balloon filled with 39.1 mol of helium has a volume of 876 L at 0.0°C and 1.00 atm pressure. At constant pressure, the temperature of the balloon is increased to 38.0°C, causing the balloon to expand to a volume of 998 L. Calculate q, w, and ΔE for the helium in the balloon. (The molar heat capacity for helium gas is 20.8 J °C^{-1} mol^{-1}.)

22. Consider a mixture of air and gasoline vapor in a cylinder with a piston. The original volume is 40. cm^3. If the combustion of this mixture releases 950. J of energy, to what volume will the gases expand against a constant pressure of 650. torr if all the energy of combustion is converted into work to push back the piston?

23. One mole of $H_2O(g)$ at 1.00 atm and 100.°C occupies a volume of 30.6 L. When one mole of $H_2O(g)$ is condensed to one mole of $H_2O(l)$ at 1.00 atm and 100.°C, 40.66 kJ of heat is released. If the density of $H_2O(l)$ at this temperature and pressure is 0.996 g/cm^3, calculate ΔE for the condensation of one mole of water at 1.00 atm and 100.°C.

24. As a system increases in volume, it absorbs 52.5 J of energy in the form of heat from the surroundings. The piston is working against a pressure of 0.500 atm. The final volume of the system is 58.0 L. What was the initial volume of the system if the internal energy of the system decreased by 102.5 J?

Properties of Enthalpy

25. What is the difference between ΔH and ΔE at constant P?

26. Are the following processes exothermic or endothermic?
    a. the combustion of gasoline in a car engine
    b. water condensing on a cold pipe
    c. $CO_2(s) \longrightarrow CO_2(g)$
    d. $F_2(g) \longrightarrow 2F(g)$

27. The reaction

$$SO_3(g) + H_2O(l) \longrightarrow H_2SO_4(aq)$$

is the last step in the commercial production of sulfuric acid. The enthalpy change for this reaction is −227 kJ. In the design of a sulfuric acid plant, is it necessary to provide for heating or cooling of the reaction mixture? Explain your answer.

28. One of the components of polluted air is NO. It is formed in the high-temperature environment of internal combustion engines by the following reaction:

$$N_2(g) + O_2(g) \longrightarrow 2NO(g) \qquad \Delta H = 180 \text{ kJ}$$

Why are high temperatures needed to convert N_2 and O_2 to NO?

29. The overall reaction in a commercial heat pack can be represented as

$$4Fe(s) + 3O_2(g) \longrightarrow 2Fe_2O_3(s) \qquad \Delta H = -1652 \text{ kJ}$$

    a. How much heat is released when 4.00 mol iron is reacted with excess O_2?
    b. How much heat is released when 1.00 mol Fe_2O_3 is produced?
    c. How much heat is released when 1.00 g iron is reacted with excess O_2?
    d. How much heat is released when 10.0 g Fe and 2.00 g O_2 are reacted?

30. Consider the combustion of propane:

$$C_3H_8(g) + 5O_2(g) \longrightarrow 3CO_2(g) + 4H_2O(l) \quad \Delta H = -2221 \text{ kJ}$$

Assume that all the heat in Example 9.1 comes from the combustion of propane. What mass of propane must be burned to furnish this amount of energy, assuming the heat transfer process is 60.% efficient?

31. For the process $H_2O(l) \longrightarrow H_2O(g)$ at 298 K and 1.0 atm, ΔH is more positive than ΔE by 2.5 kJ/mol. What does the 2.5 kJ/mol quantity represent?

32. For the following reactions at constant pressure, predict if $\Delta H > \Delta E$, $\Delta H < \Delta E$, or $\Delta H = \Delta E$.
    a. $2HF(g) \longrightarrow H_2(g) + F_2(g)$
    b. $N_2(g) + 3H_2(g) \longrightarrow 2NH_3(g)$
    c. $4NH_3(g) + 5O_2(g) \longrightarrow 4NO(g) + 6H_2O(g)$

The Thermodynamics of Ideal Gases

33. Calculate the energy required to heat 1.00 kg of ethane gas (C_2H_6) from 25.0°C to 75.0°C first under conditions of constant volume and then at a constant pressure of 2.00 atm. Calculate ΔE, ΔH, and w for these processes also. (See Table 9.1 for relevant data.)

34. Calculate q, w, ΔE, and ΔH for the process in which 88.0 g of nitrous oxide gas (N_2O) is cooled from 165°C to 55°C at a constant pressure of 5.00 atm. (See Table 9.1.)

35. Consider a sample containing 5.00 mol of a monatomic ideal gas that is taken from state A to state B by the following two pathways:

Pathway one: $P_A = 3.00$ atm, $V_A = 15.0$ L $\xrightarrow{1}$ $P_C = 3.00$ atm, $V_C = 55.0$ L $\xrightarrow{2}$ $P_B = 6.00$ atm, $V_B = 20.0$ L

Pathway two: $P_A = 3.00$ atm, $V_A = 15.0$ L $\xrightarrow{3}$ $P_D = 6.00$ atm, $V_D = 15.0$ L $\xrightarrow{4}$ $P_B = 6.00$ atm, $V_B = 20.0$ L

For each step, assume that the external pressure is constant and equals the final pressure of the gas for that step. Calculate q, w, ΔE, and ΔH for each step, and calculate overall values for each pathway. Explain how the overall values for the two pathways illustrate that ΔE and ΔH are state functions, whereas q and w are path functions.

36. Consider a sample containing 2.00 mol of a monatomic ideal gas that undergoes the following changes:

$$\begin{array}{l} P_A = 10.0 \text{ atm} \\ V_A = 10.0 \text{ L} \end{array} \xrightarrow{1} \begin{array}{l} P_B = 10.0 \text{ atm} \\ V_B = 5.0 \text{ L} \end{array} \xrightarrow{2} \begin{array}{l} P_C = 20.0 \text{ atm} \\ V_C = 5.0 \text{ L} \end{array} \xrightarrow{3} \begin{array}{l} P_D = 20.0 \text{ atm} \\ V_D = 25.0 \text{ L} \end{array}$$

For each step, assume that the external pressure is constant and equals the final pressure of the gas for that step. Calculate q, w, ΔE, and ΔH for each step and for the overall change from state A to state D.

Calorimetry and Heat Capacity

37. Explain how calorimetry works to calculate ΔH or ΔE for a reaction. Does the temperature of the calorimeter increase or decrease for an endothermic reaction? How about for an exothermic reaction? Explain why ΔH is obtained directly from a coffee cup calorimeter, whereas ΔE is obtained directly from a bomb calorimeter.

38. The specific heat capacity of silver is 0.24 J °C^{-1} g^{-1}.
    a. Calculate the energy required to raise the temperature of 150.0 g Ag from 273 K to 298 K.
    b. Calculate the energy required to raise the temperature of 1.0 mol Ag by 1.0°C (called the *molar heat capacity* of silver).
    c. It takes 1.25 kJ of energy to heat a sample of pure silver from 12.0°C to 15.2°C. Calculate the mass of the sample of silver.

39. Consider the substances in Table 9.3. Which substance requires the largest amount of energy to raise the temperature of 25.0 g of the substance from 15.0°C to 37.0°C? Calculate the energy. Which substance in Table 9.3 has the largest temperature change when 550. g of the substance absorbs 10.7 kJ of energy? Calculate the temperature change.

40. A 150.0-g sample of a metal at 75.0°C is added to 150.0 g of H_2O at 15.0°C. The temperature of the water rises to 18.3°C. Calculate the specific heat capacity of the metal, assuming that all the heat lost by the metal is gained by the water.

41. A biology experiment requires the preparation of a water bath at 37.0°C (body temperature). The temperature of the cold tap water is 22.0°C, and the temperature of the hot tap water is 55.0°C. If a student starts with 90.0 g of cold water, what mass of hot water must be added to reach 37.0°C?

42. A 5.00-g sample of aluminum pellets (specific heat capacity = 0.89 J °C^{-1} g^{-1}) and a 10.00-g sample of iron pellets (specific heat capacity = 0.45 J °C^{-1} g^{-1}) are heated to 100.0°C. The mixture of hot iron and aluminum is then dropped into 97.3 g of water at 22.0°C. Calculate the final temperature of the metal and water mixture, assuming no heat loss to the surroundings.

43. In a coffee cup calorimeter 50.0 mL of 0.100 M $AgNO_3$ and 50.0 mL of 0.100 M HCl are mixed. The following reaction occurs:

$$Ag^+(aq) + Cl^-(aq) \longrightarrow AgCl(s)$$

If the two solutions are initially at 22.60°C, and if the final temperature is 23.40°C, calculate ΔH for the reaction in kJ/mol of AgCl formed. Assume a mass of 100.0 g for the combined solution and a specific heat capacity of 4.18 J °C^{-1} g^{-1}.

44. In a coffee cup calorimeter, 1.60 g of NH_4NO_3 is mixed with 75.0 g of water at an initial temperature of 25.00°C. After dissolution of the salt, the final temperature of the calorimeter contents is 23.34°C. Assuming the solution has a heat capacity of 4.18 J °C^{-1} g^{-1} and assuming no heat loss to the calorimeter, calculate the enthalpy change for the dissolution of NH_4NO_3 in units of kJ/mol.

45. Consider the dissolution of $CaCl_2$:

$$CaCl_2(s) \longrightarrow Ca^{2+}(aq) + 2Cl^-(aq) \qquad \Delta H = -81.5 \text{ kJ}$$

An 11.0-g sample of $CaCl_2$ is dissolved in 125 g of water, with both substances at 25.0°C. Calculate the final temperature of the solution assuming no heat loss to the surroundings and assuming the solution has a specific heat capacity of 4.18 J °C^{-1} g^{-1}.

46. Consider the reaction

$$2HCl(aq) + Ba(OH)_2(aq) \longrightarrow BaCl_2(aq) + 2H_2O(l) \qquad \Delta H = -118 \text{ kJ}$$

Calculate the heat when 100.0 mL of 0.500 M HCl is mixed with 300.0 mL of 0.100 M $Ba(OH)_2$. Assuming that the temperature of both solutions was initially 25.0°C and that the final mixture has a mass of 400.0 g and a specific heat capacity of 4.18 J °C^{-1} g^{-1}, calculate the final temperature of the mixture.

47. A 0.1964-g sample of quinone ($C_6H_4O_2$) is burned in a bomb calorimeter that has a heat capacity of 1.56 kJ/°C. The temperature of the calorimeter increases by 3.2°C. Calculate the energy of combustion of quinone per gram and per mole.

48. The heat capacity of a bomb calorimeter was determined by burning 6.79 g of methane (energy of combustion = −802 kJ/mol CH_4) in the bomb. The temperature changed by 10.8°C.
    a. What is the heat capacity of the bomb?
    b. A 12.6-g sample of acetylene (C_2H_2) produced a temperature increase of 16.9°C in the same calorimeter. What is the energy of combustion of acetylene (in kJ/mol)?

49. Combustion of table sugar produces $CO_2(g)$ and $H_2O(l)$. When 1.46 g of table sugar is combusted in a constant-volume (bomb) calorimeter, 24.00 kJ of heat is liberated.

a. Assuming that table sugar is pure sucrose [$C_{12}H_{22}O_{11}(s)$], write the balanced equation for the combustion reaction.
b. Calculate ΔE in kJ/mol $C_{12}H_{22}O_{11}$ for the combustion reaction of sucrose.
c. Calculate ΔH in kJ/mol $C_{12}H_{22}O_{11}$ for the combustion reaction of sucrose at 25°C.

50. Calculate w and ΔE when one mole of a liquid is vaporized at its boiling point (80.°C) and 1.00 atm pressure. ΔH_{vap} for the liquid is 30.7 kJ mol^{-1} at 80.°C.

Hess's Law

51. Given the following data:

$$C_2H_2(g) + \tfrac{5}{2}O_2(g) \longrightarrow 2CO_2(g) + H_2O(l) \qquad \Delta H = -1300.\ \text{kJ}$$
$$C(s) + O_2(g) \longrightarrow CO_2(g) \qquad \Delta H = -394\ \text{kJ}$$
$$H_2(g) + \tfrac{1}{2}O_2(g) \longrightarrow H_2O(l) \qquad \Delta H = -286\ \text{kJ}$$

calculate ΔH for the reaction

$$2C(s) + H_2(g) \longrightarrow C_2H_2(g)$$

52. Given the following data:

$$2ClF(g) + O_2(g) \longrightarrow Cl_2O(g) + F_2O(g) \qquad \Delta H = 167.4\ \text{kJ}$$
$$2ClF_3(g) + 2O_2(g) \longrightarrow Cl_2O(g) + 3F_2O(g) \qquad \Delta H = 341.4\ \text{kJ}$$
$$2F_2(g) + O_2(g) \longrightarrow 2F_2O(g) \qquad \Delta H = -43.4\ \text{kJ}$$

calculate ΔH for the reaction

$$ClF(g) + F_2(g) \longrightarrow ClF_3(g)$$

53. Given the following data:

$$Ca(s) + 2C(\textit{graphite}) \longrightarrow CaC_2(s) \qquad \Delta H = -62.8\ \text{kJ}$$
$$Ca(s) + \tfrac{1}{2}O_2(g) \longrightarrow CaO(s) \qquad \Delta H = -635.5\ \text{kJ}$$
$$CaO(s) + H_2O(l) \longrightarrow Ca(OH)_2(aq) \qquad \Delta H = -653.1\ \text{kJ}$$
$$C_2H_2(g) + \tfrac{5}{2}O_2(g) \longrightarrow 2CO_2(g) + H_2O(l) \qquad \Delta H = -1300.\ \text{kJ}$$
$$C(\textit{graphite}) + O_2(g) \longrightarrow CO_2(g) \qquad \Delta H = -393.5\ \text{kJ}$$

calculate ΔH for the reaction

$$CaC_2(s) + 2H_2O(l) \longrightarrow Ca(OH)_2(aq) + C_2H_2(g)$$

54. Given the following data:

$$Fe_2O_3(s) + 3CO(g) \longrightarrow 2Fe(s) + 3CO_2(g) \qquad \Delta H = -23\ \text{kJ}$$
$$3Fe_2O_3(s) + CO(g) \longrightarrow 2Fe_3O_4(s) + CO_2(g) \qquad \Delta H = -39\ \text{kJ}$$
$$Fe_3O_4(s) + CO(g) \longrightarrow 3FeO(s) + CO_2(g) \qquad \Delta H = 18\ \text{kJ}$$

calculate ΔH for the reaction

$$FeO(s) + CO(g) \longrightarrow Fe(s) + CO_2(g)$$

55. Combustion reactions involve reacting a substance with oxygen. When compounds containing carbon and hydrogen are combusted, carbon dioxide and water are the products. Using the enthalpies of combustion for C_4H_4 (−2341 kJ/mol), C_4H_8 (−2755 kJ/mol), and H_2 (−286 kJ/mol), calculate ΔH for the reaction

$$C_4H_4(g) + 2H_2(g) \longrightarrow C_4H_8(g)$$

56. Given the following data:

$$2O_3(g) \longrightarrow 3O_2(g) \qquad \Delta H = -427\ \text{kJ}$$
$$O_2(g) \longrightarrow 2O(g) \qquad \Delta H = 495\ \text{kJ}$$
$$NO(g) + O_3(g) \longrightarrow NO_2(g) + O_2(g) \qquad \Delta H = -199\ \text{kJ}$$

calculate ΔH for the reaction

$$NO(g) + O(g) \longrightarrow NO_2(g)$$

57. The bombardier beetle uses an explosive discharge as a defensive measure. The chemical reaction involved is the oxidation of hydroquinone by hydrogen peroxide to produce quinone and water:

$$C_6H_4(OH)_2(aq) + H_2O_2(aq) \longrightarrow C_6H_4O_2(aq) + 2H_2O(l)$$

Calculate ΔH for this reaction from the following data:

$$C_6H_4(OH)_2(aq) \longrightarrow C_6H_4O_2(aq) + H_2(g) \qquad \Delta H = 177.4\ \text{kJ}$$
$$H_2(g) + O_2(g) \longrightarrow H_2O_2(aq) \qquad \Delta H = -191.2\ \text{kJ}$$
$$H_2(g) + \tfrac{1}{2}O_2(g) \longrightarrow H_2O(g) \qquad \Delta H = -241.8\ \text{kJ}$$
$$H_2O(g) \longrightarrow H_2O(l) \qquad \Delta H = -43.8\ \text{kJ}$$

58. Given the following data:

$$P_4(s) + 6Cl_2(g) \longrightarrow 4PCl_3(g) \qquad \Delta H = -1225.6\ \text{kJ}$$
$$P_4(s) + 5O_2(g) \longrightarrow P_4O_{10}(s) \qquad \Delta H = -2967.3\ \text{kJ}$$
$$PCl_3(g) + Cl_2(g) \longrightarrow PCl_5(g) \qquad \Delta H = -84.2\ \text{kJ}$$
$$PCl_3(g) + \tfrac{1}{2}O_2(g) \longrightarrow Cl_3PO(g) \qquad \Delta H = -285.7\ \text{kJ}$$

calculate ΔH for the reaction

$$P_4O_{10}(s) + 6PCl_5(g) \longrightarrow 10Cl_3PO(g)$$

59. Given the following data:

$$NH_3(g) \longrightarrow \tfrac{1}{2}N_2(g) + \tfrac{3}{2}H_2(g) \qquad \Delta H = 46\ \text{kJ}$$
$$2H_2(g) + O_2(g) \longrightarrow 2H_2O(g) \qquad \Delta H = -484\ \text{kJ}$$

calculate ΔH for the reaction

$$2N_2(g) + 6H_2O(g) \longrightarrow 3O_2(g) + 4NH_3(g)$$

On the basis of the enthalpy change, is this a useful reaction for the synthesis of ammonia?

Standard Enthalpies of Formation

60. Given the definition of the standard enthalpy of formation for a substance, write separate reactions for the

formation of NaCl, H_2O, $C_6H_{12}O_6$, and $PbSO_4$ that have $\Delta H°$ values equal to $\Delta H_f°$ for each compound.

61. Use the values of $\Delta H_f°$ in Appendix 4 to calculate $\Delta H°$ for the following reactions.

a.

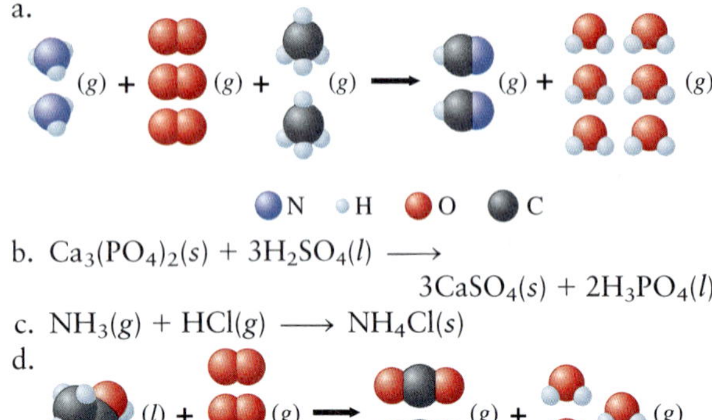

b. $Ca_3(PO_4)_2(s) + 3H_2SO_4(l) \longrightarrow 3CaSO_4(s) + 2H_3PO_4(l)$

c. $NH_3(g) + HCl(g) \longrightarrow NH_4Cl(s)$

d.

e. $SiCl_4(l) + 2H_2O(l) \longrightarrow SiO_2(s) + 4HCl(aq)$

f. $MgO(s) + H_2O(l) \longrightarrow Mg(OH)_2(s)$

62. The Ostwald process for the commercial production of nitric acid from ammonia and oxygen involves the following steps:

$$4NH_3(g) + 5O_2(g) \longrightarrow 4NO(g) + 6H_2O(g)$$

$$2NO(g) + O_2(g) \longrightarrow 2NO_2(g)$$

$$3NO_2(g) + H_2O(l) \longrightarrow 2HNO_3(aq) + NO(g)$$

a. Use the values of $\Delta H_f°$ in Appendix 4 to calculate the value of $\Delta H°$ for each of the preceding reactions.

b. Write the overall equation for the production of nitric acid by the Ostwald process by combining the preceding equations. (Water is also a product.) Is the overall reaction exothermic or endothermic?

63. Calculate $\Delta H°$ for each of the following reactions using the data in Appendix 4:

$$4Na(s) + O_2(g) \longrightarrow 2Na_2O(s)$$

$$2Na(s) + 2H_2O(l) \longrightarrow 2NaOH(aq) + H_2(g)$$

$$2Na(s) + CO_2(g) \longrightarrow Na_2O(s) + CO(g)$$

Explain why a water or carbon dioxide fire extinguisher might not be effective in putting out a sodium fire.

64. The reusable booster rockets of the space shuttle use a mixture of aluminum and ammonium perchlorate as fuel. A possible reaction is

$$3Al(s) + 3NH_4ClO_4(s) \longrightarrow Al_2O_3(s) + AlCl_3(s) + 3NO(g) + 6H_2O(g)$$

Calculate $\Delta H°$ for this reaction.

65. The space shuttle orbiter uses the oxidation of methyl hydrazine by dinitrogentetroxide for propulsion. The balanced reaction is

$$5N_2O_4(l) + 4N_2H_3CH_3(l) \longrightarrow 12H_2O(g) + 9N_2(g) + 4CO_2(g)$$

Calculate $\Delta H°$ for this reaction.

66. Does the reaction in Exercise 64 or that in Exercise 65 produce more energy per kilogram of reactant mixture (stoichiometric amounts)?

67. At 298 K, the standard enthalpies of formation for $C_2H_2(g)$ and $C_6H_6(l)$ are 227 kJ/mol and 49 kJ/mol, respectively.

a. Calculate $\Delta H°$ for

$$C_6H_6(l) \longrightarrow 3C_2H_2(g)$$

b. Both acetylene (C_2H_2) and benzene (C_6H_6) can be used as fuels. Which compound would liberate more energy per gram when combusted in air?

68. Calculate $\Delta H°$ for each of the following reactions, which occur in the atmosphere.

a. $C_2H_4(g) + O_3(g) \rightarrow CH_3CHO(g) + O_2(g)$

b. $O_3(g) + NO(g) \rightarrow NO_2(g) + O_2(g)$

c. $SO_3(g) + H_2O(l) \rightarrow H_2SO_4(aq)$

d. $2NO(g) + O_2(g) \rightarrow 2NO_2(g)$

69. Use the reaction

$$2ClF_3(g) + 2NH_3(g) \longrightarrow N_2(g) + 6HF(g) + Cl_2(g) \qquad \Delta H° = -1196 \text{ kJ}$$

to calculate $\Delta H_f°$ for $ClF_3(g)$.

70. The standard enthalpy of combustion of ethene gas $[C_2H_4(g)]$ is -1411.1 kJ/mol at 298 K. Given the following enthalpies of formation, calculate $\Delta H_f°$ for $C_2H_4(g)$.

$CO_2(g)$	-393.5 kJ/mol
$H_2O(l)$	-285.8 kJ/mol

Energy Consumption and Sources

71. The complete combustion of acetylene $[C_2H_2(g)]$ produces 1300. kJ of energy per mole of acetylene consumed. How many grams of acetylene must be burned to produce enough heat to raise the temperature of 1.00 gal of water by 10.0°C if the process is 80.0% efficient? Assume the density of water is 1.00 g/cm^3.

72. Ethanol (C_2H_5OH) has been proposed as an alternative fuel. Calculate the standard enthalpy of combustion per gram of liquid ethanol.

73. Methanol (CH_3OH) has also been proposed as an alternative fuel. Calculate the standard enthalpy of combustion per gram of liquid methanol, and compare this answer to that for ethanol in Exercise 72.

74. Some automobiles and buses have been equipped to burn propane (C_3H_8) as a fuel. Compare the amount of energy that can be obtained per gram of $C_3H_8(g)$ with that per gram of gasoline, assuming that gasoline is octane $[C_8H_{18}(l)]$. (See Example 9.8.) Look up the physical properties of propane. What disadvantages are there to using propane instead of gasoline as a fuel?

Additional Exercises

75. Consider the following cyclic process carried out in two steps on a gas:

    Step 1: 45 J of heat is added to the gas, and 10. J of expansion work is performed.

    Step 2: 60. J of heat is removed from the gas as the gas is compressed back to the initial state.

    Calculate the work for the gas compression in step 2.

76. Determine ΔE for the process $H_2O(l) \longrightarrow H_2O(g)$ at 25°C and 1 atm.

77. The standard enthalpy of formation of $H_2O(l)$ at 298 K is −285.8 kJ/mol. Calculate the change in internal energy for the following process at 298 K and 1 atm:

$$H_2O(l) \longrightarrow H_2(g) + \tfrac{1}{2}O_2(g) \qquad \Delta E° = ?$$

78. A piece of chocolate cake contains about 400 Calories. A nutritional Calorie is equal to 1000 calories (thermochemical calories). How many 8-in-high steps must a 180-lb man climb to expend the 400 Cal from the piece of cake? See Exercise 15 for the formula for potential energy.

79. In a bomb calorimeter the bomb is surrounded by water that must be added for each experiment. Since the amount of water is not constant from experiment to experiment, mass must be measured in each case. The heat capacity of the calorimeter is broken down into two parts: the water and the calorimeter components. If a calorimeter contains 1.00 kg of water and has a total heat capacity of 10.84 kJ/°C, what is the heat capacity of the calorimeter components?

80. The bomb calorimeter in Exercise 79 is filled with 987 g of water. The initial temperature of the calorimeter contents is 23.32°C. A 1.056-g sample of benzoic acid (ΔE_{comb} = −26.42 kJ/g) is combusted in the calorimeter. What is the final temperature of the calorimeter contents?

81. When 1.00 L of 2.00 M Na_2SO_4 solution at 30.0°C is added to 2.00 L of 0.750 M $Ba(NO_3)_2$ solution at 30.0°C in a calorimeter, a white solid ($BaSO_4$) forms. The temperature of the mixture increases to 42.0°C. Assuming that the specific heat capacity of the solution is 6.37 J °C^{-1} g^{-1} and that the density of the final solution is 2.00 g/mL, calculate the enthalpy change per mole of $BaSO_4$ formed.

82. If a student performs an endothermic reaction in a calorimeter, how does the calculated value of ΔH differ from the actual value if the heat exchanged with the calorimeter is not taken into account?

83. The enthalpy of neutralization for the reaction of a strong acid with a strong base is −56 kJ/mol of water produced. How much energy will be released when 200.0 mL of 0.400 M HCl is mixed with 150.0 mL of 0.500 M NaOH?

84. Three gas-phase reactions were run in a constant-pressure piston apparatus as illustrated. For each reaction, give the balanced reaction and predict the sign of w (the work done) for the reaction.

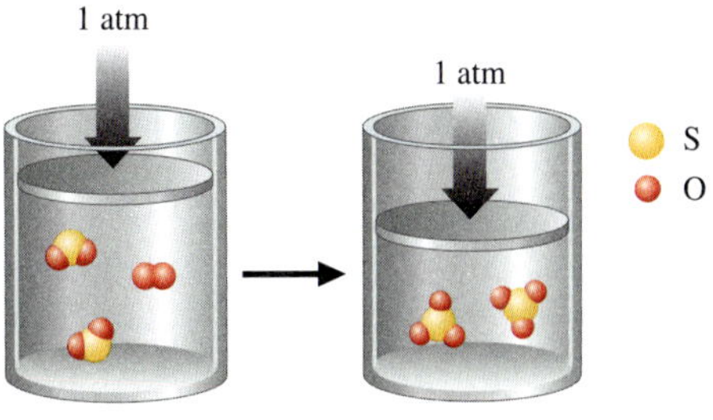

a.

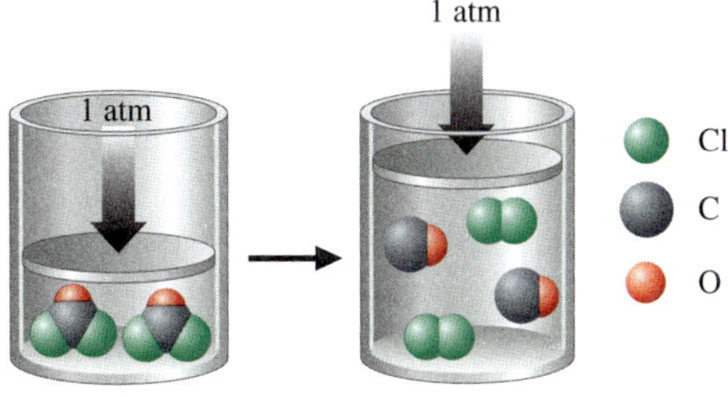

b.

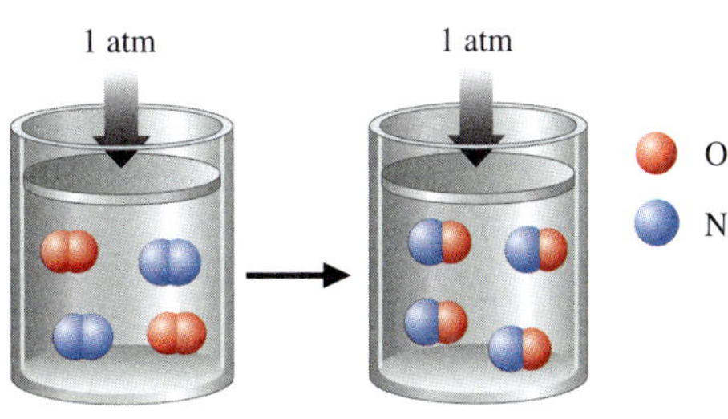

c.

If just the balanced reactions were given, how could you predict the sign of w for a reaction?

85. Consider the following changes:
    a. $N_2(g) \rightarrow N_2(l)$
    b. $CO(g) + H_2O(g) \rightarrow H_2(g) + CO_2(g)$
    c. $Ca_3P_2(s) + 6H_2O(l) \rightarrow 3Ca(OH)_2(s) + 2PH_3(g)$
    d. $2CH_3OH(l) + 3O_2(g) \rightarrow 2CO_2(g) + 4H_2O(l)$
    e. $I_2(s) \rightarrow I_2(g)$

    At constant temperature and pressure, in which of these changes is work done by the system on the surroundings? By the surroundings on the system? In which of them is no work done?

86. High-quality audio amplifiers generate large amounts of heat. To dissipate the heat and prevent damage to the electronic devices, manufacturers use heat-radiating metal fins. Would it be better to make these fins out of iron or aluminum? Why? (See Table 9.3 for specific heat capacities.)

87. Write reactions that correspond to the following enthalpy changes:
    a. ΔH_f° for solid aluminum oxide
    b. the standard enthalpy of combustion of liquid ethanol [$C_2H_5OH(l)$]

c. the standard enthalpy of neutralization of barium hydroxide solution by hydrochloric acid
d. ΔH_f° for gaseous vinyl chloride [$C_2H_3Cl(g)$]
e. the enthalpy of combustion of liquid benzene [$C_6H_6(l)$]
f. the enthalpy of solution of solid ammonium bromide

Challenge Problems

88. The heat required to raise the temperature from 300.0 K to 400.0 K for one mole of a gas at constant volume is 2079 J. The internal energy required to heat the same gas at constant pressure from 550.0 K to 600.0 K is 1305 J. The gas does 150. J of work during this expansion at constant pressure. Is this gas behaving ideally? Is the gas a monatomic gas? Explain.

89. When water is supercooled, it freezes at a temperature below 0.0°C. If 10.9 kJ of heat is released when 2.00 mol of supercooled water at −15.0°C freezes, calculate the molar enthalpy of fusion for ice at 0.0°C and 1 atm. Assume the molar heat capacities for $H_2O(s)$ and $H_2O(l)$ are 37.5 J K^{-1} mol^{-1} and 75.3 J K^{-1} mol^{-1}, respectively, and are temperature independent.

90. The sun supplies energy at a rate of about 1.0 kilowatt per square meter of surface area (1 watt = 1 J/s). The plants in an agricultural field produce the equivalent of 20. kg of sucrose ($C_{12}H_{22}O_{11}$) per hour per hectare (1 ha = 10,000 m^2). Assuming that sucrose is produced by the reaction

$$12CO_2(g) + 11H_2O(l) \longrightarrow C_{12}H_{22}O_{11}(s) + 12O_2(g) \qquad \Delta H = 5640 \text{ kJ}$$

calculate the percentage of sunlight used to produce the sucrose—that is, determine the efficiency of photosynthesis.

91. The heat of vaporization of water at the normal boiling point, 373.2 K, is 40.66 kJ/mol. The specific heat capacity of liquid water is 4.184 J K^{-1} g^{-1} and of gaseous water is 2.02 J K^{-1} g^{-1}. Assume that these values are independent of temperature. What is the heat of vaporization of water at 298.2 K? Does this result agree with Appendix 4 data?

92. Consider the following reaction at 248°C and 1.00 atm:

$$CH_3Cl(g) + H_2(g) \longrightarrow CH_4(g) + HCl(g)$$

For this reaction, the enthalpy change at 248°C is −83.3 kJ/mol. At constant pressure the molar heat capacities (C_p) for the compounds are as follows: CH_3Cl (48.5 J K^{-1} mol^{-1}), H_2 (28.9 J K^{-1} mol^{-1}), CH_4 (41.3 J K^{-1} mol^{-1}), and HCl (29.1 J K^{-1} mol^{-1}).

a. Assuming that the C_p values are independent of temperature, calculate ΔH° for this reaction at 25°C.
b. Calculate ΔH_f° for CH_3Cl using data from Appendix 4 and the result from part a.

93. The best solar panels currently available are about 13% efficient in converting sunlight to electricity. A typical home will use about 40. kWh of electricity per day (1 kWh = 1 kilowatt hour; 1 kW = 1000 J/s). Assuming 8.0 hours of useful sunlight per day, calculate the minimum solar panel surface area necessary to provide all of a typical home's electricity. (See Exercise 90 for the energy rate supplied by the sun.)

94. You have 2.4 mol of a gas contained in a 4.0-L bulb at a temperature of 32°C. This bulb is connected to a 20.0-L sealed, initially evacuated bulb via a valve. Assume the temperature remains constant.
a. What should happen to the gas when you open the valve? Calculate any changes of conditions.
b. Calculate ΔH, ΔE, q, and w for the process you described in part a.
c. Given your answer to part b, what is the driving force for the process?

95. An isothermal process is one in which the temperatures of the system and surroundings remain constant at all times. With this in mind, what is wrong with the following statement: "For an isothermal expansion of an ideal gas against a constant pressure, $\Delta T = 0$, so $q = 0$"? What is q equal to in an isothermal expansion of an ideal gas against a constant external pressure?

96. You have a 1.00-mol sample of water at −30.°C, and you heat it until you have gaseous water at 140.°C. Calculate q for the entire process. Use the following data:

Specific heat capacity of ice = 2.03 J °C^{-1} g^{-1}

Specific heat capacity of water = 4.18 J °C^{-1} g^{-1}

Specific heat capacity of steam = 2.02 J °C^{-1} g^{-1}

$H_2O(s) \longrightarrow H_2O(l) \qquad \Delta H_{fusion} = 6.01$ kJ/mol (at 0°C)

$H_2O(l) \longrightarrow H_2O(g) \qquad \Delta H_{vaporization} = 40.7$ kJ/mol (at 100.°C)

Marathon Problems

97.* A sample consisting of 22.7 g of a nongaseous, unstable compound X is placed inside a metal cylinder with a radius of 8.00 cm, and a piston is carefully placed on the surface of the compound so that, for all practical purposes, the distance between the bottom of the cylinder and the piston is zero. (A hole in the piston allows trapped air to escape as the piston is placed on the compound;

*Marathon Problems were developed by James H. Burness, Penn State University, York Campus. Used by permission.

then this hole is plugged so that nothing inside the cylinder can escape.) The piston-and-cylinder apparatus is carefully placed in 10.00 kg of water at 25.00°C. The barometric pressure is 778 torr.

When the compound spontaneously decomposes, the piston moves up, the temperature of the water reaches a maximum of 29.52°C, and then it gradually decreases as the water loses heat to the surrounding air. The distance between the piston and the bottom of the cylinder, at the maximum temperature, is 59.8 cm. Chemical analysis shows that the cylinder contains 0.300 mol carbon dioxide, 0.250 mol liquid water, 0.025 mol oxygen gas, and an undetermined amount of a gaseous element A.

It is known that the enthalpy change for the decomposition of X, according to the reaction described above, is −1893 kJ/mol X. The standard enthalpies of formation for gaseous carbon dioxide and liquid water are −393.5 kJ/mol and −286 kJ/mol, respectively. The heat capacity for water is 4.184 J $°C^{-1}$ g^{-1}. The conversion factor between L atm and J can be determined from the two values for the gas constant *R*—namely, 0.08206 L atm K^{-1} mol^{-1} and 8.3145 J K^{-1} mol^{-1}. The vapor pressure of water at 29.5°C is 31 torr. Assume that the heat capacity of the piston-and-cylinder apparatus is negligible and that the piston has negligible mass.

Given the preceding information, determine

a. the formula for X.
b. the pressure-volume work (in kJ) for the decomposition of the 22.7-g sample of X.
c. the *molar* change in internal energy for the decomposition of X and the approximate standard enthalpy of formation for X.

98. A gaseous hydrocarbon reacts completely with oxygen gas to form carbon dioxide and water vapor. Given the following data, determine ΔH_f° for the hydrocarbon:

$\Delta H_{rxn} = -2044.5$ kJ/mol

$\Delta H_f^\circ (CO_2) = -393.5$ kJ/mol

$\Delta H_f^\circ (H_2O) = -242$ kJ/mol

Density of CO_2 and H_2O mixture at 1 atm, 200.°C = 0.751 g/L

The density of the hydrocarbon is less than the density of Kr at the same conditions.

MEDIA SUMMARY

Visit the Student Website at **www.cengage.com/chemistry/zumdahl** to help prepare for class, study for quizzes and exams, understand core concepts, and visualize molecular-level interactions. The following media activities are available for this chapter:

Prepare for Class

Video Lessons *Mini-lectures from chemistry experts*

- The Nature of Energy
- Energy, Calories, and Nutrition
- The First Law of Thermodynamics
- Work
- CIA Demonstration: The Thermite Reaction
- Heats of Reaction: Enthalpy
- Heat
- CIA Demonstration: Cool Fire
- Constant-Pressure Calorimetry
- Bomb Calorimetry (Constant Volume)
- Hess's Law
- Enthalpies of Formation

Visualizations *Molecular-level animations and lab demonstration videos*

- Hess's Law
- Reduction of Iron: Thermite Reaction
- Work vs. Energy Flow

Tutorials *Animated examples and interactive activities*

- Calorimetry
- Hess's Law
- Work, Heat, and Energy Flow

Flashcards *Key terms and definitions*

Online flashcards

ACE the Test

Multiple-choice quizzes
3 ACE Practice Tests

Access these resources using your passkey, available free with new texts or for purchase separately.

10

Spontaneity, Entropy, and Free Energy

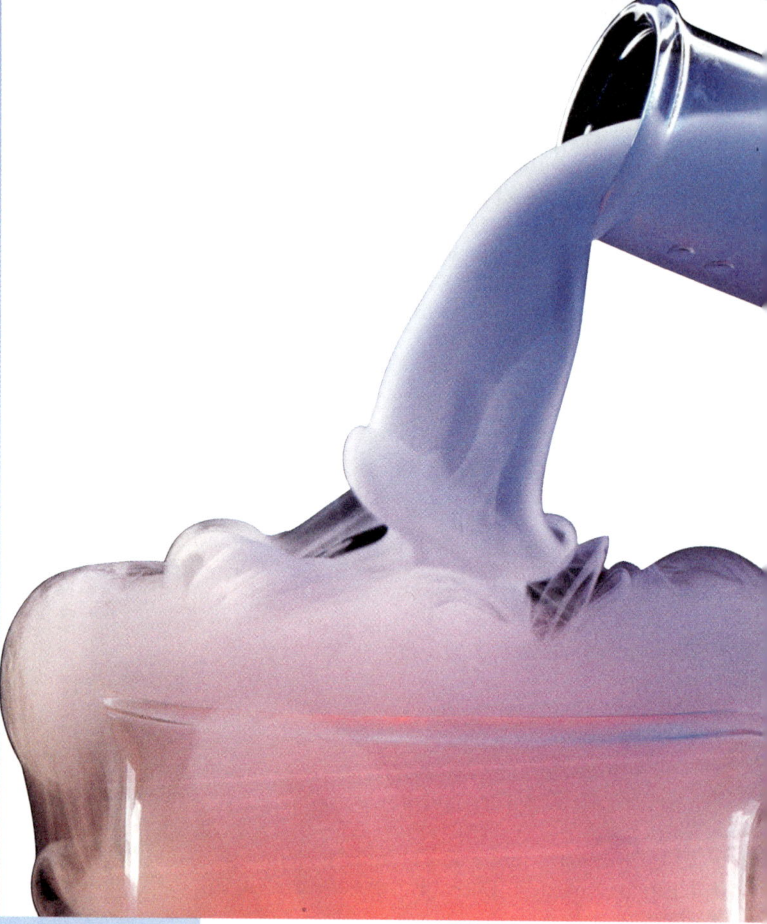

Clouds of frozen moisture are produced when dry ice is placed in water.

The *first law of thermodynamics* is a statement of the law of conservation of energy: Energy can be neither created nor destroyed. In other words, *the energy of the universe is constant.* Although the total energy remains constant, the various forms of energy can be interchanged through physical and chemical processes. For example, if you drop a book, some of the initial potential energy of the book is changed to kinetic energy, which is transferred to the atoms in the air and the floor as random motion. The net effect of this process is to change a given quantity of potential energy to exactly the same quantity of thermal energy. Energy has been converted from one form to another, but the same quantity of energy exists before and after the process.

Now we will consider a chemical example. When methane is burned in excess oxygen, the major reaction is

$$CH_4(g) + 2O_2(g) \longrightarrow CO_2(g) + 2H_2O(g) + \text{energy}$$

This reaction produces a quantity of energy that is released as heat. This energy flow results from a lowering of the potential energy stored in the bonds of CH_4 and O_2 as they react to form CO_2 and H_2O (see Fig. 10.1). Potential energy is converted to thermal energy, but the energy content of the universe remains constant.

The first law of thermodynamics is used mainly for energy bookkeeping—that is, to answer questions such as the following:

How much energy is involved in the change?
Does energy flow into or out of the system?
What form does the energy finally assume?

The first law of thermodynamics: The energy of the universe is constant.

Although the first law of thermodynamics provides the means to account for energy changes, it gives no hint as to *why* a particular process occurs in a given direction. What are the driving forces that cause a process to occur? This is the main question to be considered in this chapter.

Also you might wonder why there is an energy supply crisis in the world given that the first law of thermodynamics states that the energy of the universe is constant. It turns out that the problem is not the *quantity* of energy in the universe but the *quality* of that energy. The key question is, What happens to the usefulness of energy when we convert it from one form to another? We will see in this chapter that when we "use" energy, the amount of energy is unchanged but the new form of the energy is less useful than the original form.

10.1 Spontaneous Processes

Spontaneous does not mean fast.

A process is said to be *spontaneous* if it *occurs without outside intervention.* Spontaneous processes may be fast or slow. As we will see in this chapter, thermodynamics can tell us the *direction* in which a process will occur but can say nothing about the *speed* (rate) of the process. As we will explore in

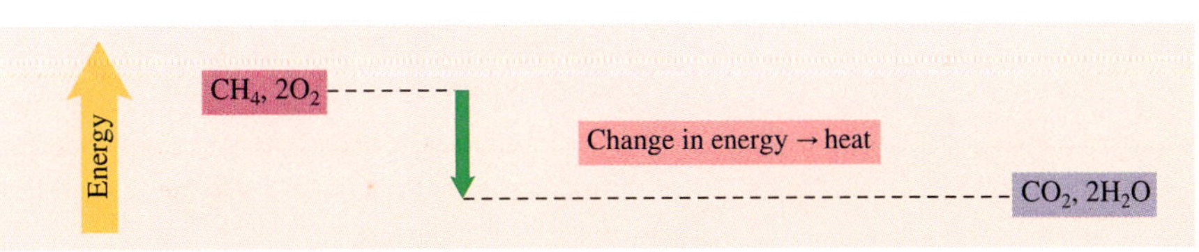

FIGURE 10.1

When methane and oxygen react to form carbon dioxide and water, the products have lower potential energy than the reactants. This change in potential energy results in energy flow (heat) to the surroundings.

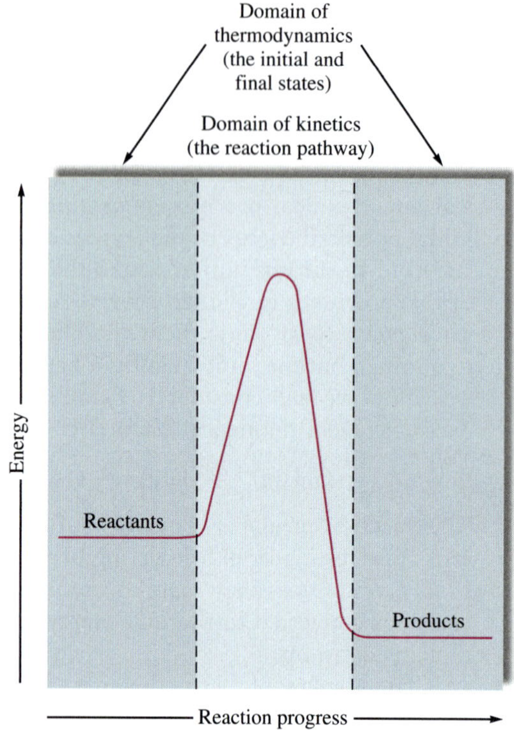

FIGURE 10.2

The rate of a reaction depends on the pathway from reactants to products; this is the domain of kinetics. Thermodynamics tells us whether a reaction is spontaneous based only on the properties of the reactants and products. The predictions of thermodynamics do not require knowledge of the pathway between reactants and products.

detail in Chapter 15, the rate of a reaction depends on many factors, including temperature and concentration. In describing a chemical reaction, the discipline of chemical kinetics (the study of reaction rates) focuses on the pathway between reactants and products; in contrast, thermodynamics considers only the initial and final states and does not require knowledge of the pathway between the reactants and products (see Fig. 10.2).

In summary, thermodynamics lets us predict whether a process will occur but gives no information about the amount of time required. For example, according to the principles of thermodynamics, a diamond should change spontaneously to graphite at 25°C and 1 atm pressure. The fact that we do not observe this process does not mean the prediction is wrong; it simply means the process is too slow to observe. Thus we need both thermodynamics and kinetics to describe reactions fully.

To explore the idea of spontaneity, consider the following physical and chemical processes:

A ball rolls down a hill but never spontaneously rolls back up the hill.

If exposed to air and moisture, steel rusts spontaneously. However, the iron oxide in rust does not spontaneously change back to iron metal and oxygen gas.

A gas fills its container uniformly. It never spontaneously collects at one end of the container.

Heat flow always occurs from a hot object to a cooler one. The reverse process never occurs spontaneously.

Wood burns spontaneously in an exothermic reaction to form carbon dioxide and water, but wood is not formed when carbon dioxide and water are heated together.

Chaos, Keep It Coming!

Can you imagine how life would be
If there were no entropy?
Or, making matters even worse,
The laws of entropy were reversed?
Books would get straighter on their shelves,
And children's rooms would clean themselves!
And every rock or stick or tree
Would form a crystal, perfectly.
There'd be no anarchy or war
For everyone would know the score.
Every thing and every face
Would have its certain time and place.
Replacing every beach would pass
An endless stretch of flawless glass.
The sea would be the brightest blue,
And every day the sky would too.
How beautiful would be our world
If order did command it.
If all were straight and never curled:
Perhaps we should demand it.

You'd think a world sans entropy
Would be a lovely place to be.
I said this recently myself,
As all my books fell off their shelf.
Yet pondering this ordered bliss,
I noticed things that I would miss,
Like rolling waves upon the sea,
Or sugar for my morning tea:
The sugar won't dissolve, it's true,
That anti-entropy holds like glue.
And after that, I saw with grief,
There'd be no fractaled maple leaf:
No beauty in the summer wood,
Should chaos disappear for good.
What a bore, to know each day
Would turn out in the same old way.
If entropy would disappear
There'd be no fortune, fate or luck
And even after many years,
Vegas wouldn't make a buck.

Heather Ryphemi Stregay

At temperatures below 0°C, water spontaneously freezes; and at temperatures above 0°C, ice spontaneously melts.

What thermodynamic principle will provide an explanation for why, under a given set of conditions, each of these diverse processes occurs in one direction and never in the reverse? In searching for an answer, we could explain the behavior of a ball on a hill in terms of gravity. But what does gravity have to do with the rusting of a nail or the freezing of water? Early developers of thermodynamics thought that exothermicity might be the key, that a process would be spontaneous if it were exothermic. Although this factor does appear to be important, since many spontaneous processes are exothermic, it is not the only factor. For example, the melting of ice, which occurs spontaneously at temperatures greater than 0°C, is an endothermic process.

What common characteristic causes the processes listed earlier to be spontaneous in one direction only? After many years of observation, scientists have concluded that the characteristic common to all spontaneous processes is an increase in a property called **entropy** (S). *The change in the entropy of the universe for a given process is a measure of the driving force behind that process.*

What is entropy? Ultimately, entropy is about how energy is distributed among the energy levels in the "particles" that constitute a given system. However, in beginning to understand how entropy operates, it is useful to think about how probability is a driving force in the macroscopic world around us. The macroscopic world illustrates that the natural progression of things is from order to disorder—that is, from lower probability to higher probability.

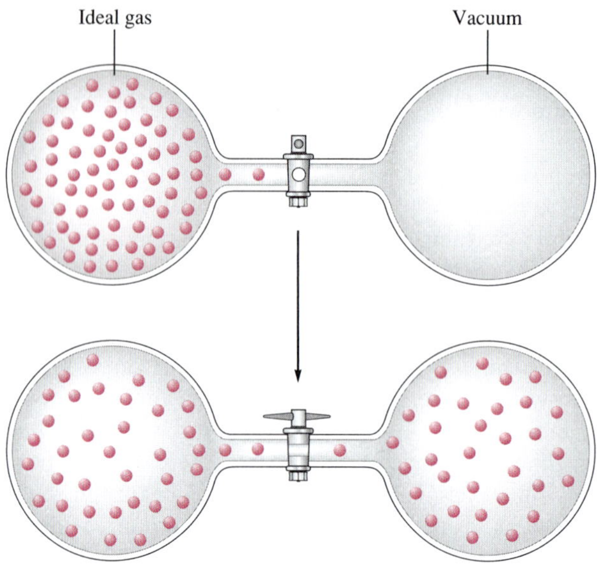

FIGURE 10.3

The expansion of an ideal gas into an evacuated bulb.

You merely have to think about the condition of your room to be convinced of this. Your room naturally tends to get messy (disordered) because an ordered room requires everything to be in its place. There are simply many more ways for things to be out of place than to be in place, as the poem by Heather Ryphemi Stregay, written while she was a student in general chemistry, observes.*

As another example, suppose you have a deck of playing cards ordered in some particular way. You throw these cards into the air and pick them all up at random. Looking at the new sequence of the cards, you would be very surprised to find that it matched the original order. Such an event would be possible but *very improbable*. There are billions of ways for the deck to be disordered but only one way to be ordered according to your definition. Thus the chances of picking the cards up out of order are much greater than of picking them up in order. It is natural for disorder to increase.

Entropy is closely associated with probability. The key concept is that the more ways a particular state can be achieved, the greater is the likelihood (probability) that that state will occur. In other words, *nature spontaneously proceeds toward the states that have the highest probabilities of existing*. This conclusion is not surprising at all. The difficulty comes in connecting this concept to real-life processes. For example, what does the spontaneous rusting of steel have to do with probability?

Understanding the connection between entropy and spontaneity will allow us to answer such questions. We will begin to explore this connection by considering a very simple process, the expansion of an ideal gas into a vacuum, as represented in Fig. 10.3. Why is this process spontaneous? What causes the gas to expand to a uniform state? The driving force can be explained in terms of simple probability. Because there are more ways of having the gas evenly spread throughout the container than there are ways for it to be in any other possible state, the gas spontaneously attains the uniform distribution.

*In her poem Ms. Stregay takes poetic license in her use of the term *entropy*.

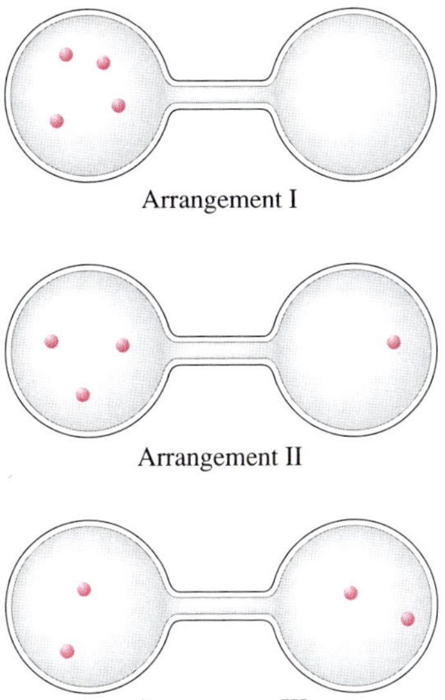

FIGURE 10.4

Three possible arrangements (states) of four molecules in a two-bulbed flask.

TABLE 10.1

The Microstates That Give a Particular Arrangement (State)

Arrangement	Microstates	
I		A B C D |
II	B D C | A	A B D | C
	A C D | B	A B C | D
III	A B | C D	B C | A D
	A C | B D	B D | A C
	A D | B C	C D | A B

To understand this conclusion, we will greatly simplify the system and consider some of the possible arrangements of only four gas molecules in a two-bulbed container* (Fig. 10.4). How many ways can each arrangement (state) be achieved? Arrangement I can be achieved in only one way—all the molecules must be in one end. Arrangement II can be achieved in four ways, as shown in Table 10.1. Each configuration that gives a particular arrangement is called a *microstate*. Arrangement I has one microstate, and arrangement II has four microstates. Arrangement III can be achieved in six ways (six microstates), as shown in Table 10.1. *Which arrangement is most likely to occur?* The one that can be achieved in the greatest number of ways is the most probable. Thus arrangement III is most probable, and the relative probabilities of arrangements III, II, and I are 6:4:1. We have discovered an important principle: *The probability of occurrence of a particular arrangement (state) depends on the number of ways (microstates) in which that arrangement can be achieved.*

The consequences of this principle are dramatic for large numbers of molecules. One gas molecule in the flask in Fig. 10.4 has one chance in two of being in the left bulb. We say that the probability of finding the molecule in the left bulb is $\frac{1}{2}$. For two molecules in the flask, there is one chance in two of finding each molecule in the left bulb, so there is one chance in four ($\frac{1}{2} \times \frac{1}{2} = \frac{1}{4}$) that *both* molecules will be in the left bulb. As the number of molecules increases, the relative probability of finding all of them in the left bulb decreases, as shown in Table 10.2. For 1 mole of gas, the probability of finding all the molecules in the left bulb is so small that this arrangement would "never" occur.

Thus a gas placed in one end of a container will spontaneously expand to fill the entire vessel evenly because for a large number of gas molecules

*Note that this treatment is oversimplified. In reality, the molecules of an ideal gas are indistinguishable—we can't really label them as in this example. The general idea, however, is correct.

TABLE 10.2

Probability of Finding All the Molecules in the Left Bulb as a Function of the Total Number of Molecules

Number of Molecules	Relative Probability of Finding All Molecules in the Left Bulb
1	$\frac{1}{2}$
2	$\frac{1}{2} \times \frac{1}{2} = \frac{1}{2^2} = \frac{1}{4}$
3	$\frac{1}{2} \times \frac{1}{2} \times \frac{1}{2} = \frac{1}{2^3} = \frac{1}{8}$
5	$\frac{1}{2} \times \frac{1}{2} \times \frac{1}{2} \times \frac{1}{2} \times \frac{1}{2} = \frac{1}{2^5} = \frac{1}{32}$
10	$\frac{1}{2^{10}} = \frac{1}{1024}$
n	$\frac{1}{2^n} = \left(\frac{1}{2}\right)^n$
6×10^{23} (1 mole)	$\left(\frac{1}{2}\right)^{6\times10^{23}} = 10^{-(2\times10^{23})}$

For two molecules in the flask, there are four possible microstates:

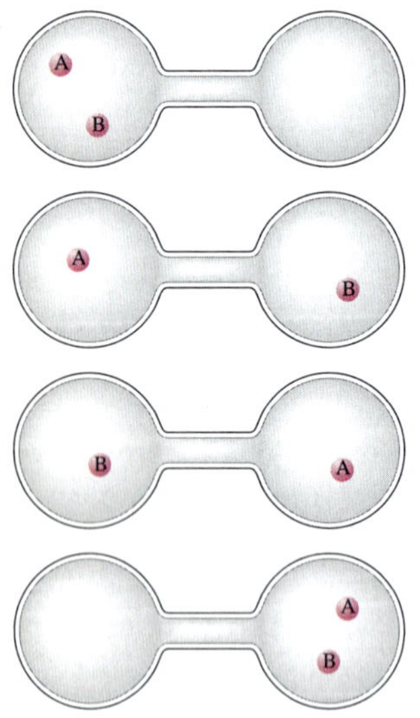

Thus there is one chance in four of finding

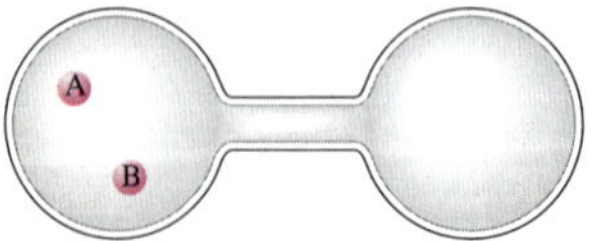

there is a huge number of microstates corresponding to equal numbers of molecules in both ends. On the other hand, the opposite process,

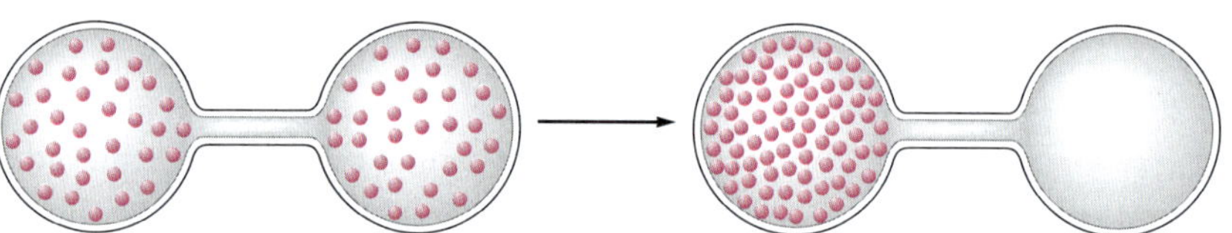

although not impossible, is *highly* improbable since only one microstate leads to this arrangement. Therefore, this process does not occur spontaneously.

The type of probability we have been considering in this example is called **positional probability** because it depends on the number of configurations in space (positional microstates) that yield a particular state. A gas expands into a vacuum to give a uniform distribution because the expanded state has the highest positional probability of the states available to the system.

Positional probability is also illustrated by changes of state. In general, positional probability increases in going from solid to liquid to gas. A mole of a substance has a much smaller volume in the solid state than in the gaseous state. In the solid state the molecules are close together, with relatively few positions available to them; in the gaseous state the molecules are far apart, with many more positions available to them.

Positional probability can also be invoked to explain the formation of solutions. The change in positional probability associated with the mixing of two pure substances is expected to be positive. There are many more microstates for the mixed condition than for the separated condition because of the increased volume available to the particles of each component of the mixture. For example, when two liquids are mixed, the molecules of each

Iodine being heated, causing it to sublime onto an evaporating dish cooled by ice.

liquid have more available space and thus more available positions. This will be discussed in detail in Chapter 17.

Example 10.1

For each of the following pairs, choose the substance with the higher positional probability (per mole) at a given temperature.

a. solid CO_2 and gaseous CO_2

b. N_2 gas at 1 atm and N_2 gas at 1.0×10^{-2} atm

Solution

a. Since a mole of gaseous CO_2 has the greater volume, the molecules have many more available positions than in a mole of solid CO_2. Thus gaseous CO_2 has the higher positional probability.

b. A mole of N_2 gas at 1×10^{-2} atm has a volume 100 times that (at a given temperature) of a mole of N_2 gas at 1 atm. Thus N_2 gas at 1×10^{-2} atm has the higher positional probability.

Example 10.2

Predict the sign of the change in positional probability for each of the following processes.

a. Solid sugar is added to water to form a solution.

b. Iodine vapor condenses on a cold surface to form crystals.

Solution

a. The sugar molecules become randomly dispersed in the water when the solution forms. The sugar molecules have access to a larger volume and therefore have more positions available to them. Thus the positional disorder increases.

b. Gaseous iodine is forming a solid. This process involves a change from a relatively large volume to a much smaller volume, which results in lower positional probability.

10.2 The Isothermal Expansion and Compression of an Ideal Gas

In this section we will lay the groundwork for several fundamental concepts of thermodynamics by considering the isothermal expansion and compression of an ideal gas. An **isothermal process** is one in which the temperatures of the system and the surroundings remain constant at all times. Recall that the energy of an ideal gas can be changed only by changing its temperature. Therefore, for any isothermal process *involving an ideal gas,*

$$\Delta E = 0$$

and since

$$\Delta E = q + w = 0$$

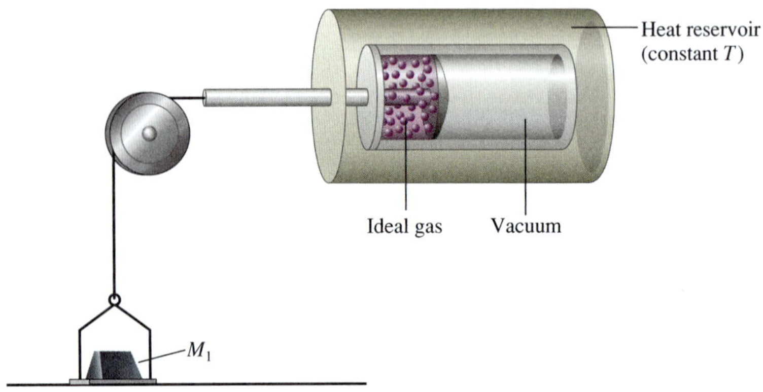

FIGURE 10.5

A device for the isothermal expansion/compression of an ideal gas.

then

$$q = -w$$

To illustrate the work and heat effects that accompany the expansion or compression of an ideal gas, consider the apparatus shown in Fig. 10.5. Assume that the pulley is frictionless and that the cable and pan have zero mass.

Initially, assume that the gas occupies a volume V_1 at pressure P_1, where P_1 is just balanced by a mass M_1 on the pan. Thus

$$P_1 = \frac{\text{force}}{\text{area}} = \frac{M_1 g}{A}$$

where A is the area of the piston and g is the gravitational constant. Thus state 1 of the gas is defined by P_1, V_1, n, and T, where n and T remain constant as the expansion occurs.

One-Step Expansion—No Work. If mass M_1 is removed from the pan, the gas will expand, moving the piston to the right end of the cylinder. After expansion, the gas occupies a volume $V_2 = 4V_1$ and pressure $P_2 = P_1/4$.

When the process goes from state 1 (P_1, V_1) to state 2 ($P_1/4$, $4V_1$) with no mass on the pan, no heat flows into or out of the gas because T is constant, and no work is done (no mass is lifted). Thus work $= w_0 = 0$. This is called a *free expansion.*

P_{external} (the pressure against which the gas expands) is zero in a free expansion.

One-Step Expansion. Now consider an experiment with the gas initially at state 1 where the mass M_1 is replaced by a mass $M_1/4$. The gas will now expand against the pressure (P_{external}):

$$P_{\text{ex}} = \frac{\left(\frac{M_1}{4}\right)g}{A} = \frac{P_1}{4}$$

The mass is lifted and the gas expands until the pressure is $P_1/4$. The new volume is $4V_1$. In this case work is performed:

$$|\text{Work}| = |w_1| = \left(\frac{M_1}{4}\right)gh$$

where h is the *change* in height of the mass.

This work can also be expressed in terms of the external pressure (P_{ex}) on the gas as it expands and the change in volume (ΔV). Recall from Chapter 9 that

$$w = -P_{\text{ex}}\Delta V$$

In this expansion the magnitude (absolute value) of the work is

$$|w_1| = P_{ex}\Delta V = \frac{P_1}{4}(V_2 - V_1) = \frac{P_1}{4}(4V_1 - V_1) = \frac{3}{4}P_1V_1$$

Since in this case work flows out of the system into the surroundings, the correct sign is

$$w_1 = -P_{ex}\Delta V = -\frac{3}{4}P_1V_1$$

Two-Step Expansion. Next, we will expand the gas in two steps by using two different weights on the pan. First, we put a weight with mass $M_1/2$ on the pan. In this case

$$P_{ex} = \frac{\left(\frac{M_1}{2}\right)g}{A} = \frac{P_1}{2}$$

and the gas expands until $P_2 = P_1/2$ and $V_2 = 2V_1$. The magnitude of the work is

$$|w_2'| = \frac{P_1}{2}(V_2 - V_1) = \frac{P_1}{2}(2V_1 - V_1) = \frac{P_1V_1}{2}$$

Next, replace the mass $M_1/2$ with a mass $M_1/4$. The gas expands again until $P_3 = P_1/4$ and $V_3 = 4V_1$. The quantity of work in this step is

$$|w_2''| = \frac{P_1}{4}(4V_1 - 2V_1) = \frac{P_1V_1}{2}$$

The total quantity of work in this two-step expansion is

$$|w_2| = \frac{P_1V_1}{2} + \frac{P_1V_1}{2} = P_1V_1$$

With its correct sign $w_2 = -P_1V_1$. This process is diagramed in Fig. 10.6. Note that $|w_2| > |w_1| > |w_0|$, even though in each case the gas is taken

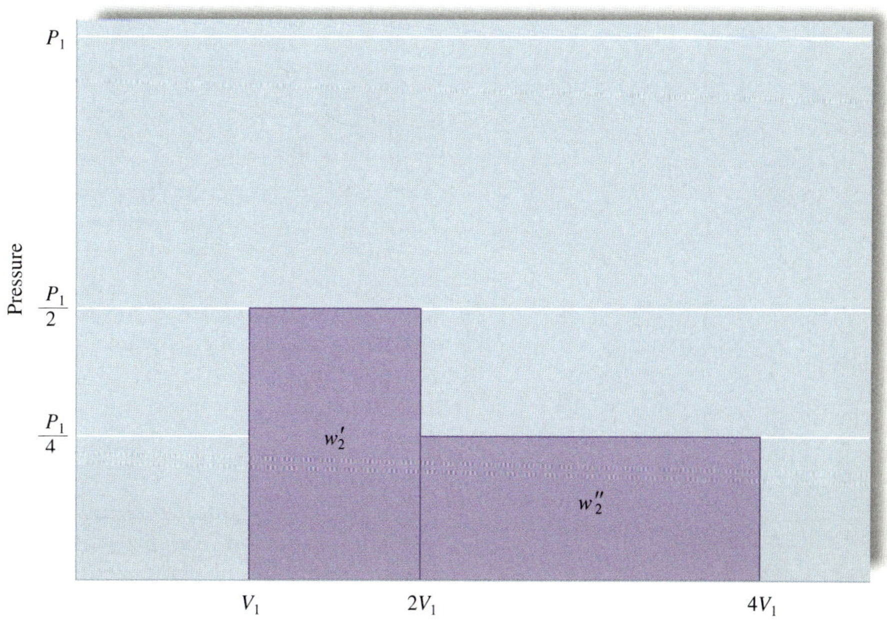

FIGURE 10.6
The *PV* diagram for a two-step expansion.

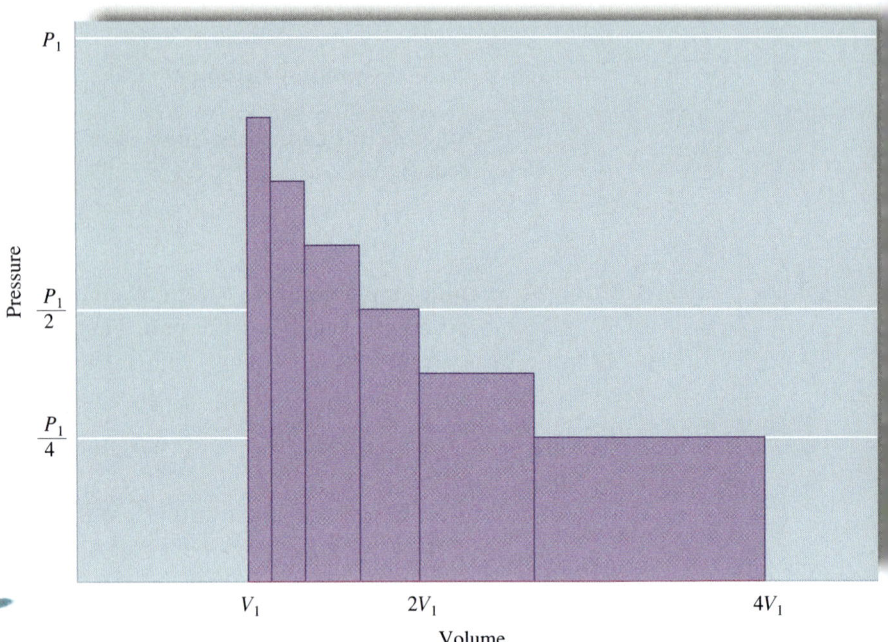

FIGURE 10.7

The *PV* diagram for a six-step expansion.

from state 1 (P_1, V_1) to state 2 ($P_1/4$, $4V_1$). This result illustrates a property of work we have discussed before: Work is pathway-dependent—it is not a state function.

Six-Step Expansion. Next, we will consider the expansion of the gas from state 1 to state 2 in six steps using several masses between M_1 and $M_1/4$. This process is summarized in Fig. 10.7. In this case $|w_6|$, which is the sum of these six steps, is clearly greater than $|w_2|$.

Infinite-Step Expansion. If one continues to increase the number of steps, the magnitude of w_n (for an n-step process),

$$|w_n| = \sum_{i=1}^{n} P_i \Delta V_i$$

continues to increase.

Now we consider the limiting case—a process in which P_{ex} is changed by infinitesimally small increments. This case corresponds to the use of an infinite number of weights, each differing from the previous one by an infinitesimally small mass. Under these conditions the successive volume changes become infinitesimally small (dV), and the process requires an infinite number of steps. The mathematical operation needed to sum the steps in this instance is the integral:

$$|\text{Work}| = \int_{V_1}^{V_2} P_{ex}\, dV$$

The diagram corresponding to this process is given in Fig. 10.8.

It is important to recognize that when the expansion of the gas is carried out in an infinite number of steps, the *external pressure is always almost exactly equal to the pressure produced by the gas.* That is, at any given time P_{ex}

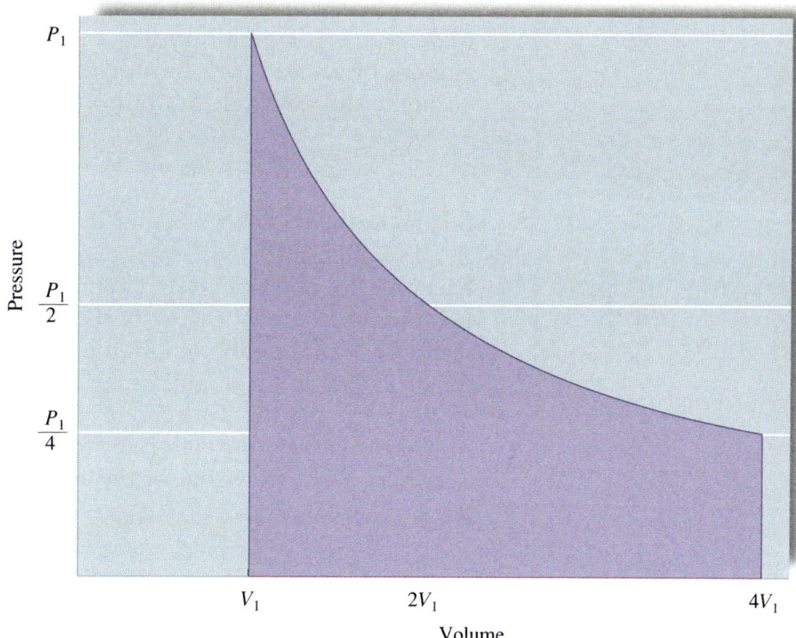

FIGURE 10.8

The *PV* diagram for the reversible expansion.

is only less than $P_{gas}(= P)$ by an infinitesimally small amount dP. Thus one can assume that at any point in the process $P \approx P_{ex}$.

A process like this one, carried out so that the system is always at equilibrium, is called a *reversible process.* The expansions described earlier that were carried out in a finite number of steps were not done reversibly because in these processes P_{ex} was smaller than P by a finite amount during each step. We will have more to say presently about what the term *reversible* means in this context.

Since $P_{ex} \approx P_{gas} = P$ in the reversible expansion, by use of the ideal gas law,

$$P_{ex} \approx P = \frac{nRT}{V}$$

and

$$|\text{Total work}|^* = |w_\infty| = |w_{rev}| = \int_{V_1}^{V_2} \frac{nRT\, dV}{V}$$

Since n and T are held constant in this experiment,

$$|w_{rev}| = nRT\int_{V_1}^{V_2} \frac{dV}{V} = nRT(\ln V_2 - \ln V_1) = nRT \ln\left(\frac{V_2}{V_1}\right)$$

In this specific experiment $V_2 = 4V_1$. Therefore,

$$|w_{rev}| = nRT \ln 4 = 1.4nRT$$

And since $P_1V_1 = nRT$,

$$|w_{rev}| = 1.4P_1V_1$$

for this particular expansion.

---

*In these calculations we deal with the magnitude for convenience. Actually, for the expansion of a gas, the work has a negative sign.

Note that as the number of steps in the expansion increases, the amount of work the gas performs also increases. The maximum work that a given amount of gas can perform in going from V_1 to V_2 at constant temperature occurs in the reversible expansion. Thus

$$|w_{\text{max}}| = |w_{\text{rev}}| = nRT \ln\left(\frac{V_2}{V_1}\right)$$

for the isothermal expansion of n moles of an ideal gas.

In doing this hypothetical experiment we have assumed that the gas behaves ideally. When a given amount of an ideal gas expands at constant temperature, the internal energy of the gas remains constant; that is, $\Delta E = 0$. As mentioned earlier, this condition means that $q = -w$. Thus a quantity of energy q (equal in magnitude to w) flows into the gas (as heat) as the expansion occurs, and the work (w) is performed. Therefore, the surroundings furnished the energy (through heat flow) necessary to perform the work.

Since work flows out of the system in the reversible expansion, it has a negative sign:

$$w_{\text{rev}} = -nRT \ln\left(\frac{V_2}{V_1}\right)$$

and

$$q_{\text{rev}} = -w_{\text{rev}} = nRT \ln\left(\frac{V_2}{V_1}\right)$$

The Isothermal Compression of an Ideal Gas

Now let's consider the opposite experiment: compressing a gas at pressure $P_1/4$ and volume $4V_1$ to pressure P_1 and volume V_1. This experiment can be done in one step or a number of steps.

One-Step Compression. Initially, the gas is at pressure $P_1/4$ and volume $4V_1$. When mass M_1 is placed on the pan, the gas will be rapidly compressed to the state described by P_1 and V_1. To see how this process works, consider the diagram for the gas at $P_1/4$ and $4V_1$ (Fig. 10.9). We raise the mass M_1 to the pan, which then causes the gas to be compressed to P_1 and V_1 as the pan returns to the original level.

The work performed to recompress the gas is equal to the work performed to lift the weight up to the pan. It also can be expressed in terms of $P_{\text{ex}}\Delta V$:

$$|w'_1| = M_1gh = P_1\Delta V = P_1(4V_1 - V_1) = 3P_1V_1$$

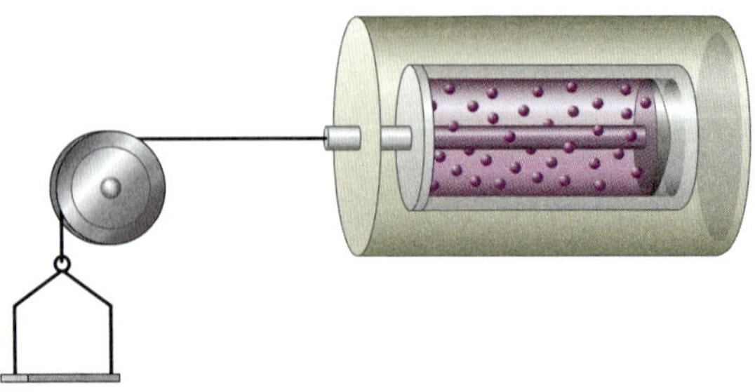

FIGURE 10.9

The situation before the one-step compression.

Two-Step Compression. We can compress the gas in two steps. For example, a mass $M_1/2$ is lifted onto the pan, and after the gas has been partially compressed, M_1 is put onto the pan in place of $M_1/2$ to finish the compression. In this case the total work required for the compression is

$$|w_2'| = \frac{P_1}{2}(4V_1 - 2V_1) + P_1(2V_1 - V_1) = 2P_1V_1$$

Infinite-Step Compression. Notice that in compressing the gas isothermally, as the number of steps increases, the work required to compress the gas decreases.

If we compress the gas in an infinite number of steps (in which, at all times, $P_{ex} \approx P$), the work required is

$$|w_\infty'| = \int_{V_1}^{V_2} P\,dV = nRT \ln\left(\frac{V_2}{V_1}\right)$$

$$= nRT \ln\left(\frac{V_1}{4V_1}\right) = 1.4P_1V_1$$

Because $P_{ex} \approx P$ throughout the process, this is a *reversible* compression. In this case w has a positive sign because we are performing work on the system. Thus, in the reversible, isothermal compression of the gas, $w_\infty' = 1.4P_1V_1$, and since $\Delta E = 0$,

$$q_\infty = -w_\infty' = -1.4P_1V_1$$

As the gas is compressed reversibly and isothermally, it releases $1.4P_1V_1$ (L atm) of energy as heat to the surroundings. In other words, the same quantity of energy flows into the gas as work and flows out of the gas as heat to produce no net change in E as the compression occurs.

In general terms, for a reversible compression from V_1 to V_2,

$$w_{rev} = -q_{rev} = -nRT \ln\left(\frac{V_2}{V_1}\right)$$

or in terms of pressures,

$$w_{rev} = -q_{rev} = -nRT \ln\left(\frac{P_1}{P_2}\right)$$

Note that since this process is a compression ($V_2 < V_1$), $\ln(V_2/V_1)$ has a negative sign, making w_{rev} positive and q_{rev} negative, as expected.

Summary

We summarize the results of these expansion and compression experiments in Table 10.3. The most important conclusion that can be drawn from these results can be stated as follows: Only when the expansion and compression *are both done reversibly* (by an infinite number of steps) is the universe the same after the cyclic process (the expansion and the subsequent compression of the gas back to its original state). That is, only for the reversible processes is the heat absorbed during expansion exactly equal to the heat released during compression. In all the processes carried out using a finite number of steps, more heat is released into the surroundings than is absorbed in the comparable expansion (same number of steps).

TABLE 10.3

Summary of the Isothermal Expansion and Compression Experiments

	Number of Steps	w	q
Expansion (constant T)	0 (no mass)	0	0
	1	$-0.75P_1V_1$	$0.75P_1V_1$
	2	$-1P_1V_1$	$1P_1V_1$
	4	$-1.16P_1V_1$	$1.16P_1V_1$
	∞	$-1.4P_1V_1$	$1.4P_1V_1$
Compression (constant T)	1	$3P_1V_1$	$-3P_1V_1$
	2	$2P_1V_1$	$-2P_1V_1$
	4	$1.67P_1V_1$	$-1.67P_1V_1$
	∞	$1.4P_1V_1$	$-1.4P_1V_1$

For example, in the one-step expansion and compression the net work done is

$$w_{\text{net}} = \underset{\substack{\uparrow \\ \text{Expansion}}}{-0.75P_1V_1} + \underset{\substack{\uparrow \\ \text{Compression}}}{3P_1V_1} = 2.25P_1V_1$$

and the net heat flow is $-2.25P_1V_1$.

In terms of the *system,* this one-step expansion/compression is cyclic. That is, the system starts at state 1 before the expansion process and ends up back at state 1 after the compression process. However, although the system is returned to the same state in this cyclic expansion–compression process, the *surroundings* are not the same. In fact, we might say that in the surroundings "work has been changed to heat." In this one-step cyclic process, $2.25P_1V_1$ (L atm) of work is changed to $2.25P_1V_1$ (L atm) of heat (thermal energy) in the surroundings. This amount represents the net extra work required to lift the weights onto the pan to compress the gas compared with the work that was obtained as the gas expanded.

This is a general result. In any finite-step cyclic expansion–compression process "work is always converted to heat":

$$\underset{\text{Ordered energy}}{\text{Work}} \longrightarrow \underset{\text{Disordered energy}}{\text{Heat}}$$

This result applies whenever the process is carried out in a nonreversible (irreversible) manner. In other words, in an *irreversible* cyclic process more work must be input to the system than the system produces. In all the finite gas compressions the work required is greater than $1.4P_1V_1$, which is the maximum work available from the expansion.

All real processes are irreversible.

Of course, all real processes are irreversible, because they cannot be carried out in an infinite number of steps without taking an infinite amount of time. In other words, *all real processes are irreversible* (in a thermodynamic sense).

The maximum work corresponds to the reversible expansion.

Another important conclusion to be drawn from the preceding example is that the *maximum work obtainable from the gas occurs when the expansion is carried out reversibly* ($w_{max} = w_{rev}$). This result is always true for PV work, as well as for any other type of work, such as electrical work performed by an electrochemical cell. We will examine this latter example in the next chapter.

The final point that this experiment reemphasizes is that work and heat are pathway-dependent and thus are not state functions. Energy, on the other hand, is a state function. In each of these isothermal expansions and compressions between (P_1, V_1) and ($P_1/4$, $4V_1$), ΔE is always zero, regardless of the number of steps, since T is constant.

At this point we can precisely define the terms *reversible* and *irreversible* in a thermodynamic sense. In a reversible cyclic process *both the system and the surroundings are returned exactly to their original conditions.* As it turns out, this process is hypothetical. On the other hand, an irreversible process is one in which, even when the system is cycled (state 1 $\rightarrow$ state 2 $\rightarrow$ state 1) and thus returned to its original state, the surroundings are changed in a permanent way. All real processes are irreversible.

10.3 The Definition of Entropy

So far we have discussed entropy in a very qualitative way. Here we want to give a precise, quantitative definition of entropy.

Ludwig Boltzmann's tomb in Vienna. Notice Boltzmann's equation on the monument.

We have said that entropy is related to probability. If a system has several states available to it, the one that can be achieved in the greatest number of ways (has the largest number of microstates) is the one most likely to occur. That is, the state with the greatest probability has the highest entropy.

Now we will connect entropy and probability quantitatively by defining the entropy function S as follows:

$$S = k_B \ln \Omega$$

where

k_B = Boltzmann's constant, the gas constant per molecule (R/N_A)

Ω = the number of microstates corresponding to a given state (including both position and energy)

This definition of entropy shows its exact relationship to probability. However, it is not useful in a practical sense for the typical types of samples used by chemists because those samples contain so many components. For example, a mole of gas contains 6.022×10^{23} individual particles. In addition, according to one estimate, describing the positions and velocities of this mole of particles would require a stack of paper 10 *light-years* tall—and this description would apply for only an instant. Clearly, we cannot deal directly with this definition of entropy for typical-sized samples. We must find a way to connect entropy to the macroscopic properties of matter. To do so, we will consider an ideal gas that expands isothermally from volume V_1 to volume $2V_1$ (see Fig. 10.10).

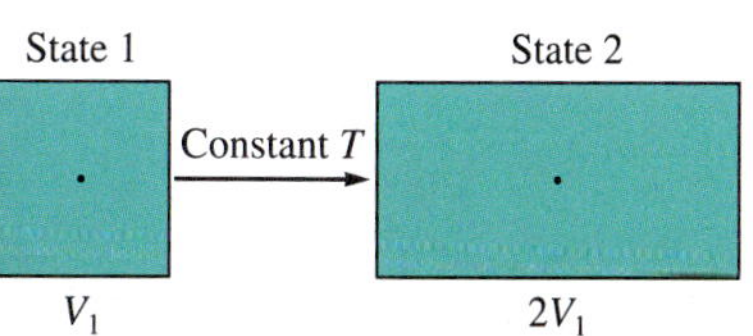

FIGURE 10.10

A particle in a gas that expands from volume V_1 to volume $2V_1$.

Focus on an individual particle in this gas. When the gas goes from volume V_1 to $2V_1$, each particle has double the number of positions* available to it. That is, $\Omega_2 = 2\Omega_1$.

*Technically, when we double the volume, we are doubling the "density" (the closeness) of energy microstates. However, this requires an understanding of concepts that are beyond the scope of this text.

Chemical Insights

Entropy: An Organizing Force?

In this text we have emphasized the meaning of the second law of thermodynamics—that the entropy of the universe is always increasing. Although the results of all our experiments support this conclusion, this does not mean that order cannot appear spontaneously in a given part of the universe. The best example of this phenomenon involves the assembly of cells in living organisms. Of course, when a process that creates an ordered system is examined in detail, it is found that other parts of the process involve an increase in disorder such that the sum of all the entropy changes is positive. In fact, scientists are now finding that the search for maximum entropy in one part of a system can be a powerful force for organization in another part of the system.

To understand how entropy can be an organizing force, look at the accompanying figure. In a system containing large and small "balls" as shown in the figure, the small balls can "herd" the large balls into clumps in the corners and near the walls. This clears out the maximum space for the small balls so that they can move more freely, thus maximizing the entropy of the system, as demanded by the second law of thermodynamics.

In essence, the ability to maximize entropy by sorting different-sized objects creates a kind of attractive force, called a *depletion,* or *excluded-volume, force.* These "entropic forces" operate for objects in the size range of approximately 10^{-8} m to approximately 10^{-6} m. For entropy-induced ordering to occur, the particles must be constantly jostling each other and must be constantly agitated by solvent molecules, thus making gravity unimportant.

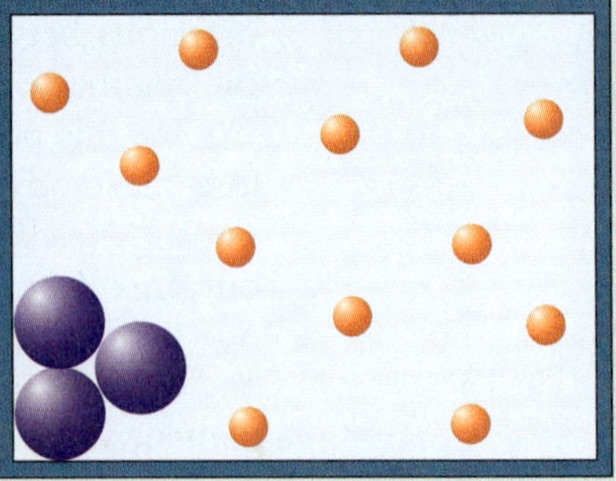

There is increasing evidence that entropic ordering is important in many biological systems. For example, this phenomenon seems to be responsible for the clumping of sickle-cell hemoglobin in the presence of much smaller proteins that act as the "smaller balls." Entropic forces also have been linked to the clustering of DNA in cells without nuclei, and Allen Minton of the National Institutes of Health in Bethesda, Maryland, is studying the role of entropic forces in the binding of proteins to cell membranes.

Entropic ordering also appears in nonbiological settings, especially in the ways polymer molecules clump together. For example, polymers added to paint to improve the flow characteristics of the paint actually caused it to coagulate because of depletion forces.

Thus, as you probably have concluded already, entropy is a complex issue. As entropy drives the universe to its ultimate death of maximum chaos, it provides some order along the way.

Now we will use the definition of entropy to calculate ΔS for this expansion. In this case $S_1 = k_B \ln \Omega_1$ and $S_2 = k_B \ln \Omega_2$, so

$$\Delta S = S_2 - S_1 = k_B \ln \Omega_2 - k_B \ln \Omega_1 = k_B \ln\left(\frac{\Omega_2}{\Omega_1}\right)$$

$$= k_B \ln\left(\frac{2\Omega_1}{\Omega_1}\right) = k_B \ln 2$$

This quantity represents ΔS for each particle in the gas. If the gas contains five particles, then the ratio of Ω_2 to Ω_1 is

$$\frac{\Omega_2}{\Omega_1} = 2 \times 2 \times 2 \times 2 \times 2 = 2^5$$

where each 2 represents the fact that each particle has twice as many positions available to it in $2V_1$ as in V_1. This argument can be readily extended to a sample with 1 mole of particles:

$$\frac{\Omega_2}{\Omega_1} = 2^{6\times 10^{23}}$$

and

$$\Delta S = k_B \ln(2^{6\times 10^{23}}) = (6 \times 10^{23})(k_B \ln 2)$$
$$= N_A k_B \ln 2 = R \ln 2$$

where R is the gas constant.

Now for a general case in which a gas is expanded for V_1 to V_2, the ratio of Ω_2 to Ω_1 is

$$\frac{\Omega_2}{\Omega_1} = \frac{V_2}{V_1}$$

and for 1 mole of gas,

$$\Delta S_{V_1 \rightarrow V_2} = R \ln\left(\frac{V_2}{V_1}\right)$$

For n moles of gas we have

$$\Delta S_{V_1 \rightarrow V_2} = nR \ln\left(\frac{V_2}{V_1}\right)$$

What we have accomplished here is to use the definition of entropy in terms of probability to derive an expression for ΔS that depends on volume, a macroscopic property of the gas. We can now relate the change in entropy to heat flow by noting the striking similarity between the preceding equation for ΔS and the one derived in Section 10.2 describing q_{rev} for the isothermal expansion–compression of an ideal gas. Compare

$$\Delta S = nR \ln\left(\frac{V_2}{V_1}\right) \qquad \text{with} \qquad q_{\text{rev}} = nRT \ln\left(\frac{V_2}{V_1}\right)$$

Combining these equations gives

$$\Delta S = \frac{q_{\text{rev}}}{T}$$

This very important relationship is the macroscopic (thermodynamic) definition of ΔS. In our treatment we started with the definition of entropy based on probability because that definition better emphasizes the fundamental character of entropy. However, it is also very important to know how entropy changes relate to changes in macroscopic properties, such as volume and heat, because these changes are relatively easy to measure.

10.4 Entropy and Physical Changes

Although chemists deal primarily with the chemical changes of matter, physical changes are also very important. In this section we will consider how the entropy of a substance depends on its temperature and on its physical state.

Temperature Dependence of Entropy

For an isothermal process we have seen that the change in entropy is defined by the relationship

$$\Delta S = \frac{q_{\text{rev}}}{T}$$

We can calculate ΔS for a change in temperature from T_1 to T_2 by summing infinitesimal increments in entropy at each temperature T:

$$dS = \frac{dq_{\text{rev}}}{T}$$

Using integration, we have

$$\Delta S_{T_1 \to T_2} = \int_{T_1}^{T_2} dS = \int_{T_1}^{T_2} \frac{dq_{\text{rev}}}{T}$$

If the process is carried out at constant pressure, then

$$dq_{\text{rev}} = nC_{\text{p}}\, dT$$

for n moles of substance. Thus

$$\Delta S_{T_1 \to T_2} = \int_{T_1}^{T_2} \frac{dq_{\text{rev}}}{T} = nC_{\text{p}} \int_{T_1}^{T_2} \frac{dT}{T}$$

assuming that C_{p} is constant between T_1 and T_2. Performing the integration gives

$$\Delta S_{T_1 \to T_2} = nC_{\text{p}} \ln\left(\frac{T_2}{T_1}\right)$$

Similarly, for a process carried out at constant volume,

$$\Delta S_{T_1 \to T_2} = nC_{\text{v}} \ln\left(\frac{T_2}{T_1}\right)$$

Entropy Changes Associated with Changes of State

At the normal melting point or boiling point of a substance the two states of matter present at that temperature and at 1 atm pressure are in equilibrium. That is, the two states can coexist indefinitely if the system is isolated (left totally undisturbed). Recall that a reversible process can occur only at equilibrium. Thus, since a change of state from solid to liquid at the substance's melting point is a reversible process, we can calculate the change in entropy for this process by using the equation

$$\Delta S = \frac{q_{\text{rev}}}{T}$$

where

$$q_{\text{rev}} = \Delta H_{\text{fusion}} = \begin{array}{l}\text{energy required to melt 1 mol}\\ \text{of solid at the melting point}\end{array}$$

$$T = \text{melting point in } K$$

The same reasoning applies to a change from liquid to gas at the boiling point, except in this case $q_{rev} = \Delta H_{vaporization}$ and T = boiling point.

Example 10.3

Calculate the change in entropy that occurs when a sample containing 2.00 mol of water is heated from 50.°C to 150.°C at 1 atm pressure. The molar heat capacities for $H_2O(l)$ and $H_2O(g)$ are 75.3 J K^{-1} mol^{-1} and 36.4 J K^{-1} mol^{-1}, respectively, and the enthalpy of vaporization for water is 40.7 kJ/mol at 100°C.

Solution Since water changes from liquid to gas at 100°C, we will do this calculation in three (reversible) steps:

1. $\Delta S(l)_{50°C \rightarrow 100°C}$ (heat the liquid)
2. $\Delta S^{100°C}_{(l)\rightarrow(g)}$ (change the liquid to a gas)
3. $\Delta S(g)_{100°C \rightarrow 150°C}$ (heat the gas)

and total the results.

1. The entropy change for heating 2.00 mol of liquid water from 50°C to 100°C is

$$\Delta S_{(1)} = (2.00 \text{ mol})\left(75.3 \frac{\text{J}}{\text{K mol}}\right) \ln\left(\frac{373}{323}\right) = 21.7 \text{ J/K}$$

2. The entropy change for the vaporization of 2.00 mol of water at 100°C can be calculated from the equation

$$\Delta S = \frac{q_{rev}}{T} = \frac{\Delta H_{vap}}{T_{bp}}$$

$$= \frac{4.07 \times 10^4 \text{ J/mol}}{373 \text{ K}} = 1.09 \times 10^2 \text{ J K}^{-1} \text{ mol}^{-1}$$

In this case 2.00 mol of water is vaporized, so

$$\Delta S_{(2)} = \left(1.09 \times 10^2 \frac{\text{J}}{\text{K mol}}\right)(2.00 \text{ mol}) = 2.18 \times 10^2 \text{ J/K}$$

3. The entropy change for heating 2.00 mol of gaseous water from 100°C to 150°C is

$$\Delta S_{(3)} = (2.00 \text{ mol})\left(36.4 \frac{\text{J}}{\text{K mol}}\right) \ln\left(\frac{423}{373}\right) = 9.16 \text{ J/K}$$

The total entropy change is

$$\Delta S_{50°C \rightarrow 150°C} = \Delta S_{(1)} + \Delta S_{(2)} + \Delta S_{(3)}$$

$$= 21.7 \text{ J/K} + 218 \text{ J/K} + 9.16 \text{ J/K} = 249 \text{ J/K}$$

10.5 Entropy and the Second Law of Thermodynamics

We have seen that processes are spontaneous when they result in an increase in disorder. Nature always moves toward the most probable state available to it. We can state this principle in terms of entropy: *In any spontaneous process there is always an increase in the entropy of the universe.* This is the **second**

The total energy of the universe is constant, but the entropy is increasing.

law of thermodynamics. Contrast this law with the first law of thermodynamics, which tells us that the energy of the universe is constant. Energy is conserved in the universe, but entropy is not. In fact, the second law can be paraphrased as follows: *The entropy of the universe is increasing.*

As in Chapter 9, we find it convenient to divide the universe into a system and the surroundings. Thus we can represent the change in the entropy of the universe as

$$\Delta S_{univ} = \Delta S_{sys} + \Delta S_{surr}$$

where ΔS_{sys} and ΔS_{surr} represent the changes in entropy that occur in the system and in the surroundings, respectively.

To predict whether a given process will be spontaneous, we must know the sign of ΔS_{univ}. If ΔS_{univ} is positive, the entropy of the universe increases, and the process is spontaneous in the direction written. If ΔS_{univ} is negative, the process is spontaneous in the *opposite* direction. If ΔS_{univ} is zero, the process has no tendency to occur, indicating that the system is at equilibrium. To predict whether a process is spontaneous, we must consider the entropy changes that occur both in the system and in the surroundings.

10.6 The Effect of Temperature on Spontaneity

To explore the interplay of ΔS_{sys} and ΔS_{surr} in determining the sign of ΔS_{univ}, we will first discuss the change of state for 1 mole of water from liquid to gas,

$$H_2O(l) \longrightarrow H_2O(g)$$

considering the water to be the system and everything else the surroundings.

What happens to the entropy of water in this process? A mole of liquid water (18 g) has a volume of approximately 18 mL. A mole of gaseous water at 1 atm and 100°C occupies a volume of approximately 31 L. Clearly, there are many more positions available to the water molecules in a volume of 31 L than in 18 mL; thus the vaporization of water is favored by this increase in positional probability. That is, for this process, the entropy of the system increases; ΔS_{sys} has a positive sign.

What about the entropy change in the surroundings? Although we will not prove it here, entropy changes in the surroundings are determined primarily by the flow of energy into or out of the system as heat. To understand this observation, suppose an exothermic process transfers 50 joules of energy as heat to the surroundings, where it becomes thermal energy—that is, kinetic energy associated with the random motions of atoms. This flow of energy into the surroundings increases the random motions of atoms there and so increases the entropy of the surroundings. The sign of ΔS_{surr} is positive. When an endothermic process occurs in the system, it produces the opposite effect. Heat flows from the surroundings to the system, causing the random motions of the atoms in the surroundings to decrease, decreasing the entropy of the surroundings. The vaporization of water is an endothermic process. Thus, for this change of state, ΔS_{surr} is negative.

In an endothermic process heat flows from the surroundings into the system. In an exothermic process heat flows into the surroundings.

Remember it is the sign of ΔS_{univ} that tells us whether the vaporization of water is spontaneous. We have seen that ΔS_{sys} is positive and favors the process and that ΔS_{surr} is negative and unfavorable. Thus the components of ΔS_{univ} are in opposition. Which one controls the situation? The answer *depends*

Steam rising from a thermal pool in Yellowstone.

on the temperature. We know that at a pressure of 1 atm water changes spontaneously from liquid to gas at all temperatures above 100°C. Below 100°C the opposite process (condensation) is spontaneous.

Since ΔS_{sys} and ΔS_{surr} are in opposition for the vaporization of water, the temperature must have an effect on the relative importance of these two terms. To understand why, we must discuss in more detail the factors that control the entropy changes in the surroundings. The central idea is that *the entropy changes in the surroundings are primarily determined by heat flow.* An exothermic process in the system increases the entropy of the surroundings because the energy flow increases the random motions in the surroundings. This means that exothermicity is an important driving force for spontaneity. In earlier chapters we have seen that a system tends to undergo changes that lower its energy. We now understand the reason for that tendency. When a system at constant temperature moves to a lower energy, the energy it gives up is transferred to the surroundings. Some of this energy is transferred as heat, thus leading to an increase in entropy in the surroundings.

The significance of exothermicity as a driving force *depends on the temperature at which the process occurs.* That is, the magnitude of ΔS_{surr} depends on the temperature at which the heat is transferred. We will not attempt to prove this fact here. Instead, we offer an analogy. Suppose that you have $50 to give away. Giving it to a millionaire will not create much of an impression—a millionaire has money to spare. However, to a poor college student, $50 represents a significant sum and will be received with considerable joy. The same principle can be applied to energy transfer via the flow of heat. If 50 joules of energy is transferred to the surroundings, the impact of that event depends greatly on the temperature. If the temperature of the surroundings is very high, the atoms there are in rapid motion. The 50 joules of energy will not make a large percentage change in these motions. On the other hand, if 50 joules of energy is transferred to the surroundings at a very low temperature, where the motions are slow, the energy will cause a large percentage change in the motions. Thus the impact of the transfer of a given quantity of energy as heat to or from the surroundings is greatest at low temperatures.

For our purposes the entropy changes that occur in the surroundings have two important characteristics.

In a process occurring at constant temperature, the tendency for the system to lower its energy is due to the resulting positive ΔS_{surr}.

1. *The sign of ΔS_{surr} depends on the direction of the heat flow.* At constant temperature an exothermic process in the system causes heat to flow into the surroundings, increasing the random motions and thereby increasing the entropy of the surroundings. For this case ΔS_{surr} is positive. This principle is often stated in terms of energy. An important driving force in nature results from the tendency of a system to achieve the lowest possible energy by transferring energy as heat to the surroundings.
2. *The magnitude of ΔS_{surr} depends on the temperature.* The transfer of a given quantity of energy as heat produces a much greater percentage change in the randomness of the surroundings at a low temperature than it does at a high temperature. Thus ΔS_{surr} depends directly on the quantity of heat transferred and inversely on temperature. In other words, the tendency for the system to lower its energy becomes a more important driving force at lower temperatures:

$$\text{Driving force provided by the energy flow (heat)} = \text{magnitude of the entropy change of the surroundings} = \frac{\text{quantity of heat (J)}}{\text{temperature (K)}}$$

These ideas are summarized as follows:

Exothermic process:

ΔS_{surr} = positive

Endothermic process:

ΔS_{surr} = negative

$$\text{Exothermic process:} \quad \Delta S_{surr} = +\frac{\text{quantity of heat (J)}}{\text{temperature (K)}}$$

$$\text{Endothermic process:} \quad \Delta S_{surr} = -\frac{\text{quantity of heat (J)}}{\text{temperature (K)}}$$

We can express ΔS_{surr} in terms of the change in enthalpy (ΔH) for a process occurring at constant pressure (where only PV work is allowed), since under these conditions

$$\text{Heat flow (constant } P) = \text{change in enthalpy} = \Delta H$$

When no subscript is present, the quantity (for example, ΔH) refers to the system.

Recall that ΔH consists of two parts: a sign and a number. The *sign* indicates the direction of flow; a plus sign means "into the system" (endothermic) and a minus sign means "out of the system" (exothermic). The *number* indicates the quantity of energy.

Combining all these concepts yields the following definition of ΔS_{surr} for a reaction that takes place under conditions of constant temperature (K) and pressure:

$$\Delta S_{surr} = -\frac{\Delta H}{T}$$

The minus sign is necessary because the sign of ΔH is determined with respect to the reaction system, and this equation expresses a property of the surroundings. This means that if the reaction is exothermic, ΔH has a negative sign, but since heat flows into the surroundings, ΔS_{surr} is positive.

Stibnite contains Sb_2S_3.

Example 10.4

In the metallurgy of antimony, the pure metal is recovered by different reactions, depending on the composition of the ore. For example, iron is used to reduce antimony in sulfide ores:

$$Sb_2S_3(s) + 3Fe(s) \longrightarrow 2Sb(s) + 3FeS(s) \qquad \Delta H = -125 \text{ kJ}$$

Carbon is used as the reducing agent in oxide ores:

$$Sb_4O_6(s) + 6C(s) \longrightarrow 4Sb(s) + 6CO(g) \qquad \Delta H = 778 \text{ kJ}$$

Calculate ΔS_{surr} for each of these reactions at 25°C and 1 atm.

Solution We use

$$\Delta S_{surr} = -\frac{\Delta H}{T}$$

where

$$T = 25 + 273 = 298 \text{ K}$$

For the sulfide ore reaction,

$$\Delta S_{surr} = -\frac{-125 \text{ kJ}}{298 \text{ K}} = 0.419 \text{ kJ/K} = 419 \text{ J/K}$$

Note that ΔS_{surr} is positive, as expected, since this reaction is exothermic; energy flows to the surroundings as heat, increasing the randomness of the surroundings.

TABLE 10.4

Interplay of ΔS_{sys} and ΔS_{surr} in Determining the Sign of ΔS_{univ}

Signs of Entropy Changes			
ΔS_{sys}	ΔS_{surr}	ΔS_{univ}	Process Spontaneous?
+	+	+	Yes
−	−	−	No (process will occur in opposite direction)
+	−	?	Yes, if ΔS_{sys} has a larger magnitude than ΔS_{surr}
−	+	?	Yes, if ΔS_{surr} has a larger magnitude than ΔS_{sys}

For the oxide ore reaction,

$$\Delta S_{surr} = -\frac{778 \text{ kJ}}{298} = -2.61 \text{ kJ/K} = -2.61 \times 10^3 \text{ J/K}$$

In this case ΔS_{surr} is negative because heat flows from the surroundings to the system.

We have seen that the spontaneity of a process is determined by the entropy change it produces in the universe. We have also seen that ΔS_{univ} has two components, ΔS_{sys} and ΔS_{surr}. If for some process both ΔS_{sys} and ΔS_{surr} are positive, then ΔS_{univ} is positive, and the process is spontaneous. If, on the other hand, both ΔS_{sys} and ΔS_{surr} are negative, the process does not occur in the direction indicated but is spontaneous in the opposite direction. Finally, if ΔS_{sys} and ΔS_{surr} have opposite signs, the spontaneity of the process depends on the sizes of the opposing terms. These cases are summarized in Table 10.4.

We can now understand why spontaneity often depends on temperature and thus why water spontaneously freezes below 0°C and melts above 0°C. The term ΔS_{surr} is temperature-dependent. Since

$$\Delta S_{surr} = -\frac{\Delta H}{T}$$

at constant pressure, the value of ΔS_{surr} changes markedly with temperature. The magnitude of ΔS_{surr} is very small at high temperatures and increases as the temperature decreases. That is, exothermicity is most important as a driving force at low temperatures.

10.7 Free Energy

So far we have used ΔS_{univ} to predict the spontaneity of a process. Now we will define another thermodynamic function that is also related to spontaneity and is especially useful in dealing with the temperature dependence of spontaneity. This function is called **free energy** (G) and is defined as

$$G = H - TS$$

where H is the enthalpy, T is the kelvin temperature, and S is the entropy.

The symbol G for free energy honors Josiah Willard Gibbs (1839–1903), who was Professor of Mathematical Physics at Yale University from 1871 to 1903. He laid the foundations of many areas of thermodynamics, particularly as they apply to chemistry.

For a process that occurs at constant temperature, the change in free energy (ΔG) is given by the equation

$$\Delta G = \Delta H - T\Delta S$$

Note that all quantities here refer to the system. From this point on we will follow the usual convention that when no subscript is included, the quantity refers to the system.

To see how this equation relates to spontaneity, we divide both sides of the equation by $-T$ to produce

$$-\frac{\Delta G}{T} = -\frac{\Delta H}{T} + \Delta S$$

Remember that at constant temperature and pressure

$$\Delta S_{surr} = -\frac{\Delta H}{T}$$

So we can write $$-\frac{\Delta G}{T} = -\frac{\Delta H}{T} + \Delta S = \Delta S_{surr} + \Delta S = \Delta S_{univ}$$

We have shown that $$\Delta S_{univ} = -\frac{\Delta G}{T} \quad \text{at constant } T \text{ and } P$$

This result is very important. It means that a process carried out at constant temperature and pressure will be spontaneous only if ΔG is negative. That is, *a process (at constant T and P) is spontaneous in the direction in which the free energy decreases* ($-\Delta G$ means $+\Delta S_{univ}$).

Now we have two functions that can be used to predict spontaneity: the entropy of the universe, which applies to all processes, and free energy, which can be used for processes carried out at constant temperature and pressure. Since so many chemical reactions occur under the latter conditions, the free energy function is more useful to chemists.

Let's use the free energy equation to predict the spontaneity of the melting of ice:

The superscript degree symbol (°) indicates that all substances are in their standard states.

$$H_2O(s) \longrightarrow H_2O(l)$$

for which $\Delta H° = 6.03 \times 10^3$ J/mol and $\Delta S° = 22.1$ J K^{-1} mol^{-1}

Results for the calculations of ΔS_{univ} and $\Delta G°$ at -10°C, 0°C, and 10°C are shown in Table 10.5. These data predict that the process is spontaneous at

TABLE 10.5

Results of the Calculations of ΔS_{univ} and $\Delta G°$ for the Process $H_2O(s) \rightarrow H_2O(l)$ at -10°C, 0°C, and 10°C*

T (°C)	T (K)	$\Delta H°$ (J/mol)	$\Delta S°$ (J K^{-1} mol^{-1})	$\Delta S_{surr} = -\frac{\Delta H°}{T}$ (J K^{-1} mol^{-1})	$\Delta S_{univ} = \Delta S° + \Delta S_{surr}$ (J K^{-1} mol^{-1})	$T\Delta S°$ (J/mol)	$\Delta G° = \Delta H° - T\Delta S°$ (J/mol)
−10	263	6.03×10^3	22.1	−22.9	−0.8	5.81×10^3	$+2.2 \times 10^2$
0	273	6.03×10^3	22.1	−22.1	0	6.03×10^3	0
10	283	6.03×10^3	22.1	−21.3	+0.8	6.25×10^3	-2.2×10^2

*Note that at 10°C, $\Delta S°$ (ΔS_{sys}) controls, and the process occurs even though it is endothermic. At -10°C, the magnitude of ΔS_{surr} is larger than that of $\Delta S°$, so the process is spontaneous in the opposite (exothermic) direction. In these calculations we are assuming that $\Delta H°$ and $\Delta S°$ are temperature-independent.

TABLE 10.6

Various Possible Combinations of ΔH and ΔS for a Process and the Resulting Dependence of Spontaneity on Temperature

Case	Result
ΔS positive, ΔH negative	Spontaneous at all temperatures
ΔS positive, ΔH positive	Spontaneous at high temperatures (where exothermicity is relatively unimportant)
ΔS negative, ΔH negative	Spontaneous at low temperatures (where exothermicity is dominant)
ΔS negative, ΔH positive	Process not spontaneous at *any* temperature (reverse process is spontaneous at *all* temperatures)

To be exact, the calculations shown in Table 10.5 should account for the temperature dependences of ΔH and ΔS. For example,

$$\begin{array}{ccc} H_2O(l) & \xrightarrow[\Delta H_2]{0°C} & H_2O(s) \\ \uparrow \Delta H_1^\circ & & \downarrow \Delta H_3^\circ \\ H_2O(l) & \xrightarrow[\Delta H_4]{-10°C} & H_2O(s) \end{array}$$

$$\Delta H_4^\circ = \Delta H_1^\circ + \Delta H_2^\circ + \Delta H_3^\circ$$

However, since ΔH_1° and ΔH_3° are small compared to ΔH_2°,

$$\Delta H_4^\circ \approx \Delta H_2^\circ$$

The same ideas hold for ΔS°.

10°C; that is, ice melts at this temperature since ΔS_{univ} is positive and ΔG° is negative. The opposite is true at −10°C, at which water freezes spontaneously.

Why is this so? The answer lies in the fact that ΔS_{sys} (ΔS°) and ΔS_{surr} oppose each other. The term ΔS° favors the melting of ice because of the increase in positional entropy, and ΔS_{surr} favors the freezing of water because it is an exothermic process. At temperatures below 0°C the change of state occurs in the exothermic direction because ΔS_{surr} is larger in magnitude than ΔS_{sys}. But above 0°C the change occurs in the direction in which ΔS_{sys} is favorable, since in this case ΔS_{sys} is larger in magnitude than ΔS_{surr}. At 0°C the *opposing tendencies just balance,* and the two states coexist; there is no driving force in either direction. An equilibrium exists between the two states of water. Note that ΔS_{univ} is equal to 0 at 0°C.

We can reach the same conclusions by examining ΔG°. At −10°C, ΔG° is positive because the ΔH° term is larger than the $T\Delta S^\circ$ term. The opposite is true at 10°C. At 0°C, ΔH° is equal to $T\Delta S^\circ$ and ΔG° is equal to 0. This means that solid H_2O and liquid H_2O have the same free energy at 0°C ($\Delta G^\circ = G_{(l)} - G_{(s)}$), and the system is at equilibrium.

We can understand the temperature dependence of spontaneity by examining the behavior of ΔG. For a process occurring at constant temperature and pressure,

$$\Delta G = \Delta H - T\Delta S$$

If ΔH and ΔS favor opposite processes, spontaneity will depend on temperature in such a way that the exothermic direction will be favored at low temperatures. For example, for the process

$$H_2O(s) \longrightarrow H_2O(l)$$

ΔH and ΔS are both positive. The natural tendency for this system to lower its energy is in opposition to its natural tendency to increase its positional randomness. At low temperatures ΔH dominates, and at high temperatures ΔS dominates. The various cases are summarized in Table 10.6.

Example 10.5

At what temperatures is the following process spontaneous at 1 atm?

$$Br_2(l) \longrightarrow Br_2(g)$$

where $\Delta H^\circ = 31.0$ kJ/mol and $\Delta S^\circ = 93.0$ J K^{-1} mol^{-1}

What is the normal boiling point of liquid Br_2?

Solution The vaporization process will be spontaneous at all temperatures at which ΔG° is negative. Note that ΔS° favors the vaporization process because of the increase in positional entropy, and ΔH° favors the *opposite* process, which is exothermic. These opposite tendencies will exactly balance at the boiling point of liquid Br_2, since at this temperature liquid and gaseous Br_2 are in equilibrium ($\Delta G^\circ = 0$). We can find this temperature by setting $\Delta G^\circ = 0$ in the equation

$$\Delta G^\circ = \Delta H^\circ - T\Delta S^\circ$$

Thus we have

$$0 = \Delta H^\circ - T\Delta S^\circ$$

$$\Delta H^\circ = T\Delta S^\circ$$

and

$$T = \frac{\Delta H^\circ}{\Delta S^\circ} = \frac{3.10 \times 10^4 \text{ J/mol}}{93.0 \text{ J K}^{-1} \text{ mol}^{-1}} = 333 \text{ K}$$

At temperatures above 333 K, $T\Delta S^\circ$ has a larger magnitude than ΔH°, and ΔG° (or $\Delta H^\circ - T\Delta S^\circ$) is negative. Above 333 K the vaporization process is spontaneous; the opposite process occurs spontaneously below this temperature. At 333 K liquid and gaseous Br_2 will coexist in equilibrium. These observations can be summarized as follows (the pressure is 1 atm in each case):

1. $T > 333$ K. The term ΔS° controls. The increase in entropy occurring when liquid Br_2 is vaporized is dominant.
2. $T < 333$ K. The process is spontaneous in the direction in which it is exothermic. The term ΔH° controls.
3. $T = 333$ K. The opposing driving forces are just balanced ($\Delta G^\circ = 0$), and the liquid and gaseous phases of Br_2 coexist. This is the normal boiling point.

10.8 Entropy Changes in Chemical Reactions

The second law of thermodynamics tells us that a process will be spontaneous if the entropy of the universe increases when the process occurs. We saw in Section 10.7 that for a process at constant temperature and pressure, we can use the change in free energy of the system to predict the sign of ΔS_{univ} and thus the direction in which it is spontaneous. So far we have applied these ideas only to physical processes, such as changes of state and the formation of solutions. However, the main business of chemistry is studying chemical reactions, and therefore, we want to apply the second law to reactions.

First, we will consider the entropy changes accompanying chemical reactions that occur under conditions of constant temperature and pressure. As for the other types of processes we have considered, the entropy changes in the *surroundings* are determined by the heat flow that occurs as the reaction takes place. However, the entropy changes in the *system* (the reactants and products of the reaction) can be predicted by considering the change in positional probability.

For example, in the ammonia synthesis reaction

$$N_2(g) + 3H_2(g) \longrightarrow 2NH_3(g)$$

four reactant molecules are changed to two product molecules, lowering the number of independent units in the system and thus leading to lower positional disorder.

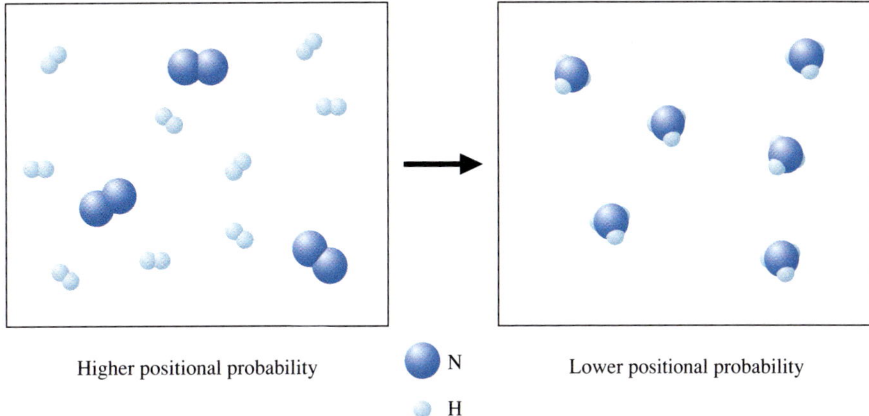

Fewer gaseous molecules means fewer possible configurations. To help clarify this result, consider a special container with a million compartments, each large enough to hold a hydrogen molecule. There are a million ways one H_2 molecule can be placed in this container. But suppose we break the H—H bond and place the two independent H atoms in the same container. A little thought will convince you that there are *many* more than a million ways to place the two separate atoms in the container. The number of arrangements possible for the two independent atoms is much greater than the number for the molecule. Thus for the process

$$H_2 \longrightarrow 2H$$

positional probability increases.

Does positional probability increase or decrease when the following reaction takes place?

$$4NH_3(g) + 5O_2(g) \longrightarrow 4NO(g) + 6H_2O(g)$$

In this case 9 gaseous molecules are changed to 10 gaseous molecules, and the positional probability increases. There are more independent units as products than as reactants. In general, when a reaction involves gaseous molecules, *the change in positional probability is dominated by the relative numbers of molecules of gaseous reactants and products.* If the number of molecules of the gaseous products is greater than the number of molecules of the gaseous reactants, positional probability typically increases, and ΔS is positive for the reaction.

Example 10.6

Predict the sign of $\Delta S°$ for each of the following reactions.

a. the thermal decomposition of solid calcium carbonate:

$$CaCO_3(s) \longrightarrow CaO(s) + CO_2(g)$$

b. the oxidation of SO_2 in air:

$$2SO_2(g) + O_2(g) \longrightarrow 2SO_3(g)$$

Solution

a. Since in this reaction a gas is produced from a solid reactant, the positional probability increases, and $\Delta S°$ is positive.

b. Here three molecules of gaseous reactants become two molecules of gaseous products. Since the number of gas molecules decreases, positional probability decreases, and ΔS° is negative.

Absolute Entropies and the Third Law of Thermodynamics

In thermodynamics it is the *change* in a certain function that is usually important. The change in enthalpy determines whether a reaction is exothermic or endothermic at constant pressure. The change in free energy determines whether a process is spontaneous at constant temperature and pressure. It is fortunate that changes in thermodynamic functions are sufficient for most purposes because absolute values for many thermodynamic characteristics of a system (such as enthalpy or free energy) cannot be determined.

However, we can assign *absolute* entropy values. Consider a solid at 0 K, at which molecular motion virtually ceases. If it is a perfect crystal, its internal arrangement is absolutely regular [see Fig. 10.11(a)]. There is only *one way* to achieve this perfect order; every particle must be in its place. For example, with N coins there is only one way to achieve the state of all heads. Thus a perfect crystal represents the lowest possible entropy; that is, *the entropy of a perfect crystal at 0 K is zero.* This is a statement of the **third law of thermodynamics.**

As the temperature of a perfect crystal is increased, the random vibrational motions increase, and disorder increases within the crystal [see Fig. 10.11(b)]. Thus the entropy of a substance increases with temperature. Since S is zero for a perfect crystal at 0 K, the entropy value for a substance at a particular temperature can be calculated if we know the temperature dependence of entropy.

We have shown that the change in entropy that accompanies a change in temperature of a substance from T_1 to T_2 can be calculated from the expression

$$\Delta S_{T_1 \to T_2} = nC \ln\left(\frac{T_2}{T_1}\right)$$

where C is C_p or C_v depending on the conditions. In using this equation, we are assuming that the heat capacity is independent of temperature. Unfortunately, this assumption is often not true, especially over a large temperature range. Thus the dependence of C on temperature must be taken into account for accurate calculations. However, we will not be concerned with these procedures here.

When a substance is heated from temperature T_1 to T_2, a change of state may occur. If it does, as we saw in Section 10.4, we use the expression

$$\Delta S = \frac{\Delta H}{T}$$

at the melting point or boiling point to account for the entropy change that accompanies the change of state.

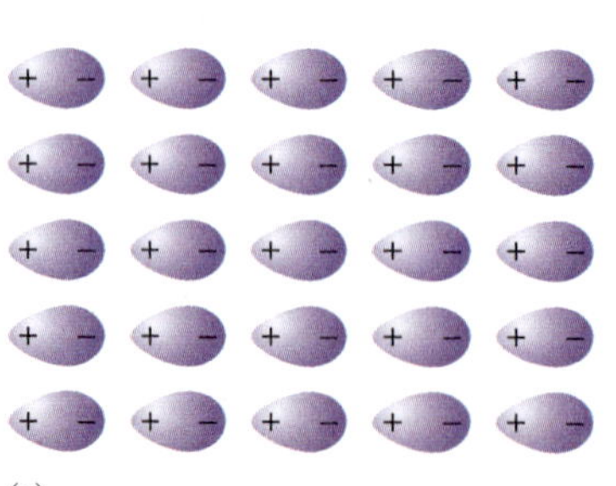
(a)

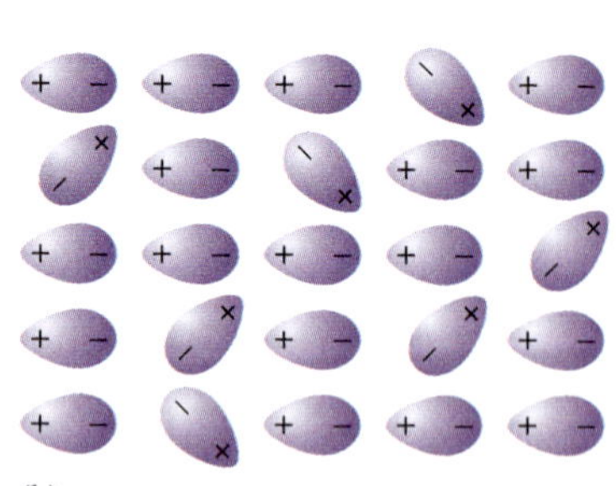
(b)

FIGURE 10.11

(a) An idealized perfect crystal of hydrogen chloride at 0 K; the dipolar HCl molecules are represented by (+ −). The entropy is zero ($S = 0$) for this crystal at 0 K. (b) As the temperature rises above 0 K, lattice vibrations allow some dipoles to change their orientations, producing some disorder and an increase in entropy ($S > 0$).

The standard entropy values represent the increase in entropy that occurs when a substance is heated from 0 K to 298 K at 1 atm.

The *standard entropy values* ($S°$) of many common substances at 298 K and 1 atm are listed in Appendix 4. From these values you will see that the entropy of a substance does indeed increase in going from solid to liquid to gas.

Because *entropy is a state function of the system* (it is not pathway-dependent), the entropy change for a given chemical reaction can be calculated by taking the difference between the standard entropy values of the products and those of the reactants:

$$\Delta S°_{\text{reaction}} = \Sigma S°_{\text{products}} - \Sigma S°_{\text{reactants}}$$

where, as usual, Σ represents the sum of the terms. It is important to note that entropy is an extensive property (it depends on the amount of substance present). This means that *the number of moles of a given reactant or product must be taken into account.*

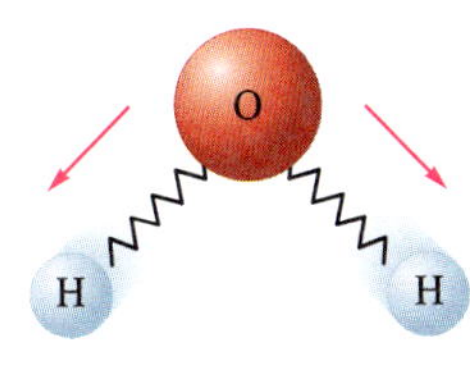

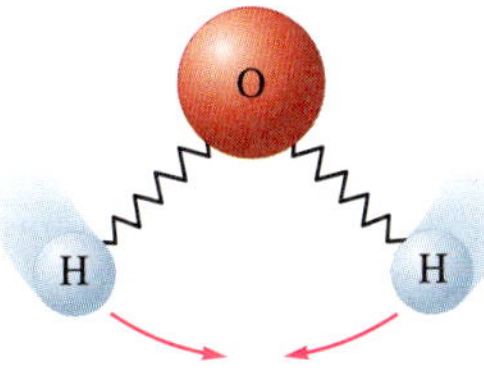

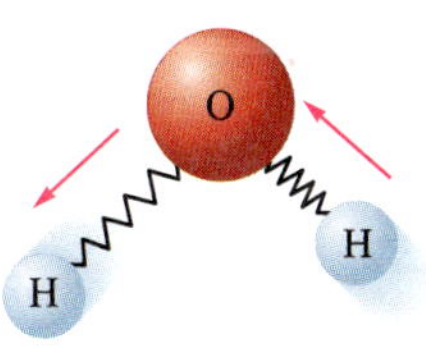

Vibrations

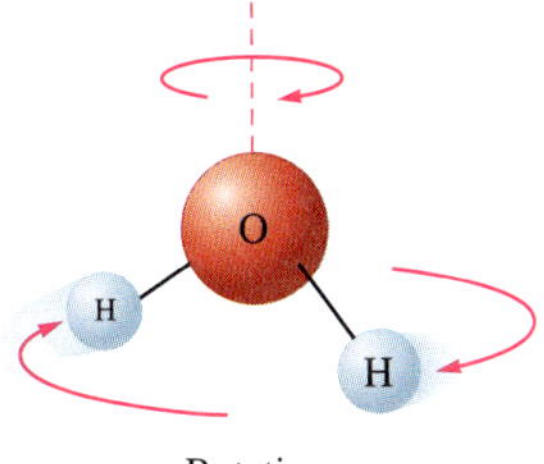

Rotation

FIGURE 10.12

The H_2O molecule can vibrate and rotate in several ways, some of which are shown here. This freedom of motion leads to a higher entropy for water than for a substance such as hydrogen, which consists of a simple diatomic molecule with fewer possible motions.

Example 10.7

Calculate $\Delta S°$ for the reduction of aluminum oxide by hydrogen gas

$$Al_2O_3(s) + 3H_2(g) \longrightarrow 2Al(s) + 3H_2O(g)$$

using the following standard entropy values.

Substance	$S°$ (J K^{-1} mol^{-1})
$Al_2O_3(s)$	51
$H_2(g)$	131
$Al(s)$	28
$H_2O(g)$	189

Solution

$$\begin{aligned}\Delta S° &= \Sigma S°_{\text{products}} - \Sigma S°_{\text{reactants}}\\ &= 2S°_{Al(s)} + 3S°_{H_2O(g)} - 3S°_{H_2(g)} - S°_{Al_2O_3(s)}\\ &= 2\text{ mol}\left(28\ \frac{\text{J}}{\text{K mol}}\right) + 3\text{ mol}\left(189\ \frac{\text{J}}{\text{K mol}}\right)\\ &\quad - 3\text{ mol}\left(131\ \frac{\text{J}}{\text{K mol}}\right) - 1\text{ mol}\left(51\ \frac{\text{J}}{\text{K mol}}\right)\\ &= 56\text{ J/K} + 567\text{ J/K} - 393\text{ J/K} - 51\text{ J/K}\\ &= 179\text{ J/K}\end{aligned}$$

The reaction considered in Example 10.7 involves 3 moles of hydrogen gas on the reactant side and 3 moles of water vapor on the product side. Would you expect ΔS to be large or small for such a case? We have assumed that ΔS depends on the relative numbers of molecules of gaseous reactants and products. On the basis of that assumption, ΔS should be near zero for the present reaction. However, ΔS is large and positive. Why? The large value for ΔS results from the difference in the entropy values for hydrogen gas and water vapor. The reason for this difference can be traced to the difference in molecular structure. Because it is a nonlinear, triatomic molecule, H_2O has more rotational and vibrational motions (see Fig. 10.12) than does the diatomic

H_2 molecule. Thus the standard entropy value for $H_2O(g)$ is greater than that for $H_2(g)$. Generally, *the more complex the molecule, the higher the standard entropy value.*

10.9 Free Energy and Chemical Reactions

For chemical reactions we are often interested in the **standard free energy change** ($\Delta G°$), *the change in free energy that occurs if the reactants in their standard states are converted to the products in their standard states.* For example, for the ammonia synthesis reaction at 25°C,

$$N_2(g) + 3H_2(g) \longrightarrow 2NH_3(g) \qquad \Delta G° = -33.3 \text{ kJ} \tag{10.1}$$

This $\Delta G°$ value represents the change in free energy that occurs when 1 mole of nitrogen gas at 1 atm reacts with 3 moles of hydrogen gas at 1 atm to produce 2 moles of gaseous ammonia at 1 atm.

It is important to recognize that the standard free energy change for a reaction is not measured directly. For example, we can measure heat flow in a calorimeter to determine $\Delta H°$, but we cannot measure $\Delta G°$ this way. The value of $\Delta G°$ for the ammonia synthesis in Equation (10.1) was *not* obtained by mixing 1 mole of N_2 with 3 moles of H_2 in a flask and measuring the change in free energy as 2 moles of NH_3 formed. For one thing, if we mixed 1 mole of N_2 and 3 moles of H_2 in a flask, the system would go to equilibrium rather than to completion. Also, we have no instrument that directly measures free energy. Although we cannot directly measure $\Delta G°$ for a reaction, we can calculate it from other measured quantities, as we will see later in this section.

Why is it useful to know $\Delta G°$ for a reaction? As we will see in more detail later, knowing the $\Delta G°$ values for several reactions allows us to compare the relative tendency of these reactions to occur. The more negative the value of $\Delta G°$, the further a reaction will go to the right to reach equilibrium. We must use standard-state free energies to make this comparison because free energy depends on pressure or concentration. Thus, to get an accurate comparison of reaction tendencies, we must compare all reactions under the same pressure or concentration conditions. We will have more to say about the significance of $\Delta G°$ later.

The value of $\Delta G°$ tells us nothing about the rate of a reaction, only its eventual equilibrium position.

There are several ways to calculate $\Delta G°$. One common method uses the equation

$$\Delta G° = \Delta H° - T\Delta S°$$

which applies to a reaction carried out at constant temperature. For example, consider the reaction

$$C(s) + O_2(g) \longrightarrow CO_2(g)$$

The values of $\Delta H°$ and $\Delta S°$ are known to be -393.5 kJ and 3.05 J/K, respectively, and $\Delta G°$ can be calculated at 298 K as follows:

$$\Delta G° = \Delta H° - T\Delta S° = -3.935 \times 10^5 \text{ J} - (298 \text{ K})(3.05 \text{ J/K})$$
$$= -3.944 \times 10^5 \text{ J} = -394.4 \text{ kJ} \qquad (\text{per mole of } CO_2)$$

Example 10.8

Consider the reaction

$$2SO_2(g) + O_2(g) \longrightarrow 2SO_3(g)$$

carried out at 25°C and 1 atm. Calculate ΔH°, ΔS°, and ΔG° using the following data:

Substance	ΔH_f° (kJ/mol)	S° (J K^{-1} mol^{-1})
$SO_2(g)$	−297	248
$SO_3(g)$	−396	257
$O_2(g)$	0	205

Solution The value of ΔH° can be calculated from the enthalpies of formation using the equation we discussed in Section 9.6:

$$\Delta H^\circ = \Sigma\Delta H_f^\circ\,(\text{products}) - \Sigma\Delta H_f^\circ\,(\text{reactants})$$

Then

$$\begin{aligned}\Delta H^\circ &= 2\Delta H^\circ_{f\,[SO_3(g)]} - 2\Delta H^\circ_{f\,[SO_2(g)]} - \Delta H^\circ_{f\,[O_2(g)]}\\ &= 2\text{ mol}(-396\text{ kJ/mol}) - 2\text{ mol}(-297\text{ kJ/mol}) - 0\\ &= -792\text{ kJ} + 594\text{ kJ} = -198\text{ kJ}\end{aligned}$$

The value of ΔS° can be calculated by using the standard entropy values and the equation discussed in Section 10.8:

$$\Delta S^\circ = \Sigma S^\circ_{\text{products}} - \Sigma S^\circ_{\text{reactants}}$$

Thus

$$\begin{aligned}\Delta S^\circ &= 2S^\circ_{SO_3(g)} - 2S^\circ_{SO_2(g)} - S^\circ_{O_2(g)}\\ &= 2\text{ mol}(257\text{ J K}^{-1}\text{ mol}^{-1}) - 2\text{ mol}(248\text{ J K}^{-1}\text{ mol}^{-1})\\ &\quad -1\text{ mol}(205\text{ J K}^{-1}\text{ mol}^{-1})\\ &= 514\text{ J/K} - 496\text{ J/K} - 205\text{ J/K} = -187\text{ J/K}\end{aligned}$$

We expect ΔS° to be negative since three molecules of gaseous reactants give two molecules of gaseous products.

The value of ΔG° can now be calculated from the equation

$$\Delta G^\circ = \Delta H^\circ - T\Delta S^\circ$$

So

$$\begin{aligned}\Delta G^\circ &= -198\text{ kJ} - (298\text{ K})\left(-187\,\frac{\text{J}}{\text{K}}\right)\left(\frac{1\text{ kJ}}{1000\text{ J}}\right)\\ &= -198\text{ kJ} + 55.7\text{ kJ} = -142\text{ kJ}\end{aligned}$$

Like enthalpy and entropy, *free energy is a state function*. Thus we can use procedures for finding ΔG that are similar to those for finding ΔH using Hess's law.

To illustrate this second method for calculating the free energy change, we will obtain ΔG° for the reaction

$$2CO(g) + O_2(g) \longrightarrow 2CO_2(g) \qquad (10.2)$$

from the following data:

$$2CH_4(g) + 3O_2(g) \longrightarrow 2CO(g) + 4H_2O(g) \qquad \Delta G^\circ = -1088\text{ kJ} \qquad (10.3)$$

$$CH_4(g) + 2O_2(g) \longrightarrow CO_2(g) + 2H_2O(g) \qquad \Delta G^\circ = -801\text{ kJ} \qquad (10.4)$$

Note that $CO(g)$ is a reactant in Equation (10.2), so Equation (10.3) must be reversed, since $CO(g)$ is a product in the reaction as written. When a reaction is reversed, the sign of $\Delta G°$ is also reversed. In Equation (10.4), $CO_2(g)$ is a product, as it is in Equation (10.2), but only one molecule of CO_2 is formed. Thus Equation (10.4) must be multiplied by 2, which means the $\Delta G°$ value for Equation (10.4) must also be multiplied by 2. Free energy is an extensive property since it is defined by two extensive properties, H and S.

Reversed Equation (10.3)

$$2CO(g) + 4H_2O(g) \longrightarrow 2CH_4(g) + 3O_2(g) \qquad \Delta G° = -(-1088\ \text{kJ})$$

$2 \times$ Equation (10.4)

$$2CH_4(g) + 4O_2(g) \longrightarrow 2CO_2(g) + 4H_2O(g) \qquad \Delta G° = 2(-801\ \text{kJ})$$

$$2CO(g) + O_2(g) \longrightarrow 2CO_2(g) \qquad \Delta G° = -(-1088\ \text{kJ}) + 2(-801\ \text{kJ}) = -514\ \text{kJ}$$

This example shows that ΔG values for reactions are manipulated in exactly the same way as ΔH values.

Example 10.9

Using the following data (at 25°C),

$$C_{(s)}^{\text{diamond}} + O_2(g) \longrightarrow CO_2(g) \qquad \Delta G° = -397\ \text{kJ} \qquad (10.5)$$

$$C_{(s)}^{\text{graphite}} + O_2(g) \longrightarrow CO_2(g) \qquad \Delta G° = -394\ \text{kJ} \qquad (10.6)$$

calculate $\Delta G°$ for the reaction

$$C_{(s)}^{\text{diamond}} \longrightarrow C_{(s)}^{\text{graphite}}$$

Solution We reverse Equation (10.6) to make graphite a product, as required, and then add the new equation to Equation (10.5):

$$C_{(s)}^{\text{diamond}} + O_2(g) \longrightarrow CO_2(g) \qquad \Delta G° = -397\ \text{kJ}$$

Reversed Equation (10.6)

$$CO_2(g) \longrightarrow C_{(s)}^{\text{graphite}} + O_2(g) \qquad \Delta G° = -(-394\ \text{kJ})$$

$$C_{(s)}^{\text{diamond}} \longrightarrow C_{(s)}^{\text{graphite}} \qquad \Delta G° = -397\ \text{kJ} + 394\ \text{kJ} = -3\ \text{kJ}$$

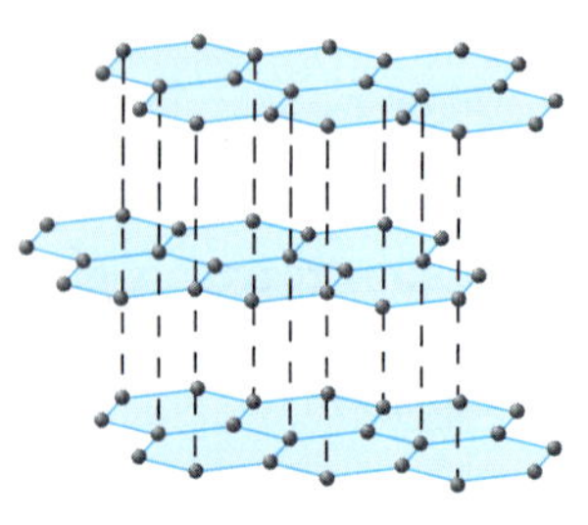

Graphite

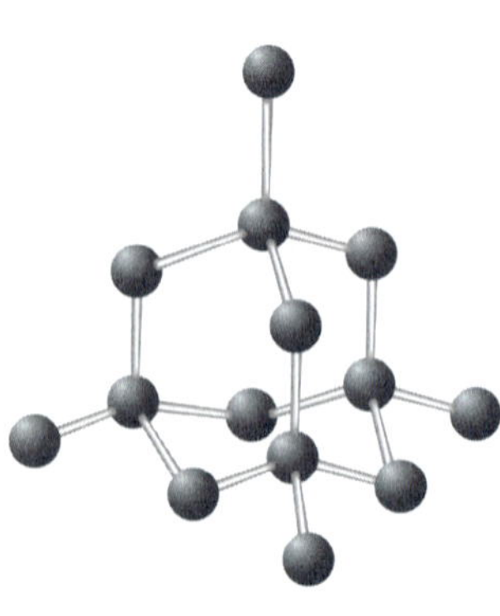

Diamond

Since $\Delta G°$ is negative for this process, diamond should spontaneously change to graphite at 25°C and 1 atm. However, the reaction is so slow under these conditions that we do not observe the process. This example shows *kinetic* rather than *thermodynamic* control of a reaction. That is, thermodynamically, diamond should change to graphite, but this spontaneous change is not observed because its rate is so slow. We say that diamond is kinetically stable with respect to graphite, even though it is thermodynamically unstable.

In Example 10.9 we saw that the process

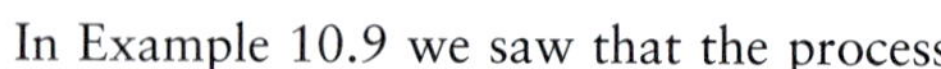

$$C_{(s)}^{\text{diamond}} \longrightarrow C_{(s)}^{\text{graphite}}$$

is spontaneous but very slow at 25°C and 1 atm. The reverse process can be forced to occur at high temperatures and pressures. Diamond has a more compact structure and thus a higher density than graphite, so the exertion of very high pressure causes it to become thermodynamically favored. If high

Synthetic diamonds.

temperatures are also used to make the process fast enough to be feasible, diamonds can be made from graphite. The conditions usually involve temperatures greater than 1000°C and pressures of about 10^5 atm. About half of all industrial diamonds are made this way. We will discuss this process in more detail in Chapter 16.

A third method for calculating the free energy change for a reaction uses standard free energies of formation. **The standard free energy of formation** (ΔG_f°) of a substance is defined as the *change in free energy that accompanies the formation of 1 mole of that substance from its constituent elements with all reactants and products in their standard states.* For example, for the formation of glucose ($C_6H_{12}O_6$), the appropriate reaction is

$$6C(s) + 6H_2(g) + 3O_2(g) \longrightarrow C_6H_{12}O_6(s)$$

The standard free energy associated with this process is called the free energy of formation of glucose. Values of the standard free energy of formation are useful in calculating ΔG° for specific chemical reactions using the equation

$$\Delta G^\circ = \Sigma\Delta G_f^\circ\,(\text{products}) - \Sigma\Delta G_f^\circ\,(\text{reactants})$$

Values of ΔG_f° for many common substances are listed in Appendix 4. Note that, analogous to the enthalpy of formation, *the standard free energy of formation of an element in its standard state is zero*.

The standard state of an element is its most stable state at the temperature of interest (usually 25°C) and 1 atm.

Calculating ΔG° from free energies of formation is very similar to calculating ΔH°, as shown in Section 9.6.

Example 10.10

Methanol is a high-octane fuel used in high-performance racing engines. Calculate ΔG° for the reaction

$$2CH_3OH(g) + 3O_2(g) \longrightarrow 2CO_2(g) + 4H_2O(g)$$

given the following free energies of formation:

Substance	ΔG_f° (kJ/mol)
$CH_3OH(g)$	−163
$O_2(g)$	0
$CO_2(g)$	−394
$H_2O(g)$	−229

Solution We use

$$\begin{aligned}\Delta G_f^\circ &= \Sigma\Delta G_f^\circ\,(\text{products}) - \Sigma\Delta G_f^\circ\,(\text{reactants})\\ &= 2\Delta G_{f\,[CO_2(g)]}^\circ + 4\Delta G_{f\,[H_2O(g)]}^\circ - 3\Delta G_{f\,[O_2(g)]}^\circ - 2\Delta G_{f\,[CH_3OH(g)]}^\circ\\ &= 2\text{ mol}(-394\text{ kJ/mol}) + 4\text{ mol}(-229\text{ kJ/mol}) - 3(0)\\ &\quad - 2\text{ mol}(-163\text{ kJ/mol})\\ &= -1378\text{ kJ}\end{aligned}$$

The large magnitude and the negative sign of ΔG° indicate that this reaction is very favorable thermodynamically.

Example 10.11

A chemical engineer wants to determine the feasibility of making ethanol (C_2H_5OH) by reacting water with ethylene (C_2H_4) according to the equation

$$C_2H_4(g) + H_2O(l) \longrightarrow C_2H_5OH(l)$$

Is this reaction spontaneous under standard conditions?

Ethylene

Ethanol

Solution To determine the spontaneity of this reaction under standard conditions, we must determine $\Delta G°$ for the reaction by using the appropriate standard free energies of formation at 25°C from Appendix 4:

$$\Delta G_{f\,[C_2H_5OH(l)]}^{\circ} = -175 \text{ kJ/mol}$$

$$\Delta G_{f\,[H_2O(l)]}^{\circ} = -237 \text{ kJ/mol}$$

$$\Delta G_{f\,[C_2H_4(g)]}^{\circ} = 68 \text{ kJ/mol}$$

Thus

$$\Delta G° = \Delta G_{f\,[C_2H_5OH(l)]}^{\circ} - \Delta G_{f\,[H_2O(l)]}^{\circ} - \Delta G_{f\,[C_2H_4(g)]}^{\circ}$$

$$= -175 \text{ kJ} - (-237 \text{ kJ}) - 68 \text{ kJ} = -6 \text{ kJ}$$

and the process is spontaneous under standard conditions at 25°C.

Although the reaction considered in Example 10.11 is spontaneous, other features of the reaction must be studied to see whether the process is feasible. For example, the chemical engineer will need to study the kinetics of the reaction to determine whether it is fast enough to be useful and, if it is not, whether a catalyst can be found to enhance the rate. In doing these studies, the engineer must remember that $\Delta G°$ depends on temperature:

$$\Delta G° = \Delta H° - T\Delta S°$$

Thus, if the process must be carried out at high temperatures to be fast enough to be feasible, $\Delta G°$ must be recalculated at that temperature using the $\Delta H°$ and $\Delta S°$ values for the reaction.

10.10 The Dependence of Free Energy on Pressure

In this chapter we have seen that a reaction system at constant temperature and pressure will proceed spontaneously in the direction that lowers its free energy. For this reason reactions proceed until they reach equilibrium. As we will see later in this section, the equilibrium position represents the lowest free energy value available to a particular reaction system. The free energy of a reaction system changes as the reaction proceeds because free energy depends on the pressure of a gas (or on the concentration of species in solution). We will deal only with the pressure dependence of the free energy of an ideal gas. The dependence of free energy on concentration can be developed using similar reasoning.

To understand the pressure dependence of free energy, we need to know how pressure affects the thermodynamic functions that constitute free energy—that is, enthalpy and entropy (recall that $G = H - TS$). For an ideal gas enthalpy is not pressure-dependent. However, entropy *does* depend on pressure because of its dependence on volume. Consider one mole of an ideal gas at a given temperature. At a volume of 10.0 L, the gas has many more positions available for the molecules than if its volume is 1.0 L. The positional probability is greater for the larger volume. In summary, at a given temperature for 1 mole of ideal gas

$$S_{\text{large volume}} > S_{\text{small volume}}$$

or since pressure and volume are inversely related,

$$S_{\text{low pressure}} > S_{\text{high pressure}}$$

We have shown qualitatively that the entropy and therefore the free energy of an ideal gas depend on its pressure. From a more detailed argument, which we will not consider here, one can show that

$$G = G^\circ + RT \ln(P)$$

where G° is the free energy of the gas at a pressure of 1 atm, G is the free energy of the gas at a pressure of P atm, R is the universal gas constant, and T is the kelvin temperature.

To see how the change in free energy for a reaction depends on pressure, we will consider the ammonia synthesis reaction

$$N_2(g) + 3H_2(g) \longrightarrow 2NH_3(g)$$

The absolute value of the free energy of a substance cannot be obtained. We use it symbolically here to show that it is the change in free energy that is really significant.

In general, $$\Delta G = \Sigma G_{\text{products}} - \Sigma G_{\text{reactants}} \quad (10.7)$$

For this reaction $$\Delta G = 2G_{NH_3} - G_{N_2} - 3G_{H_2}$$

where
$$G_{NH_3} = G^\circ_{NH_3} + RT \ln(P_{NH_3})$$
$$G_{N_2} = G^\circ_{N_2} + RT \ln(P_{N_2})$$
$$G_{H_2} = G^\circ_{H_2} + RT \ln(P_{H_2})$$

Substituting these values into Equation (10.7) gives

$$\begin{aligned}\Delta G &= 2[G^\circ_{NH_3} + RT \ln(P_{NH_3})] - [G^\circ_{N_2} + RT \ln(P_{N_2})] - 3[G^\circ_{H_2} + RT \ln(P_{H_2})]\\ &= 2G^\circ_{NH_3} - G^\circ_{N_2} - 3G^\circ_{H_2} + 2RT \ln(P_{NH_3}) - RT \ln(P_{N_2}) - 3RT \ln(P_{H_2})\\ &= \underbrace{(2G^\circ_{NH_3} - G^\circ_{N_2} - 3G^\circ_{H_2})}_{\Delta G^\circ_{\text{reaction}}} + RT[2 \ln(P_{NH_3}) - \ln(P_{N_2}) - 3 \ln(P_{H_2})]\end{aligned}$$

The first term in parentheses is ΔG° for the reaction. Thus we have

$$\Delta G = \Delta G^\circ_{\text{reaction}} + RT[2 \ln(P_{NH_3}) - \ln(P_{N_2}) - 3 \ln(P_{H_2})]$$

Since
$$2 \ln(P_{NH_3}) = \ln(P_{NH_3}{}^2)$$
$$-\ln(P_{N_2}) = \ln\left(\frac{1}{P_{N_2}}\right)$$

and
$$-3 \ln(P_{H_2}) = \ln\left(\frac{1}{P_{H_2}{}^3}\right)$$

the equation becomes

$$\Delta G = \Delta G^\circ + RT \ln\left[\frac{P_{NH_3}{}^2}{(P_{N_2})(P_{H_2}{}^3)}\right]$$

But the term
$$\frac{P_{NH_3}{}^2}{(P_{N_2})(P_{H_2}{}^3)}$$

is the reaction quotient Q discussed in Section 6.6. Therefore, we have

$$\Delta G = \Delta G^\circ + RT \ln(Q)$$

where Q is the reaction quotient (from the law of mass action), T is the temperature (K), R is the gas law constant, $\Delta G°$ is the free energy change for the reaction with all reactants and products at a pressure of 1 atm, and ΔG is the free energy change for the reaction at the specified pressures of reactants and products.

Example 10.12

One method for synthesizing methanol (CH_3OH) involves reacting gaseous carbon monoxide and hydrogen:

$$CO(g) + 2H_2(g) \longrightarrow CH_3OH(l)$$

Calculate ΔG at 25°C for this reaction, in which carbon monoxide gas at 5.0 atm and hydrogen gas at 3.0 atm are converted to liquid methanol.

Solution To calculate ΔG for this process, we use the equation

$$\Delta G = \Delta G° + RT\ln(Q)$$

We must first compute $\Delta G°$ from standard free energies of formation (see Appendix 4). Since

$$\Delta G°_{f\,[CH_3OH(l)]} = -166 \text{ kJ}$$
$$\Delta G°_{f\,[H_2(g)]} = 0$$
$$\Delta G°_{f\,[CO(g)]} = -137 \text{ kJ}$$

then

$$\Delta G° = -166 \text{ kJ} - (-137 \text{ kJ}) - 0 = -29 \text{ kJ} = -2.9 \times 10^4 \text{ J}$$

Note that this is the value of $\Delta G°$ for the reaction of 1 mol CO with 2 mol H_2 to produce 1 mol CH_3OH. We might call this the value of $\Delta G°$ for one "unit" of the reaction or for "one mole of the reaction." Thus the $\Delta G°$ value might be written more accurately as -2.9×10^4 J/mol of reaction, or -2.9×10^4 J/mol rxn.

Note that in this case ΔG is defined for 1 mol of the reaction—that is, for 1 mol $CO(g)$ reacting with 2 mol $H_2(g)$ to form 1 mol $CH_3OH(l)$. Thus ΔG, $\Delta G°$, and $RT\ln(Q)$ all have units of J/mol of reaction. In this case the units of R are actually J K^{-1} (mol of reaction)$^{-1}$, although they are usually not written this way.

We can now calculate ΔG, where

$$\Delta G° = -2.9 \times 10^4 \text{ J/mol rxn}$$
$$R = 8.3145 \text{ J K}^{-1}\text{ mol}^{-1}$$
$$T = 273 + 25 = 298 \text{ K}$$
$$Q = \frac{1}{(P_{CO})(P_{H_2}{}^2)} = \frac{1}{(5.0)(3.0)^2} = 2.2 \times 10^{-2}$$

Note that the pure liquid methanol is not included in the calculation of Q because a pure liquid has an activity of 1, as discussed in Chapter 6. Thus

$$\begin{aligned}\Delta G &= \Delta G° + RT\ln(Q)\\ &= (-2.9 \times 10^4 \text{ J/mol rxn})\\ &\quad + [8.3145 \text{ J K}^{-1}\text{ (mol rxn)}^{-1}](298 \text{ K})[\ln(2.2 \times 10^{-2})]\\ &= (-2.9 \times 10^4 \text{ J/mol rxn}) - (9.4 \times 10^3 \text{ J/mol rxn})\\ &= -3.8 \times 10^4 \text{ J/mol rxn}\\ &= -38 \text{ kJ/mol rxn}\end{aligned}$$

Note that ΔG is significantly more negative than $\Delta G°$, implying that the reaction is more spontaneous at reactant pressures greater than 1 atm. We might expect this result from Le Châtelier's principle.

The Meaning of ΔG for a Chemical Reaction

In this section we have learned to calculate ΔG for chemical reactions under various conditions. For example, in Example 10.12 the calculations show that the formation of $CH_3OH(l)$ from $CO(g)$ at 5.0 atm reacting with $H_2(g)$ at 3.0 atm is spontaneous. What does this result mean? Does it mean that if we mixed 1.0 mol of $CO(g)$ and 2.0 mol of $H_2(g)$ together at pressures of 3.0 atm and 5.0 atm, respectively, that 1.0 mol of $CH_3OH(l)$ would form in the reaction flask? The answer is no. This answer may surprise you in view of what has been said in this section. It is true that 1.0 mol of $CH_3OH(l)$ has a lower free energy than 1.0 mol of $CO(g)$ at 5.0 atm plus 2.0 mol of $H_2(g)$ at 3.0 atm. However, when $CO(g)$ and $H_2(g)$ are mixed under these conditions, there is *an even lower free energy available to this system than 1.0 mol of pure $CH_3OH(l)$*. For reasons we will discuss shortly, *the system can achieve the lowest possible free energy by going to equilibrium, not by going to completion*. At the equilibrium position some of the $CO(g)$ and $H_2(g)$ will remain in the reaction flask. So even though 1.0 mol of pure $CH_3OH(l)$ is at a lower free energy than 1.0 mol of $CO(g)$ and 2.0 mol of $H_2(g)$ at 5.0 atm and 3.0 atm, respectively, the reaction system will stop short of forming 1.0 mol of $CH_3OH(l)$. The reaction stops short of completion because the equilibrium mixture of $CH_3OH(l)$, $CO(g)$, and $H_2(g)$ exists at the lowest possible free energy available to the system.

To illustrate this point, we will explore a mechanical example. Consider balls rolling down the two hills shown in Fig. 10.13. Note that in both cases point *B* has a lower potential energy than point *A*.

In Fig. 10.13(a) the ball will roll to point *B*. This diagram is analogous to a phase change. For example, at 25°C ice will spontaneously change completely to liquid water because the latter has the lowest free energy. In this case liquid water is the only choice. There is no intermediate substance with lower free energy.

The situation is different for a chemical reaction system, as illustrated in Fig. 10.13(b). In Fig. 10.13(b) the ball will not get to point *B* because there is a lower potential energy at point C. Like the ball, a chemical system will seek the *lowest possible* free energy, which, for reasons we will discuss in the next section, is the equilibrium position.

Therefore, although the value of ΔG for a given reaction system tells us whether the products or reactants are favored under a given set of conditions,

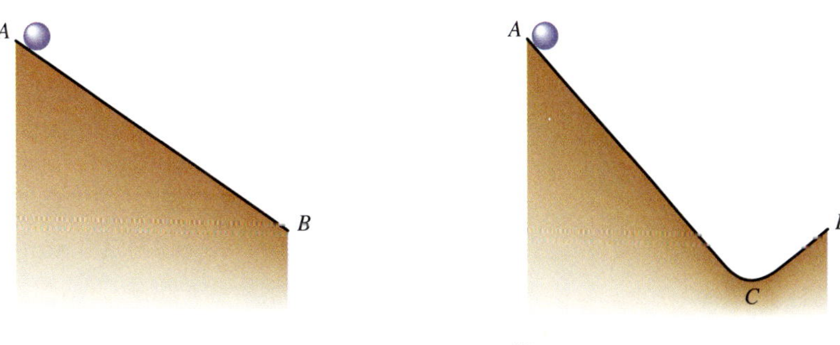

FIGURE 10.13

Schematic representations of balls rolling down two types of hills.

it does not mean that the system will proceed to pure products (if ΔG is negative) or remain at pure reactants (if ΔG is positive). Instead, the system will spontaneously go to the equilibrium position, the lowest possible free energy available to it. In the next section we will see that the value of ΔG° for a particular reaction tells us exactly where this position will be.

10.11 Free Energy and Equilibrium

When the components of a given chemical reaction are mixed, they will proceed, rapidly or slowly depending on the kinetics of the process, to the equilibrium position. In Chapter 6 we defined the equilibrium position as the point at which the forward and reverse reaction rates are equal. In this chapter we look at equilibrium from a thermodynamic point of view, and we find that *the equilibrium point occurs at the lowest value of free energy available to the reaction system.* As it turns out, the two definitions give the same equilibrium state, which must be the case for both the kinetic and thermodynamic models to be valid.

To understand the relationship between free energy and equilibrium, let's consider the following simple hypothetical reaction:

$$A(g) \rightleftharpoons B(g)$$

where 1.0 mol of gaseous A is initially placed in a reaction vessel at a pressure of 2.0 atm. The free energies for A and B are diagramed as shown in Fig. 10.14(a). As A reacts to form B, the total free energy of the system changes, yielding the following results:

$$\text{Total free energy of A} = G_A = n_A[G_A^\circ + RT\ln(P_A)]$$

$$\text{Total free energy of B} = G_B = n_B[G_B^\circ + RT\ln(P_B)]$$

where G_A and G_B are the total free energies of A and B, and n_A and n_B are the moles of A and B, respectively.

$$\text{Total free energy of system} = G = G_A + G_B$$

As A changes to B, G_A decreases because P_A is decreasing [Fig. 10.14(b)]. In contrast, G_B increases because P_B is increasing. The reaction proceeds to the right as long as the total free energy of the system decreases (as long as G_B is less than G_A). At some point the pressures of A and B reach the values P_A^e and P_B^e that make G_A equal to G_B. *The system has reached equilibrium* [Fig. 10.14(c)]. Since A at pressure P_A^e and B at pressure P_B^e have the same free energy (G_A equals G_B), ΔG is zero for A at pressure P_A^e changing to B at pressure P_B^e. *The system has reached minimum free energy.* There is no longer any driving force to change A to B or B to A, so the system remains at this position (the pressures of A and B remain constant).

Suppose that for the experiment just described, the plot of free energy versus the mole fraction of A reacted is defined as shown in Fig. 10.15(a). In this experiment minimum free energy is reached when 75% of A has been changed to B. At this point the pressure of A is 0.25 times the original pressure, or

$$(0.25)(2.0 \text{ atm}) = 0.50 \text{ atm}$$

The pressure of B is

$$(0.75)(2.0 \text{ atm}) = 1.5 \text{ atm}$$

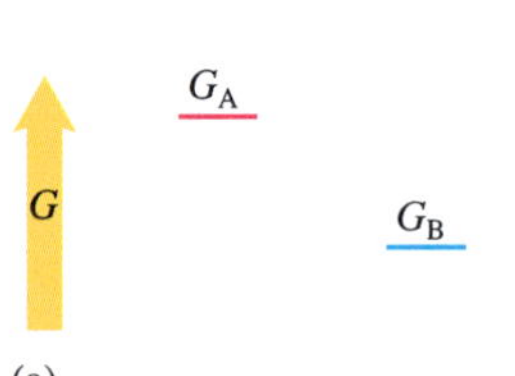

(a)

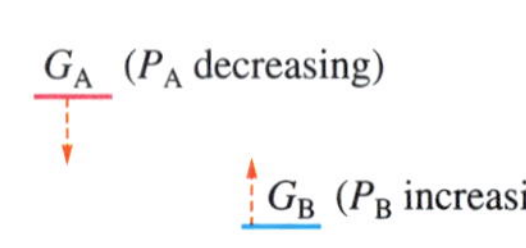

(b)

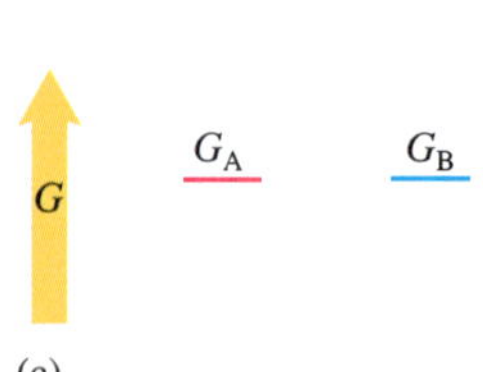

(c)

FIGURE 10.14
(a) The initial free energies of A and B. (b) As A(g) changes to B(g), the free energy of A decreases and that of B increases. (c) Eventually, pressures of A and B are achieved such that $G_A = G_B$, the equilibrium position.

Note that G_A and G_B are defined here as the total free energies of A and B and are dependent on the moles of A and B as well as the pressures of A and B.

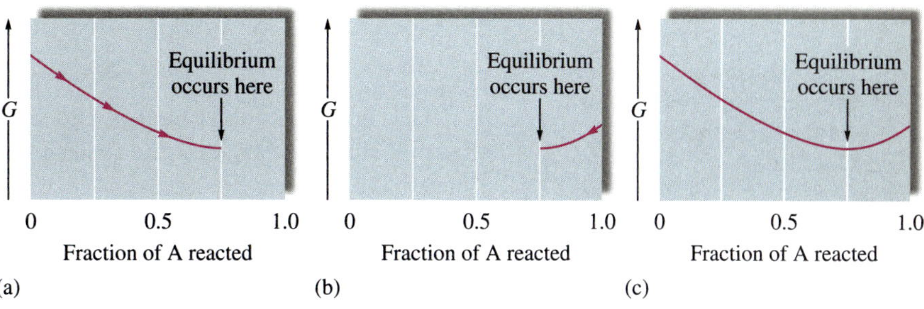

FIGURE 10.15

(a) The change in free energy to reach equilibrium, beginning with 1.0 mol of A(g) at $P_A = 2.0$ atm. (b) The change in free energy to reach equilibrium, beginning with 1.0 mol of B(g) at $P_B = 2.0$ atm. (c) The free energy profile for A(g) $\rightleftharpoons$ B(g) in a system containing 1.0 mol (A plus B) at $P_{Total} = 2.0$ atm. Each point on the curve corresponds to the total free energy for a given combination of A and B.

For the reaction A(g) $\rightleftharpoons$ B(g) the pressure is constant during the reaction, since the same number of gas molecules is always present.

Since this is the equilibrium position, we can use the equilibrium pressures to calculate a value for K for the reaction in which A is converted to B:

$$K = \frac{P_B^e}{P_A^e} = \frac{1.5 \text{ atm}}{0.50 \text{ atm}} = 3.0$$

Exactly the same equilibrium point will be achieved if we place 1.0 mol of pure B(g) in the flask at a pressure of 2.0 atm. In this case B will change to A until equilibrium ($G_B = G_A$) is reached. See Fig. 10.15(b).

The overall free energy curve for this system is shown in Fig. 10.15(c). Note that any mixture of A(g) and B(g) containing 1.0 mol of A plus B at a total pressure of 2.0 atm will react until it reaches the minimum on the curve.

In summary, when substances undergo a chemical reaction, the reaction proceeds to the minimum free energy (equilibrium), which corresponds to the point where $G_{products} = G_{reactants}$, or

Here $G_{products}$ and $G_{reactants}$ represent the sums for all products and all reactants, respectively.

$$\Delta G = G_{products} - G_{reactants} = 0$$

We can now establish a quantitative relationship between free energy and the value of the equilibrium constant. We have seen that

$$\Delta G = \Delta G^\circ + RT \ln(Q)$$

and at equilibrium ΔG equals 0 and Q equals K. So

$$\Delta G = 0 = \Delta G^\circ + RT \ln(K)$$

or

$$\Delta G^\circ = -RT \ln(K)$$

We must note the following characteristics of this very important equation:

CASE 1: $\Delta G^\circ = 0$. When ΔG° equals zero for a particular reaction, the free energies of the reactants and products are equal when all components are in the standard states (1 atm for gases). The system is at equilibrium when the pressures of all reactants and products are 1 atm, which means that *K equals 1*.

CASE 2: $\Delta G^\circ < 0$. In this case ΔG° ($G^\circ_{products} - G^\circ_{reactants}$) is negative, which means that

$$G^\circ_{products} < G^\circ_{reactants}$$

If a flask contains the reactants and products, all at 1 atm, the system is *not* at equilibrium. Since $G^\circ_{products}$ is less than $G^\circ_{reactants}$, the system adjusts to the right to reach equilibrium. In this case K is *greater than 1*, since the pressures of the products at equilibrium are greater than 1 atm and the pressures of the reactants at equilibrium are less than 1 atm.

TABLE 10.7

Qualitative Relationship Between the Change in Standard Free Energy and the Equilibrium Constant for a Given Reaction

ΔG°	K
$\Delta G^\circ = 0$	$K = 1$
$\Delta G^\circ < 0$	$K > 1$
$\Delta G^\circ > 0$	$K < 1$

CASE 3: $\Delta G^\circ > 0$. Since ΔG° ($G^\circ_{\text{products}} - G^\circ_{\text{reactants}}$) is positive,

$$G^\circ_{\text{reactants}} < G^\circ_{\text{products}}$$

If a flask contains the reactants and products, all at 1 atm, the system is *not* at equilibrium. In this case the system adjusts to the left (toward the reactants, which have a lower free energy) to reach equilibrium. The value of K is *less than 1*, since at equilibrium the pressures of the reactants are greater than 1 atm and the pressures of the products are less than 1 atm.

These results are summarized in Table 10.7. The value of K for a specific reaction can be calculated from the equation

$$\Delta G^\circ = -RT \ln(K)$$

as is shown in Examples 10.13 and 10.14.

Example 10.13

Consider the ammonia synthesis reaction

$$N_2(g) + 3H_2(g) \rightleftharpoons 2NH_3(g)$$

where $\Delta G^\circ = -33.3$ kJ per mole of N_2 consumed at 25°C. For each of the following mixtures of reactants and products at 25°C, predict the direction in which the system will shift to reach equilibrium.

a. $P_{NH_3} = 1.00$ atm, $P_{N_2} = 1.47$ atm, and $P_{H_2} = 1.00 \times 10^{-2}$ atm
b. $P_{NH_3} = 1.00$ atm, $P_{N_2} = 1.00$ atm, and $P_{H_2} = 1.00$ atm

Solution

a. We can predict the direction of the shift to equilibrium by calculating the value of ΔG using the equation

$$\Delta G = \Delta G^\circ + RT \ln(Q)$$

The units of ΔG, ΔG°, and $RT \ln(Q)$ are all per "mole of reaction," although the "per mole" is indicated only for R (as is customary).

$$\text{where } Q = \frac{P_{NH_3}{}^2}{(P_{N_2})(P_{H_2}{}^3)} = \frac{(1.00)^2}{(1.47)[(1.00 \times 10^{-2})^3]} = 6.80 \times 10^5$$

$$T = 25 + 273 = 298 \text{ K}$$

$$R = 8.3145 \text{ J K}^{-1} \text{ mol}^{-1}$$

$$\text{and } \Delta G^\circ = -33.3 \text{ kJ/mol} = -3.33 \times 10^4 \text{ J/mol}$$

Thus

$$\Delta G = (-3.33 \times 10^4 \text{ J/mol}) + (8.3145 \text{ J K}^{-1} \text{ mol}^{-1})(298 \text{ K}) \ln(6.80 \times 10^5)$$
$$= (-3.33 \times 10^4 \text{ J/mol}) + (3.33 \times 10^4 \text{ J/mol}) = 0$$

Since $\Delta G = 0$, the reactants and products have the same free energies at the given partial pressures. The system is already at equilibrium, and no shift occurs.

b. The partial pressures given here are all 1.00 atm, indicating that the system is in the standard state. That is,

$$\Delta G = \Delta G^\circ + RT \ln(Q) = \Delta G^\circ + RT \ln \frac{(1.00)^2}{(1.00)(1.00)^3}$$
$$= \Delta G^\circ + RT \ln(1.00) = \Delta G^\circ + 0 = \Delta G^\circ$$

For this reaction at 25°C,

$$\Delta G^\circ = -33.3 \text{ kJ/mol}$$

The negative value for ΔG° means that in their standard states the products have a lower free energy than the reactants. Thus the system moves to the right to reach equilibrium. That is, K is greater than 1.

Example 10.14

The overall reaction for the corrosion (rusting) of iron by oxygen is

$$4Fe(s) + 3O_2(g) \rightleftharpoons 2Fe_2O_3(s)$$

Rusted warships in Micronesia.

Using the following data, calculate the equilibrium constant for this reaction at 25°C.

Substance	ΔH_f° (kJ/mol)	S° (J K^{-1} mol^{-1})
$Fe_2O_3(s)$	−826	90
$Fe(s)$	0	27
$O_2(g)$	0	205

Solution To calculate K for this reaction, we will use the equation

$$\Delta G^\circ = -RT \ln(K)$$

We must first calculate ΔG° from

$$\Delta G^\circ = \Delta H^\circ - T\Delta S^\circ$$

where

$$\Delta H^\circ = 2\Delta H^\circ_{f\,[Fe_2O_3(s)]} - 3\Delta H^\circ_{f\,[O_2(g)]} - 4\Delta H^\circ_{f\,[Fe(s)]}$$
$$= 2 \text{ mol}(-826 \text{ kJ/mol}) - 0 - 0$$
$$= -1652 \text{ kJ} = -1.652 \times 10^6 \text{ J}$$

$$\Delta S^\circ = 2S^\circ_{Fe_2O_3} - 3S^\circ_{O_2} - 4S^\circ_{Fe}$$
$$= 2 \text{ mol}(90 \text{ J K}^{-1} \text{ mol}^{-1}) - 3 \text{ mol}(205 \text{ J K}^{-1} \text{ mol}^{-1}) - 4 \text{ mol}(27 \text{ J K}^{-1} \text{ mol}^{-1})$$
$$= -543 \text{ J/K}$$

and

$$T = 273 + 25 = 298 \text{ K}$$

The units of ΔG, ΔG°, and $RT \ln(Q)$ are all per "mole of reaction," although the "per mole" is indicated only for R (as is customary).

Then

$$\Delta G^\circ = \Delta H^\circ - T\Delta S^\circ = (-1.652 \times 10^6 \text{ J}) - (298 \text{ K})(-543 \text{ J/K})$$
$$= -1.490 \times 10^6 \text{ J}$$

and

$$\Delta G^\circ = -RT \ln(K) = -1.490 \times 10^6 \text{ J}$$
$$= -(8.3145 \text{ J K}^{-1} \text{ mol}^{-1})(298 \text{ K}) \ln(K)$$

Thus

$$\ln(K) = \frac{1.490 \times 10^6}{2.48 \times 10^3} = 601 \quad \text{and} \quad K = e^{601}$$

In terms of base = 10, $K = 10^{261}$

This is a very large equilibrium constant. The rusting of iron is clearly very favorable from a thermodynamic point of view.

The Temperature Dependence of K

In Chapter 6 we used Le Châtelier's principle to predict qualitatively how the value of K for a given reaction would change with a change in temperature. Now we can specify the quantitative dependence of the equilibrium constant on temperature from the relationship

$$\Delta G^\circ = -RT \ln(K) = \Delta H^\circ - T\Delta S^\circ$$

We can rearrange this equation to give

$$\ln(K) = -\frac{\Delta H^\circ}{RT} + \frac{\Delta S^\circ}{R} = -\frac{\Delta H^\circ}{R}\left(\frac{1}{T}\right) + \frac{\Delta S^\circ}{R}$$

Note that this is a linear equation of the form $y = mx + b$, where $y = \ln(K)$, $m = -\Delta H^\circ/R$ = slope, $x = 1/T$, and $b = \Delta S^\circ/R$ = intercept. This means that if values of K for a given reaction are determined at various temperatures, a plot of $\ln(K)$ versus $1/T$ will be linear, with slope $-\Delta H^\circ/R$ and intercept $\Delta S^\circ/R$. This result assumes that both ΔH° and ΔS° are independent of temperature over the temperature range considered. This assumption is valid only over a relatively small temperature range.

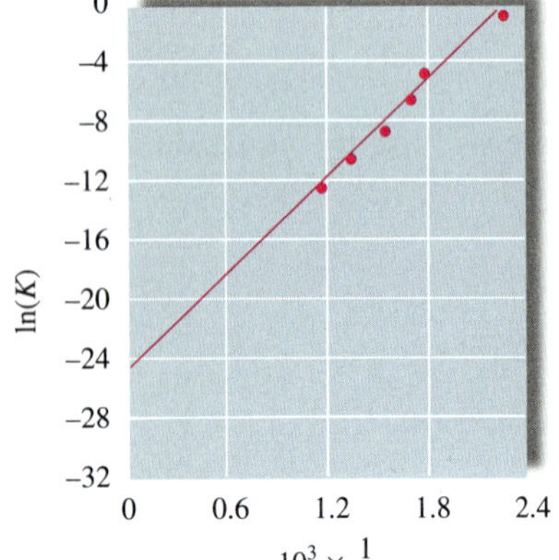

FIGURE 10.16

Experimental data showing the dependence of K on T (in kelvins) for the ammonia synthesis reaction.

An important conclusion that can be drawn from this equation is that the sign of the slope of the plot of $\ln(K)$ versus $1/T$ depends on the sign of ΔH° for the reaction. Note that an exothermic reaction ($\Delta H^\circ < 0$) will show a positive slope (ΔH° is negative, so $-\Delta H^\circ/R$ is positive) for the $\ln(K)$ versus $1/T$ plot. In this case $\ln(K)$ will increase as $1/T$ increases (T decreases). Thus K increases as T is decreased or, conversely, K decreases as T is increased. This is exactly the temperature dependence of K predicted for an exothermic reaction by Le Châtelier's principle (see Section 6.8). This effect can be shown quantitatively by examining how the value of K for the ammonia synthesis reaction

$$N_2(g) + 3H_2(g) \rightleftharpoons 2NH_3(g)$$

depends on temperature. Figure 10.16 shows a plot of $\ln(K)$ versus $1/T$ for this exothermic reaction ($\Delta H^\circ = -92$ kJ for the reaction as written). Note that $\ln(K)$ increases as $1/T$ increases, meaning that K increases as T is decreased, as expected. Of course, the reverse applies for the equilibrium constant for an endothermic reaction: The value of K increases as the temperature is increased.

Once the temperature dependence of K for a given reaction is known, this relationship can be used to predict the value of K at any temperature (assuming ΔH° is constant with T). This can be seen most easily as follows. Assuming that K_1 and K_2 are the equilibrium constants for a given reaction at temperatures T_1 and T_2, we can write

$$\ln(K_2) = \frac{-\Delta H^\circ}{RT_2} + \frac{\Delta S^\circ}{R}$$

and

$$\ln(K_1) = \frac{-\Delta H^\circ}{RT_1} + \frac{\Delta S^\circ}{R}$$

Subtracting the second equation from the first gives the combined equation:

$$\ln\left(\frac{K_2}{K_1}\right) = \frac{-\Delta H^\circ}{R}\left[\frac{1}{T_2} - \frac{1}{T_1}\right]$$

This is called the *van't Hoff equation* after Dutch chemist Jacobus van't Hoff. This equation can be used to calculate K at any temperature once ΔH° and K are known at a given temperature. The accuracy of this calculation depends on whether ΔH° and ΔS° are constant over the temperature range considered.

Example 10.15

The value of K_p is 3.7×10^{-6} at 900. K for the ammonia synthesis reaction. Assuming the value of ΔH° for this reaction is -92 kJ, calculate the value of K_p at 550. K.

Solution We use the van't Hoff equation

$$\ln\left(\frac{K_2}{K_1}\right) = \frac{-\Delta H^\circ}{R}\left(\frac{1}{T_2} - \frac{1}{T_1}\right)$$

where $K_1 = 3.7 \times 10^{-6}$, $T_1 = 900.$ K, and $T_2 = 550.$ K.

$$\ln\left(\frac{K_2}{3.7 \times 10^{-6}}\right) = -\left(\frac{-92{,}000 \text{ J/mol}}{8.3145 \dfrac{\text{J}}{\text{K mol}}}\right)\left(\frac{1}{550.} - \frac{1}{900.}\right)$$

$$\ln(K_2) - \ln(3.7 \times 10^{-6}) = 1.1 \times 10^4 \text{ K } (1.8 \times 10^{-3} - 1.1 \times 10^{-3})$$

Solving this equation gives

$$\ln(K_2) = -4.8$$

$$K_2 = 8.2 \times 10^{-3} = K_p \text{ at } 550. \text{ K}$$

Notice that the value of K_p increased as the temperature decreased, as expected for an exothermic reaction.

10.12 Free Energy and Work

One of the main reasons we are interested in physical and chemical processes is that we want to use them to do work for us, and we want this work done as efficiently and economically as possible. We have already seen that at constant temperature and pressure the sign of the change in free energy tells us whether a given process is spontaneous. This information is very useful because it prevents us from wasting effort on a process that has no inherent tendency to occur. Although a thermodynamically favorable chemical reaction may not occur to any appreciable extent at a given temperature because it is too slow, finding a catalyst to speed up the reaction makes sense in this case. On the other hand, if the reaction is prevented from occurring by its thermodynamic characteristics, we would be wasting our time looking for a catalyst.

In addition to being important qualitatively (telling us whether a process is spontaneous), the change in free energy is important quantitatively because it can tell us how much work can be done through a given process. In fact, as we will show, the *maximum possible useful work obtainable from a process at constant temperature and pressure is equal to the change in free energy:*

$$w_{\text{useful}}^{\text{max}} = \Delta G$$

This relationship explains why this function is called the *free* energy. Under certain conditions ΔG for a spontaneous process represents the energy that is *free to do useful work*. On the other hand, for a process that is not spontaneous, the value of ΔG tells us the minimum amount of work that must be *expended* to make the process occur.

Recall that the maximum work would occur only along the hypothetical reversible pathway and is thus unattainable (although it can be approached closely in some situations). In any case, knowing the maximum work for a process is still important because then we can evaluate the efficiency of any machine that might be based on the process.

We will now prove the preceding relationship between ΔG and $w_{\text{useful}}^{\text{max}}$. First, we define the total work w:

$$w = w_{\text{useful}} + \underbrace{w_{\text{useless}}}_{PV\text{ work}} = w_{\text{useful}} + w_{\text{pv}}$$

Note that PV work is related to the expansion or contraction of the system and is not counted as useful work. From the definition of ΔE, and assuming constant P and T, we have

$$\begin{aligned}\Delta E &= q_{\text{p}} + w = q_{\text{p}} + w_{\text{useful}} + w_{\text{pv}} \\ &= q_{\text{p}} + w_{\text{useful}} - P\Delta V\end{aligned}$$

From the definition of enthalpy,

$$H = E + PV$$

we have

$$\begin{aligned}\Delta H &= \Delta E + P\Delta V \\ &= \underbrace{q_{\text{p}} + w_{\text{useful}} - P\Delta V}_{\Delta E} + P\Delta V \\ &= q_{\text{p}} + w_{\text{useful}}\end{aligned}$$

Next, from the definition of free energy,

$$G = H - TS$$

we have

$$\Delta G = \Delta H - T\Delta S = \underbrace{q_{\text{p}} + w_{\text{useful}}}_{\Delta H} - T\Delta S$$

For the reversible pathway

$$w_{\text{useful}} = w_{\text{useful}}^{\text{max}} \quad \text{and} \quad q_{\text{p}} = q_{\text{p}}^{\text{rev}}$$

Thus for the reversible pathway

$$\Delta G = q_{\text{p}}^{\text{rev}} + w_{\text{useful}}^{\text{max}} - T\Delta S$$

and since

$$\Delta S = \frac{q_{\text{p}}^{\text{rev}}}{T}$$

then

$$q_{\text{p}}^{\text{rev}} = T\Delta S$$

So we have

$$\Delta G = T\Delta S + w_{\text{useful}}^{\text{max}} - T\Delta S \quad \text{or} \quad \Delta G = w_{\text{useful}}^{\text{max}}$$

Thus we have shown that at constant temperature and pressure the change in free energy for a process gives the maximum useful work available from that process.

Let's consider a few more points in connection with these relationships. If a process is carried out so that $w_{\text{useful}} = 0$, then the expression

$$\Delta G = q_{\text{P}} + w_{\text{useful}} - T\Delta S$$

becomes

$$\Delta G = q_{\text{P}} - T\Delta S$$

And since $\Delta G = \Delta H - T\Delta S$, we have

$$\Delta H - T\Delta S = q_{\text{P}} - T\Delta S$$

$$\Delta H = q_{\text{P}}$$

This relationship between ΔH and q_{P} is used frequently in thermochemical studies. We bring it up again to emphasize that $\Delta H = q_{\text{P}}$ only at constant pressure *and when no useful work is done* (only PV work is allowed). This last condition is often neglected.

If a process is carried out so that w_{useful} is at a maximum (the hypothetical reversible pathway where $\Delta G = w_{\text{useful}}$), then from the expression

$$\Delta G = q_{\text{P}} + w_{\text{useful}} - T\Delta S$$

we have

$$q_{\text{P}} = T\Delta S$$

Thus q_{P}, which is pathway-dependent, varies between ΔH (when $w_{\text{useful}} = 0$) and $T\Delta S$ (when $w_{\text{useful}} = w_{\text{useful}}^{\text{max}}$). The quantity $T\Delta S$ represents the minimum heat flow that must accompany the process under consideration. That is, $T\Delta S$ represents the minimum energy that must be "wasted" through heat flow as the process occurs.

In summary, at constant T and P,

$$q_{\text{P}} = \Delta H \quad \text{if} \quad w_{\text{useful}} = 0$$

$$q_{\text{P}} = T\Delta S \quad \text{if} \quad w_{\text{useful}} = w_{\text{useful}}^{\text{max}}$$

10.13 Reversible and Irreversible Processes: A Summary

As we demonstrated in the analysis of the isothermal expansion–compression of an ideal gas in Section 10.2, the amount of work we actually obtain from a spontaneous process is *always* less than the maximum possible amount.

To explore this idea more fully in a more realistic context than that of an ideal gas, let's consider an electric current flowing through the starter motor of a car. The current is generated from a chemical change in a battery. Since we can calculate ΔG for the battery reaction, we can determine the energy available to do work. Can we use all of this energy to do work? No, because a current flowing through a wire causes frictional heating, and the greater the current, the greater the heat. This heat represents wasted energy—it is not useful for running the starter motor. We can minimize this energy waste by running very low currents through the motor circuit. However, zero current flow would be necessary to eliminate frictional heating entirely, and we cannot derive any work from the motor if no current flows. This example shows the difficulty nature places us in. Using a process to do work requires that some of the energy be wasted, and usually, the faster we run the process, the more energy we waste.

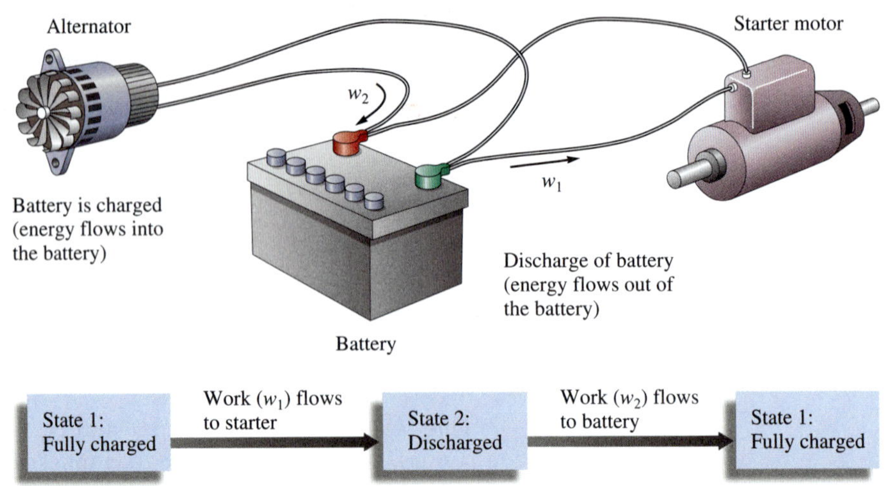

FIGURE 10.17

A battery can do work by sending current to a starter motor. The battery can then be recharged by forcing current through it. If the current flow in both processes is infinitesimally small, $|w_1| = |w_2|$. This is a *reversible process.* But if the current flow is finite, as it would be in any real case, $|w_2| > |w_1|$. This is an *irreversible process* (the *universe is different* after the cyclic process occurs). All real processes are irreversible.

Achieving the maximum work available from a spontaneous process can occur only via a hypothetical pathway. Any real pathway wastes energy in the sense that the maximum work is not obtained. If we could discharge the battery infinitely slowly by an infinitesimally small current flow, we could achieve the maximum useful work. Also, if we could then recharge the battery by using an infinitesimally small current, exactly the same amount of energy would be used to return the battery to its original state as was obtained in the infinitesimally slow discharge. After we cycle the battery in this way, the universe (the system and surroundings) is exactly the same as it was before the cyclic process. Therefore this is a reversible process (see Fig. 10.17).

However, if the battery is discharged to run the starter motor and then recharged by using a *finite* current flow, as is actually the case, *more* work will always be required to recharge the battery than the battery produces as it discharges. Thus, even though the battery (the system) has returned to its original state, the surroundings have not because the surroundings had to furnish a net amount of work as the battery was cycled. The *universe is different* after this cyclic process is performed, and this process is irreversible. *All real processes are irreversible.*

Recall that after any real cyclic process is carried out in a system, the surroundings have less ability to do work and contain more thermal energy. In other words, in any real cyclic process work is changed to heat in the surroundings, and the entropy of the universe increases. This is another way of stating the second law of thermodynamics.

Thus thermodynamics tells us the work potential of a process and then tells us that we can never achieve this potential. In this spirit, thermodynamicist Henry Bent paraphrased the first two laws of thermodynamics as follows:

First law: You can't win; you can only break even.
Second law: You can't break even.

The ideas we have discussed in this section are applicable to the energy crisis that will probably increase in severity over the next 25 years. The crisis is obviously not one of supply; the first law tells us that the universe contains a constant supply of energy. The problem is the availability of *useful* energy. *As we use energy, we degrade its usefulness.* For example, when gasoline reacts with oxygen in the combustion reaction to run our cars, the change in

When energy is used to do work, it becomes less organized and less concentrated and thus less useful.

potential energy results in heat flow. Thus the energy concentrated in the bonds of the gasoline and oxygen molecules ends up *spread* over the surroundings as thermal energy, where it is much more difficult to harness for useful work. In this way the entropy of the universe increases: Concentrated energy becomes spread out—more disordered and less useful. Therefore, the crux of the energy problem is that we are rapidly consuming the concentrated energy found in fossil fuels. It took millions of years to concentrate the sun's energy in these fuels, which we will consume in a few hundred years. Thus we must use these energy sources as wisely as possible.

10.14 Adiabatic Processes

So far in this chapter we have focused on isothermal (constant-temperature) processes for ideal gases. In this section we introduce the **adiabatic process**—*a process in which no energy as heat flows into or out of the system.* That is, an adiabatic process occurs when the system is thermally isolated (insulated) from the surroundings. For an adiabatic process

$$q = 0$$

and

$$\Delta E = q + w = w$$

Consider an ideal gas confined to a cylinder with a movable piston, as shown in Fig. 10.18. Initially, the pressure of the gas (P_{gas}) equals the external pressure (P_{ext}) and the piston is stationary. If P_{ext} is decreased, the gas will expand by doing PV work. What will happen to the temperature of the gas? (Remember that $q = 0$.) Because the expanding gas does work on the surroundings,

FIGURE 10.18

An ideal gas confined in an insulated container with a movable piston. No heat flow with the surroundings can occur.

$$w = -P_{ext}\Delta V$$

and energy flows out of the system:

$$\Delta E = w = -P_{ext}\Delta V$$

What is the source of this energy? In an isothermal expansion, energy as heat enters the system to just balance the outflow of energy as work (see Section 10.2). Since $q = 0$ for an adiabatic process, the energy to do the work must come from the thermal energy of the gas. That is, in an adiabatic expansion the temperature of the gas decreases (the average kinetic energy of sample decreases) to furnish the energy to do the work.

Recall that the energy of an ideal gas depends only on its temperature:

$$E = nC_vT$$

So for an adiabatic process,

$$\Delta E = w = -P_{ext}\Delta V = nC_v\Delta T$$

For an infinitesimal adiabatic change,

$$dE = -P_{ext}dV = nC_vdT$$

Assume that the adiabatic expansion or compression is carried out reversibly. That is, P_{ext} is only infinitesimally different from P_{gas} ($P_{ext} \sim P_{gas}$). Then

$$P_{ext} = P_{gas} = \frac{nRT}{V}$$

Thus, for a reversible, adiabatic expansion–compression, we have

$$dE = nC_v dT = -P_{ext}dV = -P_{gas}dV = -\frac{nRT}{V}\,dV$$

and

$$-\frac{nRT}{V}\,dV = nC_v dT$$

which can be rearranged to

$$\frac{C_v}{T}\,dT = -\frac{R}{V}\,dV$$

We can derive an expression for a reversible, adiabatic change from V_1 to V_2 and from T_1 to T_2 by summing (integrating) the infinitesimal changes required:

$$C_v \int_{T_1}^{T_2} \frac{1}{T}\,dT = -R \int_{V_1}^{V_2} \frac{1}{V}\,dV$$

where C_v is assumed to be independent of temperature over the interval T_1 to T_2. Evaluating the integrals gives

$$C_v \ln\left(\frac{T_2}{T_1}\right) = -R \ln\left(\frac{V_2}{V_1}\right) = R \ln\left(\frac{V_1}{V_2}\right)$$

Taking the antilog of each side we have

$$\left(\frac{T_2}{T_1}\right)^{C_v} = \left(\frac{V_1}{V_2}\right)^{R}$$

Since $C_p = C_v + R$, we can write

$$\left(\frac{T_2}{T_1}\right)^{C_v} = \left(\frac{V_1}{V_2}\right)^{(C_p - C_v)}$$

or

$$\left(\frac{T_2}{T_1}\right) = \left(\frac{V_1}{V_2}\right)^{\left(\frac{C_p}{C_v} - 1\right)} = \left(\frac{V_1}{V_2}\right)^{(\gamma - 1)}$$

where

$$\gamma = \frac{C_p}{C_v}$$

Thus

$$\frac{T_2}{T_1} = \frac{V_1^{\gamma-1}}{V_2^{\gamma-1}}$$

or

$$T_1 V_1^{\gamma-1} = T_2 V_2^{\gamma-1}$$

Using the ideal gas law we can also express this result in terms of pressure. Since in this case,

$$\frac{P_1 V_1}{T_1} = \frac{P_2 V_2}{T_2}$$

then

$$\frac{T_2}{T_1} = \frac{P_2 V_2}{P_1 V_1} = \frac{V_1^{\gamma-1}}{V_2^{\gamma-1}}$$

and

$$P_1 V_1^{\gamma} = P_2 V_2^{\gamma}$$

We can use these equations to calculate the changes in various properties of an ideal gas undergoing a reversible, adiabatic expansion or compression. This is illustrated in Example 10.16.

Example 10.16

Consider a sample containing 5.00 mol of a monatomic ideal gas at 25.0°C and an initial pressure of 10.0 atm. Suppose the external pressure is lowered to 1.00 atm in a reversible manner. Calculate the final pressure and volume of the gas sample and compute the work for the process.

Solution In this case we know the initial and final pressures, so we will use the equation

$$P_1V_1^\gamma = P_2V_2^\gamma$$

where

$$\gamma = \frac{C_p}{C_v} = \frac{\frac{5}{2}R}{\frac{3}{2}R} = \frac{5}{3}$$

for a monatomic gas. Thus we have

$$P_1V_1^{5/3} = P_2V_2^{5/3}$$

We can calculate V_1 from the ideal gas law:

$$V_1 = \frac{nRT_1}{P_1} = \frac{(5.00 \text{ mol})\left(0.08206 \frac{\text{L atm}}{\text{K mol}}\right)(298 \text{ K})}{10.0 \text{ atm}} = 12.2 \text{ L}$$

Now we can solve for V_2:

$$V_2^{5/3} = \frac{P_1V_1^{5/3}}{P_2} = \frac{(10.0 \text{ atm})(12.2 \text{ L})^{5/3}}{1.00 \text{ atm}}$$

$$V_2 = 48.6 \text{ L}$$

Thus the final volume is 48.6 L. We can calculate the work from the expression

$$\Delta E = w = nC_v\Delta T = (5.00)\left(\frac{3}{2}R\right)(T_2 - T_1)$$

but first we must compute T_2 from the ideal gas law:

$$T_2 = \frac{P_2V_2}{nR} = \frac{(1.00 \text{ atm})(48.6 \text{ L})}{(5.00 \text{ mol})\left(0.08206 \frac{\text{L atm}}{\text{K mol}}\right)} = 118 \text{ K}$$

So

$$w = \Delta E = (5.00 \text{ mol})\left(\frac{3}{2}\right)\left(8.3145 \frac{\text{J}}{\text{K mol}}\right)(118 \text{ K} - 298 \text{ K}) = -11{,}200 \text{ J}$$

Note that ΔE and w have negative signs because energy as work flows out of the system in the expansion.

Notice for the reversible adiabatic expansion considered in Example 10.16 that the temperature of the sample changed from 298 K to 118 K—a very dramatic temperature decrease. Because of this significant temperature decrease, the final volume of the gas is much smaller than if the expansion were carried out isothermally, where the temperature would remain at 298 K. For a reversible isothermal expansion at 298 K from $P_1 = 10.0$ atm and $V_1 = 12.2$ L to $P_2 = 1.00$ atm, the final volume is

$$V_2 = \frac{P_1V_1}{P_2} = \frac{(10.0 \text{ mol})(12.2 \text{ L})}{1.00 \text{ atm}} = 122 \text{ L}$$

For this expansion the work is

$$w = -nRT \ln\left(\frac{V_2}{V_1}\right) = -(5.00 \text{ mol})\left(8.3145 \frac{\text{J}}{\text{K mol}}\right)(298 \text{ K}) \ln\left(\frac{122 \text{ L}}{12.2 \text{ L}}\right)$$
$$= -28{,}500 \text{ J}$$

As expected from the much greater volume change, the work delivered to the surroundings in the reversible isothermal expansion is much greater than for the reversible adiabatic expansion. The two types of expansions starting at $P_1 = 10.0$ atm and $V_1 = 12.2$ L are compared in Fig. 10.19. Note that for reversible isothermal expansion

$$P_1V_1 = P_2V_2$$

or

$$PV = \text{constant}$$

On the other hand, for the reversible adiabatic expansion

$$P_1V_1^\gamma = P_2V_2^\gamma$$

or

$$PV^\gamma = \text{constant}$$

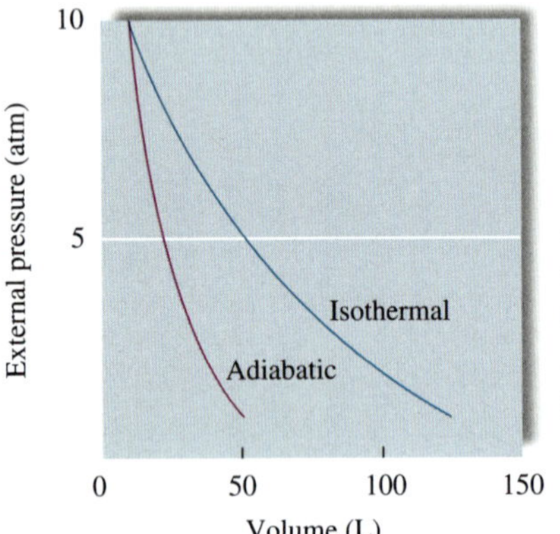

FIGURE 10.19

Comparison of the adiabatic and isothermal expansions for ideal gas samples in which $n = 5$, $P_1 = 10.0$ atm, and $V_1 = 12.2$ L.

Discussion Questions

These questions are designed to be considered by groups of students in class. Often these questions work well for introducing a particular topic in class.

1. For the process $A(l) \rightarrow A(g)$, which direction is favored by changes in energy probability? Positional probability? Explain your answer. If you wanted to favor the process as written, would you raise or lower the temperature of the system? Explain.
2. For a liquid, which would you expect to be larger: ΔS_{fusion} or $\Delta S_{\text{evaporation}}$? Why?
3. Gas A_2 reacts with gas B_2 to form gas AB at constant temperature. The bond energy of AB is much greater than that of either reactant. What can be said about the sign of ΔH? ΔS_{surr}? ΔS? Explain how potential energy changes for this process. Explain how random kinetic energy changes during the process.
4. What types of experiments can be carried out to determine if a reaction is spontaneous? Does spontaneity have any relationship to the final equilibrium position of a reaction? Explain.
5. A friend tells you "Free energy G and pressure P are directly related by the equation $G = G^\circ + RT \ln(P)$. Also, G is related to the equilibrium constant K in that when $G_{\text{products}} = G_{\text{reactants}}$, the system is at equilibrium. Therefore it must be true that a system is at equilibrium when all pressures are equal." Do you agree with this friend? Explain.
6. You remember that ΔG° is related to $RT \ln(K)$ but can't remember if it is $RT \ln(K)$ or $-RT \ln(K)$. Realizing what ΔG° and K mean, how can you figure out the correct sign?
7. Predict the sign of ΔS for each of the following and explain.
   a. the evaporation of alcohol
   b. the freezing of water
   c. compressing an ideal gas at constant temperature
   d. heating an ideal gas at constant pressure
   e. dissolving NaCl in water
8. Which is larger: ΔS at constant pressure or ΔS at constant volume? Provide a conceptual rationale.
9. Is ΔS_{surr} favorable or unfavorable for exothermic reactions? endothermic reactions? Explain.

10. At 1 atm, liquid water is heated above 100°C. For this process which of the following choices (i–iv) is correct for ΔS_{surr}? ΔS? ΔS_{univ}? Explain each answer.
    i. greater than zero
    ii. less than zero
    iii. equal to zero
    iv. cannot be determined
11. High temperatures are favorable to a reaction kinetically but may be unfavorable to a reaction thermodynamically. Explain.

EXERCISES

A blue exercise number indicates that the answer to that exercise appears at the back of this book and a solution appears in the *Solutions Guide.*

Spontaneity and Entropy

12. Define the following.
    a. spontaneous process
    b. entropy
    c. positional probability
    d. system
    e. surroundings
    f. universe
13. Consider the following energy levels, each capable of holding two objects:

$E = 2$ kJ _____

$E = 1$ kJ _____

$E = 0$ XX

Draw all the possible arrangements of the two identical particles (represented by X) in the three energy levels. What total energy is most likely—that is, occurs the greatest number of times? Assume the particles are indistinguishable from each other.

14. Do Exercise 13 with two particles A and B that can be distinguished from each other.
15. Which of the following processes require energy as they occur?
    a. Salt dissolves in H_2O.
    b. A clear solution becomes a uniform color after a few drops of dye are added.
    c. A cell produces proteins from amino acids.
    d. Iron rusts.
    e. A house is built.
    f. A satellite is launched into orbit.
    g. A satellite falls back to earth.
16. Which of the following involve an increase in the entropy of the system under consideration?
    a. melting of a solid
    b. evaporation of a liquid
    c. sublimation
    d. freezing
    e. mixing
    f. separation
    g. diffusion
17. Describe how the following changes affect the positional probability of a substance.
    a. increase in volume of a gas at constant T
    b. increase in temperature of a gas at constant V
    c. increase in pressure of a gas at constant T
18. Choose the compound with the greatest positional probability in each case.
    a. 1 mol of H_2 at STP or 1 mol of H_2 at 100°C and 0.5 atm
    b. 1 mol of N_2 at STP or 1 mol of N_2 at 100 K and 2.0 atm
    c. 1 mol of $H_2O(s)$ at 0°C or 1 mol of $H_2O(l)$ at 20°C
19. In the roll of two dice, what *total* number is the most likely to occur? Is there an energy reason why this number is favored? Would energy have to be spent to increase the probability of getting a particular number (that is, to cheat)?
20. Entropy can be calculated by a relationship proposed by Ludwig Boltzmann:

$$S = k_B \ln \Omega$$

where $k_B = 1.38 \times 10^{-23}$ J/K and Ω is the number of ways a particular state can be obtained. (This equation is engraved on Boltzmann's tombstone.) Calculate S for the three arrangements of particles in Table 10.1.

Energy, Enthalpy, and Entropy Changes Involving Ideal Gases and Physical Changes

21. Calculate the energy required to change the temperature of 1.00 kg of ethane (C_2H_6) from 25.0°C to 73.4°C in a rigid vessel. (C_v for C_2H_6 is 44.60 J K^{-1} mol^{-1}.) Calculate the energy required for this same temperature change at constant pressure. Calculate the change in internal energy of the gas in each of these processes.
22. For nitrogen gas the values of C_v and C_p at 25°C are 20.8 J K^{-1} mol^{-1} and 29.1 J K^{-1} mol^{-1}, respectively. When a sample of nitrogen is heated at constant pressure, what fraction of the energy is used to increase the internal energy of the gas? How is the remainder of the energy used? How much energy is required to raise the temperature of 100.0 g N_2 from 25.0°C to 85.0°C in a vessel having a constant volume?
23. Consider a rigid, insulated box containing 0.400 mol of He(g) at 20.0°C and 1.00 atm in one compartment and 0.600 mol of $N_2(g)$ at 100.0°C and 2.00 atm in the other compartment. These compartments are connected by a partition that transmits heat. What is the final temperature in the box at thermal equilibrium? [For He(g), C_v = 12.5 J K^{-1} mol^{-1}; for $N_2(g)$, C_v = 20.7 J K^{-1} mol^{-1}.]
24. One mole of an ideal gas is contained in a cylinder with a movable piston. The temperature is constant at 77°C.

Weights are removed suddenly from the piston to give the following sequence of three pressures:

a. $P_1 = 5.00$ atm (initial state)
b. $P_2 = 2.24$ atm
c. $P_3 = 1.00$ atm (final state)

What is the total work (in joules) in going from the initial to the final state by way of the preceding two steps? What would be the total work if the process were carried out reversibly?

25. One mole of an ideal gas with a volume of 1.0 L and a pressure of 5.0 atm is allowed to expand isothermally into an evacuated bulb to give a total volume of 2.0 L. Calculate w and q. Also calculate q_{rev} for this change of state.

26. A cylinder with an initial volume of 10.0 L is fitted with a frictionless piston and is filled with 1.00 mol of an ideal gas at 25°C. Assume that the surroundings are large enough so that if heat is withdrawn from or added to it, the temperature does not change.
    a. The gas expands isothermally and reversibly from 10.0 L to 20.0 L. Calculate the work and the heat.
    b. The gas expands isothermally and irreversibly from 10.0 L to 20.0 L as the external pressure changes instantaneously from 2.46 atm to 1.23 atm. Calculate the work and the heat.

27. The molar heat capacities for carbon dioxide at 298.0 K are

$$C_v = 28.95 \text{ J K}^{-1} \text{ mol}^{-1}$$

$$C_p = 37.27 \text{ J K}^{-1} \text{ mol}^{-1}$$

The molar entropy of carbon dioxide gas at 298.0 K and 1.000 atm is 213.64 J K^{-1} mol^{-1}.
    a. Calculate the energy required to change the temperature of 1.000 mol of carbon dioxide gas from 298.0 K to 350.0 K, both at constant volume and at constant pressure.
    b. Calculate the molar entropy of $CO_2(g)$ at 350.0 K and 1.000 atm.
    c. Calculate the molar entropy of $CO_2(g)$ at 350.0 K and 1.174 atm.

28. The molar entropy of helium gas at 25°C and 1.00 atm is 126.1 J K^{-1} mol^{-1}. Assuming ideal behavior, calculate the entropy of the following.
    a. 0.100 mol He(g) at 25°C and a volume of 5.00 L
    b. 3.00 mol He(g) at 25°C and a volume of 3000.0 L

29. Consider the process

$$\underset{75°C}{A(l)} \longrightarrow \underset{155°C}{A(g)}$$

which is carried out at constant pressure. The total ΔS for this process is known to be 75.0 J K^{-1} mol^{-1}. For A(l) and A(g), the C_p values are 75.0 J K^{-1} mol^{-1} and 29.0 J K^{-1} mol^{-1}, respectively, and are not dependent on temperature. Calculate $\Delta H_{vaporization}$ for A(l) at 125°C (its boiling point).

30. A sample of ice weighing 18.02 g, initially at −30.0°C, is heated to 140.0°C at a constant pressure of 1.00 atm. Calculate q, w, ΔE, ΔH, and ΔS for this process. The molar heat capacities (C_p) for solid, liquid, and gaseous water—37.5 J K^{-1} mol^{-1}, 75.3 J K^{-1} mol^{-1}, and 36.4 J K^{-1} mol^{-1}, respectively—are assumed to be temperature-independent. The enthalpies of fusion and vaporization are 6.01 kJ/mol and 40.7 kJ/mol, respectively. Assume ideal gas behavior.

31. Calculate the entropy change for a process in which 3.00 mol of liquid water at 0°C is mixed with 1.00 mol of water at 100.°C in a perfectly insulated container. (Assume that the molar heat capacity of water is constant at 75.3 J K^{-1} mol^{-1}.)

32. Calculate the change in entropy that occurs when 18.02 g of ice at −10.0°C is placed in 54.05 g of water at 100.0°C in a perfectly insulated vessel. Assume that the molar heat capacities for $H_2O(s)$ and $H_2O(l)$ are 37.5 J K^{-1} mol^{-1} and 75.3 J K^{-1} mol^{-1}, respectively, and the molar enthalpy of fusion for ice is 6.01 kJ/mol.

Entropy and the Second Law of Thermodynamics: Free Energy

33. The synthesis of glucose directly from CO_2 and H_2O and the synthesis of proteins directly from amino acids are both nonspontaneous processes under standard conditions. Yet these processes must occur for life to exist. In light of the second law of thermodynamics, how can life exist?

34. A green plant synthesizes glucose by photosynthesis as shown in the reaction

$$6CO_2(g) + 6H_2O(l) \longrightarrow C_6H_{12}O_6(s) + 6O_2(g)$$

Animals use glucose as a source of energy:

$$C_6H_{12}O_6(s) + 6O_2(g) \longrightarrow 6CO_2(g) + 6H_2O(l)$$

If we were to assume that both of these processes occur to the same extent in a cyclic process, what thermodynamic property must have a nonzero value?

35. What determines ΔS_{surr} for a process? To calculate ΔS_{surr} at constant pressure and temperature, we use the following equation: $\Delta S_{surr} = -\Delta H/T$. Why does a minus sign appear in the equation, and why is ΔS_{surr} inversely proportional to temperature?

36. Predict the sign of ΔS_{surr} for the following processes.
    a. $H_2O(l) \longrightarrow H_2O(g)$
    b. $CO_2(g) \longrightarrow CO_2(s)$

37. Calculate ΔS_{surr} for the following reactions at 25°C and 1 atm.
    a. $C_3H_8(g) + 5O_2(g) \longrightarrow 3CO_2(g) + 4H_2O(l)$ $\quad \Delta H° = -2221$ kJ
    b. $2NO_2(g) \longrightarrow 2NO(g) + O_2(g)$ $\quad \Delta H° = 112$ kJ

38. For each of the following pairs of substances, which substance has the greater value of $S°$ at 25°C and 1 atm?

a. $C_{graphite}(s)$ or $C_{diamond}(s)$
b. $C_2H_5OH(l)$ or $C_2H_5OH(g)$
c. $CO_2(s)$ or $CO_2(g)$
d. $N_2O(g)$ or $He(g)$
e. $HF(g)$ or $HCl(g)$

39. Predict the sign of $\Delta S°$ for each of the following changes.
a.

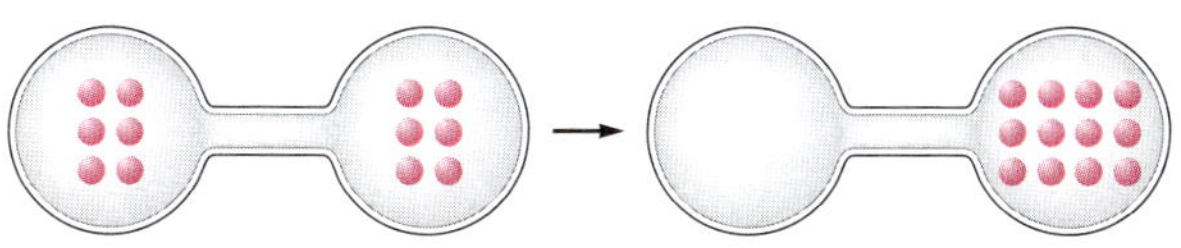

b. $AgCl(s) \longrightarrow Ag^+(aq) + Cl^-(aq)$
c. $2H_2(g) + O_2(g) \longrightarrow 2H_2O(l)$
d. $Na(s) + \frac{1}{2}Cl_2(g) \longrightarrow NaCl(s)$
e. $HCl(g) \longrightarrow H^+(aq) + Cl^-(aq)$
f. $NaCl(s) \longrightarrow Na^+(aq) + Cl^-(aq)$

40. Predict the sign of $\Delta S°$ and then calculate $\Delta S°$ for each of the following reactions.
a. $2H_2S(g) + SO_2(g) \longrightarrow 3S_{rhombic}(s) + 2H_2O(g)$
b. $2SO_3(g) \longrightarrow 2SO_2(g) + O_2(g)$
c. $Fe_2O_3(s) + 3H_2(g) \longrightarrow 2Fe(s) + 3H_2O(g)$

41. For the reaction

$$C_2H_2(g) + 4F_2(g) \longrightarrow 2CF_4(g) + H_2(g)$$

$\Delta S°$ is equal to -358 J/K. Use this value and data from Appendix 4 to calculate the value of $S°$ for $CF_4(g)$.

42. Using Appendix 4 and the following data, determine $S°$ for $Fe(CO)_5(g)$.

$Fe(s) + 5CO(g) \longrightarrow Fe(CO)_5(g)$	$\Delta S° = ?$
$Fe(CO)_5(l) \longrightarrow Fe(CO)_5(g)$	$\Delta S° = 107$ J/K
$Fe(s) + 5CO(g) \longrightarrow Fe(CO)_5(l)$	$\Delta S° = -677$ J/K

43. For the reaction

$$2Al(s) + 3Br_2(l) \longrightarrow 2AlBr_3(s)$$

$\Delta S°$ is equal to -144 J/K. Use this value and data from Appendix 4 to calculate the value of $S°$ for solid aluminum bromide.

44. Ethanethiol (C_2H_5SH; also called ethyl mercaptan) is commonly added to natural gas to provide the "rotten egg" smell of a gas leak. The boiling point of ethanethiol is 35°C and its heat of vaporization is 27.5 kJ/mol. What is the entropy of vaporization for this substance?

45. For mercury the enthalpy of vaporization is 58.51 kJ/mol and the entropy of vaporization is 92.92 J K^{-1} mol^{-1}. What is the normal boiling point of mercury?

46. The enthalpy of vaporization of ethanol is 38.7 kJ/mol at its boiling point (78°C). Determine ΔS_{sys}, ΔS_{surr}, and ΔS_{univ} when 1.00 mol of ethanol is vaporized at 78°C and 1.00 atm.

47. For ammonia (NH_3) the enthalpy of fusion is 5.65 kJ/mol, and the entropy of fusion is 28.9 J K^{-1} mol^{-1}.
a. Will $NH_3(s)$ spontaneously melt at 200. K?
b. What is the approximate melting point of ammonia?

48. It is quite common for a solid to change from one structure to another at a temperature below its melting point. For example, sulfur undergoes a phase change from the rhombic crystal structure to the monoclinic crystal form at temperatures above 95°C.
a. Predict the signs of ΔH and ΔS for the process $S_{rhombic} \longrightarrow S_{monoclinic}$.
b. Which form of sulfur has the more ordered crystalline structure (has the smaller positional probability)?

49. As $O_2(l)$ is cooled at 1 atm, it freezes at 54.5 K to form solid I. At a lower temperature, solid I rearranges to solid II, which has a different crystal structure. Thermal measurements show that ΔH for the I $\longrightarrow$ II phase transition is -743.1 J/mol, and ΔS for the same transition is -17.0 J K^{-1} mol^{-1}. At what temperature are solids I and II in equilibrium?

Free Energy and Chemical Reactions

50. From data in Appendix 4, calculate $\Delta H°$, $\Delta S°$, and $\Delta G°$ for each of the following reactions at 25°C.
a. $CH_4(g) + 2O_2(g) \longrightarrow CO_2(g) + 2H_2O(g)$
b. $6CO_2(g) + 6H_2O(l) \longrightarrow C_6H_{12}O_6(s) + 6O_2(g)$
Glucose
c. $P_4O_{10}(s) + 6H_2O(l) \longrightarrow 4H_3PO_4(s)$
d. $HCl(g) + NH_3(g) \longrightarrow NH_4Cl(s)$

51. The value of $\Delta G°$ for the reaction

$$2C_4H_{10}(g) + 13O_2(g) \longrightarrow 8CO_2(g) + 10H_2O(l)$$

is $-5490.$ kJ. Use this value and data from Appendix 4 to calculate the standard free energy of formation for $C_4H_{10}(g)$.

52. Of the functions $\Delta H°$, $\Delta S°$, and $\Delta G°$, which depends most strongly on temperature? When $\Delta G°$ is calculated at temperatures other than 25°C, what assumptions are generally made concerning $\Delta H°$ and $\Delta S°$?

53. For the reaction at 298 K,

$$2NO_2(g) \rightleftharpoons N_2O_4(g)$$

the values of $\Delta H°$ and $\Delta S°$ are -58.03 kJ and -176.6 J/K, respectively. What is the value of $\Delta G°$ at 298 K? Assuming that $\Delta H°$ and $\Delta S°$ do not depend on temperature, at what temperature is $\Delta G° = 0$? Is $\Delta G°$ negative above or below this temperature?

54. Acrylonitrile is the starting material used in the manufacture of acrylic fibers (U.S. annual production capacity is more than 2 million pounds). Three industrial processes for the production of acrylonitrile are given below. Using data from Appendix 4, calculate $\Delta S°$, $\Delta H°$, and $\Delta G°$ for each process. For part a, assume that $T = 25°C$; for part b, $T = 70.°C$; and for part c, $T = 700.°C$. Assume that $\Delta H°$ and $\Delta S°$ do not depend on temperature.

a. $CH_2{-}CH_2(g) + HCN(g)$ (with O bridging the two carbons)

Ethylene oxide

$\longrightarrow CH_2{=}CHCN(g) + H_2O(l)$

Acrylonitrile

b. $HC{\equiv}CH(g) + HCN(g) \xrightarrow[70°C–90°C]{CaCl_2 \cdot HCl} CH_2{=}CHCN(g)$

c. $4CH_2{=}CHCH_3(g) + 6NO(g)$

$\xrightarrow[Ag]{700°C} 4CH_2{=}CHCN(g) + 6H_2O(g) + N_2(g)$

55. Consider the reaction

$$2POCl_3(g) \longrightarrow 2PCl_3(g) + O_2(g)$$

a. Calculate $\Delta G°$ for this reaction. The $\Delta G_f°$ values for $POCl_3(g)$ and $PCl_3(g)$ are -502 kJ/mol and $-270.$ kJ/mol, respectively.
b. Is this reaction spontaneous under standard conditions at 298 K?
c. The value of $\Delta S°$ for this reaction is 179 J/K. At what temperatures is this reaction spontaneous at standard conditions? Assume that $\Delta H°$ and $\Delta S°$ do not depend on temperature.

56. Consider two reactions for the production of ethanol:

$$C_2H_4(g) + H_2O(g) \longrightarrow CH_3CH_2OH(l)$$
$$C_2H_6(g) + H_2O(g) \longrightarrow CH_3CH_2OH(l) + H_2(g)$$

Which would be more thermodynamically feasible? Why? Assume standard conditions and assume that $\Delta H°$ and $\Delta S°$ are temperature-independent.

57. Using data from Appendix 4, calculate $\Delta H°$, $\Delta S°$, and $\Delta G°$ for the following reactions that produce acetic acid:

$$CH_4(g) + CO_2(g) \longrightarrow CH_3\overset{O}{\overset{\|}{C}}{-}OH(l)$$

$$CH_3OH(g) + CO(g) \longrightarrow CH_3\overset{O}{\overset{\|}{C}}{-}OH(l)$$

Which reaction would you choose as a commercial method for producing acetic acid (CH_3CO_2H)? What temperature conditions would you choose for the reaction? Assume standard conditions and assume that $\Delta H°$ and $\Delta S°$ are temperature-independent.

58. Given the following data:

$$2C_6H_6(l) + 15O_2(g) \longrightarrow 12CO_2(g) + 6H_2O(l) \quad \Delta G° = -6399 \text{ kJ}$$

$$C(s) + O_2(g) \longrightarrow CO_2(g) \quad \Delta G° = -394 \text{ kJ}$$

$$H_2(g) + \tfrac{1}{2}O_2(g) \longrightarrow H_2O(l) \quad \Delta G° = -237 \text{ kJ}$$

calculate $\Delta G°$ for the reaction

$$6C(s) + 3H_2(g) \longrightarrow C_6H_6(l)$$

59. When most biological enzymes are heated, they lose their catalytic activity. The change

$$\text{Original enzyme} \longrightarrow \text{new form}$$

that occurs upon heating is endothermic and spontaneous. Is the structure of the original enzyme or its new form more ordered (has the smaller positional probability)? Explain your answer.

60. For the reaction

$$2O(g) \longrightarrow O_2(g)$$

a. predict the signs of ΔH and ΔS.
b. would the reaction be more spontaneous at high or low temperatures?

61. Hydrogen cyanide is produced industrially by the following exothermic reaction:

$2NH_3(g) + 3O_2(g) + 2CH_4(g)$

$\xrightarrow[Pt\text{-}Rh]{1000°C} 2HCN(g) + 6H_2O(g)$

Is the high temperature needed for thermodynamic or for kinetic reasons?

Free Energy: Pressure Dependence and Equilibrium

62. A reaction at constant T and P is spontaneous as long as ΔG is negative; that is, reactions always proceed as long as the products have a lower free energy than the reactants. What is so special about equilibrium? Why don't reactions move away from equilibrium?

63. ΔG predicts spontaneity for a reaction at constant T and P, whereas $\Delta G°$ predicts the equilibrium position. Explain what this statement means. Under what conditions can you use $\Delta G°$ to determine the spontaneity of a reaction?

64. Using thermodynamic data from Appendix 4, calculate $\Delta G°$ at 25°C for the process

$$2SO_2(g) + O_2(g) \longrightarrow 2SO_3(g)$$

where all gases are at 1.00 atm pressure. Also calculate $\Delta G°$ at 25°C for this same reaction but with all gases at 10.0 atm pressure.

65. Consider the reaction

$$2NO_2(g) \rightleftharpoons N_2O_4(g)$$

For each of the following mixtures of reactants and products at 25°C, predict the direction in which the reaction will shift to reach equilibrium. Use thermodynamic data in Appendix 4.

a. $P_{NO_2} = P_{N_2O_4} = 1.0$ atm
b. $P_{NO_2} = 0.21$ atm, $P_{N_2O_4} = 0.50$ atm
c. $P_{NO_2} = 0.29$ atm, $P_{N_2O_4} = 1.6$ atm

66. Using data from Appendix 4, calculate ΔG for the reaction

$$2H_2S(g) + SO_2(g) \rightleftharpoons 3S(s) + 2H_2O(g)$$

for the following conditions at 25°C:

$$P_{H_2S} = 1.0 \times 10^{-4} \text{ atm}$$
$$P_{SO_2} = 1.0 \times 10^{-2} \text{ atm}$$
$$P_{H_2O} = 3.0 \times 10^{-2} \text{ atm}$$

67. Using data from Appendix 4, calculate $\Delta H°$, $\Delta S°$, and K (at 298 K) for the synthesis of ammonia by the Haber process:

$$N_2(g) + 3H_2(g) \rightleftharpoons 2NH_3(g)$$

Calculate ΔG for this reaction under the following conditions (assume an uncertainty of ±1 in all quantities):
a. $T = 298$ K, $P_{N_2} = P_{H_2} = 200$ atm, $P_{NH_3} = 50$ atm
b. $T = 298$ K, $P_{N_2} = 200$ atm, $P_{H_2} = 600$ atm, $P_{NH_3} = 200$ atm
c. $T = 100$ K, $P_{N_2} = 50$ atm, $P_{H_2} = 200$ atm, $P_{NH_3} = 10$ atm
d. $T = 700$ K, $P_{N_2} = 50$ atm, $P_{H_2} = 200$ atm, $P_{NH_3} = 10$ atm
Assume that $\Delta H°$ and $\Delta S°$ do not depend on temperature.

68. Consider the autoionization of water at 25°C:

$$H_2O(l) \rightleftharpoons H^+(aq) + OH^-(aq) \qquad K_w = 1.00 \times 10^{-14}$$

a. Calculate $\Delta G°$ for this process at 25°C.
b. At 40.°C, $K_w = 2.92 \times 10^{-14}$. Calculate $\Delta G°$ at 40.°C.

69. How can one estimate the value of K at temperatures other than 25°C for a reaction? How can one estimate the temperature where $K = 1$ for a reaction? Do all reactions have a specific temperature where $K = 1$?

70. The Ostwald process for the commercial production of nitric acid involves three steps:

$$4NH_3(g) + 5O_2(g) \xrightarrow[825°C]{Pt} 4NO(g) + 6H_2O(g)$$
$$2NO(g) + O_2(g) \longrightarrow 2NO_2(g)$$
$$3NO_2(g) + H_2O(l) \longrightarrow 2HNO_3(l) + NO(g)$$

a. Calculate $\Delta H°$, $\Delta S°$, $\Delta G°$, and K (at 298 K) for each of the three steps in the Ostwald process (see Appendix 4).
b. Calculate the equilibrium constant for the first step at 825°C. Assume that $\Delta H°$ and $\Delta S°$ are temperature-independent.
c. Is there a thermodynamic reason for the high temperature in the first step assuming standard conditions?

71. Consider the following reaction at 800. K:

$$N_2(g) + 3F_2(g) \longrightarrow 2NF_3(g)$$

An equilibrium mixture contains the following partial pressures: $P_{N_2} = 0.021$ atm, $P_{F_2} = 0.063$ atm, and $P_{NF_3} = 0.48$ atm. Calculate $\Delta G°$ for the reaction at 800. K.

72. Consider the following reaction at 298 K:

$$2SO_2(g) + O_2(g) \longrightarrow 2SO_3(g)$$

An equilibrium mixture contains $O_2(g)$ and $SO_3(g)$ at partial pressures of 0.50 atm and 2.0 atm, respectively. Using data from Appendix 4, determine the equilibrium partial pressure of SO_2 in the mixture. Will this reaction be most favored at a high or a low temperature, assuming standard conditions?

73. For the reaction

$$A(g) + 2B(g) \rightleftharpoons C(g)$$

the initial partial pressures of gases A, B, and C are all 0.100 atm. Once equilibrium has been established, it is found that [C] = 0.040 atm. What is $\Delta G°$ for this reaction at 25°C?

74. Cells use the hydrolysis of adenosine triphosphate, abbreviated ATP, as a source of energy. Symbolically, this reaction can be represented as

$$ATP(aq) + H_2O(l) \longrightarrow ADP(aq) + H_2PO_4^-(aq)$$

where ADP represents adenosine diphosphate. For this reaction $\Delta G° = -30.5$ kJ/mol.
a. Calculate K at 25°C.
b. If all the free energy from the metabolism of glucose

$$C_6H_{12}O_6(s) + 6O_2(g) \longrightarrow 6CO_2(g) + 6H_2O(l)$$

goes into the production of ATP, how many ATP molecules can be produced for every molecule of glucose?

75. Carbon monoxide is toxic because it bonds much more strongly to the iron in hemoglobin (Hgb) than does O_2. Consider the following reactions and approximate standard free energy changes:

$$Hgb + O_2 \longrightarrow HgbO_2 \qquad \Delta G° = -70 \text{ kJ}$$
$$Hgb + CO \longrightarrow HgbCO \qquad \Delta G° = -80 \text{ kJ}$$

Using these data, estimate the equilibrium constant value at 25°C for the following reaction:

$$HgbO_2 + CO \rightleftharpoons HgbCO + O_2$$

76. One reaction that occurs in human metabolism is

$$HO_2CCH_2CH_2\underset{\underset{NH_2}{|}}{C}HCO_2H(aq) + NH_3(aq) \rightleftharpoons$$

Glutamic acid

$$H_2N\overset{\overset{O}{\|}}{C}CH_2CH_2\underset{\underset{NH_2}{|}}{C}HCO_2H(aq) + H_2O(l)$$

Glutamine

For this reaction $\Delta G° = 14$ kJ at 25°C.

a. Calculate K for this reaction at 25°C.
b. In a living cell this reaction is coupled with the hydrolysis of ATP. (See Exercise 74.) Calculate $\Delta G°$ and K at 25°C for the following reaction:

$$\text{Glutamic acid}(aq) + \text{ATP}(aq) + NH_3(aq) \rightleftharpoons \text{Glutamine}(aq) + \text{ADP}(aq) + H_2PO_4^-(aq)$$

77. At 25.0°C, for the reaction

$$2NO_2(g) \rightleftharpoons N_2O_4(g)$$

the values of $\Delta H°$ and $\Delta S°$ are −58.03 kJ/mol and −176.6 J K^{-1} mol^{-1}, respectively. Calculate the value of K at 25.0°C. Assuming $\Delta H°$ and $\Delta S°$ are temperature-independent, estimate the value of K at 100.0°C.

78. Consider the relationship

$$\ln(K) = \frac{-\Delta H°}{RT} + \frac{\Delta S°}{R}$$

The equilibrium constant for some hypothetical process was determined as a function of temperature (in kelvins) with the results plotted below.

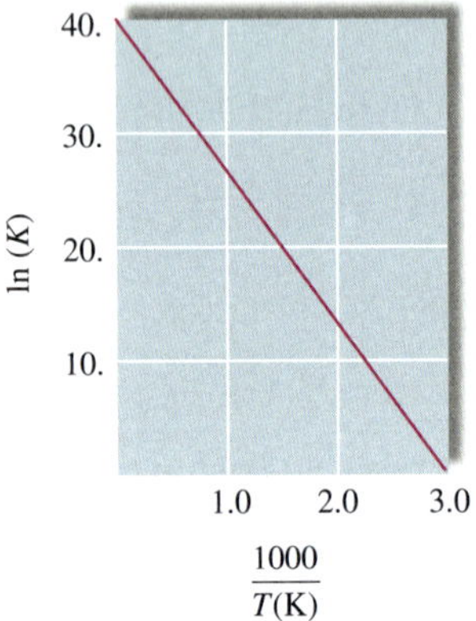

From the plot, determine the values of $\Delta H°$ and $\Delta S°$ for this process. What would be the major difference in the ln(K) versus 1/T plot for an endothermic process as compared to an exothermic process?

79. a. Use the equation in Exercise 78 to determine $\Delta H°$ and $\Delta S°$ for the autoionization of water:

$$H_2O(l) \rightleftharpoons H^+(aq) + OH^-(aq)$$

T (°C)	K
0	1.14×10^{-15}
25	1.00×10^{-14}
35	2.09×10^{-14}
40.	2.92×10^{-14}
50.	5.47×10^{-14}

b. Estimate the value of $\Delta G°$ for the autoionization of water at its critical temperature, 374°C.

80. The equilibrium constant K for the reaction

$$2Cl(g) \rightleftharpoons Cl_2(g)$$

was measured as a function of temperature (in kelvins). A graph of ln(K) versus 1/T for this reaction gives a straight line with a slope of 1.352×10^4 K and a y intercept of −14.51. Determine the values of $\Delta H°$ and $\Delta S°$ for this reaction. Reference Exercise 78.

81. The equilibrium constant for a certain reaction decreases from 8.84 to 3.25×10^{-2} when the temperature increases from 25°C to 75°C. Estimate the temperature where K = 1.00 for this reaction. Estimate the value of $\Delta S°$ for this reaction. (*Hint:* Manipulate the equation given in Exercise 78.)

Adiabatic Processes

82. A sample of a monatomic ideal gas at 1.00 atm and 25°C expands adiabatically and reversibly from 5.00 L to 12.5 L. Calculate the final temperature and pressure of the gas, the work associated with this process, and the change in internal energy.

83. A sample of 1.75 mol H_2 (C_v = 20.5 J K^{-1} mol^{-1}) at 21°C and 1.50 atm undergoes a reversible adiabatic compression until the final pressure is 4.50 atm. Calculate the final volume of the gas sample and the work associated with this process. Assume that the gas behaves ideally.

84. A 1.50-mol sample of an ideal gas is allowed to expand adiabatically and reversibly to twice its original volume. In the expansion the temperature dropped from 296 K to 239 K. Calculate ΔE and ΔH for the gas expansion.

85. Consider 1.00 mol of $CO_2(g)$ at 300. K and 5.00 atm. The gas expands until the final pressure is 1.00 atm. For each of the following conditions describing the expansion, calculate q, w, and ΔE. C_p for CO_2 is 37.1 J K^{-1} mol^{-1}, and assume that the gas behaves ideally.
a. The expansion occurs isothermally and reversibly.
b. The expansion occurs isothermally against a constant external pressure of 1.00 atm.
c. The expansion occurs adiabatically and reversibly.

86. Consider 1.00 mol of $CO_2(g)$ at 300. K and 5.00 atm. The gas expands until the final pressure is 1.00 atm. For each of the following conditions describing the expansion, calculate ΔS, ΔS_{surr}, and ΔS_{univ}. C_p for CO_2 is 37.1 J K^{-1} mol^{-1}, and assume that the gas behaves ideally.
a. The expansion occurs isothermally and reversibly.
b. The expansion occurs isothermally against a constant external pressure of 1.00 atm.
c. The expansion occurs adiabatically and reversibly.

Additional Exercises

87. Some water is placed in a coffee cup calorimeter. When 1.0 g of an ionic solid is added, the temperature of the solution increases from 21.5°C to 24.2°C as the solid dissolves. For the dissolving process, what are the signs for ΔS_{sys}, ΔS_{surr}, and ΔS_{univ}?

88. Entropy has been described as "time's arrow." Interpret this view of entropy.

89. Discuss the relationship between w_{max} and the magnitude and sign of the free energy change for a reaction. Also discuss w_{max} for real processes.

90. Human DNA contains almost twice as much information as is needed to code for all the substances produced in the body. Likewise, the digital data sent from *Voyager 2* contain one redundant bit out of every two bits of information. The Hubble space telescope transmits three redundant bits for every bit of information. How is entropy related to the transmission of information? What do you think is accomplished by having so many redundant bits of information in both DNA and the space probes?

91. In the text the equation

$$\Delta G = \Delta G° + RT \ln(Q)$$

was derived for gaseous reactions where the quantities in Q were expressed in units of pressure. We also can use units of mol/L for the quantities in Q—specifically for aqueous reactions. With this in mind, consider the reaction

$$HF(aq) \rightleftharpoons H^+(aq) + F^-(aq)$$

for which $K_a = 7.2 \times 10^{-4}$ at 25°C. Calculate ΔG for the reaction under the following conditions at 25°C:

a. $[HF] = [H^+] = [F^-] = 1.0\ M$
b. $[HF] = 0.98\ M$, $[H^+] = [F^-] = 2.7 \times 10^{-2}\ M$
c. $[HF] = [H^+] = [F^-] = 1.0 \times 10^{-5}\ M$
d. $[HF] = [F^-] = 0.27\ M$, $[H^+] = 7.2 \times 10^{-4}\ M$
e. $[HF] = 0.52\ M$, $[F^-] = 0.67\ M$, $[H^+] = 1.0 \times 10^{-3}\ M$

Based on the calculated ΔG values, in which direction will the reaction shift to reach equilibrium for each of the five sets of conditions?

92. Many biochemical reactions that occur in cells require relatively high concentrations of potassium ion (K^+). The concentration of K^+ in muscle cells is about 0.15 M. The concentration of K^+ in blood plasma is about 0.0050 M. The high internal concentration in cells is maintained by pumping K^+ from the plasma. How much work must be done to transport 1.0 mol of K^+ from the blood to the inside of a muscle cell at 37°C (normal body temperature)? When 1.0 mol of K^+ is transferred from blood to the cells, do any other ions have to be transported? Why or why not? Much of the ATP (see Exercise 74) formed from metabolic processes is used to provide energy for transport of cellular components. How much ATP must be hydrolyzed to provide the energy for the transport of 1.0 mol of K^+?

93. Consider the following system at equilibrium at 25°C:

$$PCl_3(g) + Cl_2(g) \rightleftharpoons PCl_5(g) \quad \Delta G° = -92.50 \text{ kJ}$$

What will happen to the ratio of partial pressure of PCl_5 to partial pressure of PCl_3 if the temperature is raised? Explain completely.

94. Consider the reaction

$$H_2(g) + Br_2(g) \rightleftharpoons 2HBr(g)$$

where $\Delta H° = -103.8$ kJ. In a particular experiment 1.00 atm of $H_2(g)$ and 1.00 atm of $Br_2(g)$ were mixed in a 1.00-L flask at 25°C and allowed to reach equilibrium. Then the molecules of H_2 were counted by using a very sensitive technique, and 1.10×10^{13} molecules were found. For this reaction, calculate the values of K, $\Delta G°$, and $\Delta S°$.

95. At 1500 K the process

$$\underset{10\text{ atm}}{I_2(g)} \longrightarrow \underset{10\text{ atm}}{2I(g)}$$

is not spontaneous. However, the process

$$\underset{0.10\text{ atm}}{I_2(g)} \longrightarrow \underset{0.10\text{ atm}}{2I(g)}$$

is spontaneous at 1500 K. Explain.

96. Using the following data, calculate the value of K_{sp} for $Ba(NO_3)_2$, one of the least soluble of the common nitrate salts.

Species	$\Delta G_f°$
$Ba^{2+}(aq)$	−561 kJ/mol
$NO_3^-(aq)$	−109 kJ/mol
$Ba(NO_3)_2(s)$	−797 kJ/mol

97. Sodium chloride is added to water (at 25°C) until it is saturated. Calculate the Cl^- concentration in such a solution.

Species	$\Delta G°$(kJ/mol)
$NaCl(s)$	384
$Na^+(aq)$	−262
$Cl^-(aq)$	−131

98. What is the pH of a 0.125 M solution of the weak base B if $\Delta H° = -28.0$ kJ and $\Delta S° = -175$ J/K for the following equilibrium reaction at 25°C?

$$B(aq) + H_2O(l) \rightleftharpoons BH^+(aq) + OH^-(aq)$$

99. Consider the reactions

$$Ni^{2+}(aq) + 6NH_3(aq) \longrightarrow Ni(NH_3)_6{}^{2+}(aq) \quad (1)$$

$$Ni^{2+}(aq) + 3en(aq) \longrightarrow Ni(en)_3{}^{2+}(aq) \quad (2)$$

where

$$en = H_2N{-}CH_2{-}CH_2{-}NH_2$$

The ΔH values for the two reactions are quite similar, yet $K_{reaction\ 2} > K_{reaction\ 1}$. Explain.

100. The deciding factor on why HF is a weak acid and not a strong acid like the other hydrogen halides is entropy. What occurs when HF dissociates in water as compared to the other hydrogen halides?

101. The third law of thermodynamics states that the entropy of a perfect crystal at 0 K is zero. In Appendix 4, $F^-(aq)$, $OH^-(aq)$, and $S^{2-}(aq)$ all have negative standard entropy values. How can $S°$ values be less than zero?

102. Calculate the entropy change for the vaporization of liquid methane and hexane using the following data:

	Boiling Point (1 atm)	ΔH_{vap}
Methane	112 K	8.20 kJ/mol
Hexane	342 K	28.9 kJ/mol

Compare the molar volume of gaseous methane at 112 K with that of gaseous hexane at 342 K. How do the differences in molar volume affect the values of ΔS_{vap} for these liquids?

103. The standard entropy values ($S°$) for $H_2O(l)$ and $H_2O(g)$ are 70. J K^{-1} mol^{-1} and 189 J K^{-1} mol^{-1}, respectively. Calculate the ratio of Ω_g to Ω_l for water using Boltzmann's equation. (See Exercise 20.)

104. Calculate the values of ΔS and ΔG for each of the following processes at 298 K:

$$H_2O(l, 298\ K) \longrightarrow H_2O(g, V = 1000.\ L/mol)$$

$$H_2O(l, 298\ K) \longrightarrow H_2O(g, V = 100.\ L/mol)$$

The standard enthalpy of vaporization for water at 298 K is 44.02 kJ/mol. Does either of these processes occur spontaneously?

105. Calculate the changes in free energy, enthalpy, and entropy when 1.00 mol Ar(g) at 27°C is compressed isothermally from 100.0 L to 1.00 L.

106. Consider the isothermal expansion of 1.00 mol of ideal gas at 27°C. The volume increases from 30.0 L to 40.0 L. Calculate q, w, ΔE, ΔH, ΔS, and ΔG for two situations:
   a. a free expansion
   b. a reversible expansion

107. A 1.00-mol sample of an ideal gas in a vessel with a movable piston initially occupies a volume of 5.00 L at an external pressure of 5.00 atm.
   a. If P_{ex} is suddenly lowered to 2.00 atm and the gas is allowed to expand isothermally, calculate the following quantities for the system: ΔE, ΔH, ΔS, ΔG, w, and q.
   b. Show by the second law that this process will occur spontaneously.

108. One mole of an ideal gas with a volume of 6.67 L and a pressure of 1.50 atm is contained in a vessel with a movable piston. The external pressure is suddenly increased to 5.00 atm and the gas is compressed isothermally ($T = 122$ K). Calculate ΔE, ΔH, ΔS, w, q, ΔS_{surr}, ΔS_{univ}, and ΔG.

Challenge Problems

109. Consider a 2.00-mol sample of Ar at 2.00 atm and 298 K.
   a. If the gas sample expands adiabatically and reversibly to a pressure of 1.00 atm, calculate the final temperature of the gas sample assuming ideal gas behavior.
   b. If the gas sample expands adiabatically and irreversibly against a constant 1.00 atm pressure, calculate the final temperature of the gas sample assuming ideal gas behavior.

110. Consider 1.0 mol of a monatomic ideal gas in a container fitted with a piston. The initial conditions are 5.0 L and $P = 5.0$ atm at some constant T.
   a. If the external pressure is suddenly changed to 2.0 atm, show that expansion of the gas is spontaneous.
   b. If the external pressure suddenly changes back to 5.0 atm, show that compression of the gas is spontaneous.
   c. Calculate and compare signs of ΔG for each case (parts a and b) and discuss why this sign cannot be used to predict spontaneity.

111. One mole of an ideal gas undergoes an isothermal reversible expansion at 25°C. During this process, the system absorbs 855 J of heat from the surroundings. When this gas is compressed to the original state in one step

(isothermally), *twice* as much work is done on the system as was performed on the surroundings in the expansion.

a. What is ΔS for the one-step isothermal compression?
b. What is ΔS_{univ} for the overall process (expansion and compression)?

112. At least some of what is in the following quoted statement is false. Change the incorrect statements so that they are correct and defend your answer. What is correct in the statements? What is wrong? Discuss a real-world situation that supports your position.

"The magnitude of ΔS is always larger than the magnitude of ΔS_{surr}. This is so because ΔS is related to q_{rev}, whereas ΔS_{surr} is related to q_{actual}, and the magnitude of q_{rev} is always larger than the magnitude of q_{actual}."

113. You have a 1.00-L sample of hot water (90.°C) sitting open in a 25°C room. Eventually the water cools to 25°C, whereas the temperature of the room remains unchanged. Calculate ΔS_{univ} for this process. Assume the density of water is 1.00 g/mL over this temperature range and that the heat capacity of water is constant over this temperature range and equal to 75.3 J K^{-1} mol^{-1}.

114. Consider two perfectly insulated vessels. Vessel 1 initially contains an ice cube at 0°C and water at 0°C. Vessel 2 initially contains an ice cube at 0°C and a saltwater solution at 0°C. Consider the process $H_2O(s) \rightarrow H_2O(l)$.

a. Determine the sign of ΔS, ΔS_{surr}, and ΔS_{univ} for the process in vessel 1.
b. Determine the sign of ΔS, ΔS_{surr}, and ΔS_{univ} for the process in vessel 2.

(*Hint:* Think about the effect that a salt has on the freezing point of a solvent.)

115. If wet silver carbonate is dried in a stream of hot air, the air must have a certain concentration level of carbon dioxide to prevent silver carbonate from decomposing by the reaction

$$Ag_2CO_3(s) \rightleftharpoons Ag_2O(s) + CO_2(g)$$

$\Delta H°$ for this reaction is 79.14 kJ/mol in the temperature range of 25°C–125°C. Given that the partial pressure of carbon dioxide in equilibrium with pure solid silver carbonate is 6.23×10^{-3} torr at 25°C, calculate the partial pressure of CO_2 necessary to prevent decomposition of Ag_2CO_3 at 110.°C.

116. Consider a weak acid HX. If a 0.10 *M* solution of HX has a pH of 5.83 at 25°C, what is $\Delta G°$ for the acid's dissociation reaction at 25°C?

117. Using data from Appendix 4, calculate $\Delta H°$, $\Delta G°$, and K_p (at 298 K) for the production of ozone from oxygen:

$$3O_2(g) \rightleftharpoons 2O_3(g)$$

At 30 km above the surface of the earth, the temperature is about 230. K and the partial pressure of oxygen is about 1.0×10^{-3} atm. Estimate the partial pressure of ozone in equilibrium with oxygen at 30 km above the earth's surface. Is it reasonable to assume that the equilibrium between oxygen and ozone is maintained under these conditions? Explain.

118. One mole of a monatomic ideal gas (for which $S° = 8.00$ J K^{-1} mol^{-1} at −73.0°C) was heated at a constant pressure of 2.00 atm from −73.0°C to 27.0°C. Calculate ΔH, ΔE, w, q, ΔS due to the change in volume, ΔS due to the change in temperature, and ΔG.

119. Consider the system

$$A(g) \longrightarrow B(g)$$

at 25°C.

a. Assuming that $G_A^\circ = 8996$ J/mol and $G_B^\circ = 11{,}718$ J/mol, calculate the value of the equilibrium constant for this reaction.
b. Calculate the equilibrium pressures that result if 1.00 mol A(*g*) at 1.00 atm and 1.00 mol B(*g*) at 1.00 atm are mixed at 25°C.
c. Show by calculations that $\Delta G = 0$ at equilibrium.

120. Liquid water at 25°C is introduced into an evacuated, insulated vessel. Identify the signs of the following thermodynamic functions for the process that occurs: ΔH, ΔS, ΔG, ΔT_{water}, ΔS_{surr}, ΔS_{univ}.

121. Consider 1.00 mol of an ideal gas that is expanded isothermally at 25°C from 2.45×10^{-2} atm to 2.45×10^{-3} atm in the following three irreversible steps:

Step 1: from 2.45×10^{-2} atm to 9.87×10^{-3} atm
Step 2: from 9.87×10^{-3} atm to 4.93×10^{-3} atm
Step 3: from 4.93×10^{-3} atm to 2.45×10^{-3} atm

Calculate q, w, ΔE, ΔS, ΔH, and ΔG for each step and for the overall process.

122. Consider 1.00 mol of an ideal gas at 25°C.

a. Calculate q, w, ΔE, ΔS, ΔH, and ΔG for the expansion of this gas isothermally and irreversibly from 2.45×10^{-2} atm to 2.45×10^{-3} atm in one step.
b. Calculate q, w, ΔE, ΔS, ΔH, and ΔG for the same change of pressure as in part a but performed isothermally and reversibly.
c. Calculate q, w, ΔE, ΔS, ΔH, and ΔG for the one-step isothermal, irreversible compression of 1.00 mol of an ideal gas at 25°C from 2.45×10^{-3} atm to 2.45×10^{-2} atm.
d. Construct the *PV* diagrams for the processes described in parts a, b, and c.
e. Calculate the entropy change in the surroundings for the processes described in parts a, b, and c.

123. Consider the reaction

$$2CO(g) + O_2(g) \longrightarrow 2CO_2(g)$$

a. Using data from Appendix 4, calculate *K* at 298 K.

b. What is ΔS for this reaction at $T = 298$ K if the reactants, each at 10.0 atm, are changed to products at 10.0 atm? (*Hint:* Construct a thermodynamic cycle and consider how entropy depends on pressure.)

124. Calculate ΔH° and ΔS° at 25°C for the reaction

$$2SO_2(g) + O_2(g) \longrightarrow 2SO_3(g)$$

at a constant pressure of 1.00 atm using thermodynamic data in Appendix 4. Also calculate ΔH° and ΔS° at 227°C and 1.00 atm, assuming that the constant-pressure molar heat capacities for $SO_2(g)$, $O_2(g)$, and $SO_3(g)$ are 39.9 J K^{-1} mol^{-1}, 29.4 J K^{-1} mol^{-1}, and 50.7 J K^{-1} mol^{-1}, respectively. (*Hint:* Construct a thermodynamic cycle and consider how enthalpy and entropy depend on temperature.)

125. Although we often assume that the heat capacity of a substance is not temperature-dependent, this is not strictly true, as shown by the following data for ice:

Temperature (°C)	C_p (J K^{-1} mol^{-1})
−200.	12
−180.	15
−160.	17
−140.	19
−100.	24
−60.	29
−30.	33
−10.	36
0	37

Use these data to calculate graphically the change in entropy for heating ice from −200.°C to 0°C. (*Hint:* Recall that

$$\Delta S_{T_1 \rightarrow T_2} = \int_{T_1}^{T_2} \frac{C_p\, dT}{T}$$

and that integration from T_1 to T_2 sums the area under the curve of a plot of C_p/T versus T from T_1 to T_2.)

126. Consider the following C_p values for $N_2(g)$:

C_p (J K^{-1} mol^{-1})	T (K)
28.7262	300.0
29.2937	400.0
29.8545	500.0

Assume that C_p can be expressed in the form

$$C_p = a + bT + cT^2$$

Estimate the value of C_p for $N_2(g)$ at 900. K. Assuming that C_p shows this temperature dependence over the range 100 K to 900 K, calculate ΔS for heating 1.00 mol $N_2(g)$ from 100. K to 900. K.

127. Benzene (C_6H_6) has a melting point of 5.5°C and an enthalpy of fusion of 10.04 kJ/mol at 25.0°C. The molar heat capacities at constant pressure for solid and liquid benzene are 100.4 J K^{-1} mol^{-1} and 133.0 J K^{-1} mol^{-1}, respectively. For the reaction

$$C_6H_6(l) \rightleftharpoons C_6H_6(s)$$

calculate ΔS_{sys} and ΔS_{surr} at 10.0°C.

Marathon Problems

128. Impure nickel, refined by smelting sulfide ores in a blast furnace, can be converted into metal from 99.90% to 99.99% purity by the Mond process. The primary reaction involved in the Mond process is

$$Ni(s) + 4CO(g) \rightleftharpoons Ni(CO)_4(g)$$

a. Without referring to Appendix 4, predict the sign of ΔS° for the preceding reaction. Explain.
b. The spontaneity of the preceding reaction is temperature-dependent. Predict the sign of ΔS_{surr} for this reaction. Explain.
c. For $Ni(CO)_4(g)$, $\Delta H_f^\circ = -607$ kJ/mol and $S^\circ = 417$ J K^{-1} mol^{-1} at 298 K. Using these values and data in Appendix 4, calculate ΔH° and ΔS° for the preceding reaction.
d. Calculate the temperature at which $\Delta G^\circ = 0$ ($K = 1$) for the preceding reaction, assuming that ΔH° and ΔS° do not depend on temperature.
e. The first step of the Mond process involves equilibrating impure nickel with $CO(g)$ and $Ni(CO)_4(g)$ at about 50°C. The purpose of this step is to convert as much nickel as possible into the gas phase. Calculate the equilibrium constant for the preceding reaction at 50.°C.
f. In the second step of the Mond process, the gaseous $Ni(CO)_4$ is isolated and heated at 227°C. The purpose of this step is to deposit as much nickel as possible as pure solid (the reverse of the preceding reaction). Calculate the equilibrium constant for the preceding reaction at 227°C.
g. Why is temperature increased for the second step of the Mond process?
h. The Mond process relies on the volatility of $Ni(CO)_4$ for its success. Only pressures and temperatures at which $Ni(CO)_4$ is a gas are useful. A recently developed variation of the Mond process carries out the first step at higher pressures and a temperature of 152°C. Estimate the maximum pressure of $Ni(CO)_4(g)$ that can be attained before the gas will liquefy at 152°C. The boiling point for

$Ni(CO)_4$ is 42°C, and the enthalpy of vaporization is 29.0 kJ/mol. [*Hint:* The phase-change reaction and the corresponding equilibrium expression are

$$Ni(CO)_4(l) \rightleftharpoons Ni(CO)_4(g) \qquad K_p = P_{Ni(CO)_4}$$

$Ni(CO)_4(g)$ will liquefy when the pressure of $Ni(CO)_4$ is greater than the K_p value.]

129. The initial state of an ideal gas is 2.00 atm, 2.00 L. The final state is 1.00 atm, 4.00 L. The expansion is accomplished isothermally.
   a. If the expansion is a free expansion, calculate w, q, ΔE, and ΔH.
   b. If the expansion is done in one step, calculate w, q, ΔE, and ΔH.
   c. If the expansion is done in two steps (with V = 3.00 L as the intermediate step), calculate w, q, ΔE, and ΔH.
   d. If the expansion is reversible, calculate w, q, ΔE, and ΔH.

   You have the new state of an ideal gas at 1.00 atm, 4.00 L. You take the gas back to conditions of 2.00 atm, 2.00 L. The compression is accomplished isothermally.
   e. If the compression is done in one step, calculate w, q, ΔE, and ΔH.
   f. If the compression is done in two steps (with V = 3.00 L as the intermediate step), calculate w, q, ΔE, and ΔH.
   g. If the compression is reversible, calculate w, q, ΔE, and ΔH. Explain.

   Compare your answers for the expansion and compression. Discuss the implications, especially considering the changes to the system and the changes to the surroundings that have occurred even though the system was brought back to its initial state.

MEDIA SUMMARY

Visit the Student Website at **www.cengage.com/chemistry/zumdahl** to help prepare for class, study for quizzes and exams, understand core concepts, and visualize molecular-level interactions. The following media activities are available for this chapter:

Prepare for Class

Video Lessons *Mini-lectures from chemistry experts*

- Spontaneous Processes
- Entropy and the Second Law of Thermodynamics
- Entropy and Temperature
- Gibbs Free Energy
- Standard Free Energy Changes of Formation
- Enthalpy and Entropy Contributions to K
- The Temperature Dependence of K
- Free Energy Away from Equilibrium

Improve Your Grade

Visualizations *Molecular-level animations and lab demonstration videos*

- Entropy
- Spontaneous Reactions

Flashcards *Key terms and definitions*

Online flashcards

ACE the Test

Multiple-choice quizzes
3 ACE Practice Tests

Access these resources using your passkey, available free with new texts or for purchase separately.

11 Electrochemistry

A rusted car from the 1930s.

Electrochemistry constitutes one of the most important interfaces between chemistry and everyday life. Every time you start your car, turn on your calculator, look at your digital watch, or listen to a radio at the beach, you are depending on electrochemical reactions. Our society sometimes seems to run almost entirely on batteries. Certainly, the advent of small, dependable batteries along with silicon chip technology has made possible the tiny calculators, tape recorders, and clocks that we take for granted.

Electrochemistry is important in other less obvious ways. For example, the corrosion of iron, which has tremendous economic implications, is an electrochemical process.

In addition, many important industrial materials such as aluminum, chlorine, and sodium hydroxide are prepared by electrolytic processes. In analytical chemistry, electrochemical techniques use electrodes that are specific for a given molecule or ion, including H^+ (pH meters), F^-, Cl^-, and many others. These increasingly important methods are used to analyze for trace pollutants in natural waters or for the tiny quantities of chemicals in human blood that may signal the development of a specific disease.

Electrochemistry is best defined as *the study of the interchange of chemical and electrical energy.* It is primarily concerned with two processes that involve oxidation–reduction reactions: the generation of an electric current from a chemical reaction and the opposite process, the use of a current to produce chemical change.

11.1 Galvanic Cells

Recall from Chapter 4 that an **oxidation–reduction (redox) reaction** involves a transfer of electrons from the **reducing agent** to the **oxidizing agent** and that **oxidation** involves a *loss of electrons* (an increase in oxidation number) and **reduction** involves a *gain of electrons* (a decrease in oxidation number).

To understand how a redox reaction can be used to generate a current, we will consider the reaction between MnO_4^- and Fe^{2+}:

$$8H^+(aq) + MnO_4^-(aq) + 5Fe^{2+}(aq) \longrightarrow Mn^{2+}(aq) + 5Fe^{3+}(aq) + 4H_2O(l)$$

In this reaction Fe^{2+} is oxidized and MnO_4^- is reduced; electrons are transferred from Fe^{2+} (the reducing agent) to MnO_4^- (the oxidizing agent).

Balancing half-reactions is discussed in Section 4.11.

It is useful to break a redox reaction into two **half-reactions,** one involving oxidation and the other involving reduction. For the preceding reaction, the half-reactions are

$$8H^+ + MnO_4^- + 5e^- \longrightarrow Mn^{2+} + 4H_2O$$

$$5(Fe^{2+} \longrightarrow Fe^{3+} + e^-)$$

Note that the second half-reaction must occur five times for each time the first reaction occurs. The balanced overall reaction is the sum of the half-reactions.

When MnO_4^- and Fe^{2+} are present in the same solution, the electrons are transferred directly as the reactants collide. Under these conditions, no useful work is obtained from the chemical energy associated with the reaction, which instead is released as heat. How can we harness this energy? The key is to separate the oxidizing agent from the reducing agent, thus requiring the electron transfer to occur through a wire. The current produced in the wire by the electron flow can then be directed through a device, such as an electric motor, to provide useful work.

For example, consider the system illustrated in Fig. 11.1. If our reasoning has been correct, electrons should flow through the wire from Fe^{2+} to MnO_4^-. However, when we construct the apparatus as shown, no electron flow is apparent. Why? Careful observation would show that when we connect the wires from the two compartments, current flows for an instant and then ceases. The current stops flowing because of charge buildups in the two compartments. If electrons flowed from the right to the left compartment in the apparatus as shown, the left compartment (receiving electrons) would become negatively charged, and the right compartment (losing electrons) would become positively charged. Creating a charge separation of this type requires

CHEMICAL INSIGHTS

Dental Resistance

A trip to the dentist is a necessary but often disquieting event. Because dealing with dental cavities can be expensive and unpleasant, dentists are searching for ways to find budding cavities at the very earliest stages of their development, well before they can be detected by X rays. One such method is being developed by Chris Longbottom, a dentist at the University of Dundee Dental School in Scotland. Longbottom is experimenting with measuring the electrical resistance of teeth using tiny, 10-millivolt currents. This approach takes advantage of the changes in a tooth that occur as a cavity begins to form. Bacteria in the mouth produce acids that dissolve tooth mineral, a process that starts in the pores of the teeth. As a given pore is enlarged by the action of acid, the electrical resistance of the tooth changes because the electrolyte solution that collects in the enlarged pore is a better electrical conductor than the tooth mineral. Using special electrodes that fit over the teeth at points where the teeth touch (and are thus most vulnerable to trapped food that stimulates bacterial growth) Longbottom and his colleagues have noted changes in resistance as cavities start to form. This method of detection is so sensitive that the tiny cavities can be treated with fluoride or antibiotics, thus checking the growth of cavities long before drilling is necessary. Although the technique needs further development, Longbottom hopes it will join dentists' arsenal of tools soon.

a large amount of energy. Thus sustained electron flow cannot occur under these conditions.

We can, however, solve this problem very simply. The solutions must be connected so that ions can flow to maintain the net charge of zero in each compartment. This connection might involve a **salt bridge** (a U-tube filled with an electrolyte) or a **porous disk** in a tube connecting the two solutions (see Fig. 11.2). Either of these devices allows ion flow without extensive mixing of the solutions. When we make the provision for ion flow, the circuit is complete. Electrons flow through the wire from reducing agent to oxidizing agent, and ions flow between the compartments to keep the net charge zero in each.

A galvanic cell uses a spontaneous redox reaction to produce a current that can be used to do work.

We now have covered all the essential characteristics of a **galvanic cell,** *a device in which chemical energy is changed to electrical energy.* (The opposite process, *electrolysis,* will be considered in Section 11.7.)

The reaction in an electrochemical cell occurs at the interface between an electrode and the solution where the electron transfer occurs. The electrode

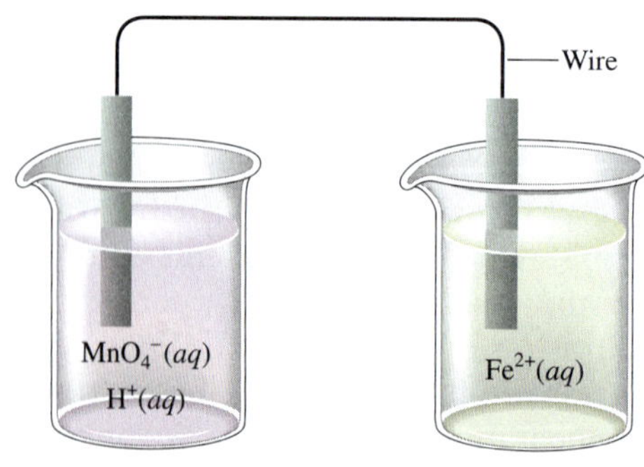

FIGURE 11.1

Schematic of a method to separate the oxidizing and reducing agents of a redox reaction. (The solutions also contain counter ions to balance the charge.)

FIGURE 11.2

Galvanic cells can contain a salt bridge as in (a) or a porous-disk connection as in (b). A salt bridge contains a strong electrolyte held in a Jello-like matrix. A porous disk contains tiny passages that allow hindered flow of ions.

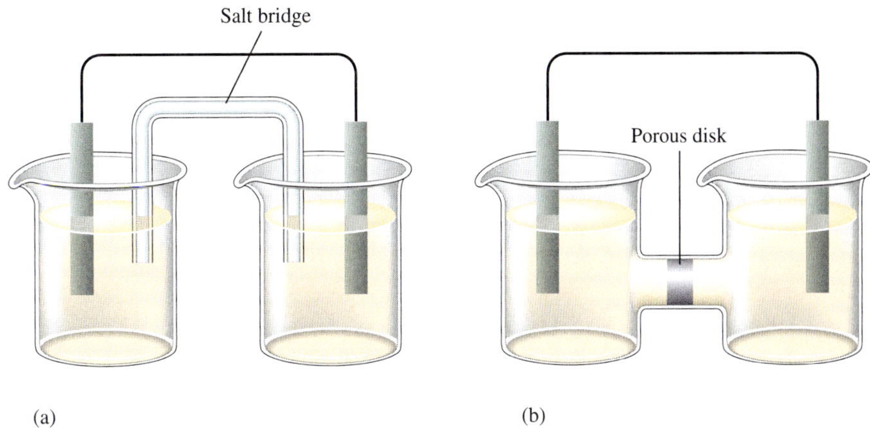

Oxidation occurs at the anode. Reduction occurs at the cathode.

at which *oxidation* occurs is called the **anode**; the electrode at which *reduction* occurs is called the **cathode** (see Fig. 11.3).

Cell Potential

A volt is 1 joule of work per coulomb of charge transferred: 1 V = 1 J/C.

A galvanic cell consists of an oxidizing agent in one compartment that pulls electrons through a wire from a reducing agent in the other compartment. The "pull," or driving force, on the electrons is called the **cell potential** ($\mathscr{E}_{cell}$), or the **electromotive force** (emf), of the cell. The unit of electrical potential is the **volt** (abbreviated V), which is defined as 1 joule of work per coulomb of charge transferred.

How can we measure the cell potential? One possible instrument is a crude **voltmeter,** which works by drawing current through a known resistance. However, when current flows through a wire, the frictional heating that occurs wastes some of the useful energy of the cell. A traditional voltmeter will therefore measure a potential that is lower than the maximum cell potential. The key to determining the maximum potential is to perform the measurement under conditions of zero current so that no energy is wasted. Traditionally, this measurement has been accomplished by inserting a variable-voltage device (powered from an external source) in *opposition* to the cell potential. The voltage on this instrument, called a **potentiometer,** is adjusted until no current flows in the cell circuit. Under such conditions the cell potential is equal in magnitude and opposite in sign to the voltage setting of the potentiometer.

FIGURE 11.3

An electrochemical process involves electron transfer at the interface between the electrode and the solution. (a) The species in the solution acting as the reducing agent supplies electrons to the anode. (b) The species in the solution acting as the oxidizing agent receives electrons from the cathode.

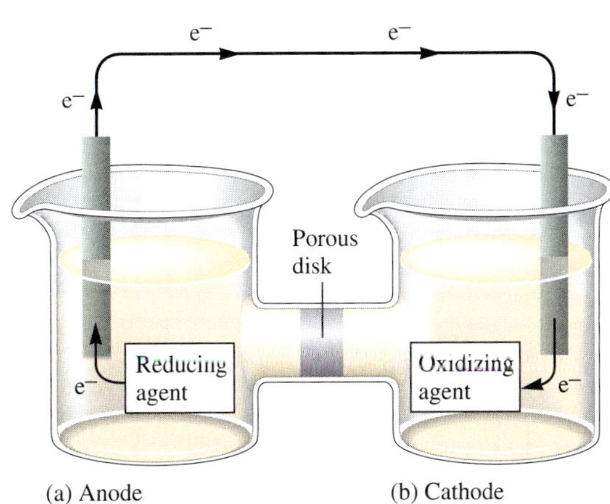

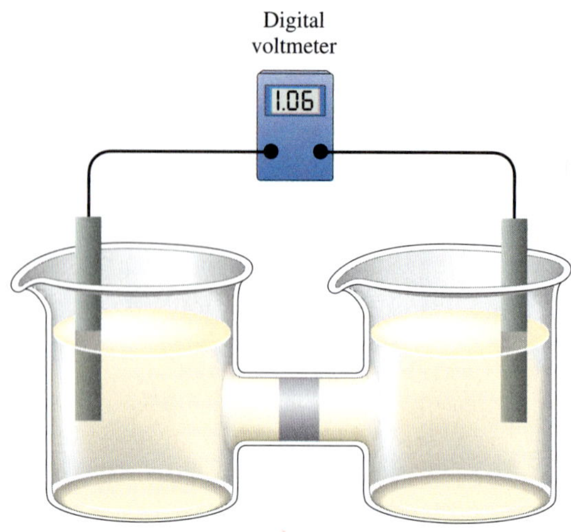

FIGURE 11.4

Digital voltmeters draw only a negligible current and are convenient to use.

This value represents the *maximum* cell potential, since no energy is wasted heating the wire. Advances in electronic technology have allowed the design of *digital voltmeters* that draw only a negligible amount of current (see Fig. 11.4). Since these instruments are more convenient to use, they have replaced potentiometers in the modern laboratory.

11.2 Standard Reduction Potentials

The name *galvanic cell* honors Luigi Galvani (1737–1798), an Italian scientist generally credited with the discovery of electricity. Galvanic cells are sometimes called *voltaic cells* after Alessandro Volta (1745–1827), another Italian, who first constructed cells of this type around 1800.

The reaction in a galvanic cell is always an oxidation–reduction reaction that can be broken down into half-reactions. It would be convenient to assign a potential to *each* half-reaction so that when we construct a cell from a given pair of half-reactions, we can obtain the cell potential by summing the half-cell potentials. For example, the observed potential for the cell shown in Fig. 11.5(a) is 0.76 volt, and the cell reaction* is

$$2H^+(aq) + Zn(s) \longrightarrow Zn^{2+}(aq) + H_2(g)$$

For this cell the anode compartment contains a zinc metal electrode with Zn^{2+} and $SO_4{}^{2-}$ ions in an aqueous solution that bathes the electrode. The anode reaction is the oxidation half-reaction:

$$Zn \longrightarrow Zn^{2+} + 2e^-$$

Each zinc atom loses two electrons to produce a Zn^{2+} ion that enters the solution. The electrons flow through the wire. For now we will assume that all cell components are in their standard states, so in this case the solution in the anode compartment will contain 1 *M* Zn^{2+}. The cathode reaction of this cell is

$$2H^+ + 2e^- \longrightarrow H_2$$

The standard hydrogen potential is the reference potential against which all half-reaction potentials are assigned.

The cathode consists of a platinum electrode (used because it is a chemically inert conductor) in contact with 1 *M* H^+ ions and bathed by hydrogen gas at

*In this text we will follow the convention of indicating the physical states of the reactants and products only in the overall redox reaction. For simplicity, half-reactions will *not* include the physical states.

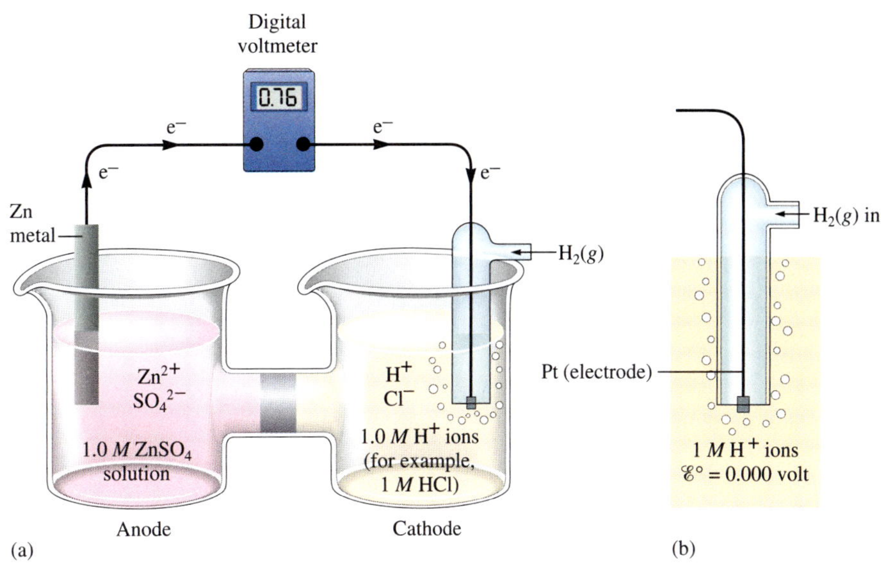

FIGURE 11.5

(a) A galvanic cell involving the reactions $Zn \longrightarrow Zn^{2+} + 2e^-$ (at the anode) and $2H^+ + 2e^- \longrightarrow H_2$ (at the cathode) has a potential of 0.76 V. (b) The standard hydrogen electrode where $H_2(g)$ at 1 atm is passed over a platinum electrode in contact with 1 M H^+ ions. This electrode process (assuming ideal behavior) is arbitrarily assigned a value of exactly zero volts.

1 atm. Such an electrode, called the **standard hydrogen electrode,** is shown in Fig. 11.5(b).

Although we can measure the *total* potential of this cell (0.76 volt), there is no way to measure the potentials of the individual electrodes. Thus, if we desire potentials for half-reactions (half-cells), we must arbitrarily divide up the total cell potential. For example, if we assign the reaction

$$2H^+ + 2e^- \longrightarrow H_2$$

where

$$[H^+] = 1\ M \qquad \text{and} \qquad P_{H_2} = 1 \text{ atm}$$

as having a potential of exactly 0 volts, then the reaction

$$Zn \longrightarrow Zn^{2+} + 2e^-$$

will have a potential of 0.76 volt, since

$$\underset{0.76\text{ V}}{\underset{\uparrow}{\mathscr{E}^\circ_{\text{cell}}}} = \underset{0\text{ V}}{\underset{\uparrow}{\mathscr{E}^\circ_{H^+\rightarrow H_2}}} + \underset{0.76\text{ V}}{\underset{\uparrow}{\mathscr{E}^\circ_{Zn\rightarrow Zn^{2+}}}}$$

Standard states were discussed in Section 9.6.

Recall that the superscript ° indicates that *standard states* are used.

By setting the standard potential for the half-reaction $2H^+ + 2e^- \rightarrow H_2$ equal to zero, we can assign values to all other half-reactions. For example, the measured potential for the cell shown in Fig. 11.6 is 1.10 volt. The cell reaction is

$$Zn(s) + Cu^{2+}(aq) \longrightarrow Zn^{2+}(aq) + Cu(s)$$

which can be divided into the half-reactions

$$\text{Anode:} \qquad Zn \longrightarrow Zn^{2+} + 2e^-$$

$$\text{Cathode:} \qquad Cu^{2+} + 2e^- \longrightarrow Cu$$

Thus

$$\mathscr{E}^\circ_{\text{cell}} = \mathscr{E}^\circ_{Zn\rightarrow Zn^{2+}} + \mathscr{E}^\circ_{Cu^{2+}\rightarrow Cu}$$

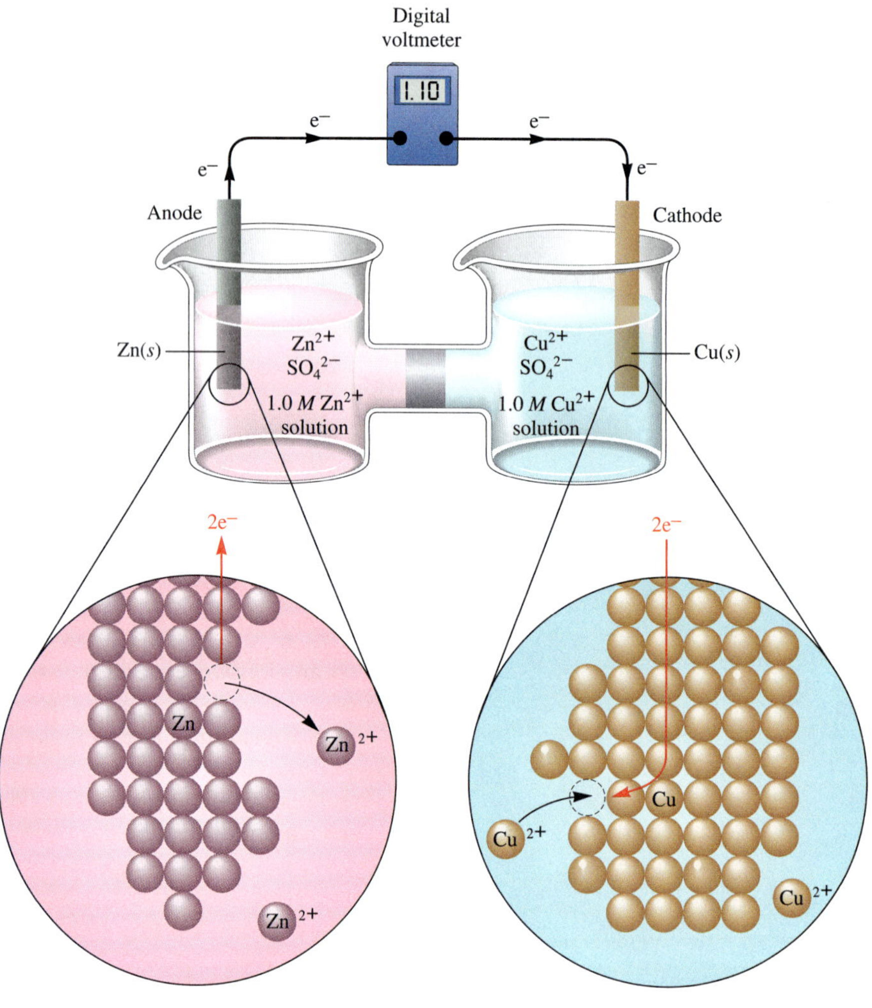

FIGURE 11.6

A galvanic cell involving the half-reactions $Zn \longrightarrow Zn^{2+} + 2e^-$ (anode) and $Cu^{2+} + 2e^- \longrightarrow Cu$ (cathode), with $\mathscr{E}^\circ_{cell} = 1.10$ V.

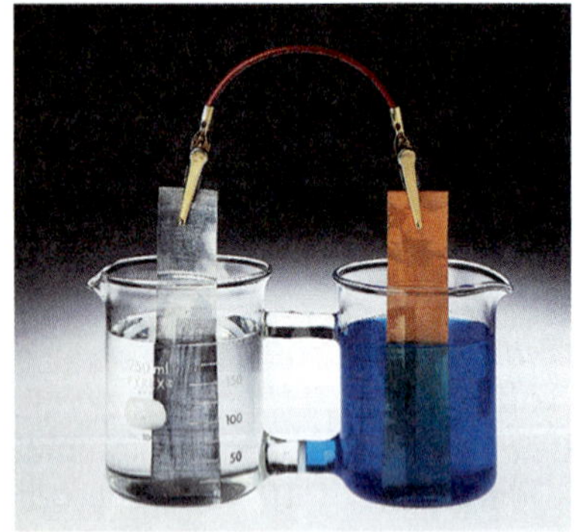

Copper being plated onto the copper metal cathode on the right, and zinc dissolving from the zinc metal anode on the left. Note the porous disk in the tube connecting the two solutions, which allows the ion flow to balance the electron flow through the wire.

All half-reactions are given as reduction processes in standard tables.

Since $\mathscr{E}^\circ_{Zn \to Zn^{2+}}$ is assigned a value of 0.76 volt, the value of $\mathscr{E}^\circ_{Cu^{2+} \to Cu}$ must be 0.34 volt:

$$1.10 \text{ V} = 0.76 \text{ V} + 0.34 \text{ V}$$

The scientific community has universally accepted the values for half-reaction potentials based on the assignment of 0 volts to the process $2H^+ + 2e^- \rightarrow H_2$ (under standard conditions where ideal behavior is assumed). However, before we can use these values, we need to understand several essential characteristics of half-cell potentials.

The accepted convention is to give the potentials of half-reactions as *reduction* processes. For example,

$$2H^+ + 2e^- \longrightarrow H_2$$

$$Cu^{2+} + 2e^- \longrightarrow Cu$$

$$Zn^{2+} + 2e^- \longrightarrow Zn$$

The $\mathscr{E}^\circ$ values corresponding to these half-reactions are called **standard reduction potentials.** Standard reduction potentials for the most common half-reactions are given in Table 11.1 and Appendix A5.5.

TABLE 11.1

Standard Reduction Potentials at 25°C (298 K) for Many Common Half-reactions

Half-reaction	$\mathscr{E}°$ (V)	Half-reaction	$\mathscr{E}°$ (V)
$F_2 + 2e^- \rightarrow 2F^-$	2.87	$O_2 + 2H_2O + 4e^- \rightarrow 4OH^-$	0.40
$Ag^{2+} + e^- \rightarrow Ag^+$	1.99	$Cu^{2+} + 2e^- \rightarrow Cu$	0.34
$Co^{3+} + e^- \rightarrow Co^{2+}$	1.82	$Hg_2Cl_2 + 2e^- \rightarrow 2Hg + 2Cl^-$	0.27
$H_2O_2 + 2H^+ + 2e^- \rightarrow 2H_2O$	1.78	$AgCl + e^- \rightarrow Ag + Cl^-$	0.22
$Ce^{4+} + e^- \rightarrow Ce^{3+}$	1.70	$SO_4^{2-} + 4H^+ + 2e^- \rightarrow H_2SO_3 + H_2O$	0.20
$PbO_2 + 4H^+ + SO_4^{2-} + 2e^- \rightarrow PbSO_4 + 2H_2O$	1.69	$Cu^{2+} + e^- \rightarrow Cu^+$	0.16
$MnO_4^- + 4H^+ + 3e^- \rightarrow MnO_2 + 2H_2O$	1.68	$2H^+ + 2e^- \rightarrow H_2$	0.00
$IO_4^- + 2H^+ + 2e^- \rightarrow IO_3^- + H_2O$	1.60	$Fe^{3+} + 3e^- \rightarrow Fe$	−0.036
$MnO_4^- + 8H^+ + 5e^- \rightarrow Mn^{2+} + 4H_2O$	1.51	$Pb^{2+} + 2e^- \rightarrow Pb$	−0.13
$Au^{3+} + 3e^- \rightarrow Au$	1.50	$Sn^{2+} + 2e^- \rightarrow Sn$	−0.14
$PbO_2 + 4H^+ + 2e^- \rightarrow Pb^{2+} + 2H_2O$	1.46	$Ni^{2+} + 2e^- \rightarrow Ni$	−0.23
$Cl_2 + 2e^- \rightarrow 2Cl^-$	1.36	$PbSO_4 + 2e^- \rightarrow Pb + SO_4^{2-}$	−0.35
$Cr_2O_7^{2-} + 14H^+ + 6e^- \rightarrow 2Cr^{3+} + 7H_2O$	1.33	$Cd^{2+} + 2e^- \rightarrow Cd$	−0.40
$O_2 + 4H^+ + 4e^- \rightarrow 2H_2O$	1.23	$Fe^{2+} + 2e^- \rightarrow Fe$	−0.44
$MnO_2 + 4H^+ + 2e^- \rightarrow Mn^{2+} + 2H_2O$	1.21	$Cr^{3+} + e^- \rightarrow Cr^{2+}$	−0.50
$IO_3^- + 6H^+ + 5e^- \rightarrow \frac{1}{2}I_2 + 3H_2O$	1.20	$Cr^{3+} + 3e^- \rightarrow Cr$	−0.73
$Br_2 + 2e^- \rightarrow 2Br^-$	1.09	$Zn^{2+} + 2e^- \rightarrow Zn$	−0.76
$VO_2^+ + 2H^+ + e^- \rightarrow VO^{2+} + H_2O$	1.00	$2H_2O + 2e^- \rightarrow H_2 + 2OH^-$	−0.83
$AuCl_4^- + 3e^- \rightarrow Au + 4Cl^-$	0.99	$Mn^{2+} + 2e^- \rightarrow Mn$	−1.18
$NO_3^- + 4H^+ + 3e^- \rightarrow NO + 2H_2O$	0.96	$Al^{3+} + 3e^- \rightarrow Al$	−1.66
$ClO_2 + e^- \rightarrow ClO_2^-$	0.954	$H_2 + 2e^- \rightarrow 2H^-$	−2.23
$2Hg^{2+} + 2e^- \rightarrow Hg_2^{2+}$	0.91	$Mg^{2+} + 2e^- \rightarrow Mg$	−2.37
$Ag^+ + e^- \rightarrow Ag$	0.80	$La^{3+} + 3e^- \rightarrow La$	−2.37
$Hg_2^{2+} + 2e^- \rightarrow 2Hg$	0.80	$Na^+ + e^- \rightarrow Na$	−2.71
$Fe^{3+} + e^- \rightarrow Fe^{2+}$	0.77	$Ca^{2+} + 2e^- \rightarrow Ca$	−2.76
$O_2 + 2H^+ + 2e^- \rightarrow H_2O_2$	0.68	$Ba^{2+} + 2e^- \rightarrow Ba$	−2.90
$MnO_4^- + e^- \rightarrow MnO_4^{2-}$	0.56	$K^+ + e^- \rightarrow K$	−2.92
$I_2 + 2e^- \rightarrow 2I^-$	0.54	$Li^+ + e^- \rightarrow Li$	−3.05
$Cu^+ + e^- \rightarrow Cu$	0.52		

Combining two half-reactions to obtain a balanced oxidation–reduction reaction often requires two manipulations:

When a half-reaction is reversed, the sign of $\mathscr{E}°$ is reversed.

1. One of the reduction half-reactions must be reversed (since redox reactions must involve a substance being oxidized and a substance being reduced). The half-reaction with the largest positive potential will run as written (as a reduction), and the other half-reaction will be forced to run in reverse (will be the oxidation reaction). The net potential of the cell will be the *difference* between the two. Since the reduction process occurs at the cathode and the oxidation process occurs at the anode, we can write

$$\mathscr{E}°_{cell} = \mathscr{E}° \text{ (cathode)} - \mathscr{E}° \text{ (anode)}$$

Because subtraction means "change the sign and add," in the examples done here we will change the sign of the oxidation (anode) reaction when we reverse it and add it to the reduction (cathode) reaction.

When a half-reaction is multiplied by an integer, $\mathscr{E}°$ remains the same.

2. Since the number of electrons lost must equal the number gained, the half-reactions must be multiplied by integers as necessary to achieve electron balance. However, the *value of $\mathscr{E}°$ is not changed* when a half-reaction is multiplied by an integer. Since a standard reduction potential is an *intensive*

property (it does not depend on how many times the reaction occurs), the potential is *not* multiplied by the integer required to balance the cell reaction.

Consider a galvanic cell based on the redox reaction

$$Fe^{3+}(aq) + Cu(s) \longrightarrow Cu^{2+}(aq) + Fe^{2+}(aq)$$

The pertinent half-reactions are

$$Fe^{3+} + e^- \longrightarrow Fe^{2+} \qquad \mathscr{E}° = 0.77\ V \qquad (1)$$

$$Cu^{2+} + 2e^- \longrightarrow Cu \qquad \mathscr{E}° = 0.34\ V \qquad (2)$$

Since the Cu^{2+}/Cu half-reaction has the lower positive $\mathscr{E}°$ value, it will be forced to run in reverse:

$$Cu \longrightarrow Cu^{2+} + 2e^-$$

It is the anode reaction.

Then, since each Cu atom produces two electrons but each Fe^{3+} ion accepts only one electron, reaction (1) must be multiplied by 2:

$$2Fe^{3+} + 2e^- \longrightarrow 2Fe^{2+}$$

Now we can obtain the balanced cell reaction by summing the appropriately modified half-reactions:

$$2Fe^{3+} + 2e^- \longrightarrow 2Fe^{2+} \qquad \mathscr{E}\ (\text{cathode}) = 0.77\ V$$

$$Cu \longrightarrow Cu^{2+} + 2e^- \qquad -\mathscr{E}\ (\text{anode}) = -0.34\ V$$

Cell reaction: $Cu(s) + 2Fe^{3+}(aq) \longrightarrow Cu^{2+}(aq) + 2Fe^{2+}(aq)$

$$\mathscr{E}°_{cell} = \mathscr{E}°\ (\text{cathode}) - \mathscr{E}°\ (\text{anode}) = 0.77\ V - 0.34\ V = 0.43\ V$$

A galvanic cell runs spontaneously in the direction that gives a positive value for $\mathscr{E}°_{cell}$.

Next, we want to consider how to describe a galvanic cell fully, given just its half-reactions. This description will include the cell reaction, the cell potential, and the physical setup of the cell. Let's consider a galvanic cell based on the following half-reactions:

$$Fe^{2+} + 2e^- \longrightarrow Fe \qquad \mathscr{E}° = -0.44\ V$$

$$MnO_4^- + 5e^- + 8H^+ \longrightarrow Mn^{2+} + 4H_2O \qquad \mathscr{E}° = 1.51\ V$$

In a working galvanic cell one of these reactions must run in reverse—the one with the least positive potential.

This will lead to a positive overall cell potential: *A cell will always run spontaneously in the direction that produces a positive cell potential.* Thus in the present case the half-reaction involving iron must be reversed.

$$Fe \longrightarrow Fe^{2+} + 2e^- \qquad \text{Anode reaction}$$

$$MnO_4^- + 5e^- + 8H^+ \longrightarrow Mn^{2+} + 4H_2O \qquad \text{Cathode reaction}$$

where

$$\mathscr{E}°_{cell} = \mathscr{E}°\ (\text{cathode}) - \mathscr{E}°\ (\text{anode}) = 1.51\ V - 0.44\ V$$

The balanced cell reaction is obtained as follows:

$$5(Fe \longrightarrow Fe^{2+} + 2e^-)$$

$$2(MnO_4^- + 5e^- + 8H^+ \longrightarrow Mn^{2+} + 4H_2O)$$

$$2MnO_4^-(aq) + 5Fe(s) + 16H^+(aq) \longrightarrow 5Fe^{2+}(aq) + 2Mn^{2+}(aq) + 8H_2O(l)$$

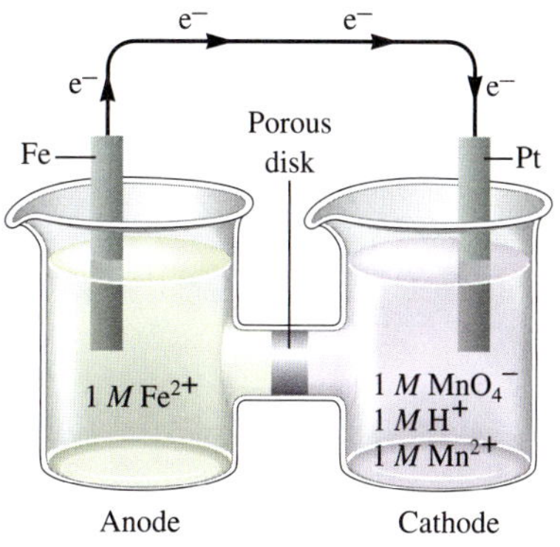

FIGURE 11.7

The schematic of a galvanic cell based on the half-reactions

$$Fe \longrightarrow Fe^{2+} + 2e^-$$

$$MnO_4^- + 5e^- + 8H^+ \longrightarrow Mn^{2+} + 4H_2O$$

Now consider the physical setup of the cell, shown schematically in Fig. 11.7. In the left compartment the active components in their standard states are pure metallic iron (Fe) and 1.0 *M* Fe^{2+}. The anion present depends on the iron salt used. In this compartment the anion does not participate in the reaction but simply balances the charge. The half-reaction that takes place at this electrode is

$$Fe \longrightarrow Fe^{2+} + 2e^-$$

which is an oxidation reaction, so this compartment is the anode. The electrode consists of pure iron metal.

In the right compartment the active components in their standard states are 1.0 *M* MnO_4^-, 1.0 *M* H^+, and 1.0 *M* Mn^{2+}, with appropriate unreacting ions (often called *counter ions*) to balance the charge. The half-reaction in this compartment is

$$MnO_4^- + 5e^- + 8H^+ \longrightarrow Mn^{2+} + 4H_2O$$

which is a reduction reaction, so this compartment is the cathode. Since neither MnO_4^- nor Mn^{2+} ion can serve as the electrode, a nonreacting conductor must be used. The usual choice is platinum.

The next step is to determine the direction of electron flow. In the left compartment the half-reaction is the oxidation of iron:

$$Fe \longrightarrow Fe^{2+} + 2e^-$$

In the right compartment the half-reaction is the reduction of MnO_4^-:

$$MnO_4^- + 5e^- + 8H^+ \longrightarrow Mn^{2+} + 4H_2O$$

Thus the electrons flow from Fe to MnO_4^- in this cell, or from the anode to the cathode, as is always the case.

Items Needed for a Description of a Galvanic Cell

1. The cell potential (always positive for a galvanic cell) and the balanced cell reaction.
2. The direction of electron flow, obtained by inspecting the half-reactions and using the directions that give a positive $\mathscr{E}^\circ_{cell}$.
3. Designation of the anode and the cathode.
4. The nature of each electrode and the ions present in each compartment. A chemically inert conductor is required if none of the substances participating in the half-reaction is a conducting solid.

Example 11.1

Describe completely the galvanic cell based on the following half-reactions under standard conditions:

$$Ag^+ + e^- \longrightarrow Ag \qquad \mathscr{E}^\circ = 0.80 \text{ V} \qquad (1)$$

$$Fe^{3+} + e^- \longrightarrow Fe^{2+} \qquad \mathscr{E}^\circ = 0.77 \text{ V} \qquad (2)$$

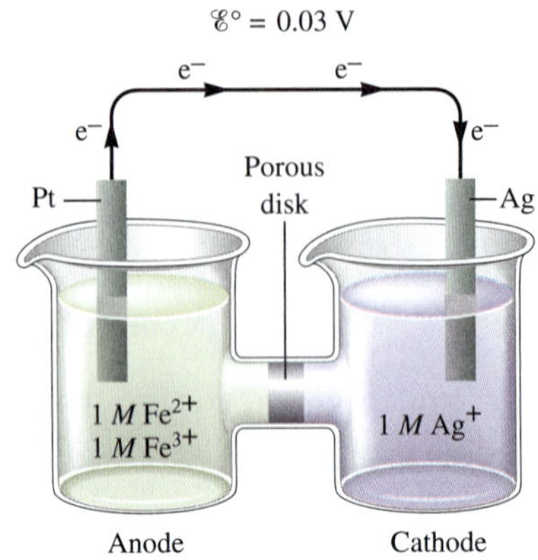

FIGURE 11.8

Schematic diagram for the galvanic cell based on the half-reactions

$$Ag^+ + e^- \longrightarrow Ag$$
$$Fe^{2+} \longrightarrow Fe^{3+} + e^-$$

Solution Since a positive $\mathscr{E}°_{cell}$ value is required, reaction (2) must run in reverse:

$$Ag^+ + e^- \longrightarrow Ag \qquad \mathscr{E}° \text{ (cathode)} = 0.80 \text{ V}$$
$$Fe^{2+} \longrightarrow Fe^{3+} + e^- \qquad -\mathscr{E}° \text{ (anode)} = -0.77 \text{ V}$$

$$\text{Cell reaction:} \quad Ag^+(aq) + Fe^{2+}(aq) \longrightarrow Fe^{3+}(aq) + Ag(s) \qquad \mathscr{E}°_{cell} = 0.03 \text{ V}$$

Since Ag^+ receives electrons and Fe^{2+} loses electrons in the cell reaction, the electrons flow from the compartment containing Fe^{2+} to the compartment containing Ag^+.

Oxidation occurs in the compartment containing Fe^{2+}. Hence this compartment functions as the anode. Reduction occurs in the compartment containing Ag^+, so this compartment is the cathode.

The electrode in the Ag/Ag^+ compartment is silver metal, and an inert conductor, such as platinum, must be used in the Fe^{2+}/Fe^{3+} compartment. Appropriate counter ions are assumed to be present. The diagram for this cell is shown in Fig. 11.8.

11.3 Cell Potential, Electrical Work, and Free Energy

So far we have considered electrochemical cells in a very practical manner without much theoretical background. The next step will be to explore the relationship between thermodynamics and electrochemistry.

The work that can be accomplished when electrons are transferred through a wire depends on the "push" (the thermodynamic driving force) behind the electrons. This driving force (the emf) is defined in terms of a *potential* difference (in volts) between two points in the circuit. Recall that a volt represents a joule of work per coulomb of charge transferred:

$$\text{emf} = \text{potential difference (V)} = \frac{\text{work (J)}}{\text{charge (C)}}$$

Thus 1 joule of work is produced or required (depending on the direction) when 1 coulomb of charge is transferred between two points in the circuit that differ by a potential of 1 volt.

In this book *work is viewed from the point of view of the system.* Thus work flowing out of the system is indicated by a minus sign. When a cell produces a current, the cell potential is positive, and the current can be used to do work—to run a motor, for instance. Thus the cell potential ($\mathscr{E}$) and the work (w) have opposite signs:

$$\mathscr{E} = \frac{-w \leftarrow \text{Work}}{q \leftarrow \text{Charge}}$$

Thus

$$-w = q\mathscr{E}$$

From this equation we can see that the maximum work in a cell is obtained at the maximum cell potential:

$$-w_{max} = q\mathscr{E}_{max} \quad \text{or} \quad w_{max} = -q\mathscr{E}_{max}$$

The work is never the maximum possible if any current is flowing.

Michael Faraday lecturing at the Royal Institution before Prince Albert and others (1855).

However, there is a problem. For the system to perform electrical work, current must flow. When current flows, some energy is inevitably wasted through frictional heating, so the maximum work is not obtained. This observation reflects the important general principle introduced in Section 10.12: *In any real, spontaneous process some energy is always wasted—the actual work realized is always less than the calculated maximum.* This principle is a consequence of the fact that the entropy of the universe must increase in any spontaneous process. Recall from Section 10.12 that the only process from which maximum work could be realized is the hypothetical reversible process. For a galvanic cell, this would involve an infinitesimally small current flow and thus an infinite amount of time to do the work. Even though we can never achieve the maximum work through the actual discharge of a galvanic cell, we can measure the maximum potential. There is negligible current flow when a cell potential is measured with a potentiometer or an efficient digital voltmeter. No current flow implies no waste of energy, so the measured potential is the maximum.

Although we can never actually realize the maximum work from a cell reaction, its value is still useful for evaluating the efficiency of a real process based on the cell reaction. For example, suppose a certain galvanic cell has a maximum potential of 2.50 V. In a particular experiment 1.33 moles of electrons passes through this cell at an average actual potential of 2.10 V. The actual work done is

$$w = -q\mathscr{E}$$

where $\mathscr{E}$ represents the actual potential difference at which the current flowed (2.10 V or 2.10 J/C) and q is the quantity of charge transferred (in coulombs). The charge on 1 mole of electrons is called the **faraday** (abbreviated F): *96,485 coulombs of charge per mole of electrons.* Thus q equals the number of moles of electrons times the charge per mole of electrons:

$$q = nF = 1.33 \text{ mol e}^- \times 96{,}485 \text{ C/mol e}^-$$

For the experiment above, the actual work is

$$w = -q\mathscr{E} = -\left(1.33 \text{ mol e}^- \times 96{,}485 \frac{\text{C}}{\text{mol e}^-}\right) \times \left(2.10 \frac{\text{J}}{\text{C}}\right)$$

$$= -2.69 \times 10^5 \text{ J}$$

For the maximum possible work, the calculation is similar, except that the maximum potential is used:

$$w_{\text{max}} = -q\mathscr{E}_{\text{max}} = -\left(1.33 \text{ mol e}^- \times 96{,}485 \frac{\text{C}}{\text{mol e}^-}\right)\left(2.50 \frac{\text{J}}{\text{C}}\right)$$

$$= -3.21 \times 10^5 \text{ J}$$

Thus, in its actual operation, the efficiency of this cell is

$$\frac{w}{w_{\text{max}}} \times 100\% = \frac{-2.69 \times 10^5 \text{ J}}{-3.21 \times 10^5 \text{ J}} \times 100\% = 83.8\%$$

Next we want to relate the potential of a galvanic cell to free energy. In Section 10.12 we saw that for a process carried out at constant temperature and pressure, the change in free energy equals the maximum useful work obtainable from that process:

$$w_{\text{max}} = \Delta G$$

For a galvanic cell,

$$w_{\text{max}} = -q\mathscr{E}_{\text{max}} = \Delta G$$

Since

$$q = nF$$

we have

$$\Delta G = -q\mathscr{E}_{\text{max}} = -nF\mathscr{E}_{\text{max}}$$

From now on the subscript on $\mathscr{E}_{\text{max}}$ will be deleted, with the understanding that any potential given in this book is the maximum potential. Thus

$$\Delta G = -nF\mathscr{E}$$

For standard conditions

$$\Delta G^\circ = -nF\mathscr{E}^\circ$$

This equation states that *the maximum cell potential is directly related to the free energy difference between the reactants and the products in the cell.* This relationship is important because it provides an experimental means of obtaining ΔG for a reaction. It also confirms that a galvanic cell runs in the direction that gives a positive value for $\mathscr{E}_{\text{cell}}$; a positive $\mathscr{E}_{\text{cell}}$ value corresponds to a negative ΔG value, which is the condition for spontaneity.

Example 11.2

Using the data in Table 11.1, calculate ΔG° for the reaction

$$Cu^{2+}(aq) + Fe(s) \longrightarrow Cu(s) + Fe^{2+}(aq)$$

Is this reaction spontaneous?

Solution The half-reactions are

$$Cu^{2+} + 2e^- \longrightarrow Cu \qquad \mathscr{E}^\circ \text{ (cathode)} = 0.34 \text{ V}$$

$$Fe \longrightarrow Fe^{2+} + 2e^- \qquad -\mathscr{E}^\circ \text{ (anode)} = 0.44 \text{ V}$$

$$Cu^{2+} + Fe \longrightarrow Fe^{2+} + Cu \qquad \mathscr{E}^\circ_{\text{cell}} = 0.78 \text{ V}$$

We can calculate ΔG° from the equation

$$\Delta G^\circ = -nF\mathscr{E}^\circ$$

Since two electrons are transferred in the reaction, 2 moles of electrons are required per mole of reactants and products. Thus $n = 2$ mol e^-, $F = 96{,}485$ C/mol e^-, and $\mathscr{E}° = 0.78$ V $= 0.78$ J/C. Therefore,

$$\Delta G° = -(2 \text{ mol e}^-)\left(96{,}485 \frac{\text{C}}{\text{mol e}^-}\right)\left(0.78 \frac{\text{J}}{\text{C}}\right)$$
$$= -1.5 \times 10^5 \text{ J}$$

The process is spontaneous, as indicated both by the negative sign of $\Delta G°$ and by the positive sign of $\mathscr{E}°_{\text{cell}}$.

This reaction is used industrially to deposit copper metal from solutions containing dissolved copper ores.

Example 11.3

Using the data from Table 11.1, predict whether 1 *M* HNO_3 will dissolve gold metal to form a 1 *M* Au^{3+} solution.

A gold ring does not dissolve in nitric acid.

Solution The half-reaction for HNO_3 acting as an oxidizing agent is

$$NO_3^- + 4H^+ + 3e^- \longrightarrow NO + 2H_2O \qquad \mathscr{E}° \text{ (cathode)} = 0.96 \text{ V}$$

The reaction for the oxidation of solid gold to Au^{3+} ions is

$$\text{Au} \longrightarrow Au^{3+} + 3e^- \qquad -\mathscr{E}° \text{ (anode)} = -1.50 \text{ V}$$

The sum of these half-reactions gives the required reaction:

$$\text{Au}(s) + NO_3^-(aq) + 4H^+(aq) \longrightarrow Au^{3+}(aq) + \text{NO}(g) + 2H_2O(l)$$

and

$$\mathscr{E}°_{\text{cell}} = 0.96 \text{ V} - 1.50 \text{ V} = -0.54 \text{ V}$$

Since the $\mathscr{E}°$ value is negative, the process will *not* occur under standard conditions. That is, gold will not dissolve in 1 *M* HNO_3 to give 1 *M* Au^{3+}. In fact, a mixture (1:3 by volume) of concentrated nitric and hydrochloric acid, called *aqua regia,* is required to dissolve gold.

11.4 Dependence of the Cell Potential on Concentration

So far we have described cells under standard conditions. In this section we consider the dependence of the cell potential on concentration. For example, under standard conditions (all concentrations 1 *M*), the cell with the reaction

$$\text{Cu}(s) + 2Ce^{4+}(aq) \longrightarrow Cu^{2+}(aq) + 2Ce^{3+}(aq)$$

has a potential of 1.36 V. What will the cell potential be if $[Ce^{4+}]$ is greater than 1.0 *M*? This question can be answered qualitatively in terms of Le Châtelier's principle. An increase in the concentration of Ce^{4+} will favor the forward reaction and thus increase the driving force on the electrons. The cell potential will increase. On the other hand, an increase in the concentration of a product (Cu^{2+} or Ce^{3+}) will oppose the forward reaction, thus decreasing the cell potential.

These ideas are illustrated in Example 11.4.

Example 11.4

For the cell reaction

$$2Al(s) + 3Mn^{2+}(aq) \longrightarrow 2Al^{3+}(aq) + 3Mn(s) \qquad \mathscr{E}^\circ_{cell} = 0.48 \text{ V}$$

predict whether $\mathscr{E}_{cell}$ will be larger or smaller than $\mathscr{E}^\circ_{cell}$ for the following cases.

a. $[Al^{3+}] = 2.0\ M$, $[Mn^{2+}] = 1.0\ M$
b. $[Al^{3+}] = 1.0\ M$, $[Mn^{2+}] = 3.0\ M$

Solution

a. A product concentration has been raised above 1.0 *M*. This will oppose the cell reaction, causing $\mathscr{E}_{cell}$ to be less than $\mathscr{E}^\circ_{cell}$ ($\mathscr{E}_{cell} < 0.48$ V).

b. A reactant concentration has been increased above 1.0 *M*, and $\mathscr{E}_{cell}$ will be greater than $\mathscr{E}^\circ_{cell}$ ($\mathscr{E}_{cell} > 0.48$ V).

The Nernst Equation

Professor Walter Hermann von Nernst (1864–1941) was one of the pioneers in the development of electrochemical theory and is generally given credit for first stating the third law of thermodynamics. He won the Nobel Prize in chemistry in 1920 for his contributions to our understanding of thermodynamics.

The dependence of the cell potential on concentration results directly from the dependence of free energy on concentration. Recall from Chapter 10 that the equation

$$\Delta G = \Delta G^\circ + RT\ln(Q)$$

where Q is the reaction quotient, was used to calculate the effect of concentration on ΔG. Since $\Delta G = -nF\mathscr{E}$ and $\Delta G^\circ = -nF\mathscr{E}^\circ$, the equation becomes

$$-nF\mathscr{E} = -nF\mathscr{E}^\circ + RT\ln(Q)$$

Dividing each side of the equation by $-nF$ gives

$$\mathscr{E} = \mathscr{E}^\circ - \frac{RT}{nF}\ln(Q) \tag{11.1}$$

Equation (11.1), which gives the relationship between the cell potential and the concentrations of the cell components, is commonly called the **Nernst equation** after German chemist Walter Hermann von Nernst.

The Nernst equation is often given in terms of a log (base-10) form that is valid at 25°C:

$$\mathscr{E} = \mathscr{E}^\circ - \frac{0.0591}{n}\log(Q)$$

Using this relationship we can calculate the potential of a cell in which some or all of the components are not in their standard states.

For example, $\mathscr{E}^\circ_{cell}$ is 0.48 V for the galvanic cell based on the reaction

$$2Al(s) + 3Mn^{2+}(aq) \longrightarrow 2Al^{3+}(aq) + 3Mn(s)$$

Consider a cell in which

$$[Mn^{2+}] = 0.50\ M \qquad \text{and} \qquad [Al^{3+}] = 1.50\ M$$

The cell potential at 25°C for these concentrations must be calculated using the Nernst equation:

$$\mathscr{E}_{cell} = \mathscr{E}^\circ_{cell} - \frac{0.0591}{n}\log(Q)$$

We know that

$$\mathscr{E}^\circ_{cell} = 0.48 \text{ V} \qquad \text{and} \qquad Q = \frac{[Al^{3+}]^2}{[Mn^{2+}]^3} = \frac{(1.50)^2}{(0.50)^3} = 18$$

Since the half-reactions are

$$2Al \longrightarrow 2Al^{3+} + 6e^- \qquad \text{and} \qquad 3Mn^{2+} + 6e^- \longrightarrow 3Mn$$

we know that

$$n = 6$$

Thus

$$\mathscr{E}_{cell} = 0.48 - \frac{0.0591}{6}\log(18) = 0.47 \text{ V}$$

Note that the cell voltage decreases slightly because of the nonstandard concentrations. This change is consistent with the predictions of Le Châtelier's principle (see Example 11.4). In this case, since the reactant concentration is lower than 1.0 *M* and the product concentration is higher than 1.0 *M*, $\mathscr{E}_{cell}$ is less than $\mathscr{E}^\circ_{cell}$.

The potential calculated from the Nernst equation is the maximum potential before any current flow occurs. As the cell discharges and current flows from anode to cathode, the concentrations change; as a result, $\mathscr{E}_{cell}$ changes. In fact, *the cell will spontaneously discharge until it reaches equilibrium,* at which point

$$Q = K \text{ (the equilibrium constant)} \qquad \text{and} \qquad \mathscr{E}_{cell} = 0$$

A "dead" battery is one in which the cell reaction has reached equilibrium; there is no longer any chemical driving force to push electrons through the wire. In other words, *at equilibrium the components in the two cell compartments have the same free energy;* that is, $\Delta G = 0$ for the cell reaction at the equilibrium concentrations. The cell no longer has the ability to do work.

Example 11.5

Describe the cell based on the following half-reactions:

$$VO_2^{+} + 2H^+ + e^- \longrightarrow VO^{2+} + H_2O \qquad \mathscr{E}^\circ = 1.00 \text{ V} \qquad (1)$$

$$Zn^{2+} + 2e^- \longrightarrow Zn \qquad \mathscr{E}^\circ = -0.76 \text{ V} \qquad (2)$$

where

$$T = 25°C \qquad [VO_2^{+}] = 2.0\ M \qquad [VO^{2+}] = 1.0 \times 10^{-2}\ M$$

$$[H^+] = 0.50\ M \qquad [Zn^{2+}] = 1.0 \times 10^{-1}\ M$$

Solution The balanced cell reaction is obtained by reversing reaction (2) and multiplying reaction (1) by 2:

2 × reaction (1)

$$2VO_2^{+} + 4H^+ + 2e^- \longrightarrow 2VO^{2+} + 2H_2O \qquad \mathscr{E}^\circ \text{ (cathode)} = 1.00 \text{ V}$$

Reaction (2) reversed

$$Zn \longrightarrow Zn^{2+} + 2e^- \qquad -\mathscr{E}^\circ \text{ (anode)} = 0.76 \text{ V}$$

Cell reaction:

$$2VO_2^{+}(aq) + 4H^+(aq) + Zn(s) \longrightarrow 2VO^{2+}(aq) + 2H_2O(l) + Zn^{2+}(aq) \qquad \mathscr{E}^\circ_{cell} = 1.76 \text{ V}$$

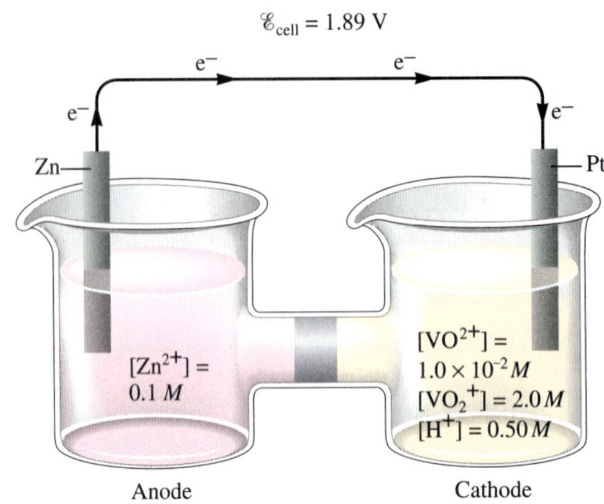

FIGURE 11.9

Schematic diagram of the cell described in Example 11.5.

Since the cell contains components at concentrations other than 1 M, we must use the Nernst equation, where $n = 2$ (since two electrons are transferred), to calculate the cell potential. At 25°C we can use the equation in the form

$$\mathscr{E} = \mathscr{E}° - \frac{0.0591}{n}\log(Q)$$

Thus

$$\mathscr{E} = 1.76 - \frac{0.0591}{2}\log\left(\frac{[Zn^{2+}][VO^{2+}]^2}{[VO_2^{+}]^2[H^+]^4}\right)$$

$$= 1.76 - \frac{0.0591}{2}\log\left[\frac{(1.0 \times 10^{-1})(1.0 \times 10^{-2})^2}{(2.0)^2(0.50)^4}\right]$$

$$= 1.76 - \frac{0.0591}{2}\log(4 \times 10^{-5}) = 1.76 + 0.13 = 1.89 \text{ V}$$

The cell diagram is given in Fig. 11.9.

Ion-Selective Electrodes

Because the cell potential is sensitive to the concentrations of the reactants and products involved in the cell reaction, measured potentials can be used to determine the concentration of an ion. A pH meter is a familiar example of an instrument that measures concentration from an observed potential. The pH meter has three main components: a standard electrode of known potential, a special **glass electrode** that changes potential depending on the concentration of H^+ ion in the solution into which it is dipped, and a potentiometer that measures the potential between the two electrodes. The potentiometer reading is automatically converted electronically to a direct reading of the pH of the solution being tested.

The glass electrode (see Fig. 11.10) contains a reference solution of dilute hydrochloric acid in contact with a thin glass membrane. The electrical potential

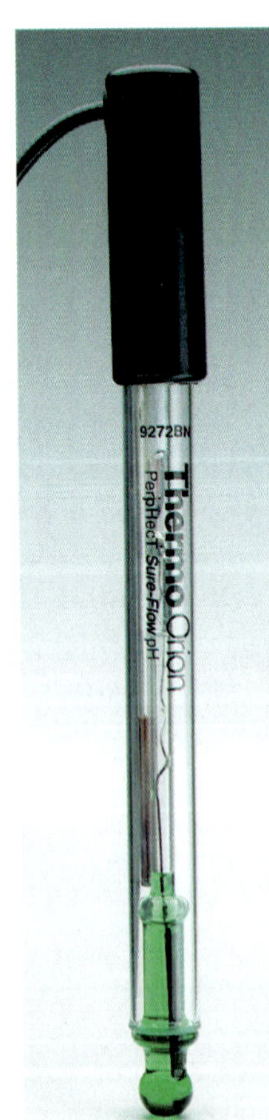

FIGURE 11.10

A glass electrode contains a reference solution of dilute hydrochloric acid in contact with a thin glass membrane, in which a silver wire coated with silver chloride has been embedded. When the electrode is dipped into a solution containing H^+ ions, the electrode potential is determined by the difference in $[H^+]$ between the two solutions.

TABLE 11.2

Some Ions Whose Concentrations Can Be Detected by Ion-Selective Electrodes

Cations	Anions
H^+	Br^-
Cd^{2+}	Cl^-
Ca^{2+}	CN^-
Cu^{2+}	F^-
K^+	NO_3^-
Ag^+	S^{2-}
Na^+	

of the glass electrode depends on the difference in $[H^+]$ between the reference solution and the solution into which the electrode is dipped. Thus the electrical potential varies with the pH of the solution being tested.

Electrodes that are sensitive to the concentration of a particular ion are called **ion-selective electrodes,** of which the glass electrode for pH measurement is just one example. Glass electrodes can be made sensitive to ions such as Na^+, K^+, or NH_4^+ by changing the composition of the glass. Other ions can be detected if an appropriate crystalline solid replaces the glass membrane. For example, a crystal of lanthanum(III) fluoride (LaF_3) can be used in an electrode to measure $[F^-]$. Solid silver sulfide (Ag_2S) can be used to measure $[Ag^+]$ and $[S^{2-}]$. Some of the ions that can be detected by ion-selective electrodes are listed in Table 11.2.

Calculation of Equilibrium Constants for Redox Reactions

The quantitative relationship between $\mathscr{E}°$ and $\Delta G°$ allows the calculation of equilibrium constants for redox reactions. For a cell at equilibrium,

$$\mathscr{E}_{\text{cell}} = 0 \qquad \text{and} \qquad Q = K$$

Applying these conditions to the form of the Nernst equation valid at 25°C,

$$\mathscr{E} = \mathscr{E}° - \frac{0.0591}{n}\log(Q)$$

gives

$$0 = \mathscr{E}° - \frac{0.0591}{n}\log(K)$$

or

$$\log(K) = \frac{n\mathscr{E}°}{0.0591} \quad \text{at } 25°\text{C}$$

Example 11.6

For the oxidation–reduction reaction

$$S_4O_6^{2-}(aq) + Cr^{2+}(aq) \longrightarrow Cr^{3+}(aq) + S_2O_3^{2-}(aq)$$

the appropriate half-reactions are

$$S_4O_6^{2-} + 2e^- \longrightarrow 2S_2O_3^{2-} \qquad \mathscr{E}° = 0.17\text{ V} \qquad (1)$$

$$Cr^{3+} + e^- \longrightarrow Cr^{2+} \qquad \mathscr{E}° = -0.50\text{ V} \qquad (2)$$

Balance the redox reaction and calculate $\mathscr{E}°$ and K (at 25°C).

Solution To obtain the balanced reaction, we must reverse reaction (2), multiply it by 2, and add it to reaction (1):

$$S_4O_6^{2-} + 2e^- \longrightarrow 2S_2O_3^{2-} \qquad \mathscr{E}°\text{ (cathode)} = 0.17\text{ V}$$

$$2(Cr^{2+} \longrightarrow Cr^{3+} + e^-) \qquad -\mathscr{E}°\text{ (anode)} = -(-0.50)\text{ V}$$

Cell reaction:

$$2Cr^{2+}(aq) + S_4O_6^{2-}(aq) \longrightarrow 2Cr^{3+}(aq) + 2S_2O_3^{2-}(aq) \qquad \mathscr{E}° = 0.67\text{ V}$$

In this reaction 2 mol of electrons is transferred for every unit of reaction—that is, for every 2 mol of Cr^{2+} reacting with 1 mol of $S_4O_6^{2-}$ to form 2 mol

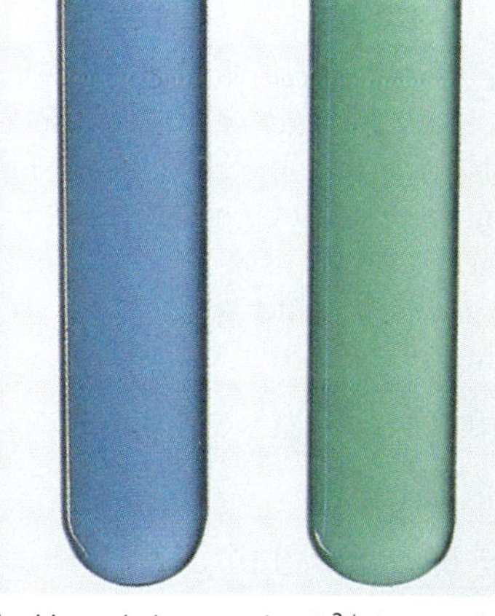

The blue solution contains Cr^{2+} ions, and the green solution contains Cr^{3+} ions.

of Cr^{3+} and 2 mol of $S_2O_3^{2-}$. Thus $n = 2$. Then

$$\log(K) = \frac{n\mathscr{E}°}{0.0591} = \frac{2(0.67)}{0.0591} = 22.6$$

$$K = 10^{22.6} = 4 \times 10^{22}$$

This very large equilibrium constant is not unusual for a redox reaction.

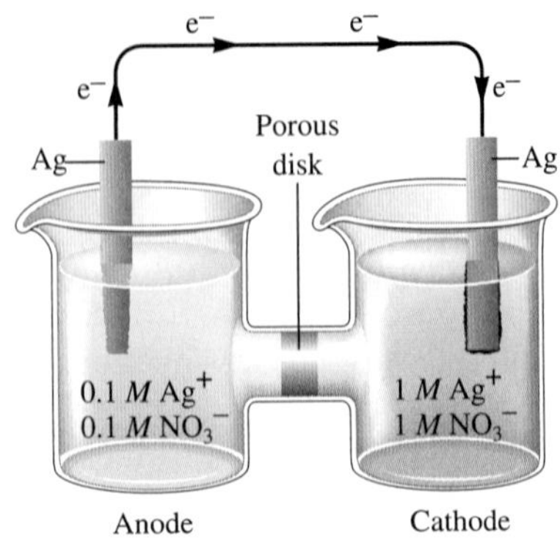

FIGURE 11.11

A concentration cell that contains a silver electrode and aqueous silver nitrate in both compartments. Because the right compartment contains 1 M Ag^+ and the left compartment contains 0.1 M Ag^+, there will be a driving force to transfer electrons from left to right. Silver will be deposited on the right electrode, thus lowering the concentration of Ag^+ in the right compartment. In the left compartment the silver electrode dissolves (producing Ag^+ ions) to raise the concentration of Ag^+ in solution.

Concentration Cells

Because cell potentials depend on concentration, we can construct galvanic cells in which both compartments contain the same components but at different concentrations. For example, in the cell in Fig. 11.11 both compartments contain aqueous $AgNO_3$, but with different molarities. Let's consider the potential of this cell and the direction of electron flow. The half-reaction relevant to both compartments of this cell is

$$Ag^+ + e^- \longrightarrow Ag \qquad \mathscr{E}° = 0.80 \text{ V}$$

If the cell had 1 M Ag^+ in both compartments,

$$\mathscr{E}°_{cell} = 0.80 \text{ V} - 0.80 \text{ V} = 0 \text{ V}$$

However, in the cell shown in Fig. 11.11 the concentrations of Ag^+ in the two compartments are 1 M and 0.1 M. Because the concentrations of Ag^+ are unequal, the actual half-cell potentials will not be identical. Thus the cell will exhibit a positive voltage. In which direction will the electrons flow in this cell? The best way to think about this question is to recognize that nature will try to equalize the concentrations of Ag^+ in the two compartments. This can be done by transferring electrons from the compartment containing 0.1 M Ag^+ to the one containing 1 M Ag^+ (left to right in Fig. 11.11). This electron transfer will produce Ag^+ in the left compartment and consume Ag^+ (to form Ag) in the right compartment.

To calculate the potential at 25°C for the cell shown in Fig. 11.11, we use the Nernst equation in the form

$$\mathscr{E} = \mathscr{E}° - \frac{0.0591}{n}\log(Q)$$

In this case $n = 1$ because the reaction in both compartments is $Ag^+ + e^- \rightarrow Ag$ but running in opposite directions. What is $\mathscr{E}°$ for this cell? Because $\mathscr{E}°$ refers to standard conditions, this corresponds to a cell like the one in Fig. 11.11, *except* that $[Ag^+] = 1$ M in both compartments. Such a cell has a potential of zero. That is, $\mathscr{E}° = 0$ for this cell, as noted previously. Next, we need to consider the form of Q. Recall that Q is a ratio of product to reactant concentrations. We can represent the process taking place in this cell as

$$Ag^+(1\ M) \longrightarrow Ag^+(0.1\ M)$$

Thus we can write the Nernst equation for this case:

$$\mathscr{E} = \mathscr{E}° - \frac{0.0591}{n}\log(Q) = 0 - \frac{0.0591}{1}\log\left(\frac{0.10}{1.0}\right)$$

$$= -\frac{0.0591}{1}(-1.00) = 0.0591 \text{ V}$$

Thus the cell potential is 0.0591 V.

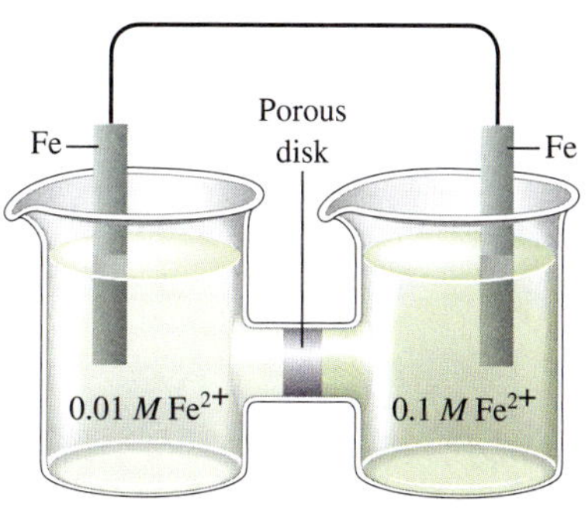

FIGURE 11.12

A concentration cell containing iron electrodes and different concentrations of Fe^{2+} ion in the two compartments.

A cell in which both compartments have the same components but at different concentrations is called a **concentration cell.** The difference in concentration is the only factor that produces a cell potential in this case, and the voltages are typically small.

Example 11.7

Determine the direction of electron flow, designate the anode and cathode, and calculate the potential at 25°C for the cell represented in Fig. 11.12.

Solution The concentrations of Fe^{2+} ion in the compartments can eventually be equalized by transferring electrons from the left compartment to the right. This will cause Fe^{2+} to be formed in the left compartment, and iron metal will be deposited on the right electrode, thus consuming Fe^{2+} ions. Therefore, electron flow is from left to right, oxidation occurs in the left compartment (the anode), and reduction occurs in the right compartment (the cathode).

To calculate the cell potential, we use the Nernst equation in the form

$$\mathscr{E} = \mathscr{E}^\circ - \frac{0.0591}{n} \log(Q)$$

where $n = 2$ because the cell half-reaction is $Fe^{2+} + 2e^- \rightarrow Fe$ or its opposite. Also, $\mathscr{E}^\circ = 0$ (as always) for a concentration cell, and $Q = 0.01/0.10$ because Fe^{2+} is being formed in the compartment with the lower concentration. Thus

$$\mathscr{E} = 0 - \frac{0.0591}{2} \log\left(\frac{0.01}{0.10}\right) = -0.0296(-1.00) = 0.0296 \text{ V}$$

Because the potential of an electrochemical cell depends on the concentrations of the participating ions, the observed potential can be used as a sensitive method for measuring ion concentrations in solution. We have already mentioned the ion-selective electrodes that work by this principle. Another application of the relationship between cell potential and concentration is the determination of equilibrium constants for reactions that are not redox reactions. For example, consider a modified version of the silver concentration cell shown in Fig. 11.11. If the 0.10 M $AgNO_3$ solution in the left-hand compartment is replaced by 1.0 M NaCl and an excess of solid AgCl is added to the cell, the observed cell potential can be used to determine the concentration of Ag^+ in equilibrium with the AgCl(s). In other words, at 25°C we can write the Nernst equation as

$$\mathscr{E} = 0 - \frac{0.0591}{1} \log\left(\frac{\overset{\substack{Ag^+ \text{ in equilibrium with AgCl}(s)\\ \downarrow}}{[Ag^+]}}{1.0}\right)$$

Note that the measurement of $\mathscr{E}$ for the cell allows the $[Ag^+]$ in equilibrium with AgCl(s) to be calculated, which, in turn, allows the K_{sp} for AgCl to be calculated:

$$K_{sp} = [Ag^+][Cl^-]$$

In this case $[Cl^-] = 1.0$ M from the 1.0 M NaCl, and $[Ag^+]$ can be obtained from the measured value for $\mathscr{E}$. We will illustrate this process in Example 11.8.

Example 11.8

A silver concentration cell similar to the one shown in Fig. 11.11 is set up at 25°C with 1.0 *M* $AgNO_3$ in the left compartment and 1.0 *M* NaCl along with excess AgCl(*s*) in the right compartment. The measured cell potential is 0.58 V. Calculate the K_{sp} value for AgCl at 25°C.

Solution In this case at 25°C

$$\mathscr{E} = 0.58\text{ V} = 0 - \frac{0.0591}{1}\log\left(\frac{[Ag^+]}{1.0}\right)$$

where $[Ag^+]$ represents the equilibrium concentration of Ag^+ in the compartment containing 1.0 *M* NaCl and AgCl(*s*). We calculate $[Ag^+]$ as follows:

$$\log[Ag^+] = -\frac{0.58}{0.0591} = -9.80 \qquad \text{and} \qquad [Ag^+] = 1.6 \times 10^{-10}\ M$$

Thus

$$K_{sp} = [Ag^+][Cl^-] = (1.6 \times 10^{-10})(1.0) = 1.6 \times 10^{-10}$$

This calculation neglects any complications from complex ions and ion pairs.

Because the measured potential of an electrochemical cell provides a very sensitive method for the experimental determination of equilibrium concentrations, the values of equilibrium constants are often determined from electrochemical measurements.

11.5 Batteries

A **battery** is a galvanic cell or, more commonly, *a group of galvanic cells connected in series*, where the potentials of the individual cells add to give the total battery potential. Batteries are a source of direct current and have become an essential source of portable power in our society. In this section we examine the most common types of batteries. Some new batteries currently being developed are described at the end of the chapter.

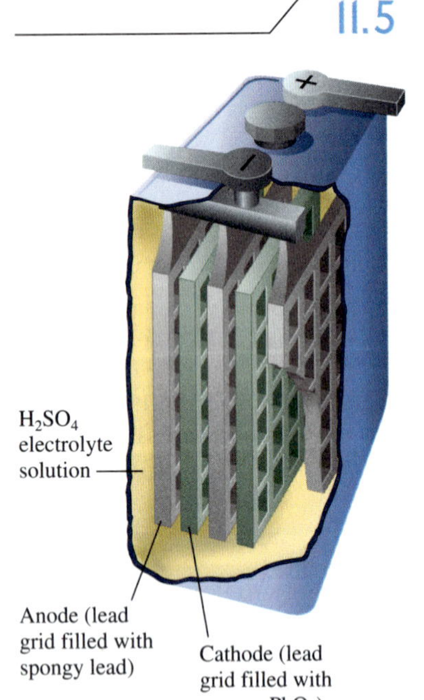

FIGURE 11.13

One of the six cells in a 12-V lead storage battery. The anode consists of a lead grid filled with spongy lead, and the cathode is a lead grid filled with lead dioxide. The cell also contains 38% (by mass) sulfuric acid.

Lead Storage Battery

Since about 1912 when self-starters were first used in automobiles, the **lead storage battery** has been a major factor in making the automobile a practical means of transportation. This type of battery can function for several years under temperature extremes from −30°F to 100°F and under incessant punishment from rough roads.

In this battery, lead serves as the anode, and lead coated with lead dioxide serves as the cathode. The electrodes dip into an electrolyte solution of sulfuric acid. The electrode reactions are as follows:

Anode reaction: $Pb + HSO_4^- \longrightarrow PbSO_4 + H^+ + 2e^-$

Cathode reaction: $PbO_2 + HSO_4^- + 3H^+ + 2e^- \longrightarrow PbSO_4 + 2H_2O$

Cell reaction: $Pb(s) + PbO_2(s) + 2H^+(aq) + 2HSO_4^-(aq) \longrightarrow 2PbSO_4(s) + 2H_2O(l)$

The typical automobile lead storage battery has six cells connected in series. Each cell contains multiple electrodes in the form of grids (Fig. 11.13) and produces about 2 volts, to give a total battery potential of about 12 volts. Note from the cell reaction that sulfuric acid is consumed as the battery discharges, which lowers the density of the electrolyte solution from its initial value of about 1.28 g/cm^3 in the fully charged battery. As a result, the condition of the battery can be monitored by measuring the density of the sulfuric acid solution. The solid lead sulfate formed in the cell reaction during discharge adheres to the grid surfaces of the electrodes. The battery is recharged by forcing current through the battery in the opposite direction to reverse the cell reaction. A car's battery is continuously charged by an alternator driven by the automobile's engine.

An automobile with a dead battery can be jump-started by connecting its battery to the battery in a running automobile. This process can be dangerous, however, because the resulting flow of current causes electrolysis of water in the dead battery, producing hydrogen and oxygen gases (see Section 11.7 for details). Disconnecting the jumper cables after the disabled car starts causes an arc that can ignite the gaseous mixture. If this happens, the battery may explode, ejecting corrosive sulfuric acid. This problem can be averted by connecting the ground jumper cable to a part of the engine remote from the battery. Any arc produced when this cable is disconnected will then be harmless.

Traditional types of storage batteries require periodic "topping off" because the water in the electrolyte solution is depleted by the electrolysis that accompanies the charging process. Recent types of batteries have electrodes made of an alloy of calcium and lead that inhibits the electrolysis of water. These batteries can be sealed since they require no addition of water.

It is rather amazing that in the 85 years that lead storage batteries have been used, no better system has been found. Although a lead storage battery does provide excellent service, it has a useful lifetime of only 3–5 years in an automobile. Although it might seem that the battery could undergo an indefinite number of discharge–charge cycles, physical damage from road shock, which tends to shake the solid $PbSO_4$ from the electrodes, and chemical side reactions eventually cause it to fail.

Dry Cell Batteries

The calculators, electronic watches, portable radios, and portable audio players that are so familiar to us are all powered by small, efficient **dry cell batteries.** The common dry cell battery was invented more than 100 years ago by George Leclanché (1839–1882), a French chemist. In its *acid version* the dry cell battery contains a zinc inner case that acts as the anode and a carbon rod in contact with a moist paste of solid MnO_2, solid NH_4Cl, and carbon that acts as the cathode (Fig. 11.14). The half-cell reactions are complex but can be approximated as follows:

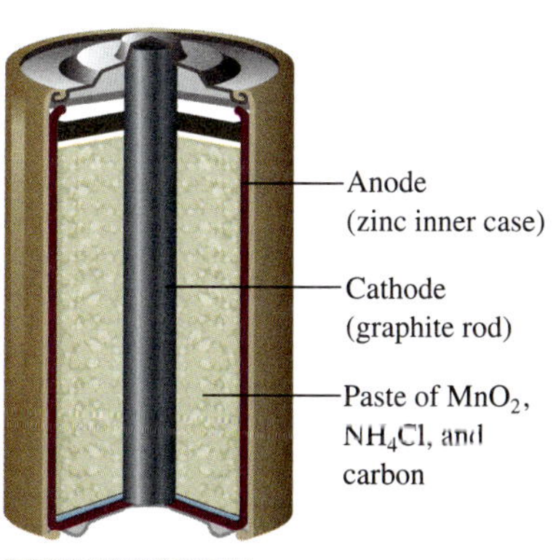

FIGURE 11.14
A common dry cell battery.

Anode reaction: $$Zn \longrightarrow Zn^{2+} + 2e^-$$

Cathode reaction: $$2NH_4^+ + 2MnO_2 + 2e^- \longrightarrow Mn_2O_3 + 2NH_3 + H_2O$$

This cell produces a potential of about 1.5 volts.

In the *alkaline version* of the dry cell battery the solid NH_4Cl is replaced with KOH or NaOH. In this case the half-reactions can be approximated as follows:

Anode reaction: $$Zn + 2OH^- \longrightarrow ZnO + H_2O + 2e^-$$

Cathode reaction: $$2MnO_2 + H_2O + 2e^- \longrightarrow Mn_2O_3 + 2OH^-$$

CHEMICAL INSIGHTS

Electrochemical Window Shades

Everybody likes large windows in their homes and workplaces. However, windows present major problems in making buildings energy efficient—they admit sunlight in the summer, increasing air-conditioning demands, and they allow heat to flow out of the building in the winter. To combat these problems, researchers from Electro-Optics Technology Center at Tufts University are developing a "smart" window that uses a short pulse of electricity to vary the transparency of the window in response to changing climate conditions.

The Tufts smart window consists of seven thin layers as shown in the accompanying illustration. Production of the incredibly thin layers (the seven layers have a total thickness of about 10^{-3} mm) has a lot in common with the manufacture of computer chips and has proved to be very difficult. The center layer of the window contains an electrolyte that allows ion flow to both of the surrounding layers. One of the sandwiching layers contains tungsten(VI) oxide (WO_3) and the other contains lithium cobalt(III) oxide ($LiCoO_2$). Because the transparency and reflectivity of these compounds change in response to electron and ion flow, they can be used to regulate the passage of light through the window. Depending on the size of the current pulse delivered across the window layers, a continuum of optical properties can be produced ranging from completely transparent to opaque.

A somewhat simpler electrochemical system is being developed by Claes-Goran Granquist and coworkers at the University of Uppsala in Sweden. This system involves two transparent polyester foils that are first coated with a conductive indium–tin oxide layer. One of the foils is then coated with tungsten oxide and the other with nickel oxide. An ion-conducting polymer is then sandwiched between the foils. The system works similar to a battery, with the foils acting as the anode and the cathode with the sandwiched polymer acting as the electrolyte. When 1.4 V is applied to the system, the direction that charges the tungsten oxide coating (turns dark) and discharges the nickel oxide coating (turns dark as it discharges) causes the system to block the passage of light. When the voltage is reversed, both oxides become transparent. The voltage can be controlled with a common switch, photo detector, or voice-sensitive switch. The product is projected to be

The alkaline dry cell lasts longer than the acidic cell mainly because the zinc anode corrodes less rapidly under basic conditions than under acidic conditions.

Other types of dry cell batteries include the *silver cell,* which has a Zn anode. Its cathode uses Ag_2O as the oxidizing agent in a basic environment. *Mercury cells,* often used in calculators, also have a Zn anode. The cathode uses HgO as the oxidizing agent in a basic medium (see Fig. 11.15).

An especially important type of dry cell is the *nickel–cadmium battery,* in which the electrode reactions are as follows:

Anode reaction: $Cd + 2OH^- \longrightarrow Cd(OH)_2 + 2e^-$

Cathode reaction: $NiO_2 + 2H_2O + 2e^- \longrightarrow Ni(OH)_2 + 2OH^-$

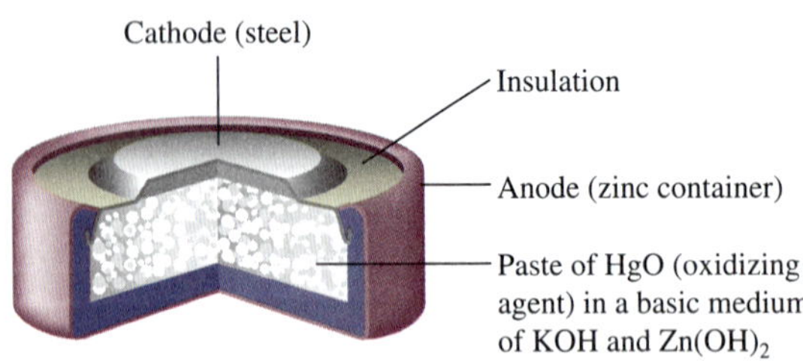

FIGURE 11.15

A mercury battery of the type used in calculators.

useful in motorcycle helmets, face shields, ski goggles, and windows in buildings.

Although many practical problems remain to be solved, there is real hope that smart windows will be available for buildings and cars in the near future. This could lead to a significant change in the energy demands of our society—it is estimated that smart windows can reduce a building's energy usage by as much as 50%.

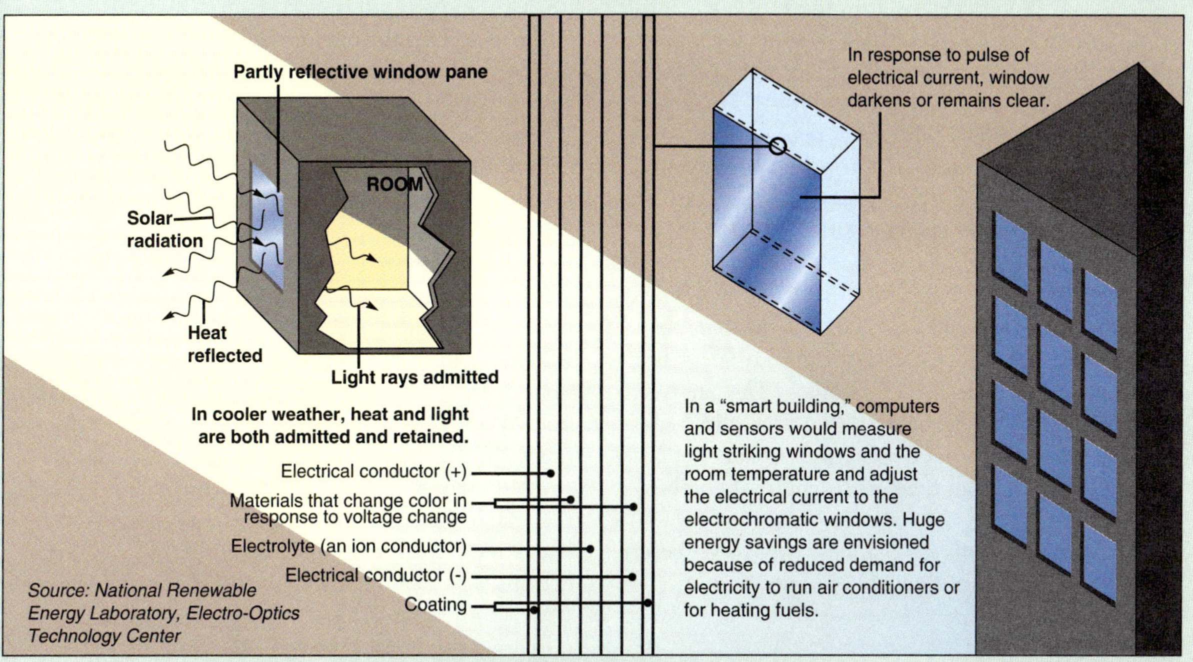

As in the lead storage battery, the products adhere to the electrodes. Therefore, a nickel–cadmium battery can be recharged an indefinite number of times.

Lithium-ion batteries involve the migration of Li^+ ions from the cathode to the anode, where they intercolate (enter the interior) as the battery is charged. At the same time, charge-balancing electrons travel to the anode through the external circuit in the charger. On discharge, the opposite process occurs. The cathode of the first successful lithium-ion batteries originally contained $LiCoO_2$ and a lithium-intercolated carbon (LiC_6) anode. More recently manufacturers have included transition metals such as nickel and manganese in the cathode in addition to cobalt. The mixed-metal cathodes have greater charge capacity and power output and shorter recharge times.

Lithium-ion batteries are used in a wide variety of applications, including cell phones, laptop computers, power tools, and even electric drive systems in automobiles and motorcycles.

Fuel Cells

A **fuel cell** is *a galvanic cell in which the reactants are continuously supplied.* To illustrate the principles of fuel cells, we will consider the exothermic redox reaction of methane with oxygen:

$$CH_4(g) + 2O_2(g) \longrightarrow CO_2(g) + 2H_2O(g) + \text{energy}$$

Fuel Cells—Portable Energy

The promise of an energy-efficient, environmentally sound source of electrical power has spurred an intense interest in fuel cells in recent years. Although fuel cells have long been used in the U.S. space program, no practical fuel cell for powering automobiles has been developed. However, we are now on the verge of practical fuel-cell–powered cars. For example, DaimlerChrysler's NECAR 5 was driven across the United States, a 3000-mile trip that took 16 days. NECAR 5 is powered by a H_2/O_2 fuel cell that generates its H_2 from decomposition of methanol (CH_3OH).

General Motors, which has also been experimenting with H_2/O_2 fuel cells, in 2005 introduced the Chevrolet Sequel, which can go from 0 to 60 mph in 10 seconds with a range of 300 miles. The car has composite tanks for storage of 8 kg of liquid hydrogen. GM's goal is to build a fuel-cell system by 2010 that can compete with the current internal combustion engine.

In reality, fuel cells have a long way to go before they can be economically viable in automobiles. The main problem is the membrane that separates the hydrogen electrode from the oxygen electrode. This membrane must prevent H_2 molecules from passing through and still allow ions to pass between the electrodes. Current membranes cost over $3000 for an automobile-size fuel cell. The hope is to reduce this by a factor of 10 in the next few years.

The Chevrolet Sequel car.

Besides providing power for automobiles, fuel cells are being considered for powering small electronic devices such as cameras, cell phones, and laptop computers. Many of these micro fuel cells currently use methanol as the fuel (reducing agent) rather than H_2. However, these direct-methanol fuel cells are rife with problems. A major difficulty is water management. Water is needed at the anode to react with the methanol and is produced at the cathode. Water is also needed to moisten the electrolyte to promote charge migration.

Although the direct-methanol fuel cell is currently the leader among micro fuel-cell designs, its drawbacks have encouraged the development of other designs. For example, Richard Masel at the University of Illinois at Urbana–Champaign has designed a micro fuel cell that uses formic acid as the fuel. Masel and others are also experimenting with mini hot chambers external to the fuel cell that break down hydrogen-rich fuels into hydrogen gas, which is then fed into the tiny fuel cells.

To replace batteries, fuel cells must be demonstrated to be economically feasible, safe, and dependable. Today, rapid progress is being made to overcome the current problems. A recent estimate indicates that by late in this decade annual sales of the little power plants may reach 200 million units per year. It appears that after years of hype about the virtues of fuel cells, we are finally going to realize their potential.

Usually, the energy from this reaction is released as heat to warm homes and to run machines. However, in a fuel cell designed to use this reaction, the energy is used to produce an electric current; the electrons flow from the reducing agent (CH_4) to the oxidizing agent (O_2) through a conductor.

The U.S. space program has supported extensive research to develop fuel cells. The space shuttle missions use a fuel cell based on the reaction of hydrogen and oxygen to form water:

$$2H_2(g) + O_2(g) \longrightarrow 2H_2O(l)$$

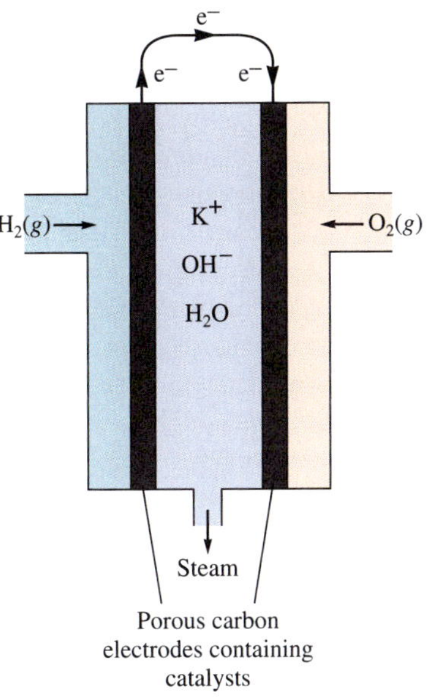

FIGURE 11.16
Schematic of the hydrogen–oxygen fuel cell.

A schematic of a fuel cell that uses this reaction is shown in Fig. 11.16. The half-cell reactions are as follows:

$$\text{Anode reaction:} \quad 2H_2 + 4OH^- \longrightarrow 4H_2O + 4e^-$$

$$\text{Cathode reaction:} \quad 4e^- + O_2 + 2H_2O \longrightarrow 4OH^-$$

Fuel cells are finding use as permanent power sources. A power plant built in New York City contains stacks of hydrogen–oxygen fuel cells, which can be rapidly put online in response to fluctuating power demands. The hydrogen gas is obtained by decomposing the methane in natural gas. A plant of this type has also been constructed in Tokyo.

Although the H_2/O_2 fuel cells developed for the U.S. space program are too expensive for commercial use, progress is being made toward achieving more widespread use of fuel cells for transportation. For example, the cities of Chicago and Vancouver (Canada) each have three city buses that are powered by H_2/O_2 fuel cells. These buses are more powerful and much quieter than their diesel-powered counterparts and emit only water in their exhausts. The major problem with these buses is their prohibitive cost—$1.4 million each. The large automobile manufacturers are also working on a fuel-cell–powered vehicle. DaimlerChrysler is now experimenting with a fourth-generation fuel-cell car and expects to have a practical version of this car available to the public very soon.

Because of the problems associated with the storage of hydrogen gas, scientists are also working to develop fuel cells that use gasoline or methanol to generate the H_2 needed for the fuel cell.

11.6 Corrosion

Some metals, such as copper, gold, silver, and platinum, are relatively difficult to oxidize. They are often called noble metals.

Corrosion can be viewed as the process of returning metals to their natural state—the ores from which they were originally obtained. Corrosion involves oxidation of the metal. Since corroded metal often loses its structural integrity and attractiveness, this spontaneous process has great economic impact. About one-fifth of the iron and steel produced annually is used to replace rusted metal.

Metals corrode because they oxidize easily. Table 11.1 shows that, with the exception of gold, metals commonly used for structural and decorative purposes all have standard reduction potentials less positive than that of oxygen gas. When any one of these half-reactions is reversed (to show oxidation of the metal) and combined with the reduction half-reaction for oxygen, the result is a positive $\mathscr{E}°$ value. Thus the oxidation of most metals by oxygen is spontaneous (although we cannot tell from the potential how fast it will occur).

In view of the large differences in the reduction potentials between oxygen and most metals, it is surprising that the problem of corrosion does not prevent the use of metals in air. However, most metals develop a thin oxide coating that tends to protect their internal atoms against further oxidation. The metal that best demonstrates this phenomenon is aluminum. With a reduction potential of −1.7 volts, aluminum should be easily oxidized by O_2. According to the apparent thermodynamics of the reaction, an aluminum airplane could dissolve in a rainstorm. This very active metal can be used as a structural material because the formation of a thin, adherent layer of aluminum oxide [Al_2O_3, more properly represented as $Al_2(OH)_6$] greatly inhibits further corrosion. The potential of the "passive," oxide-coated aluminum is −0.6 volt, a value that causes it to behave much like a noble metal.

CHEMICAL INSIGHTS

Refurbishing the Lady

The restoration of the Statue of Liberty in New York harbor more than 30 years ago represents a fascinating blend of science, technology, and art. The statue consists of copper sheets attached to a framework of iron, which had become so weakened by corrosion during its 100 years of exposure to the elements that it was in danger of collapsing.

Gustave Eiffel, the French engineer who designed the ingenious support structure, knew from experience that if the copper touched the iron framework, the more active iron would corrode very rapidly. Why does this happen? It is apparent from the reduction potentials

$$Cu^{2+} + 2e^- \longrightarrow Cu \qquad \mathscr{E}° = 0.34\ V$$

$$Fe^{2+} + 2e^- \longrightarrow Fe \qquad \mathscr{E}° = -0.44\ V$$

that Cu^{2+} will spontaneously oxidize iron. However, as with many electrochemical processes, the situation is more complex than it first appears. Since we are talking about two metal strips touching each other, the question is: Where do the Cu^{2+} ions (the oxidizing agents) come from? In fact, research on this process suggests that the copper simply acts as a conductor for electrons and that the oxidizing agent is not Cu^{2+} at all but probably O_2 or oxides of N or S.

Whatever the mechanism, Eiffel attempted to combat the corrosion problem by inserting asbestos pads between the copper sheets and the frame. However, this idea did not work, probably because copper is such a good conductor that *any* contact between the two metals anywhere on the statue totally thwarted the effect of the insulation. In fact, workers found that the iron framework was so corroded it had to be completely replaced with stainless steel, which is much more corrosion-resistant.

Stainless steel has its own problems, however. In being bent to achieve the intricate shapes needed, the steel bars often became brittle. Flexibility was restored by heating each bar to a very high temperature, using a current of 30,000 amperes, and then cooling the bar suddenly. Unfortunately, this process also destroyed the corrosion resistance of the stainless steel. That resistance had to be restored by soaking the bars in nitric acid, an oxidizing acid that re-forms the protective oxide coating removed by heating.

Another problem faced by the restorers was the removal of layers of coal tar and paint that had been applied in vain attempts to protect the statue's interior. Although the iron could be cleaned by blasting with aluminum oxide powder, the more fragile copper sheets required a gentler treatment. The restorers discovered that liquid nitrogen (77 K) cracked the paint and caused it to peel away. The coal-tar layer under the paint was removed by blasting with baking soda ($NaHCO_3$), a substance sometimes used by museum curators for polishing dinosaur bones. Although this treatment worked very well, it created another chemical problem: Where the baking soda seeped between the seams in the copper plates onto the exterior, the statue turned from green

Iron can also form a protective oxide coating. This coating is not an infallible shield against corrosion, however; when steel is exposed to oxygen in moist air, the oxide that forms tends to scale off, exposing new metal surfaces to corrosion.

The corrosion products of noble metals such as copper and silver are complex and affect the use of these metals as decorative materials. Under normal atmospheric conditions, copper forms an external layer of greenish copper carbonate called *patina*. *Silver tarnish* is silver sulfide (Ag_2S), which in thin layers gives the silver surface a richer appearance. Gold, with a positive standard reduction potential (1.50 volts) significantly larger than that for oxygen (1.23 volts), shows no appreciable corrosion in air.

Corrosion of Iron

Since steel is the main structural material for bridges, buildings, and automobiles, controlling its corrosion is extremely important. To do so, we must

The restored Statue of Liberty in New York harbor.

to light blue when it rained. After this phenomenon was observed for the first time, workers stationed on the exterior scaffolding immediately cleaned off any leaking $NaHCO_3$.

One of the most interesting aspects of the chemistry of the Statue of Liberty is the green patina on its surface. Copper metal exposed to the atmosphere changes from its bright reddish brown metallic luster first to an almost black color and then to the familiar green patina. The initial blackening is mainly caused by formation of copper oxide and copper sulfide. The green patina that forms consists of thin layers of two types of basic copper sulfates: brochantite, $CuSO_4 \cdot 3Cu(OH)_2$, and antlerite, $CuSO_4 \cdot 2Cu(OH)_2$. Crystals of these compounds seem to be cemented onto the copper surface by organic molecules from the air.

In recent years large areas of the statue's left side have been observed to darken as if the patina were being removed. Scientists speculated that this darkening may be the result of acid rain, which converts brochantite to the more soluble antlerite, which is then washed off by rainwater.

To make any new copper sheets look like the old ones, the restorers transplanted the patina from a weathered piece of copper to the new surface. Applying acetone, an organic solvent, to the weathered surface and scraping with an abrasive cloth caused tiny flakes of patina to fall off. These flakes, applied to the new copper, attached themselves permanently in one to three weeks of exposure to the atmosphere.

The restoration of the Statue of Liberty made use of the latest advances in chemistry as well as facts known to most general chemistry students. It's an example of the fascinating and varied problems faced by chemists as they pursue their profession.

Suggested Reading

Ivars Peterson, "Lessons Learned from a Lady," *Science News* 130 (1986): 392.

understand the corrosion mechanism. Instead of being a direct oxidation process, as we might expect, the corrosion of iron is an electrochemical reaction, as illustrated in Fig. 11.17.

FIGURE 11.17

The electrochemical corrosion of iron.

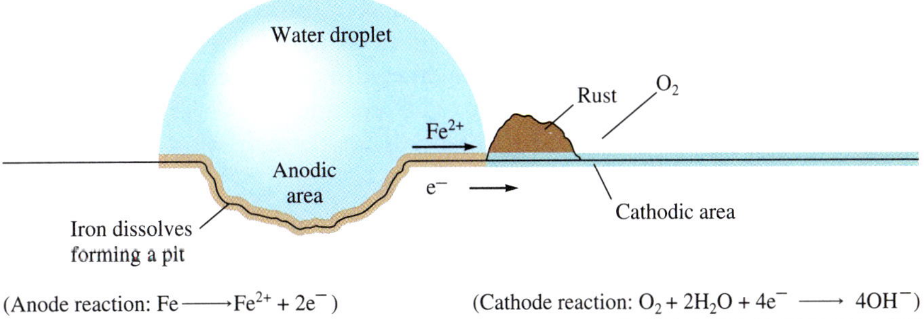

Steel has a nonuniform surface because its chemical composition is not completely homogeneous. In addition, physical strains leave stress points in the metal. These nonuniformities produce areas where the iron is more easily oxidized (*anodic regions*) than it is at others (*cathodic regions*). In the anodic regions each iron atom gives up two electrons to form the Fe^{2+} ion:

$$\text{Fe} \longrightarrow \text{Fe}^{2+} + 2\text{e}^-$$

The electrons that are released flow through the steel, as they do through the wire of a galvanic cell, to a cathodic region where they react with oxygen:

$$\text{O}_2 + 2\text{H}_2\text{O} + 4\text{e}^- \longrightarrow 4\text{OH}^-$$

The Fe^{2+} ions formed in the anodic regions travel to the cathodic regions through the moisture on the surface of the steel, just as ions travel through a salt bridge in a galvanic cell. In the cathodic regions Fe^{2+} ions react with oxygen to form rust, which is hydrated iron(III) oxide of variable composition:

$$4\text{Fe}^{2+}(aq) + \text{O}_2(g) + (4 + 2n)\text{H}_2\text{O}(l) \longrightarrow \underbrace{2\text{Fe}_2\text{O}_3 \cdot n\text{H}_2\text{O}(s)}_{\text{Rust}} + 8\text{H}^+(aq)$$

Because of the migration of ions and electrons, rust often forms at sites that are remote from those where the iron dissolved to form pits in the steel. Also the degree of hydration of the iron oxide affects the color of the rust, which may vary from black to yellow to the familiar reddish brown.

The electrochemical nature of the rusting of iron explains the importance of moisture in the corrosion process. Moisture must be present to act as a "salt bridge" between anodic and cathodic regions. Steel does not rust in dry air, a fact that explains why cars last much longer in the arid Southwest than in the relatively humid Midwest. Salt also accelerates rusting, a fact all too easily recognized by car owners in the colder parts of the United States, where salt is used on roads to melt snow and ice. The severity of rusting is greatly increased because the dissolved salt on the moist steel surface increases the conductivity of the aqueous solution formed there and thus accelerates the electrochemical corrosion process. Chloride ions also form very stable complex ions with Fe^{3+}, and this factor tends to encourage the dissolving of the iron, further accelerating the corrosion.

Prevention of Corrosion

Prevention of corrosion is an important way of conserving our natural resources of energy and metals. The primary means of protection is the application of a coating, most commonly paint or metal plating, to protect the metal from oxygen and moisture. Chromium and tin are often used to plate steel (Section 11.8) because they react with oxygen to form a durable, effective oxide coating. Zinc, also used to coat steel in a process called **galvanizing**, does not form an oxide coating. However, since it is a more active metal than iron, as the potentials for the oxidation half-reactions show,

$$\text{Fe} \longrightarrow \text{Fe}^{2+} + 2\text{e}^- \qquad -\mathscr{E}° = 0.44\ \text{V}$$

$$\text{Zn} \longrightarrow \text{Zn}^{2+} + 2\text{e}^- \qquad -\mathscr{E}° = 0.76\ \text{V}$$

any oxidation that occurs dissolves zinc rather than iron. Recall that the reaction with the most positive standard potential has the greatest thermodynamic tendency to occur. Thus zinc acts as a "sacrificial" coating on steel.

Paint That Stops Rust—Completely

Traditionally, paint has provided the most economical method for protecting steel against corrosion. However, as people who live in the Midwest know well, paint cannot prevent a car from rusting indefinitely. Eventually, flaws develop in the paint that allow the ravages of rusting to take place.

This situation may soon change. Chemists at Glidden Research Center in Ohio have developed a paint called Rustmaster Pro that worked so well to prevent rusting in its initial tests that the scientists did not believe their results. Steel coated with the new paint showed no signs of rusting after an astonishing 10,000 hours of exposure in a salt spray chamber at 38°C.

Rustmaster is a water-based polymer formulation that prevents corrosion in two different ways. First, the polymer layer that cures in air forms a barrier impenetrable to both oxygen and water vapor. Second, the chemicals in the coating react with the steel surface to produce an interlayer between the metal and the polymer coating. This interlayer is a complex mineral called pyroaurite that contains cations of the form $[M_{1-x}Z_x(OH)_2]^{x+}$, where M is a 2+ ion (Mg^{2+}, Fe^{2+}, Zn^{2+}, Co^{2+}, or Ni^{2+}), Z is a 3+ ion (Al^{3+}, Fe^{3+}, Mn^{3+}, Co^{3+}, or Ni^{3+}), and x is a number between 0 and 1. The anions in pyroaurite are typically CO_3^{2-}, Cl^-, and/or SO_4^{2-}.

This pyroaurite interlayer is the real secret of the paint's effectiveness. Because the corrosion of steel has an electrochemical mechanism, motion of ions must be possible between the cathodic and anodic areas on the surface of the steel for rusting to occur. However, the pyroaurite interlayer grows into the neighboring polymer layer, thus preventing this crucial movement of ions. In effect, this layer prevents corrosion in the same way that removing the salt bridge prevents current from flowing in a galvanic cell.

In addition to having an extraordinary corrosion-fighting ability, Rustmaster yields an unusually small quantity of volatile solvents as it dries. A typical paint can produce from 1 to 5 kg of volatiles per gallon; Rustmaster produces only 0.05 kg. This paint may signal a new era in corrosion prevention.

Alloying is also used to prevent corrosion. *Stainless steel* contains chromium and nickel, both of which form oxide coatings that change steel's reduction potential to one characteristic of the noble metals. A new technology is now being developed to create surface alloys. Instead of forming a metal alloy such as stainless steel, which has the same composition throughout, a cheaper carbon steel is treated by ion bombardment to produce a thin layer of stainless steel or other desirable alloy on the surface. In this process a "plasma" or "ion gas" of the alloying ions is formed at high temperatures and is then directed onto the surface of the metal.

Cathodic protection is a method most often used to protect steel in buried fuel tanks and pipelines. An active metal, such as magnesium, is connected by a wire to the pipeline or tank to be protected (Fig. 11.18). Because magnesium

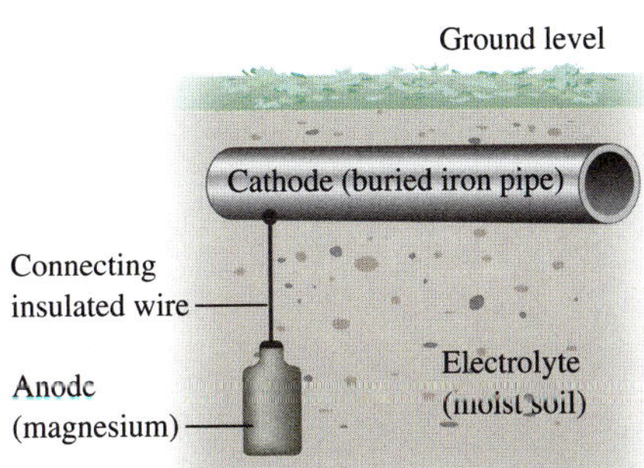

FIGURE 11.18
Cathodic protection of an underground pipe.

is a better reducing agent than iron, electrons are furnished by the magnesium, keeping the iron from being oxidized. As oxidation occurs, the magnesium anode dissolves, and so it must be replaced periodically.

11.7 Electrolysis

A galvanic cell produces current when an oxidation–reduction reaction proceeds spontaneously. A similar apparatus, an **electrolytic cell,** uses electrical energy to produce chemical change. The process of **electrolysis** involves *forcing a current through a cell to produce a chemical change for which the cell potential is negative;* that is, electrical work causes an otherwise nonspontaneous chemical reaction to occur. Electrolysis has great practical importance; for example, charging a battery, producing aluminum metal, and chrome plating an object are all done electrolytically.

To illustrate the difference between a galvanic cell and an electrolytic cell, consider the cell shown in Fig. 11.19(a) as it runs spontaneously to produce 1.10 volts. In this *galvanic cell* the reaction at the anode is

$$Zn \longrightarrow Zn^{2+} + 2e^-$$

and the cathode reaction is

$$Cu^{2+} + 2e^- \longrightarrow Cu$$

An electrolytic cell uses electrical energy to produce a chemical change that would otherwise not occur spontaneously.

Figure 11.19(b) shows an external power source forcing electrons through the cell in the *opposite* direction to that in (a). This requires an external potential greater than 1.10 V, which must be applied in opposition to the natural cell potential. This device is an *electrolytic cell.* Notice that since electron flow is opposite in the two cases, the anode and cathode are reversed in (a) and (b). Also, ion flow through the salt bridge is opposite in the two cells.

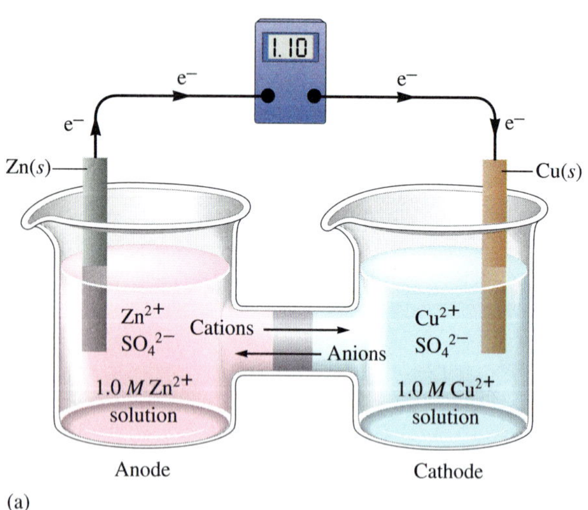

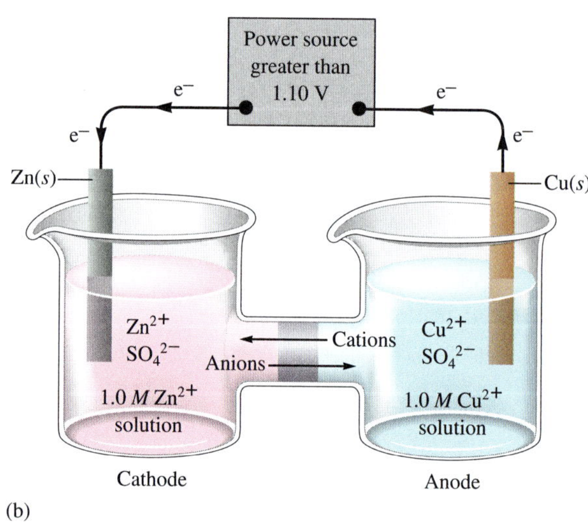

FIGURE 11.19

(a) A standard galvanic cell based on the spontaneous reaction

$$Zn + Cu^{2+} \longrightarrow Zn^{2+} + Cu$$

(b) A standard electrolytic cell. A power source forces the opposite reaction

$$Cu + Zn^{2+} \longrightarrow Cu^{2+} + Zn$$

Now we will consider the stoichiometry of electrolytic processes—that is, *how much chemical change occurs with the flow of a given current for a specified time.* Suppose we wish to determine the mass of copper that is plated out when a current of 10.0 amperes [an **ampere** (amp) is *1 coulomb of charge per second*] is passed for 30.0 minutes through a solution containing Cu^{2+}. *Plating* means depositing the neutral metal on the electrode surface by reducing the metal ions in solution. In this case each Cu^{2+} ion requires two electrons to become an atom of copper metal:

1 A = 1 C/s

$$Cu^{2+}(aq) + 2e^- \longrightarrow Cu(s)$$

This reduction process occurs at the cathode of the electrolytic cell.

To solve this stoichiometry problem, we use the following steps:

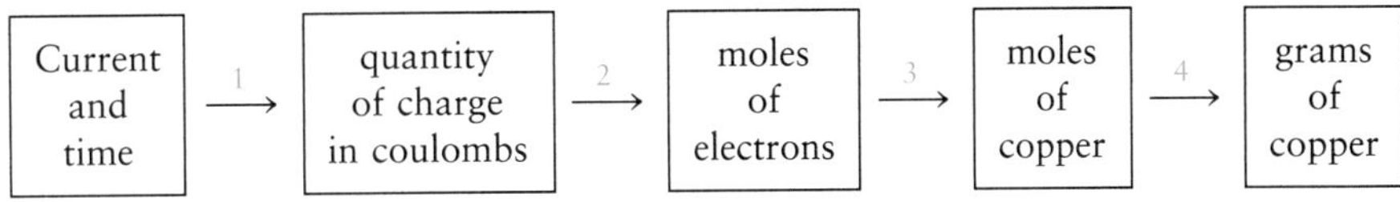

1. Since an amp is a coulomb of charge per second, we multiply the current by the time in seconds to obtain the total coulombs of charge passed into the solution at the cathode:

$$\text{Coulombs of charge} = \text{amps} \times \text{seconds} = \frac{C}{s} \times s$$
$$= 10.0 \frac{C}{s} \times 30.0 \text{ min} \times 60.0 \frac{s}{\text{min}}$$
$$= 1.80 \times 10^4 \text{ C}$$

2. Since 1 mole of electrons carries a charge of 1 faraday, or 96,485 coulombs, we can calculate the number of moles of electrons required to carry 1.80×10^4 coulombs of charge:

$$1.80 \times 10^4 \text{ C} \times \frac{1 \text{ mol e}^-}{96{,}485 \text{ C}} = 1.87 \times 10^{-1} \text{ mol e}^-$$

This means that 0.187 mole of electrons flowed into the solution containing Cu^{2+}.

3. Each Cu^{2+} ion requires two electrons to become a copper atom. Thus each mole of electrons produces $\frac{1}{2}$ mole of copper metal:

$$1.87 \times 10^{-1} \text{ mol e}^- \times \frac{1 \text{ mol Cu}}{2 \text{ mol e}^-} = 9.35 \times 10^{-2} \text{ mol Cu}$$

4. Since we now know the number of moles of copper metal plated onto the cathode, we can calculate the mass of copper formed:

$$9.35 \times 10^{-2} \text{ mol Cu} \times \frac{63.546 \text{ g}}{\text{mol Cu}} = 5.94 \text{ g Cu}$$

Electrolysis of Water

We have seen that hydrogen and oxygen combine spontaneously to form water and that the accompanying decrease in free energy can be used to run a fuel cell to produce electricity. The reverse process, which is, of course, nonspontaneous, can be forced by electrolysis:

Anode reaction:	$2H_2O \longrightarrow O_2 + 4H^+ + 4e^-$	$-\mathscr{E}° = -1.23$ V
Cathode reaction:	$4H_2O + 4e^- \longrightarrow 2H_2 + 4OH^-$	$\mathscr{E}° = -0.83$ V
Net reaction:	$6H_2O \longrightarrow 2H_2 + O_2 + \underbrace{4(H^+ + OH^-)}_{4H_2O}$	
or	$2H_2O \longrightarrow 2H_2 + O_2$	$\mathscr{E}° = -2.06$ V

Note that these potentials assume an anode chamber with 1 *M* H^+ and a cathode chamber with 1 *M* OH^-. In pure water, where $[H^+] = [OH^-] = 10^{-7}$ *M*, the potential for the overall process is -1.23 V.

In practice, however, if platinum electrodes connected to a 6-volt battery are dipped into pure water, no reaction is observed; pure water contains so few ions that only a negligible current can flow. However, addition of even a small amount of a soluble salt causes an immediate evolution of bubbles of hydrogen and oxygen, as illustrated in Fig. 11.20.

Electrolysis of Mixtures of Ions

Suppose a solution in an electrolytic cell contains the ions Cu^{2+}, Ag^+, and Zn^{2+}. If the voltage, which is initially very low, is gradually turned up, in which order will the metals be plated out onto the cathode? This question can be answered by looking at the standard reduction potentials of these ions:

$$Ag^+ + e^- \longrightarrow Ag \qquad \mathscr{E}° = 0.80 \text{ V}$$
$$Cu^{2+} + 2e^- \longrightarrow Cu \qquad \mathscr{E}° = 0.34 \text{ V}$$
$$Zn^{2+} + 2e^- \longrightarrow Zn \qquad \mathscr{E}° = -0.76 \text{ V}$$

Remember that the more *positive* the $\mathscr{E}°$ value, the more the reaction has a tendency to proceed in the direction indicated. Of the three reactions listed, the order of oxidizing ability is

$$Ag^+ > Cu^{2+} > Zn^{2+}$$

This means that silver will plate out first as the potential is increased, followed by copper, and finally zinc.

The principle described in this section is very useful, but it must be applied with some caution. For example, in the electrolysis of an aqueous solution of sodium chloride, we should be able to use $\mathscr{E}°$ values to predict which products are expected. Of the major species in the solution (Na^+, Cl^-, and H_2O), only Cl^- and H_2O can be readily oxidized. The half-reactions (written as oxidation processes) are

$$2Cl^- \longrightarrow Cl_2 + 2e^- \qquad -\mathscr{E}° = -1.36 \text{ V}$$
$$2H_2O \longrightarrow O_2 + 4H^+ + 4e^- \qquad -\mathscr{E}° = -1.23 \text{ V}$$

Since water has the more positive potential, we would expect to see O_2 produced at the anode. However, this does not happen. As the voltage is increased in the cell, the Cl^- ion is the first to be oxidized. A much higher potential than expected is required to oxidize water. The voltage required in excess of the expected value (called the *overvoltage*) is much greater for the production of O_2 than for Cl_2, which explains why chlorine is produced at the lower voltage.

The causes of overvoltage are very complex. Basically, the phenomenon is caused by difficulties in transferring electrons from the species in the solution

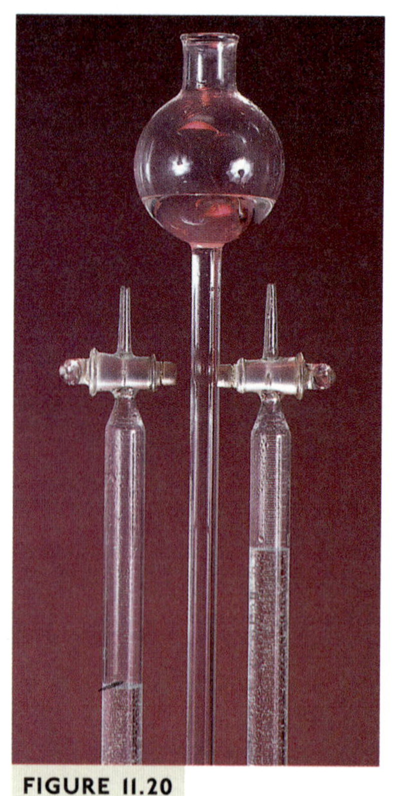

FIGURE 11.20

The electrolysis of water produces hydrogen gas at the cathode (on the left) and oxygen gas at the anode (on the right).

The Chemistry of Sunken Treasure

When the galleon *Atocha* was destroyed on a reef by a hurricane in 1622, it was bound for Spain carrying about 47 tons of copper, gold, and silver from the New World. The bulk of the treasure was silver bars and coins packed in wooden chests. When treasure hunter Mel Fisher salvaged the silver in 1985, corrosion and marine growth had transformed the shiny metal into something that looked like coral. Restoring the silver to its original condition required an understanding of the chemical changes that had occurred in 350 years of being submerged in the ocean. Much of this chemistry we have already considered at various places in this text.

As the wooden chests containing the silver decayed over the years, the oxygen supply was depleted. This favored the growth of certain bacteria that use the sulfate ion rather than oxygen as an oxidizing agent to generate energy. As these bacteria consume sulfate ions, they release hydrogen sulfide gas that reacts with silver to form black silver sulfide:

$$2Ag(s) + H_2S(aq) \longrightarrow Ag_2S(s) + H_2(g)$$

Thus over the years the surface of the silver became covered with a tightly adhering layer of corrosion, which fortunately protected the silver underneath, thus preventing total conversion of the silver to silver sulfide.

Another change that took place as the wood decomposed was the formation of carbon dioxide.

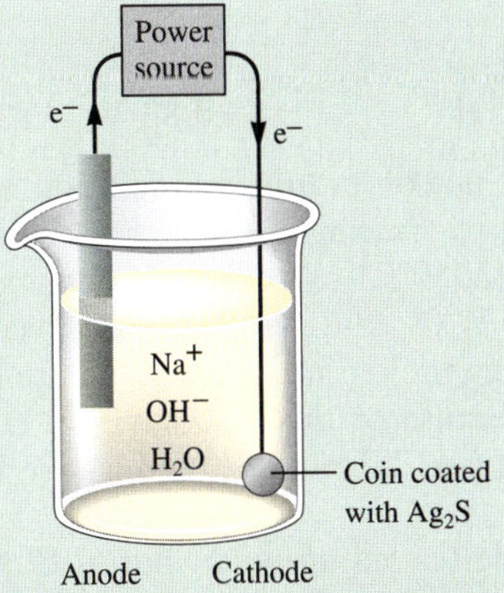

This shifted the equilibrium that is present in the ocean,

$$CO_2(aq) + H_2O(l) \rightleftharpoons HCO_3^-(aq) + H^+(aq)$$

to the right, producing higher concentrations of HCO_3^-. In turn, the HCO_3^- reacted with Ca^{2+} ions present in the seawater to form calcium carbonate:

$$Ca^{2+}(aq) + HCO_3^-(aq) \rightleftharpoons CaCO_3(s) + H^+(aq)$$

Calcium carbonate is the main component of limestone. Thus over time the corroded silver coins and bars became encased in limestone.

Both the limestone formation and the corrosion had to be dealt with. Since $CaCO_3$ contains the basic anion CO_3^{2-}, acid dissolves limestone:

$$2H^+(aq) + CaCO_3(s) \longrightarrow Ca^{2+}(aq) + CO_2(g) + H_2O(l)$$

Soaking the mass of coins in a buffered acidic bath for several hours allowed the individual pieces to be separated, and the black Ag_2S on the surfaces was revealed. An abrasive could not have been used to remove this corrosion; it would have destroyed the details of the engraving—a very valuable feature of the coins to a historian or a collector—and it would have washed away some of the silver. Instead, the corrosion reaction was reversed through electrolytic reduction. The coins were connected to the cathode of an electrolytic cell in a dilute sodium hydroxide solution, as represented in the figure.

As electrons flow, the Ag^+ ions in the silver sulfide are reduced to silver metal,

$$Ag_2S + 2e^- \longrightarrow Ag + S^{2-}$$

As a by-product, bubbles of hydrogen gas from the reduction of water,

$$2H_2O + 2e^- \longrightarrow H_2 + 2OH^-$$

form on the surface of the coins.

The agitation caused by the bubbles loosens the flakes of metal sulfide and helps clean the coins.

Using these procedures, technicians have been able to restore the treasure to very nearly the same condition it was in when the *Atocha* sailed many years ago.

to the atoms on the electrode across the electrode–solution interface. Because of this situation, $\mathscr{E}°$ values must be used cautiously in predicting the actual order of oxidation or reduction of species in an electrolytic cell.

11.8 Commercial Electrolytic Processes

FIGURE 11.21

Charles Martin Hall (1863–1914) was a student at Oberlin College in Ohio when he first became interested in aluminum. One of his professors commented that anyone who could manufacture aluminum cheaply would make a fortune, so Hall decided to give it a try. The 21-year-old Hall worked in a wooden shed near his house with an iron frying pan as a container, a blacksmith's forge as a heat source, and galvanic cells constructed from fruit jars. Using these crude galvanic cells, Hall found that he could produce aluminum by passing a current through a molten Al_2O_3/Na_3AlF_6 mixture. By a strange coincidence, Paul Heroult, a Frenchman who was born and died in the same years as Hall, made the same discovery at about the same time.

The chemistry of metals is characterized by their ability to donate electrons to form ions. Because metals are typically very good reducing agents, most are found in nature in *ores*, mixtures of ionic compounds often containing oxide, sulfide, and silicate anions. The noble metals, such as gold, silver, and platinum, are more difficult to oxidize and are often found as pure metals.

Production of Aluminum

Aluminum is one of the most abundant elements on earth, ranking third behind oxygen and silicon. Since aluminum is a very active metal, it is found in nature as its oxide in an ore called *bauxite* (named after Les Baux, France, where it was discovered in 1821). Production of aluminum metal from its ore proved to be more difficult than production of most other metals. In 1782 Lavoisier recognized aluminum as a metal "whose affinity for oxygen is so strong that it cannot be overcome by any known reducing agent." As a result, pure aluminum metal remained unknown. Finally, in 1854 a process was found for producing metallic aluminum using sodium, but aluminum remained a very expensive rarity. In fact, it is said that Napoleon III served his most honored guests with aluminum forks and spoons, while the others had to settle for gold and silver utensils.

The breakthrough came in 1886 when two men, Charles M. Hall (Fig. 11.21) in the United States and Paul Heroult in France, almost simultaneously discovered a practical electrolytic process for producing aluminum. The key factor in the *Hall–Heroult process* is the use of molten cryolite (Na_3AlF_6) as the solvent for the aluminum oxide.

Electrolysis is possible only if ions can move to the electrodes. A common method for producing ion mobility is dissolving the substance to be electrolyzed in water. This method cannot be used for aluminum because water is more easily reduced than Al^{3+}, as the following standard reduction potentials show:

$$Al^{3+} + 3e^- \longrightarrow Al \qquad \mathscr{E}° = -1.66 \text{ V}$$

$$2H_2O + 2e^- \longrightarrow H_2 + 2OH^- \qquad \mathscr{E}° = -0.83 \text{ V}$$

Thus aluminum metal cannot be plated out of an aqueous solution of Al^{3+}.

Ion mobility can also be produced by melting the salt. But the melting point of solid Al_2O_3 is much too high (2050°C) to allow practical electrolysis of the molten oxide. A mixture of Al_2O_3 and Na_3AlF_6, however, has a melting point of 1000°C, and the resulting molten mixture can be used to obtain aluminum metal electrolytically. Because of this discovery by Hall and Heroult, the price of aluminum plunged (see Table 11.3), and its use became economically feasible.

Bauxite is not pure aluminum oxide (called *alumina*) but also contains the oxides of iron, silicon, and titanium, and various silicate materials. The pure hydrated alumina ($Al_2O_3 \cdot nH_2O$) is obtained by treating the crude bauxite with aqueous sodium hydroxide. Being amphoteric, alumina dissolves in the basic solution:

$$Al_2O_3(s) + 2OH^-(aq) \longrightarrow 2AlO_2^-(aq) + H_2O(l)$$

TABLE 11.3

The Price of Aluminum Over the Past Century

Date	Price of Aluminum ($/lb)*
1855	100,000
1885	100
1890	2
1895	0.50
1970	0.30
1980	0.80
1990	0.74

*Note the precipitous drop in price after the discovery of the Hall–Heroult process.

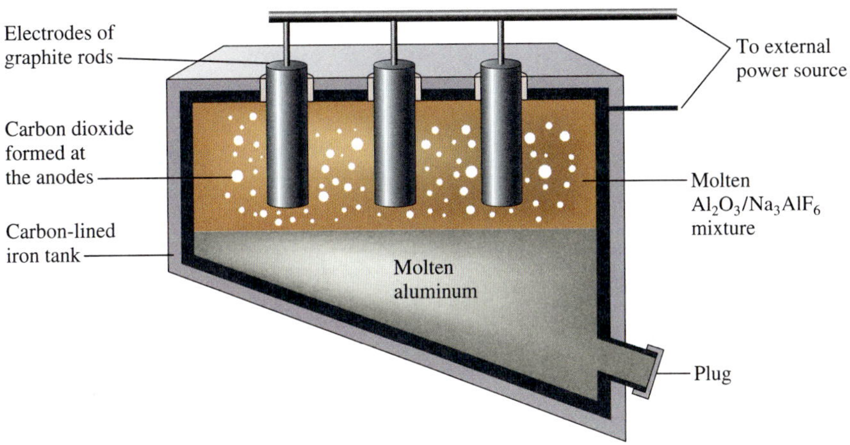

FIGURE 11.22

A schematic diagram of an electrolytic cell for producing aluminum by the Hall–Heroult process. Because molten aluminum is more dense than the mixture of molten cryolite and alumina, it settles to the bottom of the cell and is drawn off periodically. The graphite electrodes are gradually eaten away and must be replaced from time to time. The cell operates at a current flow of up to 250,000 A.

The other metal oxides, which are basic, remain as solids. The solution containing the aluminate ion (AlO_2^-) is separated from the sludge of other oxides and is acidified with carbon dioxide gas, causing the hydrated alumina to reprecipitate:

$$2CO_2(g) + 2AlO_2^-(aq) + (n + 1)H_2O(l) \longrightarrow 2HCO_3^-(aq) + Al_2O_3 \cdot nH_2O(s)$$

The purified alumina is then mixed with cryolite and melted, and the aluminum ion is reduced to aluminum metal in an electrolytic cell of the type shown in Fig. 11.22. Because the electrolyte solution contains a large number of aluminum-containing ions, the chemistry is not completely understood. However, the alumina probably reacts with the cryolite anion as follows:

$$Al_2O_3 + 4AlF_6^{3-} \longrightarrow 3Al_2OF_6^{2-} + 6F^-$$

The electrode reactions are thought to be the following:

Cathode reaction: $AlF_6^{3-} + 3e^- \longrightarrow Al + 6F^-$

Anode reaction: $2Al_2OF_6^{2-} + 12F^- + C \longrightarrow 4AlF_6^{3-} + CO_2 + 4e^-$

The overall cell reaction can be written as

$$2Al_2O_3 + 3C \longrightarrow 4Al + 3CO_2$$

The aluminum produced in this electrolytic process is 99.5% pure. To be useful as a structural material, aluminum is alloyed with metals such as zinc (used for trailer and aircraft construction) and manganese (used for cooking utensils, storage tanks, and highway signs). The production of aluminum consumes almost 5% of all electricity used in the United States.

Electrorefining of Metals

Purification of metals is another important application of electrolysis. For example, impure copper from the chemical reduction of copper ore is cast into large slabs that serve as the anodes for electrolytic cells. Aqueous copper sulfate

FIGURE 11.23

Ultrapure copper sheets (serving as cathodes) are lowered between slabs of impure copper (serving as anodes) into a tank containing an aqueous solution of copper sulfate ($CuSO_4$). It takes about four weeks for the anodes to dissolve and for the pure copper to be deposited on the cathodes.

is the electrolyte, and thin sheets of ultrapure copper function as the cathodes (see Fig. 11.23).

The main reaction at the anode is

$$Cu \longrightarrow Cu^{2+} + 2e^-$$

Other metals such as iron and zinc are also oxidized from the impure anode:

$$Zn \longrightarrow Zn^{2+} + 2e^-$$
$$Fe \longrightarrow Fe^{2+} + 2e^-$$

Noble metal impurities in the anode are not oxidized at the voltage used; they fall to the bottom of the cell to form a sludge, which is processed to remove the valuable silver, gold, and platinum.

The Cu^{2+} ions from the solution are deposited onto the cathode,

$$Cu^{2+} + 2e^- \longrightarrow Cu$$

producing copper that is 99.95% pure.

Metal Plating

Metals that readily corrode can often be protected by the application of a thin coating of a metal that resists corrosion. Examples are "tin" cans, which are actually steel cans with a thin coating of tin, and chrome-plated steel bumpers for automobiles.

An object can be plated by making it the cathode in a tank containing ions of the plating metal. The silver plating of a spoon is shown schematically in Fig. 11.24. In an actual plating process the solution also contains ligands that form complexes with the silver ion. When the concentration of Ag^+ is lowered in this way, a smooth, even coating of silver is obtained.

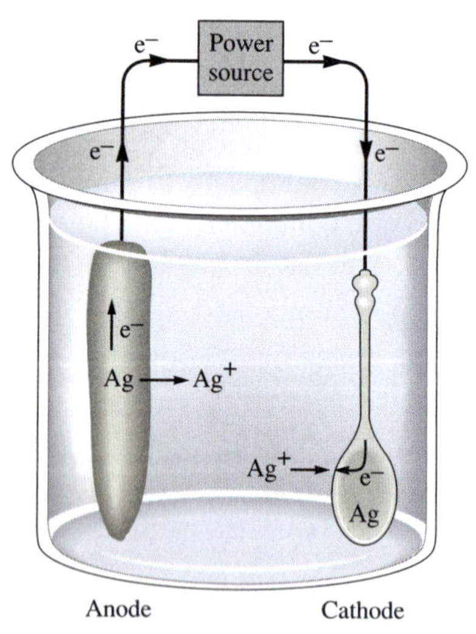

FIGURE 11.24

Schematic of the electroplating of a spoon. The item to be plated is the cathode, and the anode is a silver bar. Silver is plated out at the cathode:

$$Ag^+ + e^- \longrightarrow Ag$$

Note that a salt bridge is not needed here since Ag^+ ions are involved at both electrodes.

Electrolysis of Sodium Chloride

Sodium metal is produced mainly by the electrolysis of molten sodium chloride. Because solid NaCl has a rather high melting point (800°C), it is usually mixed with solid $CaCl_2$ to lower the melting point to about 600°C. The mixture is

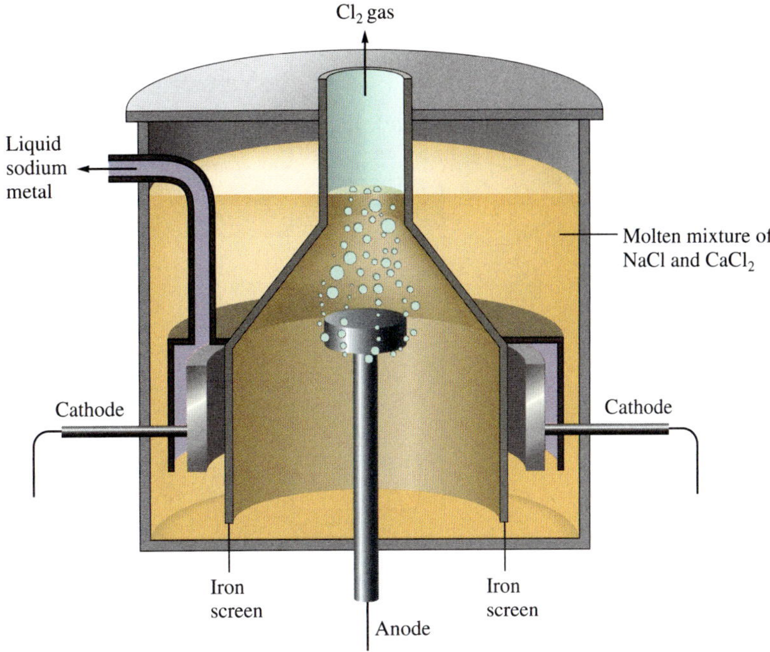

FIGURE 11.25

The Downs cell for the electrolysis of molten sodium chloride. The cell is designed so that the sodium and chlorine produced cannot come into contact with each other to re-form NaCl.

then electrolyzed in a **Downs cell,** as illustrated in Fig. 11.25, where the reactions are as follows:

$$\text{Anode reaction:} \qquad 2Cl^- \longrightarrow Cl_2 + 2e^-$$

$$\text{Cathode reaction:} \qquad Na^+ + e^- \longrightarrow Na$$

At the temperatures in the Downs cell the sodium is liquid and can be drained off, cooled, and cast into blocks. Because it is so reactive, sodium must be stored in an inert solvent, such as mineral oil, to prevent its oxidation.

Electrolysis of aqueous sodium chloride (brine) is an important industrial process for the production of chlorine and sodium hydroxide. In fact, this process is second only to the production of aluminum as a consumer of electricity in the United States. Sodium is not produced in this process under normal circumstances because H_2O is more easily reduced than Na^+, as the standard reduction potentials show:

$$Na^+ + e^- \longrightarrow Na \qquad \mathscr{E}° = -2.71 \text{ V}$$

$$2H_2O + 2e^- \longrightarrow H_2 + 2OH^- \qquad \mathscr{E}° = -0.83 \text{ V}$$

Hydrogen, not sodium, is produced at the cathode.

For the reasons we discussed in Section 11.7, chlorine gas is produced at the anode. Thus the electrolysis of brine produces hydrogen and chlorine:

$$\text{Anode reaction:} \qquad 2Cl^- \longrightarrow Cl_2 + 2e^-$$

$$\text{Cathode reaction:} \qquad 2H_2O + 2e^- \longrightarrow H_2 + 2OH^-$$

It leaves a solution containing dissolved NaOH and NaCl.

The contamination of the sodium hydroxide by NaCl can be virtually eliminated by using a special **mercury cell** for electrolyzing brine (see Fig. 11.26). In this cell mercury is the conductor at the cathode, and because hydrogen gas has an extremely high overvoltage with a mercury electrode,

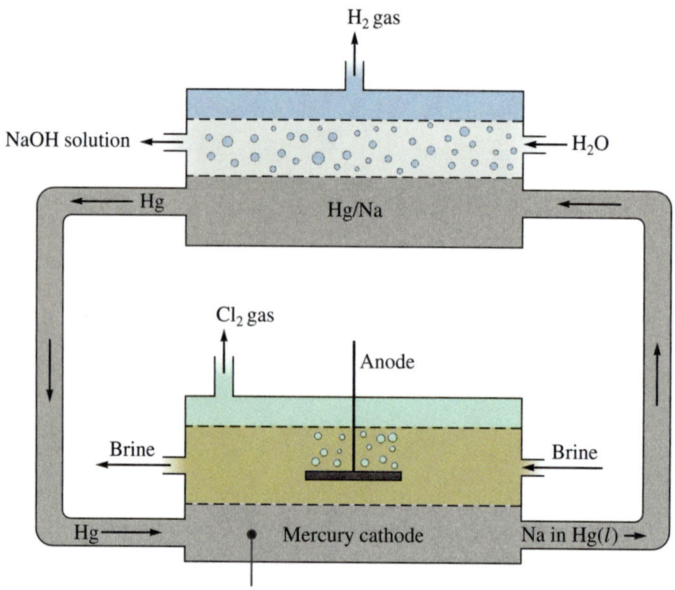

FIGURE 11.26

The mercury cell for production of chlorine and sodium hydroxide. The large overvoltage required to produce hydrogen at a mercury electrode means that Na^+ ions are reduced rather than water. The sodium formed dissolves in the liquid mercury and is then pumped to a chamber, where it reacts with water.

Na^+ is reduced instead of H_2O. The resulting sodium metal dissolves in the mercury, forming a liquid alloy, which is then pumped to a chamber where the dissolved sodium is reacted with water to produce hydrogen:

$$2Na(s) + 2H_2O(l) \longrightarrow 2Na^+(aq) + 2OH^-(aq) + H_2(g)$$

Relatively pure solid NaOH is recovered from the aqueous solution, and the regenerated mercury is then pumped back to the electrolysis cell. This process, called the **chlor-alkali process,** has often resulted in significant mercury contamination of the environment; the waste solutions from this process are now carefully treated to remove mercury.

Because of the environmental problems associated with the mercury cell, it has been primarily displaced in the chlor-alkali industry by other technologies. In the United States nearly 75% of chlor-alkali production is now carried out in diaphragm cells. In a diaphragm cell the cathode and the anode are separated by a diaphragm that allows passage of H_2O molecules, Na^+ ions, and, to a limited extent, Cl^- ions. The diaphragm does not allow OH^- ions to pass through it. Thus the H_2 and OH^- formed at the cathode are kept separate from the Cl_2 formed at the anode. The major disadvantage of this process is that the aqueous effluent pumped from the cathode compartment contains a mixture of sodium hydroxide and unreacted sodium chloride, which must be separated if pure sodium hydroxide is a desired product.

In the last twenty-five years a new process has been developed in the chlor-alkali industry that uses a membrane to separate the anode and cathode compartments in brine electrolysis cells. The membrane is superior to the diaphragm used in diaphragm cells because the membrane is impermeable to anions. Only cations can flow through the membrane. Because neither Cl^- nor OH^- ions can pass through the membrane separating the anode and cathode compartments, NaCl contamination of the NaOH formed at the cathode is not a problem. Although membrane technology is only now becoming prominent in the United States, it is already the dominant method for chlor-alkali production in Japan.

Discussion Questions

These questions are designed to be considered by groups of students in class. Often these questions work well for introducing a particular topic in class.

1. Sketch a galvanic cell, and explain how it works. Look at Figs. 11.1 and 11.2. Explain what is occurring in each container and why the cell in Fig. 11.2 "works" but the one in Fig. 11.1 does not.
2. In making a specific galvanic cell, explain how one determines which electrodes and solutions to use in the cell.
3. You want to "plate out" nickel metal from a nickel nitrate solution onto a piece of metal inserted into the solution. Should you use copper or zinc (or can you use either)? Explain.
4. A copper penny can be dissolved in nitric acid but not in hydrochloric acid. Using reduction potentials given in the book, show why this is so. What are the products of the reaction? Newer pennies contain a mixture of zinc and copper. What happens to the zinc in the penny when placed in nitric acid? Hydrochloric acid? Support your explanations with the data from the book, and include balanced equations for all reactions.
5. Sketch a cell that forms iron metal from iron(II) while changing chromium metal to chromium(III). Calculate the voltage, show the electron flow, label the anode and cathode, and balance the overall cell equation.
6. Which of the following is the best reducing agent: F_2, H^+, Na, Na^+, or F^-? Explain. Order as many of these species as possible from the best to the worst oxidizing agent. Why can't you order all of them? From Table 11.1 choose the species that is the best oxidizing agent. Choose the best reducing agent. Explain.
7. You are told that metal A is a better reducing agent than metal B. What, if anything, can be said about A^+ compared with B^+? Explain.
8. Explain the following relationships: ΔG and w, cell potential and w, cell potential and ΔG, cell potential and Q. Using these relationships, explain how you could make a cell in which both electrodes are the same metal and both solutions contain the same compound, but at different concentrations. How could this cell run spontaneously?
9. Explain why cell potentials are not multiplied by the coefficients in the balanced equation. (*Hint:* Use the relationship between ΔG and cell potential.)
10. What is the difference between $\mathscr{E}$ and $\mathscr{E}°$? When is $\mathscr{E}$ equal to zero? When is $\mathscr{E}°$ equal to zero? (Consider "regular" galvanic cells as well as concentration cells.)
11. Consider the following galvanic cell:

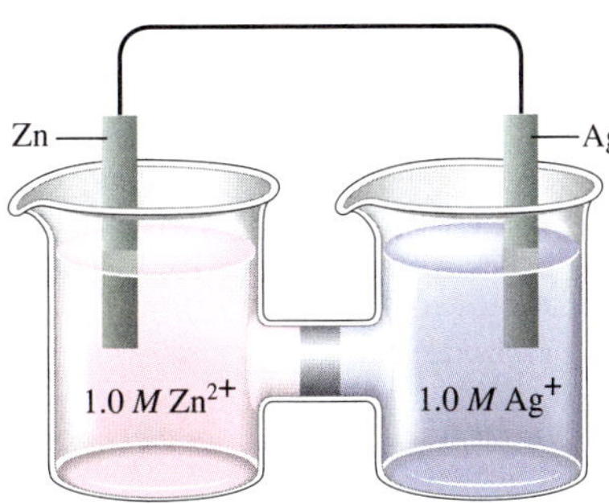

What happens to $\mathscr{E}$ as the concentration of Zn^{2+} is increased? as the concentration of Ag^+ is increased? What happens to $\mathscr{E}°$ in these cases?

12. Look up the reduction potential for Fe^{3+} to Fe^{2+}. Look up the reduction potential for Fe^{2+} to Fe. Finally, look up the reduction potential for Fe^{3+} to Fe. You should notice that adding the reduction potentials for the first two does not give you the potential for the third. Why not? Show how you can use the first two potentials to calculate the third potential.
13. If the cell potential is proportional to work and the standard reduction potential for the hydrogen ion is zero, does this mean that the reduction of the hydrogen ion requires no work?
14. Is the following statement true or false? Concentration cells work because standard reduction potentials are dependent on concentration. Explain.

Exercises

A blue exercise number indicates that the answer to that exercise appears at the back of this book and a solution appears in the *Solutions Guide*.

Galvanic Cells, Cell Potentials, and Standard Reduction Potentials

15. What is electrochemistry? What are redox reactions? Explain the difference between a galvanic cell and an electrolytic cell.
16. When magnesium metal is added to a beaker of HCl(*aq*), a gas is produced. Knowing that magnesium is oxidized and that hydrogen is reduced, write the balanced equation for the reaction. How many electrons are transferred in the balanced equation? What quantity of useful work can be obtained when Mg is added directly to the beaker of HCl? How can you harness this reaction to do useful work?
17. Sketch the galvanic cells based on the following overall reactions. Calculate $\mathscr{E}°$, show the direction of electron

flow and the direction of ion migration through the salt bridge, identify the cathode and anode, and give the overall balanced reaction. Assume that all concentrations are 1.0 *M* and that all partial pressures are 1.0 atm. Standard reduction potentials are found in Table 11.1.

a. $Cr^{3+}(aq) + Cl_2(g) \rightleftharpoons Cr_2O_7^{2-}(aq) + Cl^-(aq)$
b. $Cu^{2+}(aq) + Mg(s) \rightleftharpoons Mg^{2+}(aq) + Cu(s)$
c. $IO_3^-(aq) + Fe^{2+}(aq) \rightleftharpoons Fe^{3+}(aq) + I_2(s)$
d. $Zn(s) + Ag^+(aq) \rightleftharpoons Zn^{2+}(aq) + Ag(s)$

18. Calculate $\mathscr{E}°$ values for the following cells. Which reactions are spontaneous as written (under standard conditions)? Balance the reactions. Standard reduction potentials are found in Table 11.1.

a. $MnO_4^-(aq) + I^-(aq) \rightleftharpoons I_2(aq) + Mn^{2+}(aq)$
b. $MnO_4^-(aq) + F^-(aq) \rightleftharpoons F_2(g) + Mn^{2+}(aq)$
c. $H_2(g) \rightleftharpoons H^+(aq) + H^-(aq)$
d. $Au^{3+}(aq) + Ag(s) \rightleftharpoons Ag^+(aq) + Au(s)$

19. Sketch the galvanic cells based on the following half-reactions. Calculate $\mathscr{E}°$, show the direction of electron flow and the direction of ion migration through the salt bridge, identify the cathode and anode, and give the overall balanced reaction. Assume that all concentrations are 1.0 *M* and that all partial pressures are 1.0 atm.

a. $Cl_2 + 2e^- \longrightarrow 2Cl^-$ $\quad \mathscr{E}° = 1.36$ V
$Br_2 + 2e^- \longrightarrow 2Br^-$ $\quad \mathscr{E}° = 1.09$ V
b. $MnO_4^- + 8H^+ + 5e^- \longrightarrow Mn^{2+} + 4H_2O$ $\quad \mathscr{E}° = 1.51$ V
$IO_4^- + 2H^+ + 2e^- \longrightarrow IO_3^- + H_2O$ $\quad \mathscr{E}° = 1.60$ V
c. $H_2O_2 + 2H^+ + 2e^- \longrightarrow 2H_2O$ $\quad \mathscr{E}° = 1.78$ V
$O_2 + 2H^+ + 2e^- \longrightarrow H_2O_2$ $\quad \mathscr{E}° = 0.68$ V
d. $Mn^{2+} + 2e^- \longrightarrow Mn$ $\quad \mathscr{E}° = -1.18$ V
$Fe^{3+} + 3e^- \longrightarrow Fe$ $\quad \mathscr{E}° = -0.036$ V

20. Consider the following galvanic cells:

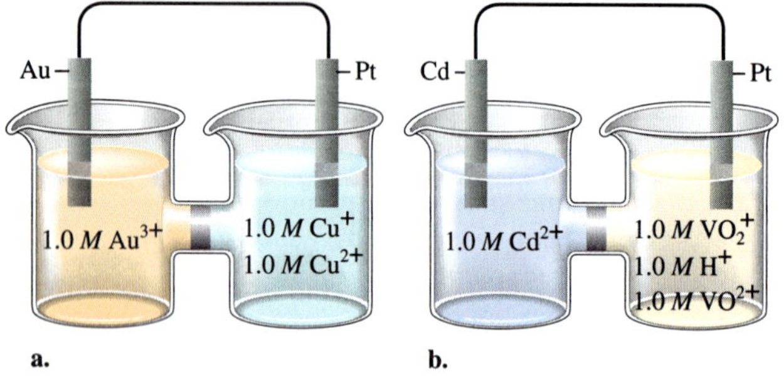

For each galvanic cell, give the balanced cell reaction and determine $\mathscr{E}°$. Standard reduction potentials are found in Table 11.1.

21. Give the standard line notation for each cell in Exercise 19.

22. The saturated calomel electrode, abbreviated SCE, is often used as a reference electrode in making electrochemical measurements. The SCE is composed of mercury in contact with a saturated solution of calomel (Hg_2Cl_2). The electrolyte solution is saturated KCl. $\mathscr{E}_{SCE}$ is +0.242 V relative to the standard hydrogen electrode. Calculate the potential for each of the following galvanic cells containing a saturated calomel electrode and the given half-cell components as standard conditions. In each case indicate whether the SCE is the cathode or the anode. Standard reduction potentials are found in Table 11.1.

a. $Cu^{2+} + 2e^- \longrightarrow Cu$
b. $Fe^{3+} + e^- \longrightarrow Fe^{2+}$
c. $AgCl + e^- \longrightarrow Ag + Cl^-$
d. $Al^{3+} + 3e^- \longrightarrow Al$
e. $Ni^{2+} + 2e^- \longrightarrow Ni$

23. Answer the following questions using data from Table 11.1 (all under standard conditions).

a. Is $H^+(aq)$ capable of oxidizing $Cu(s)$ to $Cu^{2+}(aq)$?
b. Is $Fe^{3+}(aq)$ capable of oxidizing $I^-(aq)$?
c. Is $H_2(g)$ capable of reducing $Ag^+(aq)$?
d. Is $Fe^{2+}(aq)$ capable of reducing $Cr^{3+}(aq)$ to $Cr^{2+}(aq)$?

24. Using data from Table 11.1, place the following in order of increasing strength as oxidizing agents (all under standard conditions).

$$Cd^{2+}, IO_3^-, K^+, H_2O, AuCl_4^-, \text{and } I_2$$

25. Using data from Table 11.1, place the following in order of increasing strength as reducing agents (all under standard conditions).

$$Cu^+, F^-, H^-, H_2O, I_2, \text{and } K$$

26. Consider only the species (at standard conditions)

$$Na^+, Cl^-, Ag^+, Ag, Zn^{2+}, Zn, \text{and } Pb$$

in answering the following questions. Give reasons for your answers. (Use data from Table 11.1.)

a. Which is the strongest oxidizing agent?
b. Which is the strongest reducing agent?
c. Which species can be oxidized by $SO_4^{2-}(aq)$ in acid?
d. Which species can be reduced by $Al(s)$?

27. Use the table of standard reduction potentials (Table 11.1) to pick a reagent that is capable of each of the following oxidations (under standard conditions in acidic solution).

a. oxidizes Br^- to Br_2 but does not oxidize Cl^- to Cl_2
b. oxidizes Mn to Mn^{2+} but does not oxidize Ni to Ni^{2+}

28. Use the table of standard reduction potentials (Table 11.1) to pick a reagent that is capable of each of the following reductions (under standard conditions in acidic solution).

a. reduces Cu^{2+} to Cu but does not reduce Cu^{2+} to Cu^+
b. reduces Br_2 to Br^- but does not reduce I_2 to I^-

29. Hydrazine is somewhat toxic. Use the half-reactions shown below to explain why household bleach (a highly alkaline solution of sodium hypochlorite) should not be mixed with household ammonia or glass cleansers that contain ammonia.

$$ClO^- + H_2O + 2e^- \longrightarrow 2OH^- + Cl^- \qquad \mathscr{E}° = 0.90 \text{ V}$$

$$N_2H_4 + 2H_2O + 2e^- \longrightarrow 2NH_3 + 2OH^- \quad \mathscr{E}° = -0.10 \text{ V}$$

30. A patent attorney has asked for your advice concerning the merits of a patent application claiming the invention of an aqueous single galvanic cell capable of producing a 12-V potential. Comment.

Cell Potential, Free Energy, and Equilibrium

31. The free energy change for a reaction ΔG is an extensive property. What is an extensive property? Surprisingly, one can calculate ΔG from the cell potential $\mathscr{E}$ for the reaction. This is surprising because $\mathscr{E}$ is an intensive property. How can the extensive property ΔG be calculated from the intensive property $\mathscr{E}$?

32. The equation $\Delta G° = -nF\mathscr{E}°$ also can be applied to half-reactions. Use standard reduction potentials to estimate $\Delta G_f°$ for $Fe^{2+}(aq)$ and $Fe^{3+}(aq)$. ($\Delta G_f°$ for $e^- = 0$.)

33. Calculate the maximum amount of work that can be obtained from the galvanic cells at standard conditions in Exercise 20.

34. Under standard conditions, what reaction occurs, if any, when each of the following operations are performed?
   a. Crystals of I_2 are added to a solution of NaCl.
   b. Cl_2 gas is bubbled into a solution of NaI.
   c. A silver wire is placed in a solution of $CuCl_2$.
   d. An acidic solution of $FeSO_4$ is exposed to air.

   For the reactions that occur, write a balanced equation and calculate $\mathscr{E}°$, $\Delta G°$, and K at 25°C.

35. Calculate $\Delta G°$ and K at 25°C for the galvanic cell reactions in Exercise 19.

36. Chlorine dioxide (ClO_2), which is produced by the reaction

$$2NaClO_2(aq) + Cl_2(g) \longrightarrow 2ClO_2(g) + 2NaCl(aq)$$

   has been tested as a disinfectant for municipal water treatment.
   a. Using data from Table 11.1, calculate $\mathscr{E}°$, $\Delta G°$, and K at 25°C for the production of ClO_2.
   b. One of the concerns in using ClO_2 as a disinfectant is that the carcinogenic chlorate ion (ClO_3^-) might be a by-product. It can be formed from the reaction

$$ClO_2(g) \rightleftharpoons ClO_3^-(aq) + Cl^-(aq)$$

   Balance the equation for the decomposition of ClO_2.

37. The amount of manganese in steel is determined by changing it to permanganate ion. The steel is first dissolved in nitric acid, producing Mn^{2+} ions. These ions are then oxidized to the deeply colored MnO_4^- ions by periodate ion (IO_4^-) in acid solution.
   a. Complete and balance an equation describing each of the above reactions.
   b. Calculate $\mathscr{E}°$, $\Delta G°$, and K at 25°C for each reaction.

38. The overall reaction and equilibrium constant value for a hydrogen–oxygen fuel cell at 298 K is

$$2H_2(g) + O_2(g) \longrightarrow 2H_2O(l) \quad K = 1.28 \times 10^{83}$$

   a. Calculate $\mathscr{E}°$ and $\Delta G°$ at 298 K for the fuel-cell reaction.
   b. Predict the signs of $\Delta H°$ and $\Delta S°$ for the fuel-cell reaction.
   c. As temperature increases, does the maximum amount of work obtained from the fuel-cell reaction increase, decrease, or remain the same? Explain.

39. Combine the equations

$$\Delta G° = -nF\mathscr{E}° \quad \text{and} \quad \Delta G° = \Delta H° - T\Delta S°$$

   to derive an expression for $\mathscr{E}°$ as a function of temperature. Describe how one can graphically determine $\Delta H°$ and $\Delta S°$ from measurements of $\mathscr{E}°$ at different temperatures, assuming that $\Delta H°$ and $\Delta S°$ do not depend on temperature. What property would you look for in designing a reference half-cell that would produce a potential relatively stable with respect to temperature?

40. Calculate $\mathscr{E}°$ for the reaction

$$CH_3OH(l) + \tfrac{3}{2}O_2(g) \longrightarrow CO_2(g) + 2H_2O(l)$$

   using values of $\Delta G_f°$ in Appendix 4. Will $\mathscr{E}°$ increase or decrease with an increase in temperature? (See Exercise 39 for the dependence of $\mathscr{E}°$ on temperature.)

41. A disproportionation reaction involves a substance that acts as both an oxidizing agent and a reducing agent, producing higher and lower oxidation states of the same element in the products. Which of the following disproportionation reactions are spontaneous under standard conditions? Calculate $\Delta G°$ and K at 25°C for those reactions that are spontaneous under standard conditions.
   a. $2Cu^+(aq) \longrightarrow Cu^{2+}(aq) + Cu(s)$
   b. $3Fe^{2+}(aq) \longrightarrow 2Fe^{3+}(aq) + Fe(s)$
   c. $HClO_2(aq) \longrightarrow ClO_3^-(aq) + HClO(aq)$ (unbalanced)

   Use the half-reactions:

$$ClO_3^- + 3H^+ + 2e^- \longrightarrow HClO_2 + H_2O \quad \mathscr{E}° = 1.21 \text{ V}$$

$$HClO_2 + 2H^+ + 2e^- \longrightarrow HClO + H_2O \quad \mathscr{E}° = 1.65 \text{ V}$$

42. Calculate the value of the equilibrium constant for the reaction of zinc metal in a solution of silver nitrate at 25°C.

43. For the following half-reaction, $\mathscr{E}° = -2.07$ V:

$$AlF_6^{3-} + 3e^- \longrightarrow Al + 6F^-$$

   Using data from Table 11.1, calculate the equilibrium constant at 25°C for the reaction

$$Al^{3+}(aq) + 6F^-(aq) \rightleftharpoons AlF_6^{3-}(aq)$$

44. Calculate K_{sp} for iron(II) sulfide given the following data:

$$FeS(s) + 2e^- \longrightarrow Fe(s) + S^{2-}(aq) \quad \mathscr{E}° = -1.01 \text{ V}$$

$$Fe^{2+}(aq) + 2e^- \longrightarrow Fe(s) \quad \mathscr{E}° = -0.44 \text{ V}$$

45. The solubility product for CuI(s) is 1.1×10^{-12}. Calculate the value of $\mathscr{E}°$ for the half-reaction

$$CuI + e^- \longrightarrow Cu + I^-$$

Galvanic Cells: Concentration Dependence

46. Calculate the pH of the cathode compartment for the following reaction given $\mathscr{E}_{cell} = 3.01$ V when $[Cr^{3+}] = 0.15$ *M*, $[Al^{3+}] = 0.30$ *M*, and $[Cr_2O_7^{2-}] = 0.55$ *M*.

$$2Al(s) + Cr_2O_7^{2-}(aq) + 14H^+(aq) \longrightarrow 2Al^{3+}(aq) + 2Cr^{3+}(aq) + 7H_2O(l)$$

47. Consider the galvanic cell based on the following half-reactions:

$$Au^{3+} + 3e^- \longrightarrow Au \quad \mathscr{E}° = 1.50 \text{ V}$$
$$Tl^+ + e^- \longrightarrow Tl \quad \mathscr{E}° = -0.34 \text{ V}$$

a. Determine the overall cell reaction and calculate $\mathscr{E}°_{cell}$.
b. Calculate $\Delta G°$ and K for the cell reaction at 25°C.
c. Calculate $\mathscr{E}_{cell}$ at 25°C when $[Au^{3+}] = 1.0 \times 10^{-2}$ *M* and $[Tl^+] = 1.0 \times 10^{-4}$ *M*.

48. Consider the following galvanic cell at 25°C:

$$Pt|Cr^{2+}(0.30\ M), Cr^{3+}(2.0\ M)||Co^{2+}(0.20\ M)|Co$$

The overall reaction and equilibrium constant value are

$$2Cr^{2+}(aq) + Co^{2+}(aq) \longrightarrow 2Cr^{3+}(aq) + Co(s) \quad K = 2.79 \times 10^7$$

Calculate the cell potential $\mathscr{E}$ for this galvanic cell and ΔG for the cell reaction at these conditions.

49. Consider the cell described below:

$$Al|Al^{3+}(1.00\ M)||Pb^{2+}(1.00\ M)|Pb$$

Calculate the cell potential after the reaction has operated long enough for the $[Al^{3+}]$ to have changed by 0.60 mol/L. (Assume $T = 25°C$.)

50. The Nernst equation can be applied to half-reactions. Calculate the reduction potential at 25°C of each of the following half-cells.
a. Cu/Cu^{2+} (0.10 *M*)
(The half-reaction is $Cu^{2+} + 2e^- \rightarrow Cu$.)
b. Cu/Cu^{2+} (2.0 *M*)
c. $Cu/Cu^{2+}(1.0 \times 10^{-4}$ *M*)
d. MnO_4^- (0.10 *M*)/Mn^{2+} (0.010 *M*) at pH = 3.00
(The half-reaction is $MnO_4^- + 8H^+ + 5e^- \rightarrow Mn^{2+} + 4H_2O$.)
e. MnO_4^- (0.10 *M*)/Mn^{2+} (0.010 *M*) at pH = 1.00

51. The overall reaction in the lead storage battery is

$$Pb(s) + PbO_2(s) + 2H^+(aq) + 2HSO_4^-(aq) \longrightarrow 2PbSO_4(s) + 2H_2O(l)$$

a. Calculate $\mathscr{E}$ at 25°C for this battery when $[H_2SO_4] = 4.5$ *M*; that is, $[H^+] = [HSO_4^-] = 4.5$ *M*. At 25°C, $\mathscr{E}° = 2.04$ V for the lead storage battery.
b. For the cell reaction $\Delta H° = -315.9$ kJ and $\Delta S° = 263.5$ J/K. Calculate $\mathscr{E}°$ at −20.°C. (See Exercise 39.)
c. Calculate $\mathscr{E}$ at −20.°C when $[H_2SO_4] = 4.5$ *M*.
d. Based on your previous answers, why does it seem that batteries fail more often on cold days than on warm days?

52. A chemist wishes to determine the concentration of CrO_4^{2-} electrochemically. A cell is constructed consisting of a saturated calomel electrode (SCE; see Exercise 22) and a silver wire coated with Ag_2CrO_4. The $\mathscr{E}°$ value for the following half-reaction is +0.446 V relative to the standard hydrogen electrode:

$$Ag_2CrO_4 + 2e^- \longrightarrow 2Ag + CrO_4^{2-}$$

a. Calculate $\mathscr{E}_{cell}$ and ΔG at 25°C for the cell reaction when $[CrO_4^{2-}] = 1.00$ mol/L.
b. Write the Nernst equation for the cell. Assume that the SCE concentrations are constant.
c. If the coated silver wire is placed in a solution (at 25°C) in which $[CrO_4^{2-}] = 1.00 \times 10^{-5}$ *M*, what is the expected cell potential?
d. The measured cell potential at 25°C is 0.504 V when the coated wire is dipped into a solution of unknown $[CrO_4^{2-}]$. What is the $[CrO_4^{2-}]$ for this solution?
e. Using data from this problem and from Table 11.1, calculate the solubility product (K_{sp}) for Ag_2CrO_4.

53. What are concentration cells? What is $\mathscr{E}°$ in a concentration cell? What is the driving force for a concentration cell to produce a voltage? Is the higher or the lower ion concentration solution present at the anode? When the anode ion concentration is decreased and/or the cathode ion concentration is increased, both give rise to larger cell potentials. Why?

54. Concentration cells are commonly used to calculate the value of equilibrium constants for various reactions. For example, the silver concentration cell illustrated in Fig. 11.11 can be used to determine the K_{sp} value for AgCl(*s*). To do so, NaCl is added to the anode compartment until no more precipitate forms. The $[Cl^-]$ in solution is then determined somehow. What happens to $\mathscr{E}_{cell}$ when NaCl is added to the anode compartment? To calculate the K_{sp} value, $[Ag^+]$ must be calculated. Given the value of $\mathscr{E}_{cell}$, how is $[Ag^+]$ determined at the anode?

55. Consider the concentration cell shown below. Calculate the cell potential at 25°C when the concentration of Ni^{2+} in the compartment on the right has each of the following values.

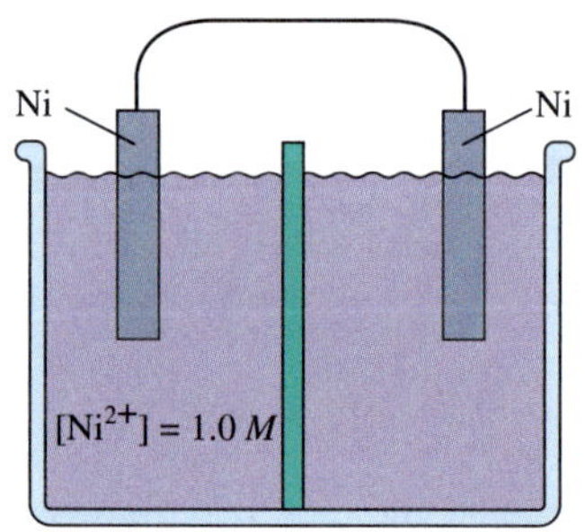

a. 1.0 *M*
b. 2.0 *M*
c. 0.10 *M*
d. 4.0×10^{-5} *M*
e. Calculate the potential when both solutions are 2.5 *M* in Ni^{2+}.

For each case, identify the cathode, the anode, and the direction in which electrons flow.

56. Consider a concentration cell that has both electrodes made of some metal M. Solution A in one compartment of the cell contains 1.0 M M^{2+}. Solution B in the other cell compartment has a volume of 1.00 L. At the beginning of the experiment 0.0100 mol of $M(NO_3)_2$ and 0.0100 mol of Na_2SO_4 are dissolved in solution B (ignore volume changes), where the reaction

$$M^{2+}(aq) + SO_4^{2-}(aq) \rightleftharpoons MSO_4(s)$$

occurs. For this reaction equilibrium is rapidly established, whereupon the cell potential is found to be +0.44 V at 25°C. Assume that the process

$$M^{2+} + 2e^- \longrightarrow M$$

has a standard reduction potential of +0.80 V and that no other redox process occurs in the cell. Calculate the value of K_{sp} for $MSO_4(s)$ at 25°C.

57. An electrochemical cell consists of a standard hydrogen electrode and a copper metal electrode.
   a. What is the potential of the cell at 25°C if the copper electrode is placed in a solution in which $[Cu^{2+}] = 2.5 \times 10^{-4}$ M?
   b. If the copper electrode is placed in a solution of 0.10 M NaOH that is saturated with $Cu(OH)_2$, what is the cell potential at 25°C? For $Cu(OH)_2$, $K_{sp} = 1.6 \times 10^{-19}$.
   c. The copper electrode is placed in a solution of unknown $[Cu^{2+}]$. The measured potential at 25°C is 0.195 V. What is $[Cu^{2+}]$? (Assume that Cu^{2+} is reduced.)
   d. If you wish to construct a calibration curve to show how the cell potential varies with $[Cu^{2+}]$, what should you plot to obtain a straight line? What will the slope of this line be?

58. An electrochemical cell consists of a nickel metal electrode immersed in a solution with $[Ni^{2+}] = 1.0$ M separated by a porous disk from an aluminum metal electrode immersed in a solution with $[Al^{3+}] = 1.0$ M. Sodium hydroxide is added to the aluminum compartment, causing $Al(OH)_3(s)$ to precipitate. After precipitation of $Al(OH)_3$ has ceased, the concentration of OH^- is 1.0×10^{-4} M, and the measured cell potential is 1.82 V. Calculate the K_{sp} value for $Al(OH)_3$.

$$Al(OH)_3(s) \rightleftharpoons Al^{3+}(aq) + 3OH^-(aq) \qquad K_{sp} = ?$$

59. You have a concentration cell in which the cathode has a silver electrode with 0.10 M Ag^+. The anode also has a silver electrode with $Ag^+(aq)$, 0.050 M $S_2O_3^{2-}$, and 1.0×10^{-3} M $Ag(S_2O_3)_2^{3-}$. You read the voltage to be 0.76 V.
   a. Calculate the concentration of Ag^+ at the anode.
   b. Determine the value of the equilibrium constant for the formation of $Ag(S_2O_3)_2^{3-}$.

$$Ag^+(aq) + 2S_2O_3^{2-}(aq) \rightleftharpoons Ag(S_2O_3)_2^{3-}(aq) \qquad K = ?$$

Electrolysis

60. How long will it take to plate out each of the following with a current of 100.0 A?
   a. 1.0 kg of Al from aqueous Al^{3+}
   b. 1.0 g of Ni from aqueous Ni^{2+}
   c. 5.0 mol of Ag from aqueous Ag^+

61. What mass of each of the following substances can be produced in 1.0 h with a current of 15 A?
   a. Co from aqueous Co^{2+}
   b. Hf from aqueous Hf^{4+}
   c. I_2 from aqueous KI
   d. Cr from molten CrO_3

62. It took 2.30 min with a current of 2.00 A to plate out all the silver from 0.250 L of a solution containing Ag^+. What was the original concentration of Ag^+ in the solution?

63. The electrolysis of BiO^+ produces pure bismuth. How long would it take to produce 10.0 g of Bi by the electrolysis of a BiO^+ solution using a current of 25.0 A?

64. Why is the electrolysis of molten salts much easier to predict in terms of what occurs at the anode and cathode than the electrolysis of aqueous dissolved salts?

65. What reactions take place at the cathode and the anode when each of the following is electrolyzed? Assume standard conditions.
   a. molten $NiBr_2$
   b. molten AlF_3
   c. molten MnI_2
   d. 1.0 M $NiBr_2$ solution
   e. 1.0 M AlF_3 solution
   f. 1.0 M MnI_2 solution

66. a. In the electrolysis of an aqueous solution of Na_2SO_4, what reactions occur at the anode and the cathode (assuming standard conditions)?

	$\mathscr{E}°$
$S_2O_8^{2-} + 2e^- \longrightarrow 2SO_4^{2-}$	2.01 V
$O_2 + 4H^+ + 4e^- \longrightarrow 2H_2O$	1.23 V
$2H_2O + 2e^- \longrightarrow H_2 + 2OH^-$	−0.83 V
$Na^+ + e^- \longrightarrow Na$	−2.71 V

   b. When water containing a small amount (~0.01 M) of sodium sulfate is electrolyzed, measurement of the volume of gases generated consistently gives a result that the volume ratio of hydrogen to oxygen is not quite 2:1. To what do you attribute this discrepancy? Predict whether the measured ratio is greater than or less than 2:1.

67. A solution at 25°C contains 1.0 M Cd^{2+}, 1.0 M Ag^+, 1.0 M Au^{3+}, and 1.0 M Ni^{2+} in the cathode compartment of an electrolytic cell. Predict the order in which the metals will plate out as the voltage is gradually increased.

68. An aqueous solution of an unknown salt of ruthenium is electrolyzed by a current of 2.50 A passing for 50.0 min. If 2.618 g Ru is produced at the cathode, what is the charge on the ruthenium ions in solution?

69. Consider the following half-reactions:

$$IrCl_6^{3-} + 3e^- \longrightarrow Ir + 6Cl^- \qquad \mathscr{E}° = 0.77\ V$$

$$PtCl_4^{2-} + 2e^- \longrightarrow Pt + 4Cl^- \qquad \mathscr{E}° = 0.73\ V$$

$$PdCl_4^{2-} + 2e^- \longrightarrow Pd + 4Cl^- \qquad \mathscr{E}° = 0.62\ V$$

A hydrochloric acid solution contains platinum, palladium, and iridium as chloro-complex ions. The solution is a constant 1.0 *M* in chloride ion and 0.020 *M* in each complex ion. Is it feasible to separate the three metals from this solution by electrolysis? (Assume that 99% of a metal must be plated out before another metal begins to plate out.)

70. An unknown metal M is electrolyzed. It took 74.1 s for a current of 2.00 A to plate out 0.107 g of the metal from a solution containing $M(NO_3)_3$. Identify the metal.

71. Electrolysis of an alkaline earth metal chloride using a current of 5.00 A for 748 seconds deposits 0.471 g of metal at the cathode. What is the identity of the alkaline earth metal chloride?

72. One of the few industrial-scale processes that produces organic compounds electrochemically is used by the Monsanto Company to produce 1,4-dicyanobutane. The reduction reaction is

$$2CH_2{=}CHCN + 2H^+ + 2e^- \longrightarrow NC{-}(CH_2)_4{-}CN$$

The $NC{-}(CH_2)_4{-}CN$ is then chemically reduced by hydrogen to $H_2N{-}(CH_2)_6{-}NH_2$, which is used in the production of nylon. What current must be used to produce 150. kg of $NC{-}(CH_2)_4{-}CN$ per hour?

73. What volume of F_2 gas, at 25°C and 1.00 atm, is produced when molten KF is electrolyzed by a current of 10.0 A for 2.00 h? What mass of potassium metal is produced? At which electrode does each reaction occur?

74. It takes 15 kWh (kilowatt hours) of electrical energy to produce 1.0 kg of aluminum metal from aluminum oxide by the Hall–Heroult process. Compare this value with the amount of energy necessary to melt 1.0 kg of aluminum metal. Why is it economically feasible to recycle aluminum cans? (The enthalpy of fusion for aluminum metal is 10.7 kJ/mol and 1 watt = 1 J/s.)

75. In the electrolysis of a sodium chloride solution, what volume of $Cl_2(g)$ is produced in the same time it takes to produce 6.00 L of $H_2(g)$, both volumes measured at 0°C and 1.00 atm?

76. What volumes of $H_2(g)$ and $O_2(g)$ at STP are produced from the electrolysis of water by a current of 2.50 A in 15.0 min?

Additional Exercises

77. An experimental fuel cell has been designed that uses carbon monoxide as fuel. The overall reaction is

$$2CO(g) + O_2(g) \longrightarrow 2CO_2(g)$$

The two half-cell reactions are

$$CO + O^{2-} \longrightarrow CO_2 + 2e^-$$

$$O_2 + 4e^- \longrightarrow 2O^{2-}$$

The two half-reactions are carried out in separate compartments connected with a solid mixture of CeO_2 and Gd_2O_3. Oxide ions can move through this solid at high temperatures (about 800°C). ΔG for the overall reaction at 800°C under certain concentration conditions is -380 kJ. Calculate the cell potential for this fuel cell at the same temperature and concentration conditions.

78. Batteries are galvanic cells. What happens to $\mathscr{E}_{cell}$ as a battery discharges? Does a battery represent a system at equilibrium? Explain. What is $\mathscr{E}_{cell}$ when a battery reaches equilibrium? How are batteries and fuel cells alike? How are they different? The U.S. space program uses hydrogen–oxygen fuel cells to produce power for its spacecraft. What is a hydrogen–oxygen fuel cell?

79. A fuel cell designed to react grain alcohol with oxygen has the following net reaction:

$$C_2H_5OH(l) + 3O_2(g) \longrightarrow 2CO_2(g) + 3H_2O(l)$$

The maximum work 1 mol of alcohol can yield by this process is 1320 kJ. What is the theoretical maximum voltage this cell can achieve?

80. What is the maximum work that can be obtained from a hydrogen–oxygen fuel cell at standard conditions that produces 1.00 kg of water at 25°C? Why do we say that this is the maximum work that can be obtained? What are the advantages and disadvantages in using fuel cells rather than the corresponding combustion reactions to produce electricity?

81. The overall reaction and standard cell potential at 25°C for the rechargeable nickel–cadmium alkaline battery is

$$Cd(s) + NiO_2(s) + 2H_2O(l) \longrightarrow Ni(OH)_2(s) + Cd(OH)_2(s) \qquad \mathscr{E}° = 1.10\ V$$

For every mole of Cd consumed in the cell, what is the maximum useful work that can be obtained at standard conditions?

82. Not all spontaneous redox reactions produce wonderful results. Corrosion is an example of a spontaneous redox process that has negative effects. What happens in the corrosion of a metal such as iron? What must be present for the corrosion of iron to take place? How can moisture and salt increase the severity of corrosion?

83. Explain how the following protect metals from corrosion.
    a. paint
    b. durable oxide coatings
    c. galvanizing
    d. sacrificial metal
    e. alloying
    f. cathodic protection

84. In theory, most metals should easily corrode in air. Why? A group of metals called the noble metals are relatively difficult to corrode in air. Some noble metals include gold, platinum, and silver. Reference Table 11.1 to come up with a possible reason why the noble metals are relatively difficult to corrode.

85. In 1973 the wreckage of the Civil War ironclad USS *Monitor* was discovered near Cape Hatteras, North Carolina. [The *Monitor* and the CSS *Virginia* (formerly the USS *Merrimack*) fought the first battle between iron-armored ships.] In 1987 investigations were begun to see whether the ship could be salvaged. *Time* reported (June 22, 1987) that scientists were considering adding sacrificial anodes of zinc to the rapidly corroding metal hull of the *Monitor*. Describe how attaching zinc to the hull would protect the *Monitor* from further corrosion.

86. A standard galvanic cell is constructed so that the overall cell reaction is

$$2Al^{3+}(aq) + 3M(s) \longrightarrow 3M^{2+}(aq) + 2Al(s)$$

where M is an unknown metal. If $\Delta G° = -411$ kJ for the overall cell reaction, identify the metal used to construct the standard cell.

87. Consider the following half-reactions:

$$Pt^{2+} + 2e^- \longrightarrow Pt \qquad \mathscr{E}° = 1.188\text{ V}$$

$$PtCl_4{}^{2-} + 2e^- \longrightarrow Pt + 4Cl^- \qquad \mathscr{E}° = 0.755\text{ V}$$

$$NO_3{}^- + 4H^+ + 3e^- \longrightarrow NO + 2H_2O \qquad \mathscr{E}° = 0.96\text{ V}$$

Explain why platinum metal will dissolve in aqua regia (a mixture of hydrochloric and nitric acids) but not in either concentrated nitric or concentrated hydrochloric acid individually.

88. Consider the following reduction potentials:

$$Co^{3+} + 3e^- \longrightarrow Co \qquad \mathscr{E}° = 1.26\text{ V}$$

$$Co^{2+} + 2e^- \longrightarrow Co \qquad \mathscr{E}° = -0.28\text{ V}$$

a. When cobalt metal dissolves in 1.0 *M* nitric acid, will Co^{3+} or Co^{2+} be the primary product (assuming standard conditions)?
b. Is it possible to change the concentration of HNO_3 to get a different result in part a?

89. The measurement of pH using a glass electrode obeys the Nernst equation. The typical response of a pH meter at 25.00°C is given by the equation

$$\mathscr{E}_{meas} = \mathscr{E}_{ref} + 0.05916\text{ pH}$$

where $\mathscr{E}_{ref}$ contains the potential of the reference electrode and all other potentials that arise in the cell that are not related to the hydrogen ion concentration. Assume that $\mathscr{E}_{ref} = 0.250$ V and that $\mathscr{E}_{meas} = 0.480$ V.
a. What is the uncertainty in the values of pH and $[H^+]$ if the uncertainty in the measured potential is ± 1 mV (± 0.001 V)?
b. To what accuracy must the potential be measured for the uncertainty in pH to be ± 0.02 pH unit?

90. The black silver sulfide discoloration of silverware can easily be removed by heating the silver article in a sodium carbonate solution in an aluminum pan. The reaction is

$$3Ag_2S(s) + 2Al(s) \rightleftharpoons 6Ag(s) + 3S^{2-}(aq) + 2Al^{3+}(aq)$$

a. Using data in Appendix 4, calculate $\Delta G°$, K, and $\mathscr{E}°$ for the above reaction. [For $Al^{3+}(aq)$, $\Delta G_f° = -480.$ kJ/mol.]
b. Calculate the value of the standard reduction potential for the following half-reaction:

$$2e^- + Ag_2S(s) \longrightarrow 2Ag(s) + S^{2-}(aq)$$

91. Consider the standard galvanic cell based on the following half-reactions

$$Cu^{2+} + 2e^- \longrightarrow Cu$$

$$Ag^+ + e^- \longrightarrow Ag$$

The electrodes in this cell are Ag(*s*) and Cu(*s*). Does the cell potential increase, decrease, or remain the same when the following changes occur to the standard cell?
a. $CuSO_4(s)$ is added to the copper half-cell compartment (assume no volume change).
b. $NH_3(aq)$ is added to the copper half-cell compartment. [*Hint:* Cu^{2+} reacts with NH_3 to form $Cu(NH_3)_4{}^{2+}(aq)$.]
c. NaCl(*s*) is added to the silver half-cell compartment. [*Hint:* Ag^+ reacts with Cl^- to form AgCl(*s*).]
d. Water is added to both half-cell compartments until the volume of solution is doubled.
e. The silver electrode is replaced with a platinum electrode.

$$Pt^{2+} + 2e^- \longrightarrow Pt \qquad \mathscr{E}° = 1.19\text{ V}$$

92. Consider a cell based on the following half-reactions:

$$Au^{3+} + 3e^- \longrightarrow Au \qquad \mathscr{E}° = 1.50\text{ V}$$

$$Fe^{3+} + e^- \longrightarrow Fe^{2+} \qquad \mathscr{E}° = 0.77\text{ V}$$

a. Draw this cell under standard conditions, labeling the anode, the cathode, the direction of electron flow, and the concentrations, as appropriate.
b. When enough NaCl(*s*) is added to the compartment containing gold to make $[Cl^-] = 0.10$ *M*, the cell potential is observed to be 0.31 V. Assume that Au^{3+} is reduced and that the reaction in the compartment containing gold is

$$Au^{3+}(aq) + 4Cl^-(aq) \rightleftharpoons AuCl_4{}^-(aq)$$

Calculate the value of *K* for this reaction at 25°C.

Challenge Problems

93. Three electrochemical cells were connected in series so that the same quantity of electrical current passes through all three cells. In the first cell, 1.15 g of chromium metal was deposited from a chromium(III) nitrate solution. In the second cell, 3.15 g of osmium was deposited from a solution made of Os^{n+} and nitrate ions. What is the name of the salt? In the third cell, the electrical charge passed through a solution containing X^{2+} ions caused deposition of 2.11 g of metallic X. Identify X.

94. An electrochemical cell is set up using the following unbalanced reaction:

$$M^{a+}(aq) + N(s) \longrightarrow N^{2+}(aq) + M(s)$$

The standard reduction potentials are

$$M^{a+} + ae^- \longrightarrow M \qquad \mathscr{E}° = +0.400 \text{ V}$$

$$N^{2+} + 2e^- \longrightarrow N \qquad \mathscr{E}° = +0.240 \text{ V}$$

The cell contains 0.10 M N^{2+} and produces a voltage of 0.180 V. If the concentration of M^{a+} is such that the value of the reaction quotient Q is 9.32×10^{-3}, calculate $[M^{a+}]$. Calculate w_{max} for this electrochemical cell.

95. A zinc–copper battery is constructed as follows:

$$Zn|Zn^{2+}(0.10\ M)||Cu^{2+}(2.50\ M)|Cu$$

The mass of each electrode is 200. g.
a. Calculate the cell potential when this battery is first connected.
b. Calculate the cell potential after 10.0 A of current has flowed for 10.0 h. (Assume each half-cell contains 1.00 L of solution.)
c. Calculate the mass of each electrode after 10.0 h.
d. How long can this battery deliver a current of 10.0 A before it goes dead?

96. The measurement of F^- ion concentration by ion-selective electrodes at 25.00°C obeys the equation

$$\mathscr{E}_{meas} = \mathscr{E}_{ref} - 0.05916 \log[F^-]$$

a. For a given solution, $\mathscr{E}_{meas}$ is 0.4462 V. If $\mathscr{E}_{ref}$ is 0.2420 V, what is the concentration of F^- in the solution?
b. Hydroxide ion interferes with the measurement of F^-. Therefore, the response of a fluoride electrode is

$$\mathscr{E}_{meas} = \mathscr{E}_{ref} - 0.05916 \log([F^-] + k[OH^-])$$

where $k = 1.00 \times 10^1$ and is called the selectivity factor for the electrode response. Calculate $[F^-]$ for the data in part a if the pH is 9.00. What is the percent error introduced in the $[F^-]$ if the hydroxide interference is ignored?
c. For the $[F^-]$ in part b, what is the maximum pH such that $[F^-]/k[OH^-] = 50.$?
d. At low pH, F^- is mostly converted to HF. The fluoride electrode does not respond to HF. What is the minimum pH at which 99% of the fluoride is present as F^- and only 1% is present as HF?
e. Buffering agents are added to solutions containing fluoride before making measurements with a fluoride-selective electrode. Why?

97. When copper reacts with nitric acid, a mixture of $NO(g)$ and $NO_2(g)$ is evolved. The volume ratio of the two product gases depends on the concentration of the nitric acid according to the equilibrium

$$2H^+(aq) + 2NO_3^-(aq) + NO(g) \rightleftharpoons 3NO_2(g) + H_2O(l)$$

Consider the following standard reduction potentials at 25°C:

$$3e^- + 4H^+(aq) + NO_3^-(aq) \longrightarrow NO(g) + 2H_2O(l) \qquad \mathscr{E}° = 0.957 \text{ V}$$

$$e^- + 2H^+(aq) + NO_3^-(aq) \longrightarrow NO_2(g) + H_2O(l) \qquad \mathscr{E}° = 0.775 \text{ V}$$

a. Calculate the equilibrium constant for this reaction.
b. What concentration of nitric acid will produce an NO and NO_2 mixture with only 0.20% NO_2 (by moles) at 25°C and 1.00 atm? Assume that no other gases are present and that the change in acid concentration can be neglected.

98. You make a galvanic cell with a piece of nickel, 1.0 M $Ni^{2+}(aq)$, a piece of silver, and 1.0 M $Ag^+(aq)$. Calculate the concentrations of $Ag^+(aq)$ and $Ni^{2+}(aq)$ once the cell is "dead."

99. A galvanic cell is based on the following half-reactions:

$$Fe^{2+} + 2e^- \longrightarrow Fe(s) \qquad \mathscr{E}° = -0.440 \text{ V}$$

$$2H^+ + 2e^- \longrightarrow H_2(g) \qquad \mathscr{E}° = 0.000 \text{ V}$$

In this cell the iron compartment contains an iron electrode and $[Fe^{2+}] = 1.00 \times 10^{-3}$ M, and the hydrogen compartment contains a platinum electrode, $P_{H_2} = 1.00$ atm and a weak acid HA at an initial concentration of 1.00 M. If the observed cell potential is 0.333 V at 25°C, calculate the K_a value for the weak acid HA at 25°C.

100. You have a concentration cell with Cu electrodes and $[Cu^{2+}] = 1.00\ M$ (right side) and $1.0 \times 10^{-4}\ M$ (left side).
a. Calculate the potential for this cell at 25°C.
b. The Cu^{2+} ion reacts with NH_3 to form $Cu(NH_3)_4^{2+}$, where the stepwise formation constants are $K_1 = 1.0 \times 10^3$, $K_2 = 1.0 \times 10^4$, $K_3 = 1.0 \times 10^3$, and $K_4 = 1.0 \times 10^3$. Calculate the new cell potential after enough NH_3 is added to the left cell compartment such that at equilibrium $[NH_3] = 2.0\ M$.

101. A galvanic cell is based on the following half-reactions:

$$Ag^+ + e^- \longrightarrow Ag(s) \qquad \mathscr{E}° = 0.80 \text{ V}$$

$$Cu^{2+} + 2e^- \longrightarrow Cu(s) \qquad \mathscr{E}° = 0.34 \text{ V}$$

In this cell the silver compartment contains a silver electrode and excess AgCl(*s*) ($K_{sp} = 1.6 \times 10^{-10}$), and the copper compartment contains a copper electrode and $[Cu^{2+}] = 2.0\ M$.

a. Calculate the potential for this cell at 25°C.
b. Assuming 1.0 L of 2.0 *M* Cu^{2+} in the copper compartment, calculate the moles of NH_3 that would have to be added to give a cell potential of 0.52 V at 25°C (assume no volume change on addition of NH_3).

$$Cu^{2+}(aq) + 4NH_3(aq) \rightleftharpoons Cu(NH_3)_4{}^{2+}(aq) \qquad K = 1.0 \times 10^{13}$$

102. Consider the following galvanic cell:

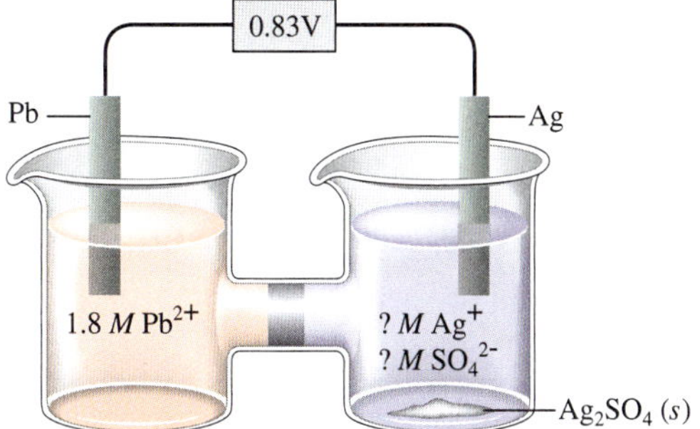

Calculate the K_{sp} value for $Ag_2SO_4(s)$. Note that to obtain silver ions in the right compartment (the cathode compartment), excess solid Ag_2SO_4 was added and some of the salt dissolved.

103. Consider the following galvanic cell:

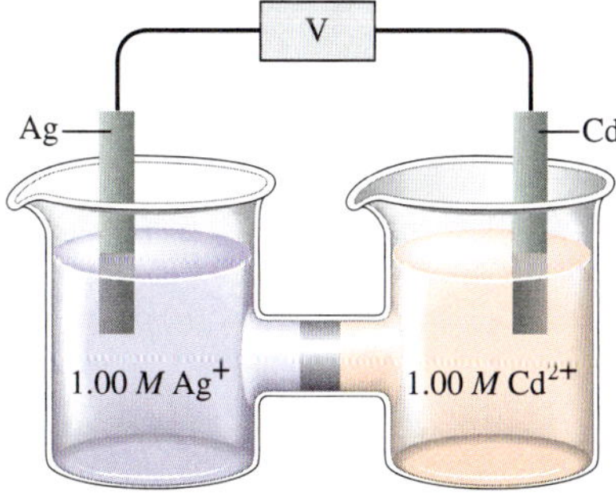

A 15.0-mol sample of NH_3 is added to the Ag compartment (assume 1.00 L of total solution after the addition). The silver ion reacts with ammonia to form complex ions as shown:

$$Ag^+(aq) + NH_3(aq) \rightleftharpoons AgNH_3{}^+(aq) \qquad K_1 = 2.1 \times 10^3$$

$$AgNH_3{}^+(aq) + NH_3(aq) \rightleftharpoons Ag(NH_3)_2{}^+(aq) \qquad K_2 = 8.2 \times 10^3$$

Calculate the cell potential after the addition of 15.0 mol NH_3.

104. Given the following two standard reduction potentials,

$$M^{3+} + 3e^- \longrightarrow M \qquad \mathscr{E}° = -0.10\ V$$

$$M^{2+} + 2e^- \longrightarrow M \qquad \mathscr{E}° = -0.50\ V$$

solve for the standard reduction potential of the half-reaction

$$M^{3+} + e^- \longrightarrow M^{2+}$$

(*Hint:* You must use the extensive property $\Delta G°$ to determine the standard reduction potential.)

105. Zirconium is one of the few metals that retains its structural integrity upon exposure to radiation. For this reason, the fuel rods in most nuclear reactors are made of zirconium. Answer the following questions about the redox properties of zirconium based on the half-reaction

$$ZrO_2 \cdot H_2O + H_2O + 4e^- \longrightarrow Zr + 4OH^- \qquad \mathscr{E}° = -2.36\ V$$

a. Is zirconium metal capable of reducing water to form hydrogen gas at standard conditions?
b. Write a balanced equation for the reduction of water by zirconium metal.
c. Calculate $\mathscr{E}°$, $\Delta G°$, and *K* for the reduction of water by zirconium metal.
d. The reduction of water by zirconium occurred during the accident at Three Mile Island, Pennsylvania, in 1979. The hydrogen produced was successfully vented and no chemical explosion occurred. If 1.00×10^3 kg of Zr reacts, what mass of H_2 is produced? What volume of H_2 at 1.0 atm and 1000.°C is produced?
e. At Chernobyl, USSR, in 1986, hydrogen was produced by the reaction of superheated steam with the graphite reactor core:

$$C(s) + H_2O(g) \longrightarrow CO(g) + H_2(g)$$

A chemical explosion involving the hydrogen gas did occur at Chernobyl. In light of this fact, do you think it was a correct decision to vent the hydrogen and other radioactive gases into the atmosphere at Three Mile Island? Explain.

Marathon Problems

106. A galvanic cell is based on the following half-reactions:

$$Cu^{2+} + 2e^- \longrightarrow Cu(s) \qquad \mathscr{E}° = 0.34\ V$$

$$V^{2+} + 2e^- \longrightarrow V(s) \qquad \mathscr{E}° = -1.20\ V$$

In this cell the copper compartment contains a copper electrode and $[Cu^{2+}] = 1.00\ M$, and the vanadium compartment contains a vanadium electrode and V^{2+} at an unknown concentration. The compartment containing the vanadium (1.00 L of solution) was titrated with 0.0800 *M* H_2EDTA^{2-}, resulting in the reaction

$$H_2EDTA^{2-}(aq) + V^{2+}(aq) \rightleftharpoons VEDTA^{2-}(aq) + 2H^+(aq) \qquad K = ?$$

The potential of the cell was monitored to determine the stoichiometric point for the process, which occurred at a volume of 500.0 mL of H_2EDTA^{2-} solution added. At the stoichiometric point, $\mathscr{E}_{cell}$ was observed to be 1.98 V. The solution was buffered at a pH of 10.00.

a. Calculate $\mathscr{E}_{cell}$ before the titration was carried out.

b. Calculate the value of the equilibrium constant *K* for the titration reaction.

c. Calculate $\mathscr{E}_{cell}$ at the halfway point in the titration.

107. The table below lists the cell potentials for the 10 possible galvanic cells assembled from the metals A, B, C, D, and E and their respective 1.00 *M* 2+ ions in solution. Using the data in the table, establish a standard reduction potential table similar to Table 11.1 in the text. Assign a reduction potential of 0.00 V to the half-reaction that falls in the middle of the series. You should get two different tables. Explain why, and discuss what you could do to determine which table is correct.

	A(*s*) in $A^{2+}(aq)$	B(*s*) in $B^{2+}(aq)$	C(*s*) in $C^{2+}(aq)$	D(*s*) in $D^{2+}(aq)$
E(*s*) in $E^{2+}(aq)$	0.28 V	0.81 V	0.13 V	1.00 V
D(*s*) in $D^{2+}(aq)$	0.72 V	0.19 V	1.13 V	—
C(*s*) in $C^{2+}(aq)$	0.41 V	0.94 V	—	—
B(*s*) in $B^{2+}(aq)$	0.53 V	—	—	—

Media Summary

Visit the Student Website at **www.cengage.com/chemistry/zumdahl** to help prepare for class, study for quizzes and exams, understand core concepts, and visualize molecular-level interactions. The following media activities are available for this chapter:

Prepare for Class

Video Lessons *Mini-lectures from chemistry experts*

- Reviewing Oxidation–Reduction Reactions
- Electrochemical Cells
- Electromotive Force
- The Activity Series of the Elements
- Standard Reduction Potentials
- Using Standard Reduction Potentials
- The Nernst Equation
- Electrochemical Determinants of Equilibria
- Batteries
- CIA Demonstration: The Fruit-Powered Clock
- Corrosion and the Prevention of Corrosion
- The Stoichiometry of Electrolysis
- Electrolytic Cells

Visualizations *Molecular-level animations and lab demonstration videos*

- Electrochemical Half-reactions in a Galvanic Cell
- Electrolysis of Water
- Galvanic (Voltaic) Cells
- Voltaic Cell: Anode Reaction
- Voltaic Cell: Cathode Reaction
- Zinc/Copper Cells (Lemon Battery)

Flashcards *Key terms and definitions*

Online flashcards

ACE the Test

Multiple-choice quizzes
3 ACE Practice Tests

Improve Your Grade

Tutorials *Animated examples and interactive activities*

Galvanic Cells

Access these resources using your passkey, available free with new texts or for purchase separately.

12 Quantum Mechanics and Atomic Theory

White light passing through two prisms.

In the past 200 years a great deal of experimental evidence has accumulated to support the atomic model. This theory has proved to be both extremely useful and physically reasonable. When atoms were first suggested by the Greek philosophers Democritus and Leucippus about 400 B.C., the concept was based mostly on intuition. In fact, for the following 20 centuries, no convincing experimental evidence was available to support the existence of atoms. The first real scientific data were gathered by Lavoisier and others from quantitative measurements of chemical reactions. The results of these stoichiometric experiments led John Dalton to propose the first systematic atomic theory. Dalton's theory, although crude, has stood the test of time extremely well.

Once we came to "believe in" atoms, it was logical to ask: What is the nature of an atom? Does an atom have parts, and if so, what are they? In Chapter 2 we considered some of the experiments most important for shedding light on the nature of the atom. Now we will see how the atomic theory has evolved to its present state.

One of the most striking things about the chemistry of the elements is the periodic repetition of properties. There are several groups of elements that show great similarities in chemical behavior. As we saw in Chapter 2, these similarities led to the development of the periodic table of the elements. In this chapter we will see that the modern theory of atomic structure accounts for periodicity in terms of the electron arrangements in atoms.

However, before we examine atomic structure, we must consider the revolution that took place in physics in the first 30 years of the twentieth century. During that time, experiments were carried out, the results of which could not be explained by the theories of classical physics developed by Isaac Newton and many others who followed him. A radical new theory called quantum mechanics was developed to account for the behavior of light and atoms. This "new physics" provides many surprises for people who are used to the macroscopic world, but it seems to account flawlessly (within the bounds of necessary approximations) for the behavior of matter.

As the first step in our exploration of this revolution in science, we will consider the properties of light, more properly called electromagnetic radiation.

12.1 Electromagnetic Radiation

One of the ways that energy travels through space is by **electromagnetic radiation.** The light from the sun, the energy used to cook food in a microwave oven, the X rays used by dentists, and the radiowaves used by physicians to make MRI maps of body tissues are all examples of electromagnetic radiation. Although these forms of radiant energy seem quite different, they all exhibit the same type of wavelike behavior and travel at the speed of light in a vacuum. Electromagnetic radiation is so-named because it has electrical and magnetic fields that simultaneously oscillate in planes mutually perpendicular to each other and to the direction of propagation through space (see Fig. 12.1).

Waves are characterized by wavelength, frequency, and speed. As shown in Fig. 12.2, **wavelength** (symbolized by the Greek letter lambda, λ) is the *distance between two consecutive peaks or troughs in a wave.* The **frequency** (symbolized by the Greek letter nu, ν) is defined as the *number of waves (cycles) per second that pass a given point in space.* Since all types of electromagnetic

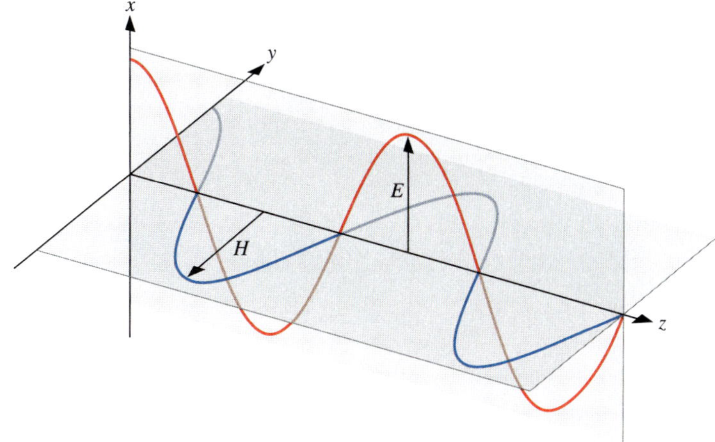

FIGURE 12.1

Electromagnetic radiation has oscillating electric (E) and magnetic (H) fields in planes perpendicular to each other and to the direction of propagation.

radiation travel at the speed of light, short-wavelength radiation must have a high frequency. You can see this in Fig. 12.2, where three waves are shown traveling between two points at constant speed. Note that the wave with the shortest wavelength (λ_3) has the highest frequency, and the wave with the longest wavelength (λ_1) has the lowest frequency. This implies an inverse relationship between wavelength and frequency; that is, $\lambda \propto 1/\nu$, or

$$\lambda\nu = c$$

Wavelength (λ) and frequency (ν) are inversely related.

In this equation λ is the wavelength in meters, ν is the frequency in cycles per second, and c is the speed of light, a defined quantity with the exact value of 2.99792458×10^8 m/s. In the SI system, *cycles* is understood, and the unit cycles per second becomes 1/s, or s^{-1}, which is called the *hertz* (abbreviated Hz).

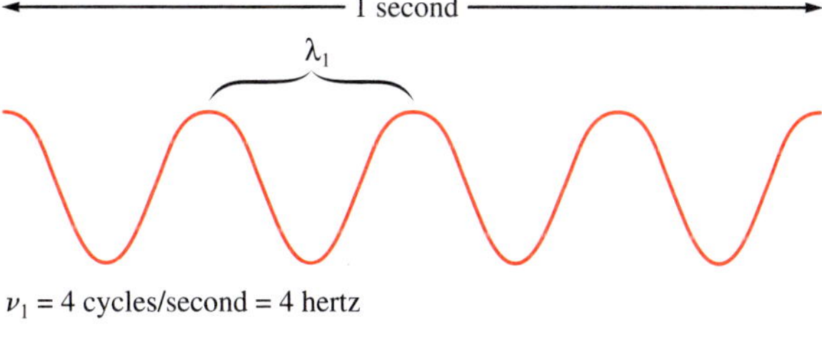

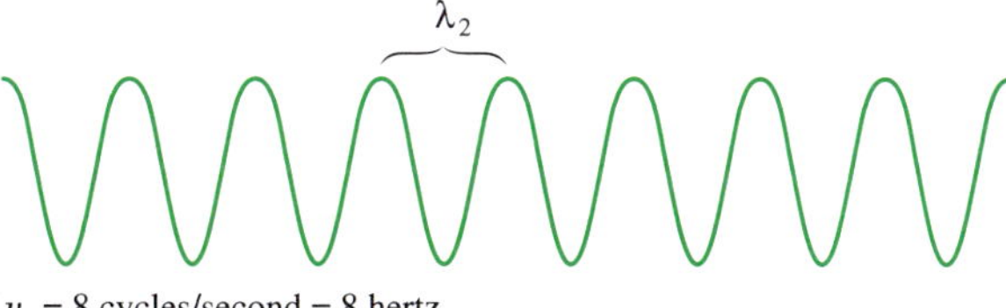

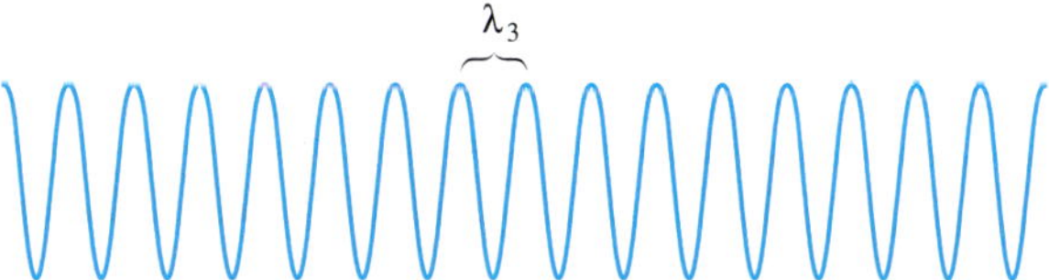

FIGURE 12.2

The nature of waves. Note that the radiation with the shortest wavelength has the highest frequency. This diagram can represent the oscillating electric or magnetic field of the wave.

New-Wave Sunscreens

Skin cancer is an increasingly important problem. Recent studies indicate that basal cell and squamous cell carcinomas have more than tripled in the last 30 years due in greatest part to overexposure to ultraviolet (UV) light. In addition to causing cancer, overexposure to UV radiation also accelerates the "aging" processes in the skin because it damages proteins, elastin, and DNA in the upper and medium layers of the skin. Ultraviolet radiation (UVR) is classified as UVB (wavelengths from 280–320 nm) and UVA (wavelengths from 320–400 nm). Studies show that UVA accounts for over 80% of the detrimental effects of exposure to UVR, including DNA damage and ultimately skin cancer. UVA rays, which penetrate deeper into the skin, can pass through clouds, windows, and even clothing.

Although it's impossible to eliminate UVR exposure to the skin, it can be minimized by wearing protective clothing and by applying sunblocks and sunscreens. Sunblocks, which are physical agents that act as barriers to UVR, often contain zinc oxide (ZnO), titanium(IV) oxide (TiO_2), or both, which are highly reflective white powders. Traditional ZnO sunblocks worn by lifeguards are opaque white, but by reducing the particle sizes of the white oxide to about 100 nm, they can be made invisible on the skin while retaining their UV-blocking ability. Traditional chemical sunscreens, most of which are based on *p*-aminobenzoic acid, act primarily by binding to skin proteins and absorbing UVB photons. Recently, the Food and Drug Administration (FDA) has approved a sunscreen formula for use in the United States that has long been used abroad. The facial moisturizer Anthelios SX, marketed by L'Oreal, contains ecamsule (trade name Mexoryl SX), a sunscreen ingredient that is particularly effective at blocking short UVA radiation (320–340 nm). The product also contains other ingredients that block longer UVA rays and UVB rays to make it a broad-spectrum sunscreen. Although there are already some sunscreens, such as avobenzene, that offer UVA protection, the advantage of ecamsule is that it does not decompose on exposure to sunlight. A major disadvantage is that ecamsule is water soluble and thus is not useful at the beach.

SO_3H SO_3H O O C H C H

Structure of ecamsule

Electromagnetic radiation is classified as shown in Fig. 12.3. Radiation provides an important means of energy transfer. For example, the energy from the sun reaches the earth mainly in the forms of visible and ultraviolet radiation, and the glowing coals of a fireplace transmit heat energy by infrared

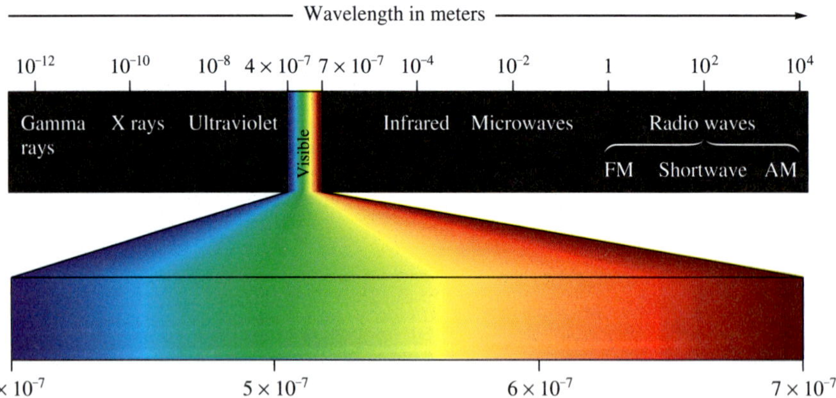

FIGURE 12.3

Classification of electromagnetic radiation.

radiation. In a microwave oven the water molecules in food absorb microwave radiation, which increases their motions. This energy is then transferred to other types of molecules via collisions, causing an increase in the food's temperature. As we proceed in the study of chemistry, we will consider many of the classes of electromagnetic radiation and the ways in which they affect matter.

12.2 The Nature of Matter

It is probably fair to say that at the end of the nineteenth century physicists were feeling rather smug. Available theories could explain phenomena ranging from the motions of the planets to the dispersion of visible light by a prism. Rumor has it that students were being discouraged from pursuing physics as a career because it was believed that all the major problems had been solved or at least described in terms of the current physical theories.

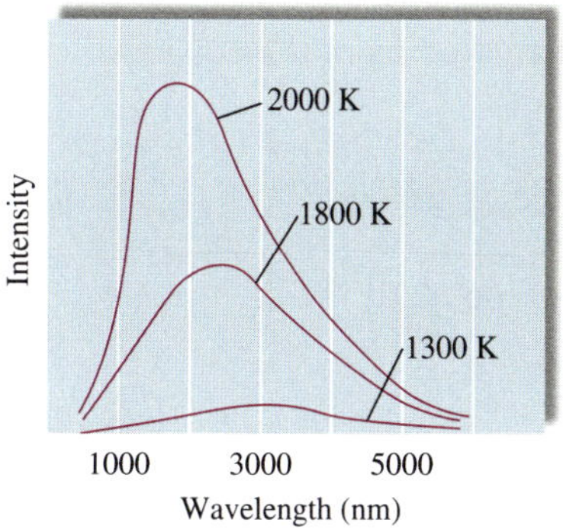

At the end of the nineteenth century the idea prevailed that matter and energy were distinct. Matter was thought to consist of particles, whereas energy in the form of light (electromagnetic radiation) was described as a wave. Particles were things that had mass and whose positions in space could be specified. Waves were described as massless and delocalized; that is, their positions in space could not be specified. It was also assumed that there was no intermingling of matter and light. Everything known before 1900 seemed to fit neatly into this view.

FIGURE 12.4

The profile of radiation emitted from a blackbody. Note that the maximum shifts to shorter wavelengths as the temperature is increased, in agreement with the observed change from a reddish to a white glow as iron is heated to higher temperatures.

At the beginning of the twentieth century, however, certain experimental results suggested that this picture was incorrect. The first important advance came in 1901 from German physicist Max Planck, who studied the profile (intensity versus wavelength) of the electromagnetic radiation emitted from a solid body heated to incandescence. (An example is a piece of iron that glows red and then white as it is heated to higher and higher temperatures.) The profiles shown in Fig. 12.4 are for so-called blackbody radiation. This may seem like a contradiction in terms, since a blackbody is an idealized object that absorbs all the radiation incident on it. The term is used in this context to mean radiation that originates from the thermal energy of the body only. It does not include radiation reflected from the object and does not depend on the material composing the object. Blackbody radiation is closely approximated by the radiation emitted through a tiny hole from a cavity inside an object. The main point to be made here is that the radiation profiles shown in Fig. 12.4 are not the ones expected from classical physics. The classical theory of matter, which assumes that matter can absorb or emit any quantity of energy, predicts a radiation profile that has no maximum and goes to infinite intensity at very short wavelengths (an effect often called the **ultraviolet catastrophe**).

Energy can be gained or lost only in integer multiples of $h\nu$.

Planck's constant = 6.626×10^{-34} J s.

Planck found that the observed profiles (with their intensity maxima) could be accounted for by postulating that energy can be gained or lost only in *whole-number multiples* of the quantity $h\nu$, where h is a constant now called **Planck's constant,** determined by experiment to have the value 6.626×10^{-34} J s. That is, the change in energy for a system ΔE can be represented by the equation

$$\Delta E = nh\nu$$

where n is an integer (1, 2, 3, . . .), h is Planck's constant, and ν is the frequency of the electromagnetic radiation absorbed or emitted.

CuCl produces a blue flame when heated in a burner.

Planck's result was a real surprise. Physicists had always assumed that the energy of matter was continuous, which meant that the transfer of any quantity of energy was possible. Now it seemed clear that energy is in fact **quantized** and can be transferred only in discrete units of size $h\nu$. Each of these small "packets" of energy is called a *quantum*. A system can transfer energy only in whole *quanta*. Thus energy seems to have particulate properties.

Example 12.1

The blue color in fireworks is often achieved by heating copper(I) chloride (CuCl) to about 1200°C. The hot compound emits blue light having a wavelength of 450 nm. What is the increment of energy (the quantum) that is emitted at 4.50×10^2 nm by CuCl?

Solution The quantum of energy can be calculated from the equation

$$\Delta E = h\nu$$

The frequency ν for this case can be calculated as follows:

$$\nu = \frac{c}{\lambda} = \frac{2.9979 \times 10^8 \text{ m/s}}{4.50 \times 10^{-7} \text{ m}} = 6.66 \times 10^{14} \text{ s}^{-1}$$

So

$$\Delta E = h\nu = (6.626 \times 10^{-34} \text{ J s})(6.66 \times 10^{14} \text{ s}^{-1}) = 4.41 \times 10^{-19} \text{ J}$$

A sample of CuCl emitting light at 450 nm can lose energy only in increments of 4.41×10^{-19} J, the size of the quantum in this case.

G. N. Lewis (see Section 13.10) actually coined the term "photon" in 1926.

The next important development in the knowledge of atomic structure came when Albert Einstein (see Fig. 12.5) proposed that electromagnetic radiation is itself quantized. Einstein suggested that electromagnetic radiation can be viewed as a stream of "particles" now called **photons.** The energy of each photon is given by the expression

$$E_{\text{photon}} = h\nu = \frac{hc}{\lambda}$$

where h is Planck's constant, ν is the frequency of the radiation, and λ is the wavelength of the radiation.

Einstein arrived at this conclusion through his analysis of the **photoelectric effect** (for which he later was awarded the Nobel Prize). The photoelectric effect refers to the phenomenon in which electrons are emitted from the surface of a metal when light strikes it. The following observations characterize the photoelectric effect.

1. Studies in which the frequency of the light is varied show that no electrons are emitted by a given metal below a specific threshold frequency ν_0.
2. For light with frequency lower than the threshold frequency, no electrons are emitted regardless of the intensity of the light.
3. For light with frequency greater than the threshold frequency, the number of electrons emitted increases with the intensity of the light.

FIGURE 12.5

Albert Einstein (1879–1955) was born in Germany. Nothing in his early development suggested genius; even at the age of 9 he did not speak clearly, and his parents feared that he might have a handicap. When asked what profession Einstein should follow, his school principal replied, "It doesn't matter; he'll never make a success of anything." When he was 10, Einstein entered the Luitpold Gymnasium (high school), which was typical of German schools of that time in being harshly disciplinarian. There he developed a deep suspicion of authority and a skepticism that encouraged him to question and doubt—valuable qualities in a scientist. In 1905, while a patent clerk in Switzerland, Einstein published a paper explaining the photoelectric effect via the quantum theory. For this revolutionary thinking he received a Nobel Prize in 1921. Highly regarded by this time, he worked in Germany until 1933, when Hitler's persecution of Jews forced him to come to the United States. He worked at the Institute for Advanced Study at Princeton University until his death in 1955.

Einstein was undoubtedly the greatest physicist of our age. Even if someone else had derived the theory of relativity, his other work would have ensured his ranking as the second greatest physicist of his time. Our concepts of space and time were radically changed by ideas he first proposed when he was 26 years old. From then until the end of his life, he attempted unsuccessfully to find a single unifying theory that would explain all physical events.

4. For light with frequency greater than the threshold frequency, the kinetic energy of the emitted electrons increases linearly with the frequency of the light.

These observations can be explained by assuming that electromagnetic radiation is quantized (consists of photons) and that the threshold frequency represents the minimum energy required to remove the electron from the metal's surface.

$$\text{Minimum energy required to remove an electron} = E_0 = h\nu_0$$

Because a photon with energy less than E_0 ($\nu < \nu_0$) cannot remove an electron, light with a frequency less than the threshold frequency produces no electrons. On the other hand, for light where $\nu > \nu_0$, the energy in excess of that required to remove the electron is given to the electron as kinetic energy (KE):

$$\text{KE}_{\text{electron}} = \frac{1}{2}mv^2 = h\nu - h\nu_0$$

- m: Mass of electron
- v: Velocity of electron
- $h\nu$: Energy of incident photon
- $h\nu_0$: Energy required to remove electron from metal's surface

Because in this picture the intensity of light is a measure of the number of photons present in a given part of the beam, a greater intensity means that more photons are available to release electrons (as long as $\nu > \nu_0$ for the radiation).

At about the same time that Einstein was performing his analysis of the photoelectric effect, he was also constructing the theory of special relativity. In connection with this work Einstein derived the famous equation

$$E = mc^2$$

which he published in 1905. This equation points out the close relationship between energy and mass. In fact, we have learned that mass is a form of

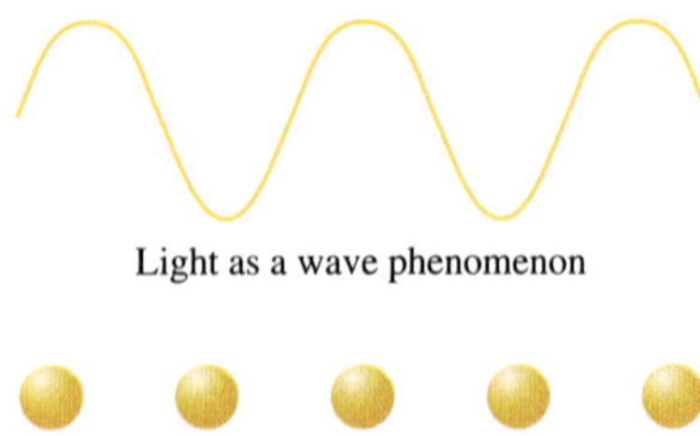

FIGURE 12.6

Electromagnetic radiation exhibits wave properties and particulate properties. The energy of each photon of the radiation is related to the wavelength and frequency by the equation $E_{\text{photon}} = h\nu = hc/\lambda$.

energy. When a system loses energy, it also loses mass. This result is more apparent if we rearrange the equation to the following form:

$$m = \frac{E}{c^2}$$

(m ← Mass; E ← Energy; c ← Speed of light)

Using this form of the equation, we can calculate the mass associated with a given quantity of energy.

If a beam of light can be considered to be a stream of "particles" (photons), do the photons have mass? The answer to this question is "no." Photons do not exhibit mass in the same way as classical particles do. Einstein's equations, however, predict that a photon has momentum, which is best thought of as an intrinsic property of the photon that does not depend separately on mass and velocity, unlike the case for a classical particle. In 1922 American physicist Arthur Compton performed experiments involving collisions of X rays with electrons. These experiments showed that photons do exhibit the momentum calculated from Einstein's equation. Also, photons do seem to be affected by gravity, as Einstein postulated in his general theory of relativity. However, it is important to recognize that the photon is in no sense a typical particle. A photon has mass only in a relativistic sense—it has no rest mass.

We can summarize the important conclusions from the work of Planck and Einstein as follows:

Energy is quantized. It can be transferred only in discrete units called quanta.

Electromagnetic radiation, which was previously thought to exhibit only wave properties, seems to show certain characteristics of particulate matter as well. This phenomenon, illustrated in Fig. 12.6, is sometimes referred to as the **dual nature of light.**

Thus light, which was previously thought to be purely wavelike, was found to have certain characteristics of particulate matter. But is the opposite also true? That is, does matter that is normally assumed to be particulate exhibit wave properties? This question was raised in 1923 by a young French physicist named Louis de Broglie (1892–1987), who derived the following relationship for the wavelength of a particle with momentum mv:

Do not confuse ν (frequency) with v (velocity).

$$\lambda = \frac{h}{mv}$$

This equation, called de Broglie's equation, allows us to calculate the wavelength for a particle, as shown in Example 12.2.

Example 12.2

Compare the wavelength for an electron (mass = 9.11×10^{-31} kg) traveling at a speed of 1.0×10^7 m/s with that for a ball (mass = 0.10 kg) traveling at 35 m/s.

Solution We use the equation $\lambda = h/mv$, where

$$h = 6.626 \times 10^{-34} \text{ J s} \quad \text{or} \quad 6.626 \times 10^{-34} \text{ kg m}^2\text{/s}$$

since

$$1\ \text{J} = 1\ \text{kg m}^2/\text{s}^2$$

For the electron,

$$\lambda_e = \frac{6.626 \times 10^{-34}\ \dfrac{\text{kg m}^2}{\text{s}}}{(9.11 \times 10^{-31}\ \text{kg})(1.0 \times 10^{7}\ \text{m/s})} = 7.3 \times 10^{-11}\ \text{m}$$

For the ball,

$$\lambda_b = \frac{6.626 \times 10^{-34}\ \dfrac{\text{kg m}^2}{\text{s}}}{(0.10\ \text{kg})(35\ \text{m/s})} = 1.9 \times 10^{-34}\ \text{m}$$

Spectrum obtained by shining a beam of light at an angle to a diffraction grating.

Notice from Example 12.2 that the wavelength associated with the ball is incredibly short. On the other hand, the wavelength of the electron, although quite small, happens to be of the same order as the spacing between the atoms in a typical crystal. This is important because, as we will see presently, it provides a means for testing de Broglie's equation.

Diffraction results when light is scattered from a regular array of points or lines. The diffraction of light from a diffraction grating is shown in the accompanying photograph. The colors result because the various wavelengths of visible light are not all scattered in the same way. The colors are "separated," giving the same effect as light passing through a prism. Just as a regular arrangement of ridges and grooves produces diffraction, so does a regular array of atoms or ions in a crystal. For example, when X rays are directed onto a crystal of sodium chloride with its regular array of Na^+ and Cl^- ions, the scattered radiation produces a **diffraction pattern** of bright areas and dark spots on a photographic plate, as shown in Fig. 12.7(a). This pattern occurs because the scattered light can interfere constructively (the peaks and troughs of the beams are in phase) to produce a bright area [Fig. 12.7(b)] or destructively (the peaks and troughs are out of phase) to produce a dark spot [Fig. 12.7(c)].

A diffraction pattern can be explained only in terms of waves. Thus this phenomenon provides a test for the postulate that particles such as electrons have wave properties. As we saw in Example 12.2, an electron with a velocity of 10^7 m/s (easily achieved by acceleration of the electron in an electric field) has a wavelength of about 10^{-10} m, roughly the distance between the components in a typical crystal. This is important because diffraction occurs most efficiently when the spacing between the scattering points is about the same as the wavelength. Thus, if electrons actually do have an associated wavelength, a crystal should diffract electrons. An experiment to test this idea was

FIGURE 12.7

(a) Diffraction occurs when electromagnetic radiation is scattered from a regular array of objects, such as the ions in a crystal of sodium chloride. The large spot in the center is from the main incident beam of X rays. (b) Bright areas in the diffraction pattern result from *constructive interference* of waves. The waves are in phase; that is, their peaks match. (c) Dark spots result from *destructive interference* of waves. The waves are out of phase; the peaks of one wave coincide with the troughs of another wave.

(a) Diffraction

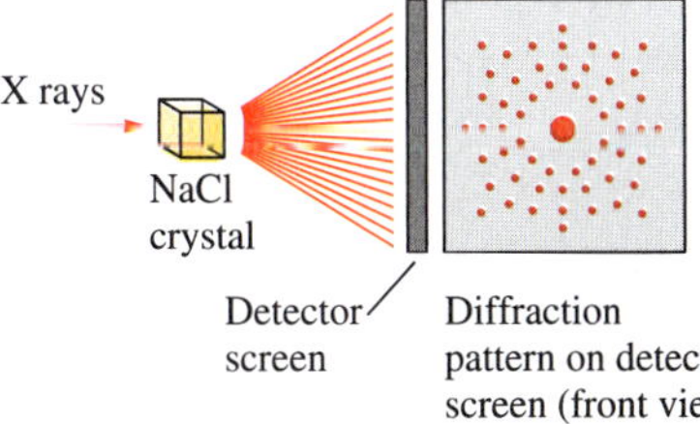

(b) Constructive interference

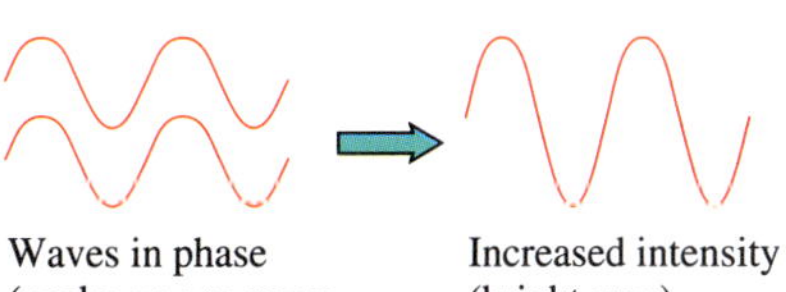

(c) Destructive interference

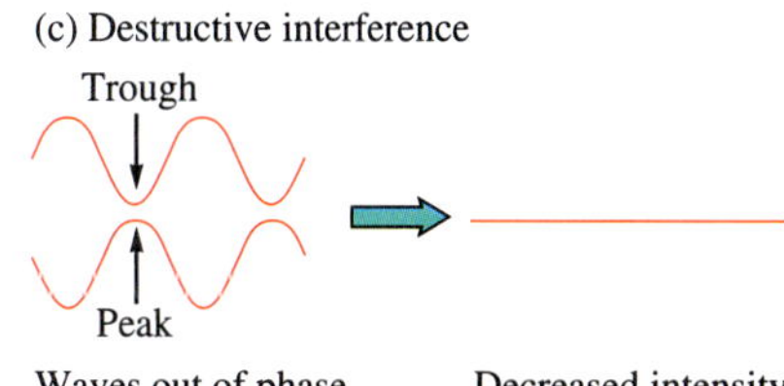

carried out in 1927 by Davisson and Germer at Bell Laboratories. When they directed a beam of electrons at a nickel crystal, they observed a diffraction pattern similar to that seen from the diffraction of X rays. This result verified de Broglie's relationship, at least for electrons. Larger chunks of matter, such as balls, have wavelengths (Example 12.2) too small to verify experimentally. However, we believe that all matter obeys de Broglie's equation.

Now we have come full circle. Electromagnetic radiation, which at the turn of the twentieth century was thought to be a pure waveform, was found to exhibit particulate properties. Conversely, electrons, which were thought to be particles, were found to have a wavelength associated with them. The significance of these results is that matter and energy are not distinct. Energy is really a form of matter, and all matter shows the same types of properties. That is, *all matter exhibits both particulate and wave properties.* Large "pieces" of matter, such as baseballs, exhibit predominantly particulate properties. The associated wavelength is so small that it is not observed. Very small "pieces" of matter, such as photons, while showing some particulate properties through relativistic effects, exhibit predominantly wave properties. Pieces of matter with intermediate mass, such as electrons, show both the particulate and wave properties of matter.

12.3 The Atomic Spectrum of Hydrogen

Recall from Chapter 2 that key information about the structure of the atom came from several experiments carried out in the early twentieth century. Particularly important were Thomson's discovery of the electron and Rutherford's discovery of the nucleus. Another important experiment concerned the study of the emission of light by excited hydrogen atoms. When a high-energy discharge is passed through a sample of hydrogen gas, the H_2 molecules absorb energy, causing some of the H—H bonds to break. The resulting hydrogen atoms are *excited;* that is, they contain excess energy, which they release by emitting light of various wavelengths to produce what is called the *emission spectrum* of the hydrogen atom.

To understand the significance of the hydrogen emission spectrum, we must first describe the **continuous spectrum** that results when white light is passed through a prism, as shown in Fig. 12.8(a). This spectrum, like the rainbow produced when sunlight is dispersed by raindrops, contains *all* the wavelengths of visible light. In contrast, when the hydrogen emission spectrum in the visible region is passed through a prism, as shown in Fig. 12.8(b), we see only a few lines, each corresponding to a discrete wavelength. The hydrogen emission spectrum is called a **line spectrum.**

The energy of the electron in the hydrogen atom is quantized.

What is the significance of the line spectrum of hydrogen? It indicates that *only certain energies are allowed for the electron in the hydrogen atom.* In other words, the energy of the electron in the hydrogen atom is *quantized.* Changes in energy between discrete energy levels in hydrogen produce only certain wavelengths of emitted light, as shown in Fig. 12.9. For example, a given change in energy from a high to a lower level gives a wavelength of light that can be calculated from Einstein's equation:

$$\Delta E = h\nu = \frac{hc}{\lambda}$$

↑ Change in energy; ↖ Frequency of light emitted; ← Wavelength of light emitted

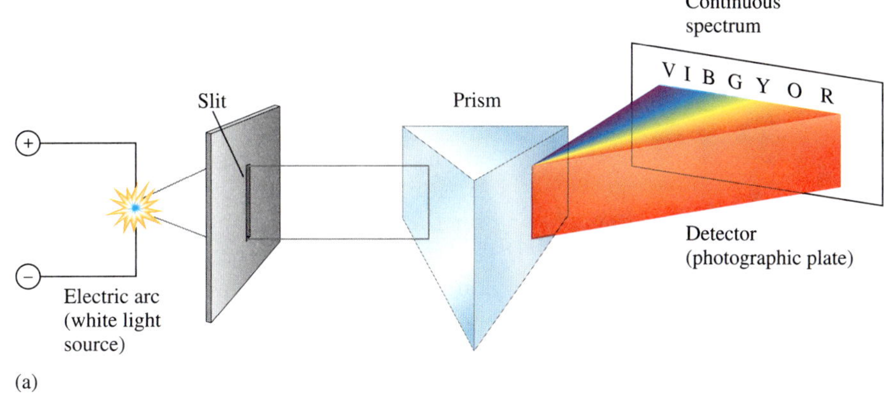

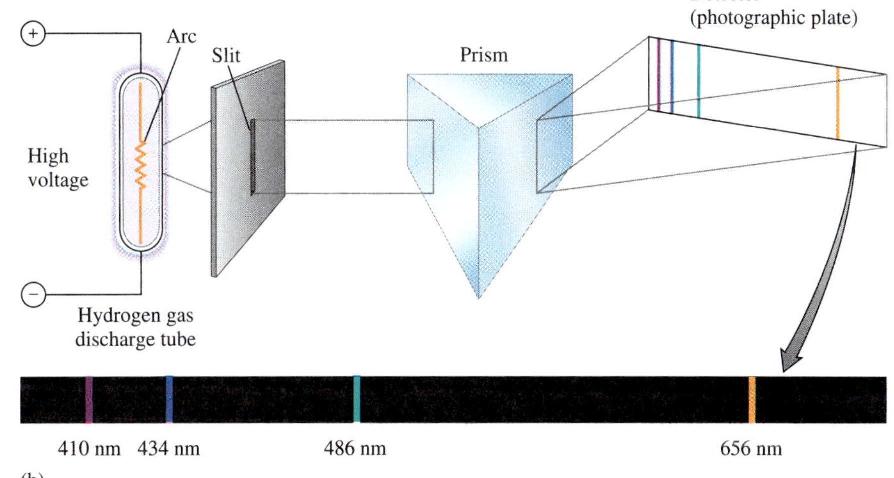

FIGURE 12.8

(a) A continuous spectrum containing all wavelengths of visible light (indicated by the first letters of the colors of the rainbow). (b) The hydrogen line spectrum contains only a few discrete wavelengths.

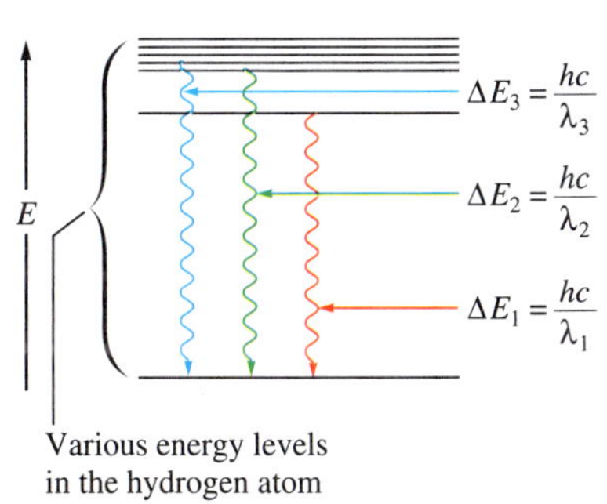

FIGURE 12.9

A change between two discrete energy levels emits a photon of light.

The discrete line spectrum of hydrogen shows that only certain energies are possible; that is, the electron energy levels are quantized. In contrast, if any energy level were allowed, the emission spectrum would be continuous.

12.4 The Bohr Model

In 1913 a Danish physicist named Niels Bohr, aware of most of the experimental results we have just discussed, developed a **quantum model** for the hydrogen atom. Bohr proposed a model that included the idea that the *electron in a hydrogen atom moves around the nucleus only in certain allowed circular orbits*. He calculated the radii for these allowed orbits by using the theories of classical physics and by making some new assumptions.

From classical physics Bohr knew that a particle in motion tends to move in a straight line and can be made to travel in a circle only by application of a force toward the center of the circle. Thus Bohr reasoned that the tendency of the revolving electron to fly off the atom must be exactly balanced by its attraction for the positively charged nucleus. But classical physics also decreed that a charged particle under acceleration should radiate energy. Since an electron revolving around the nucleus constantly changes its direction, it is

Niels Bohr at 37 years of age, photographed in 1922 when he received the Nobel Prize for physics.

constantly accelerating. Therefore, the electron should emit light and lose energy and thus be drawn into the nucleus. This conclusion, of course, does not correlate with the existence of stable atoms.

Clearly, an atomic model based solely on the theories of classical physics was untenable. Bohr dealt with the problem of the collapse of the classical atom by simply assuming that the hydrogen electron could exist only in stationary, nonradiating orbits. Ultimately, the correct model had to account for the experimental spectrum of hydrogen, which showed that only certain electron energies were allowed. The experimental data were absolutely clear on this point. Bohr's model fits the experimental results by assuming that the angular momentum of the electron could occur only in certain increments. It wasn't clear why this should be true, but with this assumption Bohr's model gave hydrogen atom energy levels consistent with the hydrogen emission spectrum. The model is represented pictorially in Fig. 12.10.

Angular momentum equals the product of mass, velocity, and orbital radius.

Although we will not discuss its origins here, the expression for the *energy levels available to the electron in the hydrogen atom is*

$$E = -2.178 \times 10^{-18}\ \text{J}\left(\frac{Z^2}{n^2}\right) \qquad (12.1)$$

where n is an integer (the larger the value of n, the larger is the orbit radius) and Z is the atomic number ($Z = 1$ for hydrogen). The negative sign in Equation (12.1) simply means that the energy of the electron bound to the nucleus is lower than it would be if the electron were at an infinite distance ($n = \infty$) from the nucleus, where there is no interaction and the energy is zero:

The "J" in Equation (12.1) stands for joules.

$$E = -2.178 \times 10^{-18}\ \text{J}\left(\frac{Z^2}{\infty}\right) = 0$$

The energy of the electron in any orbit is negative relative to this reference state.

Equation (12.1) applies to all one-electron species. In addition to being used to calculate the energy levels in the hydrogen atom, it can also be used for He^{+} ($Z = 2$), Li^{2+} ($Z = 3$), Be^{3+} ($Z = 4$), and so on.

Equation (12.1) can be used to calculate the change in energy when the electron changes orbits. For example, suppose the electron in level $n = 6$ of an excited hydrogen atom falls back to level $n = 1$ as the hydrogen atom returns to its lowest possible energy state, its **ground state.** We use Equation (12.1) with $Z = 1$ since the hydrogen nucleus contains a single proton. The energies corresponding to the two states are

For $n = 6$: $$E_6 = -2.178 \times 10^{-18}\ \text{J}\left(\frac{1^2}{6^2}\right) = -6.05 \times 10^{-20}\ \text{J}$$

For $n = 1$: $$E_1 = -2.178 \times 10^{-18}\ \text{J}\left(\frac{1^2}{1^2}\right) = -2.178 \times 10^{-18}\ \text{J}$$

Note that for $n = 1$ the electron has a more negative energy than it does for $n = 6$, which means that the electron is more tightly bound in the smallest allowed orbit.

The change in energy ΔE when the electron falls from $n = 6$ to $n = 1$ is

$$\begin{aligned}\Delta E &= \text{energy of final state} - \text{energy of initial state}\\ &= E_1 - E_6 = (-2.178 \times 10^{-18}\ \text{J}) - (-6.05 \times 10^{-20}\ \text{J})\\ &= -2.118 \times 10^{-18}\ \text{J}\end{aligned}$$

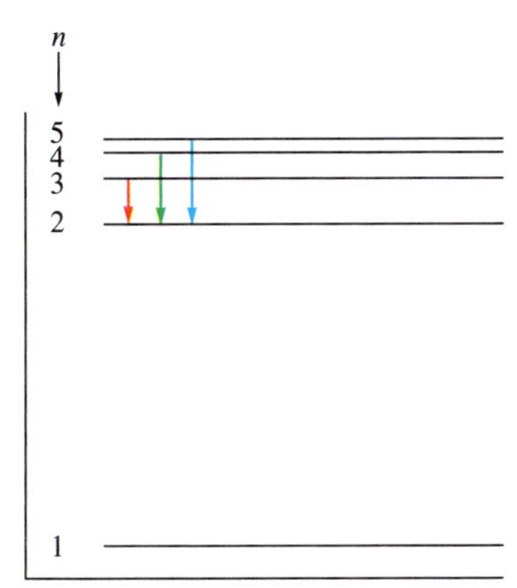

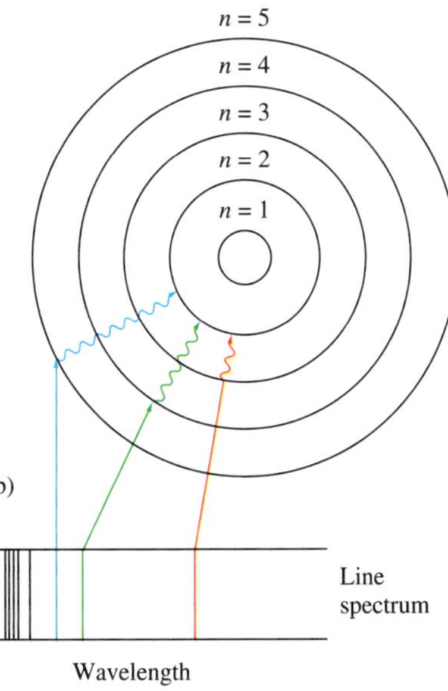

FIGURE 12.10

Electronic transitions in the Bohr model for the hydrogen atom. (a) An energy-level diagram for electronic transitions. (b) An orbit-transition diagram, which accounts for the experimental spectrum. (Note that the orbits shown are schematic. They are not drawn to scale.) (c) The resulting line spectrum on a photographic plate. Note that the lines in the visible region of the spectrum correspond to transitions from higher levels to the $n = 2$ level.

The negative sign for the *change* in energy indicates that the atom has *lost* energy and is now in a more stable state. The energy is carried away from the atom by the production (emission) of a photon.

The wavelength of the emitted photon can be calculated from the equation

$$\Delta E = h\left(\frac{c}{\lambda}\right) \quad \text{or} \quad \lambda = \frac{hc}{\Delta E}$$

where ΔE represents the change in energy of the atom and thus equals the energy of the emitted photon. We have

$$\lambda = \frac{hc}{\Delta E} = \frac{(6.626 \times 10^{-34}\ \text{J s})(2.9979 \times 10^{8}\ \text{m/s})}{2.118 \times 10^{-18}\ \text{J}} = 9.379 \times 10^{-8}\ \text{m}$$

Note that for this calculation the absolute value of ΔE is used. In this case the direction of energy flow is indicated by saying that a photon of wavelength 9.379×10^{-8} m has been *emitted* from the hydrogen atom. Simply plugging the negative value of ΔE into the equation would produce a negative value for λ, which is physically meaningless.

Example 12.3

Calculate the energy required to excite the hydrogen electron from level $n = 1$ to level $n = 2$. Also calculate the wavelength of light that must be absorbed by a hydrogen atom in its ground state to reach this excited state.

Solution Using Equation (12.1) with $Z = 1$, we have

$$E_1 = -2.178 \times 10^{-18}\ \text{J}\left(\frac{1^2}{1^2}\right) = -2.178 \times 10^{-18}\ \text{J}$$

$$E_2 = -2.178 \times 10^{-18}\ \text{J}\left(\frac{1^2}{2^2}\right) = -5.445 \times 10^{-19}\ \text{J}$$

$$\Delta E = E_2 - E_1 = (-5.445 \times 10^{-19}\ \text{J}) - (-2.178 \times 10^{-18}\ \text{J})$$
$$= 1.634 \times 10^{-18}\ \text{J}$$

The positive value for ΔE indicates that the system has gained energy. The wavelength of light that must be *absorbed* to produce this change is

$$\lambda = \frac{hc}{\Delta E} = \frac{(6.626 \times 10^{-34}\ \text{J s})(2.9979 \times 10^{8}\ \text{m/s})}{1.634 \times 10^{-18}\ \text{J}}$$
$$= 1.216 \times 10^{-7}\ \text{m}$$

Note from Fig. 12.3 that the light required to produce the transition from the $n = 1$ to $n = 2$ level in hydrogen lies in the ultraviolet region.

The New, Improved Atomic Clock

To celebrate the turn of the century, the National Institute of Standards and Technology (NIST) in Boulder, Colorado, gave the world a new precision timepiece—a clock so accurate that it will neither gain nor lose a second in 20 million years! Called NIST F-1, the new cesium atomic clock is classified as a fountain clock because it tosses spheres of cesium atoms upward inside the device.

The timepiece works as follows: Gaseous cesium atoms are introduced into a vacuum chamber where six infrared laser beams are positioned to push the cesium atoms into a spherical arrangement. In this process the cesium atoms are slowed almost to the point of absolute zero. Two vertical lasers then "toss" the sphere upward for about a meter in a microwave-filled cavity. At this point all of the lasers are turned off. The sphere of atoms then falls through the cavity under the influence of gravity, and some of the atoms are excited by the microwaves. At the bottom the cesium atoms, which were excited by the microwaves, are induced (by a laser beam) to emit light (a process called fluorescence). The entire process is repeated many times during which the frequency of the microwave energy in the cavity is adjusted until the maximum fluorescence is achieved. The microwave frequency that produces the greatest fluorescence is the natural resonance frequency for the cesium atom. This frequency (9.192631770×10^9 cycles per second) is used to define the second. The NIST F-1 clock is much more accurate than its predecessor, NIST-7, which fired heated cesium atoms at high speeds through a horizontal microwave chamber. In NIST F-1 the slower movement of the atoms gives more time for the microwaves to interact with the atoms so that the characteristic frequency can be more precisely determined.

NIST F-1, which is now the official clock in the United States, provides the extremely accurate timekeeping necessary for modern technology-based operations. NIST F-1 is also in the pool of atomic clocks used to define Coordinated Universal Time (known as UTC), the official world time.

However, the reign of the cesium-based superclocks may be over very soon. A team led by physicist John C. Berquist at NIST has developed an atomic clock, called the NIST optical clock, that is based on the interactions between ultraviolet light and a single mercury atom. Studies by the Berquist group show that the new optical clock is 10 times more precise than the cesium clock. This means that the optical clock would have an error of only 0.1 second in 70 million years of continuous operation. The search is now underway for an atom that might provide even more precise time measurements than mercury.

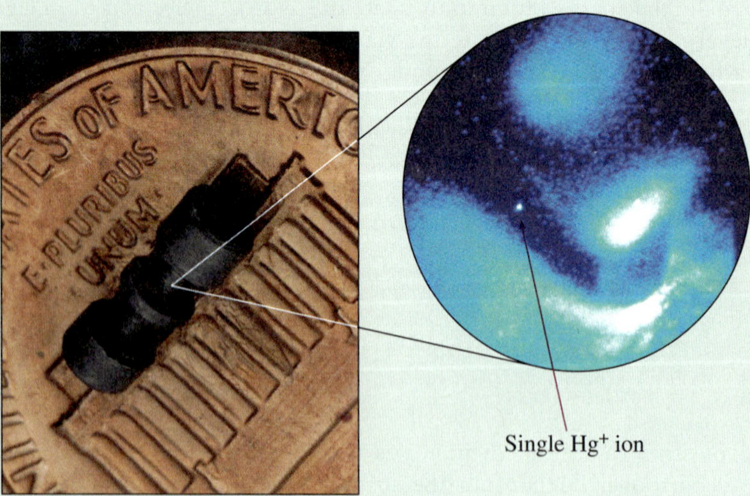

When installed in an atomic clock, this molybdenum structure (left) traps a mercury ion (right) at its center. The clock uses the ion to keep time with unprecedented precision.

At this time we must emphasize two important points about the Bohr model:

1. The model correctly fits the quantized energy levels of the hydrogen atom as inferred from its emission spectrum. These energy levels correspond to certain allowed circular orbits for the electrons.
2. As the electron becomes more tightly bound, its energy becomes more negative relative to the zero-energy reference state (corresponding to the electron being an infinite distance from the nucleus). That is, as the electron is brought closer to the nucleus, energy is released from the system.

Using Equation (12.1), we can derive a general equation for the electron moving from one level (n_{initial}) to another level (n_{final}):

$$\begin{aligned}\Delta E &= \text{energy of level } n_{\text{final}} - \text{energy of level } n_{\text{initial}} \\ &= E_{\text{final}} - E_{\text{initial}} \\ &= (-2.178 \times 10^{-18}\ \text{J})\left(\frac{1^2}{n_{\text{final}}^2}\right) - (-2.178 \times 10^{-18}\ \text{J})\left(\frac{1^2}{n_{\text{initial}}^2}\right) \\ &= -2.178 \times 10^{-18}\ \text{J}\left(\frac{1}{n_{\text{final}}^2} - \frac{1}{n_{\text{initial}}^2}\right) \qquad (12.2)\end{aligned}$$

Example 12.4

Calculate the minimum energy required to remove the electron from a hydrogen atom in its ground state.

Solution Removing the electron from a hydrogen atom in its ground state corresponds to taking the electron from $n_{\text{initial}} = 1$ to $n_{\text{final}} = \infty$. Thus

$$\begin{aligned}\Delta E &= -2.178 \times 10^{-18}\ \text{J}\left(\frac{1}{n_{\text{final}}^2} - \frac{1}{n_{\text{initial}}^2}\right) \\ &= -2.178 \times 10^{-18}\ \text{J}\left(\frac{1}{\infty} - \frac{1}{1^2}\right) \\ &= -2.178 \times 10^{-18}\ \text{J}(0 - 1) = 2.178 \times 10^{-18}\ \text{J}\end{aligned}$$

Thus the energy required to remove the electron from a hydrogen atom in its ground state is 2.178×10^{-18} J.

Equation (12.2) can be used to calculate the energy change between *any* two energy levels in a hydrogen atom, as shown in Example 12.4.

Although Bohr's model fits the energy levels for hydrogen, it is a fundamentally incorrect model for the hydrogen atom.

At first, Bohr's model appeared to be very promising. The energy levels calculated by Bohr closely agreed with the values obtained from the hydrogen emission spectrum. However, when Bohr's model was applied to atoms other than hydrogen, it did not work at all. Although some attempts were made to adapt the model using elliptical orbits, it was concluded that Bohr's model is fundamentally incorrect. The model is, however, very important historically because it shows that the observed quantization of energy in atoms can be explained by making rather simple assumptions. Bohr's model paved the way for later theories. It is important to realize, however, that the current theory of atomic structure is in no way derived from the Bohr model. Electrons do *not* move around the nucleus in circular orbits, as we will see later in this chapter.

Fireworks

The art of using mixtures of chemicals to produce explosives is an ancient one. Black powder—a mixture of potassium nitrate, charcoal, and sulfur—was being used in China well before 1000 A.D. and has been used through the centuries in military explosives, in construction blasting, and in fireworks. The DuPont Company, now a major chemical manufacturer, started out as a manufacturer of black powder. In fact, the founder, Eleuthère du Pont, learned the manufacturing technique from none other than Lavoisier.

Before the nineteenth century fireworks were mainly confined to rockets and loud bangs. Orange and yellow colors came from the presence of charcoal and iron filings. However, with the great advances in chemistry in the nineteenth century, new compounds found their way into fireworks. Salts of copper, strontium, and barium added brilliant colors. Magnesium and aluminum metals gave a dazzling white light. Fireworks, in fact, have changed very little since then.

How do fireworks produce their brilliant colors and loud bangs? Actually, only a handful of different chemicals are responsible for most of the spectacular effects. The noise and flashes are produced by an oxidizer (an oxidizing agent) and a fuel (a reducing agent). A common mixture involves potassium perchlorate ($KClO_4$) as the oxidizer and aluminum and sulfur as the fuel. The perchlorate compound oxidizes the fuel in a very exothermic reaction, which produces a brilliant flash, caused by the aluminum, and a loud report from the rapidly expanding gases produced. For a color effect an element with a colored emission spectrum is included. Recall that the electrons in atoms can be raised to higher energy levels when the atoms absorb energy. The excited atoms can then release this excess energy by emitting light of specific wavelengths, often in the visible region. In fireworks the energy to excite the electrons comes from the reaction between the oxidizer and fuel.

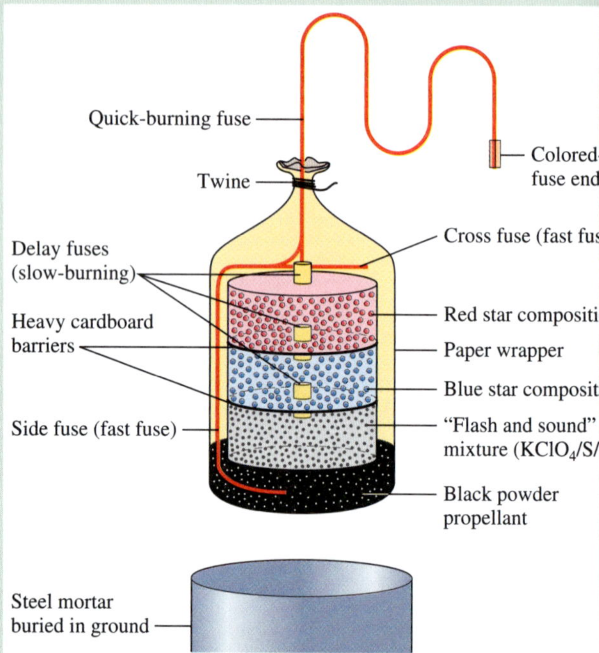

A typical aerial shell used in fireworks displays. Time-delayed fuses cause a shell to explode in stages. In this case a red starburst occurs first, followed by a blue starburst, and finally a flash and loud report. Reprinted with permission from *Chemical & Engineering News*, Vol. 59, Issue 26, June 29, 1981, p. 25. Copyright © 1981 American Chemical Society.

Yellow colors in fireworks are caused by the 589-nm emission of sodium atoms. Red colors come from strontium salts emitting at 606 nm and 636–688 nm. (This red color may be familiar from highway safety flares.) Barium salts give a green color in fireworks, caused by a series of emission lines between 505 and 535 nm. A really good blue color, however, is hard to obtain. Copper salts give a blue color, emitting in the 420–460-nm region. But

12.5 The Quantum Mechanical Description of the Atom

By the mid-1920s it had become apparent that the Bohr model was not a valid one. A totally new approach was needed. Three physicists were at the forefront of this effort: Werner Heisenberg, Louis de Broglie, and Erwin Schrödinger. The approach developed by de Broglie and Schrödinger became known as **wave mechanics** or, more commonly, **quantum mechanics.** As we

Chemicals Commonly Used in the Manufacture of Fireworks

Oxidizers	Fuels	Special Effects
Potassium nitrate	Aluminum	Red flame: strontium nitrate, strontium carbonate
Potassium chlorate	Magnesium	Green flame: barium nitrate, barium chlorate
Potassium perchlorate	Titanium	Blue flame: copper carbonate, copper sulfate, copper oxide
Ammonium perchlorate	Charcoal	Yellow flame: sodium oxalate, cryolite (Na_3AlF_6)
Barium nitrate	Sulfur	White flame: magnesium, aluminum
Barium chlorate	Antimony sulfide	Gold sparks: iron filings, charcoal
Strontium nitrate	Dextrin	White sparks: aluminum, magnesium, aluminum–magnesium alloy, titanium
	Red gum	
	Polyvinyl chloride	Whistle effect: potassium benzoate or sodium salicylate
		White smoke: mixture of potassium nitrate and sulfur
		Colored smoke: mixture of potassium chlorate, sulfur, and an organic dye

difficulties occur because another commonly used oxidizing agent, potassium chlorate ($KClO_3$), reacts with copper salts to form copper chlorate, a highly explosive compound that is dangerous to store. Paris green, a copper salt containing arsenic, was once extensively used but is now considered to be too toxic.

A typical aerial shell is shown in the figure. The shell is launched from a mortar (a steel cylinder), using black powder as the propellant. Time-delayed fuses are used to fire the shell in stages. A list of chemicals commonly used in fireworks is given in the table.

Although you might think that the chemistry of fireworks is simple, the achievement of the vivid white flashes and the brilliant colors requires complex combinations of chemicals. For example, because the white flashes produce high flame temperatures, the colors tend to wash out. Thus oxidizers, such as $KClO_4$, are sometimes used with fuels that produce relatively low flame temperatures. An added difficulty, however, is that perchlorates are very prone to accidental ignition and are therefore quite hazardous. Another problem arises from the use of sodium salts. Because sodium produces an extremely bright yellow emission, sodium salts cannot be used when other colors are desired. Carbon-based fuels also give a yellow flame that masks other colors, therefore limiting the use of organic compounds as fuels. You can see that the manufacture of fireworks that produce the desired effects and are also safe to handle requires careful selection of chemicals. And, of course, there is still the dream of that special deep blue flame.

Fireworks are a beautiful illustration of chemistry in action.

have already seen, de Broglie originated the idea that the electron, previously considered to be a particle, also shows wave properties. Pursuing this line of reasoning, Schrödinger, an Austrian physicist, decided to attack the problem of atomic structure by giving emphasis to the wave properties of the electron. To Schrödinger and de Broglie, the electron bound to the nucleus seemed similar to a standing wave, and they began research on a wave mechanical description of the atom.

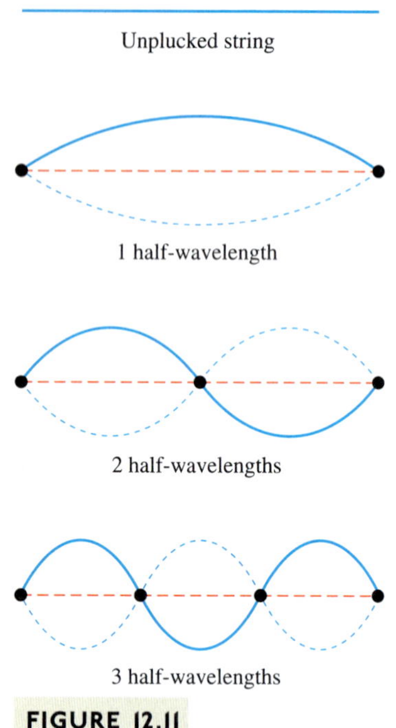

FIGURE 12.11

The standing wave produced by the vibration of a guitar string fastened at both ends. Each dot represents a node (a point of zero displacement).

The most familiar example of standing waves occurs in association with musical instruments such as guitars or violins, where a string attached at both ends vibrates to produce the musical tone. The waves are described as standing since they are stationary; the waves do not travel along the length of the string. The motions of the string can be explained as a combination of simple waves of the type shown in Fig. 12.11. The dots in this figure indicate the nodes, or points of zero lateral (sideways) displacement for a given wave. Note that there are limitations on the allowed wavelengths of the standing wave. Since each end of the string is fixed, there is always a node at each end. This means that there must be a whole number of *half*-wavelengths in any of the allowed motions of the string (see Fig. 12.11).

The wave model was applied by de Broglie to the Bohr atom by imagining the electron in the hydrogen atom to be a standing wave. As shown in Fig. 12.12, only certain circular orbits have a circumference into which a whole number of wavelengths of the standing electron wave will "fit." All other orbits would produce destructive interference of the standing electron wave and are not allowed. This seemed like a possible explanation for the observed quantization of the hydrogen atom. The mathematical formalism that Schrödinger developed in 1925 to describe the hydrogen electron as a wave was heavily based on the classical descriptions of wave phenomena. We will first give an overview of this approach before considering it in more detail.

The form of Schrödinger's equation is

$$\hat{H}\psi = E\psi$$

where ψ, called the **wave function,** is a function of the coordinates (x, y, and z) of the electron's position in three-dimensional space, and where $\hat{H}$ represents a set of mathematical instructions called an *operator.* An operator is a mathematical tool that acts on a function to produce another function. In some special cases the operator gives back the original function simply multiplied by a constant. Note that the Schrödinger equation corresponds to such a special case. In the Schrödinger equation the operator $\hat{H}$, called the Hamiltonian, acts to give back the wave function multiplied by the constant E, which represents the total energy of the atom (the sum of the potential energy due to the attraction between the proton and the electron and the kinetic energy of the moving electron). When this equation is analyzed, many solutions are found. Each solution consists of a wave function ψ that is characterized by a particular value of E. A specific wave function for a given electron is often called an **orbital.** It is important to recognize that Schrödinger could not be sure that treating an electron as a wave makes any sense: The test would be whether the model could correctly fit the experimental data for hydrogen and other atoms.

To illustrate the most important ideas of the wave mechanical (quantum mechanical) model of the atom, we will first concentrate on the wave function corresponding to the lowest energy for the hydrogen atom. This wave function is called the 1*s* orbital. The first point of interest is the meaning of the word *orbital.* One thing is clear: An orbital is *not* a Bohr orbit. The electron in the hydrogen 1*s* orbital is not moving around the nucleus in a circular orbit. How, then, is the electron moving? The answer is somewhat surprising: *We do not know.* The wave function gives us no information about the movements of the electron. This observation is somewhat disturbing. When

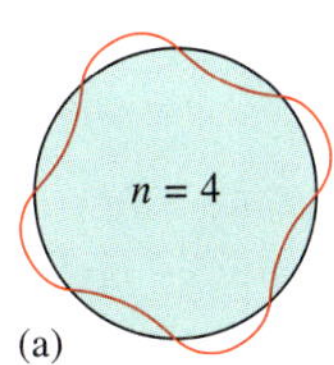

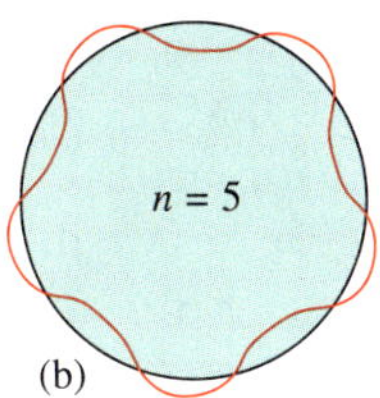

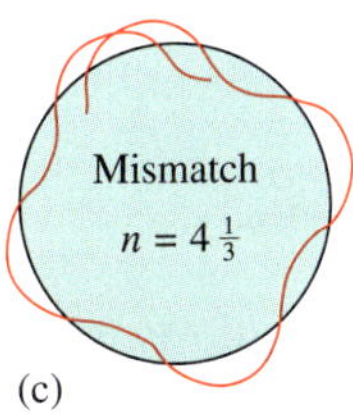

FIGURE 12.12

The hydrogen electron visualized as a standing wave around the nucleus. The circumference of a particular circular orbit has to correspond to a whole number of wavelengths, as shown in (a) and (b), or else destructive interference occurs, as shown in (c). This model is consistent with the fact that only certain electron energies are allowed; the atom is quantized. (Although this idea has encouraged scientists to use a wave theory, it does not mean that the electron really travels in circular orbits.)

The uncertainty principle is sometimes called the indeterminacy principle.

we solve problems involving the motions of particles in the macroscopic world, we are able to predict their trajectories. For example, when two billiard balls with known velocities collide, we can predict their motions after the collision. However, we cannot predict the electron's motion using the 1*s* orbital function. Does this mean that the theory is useless? Not necessarily: We have already learned that an electron does not behave much like a billiard ball, so we must examine the situation closely before we discard the theory.

Werner Heisenberg, who was also involved in the development of the quantum mechanical model for the atom, discovered a very important principle in 1927 that helps us to understand the meaning of orbitals—the **Heisenberg uncertainty principle.** Heisenberg's mathematical analysis led him to a surprising conclusion: *There is a fundamental limitation to just how precisely we can know both the position and the momentum of a particle at a given time.* Stated mathematically, the uncertainty principle is

$$\Delta x \cdot \Delta p \geq \frac{\hbar}{2}$$

where Δx is the uncertainty in a particle's position, Δp is the uncertainty in a particle's momentum, and $\hbar$ is Planck's constant divided by 2π ($\hbar = h/2\pi$). Thus the minimum uncertainty in the product $\Delta x \cdot \Delta p$ is $h/4\pi$. This relationship means that the more precisely we know a particle's position, the less precisely we can know its momentum, and vice versa. This limitation is so small for large particles such as baseballs or billiard balls that it is unnoticed. However, for a small particle such as the electron, the limitation becomes quite important. Applied to the electron, the uncertainty principle implies that we cannot know the exact path of the electron as it moves around the nucleus. It is therefore not appropriate to assume that the electron is moving around the nucleus in a well-defined orbit as in the Bohr model.

Example 12.5

The hydrogen atom has a radius on the order of 0.05 nm. Assuming that we know the position of an electron to an accuracy of 1% of the hydrogen radius, calculate the uncertainty in the velocity of the electron using the Heisenberg uncertainty principle. Then compare this value with the uncertainty in the velocity of a ball of mass 0.2 kg and radius 0.05 m whose position is known to an accuracy of 1% of its radius.

Solution From Heisenberg's uncertainty principle the smallest possible uncertainty in the product $\Delta x \cdot \Delta p$ is $\hbar/2$; that is,

$$\Delta x \cdot \Delta p = \frac{\hbar}{2} = \frac{h}{4\pi}$$

For the electron the uncertainty in position (Δx) is 1% of 0.05 nm, or

$$\Delta x = (0.01)(0.05 \text{ nm}) = 5. \times 10^{-4} \text{ nm}$$

Converting to meters gives

$$5 \times 10^{-4} \text{ nm} \times \frac{10^{-9} \text{ m}}{1 \text{ nm}} = 5 \times 10^{-13} \text{ m}$$

CHEMICAL INSIGHTS

Electrons as Waves

Although scientists talk about the dual wave and particle properties of electrons, many nonscientists still believe that electrons are only tiny particles. Rooted as we are in the macroscopic world, it can be difficult for some to picture a particle as also being a wave. One look at the accompanying picture, however, should help change that. What looks like ripples surrounding two barely submerged pebbles in a pool of water is really the surface of a copper crystal.

Although they are true believers in the wave nature of electrons, the physicists at the IBM Almaden Research Center in San Jose, California, were genuinely surprised when their scanning tunneling microscope (STM) produced this image of the copper surface. "We looked at the surface with all these waves and thought, 'Is our machine broken?'" says Michael Crommie, one of the IBM physicists. But the researchers soon realized that the waves were produced by electrons confined to the metal's surface that bounced off impurities (the two pits). Because the electrons are waves, they form interference patterns after reflecting off the impurities, producing standing waves.

To further explore this behavior, the IBM scientists constructed a "quantum corral" by using their STM to place 48 iron atoms on a copper surface in a circle approximately 14 nm in diameter. Then using the STM to study electron behavior on the copper surface inside the corral, they observed the standing electron waves shown in the photo on the right. This image provides a unique visual confirmation of what the Schrödinger equation predicts. Electrons are wavelike. Seeing is believing!

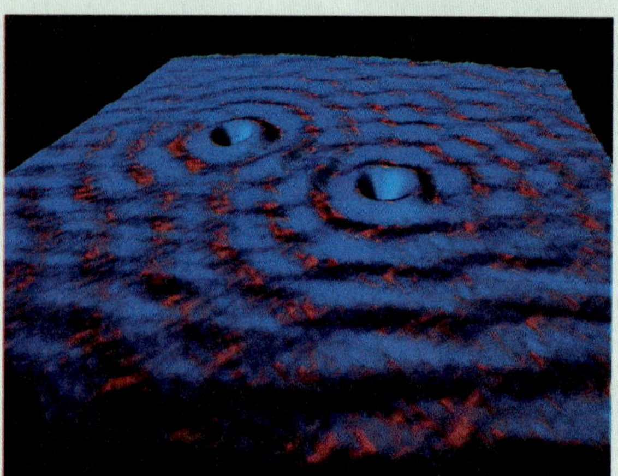

(left) The electrons form interference patterns, the ripples shown here, and produce standing waves.
(right) Iron atoms in a circular "corral" cause electron standing waves.

The values of the constants are

$$m = \text{mass of the electron} = 9.11 \times 10^{-31}\ \text{kg}$$

$$h = 6.626 \times 10^{-34}\ \text{J s} = 6.626 \times 10^{-34}\ \frac{\text{kg m}^2}{\text{s}}$$

$$\pi = 3.14$$

We can now solve for the uncertainty in momentum:

$$\Delta p = \frac{\hbar}{2 \cdot \Delta x} = \frac{h}{4\pi \cdot \Delta x} = \frac{6.626 \times 10^{-34}\ \frac{\text{kg m}^2}{\text{s}}}{4(3.14)(5 \times 10^{-13}\ \text{m})}$$

$$= 1.05 \times 10^{-22}\ \text{kg m/s}$$ (keeping extra significant figures)

Recalling that $p = mv$ and assuming that the electron mass is constant (ignoring any relativistic corrections), we have

$$\Delta p = \Delta(mv) = m\Delta v$$

and the uncertainty in velocity is

$$\Delta v = \frac{\Delta p}{m} = \frac{1.05 \times 10^{-22}\ \text{kg m/s}}{9.11 \times 10^{-31}\ \text{kg}} = 1.15 \times 10^{8}\ \text{m/s} = 1 \times 10^{8}\ \text{m/s}$$

Thus, if we know the electron's position with a minimum uncertainty of 5×10^{-13} m, the uncertainty in the electron's velocity is at least 1×10^{8} m/s. This is a very large number; in fact, it is the same magnitude as the speed of light (3×10^{8} m/s). At this level of uncertainty we have virtually no idea of the velocity of the electron.

For the ball, the uncertainty in position (Δx) is 1% of 0.05 m, or 5×10^{-4} m. Thus the minimum uncertainty in velocity is

$$\Delta v = \frac{\Delta p}{m} = \frac{h}{\Delta x \cdot m \cdot 4\pi} = \frac{6.626 \times 10^{-34}\ \frac{\text{kg m}^2}{\text{s}}}{(5 \times 10^{-4}\ \text{m})(0.2\ \text{kg})(4)(3.14)}$$
$$= 5 \times 10^{-31}\ \text{m/s}$$

This means there is a very small (undetectable) uncertainty in our measurements of the speed of a ball. Note that this uncertainty is not caused by the limitations of measuring instruments; Δv is an *inherent* uncertainty.

Thus the uncertainty principle is negligible in the world of macroscopic objects but is very important for objects with small masses, such as the electron.

12.6 The Particle in a Box

Overview

The Schrödinger wave equation, $\hat{H}\psi = E\psi$, lies at the heart of the quantum mechanical description of atoms. Recall from the preceding discussion that $\hat{H}$ represents an operator (the Hamiltonian) that "extracts" the total energy E (the sum of the potential and kinetic energies) from the wave function. The wave function ψ depends on the x, y, and z coordinates of the electron's position in space.

Note that the Schrödinger equation requires that when ψ is operated on by $\hat{H}$, the result is ψ multiplied by a constant, E, that represents the total energy of the particular state described by ψ. As we will see, there are many possible solutions to the Schrödinger equation for a given system. For example, for the hydrogen atom there are many functions that satisfy the Schrödinger equation, each one corresponding to a particular energy for hydrogen's electron. Each of these specific wave functions for the hydrogen atom is called an orbital.

Although the detailed solution of the Schrödinger equation for the hydrogen atom is not appropriate in this text, we will illustrate some of the properties of wave mechanics and wave functions by using the wave equation to describe a very simple, hypothetical system commonly called "the particle in a box," a situation in which a particle is trapped in a one-dimensional box that has infinitely high "sides." It is important to recognize that this situation is not an accurate physical model for the hydrogen atom. That is, the hydrogen atom is really not much like this particle in a box. The reasons for treating the particle in a box are that (1) it illustrates the mathematics of wave mechanics, (2) it gives an indication of the characteristics of wave functions, and (3) it shows how energy quantization arises. Thus this treatment of a particle in a box illustrates the "flavor" of the wave mechanical description of the hydrogen atom, but it should not be taken to be an accurate representation of the hydrogen atom itself.

The Particle in a Box as a Model

Consider a particle with mass m that is free to move back and forth along one dimension (we arbitrarily choose x) between the values $x = 0$ and $x = L$ (that is, we are considering a one-dimensional "box" of size L meters). We will assume that the potential energy $V(x)$ of the particle is zero at all points along its path, except at the endpoints $x = 0$ and $x = L$, where $V(x)$ is infinitely large. In effect, we have a repulsive barrier of infinite strength at each end of the box. Thus the particle is trapped in a one-dimensional box with impenetrable walls (see Fig. 12.13).

As we mentioned before, the Schrödinger equation contains the energy operator $\hat{H}$. In this case, since the potential energy is zero inside the box, the only energy possible is the kinetic energy of the particle as it moves back and forth along the x axis. The operator for this kinetic energy is

$$-\frac{\hbar^2}{2m}\frac{d^2}{dx^2}$$

where $\hbar$ is Planck's constant divided by 2π, m is the mass of the particle, and d^2/dx^2 is the second derivative with respect to x. The form of this operator comes from the description of waves in classical physics. Inserting this operator into the Schrödinger equation $\hat{H}\psi = E\psi$ gives

$$-\frac{\hbar^2}{2m}\frac{d^2\psi}{dx^2} = E\psi$$

where ψ is a function of x {[$\psi(x)$]}. We can rearrange this equation to give

$$\frac{d^2\psi}{dx^2} = -\frac{2mE}{\hbar^2}\psi$$

Our goal is to find specific functions $\psi(x)$ that satisfy this equation. Notice that the solutions to this equation are functions such that $d^2\psi/dx^2 = (\text{constant})\psi$. That is, each solution must be a function whose second derivative has the same form as the original function. One function that behaves this way is the sine function. For example, consider the function $A \sin(kx)$, where A and k are constants. We will now take the second derivative of this function with respect to x:

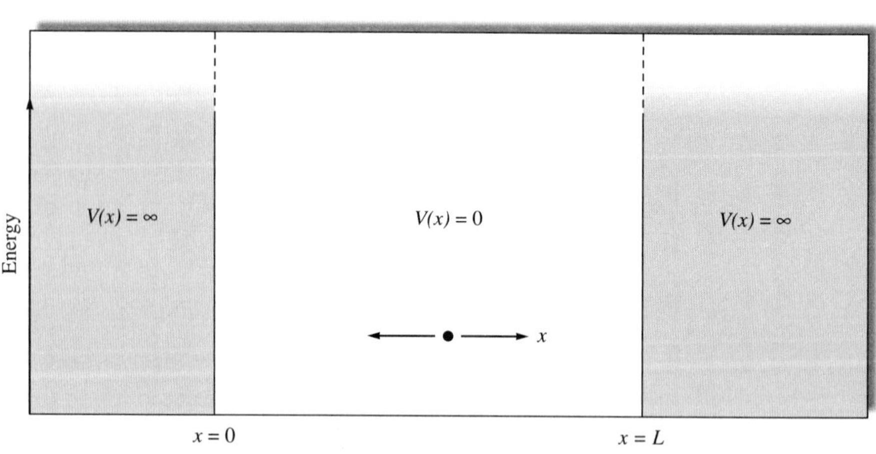

FIGURE 12.13

A schematic representation of a particle in a one-dimensional box with infinitely high potential walls.

Charles Sykes Researches Surface Architecture

Charles Sykes.

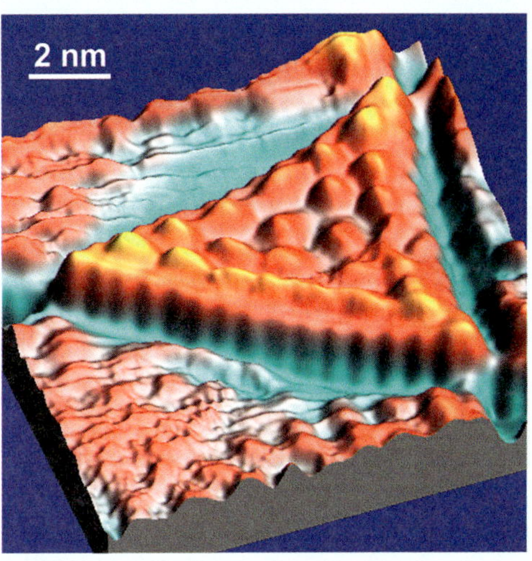

Surface architecture showing electrons forming an interference pattern.

Charles Sykes, Assistant Professor of Chemistry at Tufts University in Medford, Massachusetts, did his undergraduate work at Oxford University (United Kingdom) and obtained his Ph.D. from Cambridge University (United Kingdom). His interests focus on how atoms and molecules interact on surfaces and on building novel nanoscale surface structures using scanning tunneling microscopy (STM), which enables control of single atoms and molecules. The figure shows a "quantum corral," which is made of approximately 650 gold atoms on a metallic surface. The peaks inside the triangle represent electron density maxima that are caused by constructive interference of the electron waves as they are reflected by the edges of the box. Simple particle-in-a-box calculations can predict these maxima.

A goal of Professor Sykes's research group is to understand how molecules approaching such a surface will interact with the electron density on the surface and to explore how to tailor surfaces and better assist molecules to self-assemble in new surface architectures. Professor Sykes is also interested in studying the effect of surface structure on the ability of the surface to catalyze various chemical reactions.

$$\begin{aligned}\frac{d^2}{dx^2}(A \sin kx) &= A\frac{d}{dx}\left(\frac{d \sin kx}{dx}\right) = A\frac{d}{dx}(k \cos kx)\\ &= Ak\left(\frac{d \cos kx}{dx}\right) = Ak(-k \sin kx)\\ &= -Ak^2 \sin kx = -k^2 A \sin kx\end{aligned}$$

Thus we have shown that

$$\frac{d^2(A \sin kx)}{dx^2} = -k^2(A \sin kx)$$

This is just the type of function that will satisfy the Schrödinger equation for the particle in a box. In fact, when we compare the general form of the Schrödinger equation

$$\frac{d^2\psi}{dx^2} = -\frac{2mE}{\hbar^2}\psi$$

with

$$\frac{d^2(A \sin kx)}{dx^2} = k^2(A \sin kx)$$

we see that

$$-k^2 = -\frac{2mE}{\hbar^2}$$

which can be rearranged to give an expression for energy:

$$E = \frac{\hbar^2 k^2}{2m}$$

What does this equation mean? We have simply specified that A and k are constants. What values can these constants have? Note that if they could assume any values, this equation would lead to an infinite number of possible energies—that is, a continuous distribution of energy levels. However, this is not correct. For reasons we will discuss presently, we find that only certain energies are allowed. That is, this system is quantized. In fact, the ability of wave mechanics to account for the observed (but initially unexpected) quantization of energy in nature is one of the most important factors in convincing us that it may be a correct description of the properties of matter.

Quantization enters the wave mechanical description of the particle in a box via the boundary conditions. Boundary conditions arise from the physical requirements of natural systems. That is, we must insist that our descriptions of natural systems make physical sense. For example, assume that in describing an aqueous solution containing an acid we arrive at the expression $[H^+]^2 = 4.0 \times 10^{-8}\ M^2$. The solutions to this expression are

$$[H^+] = 2.0 \times 10^{-4}\ M \qquad \text{and} \qquad [H^+] = -2.0 \times 10^{-4}\ M$$

In doing such a problem, we automatically reject the second possibility because there is no physical meaning for a negative concentration. What we have done here is apply a type of boundary condition to this situation.

The boundary conditions for the particle in a box enforce the following facts:

1. The particle cannot be outside the box—it is bound inside the box.
2. In a given state the total probability of finding the particle in the box must be 1.
3. The wave function must be continuous.

We have seen that the function $\psi = A \sin(kx)$ satisfies the Schrödinger equation $\hat{H}\psi = E\psi$. We will now define the constants k and A so that this function also satisfies the boundary conditions based on the three constraints listed above. Because the particle must stay inside the box, and because the wave function must be continuous, the value of $\psi(x)$ must be zero at each wall. That is,

$$\psi(0) = 0 \qquad \text{and} \qquad \psi(L) = 0$$

Recall that the sine function is zero at angles of 0°, 180° (π radians), 360° (2π radians), and so on. Thus the function $A \sin(kx)$ is automatically zero when $x = 0$.

The requirement that the wave function must also be zero at the other wall, which can be stated as $\psi(L) = A \sin(kL) = 0$, means that k is limited to the values of $n\pi/L$, where n is an integer (1, 2, 3, . . .). That is,

$$\psi(x) = A \sin\left(\frac{n\pi}{L}x\right)$$

then

$$\psi(L) = A \sin\left(\frac{n\pi}{L} \cdot L\right) = A \sin(n\pi) = 0$$

To assign the value of the constant A, we need to introduce a new idea. In the application of wave mechanics to the description of matter, scientists have learned to associate the square of the wave function with probability. As we will discuss in more detail below, this means that the square of the wave function evaluated at a given point gives the relative probability of finding a particle near that point. This concept is relevant to the boundary conditions for the particle in a box because the total probability in a given state must be 1. To be more precise, the probability of finding the particle on a segment of the x axis of length dx surrounding point x is $\psi^2(x)\,dx$. Because there is one particle in the box, the sum of all those probabilities along the x axis from $x = 0$ to $x = L$ must be 1. We sum these probabilities over the length of the box (from $x = 0$ to $x = L$) by integration from $x = 0$ to $x = L$:

$$\text{Total probability of finding the particle in the box} = \int_0^L \psi^2(x)\,dx = 1$$

Substituting $\psi(x) = A\sin[(n\pi/L)x]$, we have

$$\int_0^L \psi^2(x)\,dx = \int_0^L A^2 \sin^2\left(\frac{n\pi}{L}x\right)dx = 1$$

or

$$\int_0^L \sin^2\left(\frac{n\pi}{L}x\right)dx = \frac{1}{A^2}$$

The value of the integral is $L/2$, which means that

$$\frac{L}{2} = \frac{1}{A^2} \quad \text{and} \quad A = \sqrt{\frac{2}{L}}$$

Now that we know the allowed values of k and A, we can specify the wave function for the particle in a one-dimensional box as

$$\psi(x) = \sqrt{\frac{2}{L}}\sin\left(\frac{n\pi}{L}x\right)$$

We can also substitute the value of k into the expression for energy:

$$E = \frac{\hbar^2 k^2}{2m} = \frac{\hbar^2(n\pi/L)^2}{2m}$$

Substituting $\hbar = h/2\pi$ gives

$$E = \frac{n^2h^2}{8mL^2} \quad \text{where} \quad n = 1, 2, 3, 4, \ldots$$

Note that this analysis leads to a series of solutions to the Schrödinger equation, where each function corresponds to a given energy state:

n	Function	Energy
1	$\psi_1 = \sqrt{\frac{2}{L}}\sin\left(\frac{\pi}{L}x\right)$	$E_1 = \frac{h^2}{8mL^2}$
2	$\psi_2 = \sqrt{\frac{2}{L}}\sin\left(\frac{2\pi}{L}x\right)$	$E_2 = \frac{4h^2}{8mL^2} = \frac{h^2}{2mL^2}$
3	$\psi_3 = \sqrt{\frac{2}{L}}\sin\left(\frac{3\pi}{L}x\right)$	$E_3 = \frac{9h^2}{8mL^2}$
4	$\psi_4 = \sqrt{\frac{2}{L}}\sin\left(\frac{4\pi}{L}x\right)$	$E_4 = \frac{16h^2}{8mL^2} = \frac{2h^2}{mL^2}$
⋮	⋮	⋮

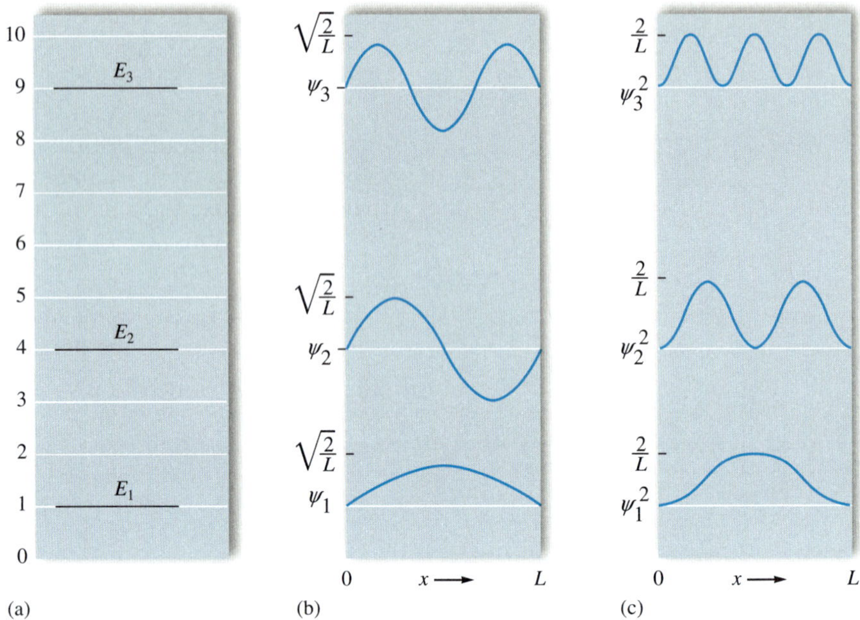

FIGURE 12.14

(a) The first three energy levels for a particle in a one-dimensional box in increments of $h^2/(8mL^2)$. (b) The wave functions for the first three levels plotted as a function of x. Note that the maximum value is $\sqrt{2/L}$ in each case. (c) The square of the wave functions for the first three levels plotted as a function of x. Note that the maximum value is $2/L$ in each case.

Notice something very important about these results. The application of the boundary conditions has led to a series of *quantized* energy levels. That is, only certain energies are allowed for the particle bound in the box. This result fits very nicely with the experimental evidence, such as the hydrogen emission spectrum, that nature does not allow continuous energy levels for *bound* systems, as classical physics had led us to expect. Note that the energies are quantized, because the boundary conditions require that n assume only integer values. Consequently, we call n the quantum number for this system.

We can diagram the solutions to the particle-in-a-box problem conveniently by showing a plot of the wave function that corresponds to each energy level. The energy level, wave function, and probability distribution are shown in Fig. 12.14 for the first three levels.

Note that each wave function goes to zero at the edges of the box, as required by the boundary conditions. Another way to say this is that the standing waves that represent the particle must have wavelengths such that an *integral number of half-wavelengths exactly equals the size of the box*. Waves with any other wavelengths could not exist because they would destructively interfere. Also note from Fig. 12.14 that the probability distribution is significantly different for the three levels. For $n = 1$ (the lowest energy or ground state) the particle is most likely to be found near the center of the box. In contrast, for $n = 2$ the particle has zero probability of being found in the center of the box. This zero point is called a node. Notice that the number of nodes increases with n.

Another interesting characteristic of the particle in a box is that the particle cannot have zero energy (that is, n cannot equal zero). For example, if n were equal to zero, ψ_0 would be zero everywhere in the box ($\sin 0 = 0$). This would mean that ψ_0^2 would also be zero. In this case there could be no particle in the box, which contradicts the boundary conditions. This fact that the particle must have a nonzero energy in its ground state is a characteristic of all particles with quantized energies. In addition, for the particle in a box a value of zero for the energy would mean that the particle was sitting still (zero kinetic energy). This condition would violate the uncertainty principle because we would simultaneously know the exact values of the momentum (zero) and the position of the particle. For similar reasons all quantized particles must possess a minimum energy, often called the *zero-point energy.*

Example 12.6

Assume that an electron is confined to a one-dimensional box 1.50 nm in length. Calculate the lowest three energy levels for this electron, and calculate the wavelength of light necessary to promote the electron from the ground state to the first excited state.

Solution To solve this problem, we need to substitute appropriate values into the general expression for energy:

$$E = \frac{n^2h^2}{8mL^2}$$

The mass of an electron (m) is 9.11×10^{-31} kg; the dimension of the box (L) is 1.50 nm, or 1.50×10^{-9} m; and the value of Planck's constant is 6.626×10^{-34} J s.

For $n = 1$ we get

$$E_1 = \frac{(1)^2(6.626 \times 10^{-34} \text{ J s})^2}{(8)(9.11 \times 10^{-31} \text{ kg})(1.50 \times 10^{-9} \text{ m})^2} = 2.68 \times 10^{-20} \text{ J}$$

Similarly, for $n = 2$ we get

$$E_2 = 1.07 \times 10^{-19} \text{ J}$$

And for $n = 3$ we get

$$E_3 = 2.41 \times 10^{-19} \text{ J}$$

Note that since

$$E_n = n^2\frac{h^2}{8mL^2} = n^2E_1$$

then

$$E_2 = (2)^2\frac{h^2}{8mL^2} - 4E_1 \qquad \text{and} \qquad E_3 - 9E_1$$

To calculate the wavelength of light necessary to excite the electron from level 1 to level 2 (the first *excited* state), we first need to obtain the energy difference between the two levels:

$$\Delta E = E_2 - E_1 = ({n_2}^2 - {n_1}^2)\frac{h^2}{8mL^2}$$

$$= (3)(2.68 \times 10^{-20} \text{ J}) = 8.04 \times 10^{-20} \text{ J}$$

Then we find the wavelength required from the equation

$$\Delta E = \frac{hc}{\lambda}$$

Inserting the appropriate values gives

$$\lambda = \frac{hc}{\Delta E} = \frac{(6.626 \times 10^{-34} \text{ J s})(2.9979 \times 10^8 \text{ m/s})}{8.04 \times 10^{-20} \text{ J}}$$

$$= 2.47 \times 10^{-6} \text{ m} = 2470 \text{ nm}$$

12.7 The Wave Equation for the Hydrogen Atom

Unlike the particle in a one-dimensional box, the electron of the hydrogen atom moves in three dimensions and has potential energy because of its attraction to the positive nucleus at the atom's center. These differences can be easily accounted for by including the second derivatives with respect to all three of the Cartesian coordinates and by inserting a term that specifies the dependence of the electron's potential energy on its position in space.

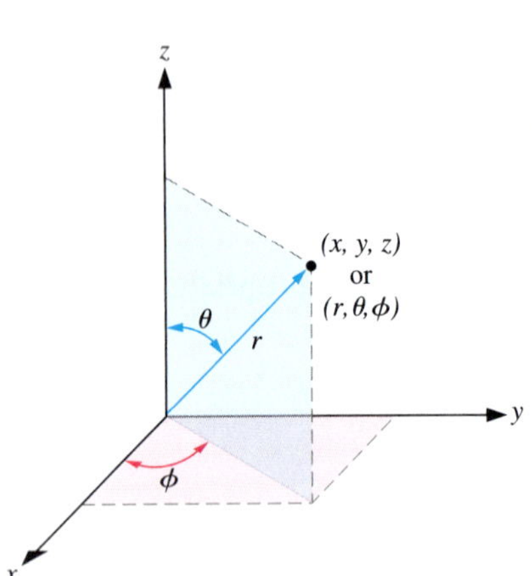

FIGURE 12.15

The spherical polar coordinate system.

Because it is more convenient mathematically, the coordinate system is changed from Cartesian to spherical polar coordinates (see Fig. 12.15) before the Schrödinger equation is solved. In the system of spherical polar coordinates a given point in space, specified by values of the Cartesian coordinates x, y, and z, is described by specific values of r, θ, and ϕ.

In the spherical polar coordinate system the wave function $\psi(r, \theta, \phi)$ can be written as a product of one function depending on r, one depending on θ, and one depending on ϕ:

$$\psi(r, \theta, \phi) = R(r)\Theta(\theta)\Phi(\phi)$$

This separation of variables allows an exact solution to the Schrödinger equation

$$\hat{H}\psi = E\psi$$

for the hydrogen atom.

In spherical polar coordinates the potential energy (in cgs units) of the electron is

$$V(r) = -\frac{(Ze)(e)}{r}$$

To convert the potential energy from cgs units to SI units (joules), the expression shown must be multiplied by $1/4\pi\epsilon_0$, where ϵ_0 (the permittivity of the vacuum) is 8.854×10^{-12} C^2/J m.

where Ze represents the nuclear charge ($Z = 1$ for the hydrogen atom). As with the particle in a box, when the Schrödinger equation for the hydrogen atom is solved and the boundary conditions are applied, a series of wave functions is obtained, each function corresponding to a particular energy. In contrast to the particle in a one-dimensional box, where one quantum number emerges from the mathematics, the three-dimensional hydrogen atom gives rise to three quantum numbers.

The boundary conditions, which differ in some important aspects from those of the particle in the box because of the very different nature of the physical system, will not be discussed here.

The conventional symbols for these quantum numbers are as follows:

n	the principal quantum number
ℓ	the angular momentum quantum number
m_ℓ	the magnetic quantum number

We will have more to say in succeeding sections about what values these quantum numbers can assume and their physical meanings.

The mathematics of wave mechanics leads to the following expression for the allowed energies of hydrogen's electron:

$$E_n = -\frac{Z^2}{n^2}\left(\frac{me^4}{8\epsilon_0{}^2h^2}\right) = -2.178 \times 10^{-18}\ \mathrm{J}\left(\frac{Z^2}{n^2}\right)$$

where $Z = 1$ for hydrogen and where n can assume only integer values (1, 2, 3, . . .). Several characteristics of this equation are worth emphasizing. First, note that the energy of the electron depends only on the principal quantum number (this is true only for one-electron species). Also note that because n is restricted to integer values, hydrogen's electron can assume only discrete

TABLE 12.1

Solutions of the Schrödinger Wave Equation for a One-Electron Atom

n	ℓ	m_ℓ	Orbital	Solution
1	0	0	1s	$\psi_{1s} = \frac{1}{\sqrt{\pi}}\left(\frac{Z}{a_0}\right)^{3/2} e^{-\sigma}$
2	0	0	2s	$\psi_{2s} = \frac{1}{4\sqrt{2\pi}}\left(\frac{Z}{a_0}\right)^{3/2} (2-\sigma)e^{-\sigma/2}$
2	1	0	$2p_z$	$\psi_{2p_z} = \frac{1}{4\sqrt{2\pi}}\left(\frac{Z}{a_0}\right)^{3/2} \sigma e^{-\sigma/2} \cos\theta$
2	1	±1	$2p_x$	$\psi_{2p_x} = \frac{1}{4\sqrt{2\pi}}\left(\frac{Z}{a_0}\right)^{3/2} \sigma e^{-\sigma/2} \sin\theta\cos\phi$
			$2p_y$	$\psi_{2p_y} = \frac{1}{4\sqrt{2\pi}}\left(\frac{Z}{a_0}\right)^{3/2} \sigma e^{-\sigma/2} \sin\theta\sin\phi$
3	0	0	3s	$\psi_{3s} = \frac{1}{81\sqrt{3\pi}}\left(\frac{Z}{a_0}\right)^{3/2} (27-18\sigma+2\sigma^2)e^{-\sigma/3}$
3	1	0	$3p_z$	$\psi_{3p_z} = \frac{\sqrt{2}}{81\sqrt{\pi}}\left(\frac{Z}{a_0}\right)^{3/2} (6\sigma-\sigma^2)e^{-\sigma/3}\cos\theta$
3	1	±1	$3p_x$	$\psi_{3p_x} = \frac{\sqrt{2}}{81\sqrt{\pi}}\left(\frac{Z}{a_0}\right)^{3/2} (6\sigma-\sigma^2)e^{-\sigma/3}\sin\theta\cos\phi$
			$3p_y$	$\psi_{3p_y} = \frac{\sqrt{2}}{81\sqrt{\pi}}\left(\frac{Z}{a_0}\right)^{3/2} (6\sigma-\sigma^2)e^{-\sigma/3}\sin\theta\sin\phi$
3	2	0	$3d_{z^2}$	$\psi_{3d_{z^2}} = \frac{1}{81\sqrt{6\pi}}\left(\frac{Z}{a_0}\right)^{3/2} \sigma^2 e^{-\sigma/3}(3\cos^2\theta-1)$
3	2	±1	$3d_{xz}$	$\psi_{3d_{xz}} = \frac{\sqrt{2}}{81\sqrt{\pi}}\left(\frac{Z}{a_0}\right)^{3/2} \sigma^2 e^{-\sigma/3}\sin\theta\cos\theta\cos\phi$
			$3d_{yz}$	$\psi_{3d_{yz}} = \frac{\sqrt{2}}{81\sqrt{\pi}}\left(\frac{Z}{a_0}\right)^{3/2} \sigma^2 e^{-\sigma/3}\sin\theta\cos\theta\sin\phi$
3	2	±2	$3d_{xy}$	$\psi_{3d_{xy}} = \frac{1}{81\sqrt{2\pi}}\left(\frac{Z}{a_0}\right)^{3/2} \sigma^2 e^{-\sigma/3}\sin^2\theta\sin 2\phi$
			$3d_{x^2-y^2}$	$\psi_{3d_{x^2-y^2}} = \frac{1}{81\sqrt{2\pi}}\left(\frac{Z}{a_0}\right)^{3/2} \sigma^2 e^{-\sigma/3}\sin^2\theta\cos 2\phi$

Note: $\sigma = Zr/a_0$, where $Z = 1$ for hydrogen; $a_0 = \epsilon_0 h^2/\pi m e^2 = 5.29 \times 10^{-11}$ m.

energy values—the energy levels are quantized. Finally, note that this is exactly the same equation for energy as obtained in the Bohr model.

So that you have an idea of what they look like, the first few wave functions for hydrogen are shown in Table 12.1, along with the three quantum numbers n, ℓ, and m_ℓ.

When we solve the Schrödinger equation for the hydrogen atom, some of the solutions contain complex numbers (that is, they contain $i = \sqrt{-1}$). Because it is more convenient physically to deal with orbitals that contain only real numbers, the complex orbitals are usually combined (added and subtracted) to remove the complex portions. For example, the p_x and p_y orbitals shown in Table 12.1 are combinations of the complex orbitals that correspond to values of m_ℓ of $+1$ and -1. These orbitals are indicated with a brace in Table 12.1. The last four d orbitals listed are also obtained by combination of complex orbitals, as indicated by braces in Table 12.1.

12.8 The Physical Meaning of a Wave Function

Now that we have examined some of the mathematical details of the quantum mechanical treatment of the hydrogen atom, we need to consider what it all means. What is a wave function, and what does it tell us about the electron to which it applies? First, a warning: There is always danger in taking a mathematical description of nature and using our human experiences to interpret it. Although our attempts to attach physical significance to mathematical descriptions are quite useful to us as we try to understand how nature operates, they must be viewed with caution. Simple pictorial models of a particular natural phenomenon always oversimplify the phenomenon and should not be taken too literally. With this caveat we will proceed to try to picture what the "quantum mechanical atom" is like.

Recall that the uncertainty principle indicates that there is no way of knowing the detailed movements of the electron in a hydrogen atom. Given this severe limitation, what then is the physical meaning of a wave function for an electron? Although the function itself has no easily visualized meaning, as we mentioned in the treatment of the particle in a box, the square of the wave function does have a physical significance. *The square of the function evaluated at a particular point in space indicates the probability of finding an electron near that point.* For example, suppose we have two positions in space: one defined by the coordinates r_1, θ_1, and ϕ_1 and the other defined by the coordinates r_2, θ_2, and ϕ_2. The relative probability of finding the electron near positions 1 and 2 is determined by substituting the values of r, θ, and ϕ for the two positions into the wave function, squaring the function value, and computing the following ratio:

The square of the function here means the square of the magnitude, $|\psi|^2$. This distinction is important when orbitals with complex numbers are being considered: $|\psi|^2 = (\text{real part})^2 + (\text{imaginary part})^2$.

$$\frac{[\psi(r_1, \theta_1, \phi_1)]^2\, dv}{[\psi(r_2, \theta_2, \phi_2)]^2\, dv} = \frac{N_1}{N_2}$$

The quotient N_1/N_2 is the ratio of the probabilities of finding the electron in the infinitesimally small volume elements dv around points 1 and 2. For example, if the value of the ratio N_1/N_2 is 100, the electron is 100 times more likely to be found at position 1 than at position 2. The model gives no information concerning when the electron will be at either position or how it moves between the positions. This vagueness is consistent with the concept of the Heisenberg uncertainty principle.

The square of the wave function is most conveniently represented as a **probability distribution,** in which the intensity of color is used to indicate the probability value at a given point in space. The probability distribution for the hydrogen 1s orbital is shown in Fig. 12.16(a). The best way to think about this diagram is as a three-dimensional time exposure, with the electron as a tiny moving light. The more times the electron visits a particular point, the darker the negative becomes. Thus the darkness (intensity) of a point indicates the probability of finding an electron at that position. This diagram is sometimes known as an *electron density map;* electron density and electron probability mean the same thing.

Another way of representing the electron probability distribution for the 1s orbital is to calculate the probability at points along a line drawn outward

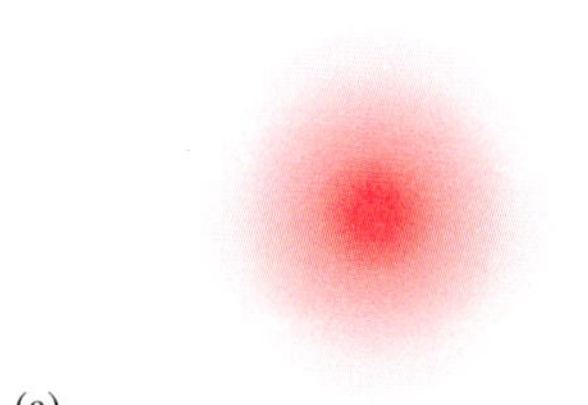

(a)

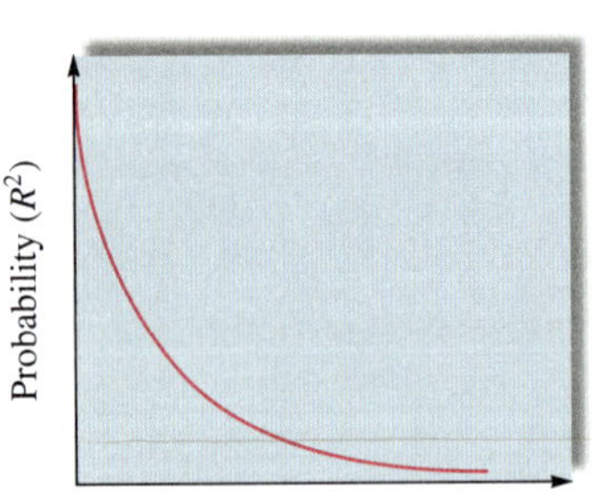

(b)

FIGURE 12.16

(a) The probability distribution for the hydrogen 1s orbital in three-dimensional space. (b) The probability density of the electron at points along a line drawn outward from the nucleus in any direction for the hydrogen 1s orbital.

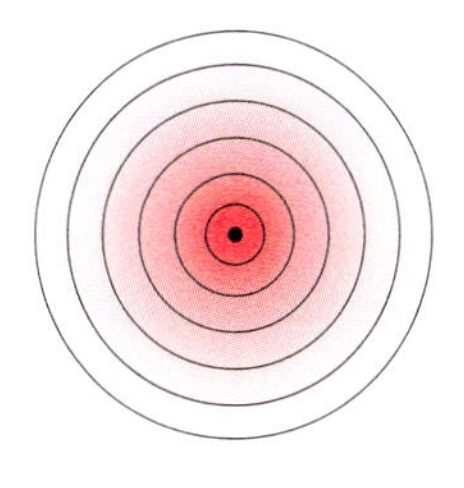

(a)

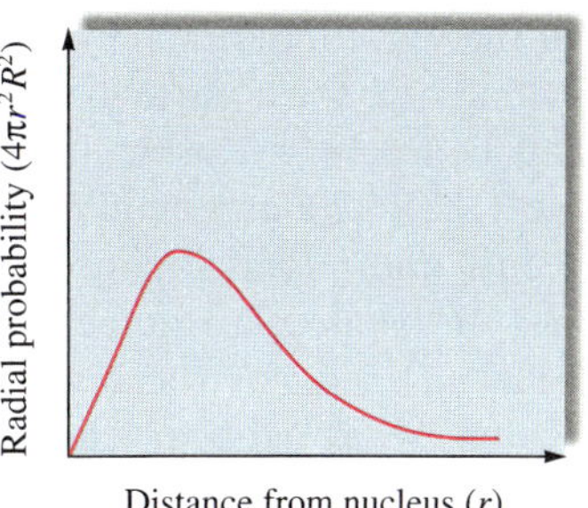

(b)

FIGURE 12.17

(a) Cross section of the hydrogen 1s orbital probability distribution divided into successive thin spherical shells. (b) The radial probability distribution. A plot of the total probability of finding the electron in each thin spherical shell as a function of distance from the nucleus.

in any direction from the nucleus. The result is shown in Fig. 12.16(b), where R^2 (the square of the radial part—the part that depends on r—of the 1s orbital) is plotted versus r. Note that the probability of finding the electron at a particular position is greatest close to the nucleus and that it drops off rapidly as the distance from the nucleus increases.

We are also interested in knowing the *total* probability of finding the electron in the hydrogen atom at a particular *distance* from the nucleus. Imagine that the space around the hydrogen nucleus is made up of a series of thin spherical shells (rather like layers in an onion), as shown in Fig. 12.17(a). When the total probability of finding the electron in each spherical shell is plotted versus the distance from the nucleus, the plot in Fig. 12.17(b) is obtained. This graph is called the **radial probability distribution,** which is a plot of $4\pi r^2R^2$ versus r, where R represents the radial part of the wave function.

The maximum in the curve occurs because of two opposing effects. The probability of finding an electron at a particular position is greatest near the nucleus, but the volume of the spherical shell increases with the distance from the nucleus. Therefore, as we move away from the nucleus, the probability of finding the electron at a given position decreases. However, we are summing more positions. Thus the total probability increases to a certain radius and then decreases as the electron probability at each position becomes very small. Mathematically, the maximum occurs because in the function $4\pi r^2R^2$, r^2 increases with r while R^2 decreases with r [see Fig. 12.16(b)]. For the hydrogen 1s orbital, the maximum radial probability (the distance at which the electron is most likely to be found) occurs at a distance of 5.29×10^{-2} nm, or 0.529 Å (angstrom), from the nucleus. Interestingly, this distance is exactly the radius of the innermost orbit in the Bohr model and thus is called the Bohr radius, denoted by a_0. Note that in Bohr's model the electron is assumed to have a circular path and so is *always* found at this distance. In the wave mechanical model the specific electron motions are unknown; therefore, this is the *most probable* distance at which the electron is found.

1 Å = 10^{-10} m; the angstrom is often used as the unit for atomic radius because of its convenient size. Another convenient unit is the picometer (1 pm = 10^{-12} m).

One more characteristic of the hydrogen 1s orbital that we must consider is its size. As we can see from Fig. 12.16, the size of this orbital cannot be precisely defined, since the probability never becomes zero (although it drops to an extremely small value at large values of r). Therefore, the hydrogen 1s orbital has no distinct size. However, it is useful to have a definition of relative orbital size. *The normally accepted arbitrary definition of the size of the hydrogen 1s orbital is the radius of the sphere that encloses 90% of the total electron probability.* That is, 90% of the time the electron is found inside this sphere. Application of this rule to the hydrogen atom 1s orbital gives a sphere with radius 2.6 a_0, or 1.4×10^{-10} m (140 pm).

So far we have described only the lowest-energy wave function in the hydrogen atom, the 1s orbital. Hydrogen has many other orbitals, which are described in the next section.

12.9 The Characteristics of Hydrogen Orbitals

Quantum Numbers

As we have seen, when we solve the Schrödinger equation for the hydrogen atom, we find many wave functions (orbitals) that satisfy it. Each of these orbitals is characterized by a set of quantum numbers that arise when the boundary conditions are applied. Now we will systematically describe these

TABLE 12.2

The Angular Momentum Quantum Numbers and Corresponding Letter Symbols

Value	Letter Used
0	*s*
1	*p*
2	*d*
3	*f*
4	*g*

quantum numbers in terms of the values they can assume and their physical meanings.

The **principal quantum number** (n), which can have integral values (1, 2, 3, . . .), is related to the size and energy of the orbital. As n increases, the orbital becomes larger and the electron spends more time farther from the nucleus. An increase in n also means higher energy because the electron is less tightly bound to the nucleus, and the energy is less negative. The **angular momentum quantum number** (ℓ) can have integral values from 0 to $n - 1$ for each value of n. This quantum number relates to the angular momentum of an electron in a given orbital. The dependence of the wave functions on ℓ determines the shapes of the atomic orbitals. The value of ℓ for a particular orbital is commonly assigned a letter: $\ell = 0$ is called *s*, $\ell = 1$ is called *p*, $\ell = 2$ is called *d*, and $\ell = 3$ is called *f* (see Table 12.2). The **magnetic quantum number** (m_ℓ) can have integral values between ℓ and $-\ell$, including zero. The value of m_ℓ relates to the orientation in space of the angular momentum associated with the orbital. As we mentioned earlier, many of the familiar atomic orbitals are actually a combination of a complex orbital characterized by m_ℓ and one characterized by $-m_\ell$.

$n = 1, 2, 3, \ldots$
$\ell = 0, 1, \ldots, (n - 1)$
$m_\ell = -\ell, \ldots, 0, \ldots, +\ell$

The labels *s*, *p*, *d*, and *f* are used for historical reasons. They originally referred to characteristics of lines observed in the atomic spectra: *s* (sharp), *p* (principal), *d* (diffuse), and *f* (fundamental). Beyond *f* the letters become alphabetic: *g*, *h*, . . . , skipping *j*, which is reserved as a symbol for angular momentum.

Example 12.7

For principal quantum level $n = 5$, determine the number of subshells (different values of ℓ) and give the designation of each.

Solution For $n = 5$ the allowed values of ℓ run from 0 to 4 ($n - 1 = 5 - 1$). Thus the subshells and their designations are

$\ell = 0$	$\ell = 1$	$\ell = 2$	$\ell = 3$	$\ell = 4$
5*s*	5*p*	5*d*	5*f*	5*g*

Number of Orbitals per Subshell

s = 1
p = 3
d = 5
f = 7
g = 9

The first four levels of orbitals in the hydrogen atom are listed with their quantum numbers in Table 12.3. Note that each set of orbitals with a given value of ℓ (sometimes called a **subshell**) is designated by giving the value of n and the letter for ℓ. Thus an orbital where $n = 2$ and $\ell = 1$ is symbolized as 2*p*. There are three 2*p* orbitals, which have different orientations in space. We will describe these orbitals in the next section.

TABLE 12.3

Quantum Numbers for the First Four Levels of Orbitals in the Hydrogen Atom

n	ℓ	Orbital Designation	m_ℓ	Number of Orbitals
1	0	1*s*	0	1
2	0	2*s*	0	1
	1	2*p*	−1, 0, +1	3
3	0	3*s*	0	1
	1	3*p*	−1, 0, 1	3
	2	3*d*	−2, −1, 0, 1, 2	5
4	0	4*s*	0	1
	1	4*p*	−1, 0, 1	3
	2	4*d*	−2, −1, 0, 1, 2	5
	3	4*f*	−3, −2, −1, 0, 1, 2, 3	7

Orbital Shapes and Energies

We have seen that the meaning of an orbital is illustrated most clearly by a probability distribution. Each orbital in the hydrogen atom has a unique probability distribution. We also have seen that another means of representing an orbital is by the surface that surrounds 90% of the total electron probability. These three types of representations for the hydrogen 1*s*, 2*s*, and 3*s* orbitals are shown in Fig. 12.18. Note the characteristic spherical shape of each of the *s* orbitals. Note also that the 2*s* and 3*s* orbitals contain areas of high probability separated by areas of zero probability. These latter areas are called **nodal surfaces,** or simply **nodes.** The number of nodes increases as n increases. For *s* orbitals, the number of nodes is given by $n - 1$. For our purposes, however, we will think of *s* orbitals only in terms of their overall spherical shape, which becomes larger as the value of n increases.

n value ↓ $2p_x$ ← orientation in space ↑ ℓ value

Two types of representations for the 2*p* orbitals (there are no 1*p* orbitals) are shown in Fig. 12.19. Note that the *p* orbitals are not spherical, like *s* orbitals, but have two **lobes** separated by a node at the nucleus. The *p* orbitals are labeled according to the axis of the Cartesian coordinate system along which the lobes lie. For example, the 2*p* orbital with lobes along the x axis is called the $2p_x$ orbital.

At this point it is useful to remember that mathematical functions have signs. For example, a simple sine wave (see Fig. 12.1) oscillates from positive to negative and repeats this pattern. Atomic orbital functions also have signs. The functions for *s* orbitals are positive everywhere in three-dimensional space. That is, when the *s* orbital function is evaluated at any point in space, it results in a positive number. In contrast, the *p* orbital functions have different signs in different regions of space. For example, the p_z orbital has a positive

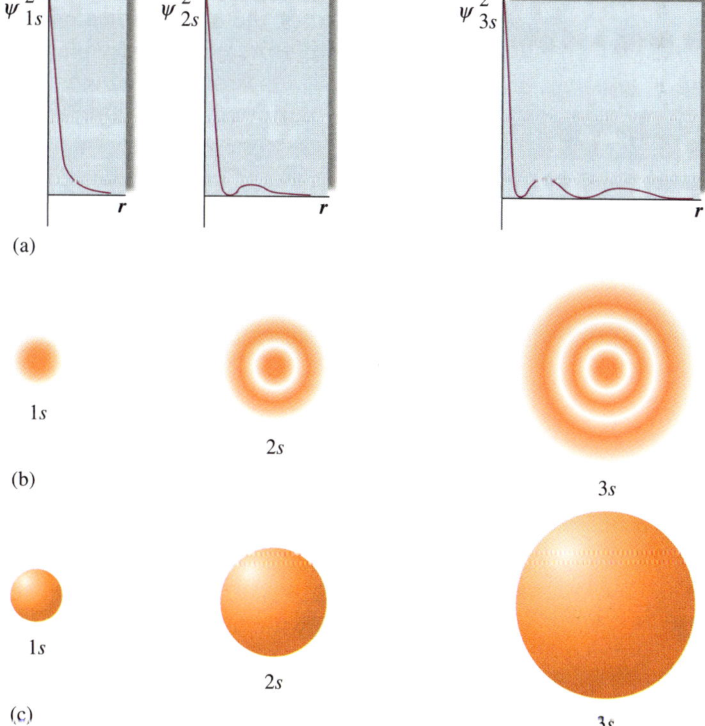

FIGURE 12.18

Three representations of the hydrogen 1*s*, 2*s*, and 3*s* orbitals. (a) The square of the wave function. (b) "Slices" of the three-dimensional electron density. (c) The surfaces that contain 90% of the total electron probability (the "sizes" of the orbitals).

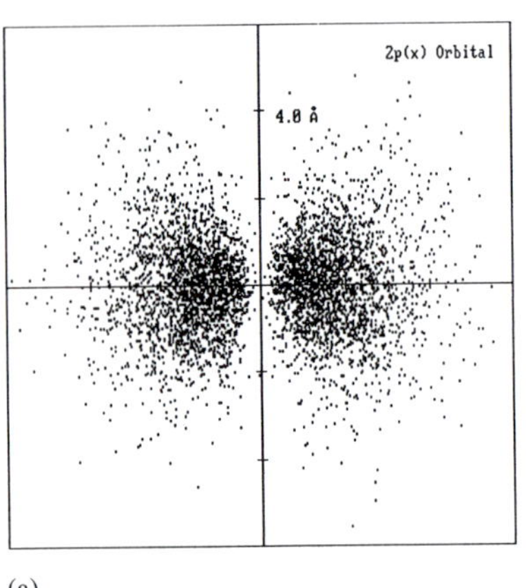

(a)

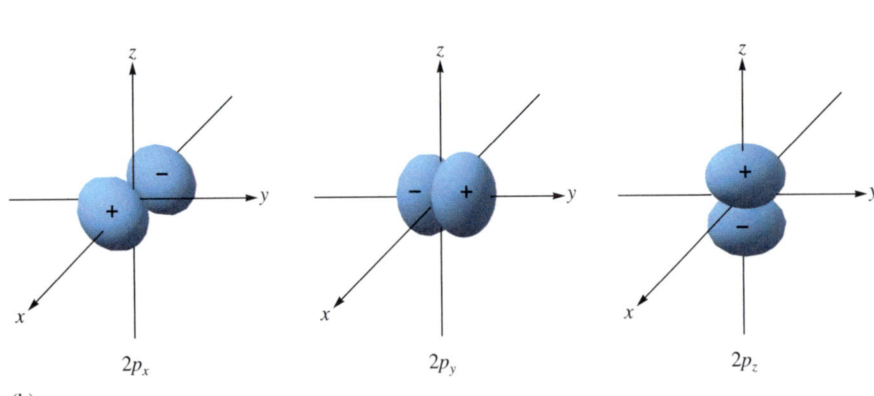

(b)

FIGURE 12.19

Representation of the $2p$ orbitals. (a) The electron probability distribution for a $2p$ orbital. Generated from a program by Robert Allendoerfer on Project SERAPHIM disk PC 2402; reprinted with permission. (b) The boundary surface representations of all three $2p$ orbitals. Note that the signs inside the surface indicate the phases (signs) of the orbital in that region of space.

sign in all the regions of space in which z is positive and has a negative sign when z is negative. This behavior is indicated in Fig. 12.19(b) by the positive and negative signs inside their boundary surfaces. It is important to understand that these are mathematical signs, not charges. Just as a sine wave has alternating positive and negative phases, p orbitals also have positive and negative phases. The phases of the p_x, p_y, and p_z orbitals are indicated in Fig. 12.19(b). As you might expect from our discussion of the s orbitals, the $3p$ orbitals have a more complex probability distribution than that of the $2p$ orbitals (see Fig. 12.20), but they can still be represented by the same boundary surface shapes. The surfaces just grow larger as the value of n increases.

There are no d orbitals that correspond to principal quantum levels $n = 1$ and $n = 2$. The d orbitals ($\ell = 2$) first occur in level $n = 3$. The five $3d$ orbitals have the shapes shown in Fig. 12.21. The d orbitals have two different fundamental shapes. Four of the orbitals (d_{xz}, d_{yz}, d_{xy}, and $d_{x^2-y^2}$) have four lobes centered in the plane indicated in the orbital label. Note that d_{xy} and $d_{x^2-y^2}$ are both centered in the xy plane; the lobes of $d_{x^2-y^2}$ lie *along* the x and y axes, but the lobes of d_{xy} lie *between* the axes. The fifth orbital, d_{z^2}, has a unique shape with two lobes along the z axis and a "belt" centered in the xy plane. The signs (phases) of the d orbital functions are indicated inside the boundary surfaces. The d orbitals for levels $n > 3$ look like the $3d$ orbitals but have larger lobes.

The f orbitals first occur in level $n = 4$, and as might be expected, they have shapes even more complex than those of the d orbitals. Figure 12.22 shows representations of the $4f$ orbitals ($\ell = 3$) along with their designations. These orbitals are not involved in the bonding in any of the compounds we will consider in this text. Their shapes and labels are included here for completeness. Because of their complexity, the phases of the f orbital functions are not represented in this diagram.

So far we have talked about the shapes of the hydrogen atomic orbitals but not about their energies. For the hydrogen atom the energy of a particular orbital is determined by its value of n. Thus *all* orbitals with the same value of n have the *same energy*—they are said to be **degenerate.** This feature is illustrated in Fig. 12.23, where the energies for the orbitals in the first three quantum levels for hydrogen are shown.

Hydrogen's single electron can occupy any of its atomic orbitals. However, in the lowest energy state, the ground state, the electron resides in the $1s$ orbital. If energy is put into the atom, the electron can be transferred to a higher-energy orbital, producing an excited state.

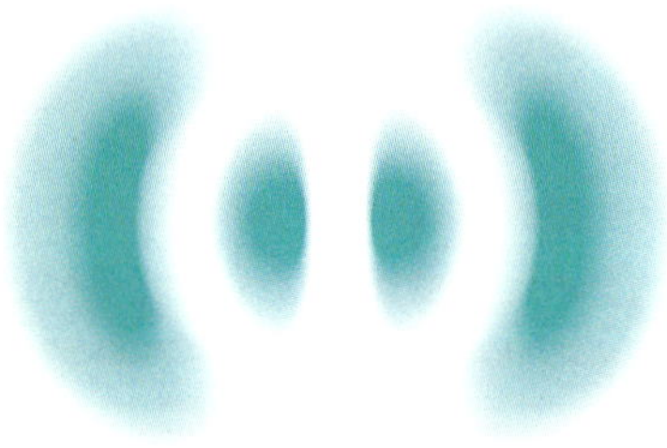

FIGURE 12.20

A cross section of the electron probability distribution for a $3p$ orbital.

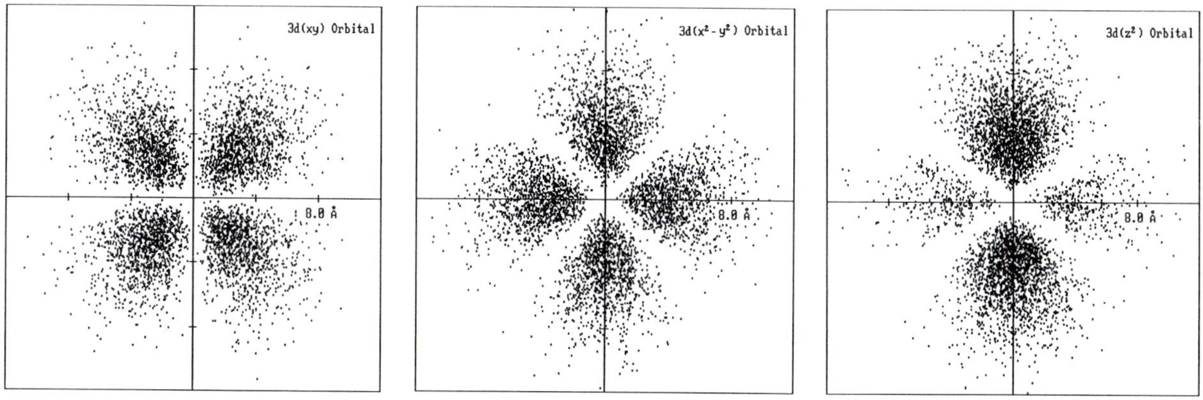

(a)

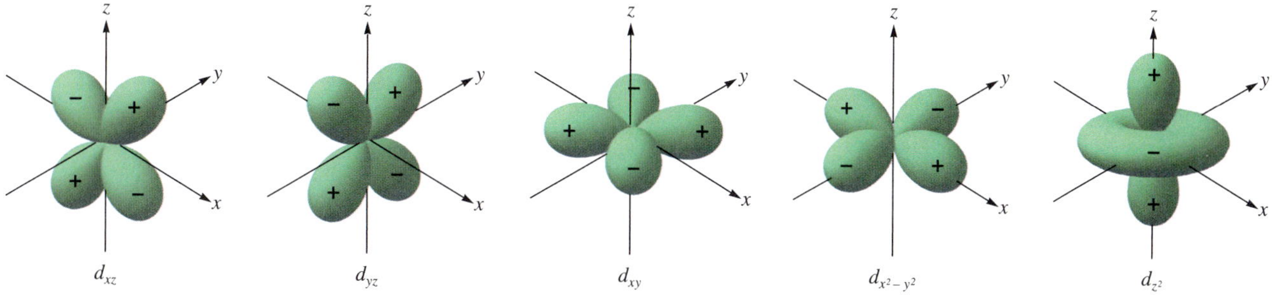

(b)

FIGURE 12.21

Representation of the 3*d* orbitals. (a) Electron density plots of selected 3*d* orbitals. Generated from a program by Robert Allendoerfer on Project SERAPHIM disk PC 2402; reprinted with permission. (b) The boundary surfaces of all five 3*d* orbitals, with the signs (phases) indicated.

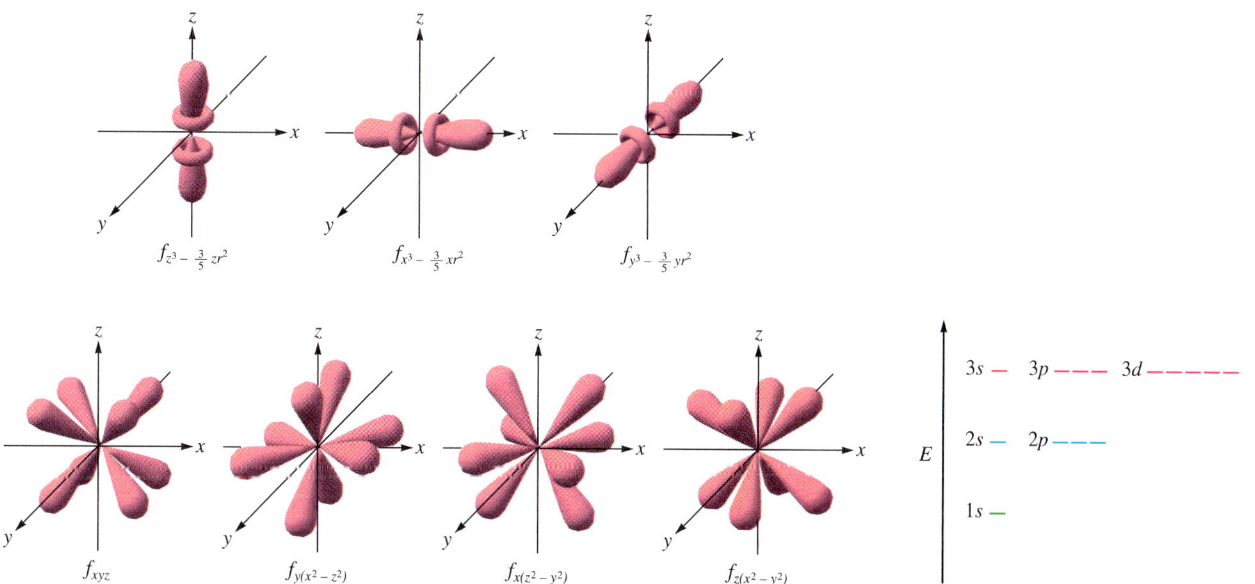

FIGURE 12.22

Representation of the 4*f* orbitals in terms of their boundary surfaces.

FIGURE 12.23

Orbital energy levels for the hydrogen atom.

The Hydrogen Atom

1. In the quantum mechanical model the electron is described as a wave. This representation leads to a series of wave functions (orbitals) that describe the possible energies and spatial distributions available to the electron.
2. In agreement with the Heisenberg uncertainty principle, the model cannot specify the detailed electron motions. Instead, the square of the wave function represents the probability distribution of the electron in that orbital. This approach allows us to picture orbitals in terms of probability distributions, or electron density maps.
3. The size of an orbital is arbitrarily defined as the surface that contains 90% of the total electron probability.
4. The hydrogen atom has many types of orbitals. In the ground state the single electron resides in the 1*s* orbital. The electron can be excited to higher-energy orbitals if the atom absorbs energy.

12.10 Electron Spin and the Pauli Principle

The electron does not literally spin. The term "spin" is just a name for the intrinsic angular momentum of the electron.

The concept of **electron spin** was developed by Samuel Goudsmit and George Uhlenbeck in 1925 while they were graduate students at the University of Leyden in the Netherlands. They found that a fourth quantum number (in addition to n, ℓ, and m_ℓ) was necessary to account for the details of the emission spectra of atoms. The new quantum number adopted to describe this phenomenon, called the **electron spin quantum number** (m_s), can have only one of two values, $+\frac{1}{2}$ and $-\frac{1}{2}$.

For our purposes the main significance of the electron spin quantum number is connected with the postulate of Austrian physicist Wolfgang Pauli (1900–1958), which is often stated as follows: *In a given atom no two electrons can have the same set of four quantum numbers* (n, ℓ, m_ℓ, and m_s). This is called the **Pauli exclusion principle.** Since electrons in the same orbital have the same values of n, ℓ, and m_ℓ, this postulate requires that they have different values of m_s. Since only two values of m_s are allowed, we might paraphrase the Pauli principle as follows: *An orbital can hold only two electrons, and they must have opposite spins.* This principle will have important consequences when we use the atomic model to relate the electron arrangement of an atom to its position in the periodic table.

$m_s = +\frac{1}{2}$ or $-\frac{1}{2}$

Each orbital can hold a maximum of two electrons.

12.11 Polyelectronic Atoms

The quantum mechanical model provides a description of the hydrogen atom that agrees very well with experimental data. However, the model would not be very useful if it did not account for the properties of the other atoms as well.

To see how the model applies to **polyelectronic atoms**—that is, atoms with more than one electron—let's consider helium, which has two protons in its nucleus and two electrons:

2+ e^- e^-

There are three energy contributions that must be considered in the description of the helium atom: (1) the kinetic energy of the electrons as they move around the nucleus, (2) the potential energy of attraction between the nucleus and the electrons, and (3) the potential energy of repulsion between the two electrons.

Although this atom can be readily described in terms of the quantum mechanical model, the Schrödinger equation that results cannot be solved exactly. The difficulty arises in dealing with the repulsion between the electrons. This so-called *electron correlation problem* refers to the fact that we cannot rigorously account for the effect a given electron has on the motions of the other electrons in an atom.

The electron correlation problem occurs with all polyelectronic atoms. To treat these systems using the quantum mechanical model, we must make approximations. The simplest approximation involves treating each electron as if it were moving in a *field of charge that is the net result of the nuclear attraction and the average repulsions of all the other electrons.* To see how this is done, let's compare the neutral helium atom and the He^+ ion:

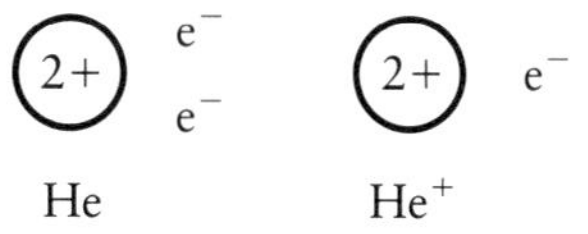

(2+) e^- e^- He   (2+) e^- He^+

What energy is required to remove an electron from each of these species? Experiments show that 2372 kJ of energy is required to remove one electron from all the atoms in a mole of helium. Removing the one electron from each ion in a mole of He^+ ions requires 5248 kJ of energy. Thus it takes more than twice as much energy to remove an electron from a He^+ ion than from a He atom.

Why is there such a large difference? In both cases the nucleus has a 2+ charge. However, in the helium atom there are two electrons that repel each other, but in the He^+ ion there is only one electron and thus no electron–electron repulsion. That is, the large difference in the energies required to remove one electron must arise from the electron–electron repulsions in the neutral atom. Each electron in the He atom is much less tightly bound to the nucleus than the electron in the He^+ ion. In other words, the effectiveness of the positively charged nucleus in binding the electrons has been decreased by the repulsions between the electrons. Thus the *effect of the electron repulsions can be thought of as reducing the nuclear charge* to an apparent value of less than 2+ toward a particular electron, as shown below:

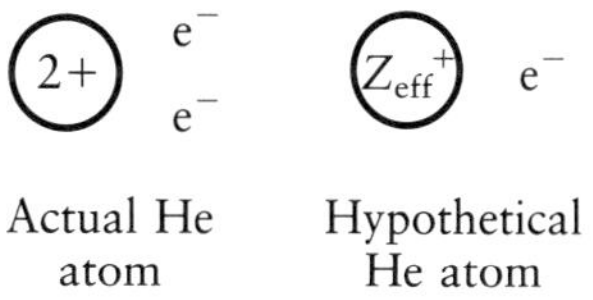

(2+) e^- e^- Actual He atom   (Z_{eff}^+) e^- Hypothetical He atom

The *apparent* nuclear charge, or the **effective nuclear charge,** is designated Z_{eff}. For a helium atom Z_{eff}, the charge "experienced" by each electron, is less than 2. In general,

$$\text{Effective nuclear charge} = Z_{eff} = Z_{actual} - (\text{effect of electron repulsions})$$

$Z_{eff} = Z -$ effect of electron repulsions

where $Z_{actual} = Z$, the atomic number (number of protons).

This simplification allows us to treat each electron individually, where each electron is viewed as moving under the influence of a positive nuclear charge Z_{eff}. This simplified atom has one electron like hydrogen, but with a positive nuclear charge of Z_{eff} instead of 1. We therefore can find the energy and wave function for each helium electron by substituting Z_{eff} in place of $Z = 1$ in the hydrogen wave mechanical equations. When we do this, we find that both helium electrons reside in a modified 1*s* orbital that is spherical, like that for the hydrogen atom, but smaller because Z_{eff} is greater than 1. The larger nuclear charge draws each of the electrons closer to the nucleus, therefore binding each more tightly than the electron in hydrogen is bound. The increased nuclear charge of the helium atom is more important than the repulsions between the two electrons, so that each of the electrons in helium is bound more tightly than the electron in the hydrogen atom.

The model we have just described so greatly oversimplifies the structure of polyelectronic atoms that, although it produces some qualitatively useful ideas about polyelectronic atoms, it is not satisfactory for the description of quantitative atomic properties. To get an accurate description of polyelectronic atoms, we must take into account the electron–electron interactions in a much more detailed manner than simply assuming that they reduce the nuclear charge.

Nothing we do will allow us to solve the problem exactly, because the electron motions are correlated. That is, because electrons repel each other, the movement of a given electron will affect the movements of all the others. This correlation problem is reflected in the Schrödinger equation for polyelectronic atoms in the following way. Because the equation contains energy terms that simultaneously involve two different electrons, it cannot be separated rigorously into equations that involve only one electron. Thus the Schrödinger equation for polyelectronic atoms cannot be solved exactly.

One approach for dealing with this problem is to solve the equation numerically. That is, a computer is used to find the numerical values of the wave functions at each point in space that produce the lowest overall energy for the atom. Although this approach allows accurate calculation of atomic properties, it suffers from two major disadvantages: It is prohibitively time-consuming for any but the simplest of atoms, and the results are very difficult to interpret physically.

A more practical approach, the **self-consistent field (SCF) method,** is now used almost universally to treat polyelectronic atoms. In this method a given electron is assumed to be moving in a potential energy field that is a result of both the nucleus and the average "electron density" of all the other electrons in the atom (residing in their various orbitals). This approximation allows the many-electron Schrödinger equation to be separated into a set of one-electron equations that can be solved by computers. The orbitals (one-electron functions) that result from this approach have angular properties exactly the same as those of the hydrogen orbitals but have radial characteristics somewhat different from those of the hydrogen orbitals. Although the quantum numbers obtained in the description of the hydrogen atom do not apply exactly to the orbitals obtained from the self-consistent field approach, we still use them as convenient labels for the atomic orbitals in polyelectronic atoms.

We will have more to say later about the self-consistent field approach, but first we will see how the atomic orbitals for polyelectronic atoms can be used to account for the form of the periodic table of the elements.

12.12 The History of the Periodic Table

The modern periodic table contains a tremendous amount of useful information. In this section we will discuss the origin of this valuable tool; later, we will see how the quantum mechanical model for the atom explains the periodicity of chemical properties. Certainly one of the greatest successes of the quantum mechanical model is its ability to account for the arrangement of the elements on the periodic table.

The periodic table was originally constructed to represent the patterns observed in the chemical properties of the elements. As chemistry progressed during the eighteenth and nineteenth centuries, it became evident that the earth is composed of a great many elements with very different properties. Things are much more complicated than the simple model of earth, air, fire, and water suggested by the ancients. At first, the array of elements and properties was bewildering. Gradually, however, patterns were noticed.

The first chemist to recognize patterns was Johann Dobereiner, who found several groups of three elements with similar properties—for example, chlorine, bromine, and iodine. However, as Dobereiner attempted to expand this model of **triads** (as he called them) to the rest of the known elements, it became clear that this concept was severely limited.

The next notable attempt was made by English chemist John Newlands, who in 1864 suggested that elements should be arranged in **octaves.** He noticed that certain properties seemed to repeat for every eighth element in a way similar to the musical scale, which repeats for every eighth tone. Although this model managed to group several elements with similar properties, it was not generally successful.

The present form of the periodic table was conceived independently by two chemists in 1869: German Julius Lothar Meyer and Russian Dmitri Ivanovich Mendeleev (Fig. 12.24). Usually, Mendeleev is given most of the credit because it was he who showed how useful the table could be in predicting the existence and properties of yet unknown elements. For example, in 1872 when Mendeleev first published his table (see Fig. 12.25), the elements gallium, scandium, and germanium were unknown. Mendeleev correctly predicted the existence and properties of these elements from gaps in his periodic table. The data for germanium (which Mendeleev called *ekasilicon*) are

FIGURE 12.24

Dmitri Ivanovich Mendeleev (1834–1907), born in Siberia as the youngest of 17 children, taught chemistry at the University of St. Petersburg. In 1860 Mendeleev heard Italian chemist Cannizzaro lecture on a reliable method for determining the correct atomic masses of the elements. This important development paved the way for Mendeleev's own brilliant contribution to chemistry—the periodic table. In 1861 Mendeleev returned to St. Petersburg, where he wrote a book on organic chemistry. Later Mendeleev also wrote a book on inorganic chemistry, and he was struck by the fact that the systematic approach characterizing organic chemistry was lacking in inorganic chemistry. In attempting to systematize inorganic chemistry, he eventually arranged the elements in the form of the periodic table.

Mendeleev was a versatile genius who was interested in many fields of science. He worked on many problems associated with Russia's natural resources, such as coal, salt, and various metals. Being particularly interested in the petroleum industry, he visited the United States in 1876 to study the Pennsylvania oil fields. His interests also included meteorology and hot-air balloons. In 1887 he made an ascent in a balloon to study a total eclipse of the sun.

Tabelle II.

Reihen	Gruppe I. — R^2O	Gruppe II. — RO	Gruppe III. — R^2O^3	Gruppe IV. RH^4 RO^2	Gruppe V. RH^3 R^2O^5	Gruppe VI. RH^2 RO^3	Gruppe VII. RH R^2O^7	Gruppe VIII. — RO^4
1	H=1							
2	Li=7	Be=9,4	B=11	C=12	N=14	O=16	F=19	
3	Na=23	Mg=24	Al=27,3	Si=28	P=31	S=32	Cl=35,5	
4	K=39	Ca=40	—=44	Ti=48	V=51	Cr=52	Mn=55	Fe=56, Co=59, Ni=59, Cu=63.
5	(Cu=63)	Zn=65	—=68	—=72	As=75	Se=78	Br=80	
6	Rb=85	Sr=87	?Yt=88	Zr=90	Nb=94	Mo=96	—=100	Ru=104, Rh=104, Pd=106, Ag=108.
7	(Ag=108)	Cd=112	In=113	Sn=118	Sb=122	Te=125	J=127	
8	Cs=133	Ba=137	?Di=138	?Ce=140	—	—	—	— — — —
9	(—)	—	—	—	—	—	—	
10	—	—	?Er=178	?La=180	Ta=182	W=184	—	Os=195, Ir=197, Pt=198, Au=199.
11	(Au=199)	Hg=200	Tl=204	Pb=207	Bi=208	—	—	
12	—	—	—	Th=231	—	U=240	—	— — — —

FIGURE 12.25

Mendeleev's early periodic table, published in 1872. Note the spaces left for missing elements with atomic masses 44, 68, 72, and 100. From *Annalen der Chemie und Pharmacie,* VIII, Supplementary Volume for 1872, page 511.

shown in Table 12.4. Note the excellent agreement between the actual values and Mendeleev's predictions, which were based on the properties of other members in the group of elements similar to germanium.

Using his table, Mendeleev was also able to correct several values of atomic masses. For example, the original atomic mass of 76 for indium was based on the assumption that indium oxide had the formula InO. This atomic mass placed indium, which has metallic properties, among the nonmetals. Mendeleev assumed that the atomic mass was probably incorrect and proposed that the formula of indium oxide was really In_2O_3. On the basis of this (correct) formula, indium has an atomic mass of about 113, placing the element among the metals. Mendeleev also corrected the atomic masses of beryllium and uranium.

Because of its obvious usefulness, Mendeleev's periodic table was almost universally adopted, and it remains one of the most valuable tools at the chemist's disposal. For example, it is still used to predict the properties of elements yet to be discovered, as shown in Table 12.5.

TABLE 12.4

Comparison of the Properties of Germanium as Predicted by Mendeleev and as Actually Observed

Properties of Germanium	Predicted in 1871	Observed in 1886
Atomic mass	72	72.3
Density	5.5 g/cm^3	5.47 g/cm^3
Specific heat	0.31 J °C^{-1} g^{-1}	0.32 J °C^{-1} g^{-1}
Melting point	Very high	960°C
Oxide formula	RO_2	GeO_2
Oxide density	4.7 g/cm^3	4.70 g/cm^3
Chloride formula	RCl_4	$GeCl_4$
Boiling point of chloride	100°C	86°C

TABLE 12.5

Predicted Properties of Elements 113 and 114

Property	Element 113	Element 114
Chemically like	Thallium	Lead
Atomic mass	297	298
Density	16 g/mL	14 g/mL
Melting point	430°C	70°C
Boiling point	1100°C	150°C

A current version of the periodic table is shown inside the front cover of this book. The only fundamental difference between this table and that of Mendeleev is that the current table lists the elements in order of atomic number rather than atomic mass. The reason for this will become clear later in this chapter as we explore the electron arrangements of the atom.

12.13 The Aufbau Principle and the Periodic Table

We can use the quantum mechanical model of the atom to show how the electron arrangements in the atomic orbitals of the various atoms account for the organization of the periodic table. Our main assumption here is that all atoms have orbitals similar to those that have been described for the hydrogen atom. *As protons are added one by one to the nucleus to build up the elements, electrons are similarly added to these atomic orbitals.* This is called the **aufbau principle.**

Aufbau is German for "building up."

Hydrogen has one electron, which occupies the 1*s* orbital in its ground state. The configuration for hydrogen is written as $1s^1$, which can be represented by the following *orbital diagram:*

		1*s*	2*s*	2*p*
H:	$1s^1$	[↑]	[ ]	[ ][ ][ ]

The arrow represents a particular electron spin state.

The next element, *helium*, has two electrons. Since two electrons with opposite spins can occupy an orbital, according to the Pauli exclusion principle, the electrons for helium are in the 1*s* orbital with opposite spins. This yields a $1s^2$ configuration:

		1*s*	2*s*	2*p*
He:	$1s^2$	[↑↓]	[ ]	[ ][ ][ ]

We will see in Section 12.14 why the 2*s* orbital is lower in energy than the 2*p* orbital.

Lithium has three electrons, two of which can go into the 1*s* orbital before the orbital is filled. Since the 1*s* orbital is the only orbital with $n = 1$, the third electron will occupy the lowest-energy orbital with $n = 2$, or the 2*s* orbital, giving a $1s^2 2s^1$ configuration:

		1*s*	2*s*	2*p*
Li:	$1s^2 2s^1$	[↑↓]	[↑]	[ ][ ][ ]

The next element, *beryllium,* has four electrons, which occupy the 1s and 2s orbitals:

		1s	2s	2p
Be:	$1s^22s^2$	↑↓	↑↓	| | |

Boron has five electrons, four of which occupy the 1s and 2s orbitals. The fifth electron goes into the second type of orbital with $n = 2$, the 2p orbitals:

		1s	2s	2p
B:	$1s^22s^22p^1$	↑↓	↑↓	↑ | |

Since all the 2p orbitals are equivalent, it does not matter which 2p orbital the electron occupies.

Carbon has six electrons: Two electrons occupy the 1s orbital, two occupy the 2s orbital, and two occupy 2p orbitals. Since there are three equivalent 2p orbitals, the electrons will occupy *separate* 2p orbitals.

For an atom with unfilled subshells, the lowest energy is achieved by electrons occupying separate orbitals, as allowed by the Pauli exclusion principle.

This behavior is summarized by **Hund's rule** (named for German physicist F. H. Hund), which states that *the lowest-energy configuration for an atom is the one having the maximum number of unpaired electrons allowed by the Pauli principle in a particular set of degenerate orbitals.*

The configuration for carbon could be written $1s^22s^22p^12p^1$ to indicate that the electrons occupy separate 2p orbitals. However, the configuration is usually given as $1s^22s^22p^2$, and it is understood that the electrons are in different 2p orbitals. The orbital diagram for carbon is

		1s	2s	2p
C:	$1s^22s^22p^2$	↑↓	↑↓	↑ | ↑ |

Note the unpaired electrons in the 2p orbitals, as required by Hund's rule.

The configuration for *nitrogen,* which has seven electrons, is $1s^22s^22p^3$. The three electrons in 2p orbitals occupy separate orbitals:

		1s	2s	2p
N:	$1s^22s^22p^3$	↑↓	↑↓	↑ | ↑ | ↑

The configuration for *oxygen,* which has eight electrons, is $1s^22s^22p^4$. One of the 2p orbitals is now occupied by a pair of electrons with opposite spins, as required by the Pauli exclusion principle:

		1s	2s	2p
O:	$1s^22s^22p^4$	↑↓	↑↓	↑↓ | ↑ | ↑

The orbital diagrams and electron configurations for *fluorine* (nine electrons) and *neon* (ten electrons) are given below:

		1s	2s	2p
F:	$1s^22s^22p^5$	↑↓	↑↓	↑↓ | ↑↓ | ↑
Ne:	$1s^22s^22p^6$	↑↓	↑↓	↑↓ | ↑↓ | ↑↓

With neon, the orbitals with $n = 1$ and $n = 2$ are now completely filled.

For *sodium* the first ten electrons occupy the 1s, 2s, and 2p orbitals, and the eleventh electron must occupy the first orbital with $n = 3$, the 3s orbital.

[Ne] is shorthand for $1s^2 2s^2 2p^6$.

The electron configuration for sodium is $1s^2 2s^2 2p^6 3s^1$. To avoid writing the inner-level electrons, we often abbreviate this configuration as $[Ne]3s^1$, where [Ne] represents the electron configuration of neon, $1s^2 2s^2 2p^6$.

The next element, *magnesium,* has the configuration $1s^2 2s^2 2p^6 3s^2$, or $[Ne]3s^2$. Then the next six elements, *aluminum* through *argon,* have configurations obtained by filling the $3p$ orbitals one electron at a time. Figure 12.26 summarizes the electron configurations of the first 18 elements by giving the number of electrons in the type of orbital occupied last.

At this point it is useful to introduce the concept of **valence electrons,** *the electrons in the outermost principal quantum level of an atom.* The valence electrons of the nitrogen atom, for example, are the $2s$ and $2p$ electrons. For the sodium atom, the valence electron is the electron in the $3s$ orbital, and so on. Valence electrons are the most important electrons to chemists because they are involved in bonding, as we will see in the next two chapters. The inner electrons are known as **core electrons.**

Note in Fig. 12.26 that a very important pattern is developing: *The elements in the same group (vertical column of the periodic table) have the same valence electron configuration.* Remember that Mendeleev originally placed the elements in groups based on similarities in chemical properties. Now we understand the reason behind these groupings. Elements with the same valence electron configuration often show similar chemical behavior.

The element after argon is *potassium.* Since the $3p$ orbitals are fully occupied in argon, we might expect the next electron to go into a $3d$ orbital (recall that for $n = 3$ the orbitals are $3s$, $3p$, and $3d$). However, the chemistry of potassium is clearly very similar to that of lithium and sodium, indicating that the last electron in potassium occupies the $4s$ orbital instead of one of the $3d$ orbitals, a conclusion confirmed by many types of experiments. The electron configuration of potassium is

[Ar] is shorthand for $1s^2 2s^2 2p^6 3s^2 3p^6$.

$$\text{K:} \quad 1s^2 2s^2 2p^6 3s^2 3p^6 4s^1 \quad \text{or} \quad [\text{Ar}]4s^1$$

The next element is *calcium:*

$$\text{Ca:} \quad [\text{Ar}]4s^2$$

The next element, *scandium,* begins a series of ten elements (scandium through zinc) called the **transition metals,** whose configurations are obtained by adding electrons to the five $3d$ orbitals. The configuration of scandium is

$$\text{Sc:} \quad [\text{Ar}]4s^2 3d^1$$

When an electron configuration is given in this text, the orbitals are listed in the order in which they fill.

That of *titanium* is Ti: $[\text{Ar}]4s^2 3d^2$

And that of *vanadium* is V: $[\text{Ar}]4s^2 3d^3$

FIGURE 12.26

The electron configurations in the type of orbital occupied last for the first 18 elements.

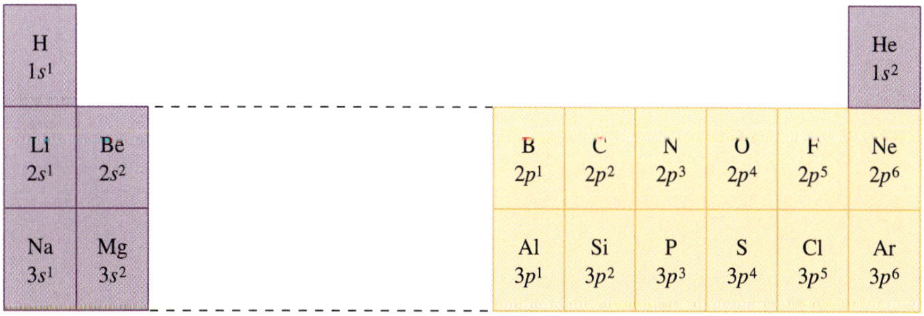

H $1s^1$							He $1s^2$
Li $2s^1$	Be $2s^2$	B $2p^1$	C $2p^2$	N $2p^3$	O $2p^4$	F $2p^5$	Ne $2p^6$
Na $3s^1$	Mg $3s^2$	Al $3p^1$	Si $3p^2$	P $3p^3$	S $3p^4$	Cl $3p^5$	Ar $3p^6$

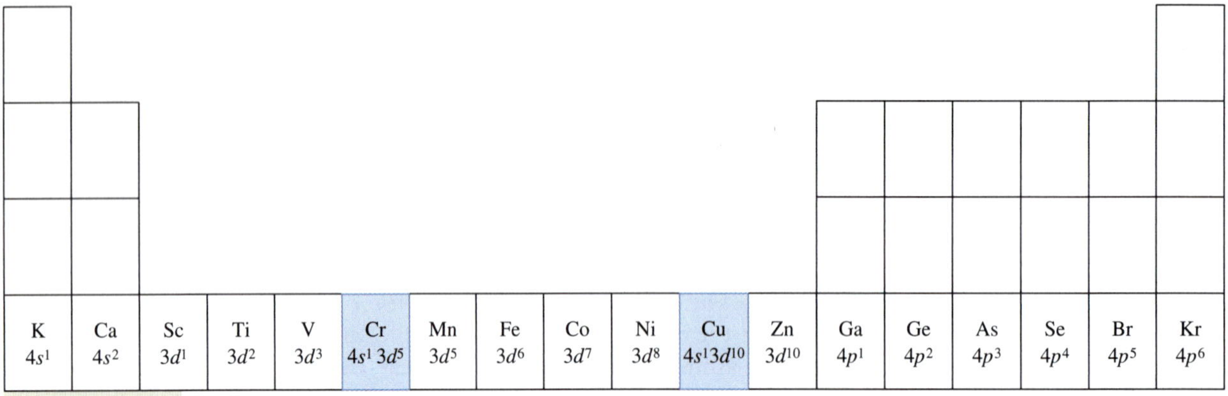

K	Ca	Sc	Ti	V	Cr	Mn	Fe	Co	Ni	Cu	Zn	Ga	Ge	As	Se	Br	Kr
$4s^1$	$4s^2$	$3d^1$	$3d^2$	$3d^3$	$4s^1 3d^5$	$3d^5$	$3d^6$	$3d^7$	$3d^8$	$4s^1 3d^{10}$	$3d^{10}$	$4p^1$	$4p^2$	$4p^3$	$4p^4$	$4p^5$	$4p^6$

FIGURE 12.27

Electron configurations for potassium through krypton. The transition metals (scandium through zinc) have the general configuration $[Ar]4s^2 3d^n$, except for chromium and copper.

Chromium is the next element. The expected configuration is $[Ar]4s^2 3d^4$. However, the observed configuration is

$$\text{Cr:} \quad [Ar]4s^1 3d^5$$

The explanation for this configuration of chromium is beyond the scope of this book. In fact, chemists are still disagreeing over the exact cause of this anomaly.* Note, however, that the observed configuration has both the $4s$ and $3d$ orbitals half-filled. This is a good way to remember the correct configuration.

The next four elements, *manganese* through *nickel,* have the expected configurations:

$$\text{Mn:} \quad [Ar]4s^2 3d^5 \qquad \text{Co:} \quad [Ar]4s^2 3d^7$$

$$\text{Fe:} \quad [Ar]4s^2 3d^6 \qquad \text{Ni:} \quad [Ar]4s^2 3d^8$$

The configuration for *copper* is expected to be $[Ar]4s^2 3d^9$. However, the observed configuration is

$$\text{Cu:} \quad [Ar]4s^1 3d^{10}$$

In this case a half-filled $4s$ orbital and a filled set of $3d$ orbitals characterize the actual configuration.

Zinc has the expected configuration:

$$\text{Zn:} \quad [Ar]4s^2 3d^{10}$$

The configurations of the transition metals are shown in Fig. 12.27. The next six elements, *gallium* through *krypton,* have configurations that correspond to filling the $4p$ orbitals (see Fig. 12.27).

The entire periodic table is represented in Fig. 12.28 in terms of which orbitals are being filled. The valence electron configurations are given in Fig. 12.29 on page 566.

From these two figures, note the following additional points:

The $(n + 1)s$ orbitals fill before the nd orbitals.

1. The $(n + 1)s$ orbitals always fill before the nd orbitals. For example, the $5s$ orbitals fill in rubidium and strontium before the $4d$ orbitals fill in the second row of transition metals (yttrium through cadmium).

*See M. P. Melrose and E. R. Scerri, *J. Chem. Educ.* 73 (1996): 498, for more information.

FIGURE 12.28

The orbitals being filled for elements in various parts of the periodic table. Note that when we move along a horizontal row (a period), the $(n + 1)s$ orbital fills before the nd orbital. The group labels indicate the number of valence electrons (ns plus np electrons) for the elements in each group.

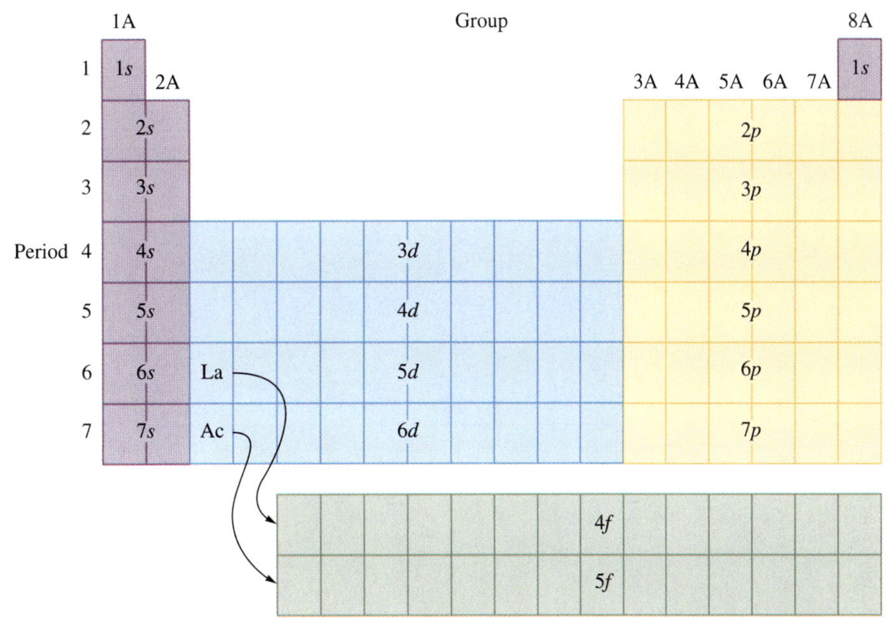

Lanthanides are elements in which the $4f$ orbitals are being filled.

Actinides are elements in which the $5f$ orbitals are being filled.

The group label tells the total number of valence electrons for that group.

2. After lanthanum, which has the configuration $[\mathrm{Xe}]6s^2 5d^1$, a group of 14 elements called the **lanthanide series,** or the lanthanides, occurs. This series of elements corresponds to the filling of the seven $4f$ orbitals. Note that sometimes one electron occupies a $5d$ instead of a $4f$ orbital. This occurs because the energies of the $4f$ and $5d$ orbitals are very similar.
3. After actinium, which has the configuration $[\mathrm{Rn}]7s^2 6d^1$, a group of 14 elements called the **actinide series,** or the actinides, occurs. This series corresponds to the filling of the seven $5f$ orbitals. Note that sometimes one or two electrons occupy the $6d$ orbitals instead of the $5f$ orbitals because these orbitals have very similar energies.
4. The group labels for the Groups 1A, 2A, 3A, 4A, 5A, 6A, 7A, and 8A indicate the *total number* of valence electrons for the atoms in these groups. For example, all the elements in Group 5A have the configuration ns^2np^3. (The d electrons fill one period late and are usually not counted as valence electrons.) The meaning of the group labels for the transition metals is not as clear as for the A group elements, so these will not be used in this text.
5. The groups labeled 1A, 2A, 3A, 4A, 5A, 6A, 7A, and 8A are often called the **main-group,** or **representative, elements.** Remember that every member of these groups has the same valence electron configuration.

In 1985 the International Union of Pure and Applied Chemistry (IUPAC), a body of scientists organized to standardize scientific conventions, recommended a new form for the periodic table, which the American Chemical Society has adopted (see Fig. 12.30 on page 567). In this new version the group number indicates the number of s, p, and d electrons added since the last noble gas. We will not use the new format in this book, but you should be aware that the familiar periodic table may soon be replaced by this or a similar format.

The results considered in this section are very important. We have seen that the wave mechanical model can be used to explain the arrangement of elements in the periodic table. This model allows us to understand that the

Representative Elements (groups 1–2) | *d*-Transition Elements (groups 3–12) | Representative Elements (groups 13–17) | Noble Gases (group 18)

Period number, highest occupied electron level	1 1A ns^1	2 2A ns^2	3	4	5	6	7	8	9	10	11	12	13 3A ns^2np^1	14 4A ns^2np^2	15 5A ns^2np^3	16 6A ns^2np^4	17 7A ns^2np^5	18 8A ns^2np^6
1	1 H $1s^1$																	2 He $1s^2$
2	3 Li $2s^1$	4 Be $2s^2$											5 B $2s^22p^1$	6 C $2s^22p^2$	7 N $2s^22p^3$	8 O $2s^22p^4$	9 F $2s^22p^5$	10 Ne $2s^22p^6$
3	11 Na $3s^1$	12 Mg $3s^2$											13 Al $3s^23p^1$	14 Si $3s^23p^2$	15 P $3s^23p^3$	16 S $3s^23p^4$	17 Cl $3s^23p^5$	18 Ar $3s^23p^6$
4	19 K $4s^1$	20 Ca $4s^2$	21 Sc $4s^23d^1$	22 Ti $4s^23d^2$	23 V $4s^23d^3$	24 Cr $4s^13d^5$	25 Mn $4s^23d^5$	26 Fe $4s^23d^6$	27 Co $4s^23d^7$	28 Ni $4s^23d^8$	29 Cu $4s^13d^{10}$	30 Zn $4s^23d^{10}$	31 Ga $4s^24p^1$	32 Ge $4s^24p^2$	33 As $4s^24p^3$	34 Se $4s^24p^4$	35 Br $4s^24p^5$	36 Kr $4s^24p^6$
5	37 Rb $5s^1$	38 Sr $5s^2$	39 Y $5s^24d^1$	40 Zr $5s^24d^2$	41 Nb $5s^14d^4$	42 Mo $5s^14d^5$	43 Tc $5s^14d^6$	44 Ru $5s^14d^7$	45 Rh $5s^14d^8$	46 Pd $4d^{10}$	47 Ag $5s^14d^{10}$	48 Cd $5s^24d^{10}$	49 In $5s^25p^1$	50 Sn $5s^25p^2$	51 Sb $5s^25p^3$	52 Te $5s^25p^4$	53 I $5s^25p^5$	54 Xe $5s^25p^6$
6	55 Cs $6s^1$	56 Ba $6s^2$	57 La* $6s^25d^1$	72 Hf $4f^{14}6s^25d^2$	73 Ta $6s^25d^3$	74 W $6s^25d^4$	75 Re $6s^25d^5$	76 Os $6s^25d^6$	77 Ir $6s^25d^7$	78 Pt $6s^15d^9$	79 Au $6s^15d^{10}$	80 Hg $6s^25d^{10}$	81 Tl $6s^26p^1$	82 Pb $6s^26p^2$	83 Bi $6s^26p^3$	84 Po $6s^26p^4$	85 At $6s^26p^5$	86 Rn $6s^26p^6$
7	87 Fr $7s^1$	88 Ra $7s^2$	89 Ac** $7s^26d^1$	104 Rf $7s^26d^2$	105 Db $7s^26d^3$	106 Sg $7s^26d^4$	107 Bh $7s^26d^5$	108 Hs $7s^26d^6$	109 Mt $7s^26d^7$	110 Ds $7s^26d^8$	111 Rg $7s^16d^{10}$	112 Uub $7s^26d^{10}$	113 Uut $7s^26d^{10}7p^1$	114 Uuq $7s^26d^{10}7p^2$	115 Uup $7s^26d^{10}7p^3$			118 Uuo $7s^27p^6$

f-Transition Elements

*Lanthanides	58 Ce $6s^24f^15d^1$	59 Pr $6s^24f^35d^0$	60 Nd $6s^24f^45d^0$	61 Pm $6s^24f^55d^0$	62 Sm $6s^24f^65d^0$	63 Eu $6s^24f^75d^0$	64 Gd $6s^24f^75d^1$	65 Tb $6s^24f^95d^0$	66 Dy $6s^24f^{10}5d^0$	67 Ho $6s^24f^{11}5d^0$	68 Er $6s^24f^{12}5d^0$	69 Tm $6s^24f^{13}5d^0$	70 Yb $6s^24f^{14}5d^0$	71 Lu $6s^24f^{14}5d^1$

**Actinides	90 Th $7s^25f^06d^2$	91 Pa $7s^25f^26d^1$	92 U $7s^25f^36d^1$	93 Np $7s^25f^46d^1$	94 Pu $7s^25f^66d^0$	95 Am $7s^25f^76d^0$	96 Cm $7s^25f^76d^1$	97 Bk $7s^25f^96d^0$	98 Cf $7s^25f^{10}6d^0$	99 Es $7s^25f^{11}6d^0$	100 Fm $7s^25f^{12}6d^0$	101 Md $7s^25f^{13}6d^0$	102 No $7s^25f^{14}6d^0$	103 Lr $7s^25f^{14}6d^1$

FIGURE 12.29

The periodic table with atomic symbols, atomic numbers, and partial electron configurations.

Cr: [Ar]$4s^13d^5$
Cu: [Ar]$4s^13d^{10}$

similar chemistry exhibited by the members of a given group arises from the fact that they all have the same valence electron configuration. Only the principal quantum number of the occupied orbitals changes in going down a particular group.

It is important to be able to give the electron configuration for each of the main-group elements. This is most easily done by using the periodic table. If you understand how the table is organized, it is not necessary to memorize the order in which the orbitals fill. Review Fig. 12.28 and Fig. 12.29 to make sure that you understand the correspondence among the orbitals and the periods and groups.

Predicting the configurations of the transition metals (3*d*, 4*d*, and 5*d* elements), the lanthanides (4*f* elements), and the actinides (5*f* elements) is somewhat more difficult because there are many exceptions of the type encountered in the first-row transition metals (the 3*d* elements). You should memorize the configurations of chromium and copper, the two exceptions in the first-row transition metals, since these elements are often encountered.

1	2	3	4	5	6	7	8	9	10	11	12	13	14	15	16	17	18
1 H																	2 He
3 Li	4 Be											5 B	6 C	7 N	8 O	9 F	10 Ne
11 Na	12 Mg											13 Al	14 Si	15 P	16 S	17 Cl	18 Ar
19 K	20 Ca	21 Sc	22 Ti	23 V	24 Cr	25 Mn	26 Fe	27 Co	28 Ni	29 Cu	30 Zn	31 Ga	32 Ge	33 As	34 Se	35 Br	36 Kr
37 Rb	38 Sr	39 Y	40 Zr	41 Nb	42 Mo	43 Tc	44 Ru	45 Rh	46 Pd	47 Ag	48 Cd	49 In	50 Sn	51 Sb	52 Te	53 I	54 Xe
55 Cs	56 Ba	57 La*	72 Hf	73 Ta	74 W	75 Re	76 Os	77 Ir	78 Pt	79 Au	80 Hg	81 Tl	82 Pb	83 Bi	84 Po	85 At	86 Rn
87 Fr	88 Ra	89 Ac**	104 Rf	105 Db	106 Sg	107 Bh	108 Hs	109 Mt	110 Ds	111 Uuu	112 Uub	113 Uut	114 Uuq	115 Uup			118 Uuo

*Lanthanide series	58 Ce	59 Pr	60 Nd	61 Pm	62 Sm	63 Eu	64 Gd	65 Tb	66 Dy	67 Ho	68 Er	69 Tm	70 Yb	71 Lu
**Actinide series	90 Th	91 Pa	92 U	93 Np	94 Pu	95 Am	96 Cm	97 Bk	98 Cf	99 Es	100 Fm	101 Md	102 No	103 Lr

FIGURE 12.30

A form of the periodic table recommended by IUPAC.

Example 12.8

Give the electron configurations for sulfur (S), cadmium (Cd), hafnium (Hf), and radium (Ra), using the periodic table inside the front cover of this book.

Solution *Sulfur,* element 16, resides in Period 3, where the $3p$ orbitals are being filled (see Fig. 12.31). Since sulfur is the fourth among the $3p$ elements, it must have four $3p$ electrons. Its configuration is

$$\text{S:} \quad 1s^2 2s^2 2p^6 3s^2 3p^4 \quad \text{or} \quad [\text{Ne}]3s^2 3p^4$$

Cadmium, element 48, is located in Period 5 at the end of the $4d$ transition metals, as shown in Fig. 12.31. It is the tenth element in the series and

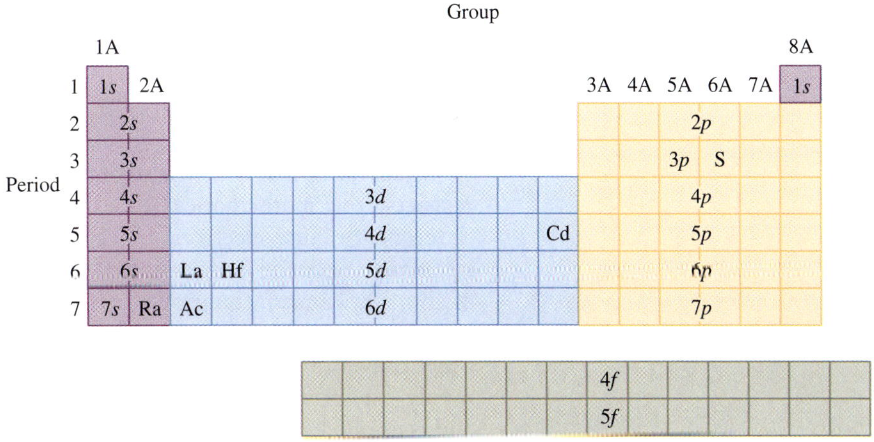

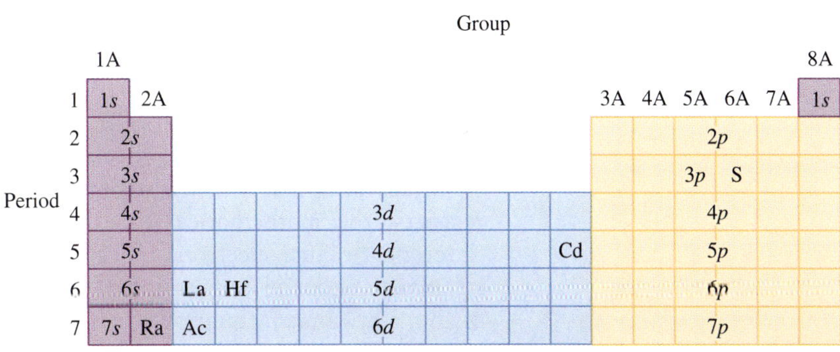

FIGURE 12.31

The positions of the elements considered in Example 12.8.

thus has ten electrons in the 4*d* orbitals (in addition to the two electrons in the 5*s* orbital). The configuration is

$$\text{Cd:}\quad 1s^22s^22p^63s^23p^64s^23d^{10}4p^65s^24d^{10} \quad \text{or} \quad [\text{Kr}]5s^24d^{10}$$

Hafnium, element 72, is found in Period 6, as shown in Fig. 12.31. Note that it occurs just after the lanthanide series. Thus the 4*f* orbitals are already filled. Hafnium is the second member of the 5*d* transition series and has two 5*d* electrons. The configuration is

$$\text{Hf:}\quad 1s^22s^22p^63s^23p^64s^23d^{10}4p^65s^24d^{10}5p^66s^24f^{14}5d^2$$
$$\text{or} \quad [\text{Xe}]6s^24f^{14}5d^2$$

Radium, element 88, is in Period 7 (and Group 2A), as shown in Fig. 12.31. Thus radium has two electrons in the 7*s* orbital, and the configuration is

$$\text{Ra:}\quad 1s^22s^22p^63s^23p^64s^23d^{10}4p^65s^24d^{10}5p^66s^24f^{14}5d^{10}6p^67s^2$$
$$\text{or} \quad [\text{Rn}]7s^2$$

12.14 Further Development of the Polyelectronic Model

Before we proceed with further discussion of polyelectronic atoms, we should summarize some of the most important things that have been said about the quantum mechanical description of atoms to this point. Most important, there is a fundamental difference between the solution of the Schrödinger equation for the hydrogen atom and the solutions for all polyelectronic atoms. The Schrödinger equation for the hydrogen atom can be solved exactly to yield the now-familiar hydrogen orbitals. These orbitals are characterized by the quantum numbers n, ℓ, m_ℓ, and m_s, and the energy levels corresponding to these orbitals depend only on n (all orbitals with the same value of n are degenerate). Recall that some of the orbitals directly obtained from the solution to the Schrödinger equation are complex (contain $\sqrt{-1}$). For example, of the three orbitals corresponding to $n = 2$ and $\ell = 1$ (2*p* orbitals), the orbital corresponding to $m_\ell = 0$ is real (the $2p_z$ orbital), but the orbitals corresponding to the values of m_ℓ of $+1$ and -1 are complex. For ease of physical interpretation these latter two orbitals are combined to produce two real orbitals ($2p_x$ and $2p_y$). These same procedures apply to all the *p* orbitals (corresponding to higher values of n). Similarly, for the 3*d* orbitals ($n = 3$, $\ell = 2$) the orbital corresponding to $m_\ell = 0$ is real (d_{z^2}), whereas the orbitals corresponding to m_ℓ values of ±1 and ±2 are complex and are used to construct the familiar real orbitals.

In contrast to the Schrödinger equation for the hydrogen atom, the Schrödinger equation for a polyelectronic atom cannot be solved exactly. For example, although the hydrogen and helium atoms are similar in many respects, the mathematical descriptions of these atoms are fundamentally different. Because electrons repel each other, the motions of the two helium electrons are correlated (coupled), and this fact prevents the exact separation of the Schrödinger equation for helium into independent, solvable equations for each electron. Thus solving the Schrödinger equation for helium (or any other polyelectronic atom) requires approximations. The approach most commonly used, the self-consistent field (SCF) method, was developed by Hartree and is applied as follows: For an atom containing N electrons, a wave function (orbital) is guessed for each electron except electron 1. For example, assume that orbitals are guessed for electrons 2, 3, 4, . . . , N: $\psi_2, \psi_3, \psi_4, \ldots, \psi_N$. The next step involves solving the Schrödinger equation for electron 1, which is moving in a potential field created by the nucleus and the electrons in orbitals $\psi_2, \psi_3, \ldots, \psi_N$. The repulsions between electron 1 and the other electrons are computed at each point in space from the sum of the average electron densities (probabilities) corresponding to $|\psi_2|^2, |\psi_3|^2, \ldots, |\psi_N|^2$ in volume element dv around that point. With the aid of a computer, the problem is solved to yield the wave function for electron 1, which we will label ψ'_1.

The next step is to do the same type of calculation to obtain a new wave function for electron 2 moving in a field of electrons described by the wave functions $\psi'_1, \psi_3, \psi_4, \ldots, \psi_N$. This step leads to a new function ψ'_2 for electron 2. Now the process is carried out for electron 3 interacting with electrons described by the wave functions $\psi'_1, \psi'_2, \psi_4, \ldots, \psi_N$ to produce a new function ψ'_3. This procedure continues until all electrons have been covered to yield the wave functions $\psi'_1, \psi'_2, \psi'_3, \ldots, \psi'_N$. Then the entire process starts again with electron 1 and continues through electron N to give the new functions $\psi''_1, \psi''_2, \psi''_3, \ldots, \psi''_N$. The procedure is diagramed in Fig. 12.32. When a given cycle produces a set of wave functions that are virtually identical to the previous set, a self-consistent field is achieved and the procedure is terminated.

The orbitals that arise from the SCF method are quite similar to hydrogen orbitals. They have the same angular characteristics (same type of boundary surfaces) as do the orbitals of hydrogen. However, the radial parts of the orbitals are different from those of the hydrogen orbitals. Although the n quantum number from the treatment of hydrogen does not apply exactly to the SCF orbitals, it is still convenient to retain it as a label. It is important to note that the energies of the SCF orbitals for polyelectronic atoms depend on both n and ℓ, not just n, as for hydrogen.

Finally, although it is not precisely correct to assume that the N electrons in an atom occupy N independent one-electron orbitals, this remains a very useful idea for understanding many atomic properties, including the organization

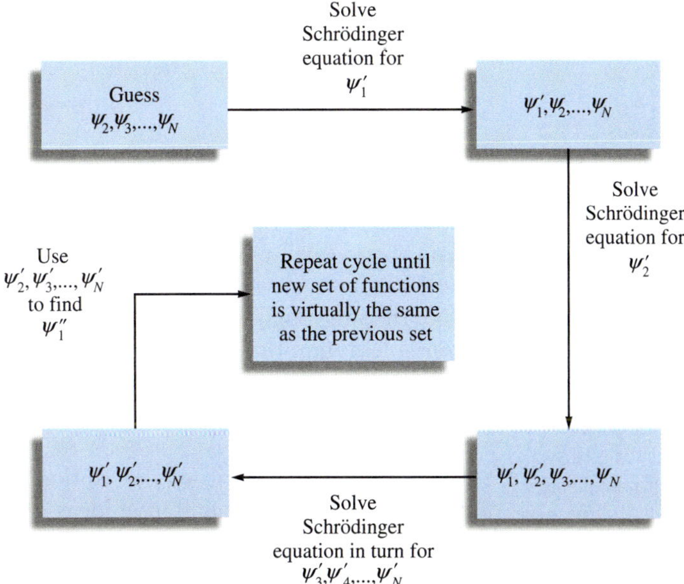

FIGURE 12.32
A schematic of the SCF method for obtaining the orbitals of a polyelectronic atom.

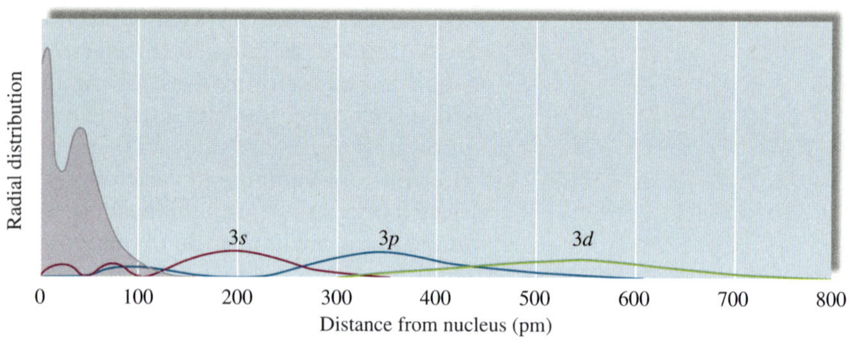

FIGURE 12.33

The radial distribution of electron probability density for the sodium atom. The shaded area represents the 10 core electrons. The radial distributions of the 3*s*, 3*p*, and 3*d* orbitals are also shown. Note the difference in the penetration effects of an electron in these three orbitals.

of the periodic table. Recall that for us to account for the arrangement of the atoms on the periodic table, the orbitals that correspond to a given value of n must fill in the order ns, then np, then nd, and finally, nf. From this observation we would expect the energies of the one-electron SCF orbitals to vary in the order

$$E_{ns} < E_{np} < E_{nd} < E_{nf}$$

and this ordering is borne out by the calculations.

We can understand the observed order of orbital energies in a qualitative sense by considering the so-called **penetration effect.** To get an appreciation for this concept, consider the radial probability plots for the 3*s*, 3*p*, and 3*d* orbitals in a sodium atom, as shown in Fig. 12.33. Note that although an electron in the 3*s* orbital spends most of its time far from the nucleus and outside the core electrons (the electrons in the 1*s*, 2*s*, and 2*p* orbitals), which **shields** it from the nuclear charge, it has a small but significant probability of being quite close to the nucleus. We say it significantly *penetrates* the shielding core electron "cloud" and thus "feels" more of the nuclear charge. On the other hand, an electron in a 3*p* orbital does not have a probability maximum close to the nucleus. Thus we can say that an electron in a 3*p* orbital penetrates the core electrons to a lesser extent than an electron in the 3*s* orbital. Similarly, an electron in a 3*d* orbital (see Fig. 12.33) shows much less penetration than a 3*p* electron does. These ideas help us to understand why an electron "prefers" the 3*s* orbital to the 3*p* or 3*d* and why, after the 3*s* orbital is filled, the next electron occupies the 3*p* rather than the 3*d* orbital. That is, the penetration effect helps us to understand qualitatively the order

$$E_{3s} < E_{3p} < E_{3d}$$

The penetration effect also helps to explain why the 4*s* orbital fills before the 3*d* orbital. Recall that potassium has the electron configuration $1s^2 2s^2 2p^6 3s^2 3p^6 4s^1$ rather than the expected $1s^2 2s^2 2p^6 3s^2 3p^6 3d^1$. We can explain this result by observing that an electron in a 4*s* orbital penetrates much more than an electron in a 3*d* orbital, as shown graphically in Fig. 12.34. Note that although the most probable distance from the nucleus for a 3*d* electron is less than that for a 4*s* electron, the 4*s* electron has a significant probability of penetrating close to the nucleus. This explains why the potassium atom in its lowest-energy state has its last electron in the 4*s* orbital rather than in the 3*d* orbital.

Although the rigorous description of polyelectronic atoms is quite complicated, our simple qualitative ideas about electrons in independent orbitals are often very useful when we try to understand why atoms behave the way they do. We will consider some specific atomic properties in the next section.

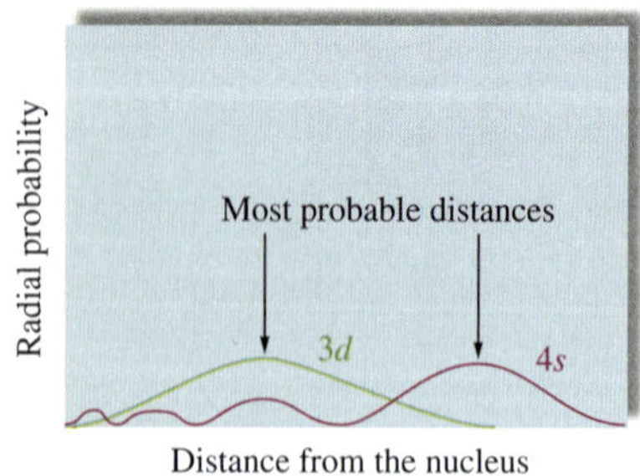

FIGURE 12.34

Radial probability distributions for the 3*d* and 4*s* orbitals. Note that the most probable distance of the electron from the nucleus for the 3*d* orbital is less than that for the 4*s* orbital. However, the 4*s* orbital allows more electron penetration close to the nucleus and thus is preferred over the 3*d* orbital.

12.15 Periodic Trends in Atomic Properties

We have developed a fairly complete picture of polyelectronic atoms that is quite successful in accounting for the periodic table of elements. We will next use the model to account for the observed trends in several important atomic properties: ionization energy, electron affinity, and atomic size.

Ionization Energy

Ionization energy is the energy required to remove an electron from a gaseous atom or ion,

$$X(g) \longrightarrow X^+(g) + e^-$$

where the atom or ion is assumed to be in its ground state. Although the values for ionization energy will be given in this text in terms of kilojoules per mole of atoms, it is quite common in other chemical literature to see values given per atom. In that context the term *ionization potential* is used, and the units are electron-volts (eV) per atom ($1\ \text{eV} = 1.602 \times 10^{-19}$ J).

The ionization energy for a particular electron in an atom is a source of information about the energy of the orbital it occupies in the atom. In fact, Koopmans' theorem states: *The ionization energy of an electron is equal to the energy of the orbital from which it came.* This rule is an approximation because, among other things, it assumes that the electrons left behind in the resulting ion will not reorganize in response to the removal of an electron. However, ionization energies do provide information that is quite useful in testing the orbital model of the atom.

To introduce some of the characteristics of ionization energy, we will consider the energy required to remove several electrons in succession from aluminum atoms in the gaseous state. The ionization energies are

$$Al(g) \longrightarrow Al^+(g) + e^- \qquad I_1 = 580\ \text{kJ/mol}$$

$$Al^+(g) \longrightarrow Al^{2+}(g) + e^- \qquad I_2 = 1815\ \text{kJ/mol}$$

$$Al^{2+}(g) \longrightarrow Al^{3+}(g) + e^- \qquad I_3 = 2740\ \text{kJ/mol}$$

$$Al^{3+}(g) \longrightarrow Al^{4+}(g) + e^- \qquad I_4 = 11{,}600\ \text{kJ/mol}$$

Several important points can be illustrated from these results. In a stepwise ionization process, it is always the highest-energy electron (the one bound least tightly) that is removed first. The energy required to remove the highest-energy electron of an atom is called the **first ionization energy** (I_1). The first electron removed from the aluminum atom comes from the $3p$ orbital (Al has the electron configuration $[\text{Ne}]3s^23p^1$). The second electron comes from the $3s$ orbital (since Al^+ has the configuration $[\text{Ne}]3s^2$).

Note that the value of I_1 is considerably smaller than the value of I_2, the **second ionization energy.** This result makes sense for several reasons. The primary factor is simply charge. Note that the first electron is removed from a neutral atom (Al), whereas the second electron is removed from a 1+ ion (Al^+). The increase in positive charge binds the electrons more firmly and the ionization energy increases. The same trend shows up in the third (I_3) and fourth (I_4) ionization energies, where the electron is removed from the Al^{2+} and Al^{3+} ions, respectively.

The increase in successive ionization energies for an atom also makes sense in terms of relative orbital energies. The increase from I_1 to I_2 is expected

TABLE 12.6

Successive Ionization Energies in Kilojoules per Mole for the Elements in Period 3

General increase →

General decrease ↑

Element	I_1	I_2	I_3	I_4	I_5	I_6	I_7
Na	495	4560					
Mg	735	1445	7730	Core electrons*			
Al	580	1815	2740	11,600			
Si	780	1575	3220	4350	16,100		
P	1060	1890	2905	4950	6270	21,200	
S	1005	2260	3375	4565	6950	8490	27,000
Cl	1255	2295	3850	5160	6560	9360	11,000
Ar	1527	2665	3945	5770	7230	8780	12,000

*Note the large jump in ionization energy in going from removal of valence electrons to removal of core electrons.

because the first electron is removed from a $3p$ orbital that is higher in energy than the $3s$ orbital from which the second electron is removed. The largest jump in ionization energy by far occurs in going from the third ionization energy (I_3) to the fourth (I_4). This large jump occurs because I_4 corresponds to removing a core electron (Al^{3+} has the configuration $1s^2 2s^2 2p^6$), and core electrons are bound much more tightly than valence electrons.

Table 12.6 gives the values of ionization energies for all the Period 3 elements. Note the large jump in energy in each case in going from the removal of valence electrons to the removal of core electrons.

The values of the first ionization energy for the elements in the first five periods of the periodic table are graphed in Fig. 12.35. Note that, in general, as we go *across a period from left to right, the first ionization energy increases.* To account for this trend qualitatively, we need to consider more fully the concept of electrons shielding each other from the nuclear charge. Shielding occurs because electrons repel each other. Our simple pictures of the atom lead us to expect that core electrons should be quite effective in shielding outer electrons from the nuclear charge, since the core electrons are between the

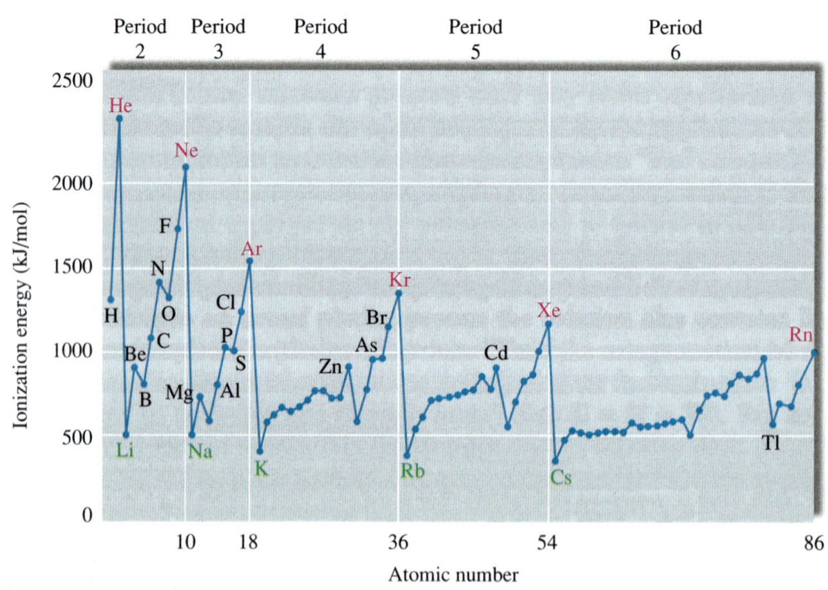

FIGURE 12.35

The values of first ionization energy for the elements in the first five periods. In general, ionization energy decreases in going down a group. For example, note the decrease in values for Group 1A and Group 8A. In general, ionization energy increases in going from left to right across a period. For example, note the sharp increase going across Period 2 from lithium through neon.

Electrons in the same principal quantum level do not shield each other as well as core electrons shield outer electrons.

nucleus and the outer electrons. On the other hand, electrons in the same principal quantum level, which on the average are all at about the same distance from the nucleus, are not expected to shield each other very well. Thus, as we move from left to right in a given period of the periodic table, we do not expect the electrons to completely shield each other from the increasing nuclear charge as the number of protons in the nucleus increases. The electrons are expected to be bound more firmly in going from left to right across a given period, which means that the ionization energy should increase.

First ionization energy increases across a period and decreases down a group.

On the other hand, *the first ionization energy values decrease in going down a group*. This can be seen most clearly by focusing on the Group 1A elements (the alkali metals) and the Group 8A elements (the noble gases), as shown in Table 12.7. The main reason for the decrease in going down a group is that the electrons being removed are, on average, farther from the nucleus. As n increases, the size of the orbital increases, and the electron is easier to remove.

TABLE 12.7

First Ionization Energies for the Alkali Metals and Noble Gases

Atom	I_1 (kJ/mol)
Group 1A	
Li	520.
Na	495
K	419
Rb	409
Cs	382
Group 8A	
He	2377
Ne	2088
Ar	1527
Kr	1356
Xe	1176
Rn	1042

In Fig. 12.35 we see that there are some discontinuities in ionization energy in going across a period. For example, discontinuities occur in Period 2, in going from beryllium to boron and from nitrogen to oxygen. These exceptions to the normal trend can be explained in terms of how electron repulsions depend on the electron configuration. We will discuss the elements in Period 2 individually to further develop the concept of shielding.

The increase in the ionization energy in going from lithium ($1s^22s^1$) to beryllium ($1s^22s^2$) is expected, since the $2s$ electrons do not shield each other completely. The decrease in the ionization energy in going from beryllium ($1s^22s^2$) to boron ($1s^22s^22p^1$) suggests that the electrons in the $2s$ orbital effectively shield the $2p$ electron. This is sensible in view of the greater penetration by the $2s$ electrons as compared with the $2p$ electron. The $2s$ electrons spend more time closer to the nucleus than the $2p$ electrons, where they provide effective shielding. The steady increase in the ionization energy in going from boron ($1s^22s^22p^1$) to carbon ($1s^22s^22p^2$) to nitrogen ($1s^22s^22p^3$) is expected because the $2p$ electrons are not very effective in shielding each other from the increasing nuclear charge. The drop in the ionization energy in going from nitrogen ($1s^22s^22p^3$) to oxygen ($1s^22s^22p^4$) is usually explained as follows: In nitrogen each $2p$ electron is in a separate orbital. When the extra electron for oxygen is added, one $2p$ orbital becomes doubly occupied. The electron repulsions between the electrons in the doubly occupied orbital make either of these electrons easier to remove.* As we move from oxygen ($1s^22s^22p^4$) to fluorine ($1s^22s^22p^5$) to neon ($1s^22s^22p^6$), the ionization energy increases because in each case the electron being considered is in a doubly occupied orbital. Note that the ionization energy values for oxygen, fluorine, and neon are different from those of boron, carbon, and nitrogen by an amount that represents the extra electron repulsions in doubly occupied orbitals, as shown in Fig. 12.35.

Example 12.9

The first ionization energy for phosphorus is 1060 kJ/mol, and that for sulfur is 1005 kJ/mol. Why?

Solution Phosphorus and sulfur are neighboring elements in Period 3 of the periodic table and have the following valence electron configurations: Phosphorus is $3s^23p^3$, and sulfur is $3s^23p^4$.

*This explanation is greatly oversimplified. A more complete explanation involves a detailed analysis of the electron interactions, which is beyond the scope of this text.

Why Is Mercury a Liquid?

The silver liquid called mercury has been known since ancient times. In fact, the symbol for mercury (Hg) comes from its Greek name *Hydrargyrum*, which means "watery silver." Although elements in the liquid state at ambient temperature and pressure are quite rare (Br_2 is another example), the liquid nature of mercury is especially confounding. For example, compare the properties of mercury and gold:

	Mercury	Gold
Melting point	−39°C	1064°C
Density	13.6 g cm^{-3}	19.3 g cm^{-3}
Enthalpy of fusion	2.30 kJ mol^{-1}	12.8 kJ mol^{-1}
Conductivity	10.4 kS m^{-1}	426 kS m^{-1}

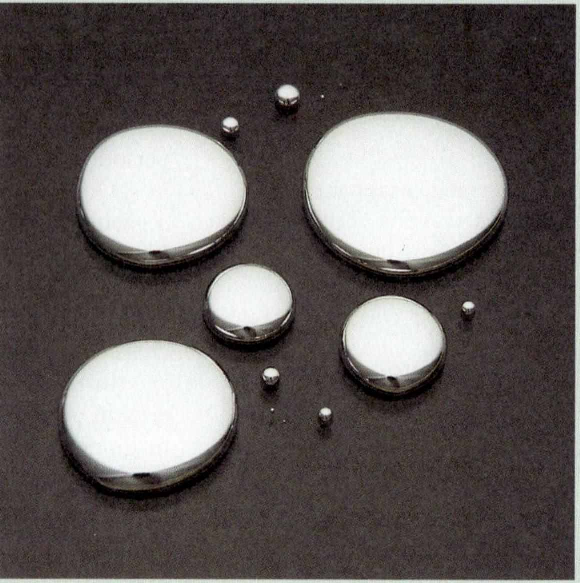

Liquid mercury forms flat drops because of its lack of surface tension.

It is quite apparent that these metals, which are neighbors on the periodic table, have strikingly different properties. Why? The answer is not at all straightforward—but very interesting. It seems to hinge on relativity.

Recall that Einstein postulated in his theory of special relativity in 1905 that the mass (m) of a moving object increases with its velocity (v):

$$m_{\text{relativistic}} = m_{\text{rest}}/\sqrt{1 - (v/c)^2}$$

where c is the speed of light. In the simple models for the atom we ignore relativistic effects on the electron mass. Although these effects are negligible for light atoms (the mass change is approximately 0.003% for the hydrogen electron), they become important for heavy elements such as gold and mercury. For example, the relativistic mass for a 1s electron in mercury is approximately 1.23 times its rest mass, and this effect leads to a very significant contraction in the radius of the 1s orbital.

It turns out that relativity has an even more profound impact on atomic theory than the preceding

Ordinarily, the first ionization energy increases as we go across a period, so we might expect sulfur to have a greater ionization energy than phosphorus. However, in this case the fourth p electron in sulfur must be placed in an already occupied orbital. The electron–electron repulsions that result cause this electron to be more easily removed than might be expected.*

Example 12.10

Consider atoms with the following electron configurations:

$$1s^2 2s^2 2p^6$$
$$1s^2 2s^2 2p^6 3s^1$$
$$1s^2 2s^2 2p^6 3s^2$$

*See page 573.

calculations suggest. A relativistic treatment of the atom fundamentally changes the way we view the electrons in atoms. In fact, as shown by British physicist Paul Dirac, the concept of electron spin occurs naturally in a relativistic treatment of the atom and does not have to be "tacked on" as in the wave mechanical treatment. The point here is not to explain these very complex ideas but to alert you to concepts you will be learning more about in higher-level courses.

How does relativity explain why mercury has a melting point of −39°C, whereas that of neighboring gold is 1064°C? The first step in answering this question involves considering the electron configurations of these atoms:

$$\text{Au: [Xe]}4f^{14}5d^{10}6s^{1}$$

$$\text{Hg: [Xe]}4f^{14}5d^{10}6s^{2}$$

Notice that gold has an unfilled 6*s* subshell, but the 6*s* level is filled in mercury. Because of its configuration, a gold atom can use its half-filled 6*s* orbital to form a bond to another gold atom. In fact, the metal–metal bond in the Au_2 molecule is an astonishingly strong 221 kJ mol^{-1}, a value very close to the bond energy of the Cl_2 molecule (239 kJ mol^{-1}) and greater than the bond energy of I_2 (149 kJ mol^{-1}). In addition, gold has an electron affinity (−220 kJ mol^{-1}) that is higher than that of oxygen and sulfur. Further, gold forms a compound with cesium (CsAu) that exhibits the CsCl crystal structure (see Fig. 16.41 and Section 16.8) in which gold atoms take the place of Cl^- ions. Thus a gold atom seems to behave a lot like a halogen atom.

What causes gold to emulate many properties of the nonmetallic halogens? The apparent answer lies in the dramatic contraction of the gold 6*s* orbital because of relativistic effects. The unexpectedly small radius of the 6*s* orbital of a gold atom results in a much lower energy than is predicted in the absence of relativistic effects. It is this very-low-energy unfilled 6*s* orbital that causes gold atoms to form very stable Au_2 molecules in the gas phase and to bind strongly to each other in the solid state, producing its high melting point. This same low-energy 6*s* orbital also leads to gold's unexpectedly high electron affinity and to its unusual color. Gold is not the silvery color exhibited by most metals because of the absorption of blue light to transfer an electron between the 5*d* and 6*s* orbitals in gold atoms.

So why are gold and mercury so different? The answer lies in the different electron configurations of the two atoms. Unlike gold, the low-energy 6*s* orbital in mercury is filled, and these two electrons are very tightly bound to the mercury atom. In fact, one can think of Hg as being analogous to He. That is, the low-energy pair of 6*s* electrons on mercury causes it to behave like a noble gas atom—it cannot bond to another mercury atom. This explains why mercury is unique among metals in that it is almost entirely monomeric in its gas phase. In contrast, the species Hg_2^{2+} is extremely stable even in aqueous solution. This fact is not surprising once it is realized that Hg^+ is isoelectronic with Au.

Thus it is the unusually low energy of the 6*s* orbitals apparently caused by relativistic effects that causes gold to behave like a halogen atom and mercury to behave like a noble gas.

Which atom has the largest first ionization energy, and which has the smallest second ionization energy? Explain your choices.

Solution The atom with the largest value of I_1 is the one with the configuration $1s^22s^22p^6$ (this is the neon atom) because this element is found at the right end of Period 2. Since the 2*p* electrons do not shield each other very effectively, I_1 will be large. The other configurations given include 3*s* electrons. These electrons are effectively shielded by the core electrons and are farther from the nucleus than the 2*p* electrons in neon. Thus I_1 for these atoms is smaller than I_1 for neon.

The atom with the smallest value of I_2 is the one with the configuration $1s^22s^22p^63s^2$ (the magnesium atom). For magnesium both I_1 and I_2 involve valence electrons. For the atom with the configuration $1s^22s^22p^63s^1$ (sodium), the second electron lost (corresponding to I_2) is a core electron from a 2*p* orbital).

Electron Affinity

Electron affinity *is the energy change associated with the addition of an electron to a gaseous atom:*

$$X(g) + e^- \longrightarrow X^-(g)$$

The sign convention for electron affinity values follows the convention for energy changes used in Chapters 9 and 10.

Because two different conventions have been used, there is a good deal of confusion in the chemical literature about the signs for electron affinity values. Electron affinity has been defined in many textbooks as the energy *released* when an electron is added to a gaseous atom. This convention requires that a positive sign be attached to an exothermic addition of an electron to an atom, which opposes normal thermodynamic conventions. Therefore, in this book we define electron affinity as a *change* in energy. This means that if the addition of the electron is exothermic, the corresponding value for electron affinity will carry a negative sign.

Figure 12.36 shows the electron affinity values for the atoms among the first 20 elements that form stable, isolated, negative ions—that is, the atoms that undergo the addition of an electron as shown above. As expected, all these elements have negative (exothermic) electron affinities. Note that the *more negative* the energy, the greater is the quantity of energy released. Although electron affinities generally become more negative from left to right across a period, there are several exceptions to this rule in each period. The dependence of electron affinity on atomic number can be explained by considering the changes in electron repulsions as a function of electron configurations. For example, the fact that the nitrogen atom does not form a stable, isolated $N^-(g)$ ion, whereas carbon forms $C^-(g)$, reflects the difference in the electron configurations of these atoms. An electron added to nitrogen ($1s^22s^22p^3$) to form the $N^-(g)$ ion ($1s^22s^22p^4$) would have to occupy a $2p$ orbital that already contains one electron. The extra repulsion between the electrons in this doubly occupied orbital causes $N^-(g)$ to be unstable. When an electron is added to carbon ($1s^22s^22p^2$) to form the $C^-(g)$ ion ($1s^22s^22p^3$), no such extra repulsions occur.

In contrast to the nitrogen atom, the oxygen atom can add one electron to form the stable $O^-(g)$ ion. Presumably, oxygen's greater nuclear charge, compared with that of nitrogen, is sufficient to overcome the repulsion associated with putting a second electron into an already occupied $2p$ orbital. However, it should be noted that a second electron *cannot* be added to an oxygen atom [$O^-(g) + e^- \not\rightarrow O^{2-}(g)$] to form an isolated oxide ion. This outcome seems strange in view of the many stable oxide compounds (MgO, Fe_2O_3, and so on) that are known. As we will discuss in detail in Chapter 13,

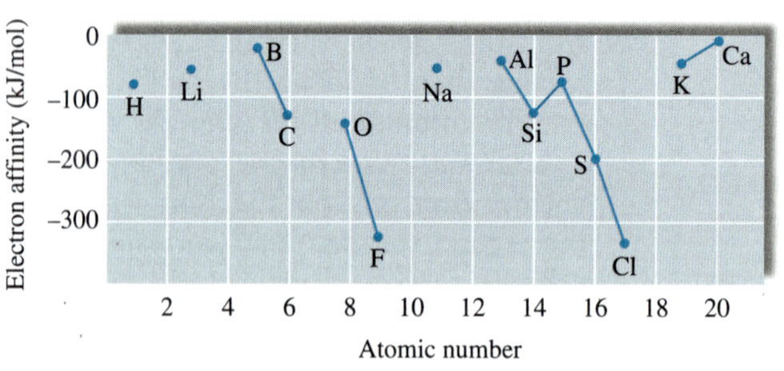

FIGURE 12.36

The electron affinity values for atoms among the first 20 elements that form stable, isolated X^- ions. The lines shown connect adjacent elements. The absence of a line indicates missing elements (He, Be, N, Ne, Mg, and Ar) whose atoms do not add an electron exothermically and thus do not form stable, isolated X^- ions.

TABLE 12.8

Electron Affinities of the Halogens

Atom	Electron Affinity (kJ/mol)
F	−327.8
Cl	−348.7
Br	−324.5
I	−295.2

the O^{2-} ion is stabilized in ionic compounds by the large attractions that occur among the positive ions and the oxide ions.

When we go down a group, electron affinity should become more positive (less energy released), since the electron is added at increasing distances from the nucleus. Although this is generally the case, the changes in electron affinity in going down most groups are relatively small and numerous exceptions occur. This behavior is demonstrated by the electron affinities of the Group 7A elements (the halogens) shown in Table 12.8. Note that the range of values is quite small compared with the changes that typically occur across a period. Also note that although chlorine, bromine, and iodine show the expected trend, the energy released when an electron is added to fluorine is smaller than might be expected. This smaller energy release has been attributed to the small size of the 2*p* orbitals. Because the electrons must be very close together in these orbitals, there are unusually large electron–electron repulsions. In the other halogens with their larger orbitals, the repulsions are not as severe.

Atomic Radius

Just as the size of an orbital cannot be specified exactly, neither can the size of an atom. We must make some arbitrary choices to obtain values for **atomic radii.** These values can be obtained by measuring the distances between atoms in chemical compounds. For example, in the bromine molecule the distance between the two nuclei is known to be 228 pm. The bromine atomic radius is assumed to be half this distance, or 114 pm, as shown in Fig. 12.37. Measurements of this type have led to the values of the atomic radii for the elements shown in Fig. 12.38. These radii are often called *covalent atomic radii* because of the way they are determined. These values are significantly smaller than might be expected from the 90% electron density volumes of isolated atoms because when atoms form bonds, their electron "clouds" interpenetrate. However, these values form a self-consistent data set that can be used to discuss the trends in atomic radii. The radii for metal atoms are obtained by halving the distance between metal atoms in solid metal crystals. Note from Fig. 12.38 that the atomic radius decreases in going from left to right across a period. This decrease can be explained in terms of the increasing effective nuclear charge (decreasing shielding) in going from left to right. This means that the valence electrons are drawn closer to the nucleus, decreasing the size of the atom.

Atomic radius decreases across a period and increases down a group.

Atomic radius increases down a group because of the increases in the orbital sizes in successive principal quantum levels.

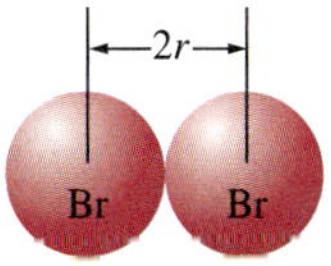

FIGURE 12.37

The radius of an atom (*r*) is defined as half the distance between the nuclei in a molecule consisting of identical atoms.

Example 12.11

Predict the trend in radius of the following ions: Be^{2+}, Mg^{2+}, Ca^{2+}, and Sr^{2+}.

Solution All these ions are formed by removing two electrons from an atom of a Group 2A element. In going from beryllium to strontium, we are going down the group, so the sizes increase.

$$Be^{2+} < Mg^{2+} < Ca^{2+} < Sr^{2+}$$

↑ Smallest radius (Be^{2+}) ↑ Largest radius (Sr^{2+})

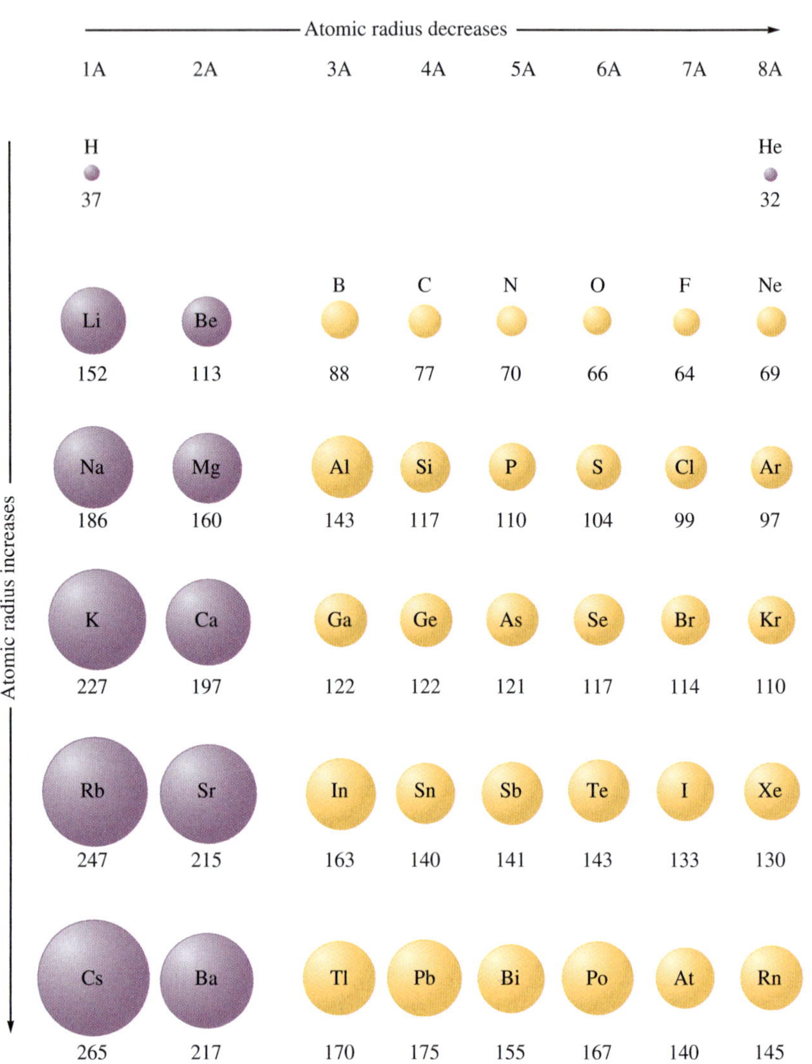

FIGURE 12.38

Atomic radii (in picometers) for selected atoms. Note that atomic radius decreases going across a period and increases going down a group. The values for the noble gases are estimated because data from bonded atoms are lacking.

12.16 The Properties of a Group: The Alkali Metals

We have seen that the periodic table originated as a way to portray the systematic properties of the elements. Mendeleev was primarily responsible for first showing its usefulness in correlating and predicting the elemental properties. In this section we will summarize much of the information available from the table. We will also illustrate the usefulness of the table by discussing the properties of a representative group, the alkali metals.

Information Contained in the Periodic Table

1. The essence of the periodic table is that members of each group of representative elements exhibit similar chemical properties that change in a regular way. The quantum mechanical model has allowed us to understand that the similarity of properties of the atoms in a group arises from the identical valence electron configurations shared by group members. *It is the number and type of valence electrons that primarily determine an atom's chemistry.*

2. One of the most valuable types of information available from the periodic table is the electron configuration of any representative element. If you understand the organization of the table, you do not need to memorize electron configurations for the elements. Although the predicted electron configurations for transition metals are sometimes incorrect, this is not a serious problem. You should, however, memorize the configuration of two exceptions, chromium and copper, since these 3*d* transition elements are found in many important compounds.
3. As we mentioned in Chapter 2, certain groups in the periodic table have special names. These names are summarized in Fig. 12.39.
4. The most fundamental classification of the elements is into metals and nonmetals. The essential chemical property of a metal is the tendency to give up electrons to form a positive ion; metals tend to have low ionization energies. The metallic elements are found on the left side of the table, as shown in Fig. 12.39. The most reactive metals are found on the lower

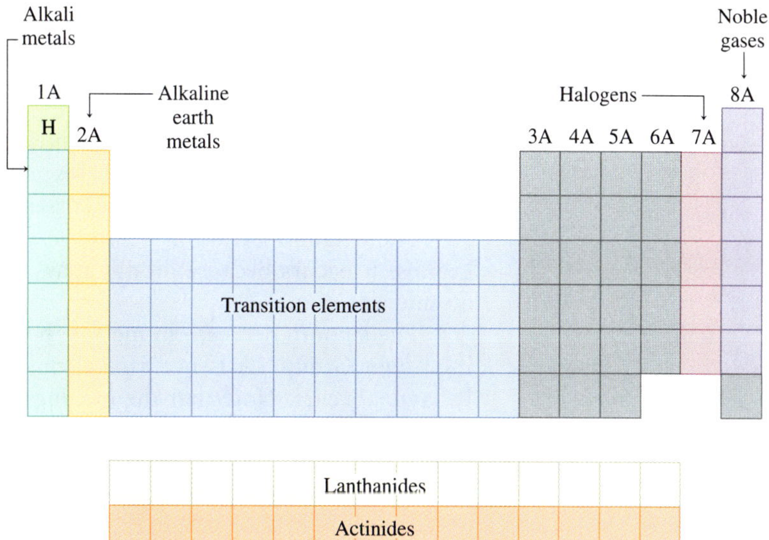

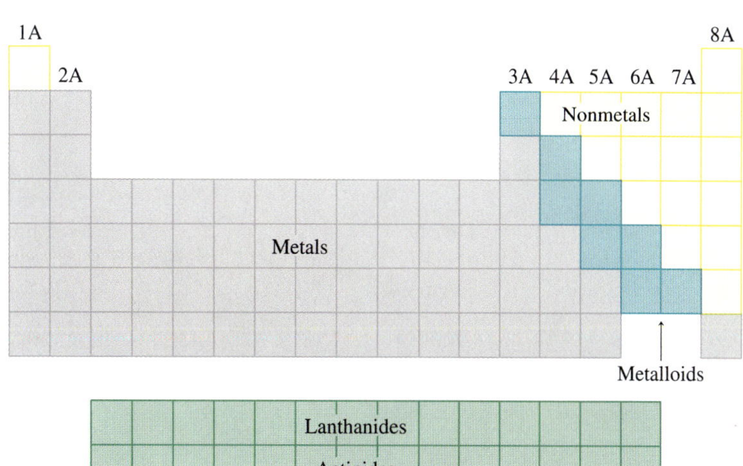

FIGURE 12.39

Special names for groups in the periodic table.

left-hand portion of the table where the ionization energies are smallest. The distinctive chemical property of a nonmetal is the ability to gain electrons to form an anion when reacting with a metal. The nonmetals have large ionization energies and most have negative electron affinities. The nonmetals are found on the right side of the table. The most reactive ones are located in the upper right-hand corner, excluding the noble gas elements, which are quite unreactive. The division between metals and nonmetals shown in Fig. 12.39 is only approximate. Many elements along the division line exhibit both metallic and nonmetallic properties under certain circumstances. These elements are called **metalloids,** or **semimetals.**

The Alkali Metals

The metals of Group 1A, the alkali metals, illustrate well the relationships among the properties of the elements in a group. Lithium, sodium, potassium, rubidium, cesium, and francium are the most reactive of the metals. We will not discuss francium here since it occurs in nature in only very small quantities. Although hydrogen is found in Group 1A, it behaves as a nonmetal, in contrast to the other members of that group. The fundamental reason for hydrogen's nonmetallic character is its very small size (see Fig. 12.38). The electron in the small 1*s* orbital is bound very tightly to the nucleus.

Some important properties of the first five alkali metals are shown in Table 12.9. The data in Table 12.9 show that when we move down the group, the first ionization energy decreases and the atomic radius increases. This agrees with the general trends discussed in Section 12.15.

The overall increase in density in going down Group 1A is typical of all groups. It occurs because atomic mass generally increases more rapidly than atomic size.

The smooth decrease in melting point (mp) and boiling point (bp) in going down Group 1A is not typical; in most other groups more complicated behavior occurs. Note that the melting point of cesium is only 29°C. Cesium can be melted readily from the heat of your hand. Cesium's low melting point is very unusual—metals typically have high melting points. For example, tungsten melts at 3410°C. The only other metals with low melting points are mercury (mp = −39°C) and gallium (mp = 30°C).

Recall that the chemical property most characteristic of a metal is the ability to lose its valence electrons. The Group 1A elements are very reactive.

TABLE 12.9

Properties of Five Alkali Metals

Element	Valence Electron Configuration	Density at 25°C (g/cm^3)	mp (°C)	bp (°C)	First Ionization Energy (kJ/mol)	Atomic (covalent) Radius (pm)	Ionic (M^+) Radius (pm)
Li	$2s^1$	0.53	180	1330	520.	152	60
Na	$3s^1$	0.97	98	892	495	186	95
K	$4s^1$	0.86	64	760	419	227	133
Rb	$5s^1$	1.53	39	688	409	247	148
Cs	$6s^1$	1.87	29	690	382	265	169

They have low ionization energies and react readily with nonmetals to form ionic solids. A typical example involves the reaction of sodium with chlorine to form sodium chloride:

$$2Na(s) + Cl_2(g) \longrightarrow 2NaCl(s)$$

This is an oxidation–reduction reaction in which chlorine oxidizes sodium to form Na^+ and Cl^- ions. In the reactions between metals and nonmetals it is typical for the nonmetal to behave as the oxidizing agent and the metal to behave as the reducing agent, as shown by the following reactions:

$$2Na(s) + S(s) \longrightarrow Na_2S(s)$$

Contains Na^+ and S^{2-} ions

$$6Li(s) + N_2(g) \longrightarrow 2Li_3N(s)$$

Contains Li^+ and N^{3-} ions

$$4Na(s) + O_2(g) \longrightarrow 2Na_2O(s)$$

Contains Na^+ and O^{2-} ions

For reactions of this type, the relative reducing powers of the alkali metals can be predicted from the first ionization energies listed in Table 12.9. Since it is much easier to remove an electron from a cesium atom than from a lithium atom, cesium should be the better reducing agent. The expected trend in reducing ability is

$$Cs > Rb > K > Na > Li$$

This order is observed experimentally for direct reactions between the solid alkali metals and nonmetals. However, this order of reducing ability is not observed when the alkali metals react in aqueous solution. For example, the reduction of water by an alkali metal is very vigorous and exothermic:

$$2M(s) + 2H_2O(l) \longrightarrow H_2(g) + 2M^+(aq) + 2OH^-(aq) + \text{energy}$$

The order of reducing abilities observed for this reaction (for the first three group members) is

$$Li > K > Na$$

which is not the order expected from the relative ionization energies of these metals.

This unexpected order occurs because the formation of the M^+ ions in aqueous solution is strongly influenced by the hydration of these ions by the polar water molecules. The hydration energy of an ion represents the change in energy that occurs when water molecules attach to the M^+ ion. The hydration energies for the Li^+, Na^+, and K^+ ions shown in Table 12.10 indicate that the process is exothermic in each case. However, nearly twice as much energy is released by the hydration of the Li^+ ion as compared with that of the K^+ ion. This difference is caused by size effects; the Li^+ ion is much smaller than the K^+ ion, and thus its *charge density* (charge per unit volume) is much greater. This means that the polar water molecules are more strongly attracted to the small Li^+ ion. Because the Li^+ ion is so strongly hydrated, its formation from the lithium atom occurs more readily than the formation of the K^+ ion from the potassium atom. Although a potassium atom in the gas phase loses its valence electron more readily than a lithium atom

TABLE 12.10

Hydration Energies for Li^+, Na^+, and K^+ Ions

Ion	Hydration Energy (kJ/mol)
Li^+	−500
Na^+	−400
K^+	−300

Chemical Insights

Lithium: Behavior Medicine

More and more people in our society seem to be suffering from the debilitating effects of mania and depression, but the alkali metal lithium can provide help for many. In fact, over 3 million prescriptions for lithium carbonate are filled annually by retail pharmacies.

Although the details are not well understood, the lithium ion seems to alleviate mood disorders by affecting the way that brain cells respond to neurotransmitters, a class of molecules that facilitate the transmission of nerve impulses.

Specifically, physiologists think that the lithium ion may interfere with a complex cycle of reactions that relays and amplifies messages carried to the cells by neurotransmitters and hormones. They theorize that exaggerated forms of behavior, such as mania or depression, arise from the overactivity of this cycle. Thus the fact that lithium inhibits this cycle may be responsible for its moderating effect on behavior.

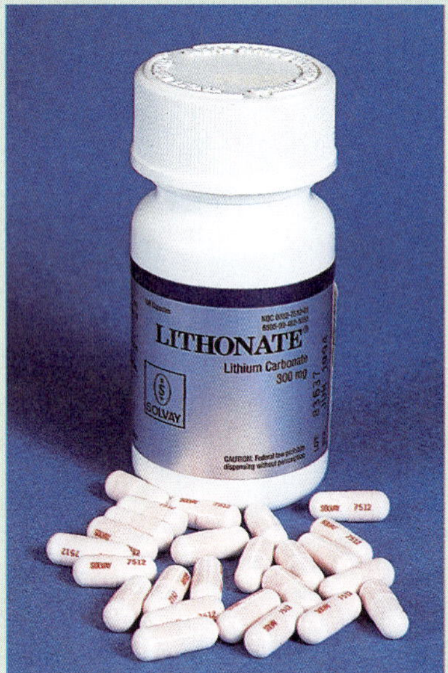

Capsules of lithium carbonate.

There is a growing collection of evidence that violent behavior may result at least partially from the improper regulation of neurotransmitters and hormones. For example, a study in Finland showed that violent criminals, especially arsonists, often had low levels of serotonin, a common neurotransmitter. Studies are now under way to determine whether lithium might also be effective for treating these and other aberrant forms of behavior.

in the gas phase, the opposite is true in aqueous solution. This anomaly is an example of the importance of the polarity of the water molecule in aqueous reactions.

There is one more surprise involving the highly exothermic reactions of the alkali metals with water. Experiments show that lithium is the best reducing agent in water, so we might expect lithium to react most violently with water. However, it does not. Sodium and potassium react much more vigorously. Why? The answer lies in the relatively high melting point for lithium. When sodium and potassium react with water, the heat evolved causes them to melt, giving a larger area of contact with water. Lithium, on the other hand, does not melt under these conditions and thus reacts more slowly. This example illustrates the important principle (which we will discuss in detail in Chapter 15) that the energy of a reaction and the rate at which it occurs are not necessarily related.

In this section we have seen that the trends in atomic properties summarized by the periodic table can be a great help in understanding the chemical behavior of the elements. This fact will be emphasized over and over as we proceed in our study of chemistry.

Discussion Questions

These questions are designed to be considered by groups of students in class. Often these questions work well for introducing a particular topic in class.

1. Explain what it means for something to have wavelike properties; for something to have particulate properties. Electromagnetic radiation can be discussed in terms of both particles and waves. Explain the experimental verification for each of these views.
2. Defend and criticize Bohr's model. Why was it reasonable that such a model was proposed, and what evidence was there that it "works"? Why do we no longer "believe" in it?
3. The first ionization energy for magnesium is 735 kJ/mol. Which electron is this for? Estimate Z_{eff} for this electron, and explain your reasoning. Calculate Z_{eff} for this electron, and compare it to your estimate.
4. The first four ionization energies for elements X and Y are shown below. The units are not kJ/mol.

	X	Y
First	170	200
Second	350	400
Third	1800	3500
Fourth	2500	5000

   Identify the elements X and Y. There may be more than one answer, so explain completely.
5. Compare the first ionization energy of helium with its second ionization energy, remembering that both electrons come from the 1*s* orbital. Explain the difference without using actual numbers from the text.
6. Which has a larger second ionization energy, lithium or beryllium? Why?
7. Explain why a graph of ionization energy versus atomic number (across a row) is not linear. Where are the exceptions? Explain why they occur.
8. Without referring to your text, predict the trend of second ionization energies for the elements sodium through argon. Compare your answer with the data in Table 12.6. Explain any differences.
9. Account for the fact that the line that separates the metals from the nonmetals on the periodic table is diagonal downward to the right instead of horizontal or vertical.
10. Explain the term *electron* from a quantum mechanical perspective, including a discussion of atomic radii, probabilities, and orbitals.
11. Choose the best response for the following. The ionization energy for the chlorine atom is equal in magnitude to the electron affinity for
    a. the Cl atom
    b. the Cl^- ion
    c. the Cl^+ ion
    d. the F atom
    e. none of these

    Explain.
12. Consider the following statement: "The ionization energy for the potassium atom is negative because when K loses an electron to become K^+ it achieves a noble gas electron configuration." Indicate what is incorrect. Explain.
13. What is the difference between Z_{eff} and Z? When are they the same? Explain.
14. In going across a row of the periodic table, electrons are added and ionization energy generally increases. In going down a column of the periodic table, electrons are also being added but ionization energy generally decreases. Explain.
15. Explain the difference between the probability density distribution for an orbital and its radial probability.
16. How does the energy of a hydrogen 1*s* orbital compare with that of a lithium 1*s* orbital? Why? What is meant by the term *energy of the orbital?* What is its sign? Why? What is meant by the term *lower in energy?*
17. Which is larger, the hydrogen 1*s* orbital or the lithium 1*s* orbital? Why? Which has the larger radius, the hydrogen atom or the lithium atom? Why?
18. Is the following statement true or false: The hydrogen atom has a 3*s* orbital. Explain.
19. Which is higher in energy: the 2*s* or 2*p* orbital in hydrogen? Is this also true for helium? Explain.
20. Prove mathematically that it is more energetically favorable for a fluorine atom to take an electron from a sodium atom than for a fluorine atom to take an electron from another fluorine atom.

Exercises

A blue exercise number indicates that the answer to that exercise appears at the end of this book and a solution appears in the *Solutions Guide*.

Light and Matter

21. Microwave radiation has a wavelength on the order of 1.0 cm. Calculate the frequency and the energy of a single

photon of this radiation. Calculate the energy of an Avogadro's number of photons (called an einstein) of this electromagnetic radiation.

22. Consider the following waves representing electromagnetic radiation:

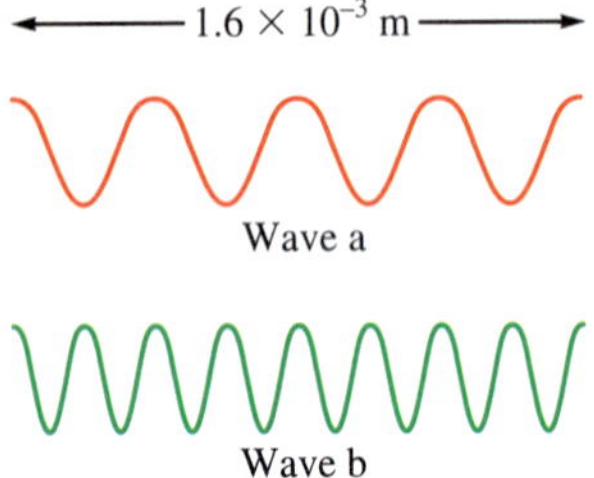

Which wave has the longer wavelength? Calculate the wavelength. Which wave has the higher frequency and larger photon energy? Calculate these values. Which wave has the faster velocity? What type of electromagnetic radiation is illustrated?

23. A carbon–oxygen double bond in a certain organic molecule absorbs radiation that has a frequency of 6.0×10^{13} s^{-1}.
    a. What is the wavelength of this radiation?
    b. To what region of the spectrum does this radiation belong?
    c. What is the energy of this radiation per photon? Per mole of photons?
    d. A carbon–oxygen bond in a different molecule absorbs radiation with frequency equal to 5.4×10^{13} s^{-1}. Is this radiation more or less energetic?

24. A photon of ultraviolet (UV) light possesses enough energy to mutate a strand of human DNA. What is the energy of a single UV photon and a mole of UV photons having a wavelength of 25 nm?

25. The work function of an element is the energy required to remove an electron from the surface of the solid. The work function for lithium is 279.7 kJ/mol (that is, it takes 279.7 kJ of energy to remove one mole of electrons from one mole of Li atoms on the surface of Li metal). What is the maximum wavelength of light that can remove an electron from an atom in lithium metal?

26. One type of electromagnetic radiation has a frequency of 107.1 MHz, another type has a wavelength of 2.12×10^{-10} m, and another type of electromagnetic radiation has photons with energy equal to 3.97×10^{-19} J/photon. Identify each type of electromagnetic radiation, and place them in order of increasing photon energy and increasing frequency.

27. It takes 208.4 kJ of energy to remove one mole of electrons from the atoms on the surface of rubidium metal. If rubidium metal is irradiated with 254-nm light, what is the maximum kinetic energy the released electrons can have?

28. What experimental evidence supports the quantum theory of light? Explain the wave–particle duality of all matter. For what size particle must one consider both the wave and the particle properties?

29. Explain the photoelectric effect.

30. Calculate the de Broglie wavelength for each of the following.
    a. an electron with a velocity 10.% of the speed of light
    b. a tennis ball (55 g) served at 35 m/s (~80 mi/h)

31. Neutron diffraction is used in determining the structures of molecules.
    a. Calculate the de Broglie wavelength of a neutron moving at 1.00% of the speed of light.
    b. Calculate the velocity of a neutron with a wavelength of 75 pm (1 pm = 10^{-12} m).

32. Calculate the velocities of electrons with de Broglie wavelengths of 1.0×10^2 nm and 1.0 nm, respectively.

33. An atom of a particular element is traveling at 1% of the speed of light. The de Broglie wavelength is found to be 3.31×10^{-3} pm. Which element is this?

Hydrogen Atom: The Bohr Model

34. Characterize the Bohr model of the atom. In the Bohr model, what do we mean when we say something is quantized? How does the Bohr model of the hydrogen atom explain the hydrogen emission spectrum? Why is the Bohr model fundamentally incorrect?

35. Calculate the wavelength of light emitted in each of the following spectral transitions in the hydrogen atom. What type of electromagnetic radiation is emitted in each transition?
    a. $n = 3 \longrightarrow n = 2$
    b. $n = 4 \longrightarrow n = 2$
    c. $n = 2 \longrightarrow n = 1$
    d. $n = 4 \longrightarrow n = 3$

36. What is the maximum wavelength of light capable of removing an electron from a hydrogen atom in the energy states characterized by $n = 1$ and $n = 3$?

37. An electron is excited from the ground state to the $n = 3$ state in a hydrogen atom. Which of the following statements are true? Correct any false statements.
    a. It takes more energy to ionize (remove) the electron from $n = 3$ than from the ground state.
    b. The electron is farther from the nucleus on average in the $n = 3$ state than in the ground state.
    c. The wavelength of light emitted if the electron drops from $n = 3$ to $n = 2$ is shorter than the wavelength of light emitted if the electron falls from $n = 3$ to $n = 1$.
    d. The wavelength of light emitted when the electron returns to the ground state from $n = 3$ is the same as the wavelength of light absorbed to go from $n = 1$ to $n = 3$.
    e. The first excited state corresponds to $n = 3$.

38. Does a photon of visible light (λ = 400–700 nm) have sufficient energy to excite an electron in a hydrogen atom from the $n = 1$ to the $n = 5$ energy state? From the $n = 2$ to the $n = 6$ energy state?

39. An excited hydrogen atom emits light with a wavelength of 397.2 nm to reach the energy level for which $n = 2$. In which principal quantum level did the electron begin?

40. An excited hydrogen atom with an electron in the $n = 5$ state emits light having a frequency of $6.90 \times 10^{14}\ s^{-1}$. Determine the principal quantum level for the final state in this electronic transition.

41. Consider an electron for a hydrogen atom in an excited state. The maximum wavelength of electromagnetic radiation that can completely remove (ionize) the electron from the H atom is 1460 nm. Determine the initial excited state for the electron ($n = ?$).

42. Calculate the energy (in kJ/mol) required to remove the electron in the ground state for each of the following one-electron species using the Bohr model.
a. H b. He^+ c. Li^{2+} d. C^{5+} e. Fe^{25+}

43. One of the emission spectral lines for Be^{3+} has a wavelength of 253.4 nm for an electronic transition that begins in the state with $n = 5$. What is the principal quantum number of the lower-energy state corresponding to this emission?

Wave Mechanics and Particle in a Box

44. The Heisenberg uncertainty principle can be expressed in the form

$$\Delta E \cdot \Delta t \geq \frac{\hbar}{2}$$

where E represents energy and t represents time. Show that the units for this form are the same as the units for the form used in this chapter:

$$\Delta x \cdot \Delta p \geq \frac{\hbar}{2}$$

45. Using the Heisenberg uncertainty principle, calculate Δx for each of the following.
a. an electron with $\Delta \nu = 0.100$ m/s
b. a baseball (mass = 145 g) with $\Delta \nu = 0.100$ m/s
c. How does the answer in part a compare with the size of a hydrogen atom?
d. How does the answer in part b correspond to the size of a baseball?

46. Calculate the wavelength of the electromagnetic radiation required to excite an electron from the ground state to the level with $n = 5$ in a one-dimensional box 40.0 pm in length.

47. An electron in a one-dimensional box requires a wavelength of 8080 nm to excite an electron from the $n = 2$ to the $n = 3$ energy level. Calculate the length of this box.

48. An electron in a 10.0-nm one-dimensional box is excited from the ground state into a higher-energy state by absorbing a photon of electromagnetic radiation with a wavelength of 1.374×10^{-5} m. Determine the final energy state for this transition.

49. Discuss what happens to the energy levels for an electron trapped in a one-dimensional box as the length of the box increases.

50. What is the total probability of finding a particle in a one-dimensional box in level $n = 3$ between $x = 0$ and $x = L/6$?

51. Which has the lowest (ground-state) energy, an electron trapped in a one-dimensional box of length 10^{-6} m or one with length 10^{-10} m?

Orbitals and Quantum Numbers

52. What are quantum numbers? What information do we get from the quantum numbers n, ℓ, and m_ℓ? We define a spin quantum number (m_s), but do we know that an electron literally spins?

53. How do $2p$ orbitals differ from each other? How do $2p$ and $3p$ orbitals differ from each other? What is a nodal surface in an atomic orbital? What is wrong with $1p$, $1d$, $2d$, $1f$, $2f$, and $3f$ orbitals? Explain what we mean when we say that a $4s$ electron is more penetrating than a $3d$ electron.

54. Which of the following sets of quantum numbers are not allowed in the hydrogen atom? For the sets of quantum numbers that are incorrect, state what is wrong in each set.
a. $n = 3, \ell = 2, m_\ell = 2$
b. $n = 4, \ell = 3, m_\ell = 4$
c. $n = 0, \ell = 0, m_\ell = 0$
d. $n = 2, \ell = -1, m_\ell = 1$

55. Which of the following sets of quantum numbers are not allowed? For each incorrect set, state why it is incorrect.
a. $n = 3, \ell = 3, m_\ell = 0, m_s = -\frac{1}{2}$
b. $n = 4, \ell = 3, m_\ell = 2, m_s = -\frac{1}{2}$
c. $n = 4, \ell = 1, m_\ell = 1, m_s = +\frac{1}{2}$
d. $n = 2, \ell = 1, m_\ell = -1, m_s = -1$
e. $n = 5, \ell = -4, m_\ell = 2, m_s = +\frac{1}{2}$
f. $n = 3, \ell = 1, m_\ell = 2, m_s = -\frac{1}{2}$

56. How many orbitals can have the designation $5p$, $3d_{z^2}$, $4d$, $n = 5$, and $n = 4$?

57. How many electrons in an atom can have the designation $1p$, $6d_{x^2-y^2}$, $4f$, $7p_y$, $2s$, and $n = 3$?

58. What is the physical significance of the value of ψ^2 at a particular point in an atomic orbital?

59. In defining the sizes of orbitals, why must we use an arbitrary value, such as 90% of the probability of finding an electron in that region?

60. From the diagrams of $2p$ and $3p$ orbitals in Fig. 12.19 and Fig. 12.20, respectively, draw a rough graph of the square of the wave function for these orbitals in the direction of one of the lobes.

61. The wave function for the $2p_z$ orbital in the hydrogen atom is

$$\psi_{2p_z} = \frac{1}{4\sqrt{2\pi}}\left(\frac{Z}{a_0}\right)^{3/2} \sigma e^{-\sigma/2} \cos\theta$$

where a_0 is the value for the radius of the first Bohr orbit in meters (5.29×10^{-11}), σ is Zr/a_0, r is the value for the distance from the nucleus in meters, and θ is an angle. Calculate the value of $\psi_{2p_z}{}^2$ at $r = a_0$ for $\theta = 0$ (z axis) and for $\theta = 90°$ (xy plane).

62. For hydrogen atoms, the wave function for the state $n = 3$, $\ell = 0$, and $m_\ell = 0$ is

$$\psi_{300} = \frac{1}{81\sqrt{3\pi}}\left(\frac{1}{a_0}\right)^{3/2}(27 - 18\sigma + 2\sigma^2)e^{-\sigma/3}$$

where $\sigma = r/a_0$ and a_0 is the Bohr radius (5.29×10^{-11} m). Calculate the position of the nodes for this wave function.

Polyelectronic Atoms

63. What is the difference between core electrons and valence electrons? Why do we emphasize the valence electrons in an atom when discussing atomic properties? What is the relationship between valence electrons and elements in the same group of the periodic table?

64. The periodic table consists of four blocks of elements that correspond to s, p, d, and f orbitals being filled. After f orbitals come g and h orbitals. In theory, if a g block and an h block of elements existed, how long would the rows of g and h elements be in this theoretical periodic table?

65. What is the maximum number of electrons in an atom that can have these quantum numbers?
    a. $n = 4$
    b. $n = 5$, $m_\ell = +1$
    c. $n = 5$, $m_s = +\frac{1}{2}$
    d. $n = 3$, $\ell = 2$
    e. $n = 2$, $\ell = 1$
    f. $n = 0$, $\ell = 0$, $m_\ell = 0$
    g. $n = 2$, $\ell = 1$, $m_\ell = -1$, $m_s = -\frac{1}{2}$
    h. $n = 3$, $m_s = +\frac{1}{2}$
    i. $n = 2$, $\ell = 2$
    j. $n = 1$, $\ell = 0$, $m_\ell = 0$

66. The elements of Si, Ga, As, Ge, Al, Cd, S, and Se are all used in the manufacture of various semiconductor devices. Write the expected electron configurations for these atoms.

67. Write the expected electron configurations for the following atoms: Sc, Fe, P, Cs, Eu, Pt, Xe, and Br.

68. Write the expected electron configurations for each of the following atoms: Cl, As, Sr, W, Pb, and Cf.

69. Using Fig. 12.29, list elements (ignore the lanthanides and actinides) that have ground-state electron configurations that differ from those we would expect from their positions in the periodic table.

70. Write the expected ground-state electron configuration for the following.
    a. the element with one unpaired $5p$ electron that forms a covalent compound with fluorine
    b. the (as yet undiscovered) alkaline earth metal after radium
    c. the noble gas with electrons occupying $4f$ orbitals
    d. the first-row transition metal with the most unpaired electrons

71. For elements 1–36, there are two exceptions to the filling order as predicted from the periodic table. Draw the atomic orbital diagrams for the two exceptions and indicate how many unpaired electrons are present.

72. In the ground state of mercury (Hg),
    a. how many electrons occupy atomic orbitals with $n = 3$?
    b. how many electrons occupy d atomic orbitals?
    c. how many electrons occupy p_z atomic orbitals?
    d. how many electrons have spin "up" ($m_s = +\frac{1}{2}$)?

73. In the ground state of element 115, Uup,
    a. how many electrons have $n = 5$ as one of their quantum numbers?
    b. how many electrons have $\ell = 3$ as one of their quantum numbers?
    c. how many electrons have $m_\ell = 1$ as one of their quantum numbers?
    d. how many electrons have $m_s = -\frac{1}{2}$ as one of their quantum numbers?

74. Give possible values for the quantum numbers of the valence electrons in an atom of titanium (Ti).

75. One bit of evidence that the quantum mechanical model is "correct" lies in the magnetic properties of matter. Atoms with unpaired electrons are attracted by magnetic fields and thus are said to exhibit *paramagnetism*. The degree to which this effect is observed is directly related to the number of unpaired electrons present in the atom. Consider the ground-state electron configurations for Li, N, Ni, Te, Ba, and Hg. Which of these atoms would be expected to be paramagnetic, and how many unpaired electrons are present in each paramagnetic atom?

76. Which of elements 1–36 have two unpaired electrons in the ground state?

77. Which of elements 1–36 have one unpaired electron in the ground state?

78. A certain oxygen atom has the electron configuration $1s^22s^22p_x{}^22p_y{}^2$. How many unpaired electrons are present? Is this an excited state for oxygen? In going from this state to the ground state, would energy be released or absorbed?

79. How many unpaired electrons are present in each of the following in the ground state: O, O^+, O^-, Os, Zr, S, F, Ar?

80. Which of the following electron configurations correspond to an excited state? Identify the atoms and write the ground-state electron configuration where appropriate.
    a. $1s^22s^23p^1$
    b. $1s^22s^22p^6$
    c. $1s^22s^22p^43s^1$
    d. $[Ar]4s^23d^54p^1$

The Periodic Table and Periodic Properties

81. Using the element phosphorus as an example, write equations for the processes in which the energy change will

correspond to the ionization energy and to the electron affinity.

82. Explain why the first ionization energy tends to increase as one proceeds from left to right across a period. Why is the first ionization energy of aluminum lower than that of magnesium and the first ionization energy of sulfur lower than that of phosphorus?

83. Why do the successive ionization energies of an atom always increase? Note the successive ionization energies for silicon given in Table 12.6. Would you expect to see any large jumps between successive ionization energies of silicon as you removed all the electrons, one by one, beyond those shown in the table?

84. The radius trend and the ionization energy trend are exact opposites. Does this make sense? Define electron affinity. Electron affinity values are both exothermic (negative) and endothermic (positive). However, ionization energy values are always endothermic (positive). Explain.

85. Arrange the following groups of atoms in order of increasing size.
a. Te, S, Se
b. K, Br, Ni
c. Ba, Si, F
d. Rb, Na, Be
e. Sr, Se, Ne
f. Fe, P, O

86. Arrange the atoms in Exercise 85 in order of increasing first ionization energy.

87. In each of the following sets, which atom or ion has the smallest ionization energy?
a. Ca, Sr, Ba
b. K, Mn, Ga
c. N, O, F
d. S^{2-}, S, S^{2+}
e. Cs, Ge, Ar

88. In each of the following sets, which atom or ion has the smallest radius?
a. H, He
b. Cl, In, Se
c. element 120, element 119, element 117
d. Nb, Zn, Si
e. Na^-, Na, Na^+

89. The first ionization energies of As and Se are 0.947 MJ/mol and 0.941 MJ/mol, respectively. Rationalize these values in terms of electron configurations.

90. Rank the elements Be, B, C, N, and O in order of increasing first ionization energy. Explain your reasoning.

91. We expect the atomic radius to increase down a group in the periodic table. Can you suggest why the atomic radius of hafnium breaks this rule? (See the following data.)

Element	Atomic Radius (Å)	Element	Atomic Radius (Å)
Sc	1.57	Ti	1.477
Y	1.693	Zr	1.593
La	1.915	Hf	1.476

92. Element 106 is named seaborgium (Sg) in honor of Glenn Seaborg, discoverer of the first transuranium element.
a. Write the expected electron configuration for Sg.
b. What other element would be most like Sg in its properties?
c. Write the formula for a possible oxide and a possible oxyanion of Sg.

93. Predict some of the properties of element 117 (the symbol is Uus following conventions proposed by the International Union of Pure and Applied Chemistry, or IUPAC).
a. What will be its electron configuration?
b. What element will it most resemble chemically?
c. What will be the formulas of the neutral binary compounds it forms with sodium, magnesium, carbon, and oxygen?
d. What oxyanions would you expect Uus to form?

94. Order each of the following sets from the least exothermic electron affinity to the most.
a. F, Cl, Br, I
b. N, O, F

95. The changes in electron affinity as one goes down a group in the periodic table are not nearly as large as the variations in ionization energies. Why?

96. Which has the more negative electron affinity, the oxygen atom or the O^- ion? Explain your answer.

97. The electron affinity for sulfur is more exothermic than that for oxygen. How do you account for this?

98. The electron affinities of the elements from aluminum to chlorine are −44 kJ/mol, −120 kJ/mol, −74 kJ/mol, −200.4 kJ/mol, and −348.7 kJ/mol, respectively. Rationalize the trend in these values.

99. Use data in this chapter to determine the following.
a. the electron affinity of Mg^{2+}
b. the electron affinity of Al^+
c. the ionization energy of Cl^-
d. the ionization energy of Cl
e. the electron affinity of Cl^+

100. For each of the following pairs of elements,

(C and N) (Ar and Br) (Mg and K) (F and Cl)

pick the one with
a. the more favorable (exothermic) electron affinity
b. the higher ionization energy
c. the larger size

The Alkali Metals

101. Does the information on alkali metals in Table 12.9 of the text confirm the general periodic trends in ionization energy and atomic radius? Explain.

102. An ionic compound of potassium and oxygen has the empirical formula KO. Would you expect this compound to be potassium(II) oxide or potassium peroxide? Explain.

103. Complete and balance the equations for the following reactions.
   a. $Li(s) + N_2(g) \longrightarrow$
   b. $Rb(s) + S(s) \longrightarrow$
   c. $Cs(s) + H_2O(l) \longrightarrow$
   d. $Na(s) + Cl_2(g) \longrightarrow$

104. Cesium was discovered in natural mineral waters in 1860 by R. W. Bunsen and G. R. Kirchhoff, using the spectroscope they invented in 1859. The name comes from the Latin word *caesius*, meaning "sky blue," which describes the prominent blue line observed for this element at 455.5 nm. Calculate the frequency and energy of a photon of this light.

105. The bright yellow light emitted by a sodium vapor lamp consists of two emission lines at 589.0 nm and 589.6 nm. What are the frequency and the energy of a photon of light at each of these wavelengths? What are the energies in kJ/mol?

106. Give the name and formula of the binary compound formed by each of the following pairs of elements.
   a. Li and N
   b. Na and Br
   c. K and S
   d. Li and P
   e. Rb and H
   f. Na and H

107. Predict the atomic number of the next alkali metal after francium, and give its ground-state electron configuration.

Additional Exercises

108. Spectroscopists use emission spectra to confirm the presence of an element in materials of unknown composition. How is this possible?

109. On which quantum number(s) does the energy of an electron depend in each of the following?
   a. a one-electron atom or ion
   b. an atom or ion with more than one electron

110. Elements with very large ionization energies also tend to have highly exothermic electron affinities. Explain. Which group of elements would you expect to be an exception to this statement?

111. Diagonal relationships in the periodic table exist as well as vertical relationships. For example, Be and Al are similar in some of their properties, as are B and Si. Rationalize why these diagonal relationships hold for properties such as size, ionization energy, and electron affinity.

112. A certain microwave oven delivers 750. watts (J/s) of power to a coffee cup containing 50.0 g of water at 25.0°C. If the wavelength of microwaves in the oven is 9.75 cm, how long does it take, and how many photons must be absorbed, to make the water boil? The specific heat capacity of water is 4.18 J $°C^{-1}$ g^{-1}. Assume that only the water absorbs the energy of the microwaves.

113. Mars is roughly 60 million km from earth. How long does it take for a radio signal originating from earth to reach Mars?

114. Photogray lenses incorporate small amounts of silver chloride in the glass of the lens. When light hits the AgCl particles, the following reaction occurs:

$$AgCl \xrightarrow{h\nu} Ag + Cl$$

The silver metal formed causes the lenses to darken. The enthalpy change for this reaction is 3.10×10^2 kJ/mol. Assuming that all this energy must be supplied by light, what is the maximum wavelength of light that can cause this reaction?

115. Consider the following approximate visible light spectrum:

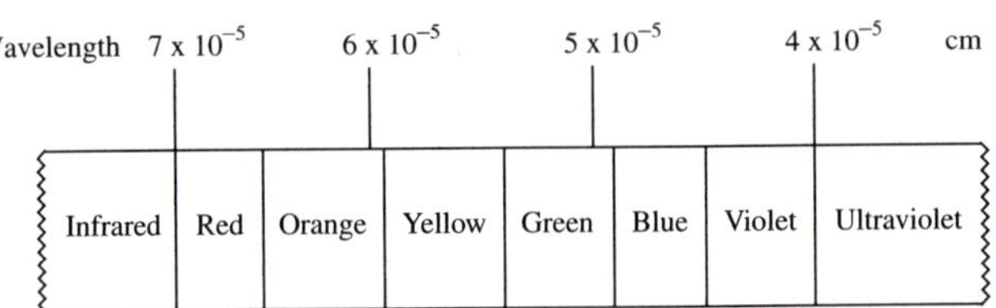

Barium emits light in the visible region of the spectrum. If each photon of light emitted from barium has an energy of 3.59×10^{-19} J, what color of visible light is emitted?

116. One of the visible lines in the hydrogen emission spectrum corresponds to the $n = 6$ to $n = 2$ electronic transition. What color light is this transition? See Exercise 115.

117. Consider the representations of the p and d atomic orbitals in Figs. 12.19 and 12.21. What do the + and − signs indicate?

118. The following graph plots the first, second, and third ionization energies for Mg, Al, and Si. Without referencing the text, which plot corresponds to which element? In one of the plots, there is a huge jump in energy between I_2 and I_3, unlike in the other two plots. Explain this phenomenon.

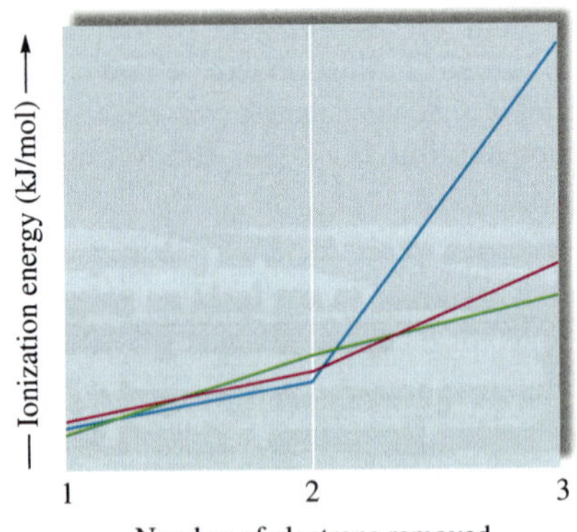

119. Using data from this chapter, calculate the change in energy expected for each of the following processes.
a. $Na(g) + Cl(g) \longrightarrow Na^+(g) + Cl^-(g)$
b. $Mg(g) + F(g) \longrightarrow Mg^+(g) + F^-(g)$
c. $Mg^+(g) + F(g) \longrightarrow Mg^{2+}(g) + F^-(g)$
d. $Mg(g) + 2F(g) \longrightarrow Mg^{2+}(g) + 2F^-(g)$

120. Write equations corresponding to the following energy terms.
a. the fourth ionization energy of Se
b. the electron affinity of S^-
c. the electron affinity of Fe^{3+}
d. the ionization energy of Mg
e. the work function of Mg (see Exercise 25)

121. The successive ionization energies for an unknown element are

$$I_1 = 896 \text{ kJ/mol}$$
$$I_2 = 1752 \text{ kJ/mol}$$
$$I_3 = 14{,}807 \text{ kJ/mol}$$
$$I_4 = 17{,}948 \text{ kJ/mol}$$

To which family in the periodic table does the unknown element most likely belong?

122. An unknown element is a nonmetal and has a valence electron configuration of ns^2np^4.
a. How many valence electrons does this element have?
b. What are some possible identities for this element?
c. What is the formula of the compound this element would form with potassium?
d. Would this element have a larger or smaller radius than barium?
e. Would this element have a greater or smaller ionization energy than fluorine?

123. An ion having a 4+ charge and a mass of 49.9 amu has two electrons with $n = 1$, eight electrons with $n = 2$, and ten electrons with $n = 3$. Supply the following properties for the ion. (*Hint:* In forming ions, the 4*s* electrons are lost before the 3*d* electrons.)
a. the atomic number
b. total number of *s* electrons
c. total number of *p* electrons
d. total number of *d* electrons
e. the number of neutrons in the nucleus
f. the mass of 3.01×10^{23} atoms
g. the ground-state electron configuration of the neutral atom

124. Consider the following ionization energies for aluminum.

$$Al(g) \longrightarrow Al^+(g) + e^- \qquad I_1 = 580 \text{ kJ/mol}$$
$$Al^+(g) \longrightarrow Al^{2+}(g) + e^- \qquad I_2 = 1815 \text{ kJ/mol}$$
$$Al^{2+}(g) \longrightarrow Al^{3+}(g) + e^- \qquad I_3 = 2740 \text{ kJ/mol}$$
$$Al^{3+}(g) \longrightarrow Al^{4+}(g) + e^- \qquad I_4 = 11{,}600 \text{ kJ/mol}$$

a. Account for the increasing trend in the values of the ionization energies.
b. Explain the large increase between I_3 and I_4.
c. Which one of the four ions has the greatest electron affinity? Explain.
d. List the four aluminum ions given in the preceding reactions in order of increasing size, and explain your ordering. (*Hint:* Remember that most of the size of an atom or ion is due to its electrons.)

125. Answer the following questions, assuming that m_s has four values rather than two and that the normal rules apply for n, ℓ, and m_ℓ.
a. How many electrons could an orbital hold?
b. How many elements would be contained in the first and second periods of the periodic table?
c. How many elements would be contained in the first transition metal series?
d. How many electrons would the set of 4*f* orbitals be able to hold?

126. Although Mendeleev predicted the existence of several undiscovered elements, he did not predict the existence of the noble gases, the lanthanides, or the actinides. Propose reasons why Mendeleev was not able to predict the existence of the noble gases.

127. Discuss why a function of the type $A \cos(Lx)$ is not an appropriate solution for the particle in a one-dimensional box.

128. Assume that four electrons are confined to a one-dimensional box 5.64×10^{-10} m in length. If two electrons can occupy each allowed energy level, calculate the wavelength of electromagnetic radiation necessary to promote the highest-energy electron into the first excited state.

129. The figure below represents part of the emission spectrum for a one-electron ion in the gas phase. All the lines result from electronic transitions from excited states to the $n = 3$ state.

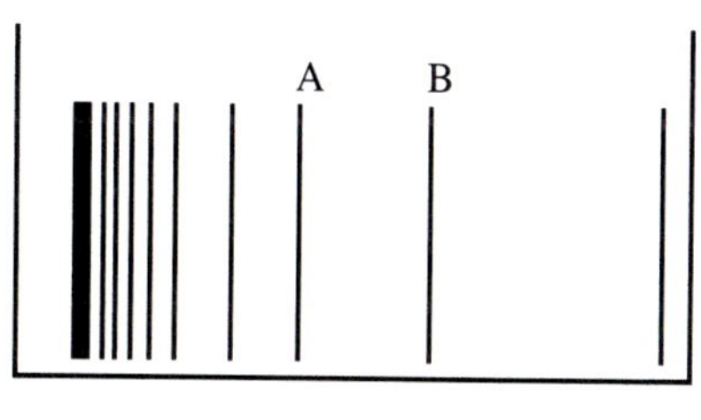

a. What electronic transitions correspond to lines A and B?
b. If the wavelength of line B is 142.5 nm, calculate the wavelength of line A.

Challenge Problems

130. An atom moving at its root mean square velocity at 100.°C has a wavelength of 2.31×10^{-11} m. Which atom is it?

131. The ground state ionization energy for the one electron ion X^{m+} is 4.72×10^4 kJ/mol. Identify X and m.

132. When the excited electron in a hydrogen atom falls from $n = 5$ to $n = 2$, a photon of blue light is emitted. If an excited electron in He^+ falls from $n = 4$, which energy level must it fall to so that a similar blue light (as with hydrogen) is emitted? Prove it.

133. The treatment of a particle in a one-dimensional box can be extended to a two-dimensional box of dimensions L_x and L_y yielding the following expression for energy:

$$E = \frac{h^2}{8m}\left(\frac{n_x^2}{L_x^2} + \frac{n_y^2}{L_y^2}\right)$$

The two quantum numbers independently can assume only integer values. Consider an electron confined to a two-dimensional box that is 8.00 nm in the x direction and 5.00 nm in the y direction.
   a. What are the quantum numbers for the first three allowed energy levels?
   b. Calculate the wavelength of light necessary to promote an electron from the first excited state to the second excited state.

134. The following numbers are the ratios of second ionization energy to first ionization energy:

Na:	9.2	P:	1.8
Mg:	2.0	S:	2.3
Al:	3.1	Cl:	1.8
Si:	2.0	Ar:	1.8

Explain these relative numbers.

135. For a hydrogen atom in its ground state, calculate the relative probability of finding the electron in the area described.
   a. in a sphere of volume 1.0×10^{-3} pm^3 centered at the nucleus
   b. in a sphere of volume 1.0×10^{-3} pm^3 centered on a point 1.0×10^{-11} m from the nucleus
   c. in a sphere of volume 1.0×10^{-3} pm^3 centered on a point 53 pm from the nucleus
   d. in a shell between two concentric spheres, one with radius 9.95 pm and the other with radius 10.05 pm
   e. in a shell between two concentric spheres, one with radius 52.85 pm and one with radius 52.95 pm

136. The treatment of a particle in a one-dimensional box can be extended to a rectangular box of dimensions L_x, L_y, and L_z, yielding the following expression for energy:

$$E = \frac{h^2}{8m}\left(\frac{n_x^2}{L_x^2} + \frac{n_y^2}{L_y^2} + \frac{n_z^2}{L_z^2}\right)$$

The three quantum numbers n_x, n_y, and n_z independently can assume only integer values.
   a. Determine the energies of the three lowest levels, assuming that the box is cubic.
   b. Describe the degeneracies of all the levels that correspond to quantum numbers having values of 1 or 2. How will these degeneracies change in a box where $L_x \neq L_y \neq L_z$?

137. Assume that eight electrons are placed into the allowed energy levels of a cubic box where two electrons can occupy each allowed energy level. (See Exercise 136 for the appropriate energy equation.) Calculate the wavelength of light necessary to promote the highest-energy ground-state electron into the lowest-energy excited state assuming a cubic box with dimensions 1.50 nm $\times$ 1.50 nm $\times$ 1.50 nm.

138. Assume that we are in another universe with different physical laws. Electrons in this universe are described by four quantum numbers with meanings similar to those we use. We will call these quantum numbers p, q, r, and s. The rules for these quantum numbers are as follows:

$p = 1, 2, 3, 4, 5, \ldots.$

q takes on positive odd integer values and $q \leq p$.

r takes on all even integer values from $-q$ to $+q$. (Zero is considered an even number.)

$s = +\frac{1}{2}$ or $-\frac{1}{2}$.

   a. Sketch what the first four periods of the periodic table will look like in this universe.
   b. What are the atomic numbers of the first four elements you would expect to be least reactive?
   c. Give an example, using elements in the first four rows, of ionic compounds with the formulas XY, XY_2, X_2Y, XY_3, and X_2Y_3.
   d. How many electrons can have $p = 3$?
   e. How many electrons can have $p = 4$, $q = 3$, $r = 2$?
   f. How many electrons can have $p = 4$, $q = 3$?
   g. How many electrons can have $p = 3$, $q = 0$, $r = 0$?
   h. What are the possible values of q and r for $p = 5$?
   i. How many electrons can have $p = 6$?

139. The ionization energy for a 1s electron in a silver atom is 2.462×10^6 kJ/mol.
   a. Determine an approximate value for Z_{eff} for the Ag 1s electron. You will first have to derive an equation that relates Z_{eff} to ionization energy.
   b. How does Z_{eff} from part a compare to Z for Ag? Rationalize the relative numbers.

140. Without looking at data in the text, sketch a qualitative graph of the third ionization energy versus atomic number for the elements Na through Ar, and explain your graph.

Marathon Problem

141.* From the information below, identify element X.

a. The wavelength of the radiowaves sent by an FM station broadcasting at 97.1 MHz is 30 million (3.00×10^7) times greater than the wavelength corresponding to the energy difference between a particular excited state of the hydrogen atom and the ground state.

b. Let V represent the principal quantum number for the valence shell of element X. If an electron in the hydrogen atom falls from shell V to the inner shell corresponding to the excited state mentioned in part a, the wavelength of light emitted is the same as the wavelength of an electron moving at a speed of 570. m/s.

c. The number of unpaired electrons for element X in the ground state is the same as the maximum number of electrons in an atom that can have the quantum number designations $n = 2$, $m_\ell = -1$, and $m_s = -\frac{1}{2}$.

d. Let A represent the principal quantum number for the electron in an excited He^+ ion in which the single electron has the same energy as the electron in the ground state of a hydrogen atom. This value of A also represents the angular momentum quantum number for the subshell containing the unpaired electron(s) for element X.

*Marathon Problems were developed by James H. Burness, Penn State University, York Campus. Used by permission.

Media Summary

Visit the Student Website at **www.cengage.com/chemistry/zumdahl** to help prepare for class, study for quizzes and exams, understand core concepts, and visualize molecular-level interactions. The following media activities are available for this chapter:

Prepare for Class

Video Lessons *Mini-lectures from chemistry experts*

- CIA Demonstration: Luminol
- Understanding Electron Spin
- The Wave Nature of Light
- The Ultraviolet Catastrophe
- The Photoelectric Effect
- Absorption and Emission
- CIA Demonstration: Flame Colors
- The Bohr Model
- The Heisenberg Uncertainty Principle
- Radial Solution to the Schrödinger Equation
- The Wave Nature of Matter
- Atomic Orbital Size
- Angular Solutions to the Schrödinger Equation
- Atomic Orbital Shapes and Quantum Numbers
- Atomic Orbital Energy
- Electron Shielding
- Creating the Periodic Table
- Electron Configurations Through Neon
- Electron Configurations Beyond Neon
- Periods and Atomic Size
- Ionization Energy
- Electron Affinity
- An Introduction to Electronegativity
- Periodic Relationships
- Hydrogen, Alkali Metals, and Alkaline Earth Metals
- Transition Metals and Nonmetals

Improve Your Grade

Visualizations *Molecular-level animations and lab demonstration videos*

- $1s$ Orbital
- $2p_x$ Orbital, $2p_y$ Orbital, $2p_z$ Orbital
- $3d_{x^2-y^2}$ Orbital, $3d_{xy}$ Orbital, $3d_{xz}$ Orbital, $3d_{yz}$ Orbital, $3d_{z^2}$ Orbital
- Cathode-Ray Tube
- Electrified Pickle
- Electromagnetic Waves
- Flame Tests
- Gold Foil Experiment
- Millikan's Oil Drop Experiment
- Orbital Energies
- Periodic Table Trends
- Photoelectric Effect
- Refraction of White Light
- The Line Spectrum of Hydrogen

Tutorials *Animated examples and interactive activities*

Tutorial: Aufbau Principle

Flashcards *Key terms and definitions*

Online flashcards

ACE the Test

Multiple-choice quizzes
3 ACE Practice Tests

Access these resources using your passkey, available free with new texts or for purchase separately.

13

Bonding: General Concepts

Model of a buckyball with potassium ion.

The world around us is composed almost entirely of compounds and mixtures of compounds: Rocks, coal, soil, petroleum, trees, and human bodies are all complex mixtures of chemical compounds in which different kinds of atoms are bound together. Substances composed of unbound atoms do exist in nature, but they are very rare. Examples are the argon in the atmosphere and the helium mixed with natural gas reserves.

The manner in which atoms are bound together in a given substance has a profound effect on its chemical and physical properties. For example, graphite is a soft, slippery material used as a lubricant in locks, and diamond is one of the hardest materials known, valuable both as a gemstone and in industrial cutting tools. Why do these materials, both composed solely of carbon atoms, have such different properties? The answer, as we will see, lies in the bonding within these substances.

Silicon and carbon are next to each other in Group 4A on the periodic table. From our knowledge of periodic trends, we might expect SiO_2 and CO_2 to be very similar. But SiO_2 is the empirical formula of silica, which is found in sand and quartz, whereas carbon dioxide is a gas, a product of respiration. Why are they so different? We will be able to answer this question after we have developed models for bonding.

(top) Quartz grows in beautiful, regular crystals. (bottom) Two forms of carbon: graphite and diamond.

Bonding and structure play a central role in determining the course of all chemical reactions, many of which are vital to our survival. Later in this book we will demonstrate the importance of bonding and structure by showing how enzymes facilitate complex chemical reactions, how genetic characteristics are transferred, and how hemoglobin in the blood carries oxygen throughout the body. All these fundamental biological reactions hinge on the geometric structures of molecules, sometimes depending on very subtle differences in molecular shape to channel the chemical reaction one way rather than another.

Many of the world's current problems require fundamentally chemical answers: disease and pollution control, the search for new energy sources, the development of new fertilizers to increase crop yields, the improvement of the protein content in various staple grains, and many more. To understand the behavior of natural materials, we must understand the nature of chemical bonding and the factors that control the structures of compounds. In this chapter we will present various classes of compounds that illustrate the different types of bonds and then develop models to describe the structure and bonding that characterize materials found in nature. Later these models will prove useful in understanding chemical reactions.

13.1 Types of Chemical Bonds

What is a chemical bond? There is no simple and yet complete answer to this question. In Chapter 2 we defined bonds as forces that hold groups of atoms together and make the atoms function as a unit.

There are many types of experiments we can perform to determine the fundamental nature of materials. For example, we can study physical properties such as melting point, hardness, and electrical and thermal conductivity. We can also study solubility characteristics and the properties of the resulting solutions. To determine the charge distribution in a molecule, we can study its behavior in an electric field. We can obtain information about the strength of a bonding interaction by measuring the energy required to break the bond, the **bond energy.** Spectroscopy, the study of the interactions of electromagnetic radiation with matter, gives a wealth of information about molecular structure and energy level spacings.

There are several ways atoms can interact with one another to form aggregates. We will consider several specific examples to illustrate the various types of chemical bonds.

When solid sodium chloride is melted, it conducts electricity, a fact that convinces us that sodium chloride contains Na^+ and Cl^- ions. Thus, when sodium and chlorine react to form sodium chloride, electrons must be transferred from the sodium atoms to the chlorine atoms to form Na^+ and Cl^- ions, which then aggregate to form solid sodium chloride. Why does this happen? The best simple answer is that *the system can achieve the lowest possible energy by behaving in this way.* Part of the favorable energy change results from the attraction of a chlorine atom for an extra electron. Even more important are the very strong attractions between the oppositely charged ions. The resulting solid sodium chloride is a very sturdy material; it has a melting point of about 800°C. The bonding forces that produce this great thermal stability result from the electrostatic attractions of the closely packed, oppositely charged ions. This is an example of *ionic bonding.* Ionic substances are formed when an atom that loses electrons relatively easily reacts with an atom that has a high affinity for electrons. That is, an **ionic compound** results when a metal reacts with a nonmetal.

The energy of interaction between a pair of ions can be calculated by using **Coulomb's law:**

$$V = \frac{Q_1Q_2}{4\pi\epsilon_0 r} = 2.31 \times 10^{-19}\text{ J nm}\left(\frac{Q_1Q_2}{r}\right)$$

where V has units of joules, r is the distance between the ion centers in nanometers, Q_1 and Q_2 are the numerical ion charges, and ϵ_0 is the permittivity of the vacuum. For example, in solid sodium chloride, where the distance between the centers of the Na^+ and Cl^- ions is 276 picometers (0.276 nm), the ionic energy per pair of ions is

$$V = 2.31 \times 10^{-19}\text{ J nm}\left[\frac{(+1)(-1)}{0.276\text{ nm}}\right] = -8.37 \times 10^{-19}\text{ J}$$

The negative sign indicates an attractive force. That is, *the ion pair has lower energy than the separated ions.* For a mole of pairs of Na^+ and Cl^- ions, the energy of interaction is

$$V = \left(-8.37 \times 10^{-19}\ \frac{\text{J}}{\text{ion pair}}\right)\left(6.022 \times 10^{23}\ \frac{\text{ion pair}}{\text{mol}}\right)$$
$$= -504\text{ kJ/mol}$$

Note that this energy refers to a mole of $Na^+ \cdot\cdot Cl^-$ ion pairs in the gas phase where a given pair is far from any other pair. In solid sodium chloride, which contains a large array of closely packed Na^+ and Cl^- ions, where a given ion is close to many oppositely charged ions, the energy associated with ionic bonding is much greater than 504 kJ/mol because of the larger numbers of interacting ions.

Coulomb's law can also be used to calculate the repulsive energy when two like-charged ions are brought together. In this case the calculated energy value will have a positive sign.

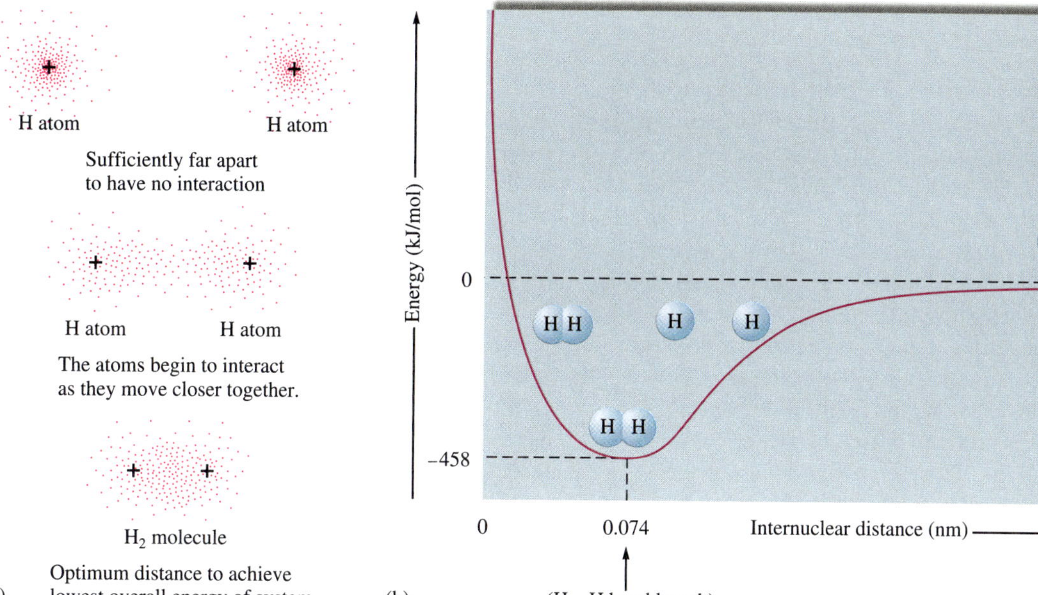

FIGURE 13.1

(a) The interaction of two hydrogen atoms. (b) Energy profile as a function of the distance between the nuclei of the hydrogen atoms. As the atoms approach each other, the energy decreases until the distance reaches 0.074 nm (0.74 Å) and then begins to increase again because of repulsions.

Bonding occurs if the energy of the aggregate is lower than that of the separated atoms.

We have seen that a bonding force develops when two very different atoms react to form oppositely charged ions. But how does a bonding force develop between two identical atoms? Let's explore this situation from a very simple point of view by considering the energy terms that result when two hydrogen atoms are brought close together, as shown in Fig. 13.1(a). For two closely spaced hydrogen atoms, there are two unfavorable energy terms, proton–proton repulsion and electron–electron repulsion, and one favorable term, proton–electron attraction. Under what conditions will the H_2 molecule be favored over the separated hydrogen atoms? That is, what conditions will favor bond formation? The answer lies in nature's strong tendency to achieve the lowest possible energy. A bond will form—that is, the two hydrogen atoms will exist as a molecular unit—if the system can lower its total energy in the process.

Therefore, the hydrogen atoms will assume the positions that give the lowest possible energy; the system will act to minimize the sum of the positive (repulsive) energy terms and the negative (attractive) energy terms. The distance at which the energy is minimum is called the equilibrium internuclear distance or, more commonly, the **bond length.** The total energy of this system as a function of distance between the hydrogen nuclei is shown in Fig. 13.1(b). Note four important features of this diagram:

1. The energy terms involved are the potential energy that results from the attractions and repulsions among the charged particles and the kinetic energy caused by the motions of the electrons.
2. The zero reference point for energy is defined for the atoms at infinite separation.
3. At very short distances the energy rises steeply because of the great importance of the internuclear repulsive forces at these distances.
4. The bond length is the distance at which the system has minimum energy, and the bond energy corresponds to the depth of the "well" at this distance.

In the H_2 molecule the electrons reside primarily in the space between the two nuclei, where they are attracted simultaneously by both protons. This

CHEMICAL INSIGHTS

No Lead Pencils

Did you ever wonder why the part of a pencil that makes the mark is called the "lead"? Pencils have no lead in them now—and they never have. Apparently the association between writing and the element lead arose during the Roman Empire, when lead rods were used as writing utensils because they leave a gray mark on paper. Many centuries later, in 1564, a deposit of a black substance found to be very useful for writing was discovered in Borrowdale, England. This substance, originally called "black lead," was shown in 1879 by Swedish chemist Carl Sheele to be a form of carbon and was subsequently named graphite (after the Greek *graphein,* meaning "to write").

Originally, chunks of graphite from Borrowdale, called "marking stones," were used as writing instruments. Later, sticks of graphite were used. Because graphite is brittle, the sticks needed reinforcement. At first they were wrapped in string, which was unwound as the core wore down. Eventually, graphite rods were tied between two wooden slats or inserted into hollowed-out wooden sticks to form the first crude pencils.

Although Borrowdale graphite was pure enough to use directly, most graphite must be mixed with other materials to be useful for writing instruments. In 1795, French chemist Nicolas-Jaques Conté invented a process in which graphite is mixed with clay and water to produce pencil "lead," a recipe that is still used today. In modern pencil manufacture, graphite and clay are mixed and crushed into a fine powder to which water is added. After the gray sludge is blended for several days, it is dried, ground up again, and mixed with more water to give a gray paste. The paste is extruded through a metal tube to form thin rods, which are then cut into pencil-length pieces called "leads." These leads are heated in an oven to 1000°C until they are smooth and hard. The ratio of clay to graphite is adjusted to vary the hardness of the lead—the more clay in the mix, the harder the lead and the lighter the line it makes.

Pencils are made from a slat of wood with several grooves cut in it to hold the leads. A similar grooved slat is then placed on top and glued to form a "sandwich" from which individual pencils are cut, sanded smooth, and painted. Although many types of wood have been used over the years to make pencils, the current favorite is incense cedar from the Sierra Nevada Mountains of California.

Modern pencils are simple but amazing instruments. The average pencil can write approximately 45,000 words, which is equivalent to a line 35 miles long. The graphite in a pencil is easily transferred to paper because graphite contains layers of carbon atoms bound together in a "chicken-wire" structure. Although the bonding *within* each layer is very strong, the bonding *between* layers is weak, giving graphite its slippery, soft nature. In this way, graphite is much different from diamond, the other common elemental form of carbon. In diamond the carbon atoms are bound tightly in all three dimensions, making it extremely hard—the hardest natural substance.

Pencils are very useful—especially for doing chemistry problems—because we can erase our mistakes. Most pencils used in the United States have erasers (first attached to pencils in 1858), although most European pencils do not. Laid end to end, the number of pencils made in the United States each year would circle the earth about 15 times. Pencils illustrate how useful a simple substance like graphite can be.

The atoms in H_2 (and all other molecules) actually vibrate back and forth around the equilibrium internuclear distance.

positioning is precisely what leads to the stability of the H_2 molecule relative to two separated hydrogen atoms. The potential energy of each electron is lowered because of the increased attractive forces in this area. Although it is not usually discussed in connection with simple models of bonding, we should note that the kinetic energy of the electrons also changes when the individual atoms form the molecule. Thus the energy plotted in Fig. 13.1(b) is the total energy of the system, not just the potential energy. When we say that a bond is formed between the hydrogen atoms, we mean that the H_2 molecule is more stable than two separated hydrogen atoms by a certain quantity of energy (the bond energy).

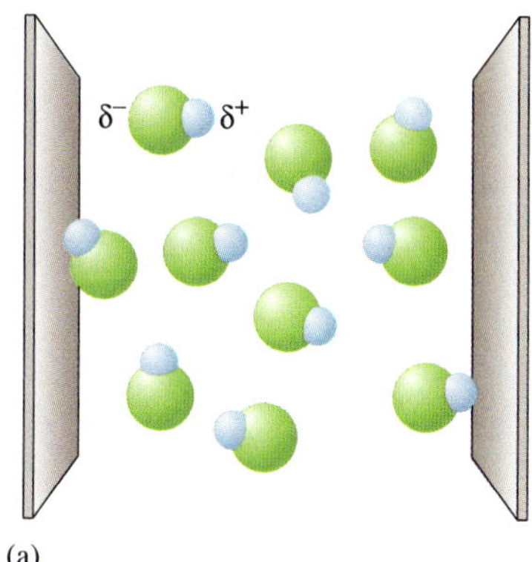

(a)

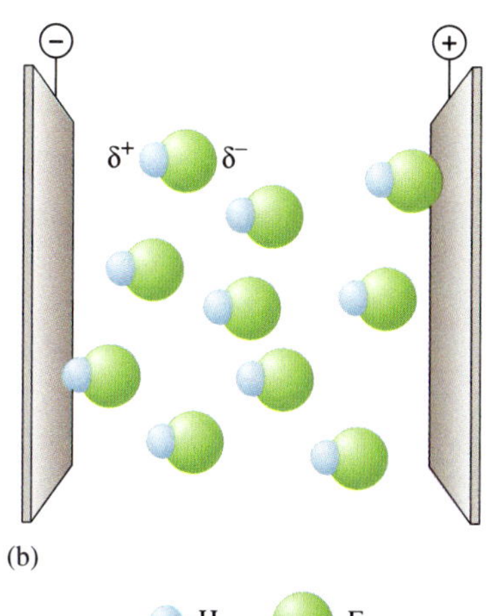

(b)

H F

FIGURE 13.2

The effect of an electric field on hydrogen fluoride molecules. (a) When no electric field is present, the molecules are randomly oriented. (b) When the field is turned on, the molecules tend to line up with their negative ends toward the positive pole and their positive ends toward the negative pole. (This illustration exaggerates the effect. Actually, only a small fraction of the molecules are lined up with the field at a given instant.)

We can also think of a bond in terms of forces. The simultaneous attraction for each electron by the two protons generates a force that pulls the protons toward each other. This attractive force just balances the proton–proton and electron–electron repulsive forces at the distance corresponding to the bond length.

The type of bonding we encounter in the hydrogen molecule and in many other molecules in which *electrons are shared by nuclei* is called **covalent bonding.**

So far we have considered two extreme types of bonding. In ionic bonding the participating atoms are so different that one or more electrons are transferred to form oppositely charged ions. The bonding results from electrostatic interactions among the resulting ions. In covalent bonding two identical atoms share electrons equally. The bonding results from the mutual attraction of the two nuclei for the shared electrons. Between these extremes lie intermediate cases in which the atoms are not so different that electrons are completely transferred but are different enough so that unequal sharing results. These are called **polar covalent bonds.** An example of this type of bond occurs in the hydrogen fluoride (HF) molecule. When a sample of hydrogen fluoride gas is placed in an electric field, the molecules tend to orient themselves as shown in Fig. 13.2, with the fluoride end closest to the positive pole and the hydrogen end closest to the negative pole. This result implies that the HF molecule has the following charge distribution:

$$\underset{\delta+}{\mathrm{H}}\!-\!\underset{\delta-}{\mathrm{F}}$$

where δ (delta) is used to indicate a fractional charge. This same effect was noted in Chapter 4, where many of water's unusual properties were attributed to the polar O—H bonds in the H_2O molecule.

The most logical explanation for the development of the partial positive and negative charges on the atoms (bond polarity) in such molecules as HF and H_2O is that the electrons in the bonds are not shared equally. For example, we can account for the polarity of the HF molecule by assuming that the fluorine atom has a stronger attraction for the shared electrons than the hydrogen atom. Similarly, in the H_2O molecule the oxygen atom appears to attract the shared electrons more strongly than the hydrogen atoms. Because bond polarity has important chemical implications, we find it useful to quantify the ability of an atom to attract shared electrons. In the next section we show how this is done.

13.2 Electronegativity

The different affinities of atoms for the electrons in a bond are described by a property called **electronegativity:** *the ability of an atom in a molecule to attract shared electrons to itself.*

The most widely accepted method for determining electronegativity values is that of Linus Pauling (1901–1995), an American scientist who won Nobel Prizes for both chemistry and peace. To understand Pauling's model, consider a hypothetical molecule HX. The relative electronegativities of the H and X atoms are determined by comparing the measured H—X bond energy with the "expected" H—X bond energy. The expected bond energy

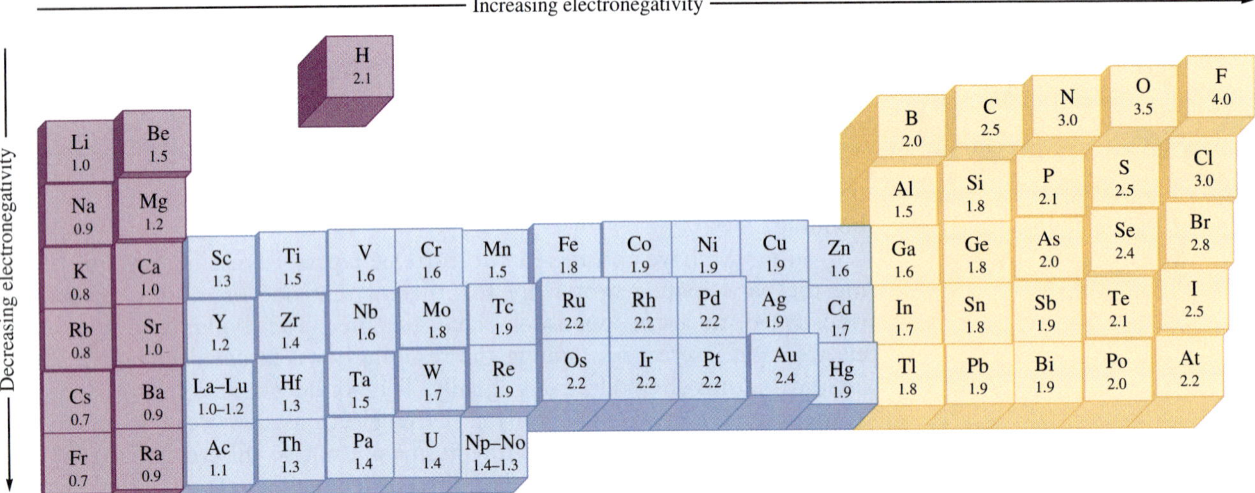

FIGURE 13.3

The Pauling electronegativity values. Electronegativity generally increases across a period and decreases down a group.

is an "average" (actually the geometric mean) of the H—H and X—X bond energies:

$$\text{Expected H—X bond energy} = [(\text{H—H bond energy})(\text{X—X bond energy})]^{1/2}$$

The difference (Δ) between the actual (measured) and expected bond energies is

$$\Delta = (\text{H—X})_{\text{act}} - (\text{H—X})_{\text{exp}}$$

If H and X have identical electronegativities, $(\text{H—X})_{\text{act}}$ and $(\text{H—X})_{\text{exp}}$ are the same and Δ is 0. On the other hand, if X has a greater electronegativity than H, the shared electron(s) will tend to be closer to the X atom. The molecule will be polar, with the following charge distribution:

$$\underset{\delta+}{\text{H}}\text{—}\underset{\delta-}{\text{X}}$$

Note that this bond can be viewed as having an ionic as well as a covalent component. The electrostatic attraction between the partially charged H and X atoms will lead to a greater bond strength. Thus $(\text{H—X})_{\text{act}}$ will be larger than $(\text{H—X})_{\text{exp}}$. The greater the difference in the electronegativities of the atoms, the greater is the ionic component of the bond and the greater is the value of Δ. Thus the relative electronegativities of H and X can be assigned from the Δ values.

The factor of 0.102 is a conversion factor between kJ and eV (the units originally used by Pauling).

The actual formula Pauling used to calculate electronegativity (EN) differences is

$$\text{EN(X)} - \text{EN(H)} = 0.102\sqrt{\Delta}$$

where all bond energies are in units of kJ/mol. Pauling then obtained absolute electronegativity values for the elements by assigning a value of 4.0 to fluorine (the element with the highest electronegativity).

Electronegativity values have been determined by this process for virtually all the elements; the results are given in Fig. 13.3. Note that for the representative elements, electronegativity generally increases from left to right across a period and decreases down a group. The range of electronegativity values is from 4.0 for fluorine to 0.7 for francium.

The relationship between electronegativity and bond type is shown in Table 13.1. For identical atoms (electronegativity difference of zero) the electrons in the bond are shared equally and no polarity occurs. When two atoms with widely differing electronegativities interact, electron transfer usually

TABLE 13.1

The Relationship Between Electronegativity and Bond Type

Electronegativity Difference in the Bonding Atoms	Bond Type
Zero	Covalent
↓ Intermediate	↓ Polar covalent
↓ Large	↓ Ionic

occurs, producing ions—an ionic substance is formed. Intermediate cases give polar covalent bonds with unequal electron sharing.

Example 13.1

Arrange the following bonds according to increasing polarity: H—H, O—H, Cl—H, S—H, and F—H.

Solution The polarity of the bond increases as the difference in electronegativity increases. From the electronegativity values in Fig. 13.3, the following variation in bond polarity is expected (the Pauling electronegativity value appears in parentheses below each element):

	H—H	< S—H	< Cl—H	< O—H	< F—H
	(2.2)(2.2)	(2.6)(2.2)	(3.2)(2.2)	(3.4)(2.2)	(4.0)(2.2)
Electronegativity difference	0	0.4	1.1	1.2	1.8

Covalent bond $\xrightarrow{\text{Polarity increases}}$ polar covalent bond

TABLE 13.2

A Comparison of Pauling's and Allen's Electronegativity Values for Selected Representative Elements

Atom	Pauling	Allen
H	2.20	2.300
Li	0.98	0.912
Be	1.57	1.576
B	2.04	2.051
C	2.55	2.544
N	3.04	3.066
O	3.44	3.610
F	3.98	4.193
Ne	—	4.787
Cl	3.16	2.869
Br	2.96	2.685
I	2.66	2.359

Although Pauling's electronegativity values are the ones most commonly found in textbooks, several other systems for obtaining electronegativity values have been proposed. Recently, Leland C. Allen of Princeton University has pioneered an electronegativity scale based on the average ionization energies of the valence electrons for a given atom. Allen's system allows calculation of an electronegativity value for an atom that is independent of its bonding environment. In this system electronegativity becomes a property of the isolated atom. This approach is fundamentally different from Pauling's system, which is derived from the bond energies of atoms attached to one another.

Although we will not consider the details of Allen's quantum mechanical analysis, it is useful to compare Allen's and Pauling's values for some of the more important representative elements (Table 13.2). While many values are similar, some significant differences appear.

Allen has made a strong case that his system for obtaining electronegativities is more meaningful than that created by Pauling. Allen's values are now becoming accepted by the chemical community.

13.3 Bond Polarity and Dipole Moments

We have seen that when hydrogen fluoride is placed in an electric field, the molecules have a preferential orientation (Fig. 13.2). This follows from the charge distribution in the HF molecule, which has a positive end and a negative end. A molecule such as HF that has a center of positive charge and a center of negative charge is said to be *dipolar,* or to have a **dipole moment.** A molecule that has a positive center of charge of magnitude Q and a negative center of charge of magnitude Q separated by a distance R has a dipole moment given by the expression

$$\text{Dipole moment} = \mu = QR$$

which has SI units of coulomb meter (C m) but is most often given in units of debye [1 debye (D) = 3.336×10^{-30} C m]. The dipolar character of a molecule is often represented by an arrow pointing to the negative charge center, with the tail of the arrow indicating the positive center of charge:

The debye is named after Peter Debye, who pioneered in the measurement of dipole moments.

$$\underset{\delta+\quad\quad\delta-}{+\!\!\longrightarrow}$$

The dipole moment of a molecule gives useful information about its bonding and electron distribution. For example, the observed dipole moment for HF is 1.83 D. If HF were totally ionic (H^+F^-), the expected dipole moment (symbolized by μ) would be

$$\mu = (\underbrace{1.60 \times 10^{-19}\ \text{C}}_{\text{Electron charge}})(\underbrace{9.17 \times 10^{-11}\ \text{m}}_{\text{H—F bond distance}})$$

$$= 1.47 \times 10^{-29}\ \text{C m} = 4.40\ \text{D}$$

This calculation shows that HF is not fully ionic, since the measured dipole moment is much less than 4.40 D. We can estimate the ionic character of HF by assuming that the hydrogen has a charge $\delta+$ and the fluorine has a charge $\delta-$. Using the measured dipole moment, we have

$$1.83\ \text{D} = (\delta)(9.17 \times 10^{-11}\ \text{m}) \times \frac{1\ \text{D}}{3.336 \times 10^{-30}\ \text{C m}}$$

Solving for δ gives 6.66×10^{-20} C. Since the charge on an electron is 1.60×10^{-19} C, each atom in HF has a fractional charge of

$$\frac{6.66 \times 10^{-20}\ \text{C}}{1.60 \times 10^{-19}\ \text{C}} = 0.416$$

From this argument we might say that HF has 42% ionic bonding. Although this analysis is somewhat oversimplified (it assumes that charge distributions can be represented as point charges, for example), it does provide useful information about bonding. Recall that the dipolar character of a molecule is often represented by an arrow pointing to the negative charge center with the tail of the arrow indicating the positive center of charge:

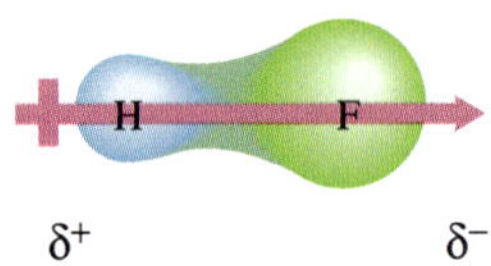

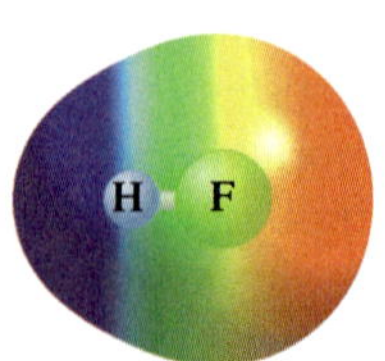

FIGURE 13.4

An electrostatic potential diagram of HF. Red indicates the most electron-rich area (the fluorine atom) and blue indicates the most electron-poor region (the hydrogen atom).

Another way to represent the charge distribution in HF is by an electrostatic potential diagram (see Fig. 13.4). For this representation the colors of visible light are used to show the variation in charge distribution. Red indicates the most electron-rich region of the molecule, and blue indicates the most electron-poor region.

From what has been said so far, we would expect any diatomic molecule with a polar bond (between atoms with different electronegativities) to exhibit a dipole moment. Although this is generally true, the observed dipole moments of diatomic molecules are sometimes smaller than expected. For example, the carbon monoxide molecule CO has a dipole moment of only 0.11 D, much smaller than expected from the polarity of the CO bond. This discrepancy is most likely caused by the lone pairs of electrons on the atoms, which make large contributions to the dipole moment in opposition to that from the bond polarity. We will not explore the details of this situation here. The dipole moments of some representative diatomic molecules are listed in Table 13.3.

TABLE 13.3

The Dipole Moments of Some Diatomic Molecules (gas phase)

Molecule	Dipole Moment (D)
CO	0.112
HF	1.83
HCl	1.11
HBr	0.78
HI	0.38
NaCl	9.00
LiF	6.33
KF	8.60
KBr	10.41

Polyatomic molecules can also exhibit dipolar behavior. For example, because the oxygen atom in the water molecule has a greater electronegativity than the hydrogen atoms, the molecular charge distribution is that shown in

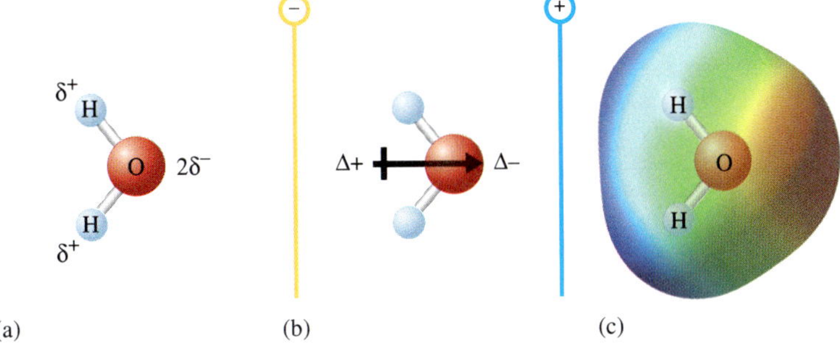

FIGURE 13.5

(a) The charge distribution in the water molecule. (b) The water molecule in an electric field. (c) The electrostatic potential diagram of the water molecule.

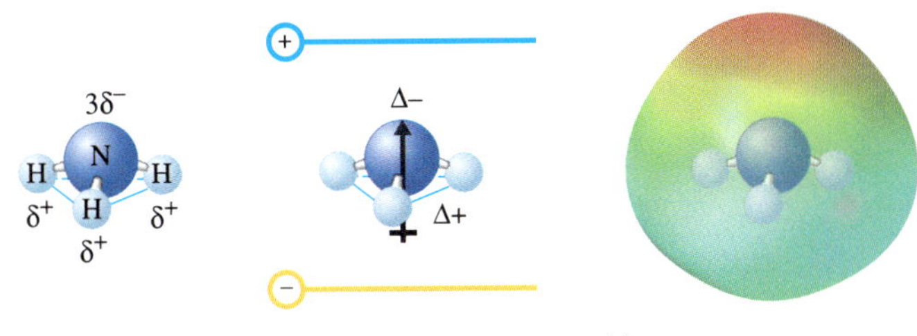

FIGURE 13.6

(a) The structure and charge distribution of the ammonia molecule. The polarity of the N—H bonds occurs because nitrogen has a greater electronegativity than hydrogen. (b) The dipole moment of the ammonia molecule oriented in an electric field. (c) The electrostatic potential diagram for ammonia.

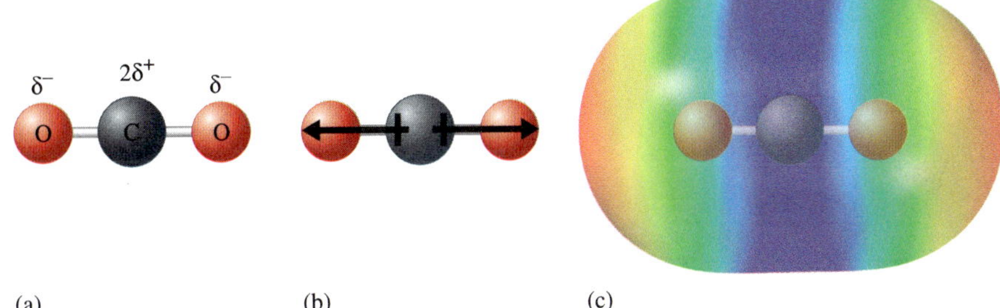

FIGURE 13.7

(a) The carbon dioxide molecule. (b) The opposed bond polarities cancel out, and the carbon dioxide molecule has no dipole moment. (c) The electrostatic potential diagram for carbon dioxide.

Fig. 13.5(a). This charge distribution causes the water molecule to behave in an electric field as if it had two centers of charge—one positive and one negative—as shown in Fig. 13.5(b). Thus the water molecule has a dipole moment. Similar behavior is observed for the NH_3 molecule (Fig. 13.6). Some molecules have polar bonds but do not have a dipole moment. This occurs when the individual bond polarities are arranged in such a way that they cancel. An example is the CO_2 molecule, a linear molecule that has the charge distribution shown in Fig. 13.7. In this case, since the opposing bond polarities cancel, the carbon dioxide molecule does not have a dipole moment. There is no preferential way for this molecule to line up in an electric field. (Try to find a preferred orientation.)

There are many cases where the bond polarities in molecules oppose and exactly cancel each other. Some common types of molecules with polar bonds but without dipole moments are shown in Table 13.4.

Example 13.2

For each of the following molecules, show the direction of the bond polarities. Also indicate which ones have dipole moments: HCl, Cl_2, SO_3 (planar), CH_4 (tetrahedral), and H_2S (V-shaped).

TABLE 13.4

Types of Molecules with Polar Bonds but No Resulting Dipole Moment

Type		Cancellation of Polar Bonds	Example	Ball-and-Stick Model
Linear molecules with two identical bonds	B—A—B		CO_2	
Planar molecules with three identical bonds 120 degrees apart	B, A, B, B, 120°		SO_3	
Tetrahedral molecules with four identical bonds 109.5 degrees apart	B, A, B, B, B		CCl_4	

Solution **The HCl molecule:** Because the electronegativity of chlorine (3.2) is greater than that of hydrogen (2.2), the chlorine is partially negative, and the hydrogen is partially positive. The HCl molecule has a dipole moment oriented as follows:

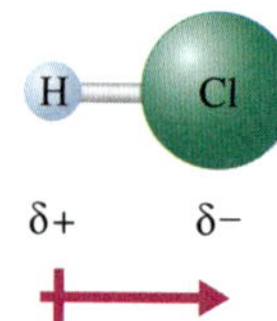

The Cl_2 molecule: Because the two chlorine atoms share the electrons equally, no bond polarity occurs. The Cl_2 molecule has no dipole moment.

The SO_3 molecule: Because the electronegativity of oxygen (3.4) is greater than that of sulfur (2.6), each oxygen has a partial negative charge, and the sulfur has a partial positive charge:

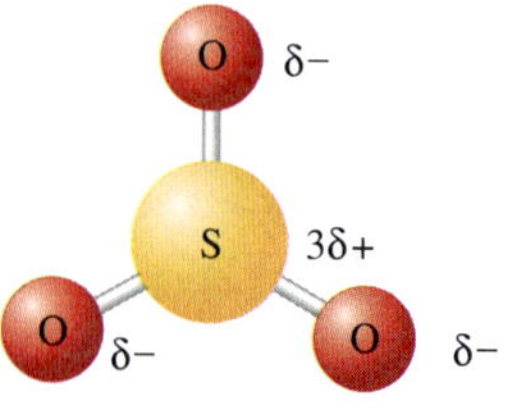

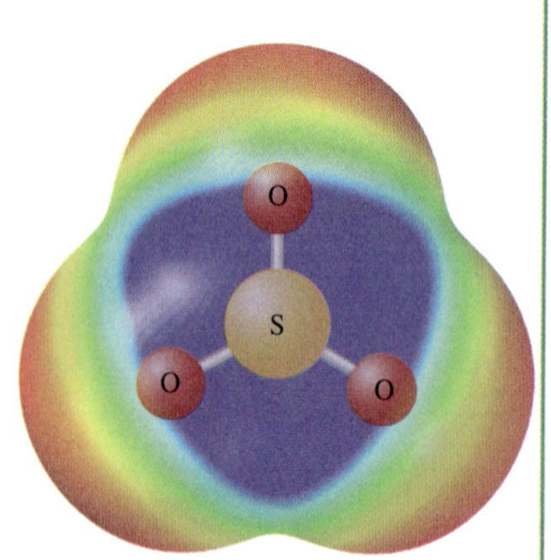

However, the bond polarities cancel, and the molecule has no dipole moment.

The CH_4 molecule: Carbon has a slightly higher electronegativity (2.6) than hydrogen (2.2). This leads to small partial positive charges on the hydrogen atoms and a small partial negative charge on the carbon:

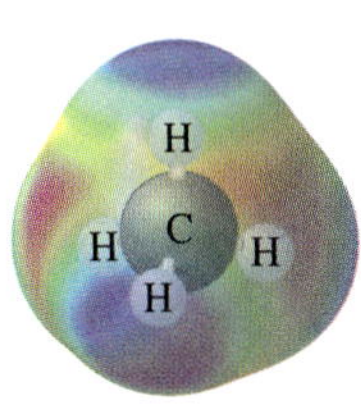

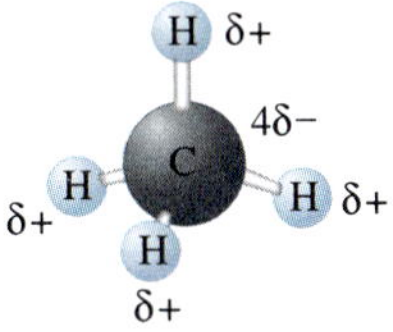

This case is similar to the third type in Table 13.4. Since the bond polarities cancel, the molecule has no dipole moment.

The H_2S molecule: Since the electronegativity of sulfur (2.6) is greater than that of hydrogen (2.2), the sulfur has a partial negative charge, and the hydrogen atoms have a partial positive charge, which can be represented as follows:

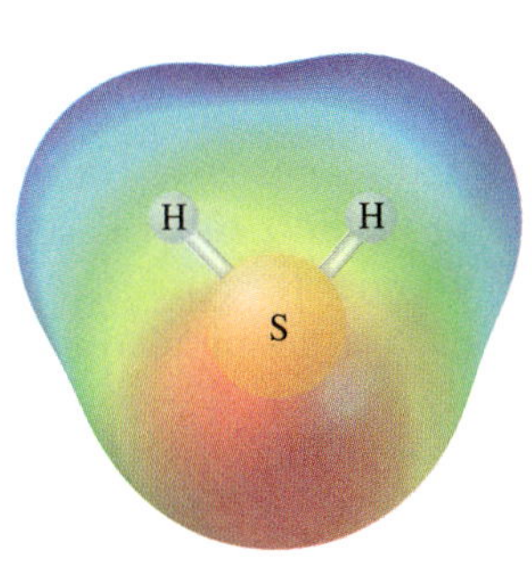

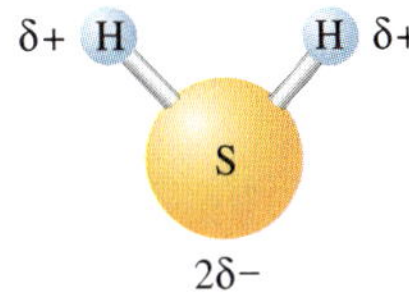

This case is analogous to the water molecule. The polar bonds result in a dipole moment oriented as shown.

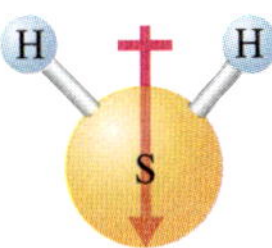

13.4 Ions: Electron Configurations and Sizes

Atoms in stable compounds usually have a noble gas electron configuration.

The description of the electron arrangements in atoms that emerged from the quantum mechanical model has helped a great deal in our understanding of what constitutes a stable compound. For example, in a very large number of stable compounds the atoms have noble gas arrangements of electrons. Nonmetallic elements achieve a noble gas electron configuration either by sharing electrons with other nonmetals to form covalent bonds or by taking electrons from metals to form ions. In the latter case the nonmetals form anions and the metals form cations. The following generalizations can be applied to the electron configurations in most stable compounds:

As we will see later, there are exceptions to these rules, but they remain a useful place to start.

- When *two nonmetals* react to form a covalent bond, they share electrons in a way that completes the valence electron configurations of both atoms. That is, both nonmetals attain noble gas electron configurations.
- When *a nonmetal and a representative group metal* react to form a binary ionic compound, the ions form so that the valence electron configuration of the nonmetal is completed and the valence orbitals of the metal are emptied. In this way both ions achieve noble gas electron configurations.

Although there are some important exceptions, these generalizations apply to the vast majority of compounds and are important to remember. We will deal with covalent bonds more thoroughly later. Next, we will consider what implications these rules hold for ionic compounds.

Predicting Formulas of Ionic Compounds

In the solid state of an ionic compound the ions are relatively close together, and many ions are simultaneously interacting:

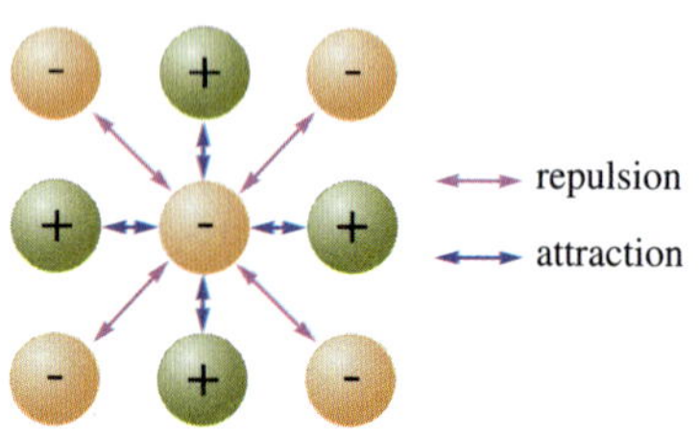

In the gas phase of an ionic substance the ions would be relatively far apart and would not contain large groups of ions:

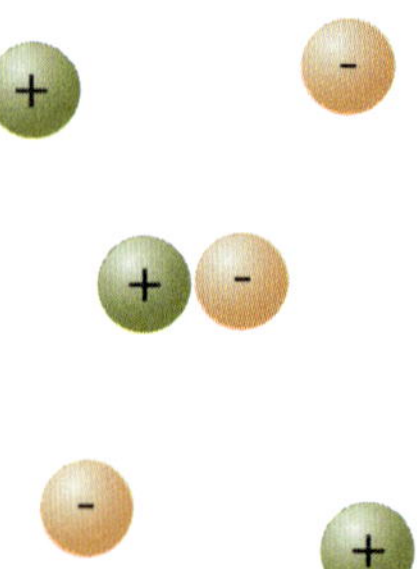

At the beginning of this discussion we should emphasize that when chemists use the term *ionic compound,* they are usually referring to the solid state of that compound. Solid ionic compounds contain a large collection of positive and negative ions packed together in a way that minimizes the $\ominus \cdot\cdot \ominus$ and $\oplus \cdot\cdot \oplus$ repulsions and maximizes the $\oplus \cdot\cdot \ominus$ attractions. This situation is in contrast to the gas phase of an ionic substance, where discrete ion pairs exist. Thus, when we speak in this text of the stability of an ionic compound, we are referring to the solid state, where the large attractive forces present among the oppositely charged ions tend to stabilize (favor the formation of) the ions. For example, as we mentioned in the previous chapter, the O^{2-} ion is not stable as an isolated gas-phase species but, of course, is very stable in many solid ionic compounds. That is, $MgO(s)$, which contains Mg^{2+} and O^{2-} ions, is very stable, but the isolated gas-phase ion pair $Mg^{2+} \cdot\cdot O^{2-}$ is not energetically favorable in comparison with the separate neutral gaseous atoms. Thus you should keep in mind that in this section, and in most other cases in which we are describing the nature of ionic compounds, the discussion usually refers to the solid state, in which many ions are simultaneously interacting.

To illustrate the principles of electron configurations in stable, solid ionic compounds, we will consider the formation of an ionic compound from calcium and oxygen. We can predict what compound will form by considering the valence electron configurations of the two atoms:

$$\text{Ca:} \quad [\text{Ar}]4s^2$$

$$\text{O:} \quad [\text{He}]2s^22p^4$$

From Fig. 13.3 we see that the electronegativity of oxygen (3.4) is much greater than that of calcium (1.0). Because of this large difference, electrons will be transferred from calcium to oxygen to form oxygen anions and calcium cations in the compound. How many electrons are transferred? We can base our prediction on the observation that noble gas configurations are generally the most stable. Note that oxygen needs two electrons to fill its $2s$ and $2p$ valence orbitals and to achieve the configuration of neon ($1s^22s^22p^6$). And by losing two electrons, calcium can achieve the configuration of argon. Two electrons are therefore transferred:

$$\underbrace{\text{Ca} + \text{O}}_{2e^-} \longrightarrow \text{Ca}^{2+} + \text{O}^{2-}$$

To predict the formula of the ionic compound, we simply recognize that chemical compounds are always electrically neutral—they have the same quantities of positive and negative charges. In this case we must have equal numbers of Ca^{2+} and O^{2-} ions, and the empirical formula of the compound is CaO.

The same principles can be applied to many other cases. For example, consider the compound formed between aluminum and oxygen. Because aluminum has the configuration $[\text{Ne}]3s^23p^1$, it must lose three electrons to form the Al^{3+} ion and thus achieve the neon configuration. Therefore, the Al^{3+} and O^{2-} ions form in this case. Since the compound must be electrically neutral,

TABLE 13.5

Common Ions with Noble Gas Electron Configurations in Ionic Compounds

Group 1A	Group 2A	Group 3A	Group 6A	Group 7A	Electron Configuration
H^-, Li^+	Be^{2+}				[He]
Na^+	Mg^{2+}	Al^{3+}	O^{2-}	F^-	[Ne]
K^+	Ca^{2+}		S^{2-}	Cl^-	[Ar]
Rb^+	Sr^{2+}		Se^{2-}	Br^-	[Kr]
Cs^+	Ba^{2+}		Te^{2-}	I^-	[Xe]

there must be three O^{2-} ions for every two Al^{3+} ions, and the compound has the empirical formula Al_2O_3.

Table 13.5 shows common elements that form ions with noble gas electron configurations in ionic compounds. In losing electrons to form cations, metals in Group 1A lose one electron, those in Group 2A lose two electrons, and those in Group 3A lose three electrons. In gaining electrons to form anions, nonmetals in Group 7A (the halogens) gain one electron, and those in Group 6A gain two electrons. Hydrogen typically behaves as a nonmetal and can gain one electron to form the hydride ion (H^-), which has the electron configuration of helium.

There are some important exceptions to the rules discussed here. For example, tin forms both Sn^{2+} and Sn^{4+} ions, and lead forms both Pb^{2+} and Pb^{4+} ions. Also, bismuth forms Bi^{3+} and Bi^{5+} ions, and thallium forms Tl^+ and Tl^{3+} ions. There are no simple explanations for the behavior of these ions. For now, just note them as exceptions to the very useful rule that ions generally adopt noble gas electron configurations in ionic compounds. Our discussion here refers to representative metals. The transition metals exhibit more complicated behavior, forming a variety of ions that will be considered in Chapter 19.

Sizes of Ions

Ion size plays an important role in determining the structure and stability of ionic solids, the properties of ions in aqueous solution, and the biological effects of ions. As with atoms, it is impossible to define precisely the sizes of ions. Most often, ionic radii are determined from the measured distances between ion centers in ionic compounds. This method, of course, involves an assumption about how the distance should be divided up between the two ions. Thus you will note considerable disagreement among ionic sizes given in various sources. Here we are mainly interested in trends and will be less concerned with absolute ion sizes.

Various factors influence ionic size. We will first consider the relative sizes of an ion and its parent atom. Since a positive ion is formed by removing electrons from a neutral atom, the resulting cation is smaller than its parent atom. The opposite is true for negative ions; the addition of electrons to a neutral atom produces an anion significantly larger than its parent atom.

It is also important to know how the sizes of ions vary depending on the positions of the parent elements in the periodic table. Figure 13.8 shows the sizes of the most important ions (each with a noble gas configuration) and their position in the periodic table. Note that ion size increases down a group. The changes that occur horizontally are complicated because of the change

FIGURE 13.8

Sizes of ions related to positions of elements in the periodic table. Note that size generally increases down a group. Also note that in a series of isoelectronic ions, size decreases with increasing atomic number. The ionic radii are given in units of picometers.

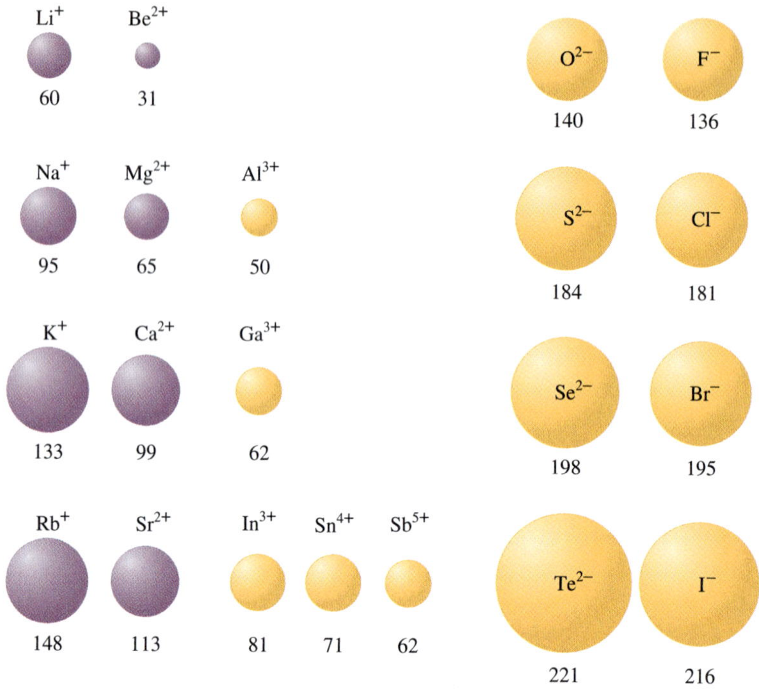

from predominantly metals on the left-hand side of the periodic table to nonmetals on the right-hand side. A given period thus contains both elements that give up electrons to form cations and ones that accept electrons to form anions.

One trend worth noting involves the relative sizes of a set of **isoelectronic ions**—*ions containing the same number of electrons.* Consider the ions O^{2-}, F^-, Na^+, Mg^{2+}, and Al^{3+}. Each of these ions has the neon electron configuration. How do the sizes of these ions vary? In general, there are two important facts to consider in predicting the relative sizes of ions: the number of electrons and the number of protons. Since these ions are isoelectronic, the number of electrons is 10 in each case. Electron repulsions should therefore be about the same in all cases. However, the number of protons increases from 8 to 13 as we go from the O^{2-} ion to the Al^{3+} ion. Thus, in going from O^{2-} to Al^{3+}, the 10 electrons experience a greater attraction as the positive charge on the nucleus increases. This causes the ions to become smaller. You can confirm this by looking at Fig. 13.8. In general, for a series of isoelectronic ions, the size decreases as the nuclear charge (Z) increases.

For isoelectronic ions, size generally decreases as *Z* increases.

Example 13.3

Arrange the ions Se^{2-}, Br^-, Rb^+, and Sr^{2+} in order of decreasing size.

Solution This is an isoelectronic series of ions with the krypton electron configuration. Since these ions all have the same number of electrons, their sizes will depend on nuclear charge. The Z values are 34 for Se^{2-}, 35 for Br^-, 37 for Rb^+, and 38 for Sr^{2+}. Since the nuclear charge is greatest for Sr^{2+}, it is the smallest of these ions. The Se^{2-} ion is largest.

$$\underset{\text{Largest}}{\underset{\uparrow}{Se^{2-}}} > Br^- > Rb^+ > \underset{\text{Smallest}}{\underset{\uparrow}{Sr^{2+}}}$$

Example 13.4

Choose the largest ion in each of the following groups.

a. Li^+, Na^+, K^+, Rb^+, Cs^+ **b.** Ba^{2+}, Cs^+, I^-, Te^{2-}

Solution

Ion size generally increases down a group.

a. The ions are all from Group 1A elements. Since size increases down a group (the ion with the greatest number of electrons is the largest), Cs^+ is the largest ion.

b. This is an isoelectronic series of ions, all of which have the xenon electron configuration. The ion with the smallest nuclear charge is the largest ion.

$$\underset{Z=52}{Te^{2-}} > \underset{Z=53}{I^-} > \underset{Z=55}{Cs^+} > \underset{Z=56}{Ba^{2+}}$$

13.5 Formation of Binary Ionic Compounds

In this section we will introduce the factors that influence the stability and the structures of solid binary ionic compounds. We know that metals and nonmetals react by transferring electrons to form cations and anions that are mutually attractive. The resulting ionic solid forms because the aggregated oppositely charged ions have a lower energy than the original elements. Just how strongly the ions attract each other in the solid state is indicated by the **lattice energy**—*the change in energy that takes place when separated gaseous ions are packed together to form an ionic solid:*

$$M^+(g) + X^-(g) \longrightarrow MX(s)$$

The structures of ionic solids will be discussed in detail in Chapter 16.

The lattice energy is often defined as the energy *released* when an ionic solid forms from its ions. However, in this book the sign of an energy term is always determined from the system's point of view: negative if the process is exothermic; positive if endothermic. Thus lattice energy has a negative sign.

We can illustrate the energy changes involved in the formation of an ionic solid by considering the formation of solid lithium fluoride from its elements:

$$Li(s) + \tfrac{1}{2}F_2(g) \longrightarrow LiF(s)$$

To see the energy terms associated with this process, we take advantage of the fact that energy is a state function and break this reaction into steps, the sum of which gives the overall reaction.

Step 1

Sublimation of solid lithium. Sublimation involves taking a substance from the solid state to the gaseous state:

$$Li(s) \longrightarrow Li(g)$$

The enthalpy of sublimation for $Li(s)$ is 161 kJ/mol.

Step 2

Ionization of lithium atoms to form Li^+ ions in the gas phase:

$$Li(g) \longrightarrow Li^+(g) + e^-$$

This process corresponds to the first ionization energy for lithium, which is 520 kJ/mol.

Step 3

Dissociation of fluorine molecules. We need to form 1 mole of fluorine atoms by breaking the F—F bonds in 0.5 mole of F_2 molecules:

$$\tfrac{1}{2}F_2(g) \longrightarrow F(g)$$

The energy required to break this bond is 154 kJ/mol. In this case we are breaking the bonds in a half mole of fluorine, so the energy required for this step is 154 kJ/2, or 77 kJ.

Step 4

Formation of F^- ions from fluorine atoms in the gas phase:

$$F(g) + e^- \longrightarrow F^-(g)$$

The energy change for this process corresponds to the electron affinity of fluorine, which is −328 kJ/mol.

Step 5

Formation of solid lithium fluoride from the gaseous Li^+ and F^- ions:

$$Li^+(g) + F^-(g) \longrightarrow LiF(s)$$

This corresponds to the lattice energy for LiF, which is −1047 kJ/mol.

Since the sum of these five processes yields the desired overall reaction, the sum of the individual energy changes gives the overall energy change:

	Process	Energy Change (kJ)
	$Li(s) \rightarrow Li(g)$	161
	$Li(g) \rightarrow Li^+(g) + e^-$	520
	$\frac{1}{2}F_2(g) \rightarrow F(g)$	77
	$F(g) + e^- \rightarrow F^-(g)$	−328
	$Li^+(g) + F^-(g) \rightarrow LiF(s)$	−1047
Overall:	$Li(s) + \frac{1}{2}F_2(g) \rightarrow LiF(s)$	−617 kJ (per mole of LiF)

In doing this calculation, we have ignored the small difference between ΔH_{sub} and ΔE_{sub}.

This process is summarized by the energy diagram in Fig. 13.9. Note that the formation of solid lithium fluoride from its elements is highly exothermic

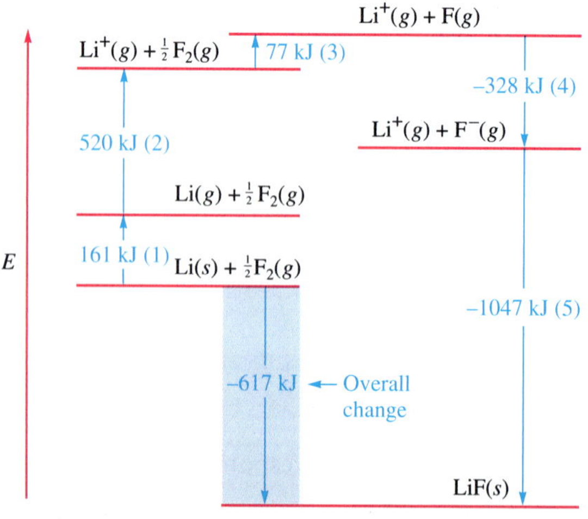

FIGURE 13.9

The energy changes involved in the formation of solid lithium fluoride from its elements. The numbers in parentheses refer to the reaction steps discussed in the text.

mainly because of the very large negative lattice energy. A great deal of energy is released when the ions combine to form the solid. In fact, note that the energy released when an electron is added to a fluorine atom to form the F^- ion (328 kJ/mol) is not enough to remove an electron from lithium (520 kJ/mol). That is, when a metallic lithium atom reacts with a nonmetallic fluorine atom to form *separated* ions,

$$Li(g) + F(g) \longrightarrow Li^+(g) + F^-(g)$$

the process is endothermic and thus unfavorable. Clearly, then, the main impetus for the formation of the ionic compound rather than a covalent compound results from the strong mutual attractions among the Li^+ and F^- ions in the solid. The lattice energy is the dominant energy term.

The structure of the solid lithium fluoride is represented in Fig. 13.10. Note the alternating arrangement of the Li^+ and F^- ions. Also note that each Li^+ is surrounded by six F^- ions and each F^- ion is surrounded by six Li^+ ions. This structure can be rationalized by assuming that the ions behave as hard spheres that pack together in a way that both maximizes the attractions among the oppositely charged ions and minimizes the repulsions among the identically charged ions.

All the binary ionic compounds formed by an alkali metal and a halogen have the structure shown in Fig. 13.10, except for the cesium salts. The arrangement of ions shown in Fig. 13.10 is often called the *sodium chloride structure*, after the most common substance that possesses it.

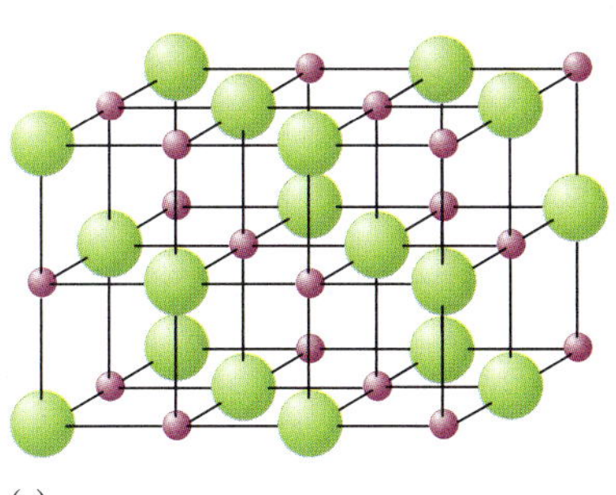

(a)

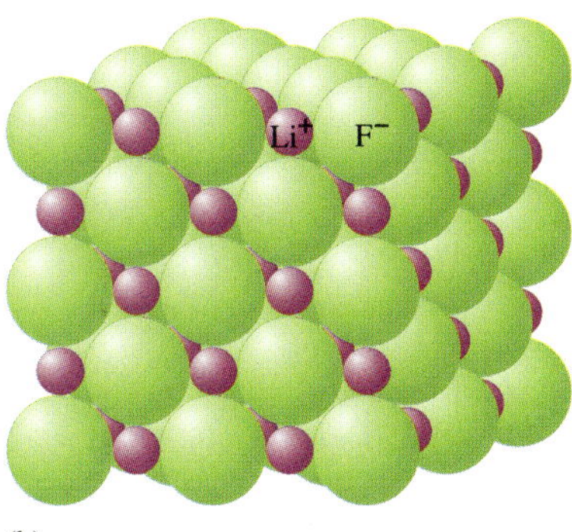

(b)

FIGURE 13.10

The structure of lithium fluoride. (a) Represented by a ball-and-stick model. Note that each Li^+ ion is surrounded by six F^- ions, and each F^- ion is surrounded by six Li^+ ions. (b) Represented with the ions shown as spheres. The structure is determined by packing the spherical ions in a way that both maximizes the ionic attractions and minimizes the ionic repulsions.

Lattice Energy Calculations

In discussing the energetics of the formation of solid lithium fluoride, we emphasized the importance of lattice energy in contributing to the stability of the ionic solid. Lattice energy can be represented by a modified form of Coulomb's law,

$$\text{Lattice energy} = k\left(\frac{Q_1Q_2}{r}\right)$$

where k is a proportionality constant that depends on the structure of the solid and the electron configurations of the ions, Q_1 and Q_2 are the charges on the ions, and r is the shortest distance between the centers of the cations and anions. Note that the lattice energy has a negative sign when Q_1 and Q_2 have opposite signs. This result is expected, since bringing cations and anions together is an exothermic process. Also note that the process becomes more exothermic as the ionic charges increase and as the distances between the ions in the solid decrease.

The importance of the charges in ionic solids can be illustrated by comparing the energies involved in the formation of NaF(s) and MgO(s). These solids contain the isoelectric ions Na^+, F^-, Mg^{2+}, and O^{2-}. The energy diagram for the formation of the two solids is given in Fig. 13.11. Note several important features:

The energy released when the gaseous Mg^{2+} and O^{2-} ions combine to form solid MgO is much greater (more than four times greater) than that released when the gaseous Na^+ and F^- ions combine to form solid NaF.

The energy required to remove two electrons from the magnesium atom (735 kJ/mol for the first and 1445 kJ/mol for the second, yielding a total

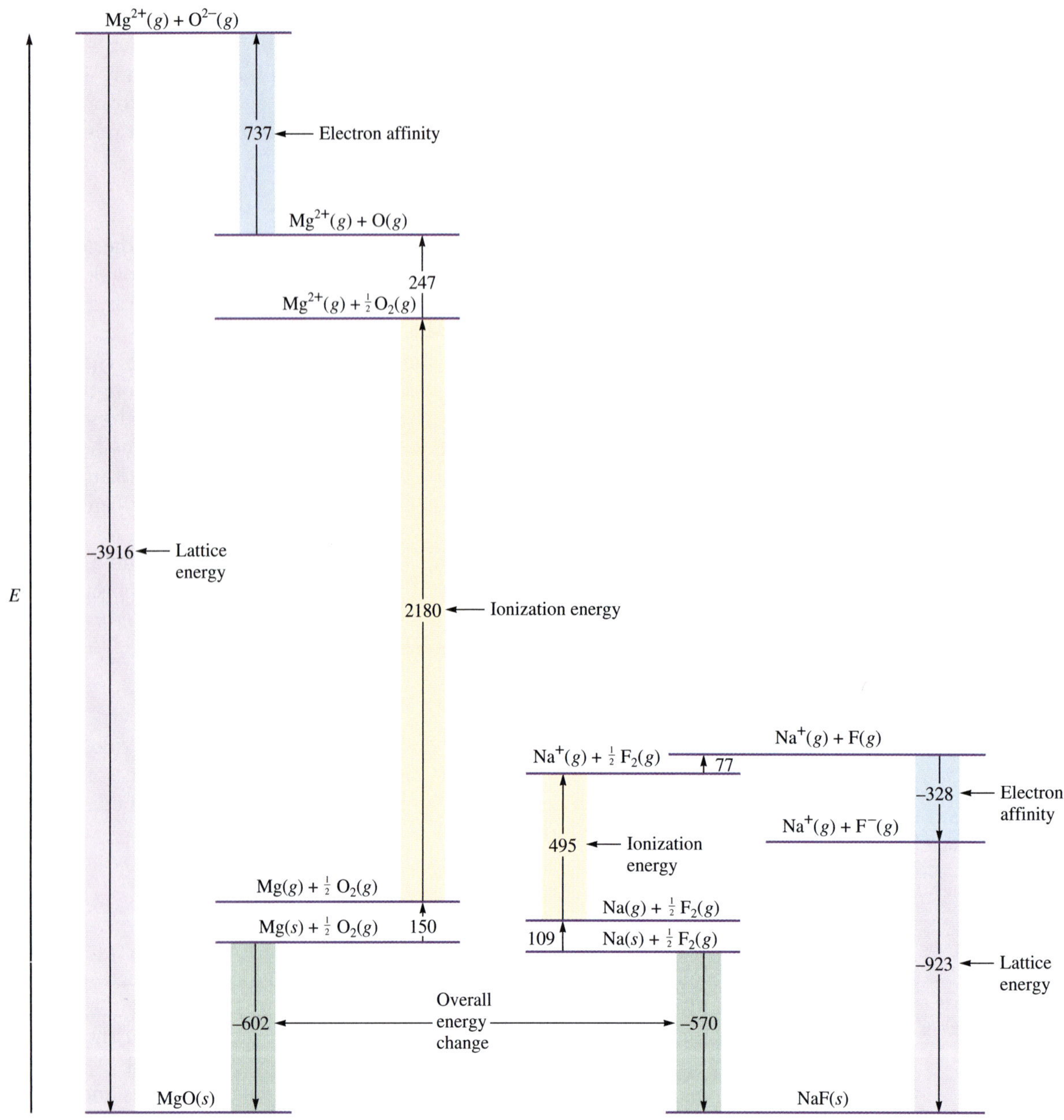

FIGURE 13.11

Comparison of the energy changes involved in the formation of solid sodium fluoride and solid magnesium oxide. Note the large lattice energy for magnesium oxide (where doubly charged ions are combining) compared with that for sodium fluoride (where singly charged ions are combining).

of 2180 kJ/mol) is much greater than the energy required to remove an electron from a sodium atom (495 kJ/mol).

Energy (737 kJ/mol) is required to add two electrons to the oxygen atom in the gas phase. Addition of the first electron is exothermic (−141 kJ/mol), but addition of the second electron is quite endothermic (878 kJ/mol). This latter energy must be obtained indirectly, since the $O^{2-}(g)$ is not stable.

Because twice as much energy is required to remove the second electron from magnesium as to remove the first, and because addition of an electron to the gaseous O^- ion is quite endothermic, it seems puzzling that magnesium oxide contains Mg^{2+} and O^{2-} ions rather than Mg^+ and O^- ions. The answer

Since the equation for lattice energy contains the product Q_1Q_2, the lattice energy for a solid with 2+ and 2− ions should be four times that for a solid with 1+ and 1− ions. That is,

$$\frac{(+2)(-2)}{(+1)(-1)} = 4$$

For MgO and NaF the observed ratio of lattice energies (see Fig. 13.11) is

$$\frac{-3916 \text{ kJ}}{-923 \text{ kJ}} = 4.24$$

lies in the lattice energy. Note that the lattice energy for combining gaseous Mg^{2+} and O^{2-} ions to form MgO(s) is 3000 kJ/mol more negative than that for combining gaseous Na^+ and F^- ions to form NaF(s). Thus the energy released in forming a solid containing Mg^{2+} and O^{2-} ions rather than Mg^+ and O^- ions more than compensates for the energies required for the processes that produce the Mg^{2+} and O^{2-} ions.

If there is so much lattice energy to be gained in going from singly charged to doubly charged ions in the case of magnesium oxide, why then does solid sodium fluoride contain Na^+ and F^- ions rather than Na^{2+} and F^{2-} ions? We can answer this question by recognizing that both Na^+ and F^- ions have the neon electron configuration. Removal of an electron from Na^+ requires an extremely large quantity of energy (4560 kJ/mol) because a $2p$ electron must be removed. Conversely, the addition of an electron to F^- would require use of the relatively high-energy $3s$ orbital, which is also an unfavorable process. Thus we can say that for sodium fluoride the extra energy required to form the doubly charged ions is greater than the gain in lattice energy that would result.

This discussion of the energies involved in the formation of solid ionic compounds illustrates that a variety of factors operate to determine the composition and structure of these compounds. The most important of these factors involve the balancing of the energies required to form highly charged ions and the energy released when highly charged ions combine to form the solid.

13.6 Partial Ionic Character of Covalent Bonds

Recall that when atoms with different electronegativities react to form molecules, the electrons are not shared equally. The possible result is a polar covalent bond or, in the case of a large electronegativity difference, a complete transfer of one or more electrons to form ions. The cases are summarized in Fig. 13.12.

(a)

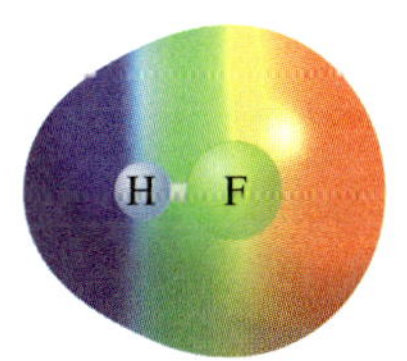

(b)

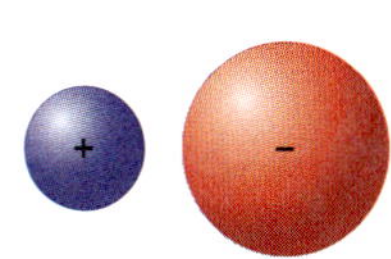

(c)

FIGURE 13.12

The three possible types of bonds: (a) a covalent bond formed between identical F atoms; (b) the polar covalent bond of HF, with both ionic and covalent components; and (c) an ionic bond with no electron sharing.

How well can we tell the difference between an ionic bond and a polar covalent bond? The only honest answer to this question is that there are probably no totally ionic bonds between *discrete pairs of atoms*. The evidence for this statement comes from calculations of the percent ionic character for the bonds of various binary compounds in the gas phase. These calculations are based on comparisons of the measured dipole moments for molecules of the type X—Y with the calculated dipole moments for the completely ionic case, X^+Y^-. We performed a calculation of this type for HF in Section 13.3. The percent ionic character of a bond can be defined as

$$\text{Percent ionic character of a bond} = \left(\frac{\text{measured dipole moment of X—Y}}{\text{calculated dipole moment of } X^+Y^-}\right) \times 100\%$$

Application of this definition to various compounds (in the gas phase) gives the results shown in Fig. 13.13, where percent ionic character is plotted versus the difference in the electronegativity values of X and Y. Note from this plot that ionic character increases with electronegativity difference, as expected. However, none of the bonds reaches 100% ionic character, even though compounds with the maximum possible electronegativity differences are considered. Thus, according to this definition, no individual bonds are completely ionic. This conclusion is in contrast to the usual classification of many of

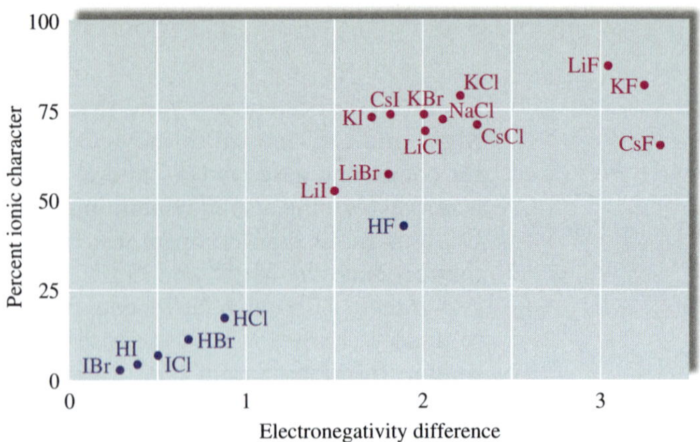

FIGURE 13.13

The relationship between the ionic character of a covalent bond and the electronegativity difference of the bonded atoms. The compounds normally considered to be ionic in the solid phase are shown in red.

Molten NaCl conducts an electric current, indicating the presence of mobile Na^+ and Cl^- ions.

these compounds (as solids). All the compounds shown in Fig. 13.13 with more than 50% ionic character are normally considered to be ionic solids. Recall, however, that the results in Fig. 13.13 are for the gas phase, where individual XY molecules exist. These results cannot necessarily be assumed to apply to the solid state, where the existence of ions is favored by the multiple ion interactions.

Another complication in identifying ionic compounds is that many substances contain polyatomic ions. For example, NH_4Cl contains NH_4^+ and Cl^- ions, and Na_2SO_4 contains Na^+ and SO_4^{2-} ions. The ammonium and sulfate ions are held together by covalent bonds. Thus, calling NH_4Cl and Na_2SO_4 ionic compounds is somewhat ambiguous.

We will avoid these problems by adopting an operational definition of ionic compounds: *Any compound that conducts an electric current when melted will be classified as ionic.* Also, the generic term *salt* will be used interchangeably with *ionic compound* in this book.

13.7 The Covalent Chemical Bond: A Model

Before we develop specific models for covalent chemical bonding, it will be helpful to summarize some of the concepts introduced in this chapter.

What is a chemical bond? Chemical bonds can be viewed as forces that cause a group of atoms to behave as a unit.

Why do chemical bonds occur? There is no principle of nature that states that bonds are favored or disfavored. Bonds are neither inherently "good" nor inherently "bad" as far as nature is concerned; they result from the tendency of a system to seek its lowest possible energy. From a simplistic point of view, bonds occur when collections of atoms are more stable (lower in energy) than the separate atoms. For example, about 1652 kJ of energy is required to break a mole of methane (CH_4) molecules into separate C and H atoms. Or, taking the opposite view, 1652 kJ of energy is released when 1 mole of methane is formed from 1 mole of gaseous C atoms and 4 moles of gaseous H atoms. Thus we can say that 1 mole of CH_4 molecules in the gas phase is 1652 kJ lower in energy than 1 mole of carbon atoms plus 4 moles of hydrogen atoms. Methane is therefore a stable molecule relative to its separated atoms.

We find it useful to interpret molecular stability in terms of a model called a chemical bond. To help understand why this model was invented, let's continue with methane, which consists of four hydrogen atoms arranged at the corners of a tetrahedron around a carbon atom:

A tetrahedron has four equal triangular faces.

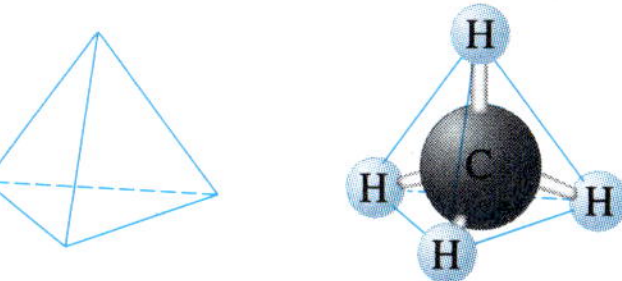

Given this structure, it is natural to envision four individual C—H interactions (we call them bonds). The energy of stabilization of CH_4 is divided equally among the four bonds to give an average C—H bond energy per mole of C—H bonds:

$$\frac{1652 \text{ kJ}}{4} = 413 \text{ kJ}$$

Next, consider methyl chloride, which consists of CH_3Cl molecules having the structure

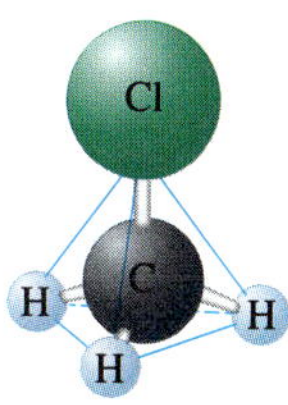

Experiments have shown that about 1578 kJ of energy is required to break down 1 mole of gaseous CH_3Cl molecules into gaseous carbon, chlorine, and hydrogen atoms. The reverse process can be represented as

$$C(g) + Cl(g) + 3H(g) \longrightarrow CH_3Cl(g) + 1578 \text{ kJ/mol}$$

A mole of gaseous methyl chloride is lower in energy by 1578 kJ than its separate gaseous atoms. Thus a mole of methyl chloride is held together by 1578 kJ of energy. Again, it is very useful to divide this energy into individual bonds. Methyl chloride can be visualized as containing one C—Cl bond and three C—H bonds. If we assume arbitrarily that a C—H interaction represents the same quantity of energy in any situation (that is, the strength of a C—H bond is independent of its molecular environment), we can do the following bookkeeping:

$$1 \text{ mol of C—Cl bonds plus 3 mol of C—H bonds} = 1578 \text{ kJ}$$

$$\text{C—Cl bond energy} + 3(\text{average C—H bond energy}) = 1578 \text{ kJ}$$

$$\text{C—Cl bond energy} + 3(413 \text{ kJ/mol}) = 1578 \text{ kJ}$$

$$\text{C—Cl bond energy} = 1578 - 1239 = 339 \text{ kJ/mol}$$

These assumptions allow us to associate given quantities of energy with C—H and C—Cl bonds.

It is important to note that the bond concept is a human invention. Bonds provide a method for dividing up the energy evolved when a stable molecule is formed from its component atoms. Thus in this context *a bond represents a quantity of energy* obtained from the molecular energy of stabilization in a rather arbitrary way. This is not to say that the concept of individual bonds is a bad idea. In fact, the modern concept of the chemical bond, conceived by American chemists G. N. Lewis and Linus Pauling, is one of the most useful ideas chemists have ever developed.

Models: An Overview

Bonding is a model proposed to explain molecular stability.

The framework of chemistry, like that of any science, consists of models—attempts to explain how nature operates on the microscopic level based on experiences in the macroscopic world. To understand chemistry, one must understand its models and how they are used. We will use the concept of bonding to reemphasize the important characteristics of models, including their origin, structure, and uses.

Models originate from our observations of the properties of nature. For example, the concept of bonds arose from the observations that most chemical processes involve collections of atoms and that chemical reactions involve rearrangements of the ways in which the atoms are grouped. So to understand reactions, we must understand the forces that bind atoms together.

In natural processes there is a tendency toward lower energy. Collections of atoms therefore occur because the aggregated state has lower energy than the separated atoms. Why? As we have seen earlier in this chapter, the best explanations for the energy change involve atoms sharing electrons or atoms transferring electrons to become ions. In the case of electron sharing we find it convenient to assume that individual bonds occur between pairs of atoms. Let's explore the validity of this assumption and see how it is useful.

In a diatomic molecule such as H_2, it is natural to assume that a bond exists between the atoms, holding them together. It is also useful to assume that individual bonds are present in polyatomic molecules such as CH_4. So instead of thinking of CH_4 as a unit with a stabilization energy of 1652 kJ

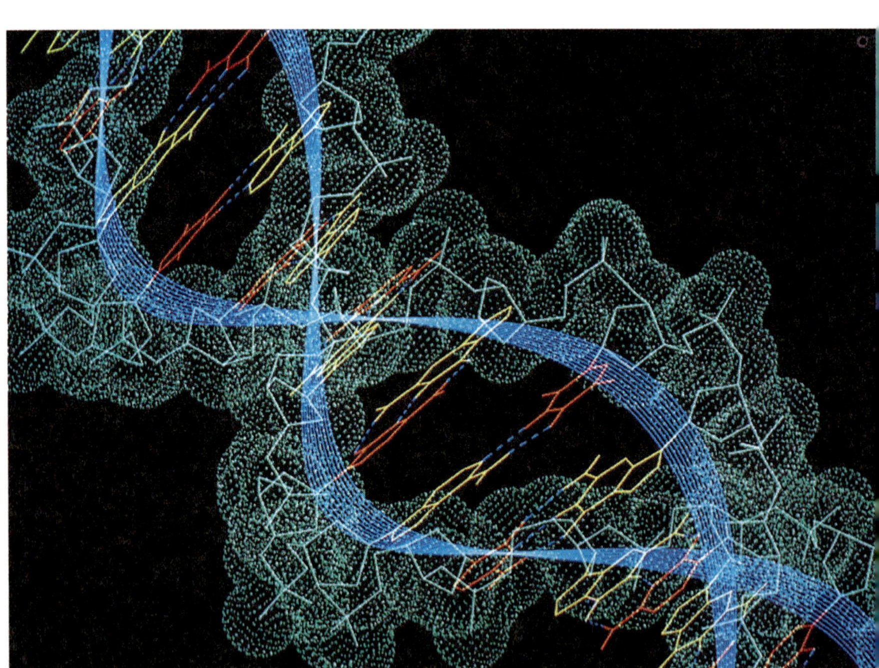

The concept of individual bonds makes it much easier to deal with complex molecules such as DNA. A small segment of a DNA molecule is shown here.

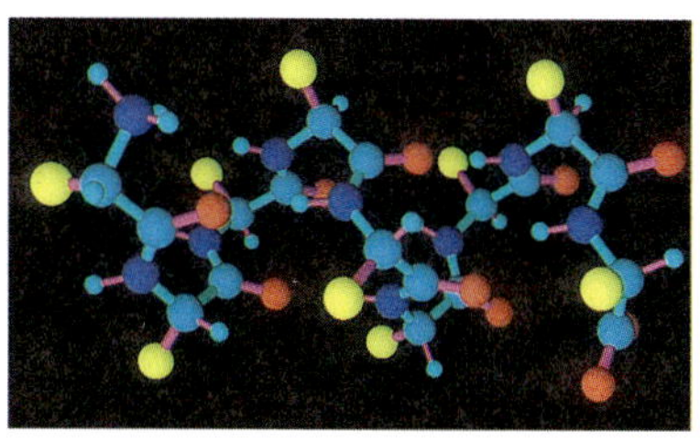
A ball-and-stick model of a protein segment illustrating the alpha helix.

per mole, we choose to think of CH_4 as containing four C—H bonds, each worth 413 kJ of energy per mole of bonds. Without this concept of individual bonds in molecules, chemistry would be hopelessly complicated. There are millions of different chemical compounds, and if each of these compounds had to be considered as an entirely new entity, the task of understanding chemical behavior would be overwhelming.

The bonding model provides a framework to systematize chemical behavior by enabling us to think of molecules as collections of common fundamental components. For example, a typical biomolecule, such as a protein, contains hundreds of atoms and might seem discouragingly complex. However, if we think of a protein as constructed of individual bonds, C—C, C—H, C—N, C—O, N—H, and so on, it helps tremendously in predicting and understanding the protein's behavior. The essential idea is that we expect a given bond to behave about the same in any molecular environment. Used in this way, the model of the chemical bond has helped chemists to systematize the reactions of the millions of existing compounds.

In addition to being very useful, the bonding model is physically sensible. It makes sense that atoms can form stable groups by sharing electrons; shared electrons give a lower energy state because they are simultaneously attracted by two nuclei.

Also, as we will see in the next section, bond energy data support the existence of discrete bonds that are relatively independent of the molecular environment. It is very important to remember, however, that the chemical bond is only a model. Although our concept of discrete bonds in molecules agrees with many of our observations, some molecular properties require that we think of a molecule as a whole, with the electrons free to move through the entire molecule. This is called *delocalization* of the electrons, a concept that will be discussed more completely in the next chapter.

Fundamental Properties of Models

1. Models are human inventions, always based on an incomplete understanding of how nature works. *A model does not equal reality.*
2. Models are often wrong. This property derives from the first property. Models are based on speculation and are always oversimplifications.
3. Models tend to become more complicated as they age. As flaws are discovered in our models, we "patch" them and thus add more detail.
4. It is very important to understand the assumptions that are inherent in a particular model before you use it to interpret observations or to make predictions. Simple methods usually involve very restrictive assumptions and can be expected to yield only qualitative information. Asking for a sophisticated explanation from a simple model is like expecting to get an accurate mass for a diamond by using a bathroom scale.

   For a model to be used effectively, we must understand its strengths and weaknesses and ask only appropriate questions. An illustration of this point is the simple aufbau principle used to explain the electron configurations of the elements. Although this model correctly predicts the configuration for most atoms, chromium and copper do not agree with the predictions. Detailed studies show that the configurations of chromium and copper result from complex electron interactions that are not taken into account in the simple model.

(continued)

However, this does not mean that we should discard the simple model that is so useful for most atoms. Instead, we must apply it with caution and not expect it to be correct in every case.

5. When a model is wrong, we often learn much more than when it is right. If a model makes a wrong prediction, it usually means we do not understand some fundamental characteristics of nature. We often learn by making mistakes. (Try to remember that when you get back your next chemistry test.)

13.8 Covalent Bond Energies and Chemical Reactions

In this section we will consider the energies associated with various types of bonds and see how the bonding concept is useful in dealing with the energies of chemical reactions. One important consideration is to establish the sensitivity of a particular type of bond to its molecular environment. For example, consider the stepwise decomposition of methane:

Process	*Energy Required (kJ/mol)*
$CH_4(g) \rightarrow CH_3(g) + H(g)$	435
$CH_3(g) \rightarrow CH_2(g) + H(g)$	453
$CH_2(g) \rightarrow CH(g) + H(g)$	425
$CH(g) \rightarrow C(g) + H(g)$	339
	Total = 1652

$$\text{Average} = \frac{1652}{4} = 413$$

Although a C—H bond is broken in each case, the energy required varies in a nonsystematic way. This example shows that the C—H bond is somewhat sensitive to its environment. We use the *average* of these individual bond dissociation energies even though this quantity only approximates the energy associated with a C—H bond in a particular molecule. The degree of sensitivity of a bond to its environment can also be seen from experimental measurements of the energy required to break the C—H bond in the following molecules:

Molecule	Measured C—H Bond Energy (kJ/mol)
$HCBr_3$	380
$HCCl_3$	380
HCF_3	430
C_2H_6	410

These data show that the C—H bond strength varies significantly with its environment, but the concept of an average C—H bond strength remains useful to chemists. The average values of bond energies for various types of bonds are listed in Table 13.6.

So far we have discussed bonds in which one pair of electrons is shared. This type of bond is called a **single bond.** As we will see in more detail later,

TABLE 13.6

Average Bond Energies (kJ/mol)

Single Bonds						Multiple Bonds	
H—H	432	N—H	391	I—I	149	C=C	614
H—F	565	N—N	160	I—Cl	208	C≡C	839
H—Cl	427	N—F	272	I—Br	175	O=O	495
H—Br	363	N—Cl	200			C=O*	745
H—I	295	N—Br	243	S—H	347	C≡O	1072
		N—O	201	S—F	327	N=O	607
C—H	413	O—H	467	S—Cl	253	N=N	418
C—C	347	O—O	146	S—Br	218	N≡N	941
C—N	305	O—F	190	S—S	266	C=N	615
C—O	358	O—Cl	203			C≡N	891
C—F	485	O—I	234	Si—Si	340		
C—Cl	339			Si—H	393		
C—Br	276	F—F	154	Si—C	360		
C—I	240	F—Cl	253	Si—O	452		
C—S	259	F—Br	237				
		Cl—Cl	239				
		Cl—Br	218				
		Br—Br	193				

*C=O (CO_2) = 799

atoms sometimes share two pairs of electrons, forming a **double bond,** or share three pairs of electrons, forming a **triple bond.** The bond energies for these *multiple bonds* are also given in Table 13.6.

A relationship also exists between the number of shared electron pairs and the bond length. As the number of shared electrons increases, the bond length shortens. This relationship is shown for selected bonds in Table 13.7.

Bond Energy and Enthalpy

Bond energy values can be used to calculate approximate energies for reactions. To illustrate how this is done, we will calculate the change in energy that accompanies the following reaction:

$$H_2(g) + F_2(g) \longrightarrow 2HF(g)$$

TABLE 13.7

Bond Lengths for Selected Bonds

Bond	Bond Type	Bond Length (pm)	Bond Energy (kJ/mol)
C—C	Single	154	347
C=C	Double	134	614
C≡C	Triple	120	839
C—O	Single	143	358
C=O	Double	123	745
C—N	Single	143	305
C=N	Double	138	615
C≡N	Triple	116	891

This reaction involves breaking one H—H and one F—F bond and forming two H—F bonds. For bonds to be broken, energy must be *added* to the system—an endothermic process. Consequently, the energy terms associated with bond breaking have *positive* signs. The formation of a bond *releases* energy, an exothermic process, and the energy terms associated with bond making carry a *negative* sign. We can write the enthalpy change for a reaction as follows:

ΔH = sum of the energies required to break old bonds (positive signs) plus the sum of the energies released in the formation of new bonds (negative signs)

This leads to the expression

$$\Delta H = \underbrace{\Sigma D \text{ (bonds broken)}}_{\text{Energy required}} - \underbrace{\Sigma D \text{ (bonds formed)}}_{\text{Energy released}}$$

where Σ represents the sum of terms and D represents the bond energy per mole of bonds. (D *always* has a positive sign.)

In the case of the formation of HF,

$$\Delta H = D_{\text{H—H}} + D_{\text{F—F}} - 2D_{\text{H—F}}$$

$$= 1 \text{ mol} \times \frac{432 \text{ kJ}}{\text{mol}} + 1 \text{ mol} \times \frac{154 \text{ kJ}}{\text{mol}} - 2 \text{ mol} \times \frac{565 \text{ kJ}}{\text{mol}}$$

$$= -544 \text{ kJ}$$

Thus, when 1 mole of $H_2(g)$ and 1 mole of $F_2(g)$ react to form 2 moles of $HF(g)$, 544 kJ of energy should be released.

Since bond energies are typically averages taken from several compounds, the ΔH calculated from bond energies is not expected to agree exactly with that calculated from enthalpies of formation.

This result can be compared with the calculation of ΔH for this reaction from the standard enthalpy of formation for HF (−271 kJ/mol):

$$\Delta H = 2 \text{ mol} \times (-271 \text{ kJ/mol}) = -542 \text{ kJ}$$

Thus the use of bond energies to calculate ΔH works quite well in this case.

Example 13.5

Using the bond energies listed in Table 13.6, calculate ΔH for the reaction of methane with chlorine and fluorine to give Freon-12 (CF_2Cl_2).

$$CH_4(g) + 2Cl_2(g) + 2F_2(g) \longrightarrow CF_2Cl_2(g) + 2HF(g) + 2HCl(g)$$

Solution The idea here is to break the bonds in the reactants to give individual atoms and then assemble these atoms into the products by forming new bonds:

$$\text{Reactants} \xrightarrow[\text{required}]{\text{Energy}} \text{atoms} \xrightarrow[\text{released}]{\text{Energy}} \text{products}$$

We then combine the energy changes to calculate ΔH:

$$\Delta H = \begin{matrix}\text{energy required}\\ \text{to break bonds}\end{matrix} - \begin{matrix}\text{energy released}\\ \text{when bonds form}\end{matrix}$$

where the minus sign gives the correct sign to the energy terms for the exothermic processes.

Reactant bonds broken:

CH_4	4 mol C—H	$4\text{ mol} \times \dfrac{413\text{ kJ}}{\text{mol}}$	$= 1652\text{ kJ}$
$2Cl_2$	2 mol Cl—Cl	$2\text{ mol} \times \dfrac{239\text{ kJ}}{\text{mol}}$	$= 478\text{ kJ}$
$2F_2$	2 mol F—F	$2\text{ mol} \times \dfrac{154\text{ kJ}}{\text{mol}}$	$= 308\text{ kJ}$

Total energy required = 2438 kJ

Product bonds formed:

CF_2Cl_2	2 mol C—F	$2\text{ mol} \times \dfrac{485\text{ kJ}}{\text{mol}}$	$= 970\text{ kJ}$
	and		
	2 mol C—Cl	$2\text{ mol} \times \dfrac{339\text{ kJ}}{\text{mol}}$	$= 678\text{ kJ}$
HF	2 mol H—F	$2\text{ mol} \times \dfrac{565\text{ kJ}}{\text{mol}}$	$= 1130\text{ kJ}$
HCl	2 mol H—Cl	$2\text{ mol} \times \dfrac{427\text{ kJ}}{\text{mol}}$	$= 854\text{ kJ}$

Total energy released = 3632 kJ

We now can calculate ΔH:

$$\Delta H = \begin{array}{c}\text{energy required}\\\text{to break bonds}\end{array} - \begin{array}{c}\text{energy released}\\\text{when bonds form}\end{array}$$

$$= 2438\text{ kJ} - 3632\text{ kJ} = -1194\text{ kJ}$$

Since the sign of the value for the enthalpy change is negative, this means that 1194 kJ of energy is released per mole of CF_2Cl_2 formed. The value of ΔH calculated for this reaction using enthalpies of formation is -1126 kJ.

In performing the calculations in this section, we have made several approximations. Of course, we made the usual assumption that average bond energies apply regardless of the specific molecular environment. We also ignored the difference between enthalpy and internal energy in these calculations. Recall from Chapter 9 that at constant pressure $\Delta E = \Delta H - P\Delta V$. Thus, if a reaction involves a change in volume, a correction should be applied to the calculations of reaction enthalpies from bond energies. However, this correction is usually small compared with the uncertainties inherent in this method.

It should also be noted that the use of bond energies to estimate enthalpies of reaction should be limited to cases in which all the reactants and products are in the gas phase, where intermolecular interactions are expected to be minimal. Bond energies do not take these interactions into account.

13.9 The Localized Electron Bonding Model

So far we have discussed the general characteristics of the chemical bonding model and have seen that properties such as bond strength and polarity can be assigned to individual bonds. In this section we introduce a specific model used to describe covalent bonds. We need a simple model that can be easily applied even to very complicated molecules and that can be used routinely by chemists to interpret and organize the wide variety of chemical phenomena. The model that serves this purpose is often called the **localized electron (LE) model.** This model assumes that *a molecule is composed of atoms that are bound together by using atomic orbitals to share electron pairs.* The electron pairs in the molecule are assumed to be localized on a particular atom or in the space between two atoms. Those pairs of electrons localized on an atom are called **lone pairs,** and those found in the space between the atoms are called **bonding pairs.**

As we will apply it, the LE model has three parts:

1. Description of the valence electron arrangement in the molecule using Lewis structures (will be discussed in the next section).
2. Prediction of the geometry of the molecule, using the valence shell electron-pair repulsion (VSEPR) model (will be discussed in Section 13.13).
3. Description of the types of atomic orbitals used by the atoms to share electrons or hold lone pairs (will be discussed in Chapter 14).

13.10 Lewis Structures

The **Lewis structure** of a molecule represents the arrangement of valence electrons among the atoms in the molecule. These representations are named after G. N. Lewis (Fig. 13.14). The rules for writing Lewis structures are based on the

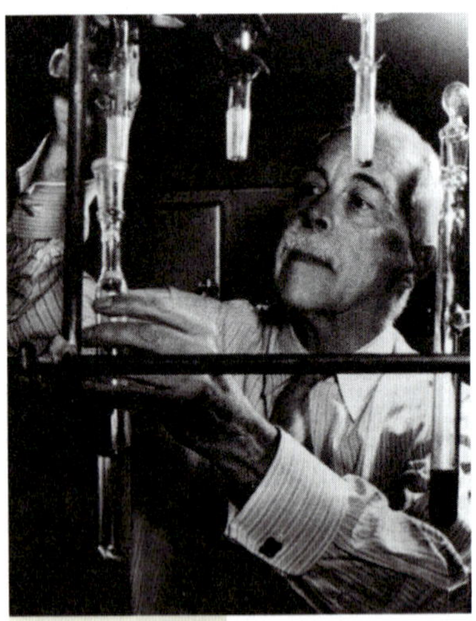

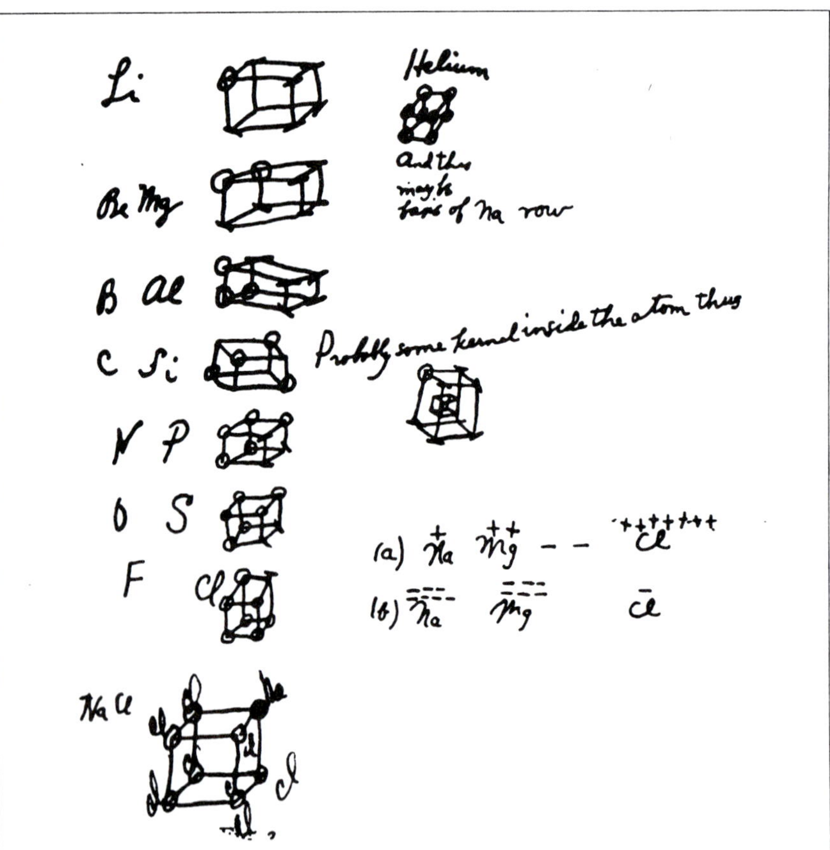

FIGURE 13.14

G. N. **Lewis** (above) conceived the octet rule while lecturing to a class of general chemistry students in 1902. He was also one of the two authors of a now classic work on thermodynamics, Lewis and Randall, *Thermodynamics and the Free Energy of Chemical Substances* (1923). (right) This is his original sketch. From G. N. Lewis, *Valence,* Dover Publications, Inc., New York, 1966.

observations of thousands of molecules, which show that *in most stable compounds the atoms achieve noble gas electron configurations.* Although this is not always the case, it is so common that it provides a very useful place to start.

Lewis structures show only valence electrons.

We have already seen that when metals and nonmetals react to form solid binary ionic compounds, electrons are transferred, and the resulting ions typically have noble gas electron configurations. An example is the formation of KBr, where the K^+ ion has the [Ar] electron configuration and the Br^- ion has the [Kr] electron configuration. In writing Lewis structures, the rule is that *only the valence electrons are included.* Using dots to represent electrons, the Lewis structure for KBr is

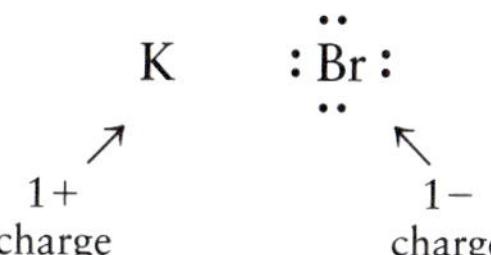

No dots are shown in the K^+ ion since it has no valence electrons. The Br^- ion is shown with eight electrons since it has a filled valence shell.

Next, we will consider Lewis structures for molecules with covalent bonds involving elements in the first and second periods. The principle of achieving a noble gas electron configuration applies to these elements as follows:

Hydrogen forms stable molecules where it shares two electrons. That is, it follows a **duet rule.** For example, when two hydrogen atoms, each with one electron, combine to form the H_2 molecule, we have

H· ·H → H:H

By sharing electrons, each hydrogen in H_2 has two electrons. This gives each hydrogen a filled valence shell.

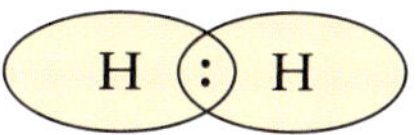

Helium does not form bonds because its valence orbitals are already filled; it is a noble gas. Helium has the electron configuration $1s^2$ and can be represented by the Lewis structure

He:

Carbon, nitrogen, oxygen, and fluorine almost always obey the octet rule in stable molecules.

The second-row nonmetals (carbon through fluorine) form stable molecules when they are surrounded by enough electrons to fill the valence orbitals—the 2*s* and the three 2*p* orbitals. Since eight electrons are required to fill the 2*s* and 2*p* orbitals, these elements typically obey the **octet rule;** they are surrounded by eight electrons. An example is the F_2 molecule, which has the following Lewis structure:

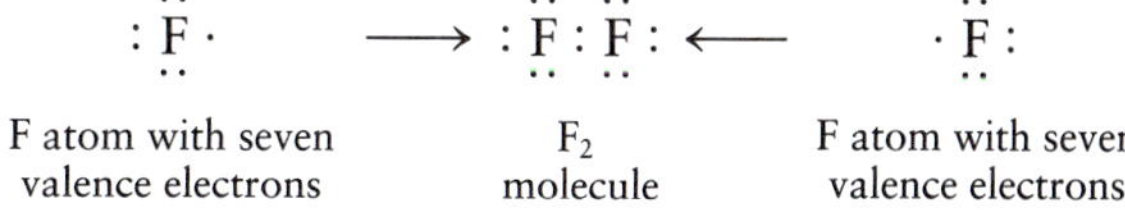

Note that each fluorine atom in F_2 is, in effect, surrounded by eight electrons, two of which are shared with the other atom. Recall that the shared pair of

electrons is called a *bonding pair.* Each fluorine atom also has three pairs of electrons not involved in bonding. These are the *lone pairs.*

Neon does not form bonds since it already has an octet of valence electrons (it is a noble gas). The Lewis structure is

$$:\ddot{\underset{\cdot\cdot}{\text{Ne}}}:$$

Note that only the valence electrons of the neon atom ($2s^2 2p^6$) are represented by the Lewis structure. The $1s^2$ electrons are core electrons and take no part in chemical reactions.

From the discussion above we can formulate the following rules for writing Lewis structures of molecules containing atoms from the first two periods.

STEPS

Writing Lewis Structures

1. Sum the valence electrons from all the atoms. Do not worry about keeping track of which electrons come from which atoms. It is the *total* number of electrons that is important.
2. Use a pair of electrons to form a bond between each pair of bound atoms.
3. Arrange the remaining electrons to satisfy the duet rule for hydrogen and the octet rule for the second-row elements.

To see how these rules are applied, we will construct the Lewis structures for a few molecules. We will first consider the water molecule and follow the rules above.

Step 1

We sum the *valence* electrons for H_2O as shown:

$$\underset{\text{H}}{1} + \underset{\text{H}}{1} + \underset{\text{O}}{6} = 8 \text{ valence electrons}$$

Step 2

Using one pair of electrons per bond, we draw in the two O—H single bonds:

H—O—H

Note that *a line instead of a pair of dots is used to indicate each pair of bonding electrons.* This is the standard notation.

Step 3

We distribute the remaining electrons around the atoms to achieve a noble gas electron configuration for each atom. Since four electrons have been used in forming the two bonds, four electrons (8 − 4) remain to be distributed. Hydrogen is satisfied with two electrons (duet rule), but oxygen needs eight electrons to achieve a noble gas configuration. Thus the remaining four electrons are added to oxygen as two lone pairs. Dots are used to represent the lone pairs:

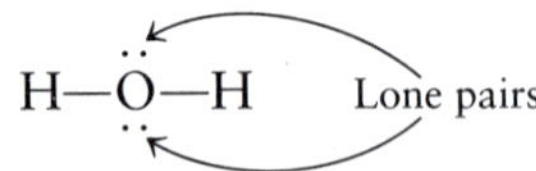

This is the correct Lewis structure for the water molecule. Each hydrogen has two electrons and the oxygen has eight:

H—O—H represents H:O:H

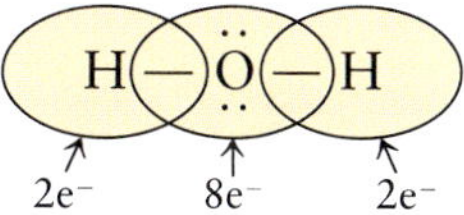

As a second example, we will write the Lewis structure for carbon dioxide. Summing the valence electrons gives

$$\underset{\text{C}}{4} + \underset{\text{O}}{6} + \underset{\text{O}}{6} = 16$$

After forming a bond between the carbon and each oxygen,

O—C—O

the remaining electrons are distributed to achieve noble gas configurations on each atom. In this case we have 12 electrons (16 − 4) remaining after the bonds are drawn. The distribution of these electrons is determined by a trial-and-error process. We have six pairs of electrons to distribute. Suppose we try three pairs on each oxygen to give

:Ö—C—Ö:

To see if this structure is correct, we need to check two things:

1. The total number of electrons. There are 16 valence electrons in this structure, which is the correct number.
2. The octet rule for each atom. Each oxygen has eight electrons, but the carbon has only four. This cannot be the correct Lewis structure.

How can we arrange the 16 available electrons to achieve an octet for each atom? Suppose that there are two shared pairs between the carbon and each oxygen:

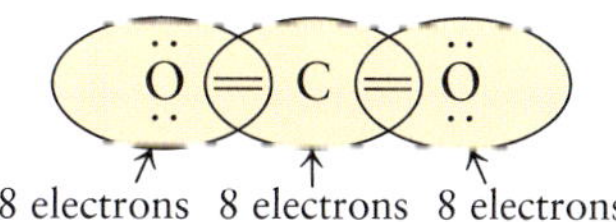

Now each atom is surrounded by 8 electrons, and the total number of electrons is 16, as required. Thus the correct Lewis structure for carbon dioxide has two double bonds.

Finally, let's consider the Lewis structure of the CN^- (cyanide) ion. Summing the valence electrons, we have

$$\underset{4}{\text{C}}\ \underset{+\,5}{\text{N}}\ \underset{+\,1\,=\,10}{^{-}}$$

Note that the negative charge requires that an extra electron must be added. After drawing a single bond (C—N), we distribute the remaining electrons to achieve a noble gas configuration for each atom. Eight electrons remain to be distributed. We can try various possibilities, for example:

C̤̈—N̤̈

This structure is incorrect because C and N have only six electrons each instead of eight. The correct arrangement is

$$[:\text{C}\equiv\text{N}:]^-$$

(Satisfy yourself that both carbon and nitrogen have eight electrons.)

Example 13.6

Give the Lewis structure for each of the following.

a. HF **b.** N_2 **c.** NH_3 **d.** CH_4 **e.** CF_4 **f.** NO^+

Solution In each case we apply the three rules for writing Lewis structures. Recall that lines are used to indicate shared electron pairs and that dots are used to indicate nonbonding pairs (lone pairs). We have the following tabulated results:

	Total Valence Electrons	Draw Single Bonds	Calculate Number of Electrons Remaining	Use Remaining Electrons to Achieve Noble Gas Configurations
a. HF	1 + 7 = 8	H—F	6	H—$\ddot{\underset{..}{\text{F}}}$:
b. N_2	5 + 5 = 10	N—N	8	:N≡N:
c. NH_3	5 + 3(1) = 8	H—N—H | H	2	H—$\ddot{\text{N}}$—H | H
d. CH_4	4 + 4(1) = 8	H | H—C—H | H	0	H | H—C—H | H
e. CF_4	4 + 4(7) = 32	F | F—C—F | F	24	:$\ddot{\text{F}}$: | :$\ddot{\underset{..}{\text{F}}}$—C—$\ddot{\underset{..}{\text{F}}}$: | :$\underset{..}{\text{F}}$:
f. NO^+	5 + 6 − 1 = 10	N—O	8	[:N≡O:]$^+$

When writing Lewis structures, don't worry about which electrons came from which atoms. The best way to look at a molecule is to regard it as a new entity that uses all of the available valence electrons of the atoms to achieve the lowest possible energy.* The valence electrons belong to the molecule rather than to the individual atoms. Simply distribute all valence electrons so that the various rules are satisfied, without regard to the origin of each particular electron.

*In a sense this approach corrects for the fact that the LE model overemphasizes the point that a molecule is simply a sum of its parts—that is, that the atoms retain their individual identities in the molecule.

13.11 Resonance

A valid Lewis structure is one that obeys the rules we have outlined.

Sometimes more than one valid Lewis structure is possible for a given molecule. For example, consider the Lewis structure for the nitrate ion (NO_3^-), which has 24 valence electrons. So that an octet of electrons surrounds each atom, a structure like the following is required:

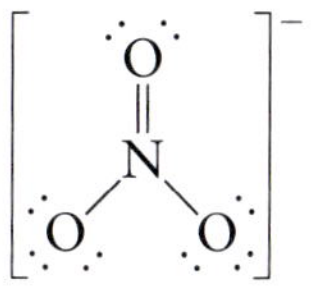

If this structure accurately represents the bonding in NO_3^-, there should be two types of N—O bonds observed in the molecule: one shorter bond (the double bond) and two identical longer ones (the two single bonds). However, experiments clearly show that NO_3^- exhibits only *one* type of N—O bond with a length and strength between those expected for a single bond and a double bond. Thus, although the structure we have shown above is a valid Lewis structure, it does *not* correctly represent the bonding in NO_3^-. This is a serious problem, and it means that the model must be modified.

Look again at the proposed Lewis structure for NO_3^-. Because there is no reason for choosing a particular oxygen atom to have the double bond, there are really three valid Lewis structures:

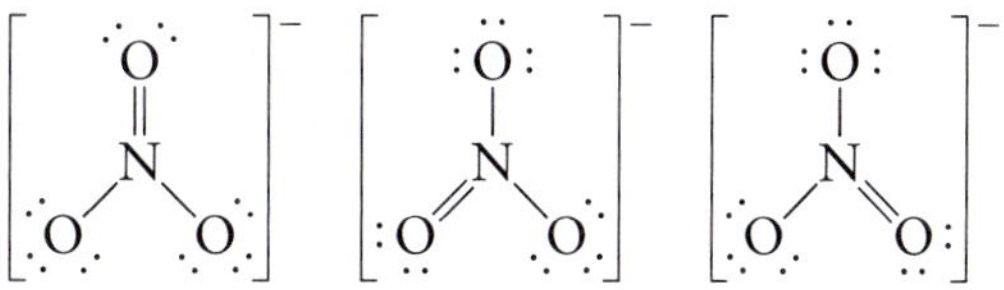

Is any of these structures a correct description of the bonding in NO_3^-? No, because NO_3^- does not have one double and two single bonds—it has three equivalent bonds. We can solve this problem by making the following assumption: The correct description of NO_3^- is *not given by any one* of the three Lewis structures individually but is given only by the *superposition of all three.*

The nitrate ion does not exist as any of the three extreme forms indicated by the individual Lewis structures but instead exists as an average of all three. **Resonance** *is invoked when more than one valid Lewis structure can be written for a particular molecule.* The resulting electron structure of the molecule is given by the average of these **resonance structures.** This situation is usually represented by double-headed arrows as follows:

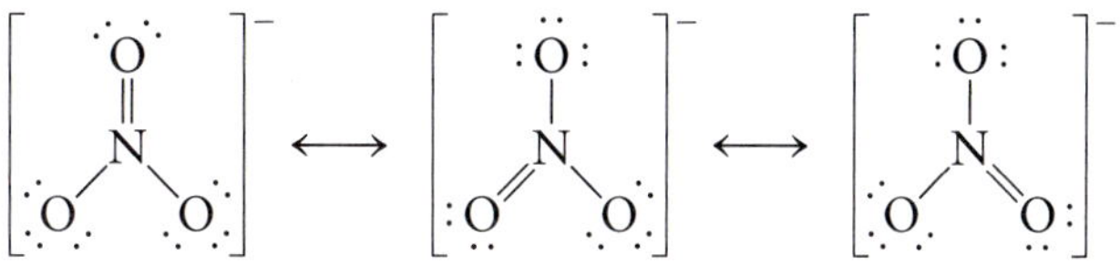

Note that in all these resonance structures the arrangement of the nuclei is the same. Only the placement of the electrons differs. The arrows do not indicate that the molecule "flips" from one resonance structure to another. They simply show that the *actual structure is an average of the three resonance structures.*

The concept of resonance is necessary because the LE model postulates that electrons are localized between a given pair of atoms. However, nature doesn't really operate this way. Electrons are actually delocalized—they can move around the entire molecule. The valence electrons in the NO_3^- molecule distribute themselves to provide equivalent N—O bonds. Resonance is necessary to compensate for this defective assumption of the LE model. However, because this model is so useful, we retain the concept of localized electrons and add resonance to accommodate species like NO_3^-.

Example 13.7

Describe the electron arrangement in the nitrite anion (NO_2^-) using the LE model.

Solution We will follow the usual procedure for obtaining the Lewis structure for the NO_2^- ion.

In NO_2^- there are 5 + 2(6) + 1 = 18 valence electrons.
Indicating the single bonds gives the structure

O—N—O

The remaining 14 electrons (18 − 4) can be distributed to produce these structures:

[:O=N—O:]$^-$ ⟷ [:O—N=O:]$^-$

This is a resonance situation. Two equivalent Lewis structures can be drawn. *The electronic structure of the molecule is not correctly represented by either resonance structure but by the average of the two.* There are two equivalent N—O bonds, each one intermediate between a single and a double bond.

Equivalent Lewis structures contain the same numbers of single and multiple bonds. For example, the resonance structures for O_3,

O—O=O

and

O=O—O

are equivalent Lewis structures. They are equally important in describing the bonding in O_3. Nonequivalent Lewis structures contain different numbers of single and multiple bonds.

13.12 Exceptions to the Octet Rule

The LE model is a simple but very successful model, and the rules we have used for Lewis structures apply to most molecules. To implement this model, we have relied heavily on the octet rule. So far we have treated molecules for which this rule is easily applied. However, inevitably, cases arise where the importance of an octet of electrons is called into question. Boron, for example, tends to form compounds in which the boron atom has fewer than eight electrons around it—it does not have a complete octet. Boron trifluoride (BF_3), a gas at normal temperatures and pressures, reacts very energetically with molecules such as water and ammonia that have available lone pairs. The violent reactivity of BF_3 with electron-rich molecules occurs because the boron atom is electron-deficient. Boron trifluoride has 24 valence electrons. The Lewis structure that seems most consistent with the properties of BF_3 is

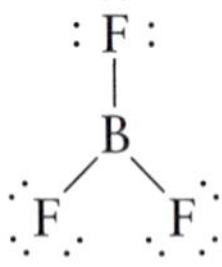

Note that in this structure boron has only six electrons around it. The octet rule for boron can be satisfied by drawing a structure with a double bond, such as

$$\begin{array}{c} \ddot{\text{F}} \\ \| \\ \text{B} \\ \diagup \quad \diagdown \\ :\ddot{\text{F}}: \quad :\ddot{\text{F}}: \end{array}$$

and there are some theoretical studies that support such a structure for BF_3. However, since fluorine is so much more electronegative than boron, this structure seems questionable. In fact, some experiments indicate that each B—F bond is probably best described by the first Lewis structure, which is also consistent with the reactivity of BF_3 toward electron-rich molecules such as NH_3 with which it reacts to form H_3NBF_3:

$$H_3N: + BF_3 \longrightarrow H_3N{-}BF_3$$

In this stable compound boron has an octet of electrons.

It is characteristic of boron to form molecules in which the boron atom is electron-deficient. On the other hand, carbon, nitrogen, oxygen, and fluorine do obey the octet rule.

A computer-generated representation of sulfur hexafluoride.

Some atoms appear to exceed the octet rule. This behavior is observed only for those elements in Period 3 of the periodic table and beyond. To see how this arises, we will consider the Lewis structure for sulfur hexafluoride (SF_6). The sum of the valence electrons for SF_6 is

$$6 + 6(7) = 48 \text{ electrons}$$

Indicating the single bonds gives the structure on the left below:

$$SF_6 \text{ (S bonded to six F by single bonds)} \qquad SF_6 \text{ (each F with three lone pairs)}$$

We have used 12 electrons to form the S—F bonds, which leaves 36 electrons. Since fluorine always follows the octet rule, we complete the six fluorine octets to give the structure on the right above. This structure uses all 48 valence electrons for SF_6, but sulfur has 12 electrons around it; that is, sulfur *exceeds* the octet rule. How can this happen? There are several ways to approach this situation. The classical explanation for molecules like SF_6 involves using the empty 3*d* orbitals on the third-period elements. Recall that the second-row elements have only 2*s* and 2*p* valence orbitals, whereas the third-row elements have 3*s*, 3*p*, and 3*d* orbitals. The 3*s* and 3*p* orbitals fill with electrons in going from sodium to argon, but the 3*d* orbitals remain empty. For example, the valence-orbital diagram for a sulfur atom is

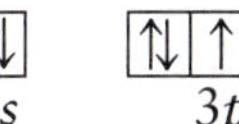

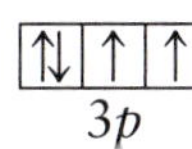

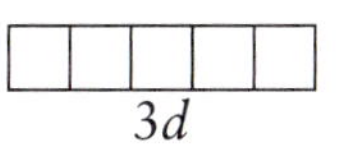

3*s* 3*p* 3*d*

Thus one way to account for the electronic structure of SF_6 is to assume that the empty 3*d* orbitals on sulfur can be used to accommodate extra electrons: The sulfur atom can have 12 electrons around it by using the 3*s* and 3*p* orbitals to hold 8 electrons, with the extra 4 electrons in the formerly empty 3*d* orbitals.

Lewis Structures and the Octet Rule

1. The second-row elements C, N, O, and F should always be assumed to obey the octet rule.
2. The second-row elements B and Be often have fewer than eight electrons around them in their compounds. These electron-deficient compounds are very reactive.
3. The second-row elements never exceed the octet rule, since their valence orbitals (2*s* and 2*p*) can accommodate only eight electrons.
4. Third-row and heavier elements often satisfy the octet rule but are assumed in the simplest model to exceed the octet rule by using their empty valence *d* orbitals.
5. When writing the Lewis structure for a molecule, first draw single bonds between all bonded atoms, and then satisfy the octet rule for all the atoms. If electrons remain after the octet rule has been satisfied, place them on the elements having available *d* orbitals (elements in the third period or beyond).

Example 13.8

Write the Lewis structure for PCl_5.

Solution We can follow the same stepwise procedure we used previously for sulfur hexafluoride.

Step 1

Sum the valence electrons.

$$\underset{\text{P}}{\underset{\uparrow}{5}} + 5(\underset{\text{Cl}}{\underset{\uparrow}{7}}) = 40 \text{ electrons}$$

Step 2

Indicate single bonds between bound atoms.

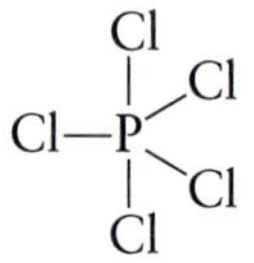

```
     Cl
     |  Cl
     | /
Cl—P
     | \
     |  Cl
     Cl
```

Step 3

Distribute the remaining electrons. In this case, 30 electrons (40 − 10) remain. These are used to satisfy the octet rule for each chlorine atom. The final Lewis

structure is

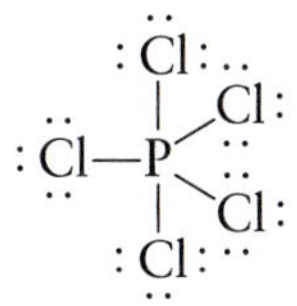

$$PCl_5$$

Note that phosphorus, a third-row element, exceeds the octet rule by two electrons.

In the PCl_5 and SF_6 molecules, the third-row central atoms (P and S, respectively) are assigned the extra electrons. However, in molecules having more than one atom that can exceed the octet rule, it is not always clear which atom should have the extra electrons. Consider the Lewis structure for the triiodide ion (I_3^-), which has

$$\underset{\substack{\uparrow\\ \text{I}}}{3(7)} + \underset{\substack{\uparrow\\ -1\ \text{charge}}}{1} = 22 \text{ valence electrons}$$

Indicating the single bonds gives I—I—I. At this point 18 electrons (22 − 4) remain. Trial and error will convince you that one of the iodine atoms must exceed the octet rule, but *which* one?

The rule we will follow is that *when it is necessary to exceed the octet rule for one of several third-row (or higher) elements, assume that the extra electrons are placed on the central atom.*

Thus for I_3^- the Lewis structure is

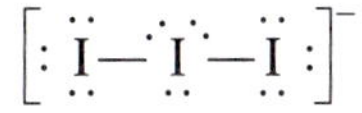

$$[:\ddot{I}—\ddot{I}—\ddot{I}:]^-$$

where the central iodine exceeds the octet rule. This structure agrees with known properties of I_3^-.

Example 13.9

Write the Lewis structure for each molecule or ion.

a. ClF_3 **b.** XeO_3 **c.** $RnCl_2$ **d.** $BeCl_2$ **e.** ICl_4^-

Solution

a. The chlorine atom (third row) accepts the extra electrons.

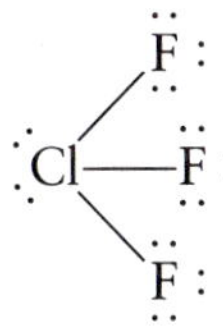

$$ClF_3$$

b. All atoms obey the octet rule.

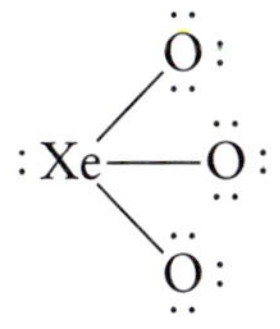

$$XeO_3$$

c. Radon, a noble gas in Period 6, accepts the extra electrons.

$$:\ddot{\underset{..}{Cl}}-\ddot{\underset{..}{Rn}}-\ddot{\underset{..}{Cl}}:$$

d. Beryllium is electron-deficient.

$$:\ddot{\underset{..}{Cl}}-Be-\ddot{\underset{..}{Cl}}:$$

e. Iodine exceeds the octet rule.

$$\left[\begin{array}{ccc} :\ddot{\underset{..}{Cl}} & & \ddot{\underset{..}{Cl}}: \\ & \ddot{\underset{..}{I}} & \\ :\ddot{\underset{..}{Cl}} & & \ddot{\underset{..}{Cl}}: \end{array}\right]^-$$

So far we have assumed that third-row and heavier atoms can exceed the octet rule by using their valence *d* orbitals to accommodate the extra electrons. However, recent calculations* indicate that because the 3*d* orbitals are so much higher in energy than the 3*s* and 3*p* orbitals for a given atom, it is not feasible to use them. Researchers argue that satisfying the octet rule is a high priority for molecules and that there are ways to describe the bonding in molecules like SF_6 and PCl_5 without formally exceeding the octet rule. One such explanation depends on a concept called *hyperconjugation,* which is explained in the special feature "Hyperconjugation—The Octet Rules" on p. 632.

The main point that should be made here is that chemists disagree on the best way to explain molecules such as SF_6 and PCl_5. Researchers continue to try to find the best answers.

Odd-Electron Molecules

In addition to the question about the use of *d* orbitals in third-row and heavier atoms, another problem for the simple LE model involves molecules with odd numbers of electrons. Although relatively few molecules formed from nonmetals contain odd numbers of electrons, there are some notable examples. One such molecule is nitric oxide (NO), which is formed when nitrogen and oxygen gases react at the high temperatures present in automobile engines. Nitric oxide is emitted into the air, where it reacts with oxygen to form gaseous nitrogen dioxide (NO_2), another odd-electron molecule.

Since the LE model is based on pairs of electrons, it does not handle odd-electron cases in a natural way, although Lewis structures are sometimes written for these species. To treat odd-electron molecules accurately, we need a more sophisticated model.

Formal Charge

Molecules or polyatomic ions often have many nonequivalent Lewis structures, all which obey the rules for writing Lewis structures. For example, as we will see in detail, the sulfate ion has a Lewis structure with all single bonds and several Lewis structures that contain double bonds. How do we decide which of the many possible Lewis structures best describes the

---

*L. Suidan, J. K. Bodenhopp, E. D. Glendening, and F. Weinhold, *J. Chem. Ed.* 72 (1995): 583.

actual bonding in sulfate? One method involves estimating the charge on each atom in the various possible Lewis structures and using these charges to select the most appropriate structure(s). We will see below how this is done, but first we must decide on a method to evaluate atomic charges in molecules.

In Chapter 4 we discussed one system for obtaining charges for atoms in molecules—oxidation states. However, in assigning oxidation states we always count *both* of the shared electrons as belonging to the more electronegative atom in a bond. This practice leads to highly exaggerated estimates of charge. In other words, although oxidation states are useful for bookkeeping electrons in redox reactions, they are not realistic estimates of the actual charges on individual atoms in a molecule, and so they are not suitable for judging the appropriateness of Lewis structures. A second definition of atomic charges in a molecule, the **formal charge,** is more suitable for evaluating Lewis structures.

The concept of formal charge requires that we compare

1. the number of valence electrons on the free neutral atom (which has a charge of zero because the number of electrons equals the number of protons) and
2. the number of valence electrons "belonging" to a given atom in a molecule.

If an atom in a molecule has the same number of valence electrons as it does in the free state, the positive and negative charges just balance, and the atom has a formal charge of 0. If an atom has one more valence electron in a molecule than it has on a free atom, it has a formal charge of -1, and so on. Thus the formal charge on an atom in a molecule is defined as

$$\text{Formal charge} = \begin{pmatrix}\text{number of valence} \\ \text{electrons on a free atom}\end{pmatrix} - \begin{pmatrix}\text{number of valence electrons} \\ \text{assigned to the atom in the} \\ \text{molecule}\end{pmatrix}$$

To compute the formal charge of an atom in a molecule, we assign the valence electrons to the various atoms by making the following assumptions:

1. Lone pair electrons belong entirely to the atom in question.
2. Shared electrons are *divided equally* between the two sharing atoms.

Thus the number of valence electrons assigned to a given atom is calculated as follows:

$$(\text{Valence electrons})_{\text{assigned}} = \begin{pmatrix}\text{number of lone} \\ \text{pair electrons}\end{pmatrix} + \tfrac{1}{2}\begin{pmatrix}\text{number of} \\ \text{shared electrons}\end{pmatrix}$$

We will illustrate the procedures for calculating formal charges by considering two of the possible Lewis structures for the sulfate ion, which has 32 valence electrons. For the Lewis structure

```
        ..
       :O:
  ..    |    ..
 :O  —  S  — O:
  ..    |    ..
       :O:
        ..         2−
```

Hyperconjugation—The Octet Rules

All our observations tell us that molecules such as SF_6 and PCl_5 are very stable, but what is the bonding like in these molecules? There are two extreme points of view: (1) The S and P atoms in these molecules exceed the octet rule, placing the extra electrons in 3*d* orbitals, and (2) the S and P atoms in these molecules obey the octet rule, resorting to hyperconjugation. *Hyperconjugation* involves binding *n* atoms to a central atom using fewer than *n* electron pairs around the central atom. To illustrate this model, consider SF_6. We know that there are six fluorine atoms around the central sulfur atom in this molecule. Can we explain the bonding without exceeding the octet rule? The answer is yes, by using Lewis structures such as the following:

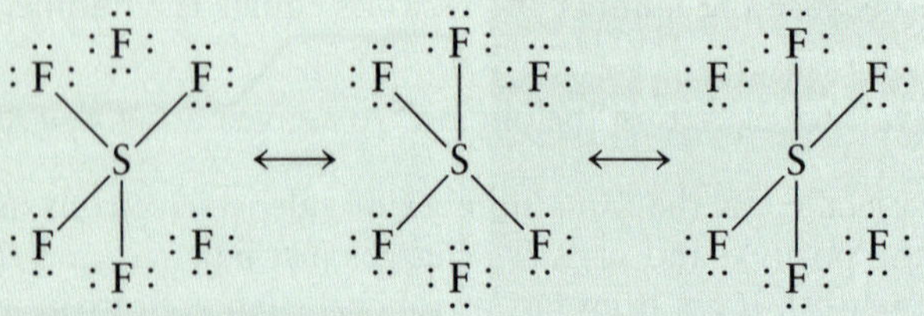

Note that in each of these resonance structures (and the many other similar ones that can be drawn) the sulfur atom always has an octet of electrons around it. The "true" structure is a composite of these equivalent resonance structures. The overall bonding is described as a combination of covalent and ionic contributions. The covalent contribution to the bonding involves four electron pairs "spread out" over the six sulfur–fluorine bonds. The ionic contribution to the bonding arises as follows. Note that for each Lewis structure two of the fluorines have −1 formal charges and the sulfur has a +2 formal charge,

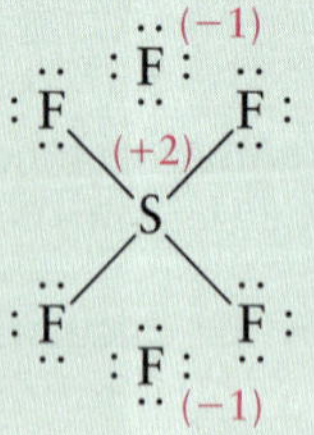

(−1)
(+2)
(−1)

leading to ionic attractive forces. It is important to recall at this point that the description of the bonding in SF_6 by this model involves *all* the resonance structures similar to the one shown. Thus the F atoms are bound to the S atom by a combination of covalent bonding,

$$\frac{\text{4 electron-pair bonds}}{\text{6 S—F interactions}} = \frac{2}{3} \text{ covalent bond per S—F interaction}$$

plus the ionic bonding described above.

A similar treatment can be given for PCl_5, using the following resonance structures,

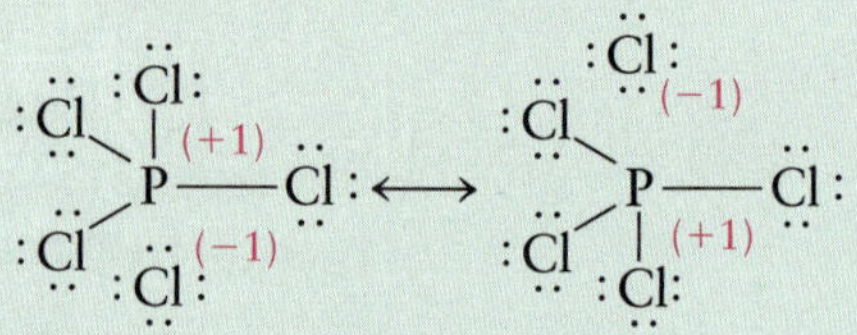

(+1) (−1) ⟷ (−1) (+1)

each of which satisfies the octet rule.

Thus the use of hyperconjugation preserves the octet rule and does not require the central atom to use *d* orbitals, which this model's proponents argue are too high in energy to participate in the bonding of these types of molecules.

So honest disagreements exist among chemists as to the best Lewis structures for molecules that, at least at first glance, appear to exceed the octet rule. This uncertainty shows the limitations of the Lewis model with its localized electron pairs. Note, however, that even with its limitations, it is still very useful because of its simplicity. The ability to obtain a reasonable bonding picture with a "back-of-the-envelope" model has led to the enduring influence of the Lewis model.

each oxygen atom has six lone pair electrons and shares two electrons with the sulfur atom. Thus, according to the preceding assumptions, each oxygen is assigned seven valence electrons:

$$\text{Valence electrons assigned to each oxygen} = \underset{\substack{\text{Lone pair}\\\text{electrons}}}{6} \text{ plus } \tfrac{1}{2}(\underset{\substack{\text{Shared}\\\text{electrons}}}{2}) = 7$$

$$\text{Formal charge on oxygen} = \underset{\substack{\text{Valence electrons on}\\\text{a free O atom}}}{6} - \underset{\substack{\text{Valence electrons}\\\text{assigned to each O}\\\text{in } SO_4^{2-}}}{7} = -1$$

The formal charge on each oxygen is -1.

For the sulfur atom there are no lone pair electrons and eight electrons are shared with the oxygen atoms. Thus for sulfur,

$$\text{Valence electrons assigned to sulfur} = \underset{\substack{\text{Lone pair}\\\text{electrons}}}{0} + \tfrac{1}{2}(\underset{\substack{\text{Shared}\\\text{electrons}}}{8}) = 4$$

$$\text{Formal charge on sulfur} = \underset{\substack{\text{Valence electrons on}\\\text{free S atom}}}{6} - \underset{\substack{\text{Valence electrons}\\\text{assigned to S}\\\text{in } SO_4^{2-}}}{4} = 2$$

A second possible Lewis structure is

$$\left[\begin{array}{c} \ddot{\text{O}} \\ \| \\ :\ddot{\text{O}}-\text{S}-\ddot{\text{O}}: \\ \| \\ \ddot{\text{O}} \end{array}\right]^{2-}$$

In this case the formal charges are as calculated below.

For oxygen atoms with single bonds:

$$\text{Valence electrons assigned} = 6 + \tfrac{1}{2}(2) = 7$$

$$\text{Formal charge} = 6 - 7 = -1$$

For oxygen atoms with double bonds:

$$\text{Valence electrons assigned} = 4 + \tfrac{1}{2}(\underset{\substack{\text{Each double bond}\\\text{has 4 electrons}}}{4}) = 6$$

$$\text{Formal charge} = 6 - 6 = 0$$

For the sulfur atom:

$$\text{Valence electrons assigned} = 0 + \tfrac{1}{2}(12) = 6$$

$$\text{Formal charge} = 6 - 6 = 0$$

Now, having determined the formal charges for the various Lewis structures of the sulfate ion, can we use them to identify which of the resonance structures is closest to the actual electronic structure of the ion? There are two schools of thought on this issue. One position assumes that the atoms in a molecule or ion will try to achieve minimum formal charges. In other words, the assumption is that electrons will naturally flow from negatively charged parts of the molecule to positively charged parts, thus minimizing the charges on the atoms. From this point of view the resonance structures of SO_4^{2-} with two double bonds (and minimum formal charges)

```
 [        O (0)          ]2-          [        O (-1)       ]2-
 [        ||(0)          ]            [        | (0)        ]
 [   :O — S — O:         ]    <-->    [    O == S == O      ]
 [  (-1)  ||   (-1)      ]            [   (0)  |    (0)     ]
 [        O              ]            [        :O:          ]
 [       (0)             ]            [           (-1)      ]
```

would be favored. Note that these structures exceed the octet rule, thus requiring the sulfur to use its 3*d* orbitals to hold electrons.

The other school of thought argues for the primacy of the octet rule and against the use of 3*d* orbitals. This position favors the single-bonded resonance structure of SO_4^{2-}

```
 [        :O: (-1)       ]2-
 [         |             ]
 [   :O — S — O:         ]
 [  (-1)   |(+2) (-1)    ]
 [        :O:            ]
 [          (-1)         ]
```

with its octet of electrons around the sulfur and its high formal charges.

Which point of view is correct? What do we know about the sulfate ion that might help us decide? One pertinent fact is that the sulfur–oxygen bonds in SO_4^{2-} are known by experiment to be shorter than expected for normal single bonds. This would seem to favor the resonance structures with double bonds. However, one can also argue that the high formal charges on the atoms in the single-bonded structure cause ionic attractions that pull the atoms closer together. Thus both schools of thought can adequately explain the short sulfur–oxygen bond lengths.

At this point the dispute continues about which position is correct. For second-row elements where the octet rule is never exceeded it seems clear that the "best" Lewis structures conform to the following rules:

1. Atoms in molecules try to achieve formal charges as close to zero as possible.
2. Any negative formal charges are expected to reside on the most electronegative atoms.

However, for molecules or ions containing third-row or heavier atoms the situation is still unclear.

Rules Governing Formal Charge

1. To calculate the formal charge on an atom:
   - Take the sum of the lone pair electrons and one-half of the shared electrons. This is the number of valence electrons assigned to a given atom in the molecule.
   - Subtract the number of assigned electrons from the number of valence electrons on the free, neutral atom to obtain the formal charge.
2. The sum of the formal charges of all atoms in a given molecule or ion must equal the overall charge on that species.
3. If nonequivalent Lewis structures exist for a species containing second-row atoms, those with formal charges closest to zero and with any negative formal charges on the most electronegative atoms are considered to best describe the bonding in the molecule or ion.

Example 13.10

Give possible Lewis structures for XeO_3, an explosive compound of xenon. Determine the formal charges of each atom in the various Lewis structures.

Solution For XeO_3 (26 valence electrons) we can draw the following possible Lewis structures (formal charges are indicated in parentheses):

Structure	Xe	O (left)	O (bottom)	O (right)
Xe—O, Xe—O, Xe—O	(+3)	(−1)	(−1)	(−1)
Xe=O (left), Xe—O, Xe—O	(+2)	(0)	(−1)	(−1)
Xe=O (bottom), Xe—O, Xe—O	(+2)	(−1)	(0)	(−1)
Xe=O (right), Xe—O, Xe—O	(+2)	(−1)	(−1)	(0)
Xe=O (bottom), Xe=O (right), Xe—O	(+1)	(−1)	(0)	(0)
Xe=O (left), Xe=O (right), Xe—O	(+1)	(0)	(−1)	(0)
Xe=O (left), Xe=O (bottom), Xe—O	(+1)	(0)	(0)	(−1)
Xe=O, Xe=O, Xe=O	(0)	(0)	(0)	(0)

The concept of formal charge is most often used to evaluate the importance of various Lewis structures for molecules that exhibit resonance. However, formal charge arguments also can be helpful in predicting which, among a given group of atoms, is the central atom in a simple molecule. For example, why is carbon dioxide O—C—O rather than C—O—O? Although this question can be pursued at many different levels of sophistication, the simplest approach involves considering the formal charges in the two possible structures. Note that in the Lewis structure given previously for carbon dioxide,

$$\ddot{O}=C=\ddot{O}$$

all atoms have formal charges of 0. However, if the atoms are arranged as follows,

$$C—O—O$$

all the Lewis structures give unreasonable formal charges. Consider the following possibilities, where the formal charges are listed below each atom:

$$:\mathrm{C}\equiv\mathrm{O}-\ddot{\mathrm{O}}: \qquad \ddot{\mathrm{C}}=\mathrm{O}=\ddot{\mathrm{O}} \qquad :\ddot{\mathrm{C}}-\mathrm{O}\equiv\mathrm{O}:$$

$$-1 \quad +2 \quad -1 \qquad -2 \quad +2 \quad 0 \qquad -3 \quad +2 \quad +1$$

None of these Lewis structures (with their resulting formal charges) agrees with our observation that oxygen has a significantly greater electronegativity than carbon. That is, it doesn't make sense that a compound would contain a negatively charged carbon atom next to a positively charged oxygen atom.

As a final note, there are several cautions about formal charge to keep in mind. First, although formal charges are closer to actual atomic charges in molecules than are oxidation states, formal charges still are only estimates of charge—they should not be taken as actual atomic charges. Second, the evaluation of Lewis structures using formal charge ideas can lead to erroneous predictions.

In this same vein, note the difference between a "correct," or valid, Lewis structure and an electronic structure that accurately accounts for a molecule's observed properties. A valid Lewis structure is one that obeys the rules we have established for Lewis structures. However, this Lewis structure may or may not give an accurate picture of the molecule and its properties. Experiments must be carried out to make the final decisions on the correct description of the bonding in a molecule or polyatomic ion.*

13.13 Molecular Structure: The VSEPR Model

The structures of molecules play a very important role in determining their chemical properties. As we will see later, structure is particularly important for biological molecules; a slight change in the structure of a large biomolecule can completely destroy its usefulness to a cell or may even change the cell from normal to cancerous.

The origin of the "repulsions" among electron pairs probably results more from the operation of the Pauli exclusion principle than from electrostatic effects, but we will not be concerned with that in this text.

Many accurate methods now exist for determining **molecular structure,** the three-dimensional arrangement of the atoms in a molecule. These methods must be used if precise information about structure is required. However, it is often useful to predict the approximate molecular structure of a molecule. In this section we consider a simple model that allows us to do this. This model, called the **valence shell electron-pair repulsion (VSEPR) model,** is useful in predicting the geometries of molecules formed from nonmetals. The main postulate of this model is that *the structure around a given atom is determined principally by minimizing electron-pair repulsions.* The idea is that the bonding and nonbonding pairs around a given atom should be positioned as far apart as possible. To see how this model works, we will first consider the molecule $BeCl_2$, which has the Lewis structure

$$:\ddot{\mathrm{Cl}}-\mathrm{Be}-\ddot{\mathrm{Cl}}:$$

Note that there are two pairs of electrons around the beryllium atom. What arrangement of these electron pairs allows them to be as far apart as possi-

*For a discussion of this issue, see Gordon H. Purser, *J. Chem. Ed.* **76** (1999): 1013.

ble to minimize the repulsions? Clearly, the best arrangement places the pairs on opposite sides of the beryllium atom at 180 degrees:

This is the maximum possible separation for two electron pairs. Once we have determined the optimal arrangement of the electron pairs around the central atom, we can specify the molecular structure of $BeCl_2$—that is, the positions of the atoms. Since each electron pair on beryllium is shared with a chlorine atom, the molecule has a **linear structure** with a bond angle of 180 degrees:

Next, let's consider BF_3, which has the Lewis structure

```
      ..
     :F:
  ..  |  ..
 :F — B — F:
  ..     ..
```

Here the boron atom is surrounded by three pairs of electrons. What arrangement will minimize the repulsions? The electron pairs are farthest apart at angles of 120 degrees:

Since each electron pair is shared with a fluorine atom, the molecular structure is

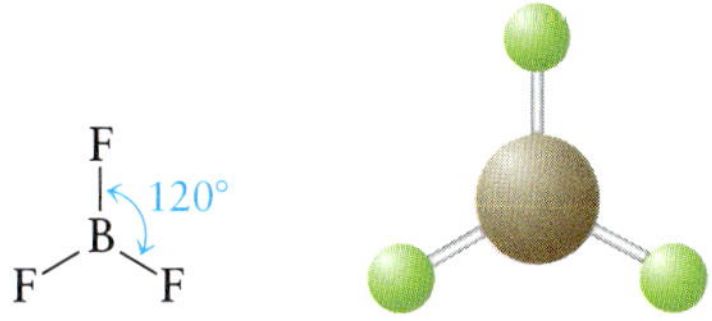

This is a planar (flat) and triangular molecule, which is commonly described as **trigonal planar.**

Next, let's consider the methane molecule, which has the Lewis structure

```
    H
    |
H — C — H
    |
    H
```

There are four pairs of electrons around the central carbon atom. What arrangement of these electron pairs best minimizes the repulsions? First, let's try a square planar arrangement:

The carbon atom and the electron pairs are centered in the plane of the paper, and the angles between the pairs are all 90 degrees.

Is there another arrangement with angles greater than 90 degrees that would put the electron pairs even farther away from each other? The answer is yes. The **tetrahedral arrangement** has angles of 109.5 degrees:

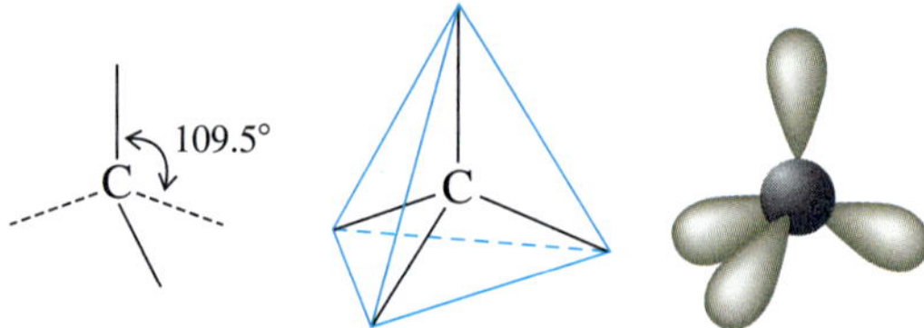

When four uniform balloons are tied together, they naturally form a tetrahedral shape.

It can be shown that this is the maximum possible separation of four pairs around a given atom. This means that *whenever four pairs of electrons are present around an atom, they should always be arranged tetrahedrally.*

Now that we have obtained the electron-pair arrangement that gives the least repulsions, we can determine the positions of the atoms and thus the molecular structure of CH_4. In methane each of the four electron pairs is shared between the carbon atom and a hydrogen atom. Thus the hydrogen atoms are placed as shown in Fig. 13.15, giving the molecule a tetrahedral structure with the carbon atom at the center.

Recall that the fundamental idea of the VSEPR model is to find the arrangement of electron pairs around the central atom that minimizes the electron repulsions. Then we can determine the molecular structure from knowing how the electron pairs are shared with the peripheral atoms.

STEPS

Steps for Using the VSEPR Model

1 Draw the Lewis structure for the molecule.

2 Count the electron pairs around the central atom, and arrange them in the way that minimizes repulsions (that is, put the pairs as far apart as possible).

3 Determine the positions of the atoms from the ways the electron pairs are shared.

4 Name the molecular structure from the positions of the *atoms*.

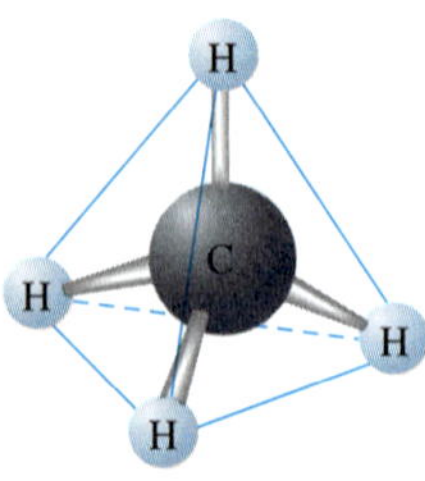

FIGURE 13.15

The molecular structure of methane. The tetrahedral arrangement of electron pairs produces a tetrahedral arrangement of hydrogen atoms.

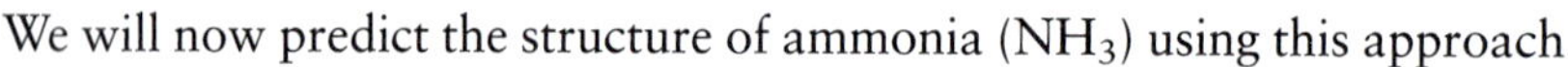

We will now predict the structure of ammonia (NH_3) using this approach.

Draw the Lewis structure.

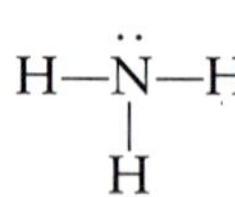

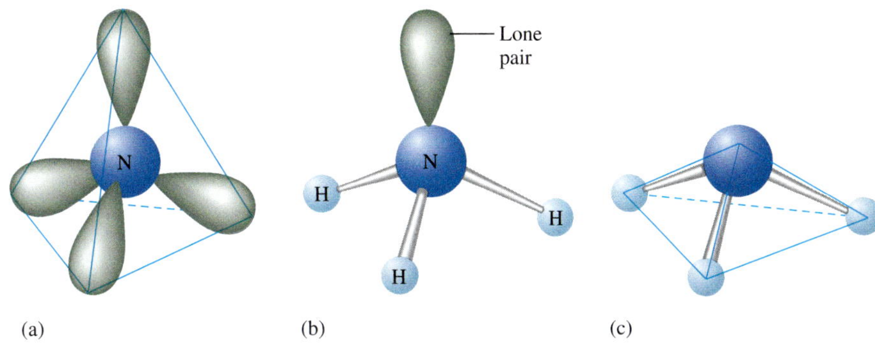

FIGURE 13.16

(a) The tetrahedral arrangement of electron pairs around the nitrogen atom in the ammonia molecule. (b) Three of the electron pairs around nitrogen are shared with hydrogen atoms, as shown, and the fourth is a lone pair. Although the arrangement of *electron pairs* is tetrahedral, as in the methane molecule, the hydrogen atoms in the ammonia molecule occupy only three corners of the tetrahedron. A lone pair occupies the fourth corner. (c) Note that molecular geometry is trigonal pyramidal, not tetrahedral.

- **Count the pairs of electrons and arrange them to minimize repulsions.**

  The NH_3 molecule has four pairs of electrons: three bonding pairs and one nonbonding pair. From the discussion of the methane molecule, we know that the best arrangement of four electron pairs is a tetrahedral array, as shown in Fig. 13.16(a).

- **Determine the positions of the atoms.**

  The three H atoms share electron pairs, as shown in Fig. 13.16(b).

- **Name the molecular structure.**

  It is very important to recognize that the *name* of the molecular structure is always based on the *positions of the atoms*. The placement of the electron pairs determines the structure, but the name is based on the positions of the atoms. Thus it is incorrect to say that the NH_3 molecule is tetrahedral. It has a tetrahedral arrangement of electron pairs but not a tetrahedral arrangement of atoms. The molecular structure of ammonia is a **trigonal pyramid** (one triangular side is different from the other three), rather than a tetrahedron, as shown in Fig. 13.16(c).

Example 13.11

Describe the molecular structure of the water molecule.

Solution The Lewis structure for water is

$$\mathrm{H{-}\ddot{\underset{\cdot\cdot}{O}}{-}H}$$

There are four pairs of electrons: two bonding pairs and two nonbonding pairs. To minimize repulsions, these are best arranged in a tetrahedral array, as shown in Fig. 13.17(a). Although H_2O has a tetrahedral arrangement of electron pairs, it is not a tetrahedral molecule. The atoms in the H_2O molecule form a V-shape, as shown in Fig. 13.17(b) and (c).

From Example 13.11 we see that the H_2O molecule is V-shaped, or bent, because of the presence of the lone pairs. If no lone pairs were present, the molecule would be linear, the polar bonds would cancel, and the molecule would have no dipole moment. This would make water very different from the polar substance so familiar to us.

From the previous discussion we would predict that the H—X—H bond angle (where X is the central atom) in CH_4, NH_3, and H_2O should be the

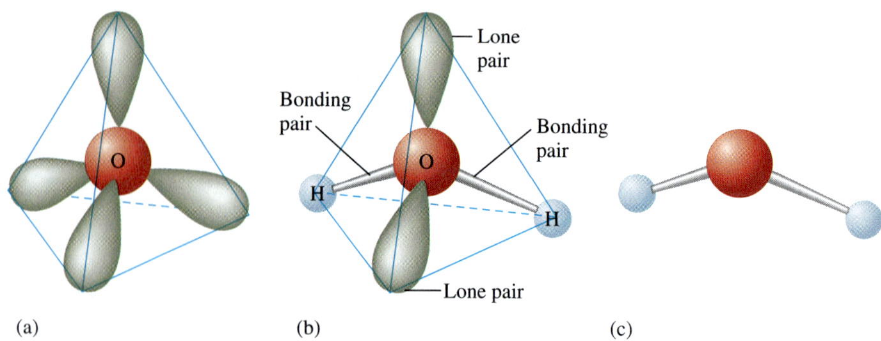

FIGURE 13.17

(a) The tetrahedral arrangement of the four electron pairs around oxygen in the water molecule. (b) Two of the electron pairs are shared between oxygen and the hydrogen atoms, and two are lone pairs. (c) The V-shaped molecular structure of the water molecule.

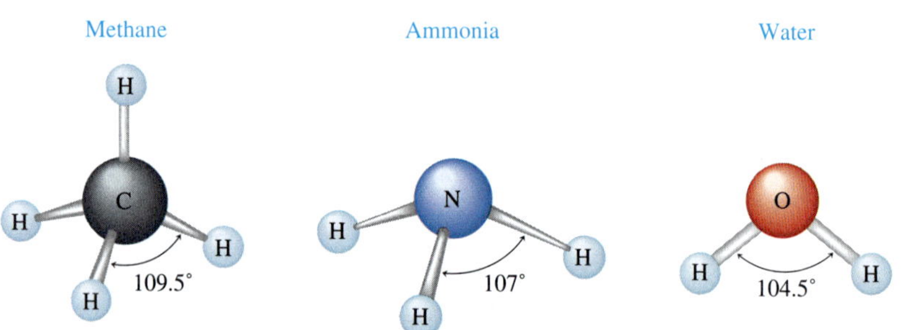

FIGURE 13.18

The bond angles in the CH_4, NH_3, and H_2O molecules. Note that the bond angle between bonding pairs decreases as the number of lone pairs increases.

tetrahedral angle (109.5 degrees). Experiments, however, show that the actual bond angles are those given in Fig. 13.18. What significance do these results have for the VSEPR model? One possible point of view is that the observed angles are close enough to the tetrahedral angle to be satisfactory. The opposite view is that the deviations are significant enough to require modification of the simple model so that it can more accurately handle similar cases. We will take the latter view.

Let's examine the following data:

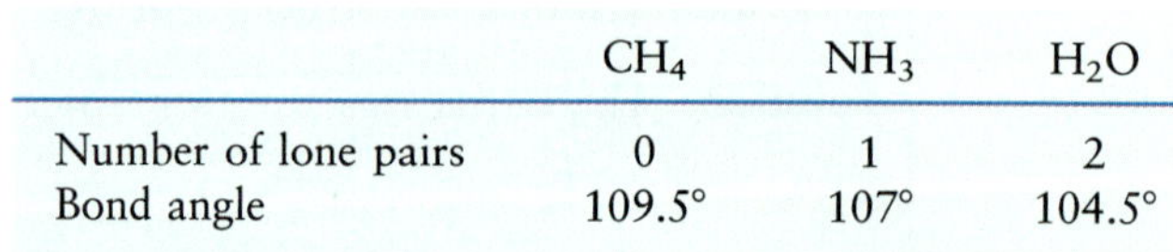

	CH_4	NH_3	H_2O
Number of lone pairs	0	1	2
Bond angle	109.5°	107°	104.5°

(a)

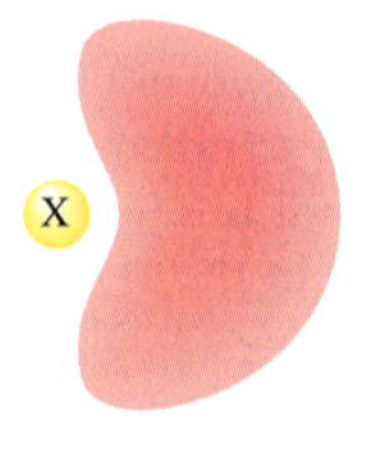

(b)

FIGURE 13.19

(a) In a bonding pair of electrons the electrons are shared by two nuclei. (b) In a lone pair, since both electrons must be close to a single nucleus, they tend to take up more of the space around that atom.

One interpretation of the trend observed here is that lone pairs require more space than bonding pairs; in other words, as the number of lone pairs increases, the bonding pairs are increasingly squeezed together.

This interpretation seems to make physical sense if we think in the following terms: A bonding pair is shared between two nuclei, and the electrons can be close to either nucleus. Therefore they are relatively confined between the two nuclei. A lone pair is localized on only one nucleus, so both electrons are close to that nucleus only, as shown schematically in Fig. 13.19. These pictures help us to understand why a lone pair may require more space near an atom than a bonding pair.

As a result of these observations, we make the following addition to the original postulate of the VSEPR model: *Lone pairs require more room than bonding pairs and tend to compress the angles between the bonding pairs.*

TABLE 13.8

Arrangements of Electron Pairs Around an Atom Yielding Minimum Repulsion

Number of Electron Pairs	Arrangement of Electron Pairs		Example
2	Linear		
3	Trigonal planar	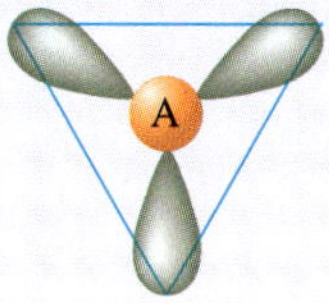	
4	Tetrahedral	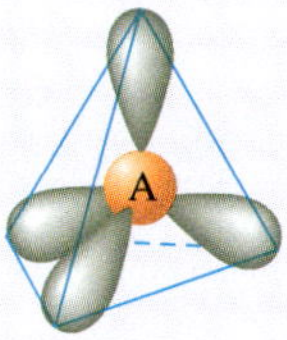	
5	Trigonal bipyramidal	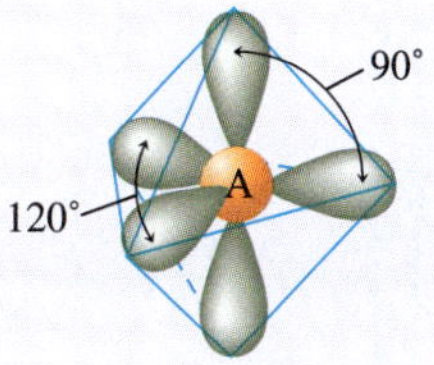 120°, 90°	
6	Octahedral	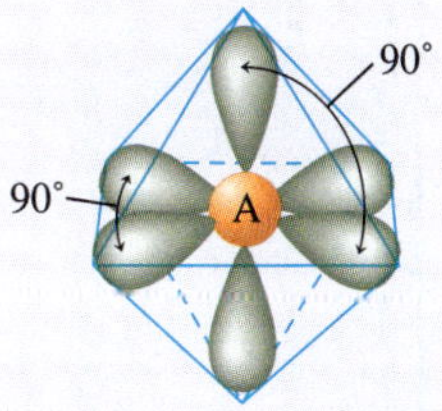 90°, 90°	

So far we have considered cases with two, three, and four electron pairs around the central atom. These are summarized in Table 13.8. Table 13.9 summarizes the structures possible for molecules in which there are four electron pairs around the central atom with various numbers of atoms bonded to it. Note that molecules with four pairs of electrons around the central atom can be tetrahedral (AB_4), trigonal pyramidal (AB_3), and V-shaped (AB_2). For five pairs of electrons there are several possible electron-pair arrangements. The one that produces minimum repulsion is a **trigonal bipyramid.** Note from Table 13.8 that this arrangement has two different angles, 90 degrees and 120 degrees. As the name suggests, the structure formed by this arrangement of pairs consists of two trigonal-based pyramids that share a common base. Table 13.10 summarizes the structures possible for molecules in which there are five electron pairs around the central atom with various numbers of atoms bonded to it. Note that molecules with five pairs of electrons around the central atom

TABLE 13.9

Structures of Molecules That Have Four Electron Pairs Around the Central Atom

Electron-Pair Arrangement | Molecular Structure

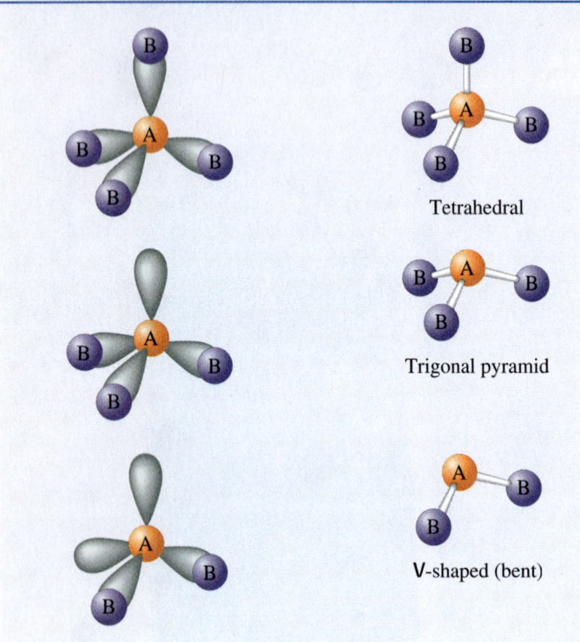

TABLE 13.10

Structures of Molecules with Five Electron Pairs Around the Central Atom

Electron-Pair Arrangement | Molecular Structure

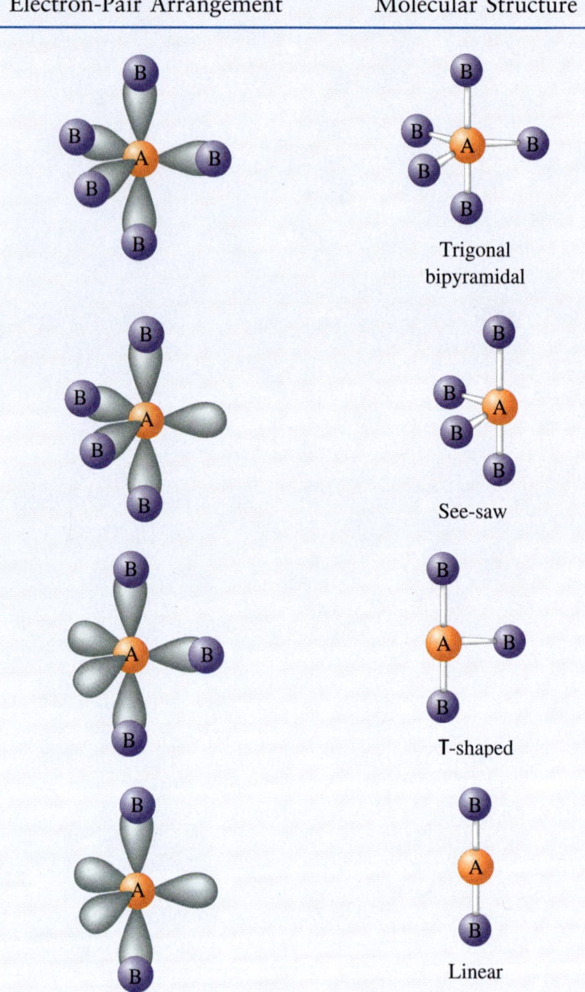

can be trigonal bipyramidal (AB_5), see-saw (AB_4), T-shaped (AB_3), and linear (AB_2). Six pairs of electrons can best be arranged around a given atom to form an **octahedral structure** with 90-degree angles, as shown in Table 13.8.

To use the VSEPR model to determine the geometric structures of molecules, you should memorize the relationships between the number of electron pairs and their best arrangement.

Example 13.12

When phosphorus reacts with excess chlorine gas, the compound phosphorus pentachloride (PCl_5) is formed. In the gaseous and liquid states this substance consists of PCl_5 molecules, but in the solid state it consists of a 1:1 mixture of PCl_4^+ and PCl_6^- ions. Predict the geometric structures of PCl_5, PCl_4^+, and PCl_6^-.

Lewis structure for PCl_5

$$:\ddot{Cl}: \quad :\ddot{Cl}-P-\ddot{Cl}: \quad (P\ \text{bonded to five}\ Cl)$$

Solution The traditional Lewis structure for PCl_5 is shown in the margin. Five pairs of electrons around the phosphorus atom require a trigonal bipyramidal arrangement (see Table 13.8). When the chlorine atoms are included, a trigonal bipyramidal molecule results.

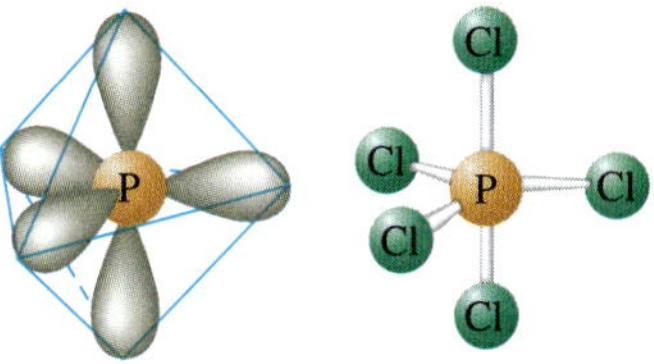

Lewis structure for PCl_4^+

$$:\ddot{Cl}-P-\ddot{Cl}: \quad (P\ \text{bonded to four}\ Cl)$$

The Lewis structure for the PCl_4^+ ion [5 + 4(7) − 1 = 32 valence electrons] is shown in the margin. There are four pairs of electrons surrounding the phosphorus atom in the PCl_4^+ ion. This requires a tetrahedral arrangement of the pairs, as shown in the figure in the margin. Since each pair is shared with a chlorine atom, a tetrahedral PCl_4^+ cation results.

The traditional Lewis structure for PCl_6^- [5 + 6(7) + 1 = 48 valence electrons] is

$$\left[P\ \text{bonded to six}\ :\ddot{Cl}: \right]^-$$

Since phosphorus is surrounded by six pairs of electrons, an octahedral arrangement is required to minimize repulsions, as shown below on the left. Since each electron pair is shared with a chlorine atom, an octahedral PCl_6^- anion is predicted.

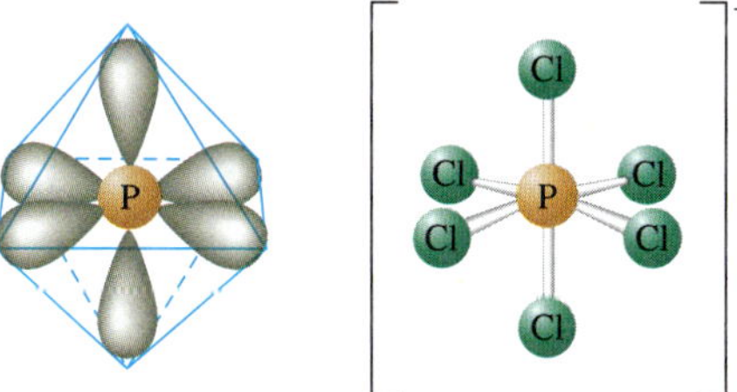

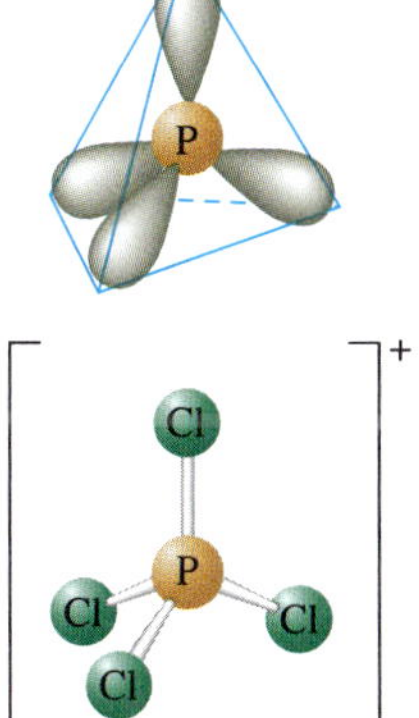

Tetrahedral PCl_4^+ cation

Example 13.13

Because the noble gases have filled *s* and *p* valence orbitals, they are not expected to be chemically reactive. In fact, for many years these elements were called *inert gases* because of this supposed inability to form any compounds. However, in the early 1960s several compounds of krypton, xenon, and radon were synthesized. For example, a team at the Argonne National Laboratory produced the stable colorless compound xenon tetrafluoride (XeF_4). Predict its structure and determine whether it has a dipole moment.

Solution The traditional Lewis structure for XeF_4 is

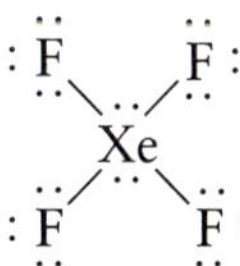

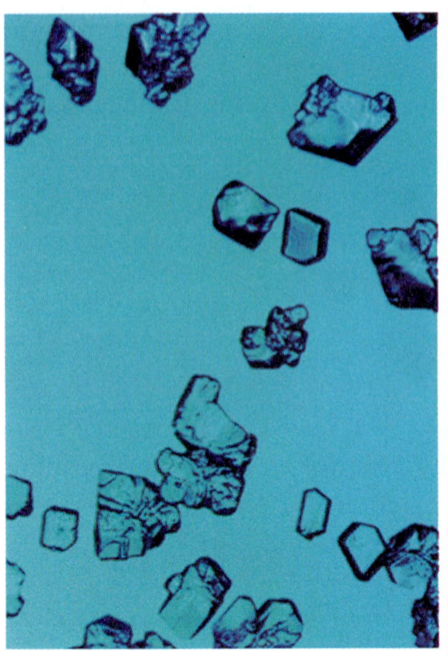

Xenon tetrafluoride crystals.

The xenon atom in this molecule is surrounded by six pairs of electrons, requiring an octahedral arrangement:

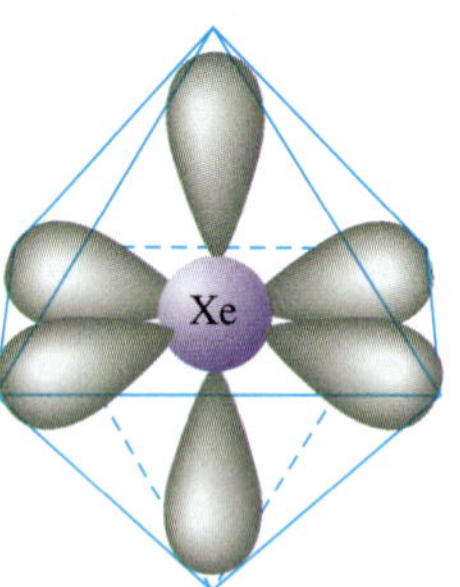

The structure predicted for this molecule depends on how the lone pairs and bonding pairs are arranged. Consider the two possibilities shown in Fig. 13.20. The bonding pairs are indicated by the presence of fluorine atoms. Since the structure predicted differs in the two cases, we must decide which of these arrangements is preferable. The key is to look at the lone pairs. In the structure in part (a) the lone pair–lone pair angle is 90 degrees; in the structure in part (b) the lone pairs are separated by 180 degrees. Since lone pairs require more room than bonding pairs, a structure with two lone pairs at 90 degrees is unfavorable. Thus the arrangement in Fig. 13.20(b) is preferred, and the molecular structure is predicted to be square planar. Note that this molecule is *not* described as being octahedral. There is an *octahedral arrangement of electron pairs*, but the *atoms* form a **square planar** structure.

Although each Xe—F bond is polar (fluorine has a greater electronegativity than xenon), the square planar arrangement of these bonds causes the polarities to cancel.

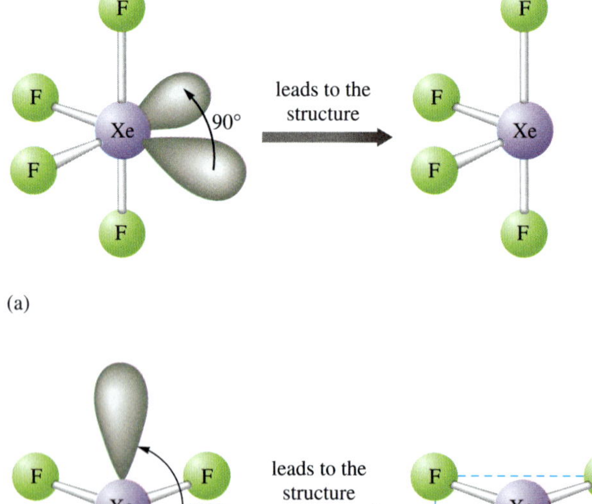

FIGURE 13.20

Possible electron-pair arrangement for XeF_4. Since arrangement (a) has lone pairs 90 degrees apart, it is less favorable than arrangement (b), where the lone pairs are 180 degrees apart.

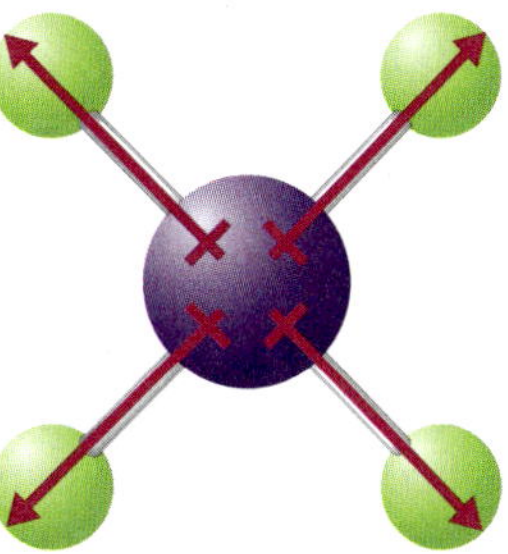

Thus XeF_4 has no dipole moment.

We can further illustrate the use of the VSEPR model for molecules or ions with lone pairs by considering the triiodide ion (I_3^-). The Lewis structure for I_3^- is

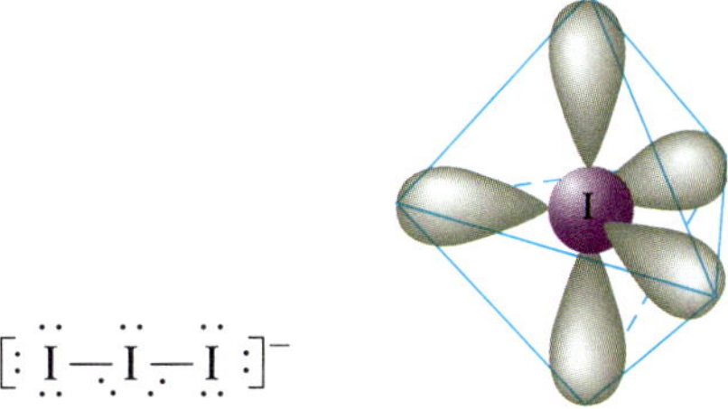

The central iodine atom has five pairs around it, requiring a trigonal bipyramidal arrangement. Several possible arrangements of the lone pairs are shown in Fig. 13.21. Note that structures (a) and (b) have lone pairs at 90 degrees, whereas in (c) all lone pairs are at 120 degrees. Thus structure (c) is preferred. The resulting molecular structure for I_3^- is linear.

$$[I—I—I]^-$$

The VSEPR Model and Multiple Bonds

So far in our treatment of the VSEPR model we have not considered any molecules with multiple bonds. To see how these molecules are handled by this model, let's consider the NO_3^- ion, which requires three resonance structures to describe its electronic structure:

$$[O{=}N(O)O]^- \longleftrightarrow [O{-}N({=}O)O]^- \longleftrightarrow [O{-}N(O){=}O]^-$$

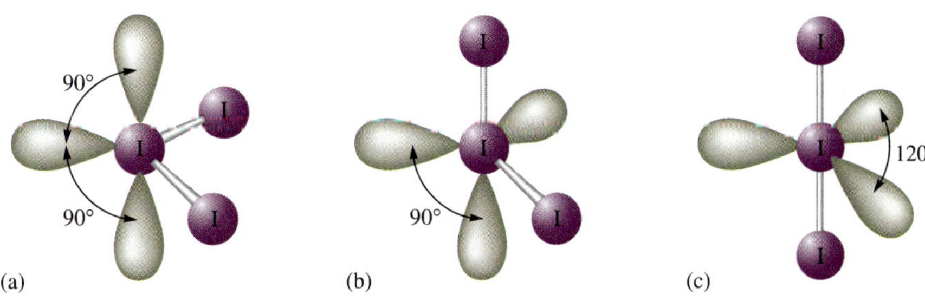

FIGURE 13.21
Three possible arrangements of the electron pairs in the I_3^- ion. Arrangement (c) is preferred because there are no 90-degree lone pair–lone pair interactions.

Chemical Insights

Chemical Structure and Communication: Semiochemicals

In this chapter we have stressed the importance of being able to predict the three-dimensional structure of a molecule. Molecular structure is important because of its effect on chemical reactivity. This is especially true in biological systems, where reactions must be efficient and highly specific. Among the hundreds of types of molecules in the fluids of a typical biological system, the appropriate reactants must find and react only with each other—they must be very discriminating. This specificity depends primarily on structure. The molecules are constructed so that only the appropriate partners can approach each other in a way that allows reaction.

Molecular structure is also central for those molecules used as a means of communication. Examples of chemical communication occurring in humans are the conduction of nerve impulses across synapses, the control of the manufacture and storage of key chemicals in cells, and the senses of smell and taste. Plants and animals also use chemical communication. For example, ants lay down a chemical trail so that other ants can find a certain food supply. Ants also warn their fellow workers of approaching danger by emitting certain chemicals.

Molecules convey messages by fitting into appropriate receptor sites in a very specific way, which is determined by their structure. When a molecule occupies a receptor site, chemical processes that produce the appropriate response are stimulated. Sometimes receptors can be fooled, as in the use of artificial sweeteners—molecules fit the sites on the taste buds that stimulate a "sweet" response in the brain, but they are not metabolized in the same way as natural sugars. Similar deception is useful in insect control. If an area is sprayed with synthetic female sex attractant molecules, the males of that species become so confused that mating does not occur.

A *semiochemical* is a molecule that delivers a message between members of the same or different species of plant or animal. There are three groups of these chemical messengers: allomones, kairomones, and pheromones. Each is of great ecological importance.

An *allomone* is defined as a chemical that gives adaptive advantage to the producer. For example, leaves of the black walnut tree contain an herbicide, juglone, that appears after the leaves fall to the ground. Juglone is not toxic to grass or certain grains, but it is effective against plants such as apple trees that would compete for the available water and food supplies.

Antibiotics are also allomones, since the microorganisms produce them to inhibit other species from growing near them.

Many plants produce bad-tasting chemicals to protect themselves from plant-eating insects and animals. The familiar compound nicotine deters animals from eating the tobacco plant. The millipede sends an unmistakable "back off" message by squirting a predator with benzaldehyde and hydrogen cyanide.

Defense is not the only use of allomones, however. Flowers use scent to attract pollinating insects.

The NO_3^- ion is known to be planar with 120-degree bond angles:

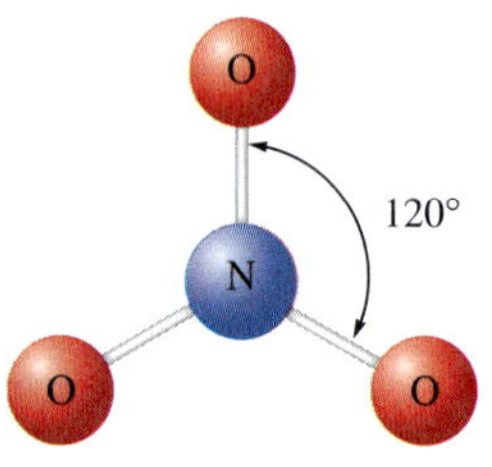

Honeybees, for instance, are guided to alfalfa flowers by a series of sweet-scented compounds.

Kairomones are chemical messengers that bring advantageous news to the receiver. For example, the floral scents are kairomones from the honeybees' viewpoint. Many predators are guided by kairomones emitted by their food. For example, apple skins exude a chemical that attracts the codling moth larva. In some cases kairomones help the underdog. Certain marine mollusks can pick up the "scent" of their predators, the sea stars, and make their escape.

Pheromones are chemicals that affect receptors of the same species as the donor. That is, they are specific within a species. *Releaser pheromones* cause an immediate reaction in the receptor, whereas *primer pheromones* cause long-term effects. Examples of releaser pheromones are the sex attractants of insects, generated in some species by the males and in others by the females. Sex pheromones have also been found in plants and mammals.

Alarm pheromones are highly volatile compounds (ones easily changed to a gas) released to warn of danger. Honeybees produce isoamyl acetate ($C_7H_{14}O_2$) in their sting glands. Because of its high volatility, this compound does not linger after the state of alert is over. Social behavior in insects is characterized by the use of *trail pheromones,* which are used to indicate a food source. Social insects such as bees, ants, wasps, and termites use these substances. Since trail pheromones are less volatile compounds, the indicators persist for some time.

Primer pheromones, which cause long-term behavioral changes, are harder to isolate and identify. One example, however, is the "queen substance" produced by queen honeybees. All the eggs in a colony are laid by one queen bee. If she is removed from the hive or dies, the worker bees are activated by the absence of the queen substance and begin to feed royal jelly to bee larvae to raise a new queen. The queen substance also prevents the development of the workers' ovaries so that only the queen herself can produce eggs.

Many studies of insect pheromones are now under way in the hope that they will provide a method of controlling insects that is more efficient and safer than the current chemical pesticides.

Honeybees are attracted to these alfalfa flowers by sweet-scented compounds the flowers emit.

This planar structure is the one expected for three pairs of electrons around a central atom, which means that *a double bond should be counted as one effective pair* in using the VSEPR model. This makes sense because the two pairs of electrons involved in the double bond are *not* independent pairs. Both of the electron pairs must be in the space between the nuclei of the two atoms to form the double bond. In other words, the double bond acts as one center of electron density to repel the other pairs of electrons. The same holds true for triple bonds. This leads us to another general rule: *For the VSEPR model, multiple bonds count as one effective electron pair.*

Smelling and Tasting Electronically

The human nose and tongue are excellent quality-control sensors. For example, we can tell whether food is spoiled by its disagreeable odor and taste. Because it's impractical to use humans as sensors in industrial settings, several companies are now developing electronic noses.

Cyano Sciences has developed a hand-held electronic nose (called a Cyranose) that can identify specific odors by the "smell print" they create in the 32 sensors of the instrument. Each sensor is composed of conductive carbon black combined with a nonconducting polymer. In the presence of the chemicals associated with an odor, the polymer swells, thereby disrupting the conductive pathways and increasing the resistance. The device is "trained" to detect particular odors by subjecting it to known sources of those odors and storing the electrical signals in the machine's database. When the "nose" is then exposed to an odor in an industrial setting, it identifies the odor by comparing the input signals with its database. The Cyranose is used in the food and beverage industries to monitor the condition of raw materials and the quality of the finished product. In the chemical industry it is used to pinpoint leaks and to identify manufacturing odors.

A Cyranose sensor (an "electronic nose").

Other artificial noses are now being developed that can furnish early detection of diseases. Various researchers are now developing "noses" that can detect different types of bacteria and can provide early diagnosis of lung cancer.

A company called Alpha M.O.S. has recently introduced what it calls the "world's first commercial electronic tongue," a device used to identify various tastes associated with liquids. Like the Cyranose, the electronic tongue must first be "trained" by exposing its probe to known liquid characteristics to record their "fingerprints." Once programmed, the "tongue" can recognize tastes such as sweetness, sourness, bitterness, and saltiness and can detect rancidity and various types of contamination. The electronic tongue is of particular interest to the pharmaceutical industry, where it is hoped it can reduce the need for human tasting panels.

Artificial tasters are also being developed for sensing diseases. For example, Eric V. Anslyn and his coworkers at the University of Texas at Austin have reported a system that can recognize the "taste of heart disease" by analyzing the "taste" of blood samples.

Humans and animals are still the best "smellers" and tasters, but computers are rapidly catching up.

The molecular structure of nitrate also illustrates one more important point: *When a molecule exhibits resonance, any one of the resonance structures can be used to predict the molecular structure using the VSEPR model.* These rules are illustrated in Example 13.14.

Example 13.14

Predict the molecular structure of the sulfur dioxide molecule. Is this molecule expected to have a dipole moment?

Solution First, we must determine the Lewis structure for the SO_2 molecule, which has 18 valence electrons. The expected resonance structures are

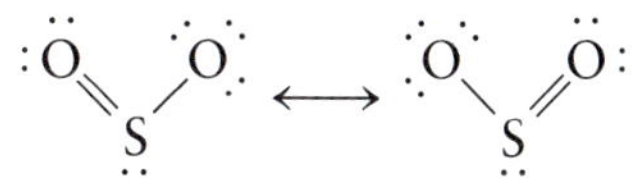

$$:\ddot{O}=\ddot{S}-\ddot{\underset{..}{O}}: \longleftrightarrow :\ddot{\underset{..}{O}}-\ddot{S}=\ddot{O}:$$

To determine the molecular structure, we must count the electron pairs around the sulfur atom. In each resonance structure the sulfur has one lone pair, one pair in a single bond, and one double bond. Counting the double bond as one pair yields three effective pairs around the sulfur. According to Table 13.8, a trigonal planar arrangement is required, yielding a V-shaped molecule:

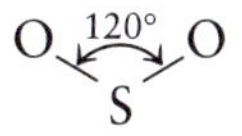

O (120°) O
S

Thus the structure of the SO_2 molecule is expected to be V-shaped with a 120-degree bond angle. The molecule has a dipole moment directed as shown:

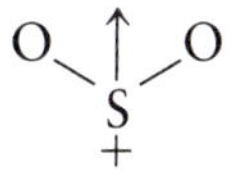

O ↑ O
S
+

Since the molecule is V-shaped, the polar bonds do not cancel.

It should be noted at this point that lone pairs oriented at least 120 degrees from other pairs do not produce significant distortions of bond angles. For example, the angle in the SO_2 molecule is actually quite close to 120 degrees. We will follow the general principle that *an angle of 120 degrees provides lone pairs with enough space so that distortions do not occur. Angles less than 120 degrees are distorted when lone pairs are present.*

Molecules Containing No Single Central Atom

So far we have considered molecules consisting of one central atom surrounded by other atoms. The VSEPR model can be readily extended to more complicated molecules, such as methanol (CH_3OH). This molecule is represented by the following Lewis structure:

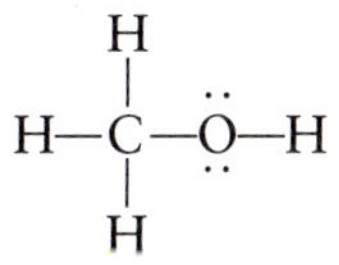

```
    H
    |    ..
H — C — O — H
    |    ..
    H
```

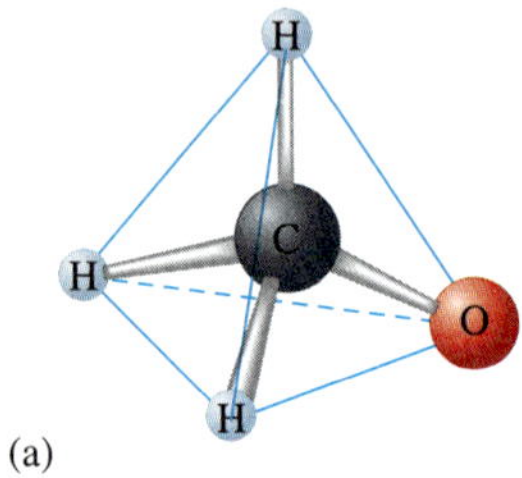

(a)

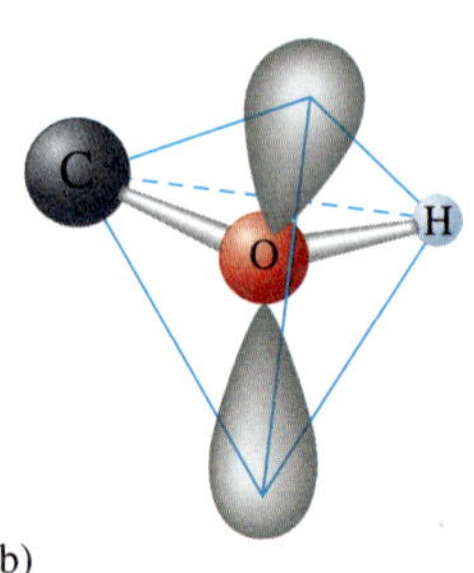

(b)

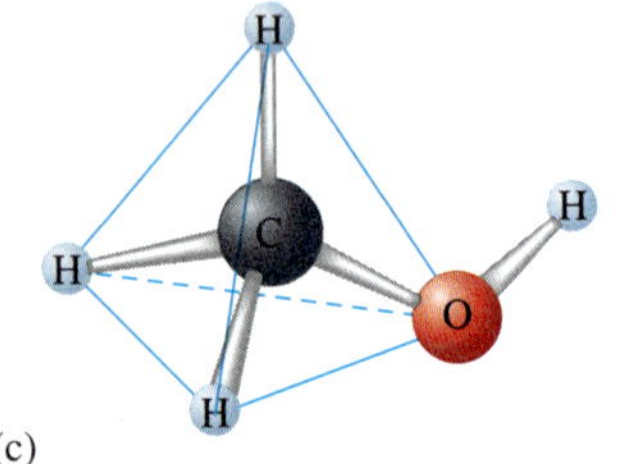

(c)

FIGURE 13.22

The molecular structure of methanol. (a) The arrangement of electron pairs and atoms around the carbon atom. (b) The arrangement of bonding and lone pairs around the oxygen atom. (c) The molecular structure.

The VSEPR Model

The following rules are helpful in using the VSEPR model to predict molecular structure.

1. Determine the Lewis structure(s) for the molecule.
2. For molecules with resonance structures, use any of the structures to predict the molecular structure.
3. Sum the electron pairs around the central atom.
4. When counting pairs, count each multiple bond as a single effective pair.
5. Determine the arrangement of the pairs that minimizes electron-pair repulsions. These arrangements are shown in Table 13.8.
6. Lone pairs require more space than bonding pairs. Choose an arrangement that gives the lone pairs as much room as possible, although it appears that an angle of at least 120 degrees between lone pairs provides enough space. Recognize that lone pairs at angles less than 120 degrees may produce distortions from the idealized structure.

The molecular structure can be predicted from the arrangement of pairs around the carbon and oxygen atoms. Note that there are four pairs of electrons around the carbon, which calls for a tetrahedral arrangement, as shown in Fig. 13.22(a). The oxygen also has four pairs, requiring a tetrahedral arrangement. However, in this case the tetrahedron will be slightly distorted by the space requirements of the lone pairs [Fig. 13.22(b)]. The overall geometric arrangement for the molecule is shown in Fig. 13.22(c).

The VSEPR Model—How Well Does It Work?

The VSEPR model is very simple. There are only a few rules to remember, yet the model correctly predicts the molecular structures of most molecules formed from nonmetallic elements. Molecules of any size can be treated by applying the VSEPR model to each appropriate atom (those bonded to at least two other atoms) in the molecule. Thus we can use this model to predict the structures of molecules with hundreds of atoms. It does, however, fail in a few instances. For example, phosphine (PH_3), which has a Lewis structure analogous to that of ammonia,

```
   ..          ..
H—P—H     H—N—H
   |           |
   H           H
```

would be predicted to have a molecular structure similar to that for NH_3 with bond angles of about 107 degrees. However, the bond angles of phosphine are actually 94 degrees. There are ways of explaining this structure, but more rules have to be added to the model.

This example again illustrates the point that simple models will certainly have exceptions. In introductory chemistry we want to use simple models that fit the majority of cases; we are willing to accept a few failures rather than complicate the model. The amazing thing about the VSEPR model is that such a simple model correctly predicts the structures of so many molecules.

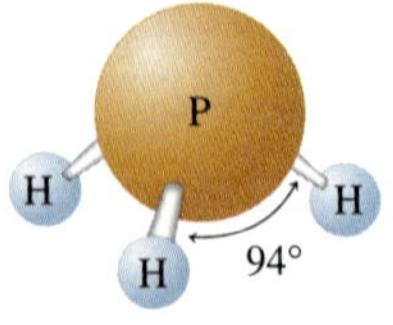

PH_3

Discussion Questions

These questions are designed to be considered by groups of students in class. Often these questions work well for introducing a particular topic in class.

1. Explain the electronegativity trends across a row and down a column of the periodic table. Compare these trends with those of ionization energy and atomic radii. How are they all related?
2. The ionic compound AB is formed. The charges on the ions may be +1, −1; +2, −2; +3, −3; or even larger. What are the factors that determine the charge for an ion in an ionic compound?
3. Using only the periodic table, predict the most stable ion for Na, Mg, Al, S, Cl, K, Ca, and Ga. Arrange these from largest to smallest radius and explain why the radius varies as it does. Compare your predictions with Fig. 13.8.
4. The bond energy for the C—H bond is about 413 kJ/mol in CH_4 but 380 kJ/mol in $CHBr_3$. Although these values are relatively close in magnitude, they are different. Explain why they are different. Does the fact that the C—H bond energy in $CHBr_3$ is lower make any sense? Why?
5. Consider the following statement: "Because oxygen seems to prefer a negative two charge, the second electron affinity is more negative than the first." Indicate everything that is correct in this statement. Indicate everything that is incorrect. Correct the incorrect information, and explain.
6. Which has the greater bond lengths: NO_2^- or NO_3^-? Explain.
7. The following ions are best described with resonance structures. Draw the resonance structures, and using formal charge arguments, predict the "best" Lewis structure for each ion.
   a. NCO^-
   b. CNO^-
8. Would you expect the electronegativity of titanium to be the same in the species Ti, Ti^{2+}, Ti^{3+}, and Ti^{4+}? Explain.
9. The second electron affinity values for both oxygen and sulfur are unfavorable (endothermic). Explain.
10. Arrange the following molecules from most to least polar and explain your order: CH_4, CF_2Cl_2, CF_2H_2, CCl_4, and CCl_2H_2.

Exercises

A blue exercise number indicates that the answer to that exercise appears at the back of this book and a solution appears in the *Solutions Guide*.

Chemical Bonds and Electronegativity

11. Distinguish between the terms *electronegativity* and *electron affinity*, *covalent bond* and *ionic bond*, and *pure covalent bond* and *polar covalent bond*. Characterize the types of bonds in terms of electronegativity difference. Energetically, why do ionic and covalent bonds form?
12. Use Coulomb's law,

$$V = \frac{Q_1Q_2}{4\pi\epsilon_0 r} = 2.31 \times 10^{-19}\ \text{J nm}\left(\frac{Q_1Q_2}{r}\right)$$

to calculate the energy of interaction for the following two arrangements of charges, each having a magnitude equal to the electron charge.

a. $(+1) \xleftrightarrow{1 \times 10^{-10}\ \text{m}} (-1) \longleftarrow \infty \longrightarrow (+1) \xleftrightarrow{1 \times 10^{-10}\ \text{m}} (-1)$

b. Four charges arranged in a square: (−1) at top, (+1) at left, (+1) at right, (−1) at bottom; each side 1×10^{-10} m.

13. Without using Fig. 13.3, predict the order of increasing electronegativity in each of the following groups of elements.
    a. C, N, O
    b. S, Se, Cl
    c. Si, Ge, Sn
    d. Tl, S, Ge
    e. Na, K, Rb
    f. B, O, Ga
14. Without using Fig. 13.3, predict which bond in each of the following groups is the most polar.
    a. C—F, Si—F, Ge—F
    b. P—Cl, S—Cl
    c. S—F, S—Cl, S—Br
    d. Ti—Cl, Si—Cl, Ge—Cl
    e. C—H, Si—H, Sn—H
    f. Al—Br, Ga—Br, In—Br, Tl—Br
15. Repeat Exercises 13 and 14. This time use the values of the electronegativities of the elements given in Fig. 13.3. Are there any differences among your answers?
16. Hydrogen has an electronegativity value between boron and carbon and identical to phosphorus. With this in mind, rank the following bonds in order of decreasing polarity: P—H, O—H, N—H, F—H, C—H.
17. Rank the following bonds in order of increasing ionic character: N—O, Ca—O, C—F, Br—Br, K—F.
18. An alternative definition of electronegativity is

$$\text{Electronegativity} = \text{constant (IE} - \text{EA)}$$

where IE is the ionization energy and EA is the electron affinity using the sign conventions of this book. Use data in Chapter 12 to calculate the (IE − EA) term for F, Cl, Br, and I. Do these values show the same trend as the electronegativity values given in this chapter? The first ionization energies of the halogens are 1678 kJ/mol, 1255 kJ/mol, 1138 kJ/mol, and 1007 kJ/mol, respectively. (*Hint:* Choose a constant so that the electronegativity of fluorine equals 4.0. Using this constant, calculate relative electronegativities for the other halogens and compare with values given in the text.)

Ionic Compounds

19. When an element forms an anion, what happens to the radius? When an element forms a cation, what happens to the radius? Why? Define the term *isoelectronic*. When comparing sizes of ions, which ion has the largest radius and which ion has the smallest radius in an isoelectronic series? Why?

20. Consider the ions Sc^{3+}, Cl^-, K^+, Ca^{2+}, and S^{2-}. Match these ions to the following pictures that represent the relative sizes of the ions.

21. For each of the following groups, place the atoms and ions in order of decreasing size.
    a. Cu, Cu^+, Cu^{2+}
    b. Ni^{2+}, Pd^{2+}, Pt^{2+}
    c. O, O^-, O^{2-}
    d. La^{3+}, Eu^{3+}, Gd^{3+}, Yb^{3+}
    e. Te^{2-}, I^-, Cs^+, Ba^{2+}, La^{3+}

22. Write electron configurations for each of the following.
    a. the cations: Mg^{2+}, Sn^{2+}, K^+, Al^{3+}, Tl^+, As^{3+}
    b. the anions: N^{3-}, O^{2-}, F^-, Te^{2-}
    c. the most stable ion formed by: Be, Rb, Ba, Se, I

23. What noble gas has the same electron configuration as each of the ions in the following compounds?
    a. cesium sulfide
    b. strontium fluoride
    c. calcium nitride
    d. aluminum bromide

24. Which of the following ions have noble gas electron configurations?
    a. Fe^{2+}, Fe^{3+}, Sc^{3+}, Co^{3+}
    b. Tl^+, Te^{2-}, Cr^{3+}
    c. Pu^{4+}, Ce^{4+}, Ti^{4+}
    d. Ba^{2+}, Pt^{2+}, Mn^{2+}

25. Give three ions that are isoelectronic with krypton. Place these ions in order of increasing size.

26. Which compound in each of the following pairs of ionic substances has the most exothermic lattice energy? Justify your answers.
    a. LiF, CsF
    b. NaBr, NaI
    c. $BaCl_2$, BaO
    d. Na_2SO_4, $CaSO_4$
    e. KF, K_2O
    f. Li_2O, Na_2S

27. Predict the empirical formulas of the ionic compounds formed from the following pairs of elements. Name each compound.
    a. Al and S
    b. K and N
    c. Mg and Cl
    d. Cs and Br

28. Following are some important properties of ionic compounds:
    i. low electrical conductivity as solids, and high conductivity in solution or when molten
    ii. relatively high melting and boiling points
    iii. brittleness

    How does the concept of ionic bonding discussed in this chapter account for these properties?

29. Use the following data to estimate ΔH_f° for potassium chloride.

$$K(s) + \tfrac{1}{2}Cl_2(g) \longrightarrow KCl(s)$$

Lattice energy	−690. kJ/mol
Ionization energy for K	419 kJ/mol
Electron affinity of Cl	−349 kJ/mol
Bond energy of Cl_2	239 kJ/mol
Enthalpy of sublimation for K	64 kJ/mol

30. Use the following data to estimate ΔH_f° for magnesium fluoride.

$$Mg(s) + F_2(g) \longrightarrow MgF_2(s)$$

Lattice energy	−2913 kJ/mol
First ionization energy of Mg	735 kJ/mol
Second ionization energy of Mg	1445 kJ/mol
Electron affinity of F	−328 kJ/mol
Bond energy of F_2	154 kJ/mol
Enthalpy of sublimation of Mg	150. kJ/mol

31. LiI(*s*) has a heat of formation of −272 kJ/mol and a lattice energy of −753 kJ/mol. The ionization energy of Li(*g*) is 520. kJ/mol, the bond energy of $I_2(g)$ is 151 kJ/mol, and the electron affinity of I(*g*) is −295 kJ/mol. Use these data to determine the heat of sublimation of Li(*s*).

32. In general, the higher the charge on the ions in an ionic compound, the more favorable is the lattice energy. Why do some stable ionic compounds have +1 charged ions even though +4, +5, and +6 charged ions would have a more favorable lattice energy?

33. Consider the following energy changes:

	ΔE (kJ/mol)
$Mg(g) \longrightarrow Mg^+(g) + e^-$	735
$Mg^+(g) \longrightarrow Mg^{2+}(g) + e^-$	1445
$O(g) + e^- \longrightarrow O^-(g)$	−141
$O^-(g) + e^- \longrightarrow O^{2-}(g)$	878

    a. Magnesium oxide exists as $Mg^{2+}O^{2-}$, not as Mg^+O^-. Explain.
    b. What experiment could be done to confirm that magnesium oxide does not exist as Mg^+O^-?

34. Use the following data (in kJ/mol) to estimate ΔH for the reaction $S^-(g) + e^- \rightarrow S^{2-}(g)$. Include an estimate of uncertainty.

	ΔH_f°	Lattice Energy	IE of M	ΔH_{sub} of M
Na_2S	−365	−2203	495	109
K_2S	−381	−2052	419	90.
Rb_2S	−361	−1949	409	82
Cs_2S	−360.	−1850.	382	78

$$S(s) \longrightarrow S(g) \qquad \Delta H = 277 \text{ kJ/mol}$$

$$S(g) + e^- \longrightarrow S^-(g) \qquad \Delta H = -200. \text{ kJ/mol}$$

35. Rationalize the following lattice energy values:

Compound	Lattice Energy (kJ/mol)
CaSe	−2862
Na_2Se	−2130
CaTe	−2721
Na_2Te	−2095

36. The lattice energies of $FeCl_3$, $FeCl_2$, and Fe_2O_3 are (in no particular order) −2631 kJ/mol, −5339 kJ/mol, and −14,774 kJ/mol. Match the appropriate formula to each lattice energy.

Bond Energies

37. Use bond energy values in Table 13.6 to estimate ΔH for each of the following reactions in the gas phase.
   a. $H_2(g) + Cl_2(g) \rightarrow 2HCl(g)$
   b. $N{\equiv}N(g) + 3H_2(g) \rightarrow 2NH_3(g)$
   c. 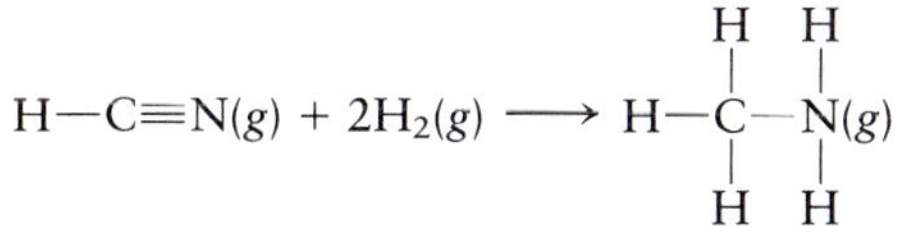 $H{-}C{\equiv}N(g) + 2H_2(g) \longrightarrow H{-}CH_2{-}NH_2(g)$ (structure: H–C(H)(H)–N(H)(H))
   d. 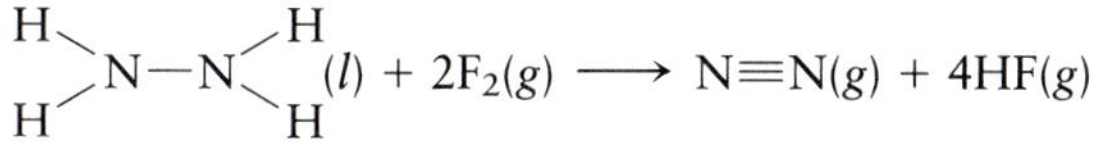 $H_2N{-}NH_2(l) + 2F_2(g) \longrightarrow N{\equiv}N(g) + 4HF(g)$

38. Compare your answers from parts a and b of Exercise 37 with ΔH values calculated for each reaction using standard enthalpies of formation in Appendix 4. Do enthalpy changes calculated from bond energies give a reasonable estimate of the actual values?

39. Use bond energies to predict ΔH for the isomerization of methyl isocyanide to acetonitrile.

$$CH_3N{\equiv}C(g) \longrightarrow CH_3C{\equiv}N(g)$$

40. Use bond energies to predict ΔH for the combustion of ethanol:

$$C_2H_5OH(l) + 3O_2(g) \longrightarrow 2CO_2(g) + 3H_2O(g)$$

41. Use bond energies to estimate ΔH for the combustion of 1 mole of acetylene:

$$C_2H_2(g) + \tfrac{5}{2}O_2(g) \longrightarrow 2CO_2(g) + H_2O(g)$$

42. Consider the following reaction:

$$A_2 + B_2 \longrightarrow 2AB \qquad \Delta H = -285 \text{ kJ}$$

The bond energy for A_2 is one-half the amount of the AB bond energy. The bond energy of B_2 = 432 kJ/mol. What is the bond energy of A_2?

43. The space shuttle orbiter uses the oxidation of methyl hydrazine by dinitrogen tetroxide for propulsion:

$$5N_2O_4(l) + 4N_2H_3CH_3(l) \longrightarrow 12H_2O(g) + 9N_2(g) + 4CO_2(g)$$

Use bond energies to estimate ΔH for this reaction. The structures for the reactants are

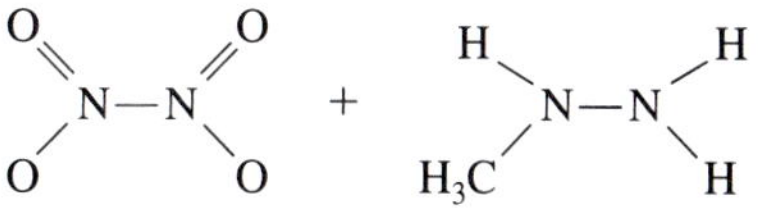

$O_2N{-}NO_2$ (each N bonded to one O by a double bond and one O by a single bond) + $(H)(H_3C)N{-}N(H)(H)$

44. Following are three processes that have been used for the industrial manufacture of acrylonitrile—an important chemical used in the manufacture of plastics, synthetic rubber, and fibers. Use bond energy values (Tables 13.6 and 13.7) to estimate ΔH for each of the reactions.

   a. 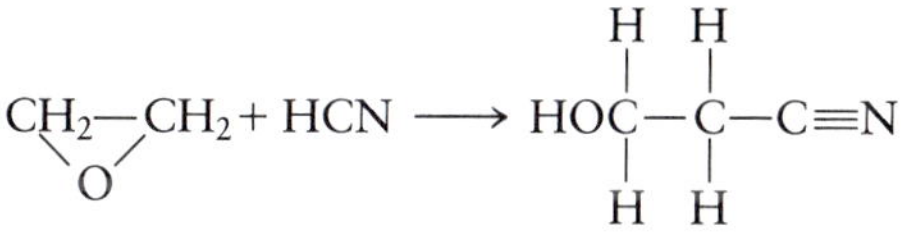 $CH_2{-}CH_2$ (ring closed by O) $+ HCN \longrightarrow HOC(H)(H){-}C(H)(H){-}C{\equiv}N$

   $HOCH_2CH_2CN \longrightarrow (H)(H)C{=}C(H)(C{\equiv}N) + H_2O$

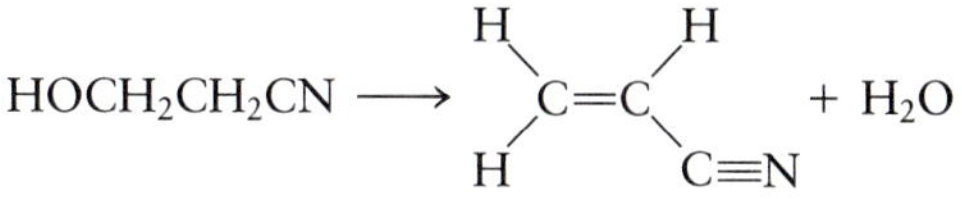

   b. $4CH_2{=}CHCH_3 + 6NO \xrightarrow[\text{Ag}]{700°C} 4CH_2{=}CHCN + 6H_2O + N_2$

   The nitrogen–oxygen bond energy in nitric oxide (NO) is 630. kJ/mol.

   c. $2CH_2{=}CHCH_3 + 2NH_3 + 3O_2 \xrightarrow[425\text{–}510°C]{\text{Catalyst}} 2CH_2{=}CHCN + 6H_2O$

45. Is the elevated temperature noted in parts b and c of Exercise 44 needed to provide energy to endothermic reactions?

46. Acetic acid is responsible for the sour taste of vinegar. It can be manufactured using the following reaction:

$$CH_3OH(g) + C{\equiv}O(g) \longrightarrow CH_3C(=O){-}OH(l)$$

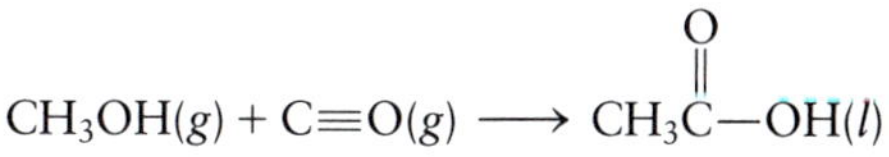

Use tabulated values of bond energies (Table 13.6) to estimate ΔH for this reaction. Compare this result to the ΔH value calculated using standard enthalpies of formation in Appendix 4. Explain any discrepancies.

47. Use bond energies (Table 13.6), values of electron affinities (Table 12.8), and the ionization energy of hydrogen (1312 kJ/mol) to estimate ΔH for each of the following reactions.
    a. $HF(g) \longrightarrow H^+(g) + F^-(g)$
    b. $HCl(g) \longrightarrow H^+(g) + Cl^-(g)$
    c. $HI(g) \longrightarrow H^+(g) + I^-(g)$
    d. $H_2O(g) \longrightarrow H^+(g) + OH^-(g)$
    (Electron affinity of $OH(g) = -180.$ kJ/mol.)

48. The standard enthalpies of formation of $S(g)$, $F(g)$, $SF_4(g)$, and $SF_6(g)$ are +278.8 kJ/mol, +79.0 kJ/mol, −775 kJ/mol, and −1209 kJ/mol, respectively.
    a. Use these data to estimate the energy of an S—F bond.
    b. Compare the value that you calculated in part a with the value given in Table 13.6. What conclusions can you draw?
    c. Why are the ΔH_f° values for $S(g)$ and $F(g)$ not equal to zero, even though sulfur and fluorine are elements?

49. Use the following standard enthalpies of formation to estimate the N—H bond energy in ammonia. Compare this with the value in Table 13.6.

$N(g)$	472.7 kJ/mol
$H(g)$	216.0 kJ/mol
$NH_3(g)$	−46.1 kJ/mol

50. What is the relationship between ΔH_f° for $H(g)$ given in Exercise 49 and the H—H bond energy listed in Table 13.6?

Lewis Structures and Resonance

51. Write Lewis structures that obey the octet rule for each of the following. Except for HCN and H_2CO, the first atom listed is the central atom. For HCN and H_2CO, carbon is the central atom.

a. HCN	d. NH_4^+	g. CO_2
b. PH_3	e. H_2CO	h. O_2
c. $CHCl_3$	f. SeF_2	i. HBr

52. Draw a Lewis structure that obeys the octet rule for each of the following molecules and ions. In each case the first atom listed is the central atom.
    a. $POCl_3$, SO_4^{2-}, XeO_4, PO_4^{3-}, ClO_4^-
    b. NF_3, SO_3^{2-}, PO_3^{3-}, ClO_3^-
    c. ClO_2^-, SCl_2, PCl_2^-

53. Considering your answers to Exercise 52, what conclusions can you draw concerning the structures of species containing the same number of atoms and the same number of valence electrons?

54. Draw Lewis structures for the following. Show all resonance structures, where applicable. Carbon is the central atom in OCN^- and SCN^-.
    a. NO_2^-, NO_3^-, N_2O_4(N_2O_4 exists as $O_2N—NO_2$.)
    b. OCN^-, SCN^-, N_3^-

55. Some of the pollutants in the atmosphere are ozone, sulfur dioxide, and sulfur trioxide. Draw Lewis structures for these three molecules. Show all resonance structures.

56. Peroxyacetyl nitrate, or PAN, is present in photochemical smog. Draw Lewis structures (including resonance forms) for PAN. The skeletal arrangement is

```
   H  O            O
   |  |           /
H—C—C—O—O—N
   |              \
   H               O
```

57. A toxic cloud covered Bhopal, India, in December 1984 when water leaked into a tank of methyl isocyanate, and the product escaped into the atmosphere. Methyl isocyanate is used in the production of many pesticides. Draw the Lewis structures for methyl isocyanate, CH_3NCO, including resonance forms.

58. Explain the terms *resonance* and *delocalized electrons.* When a substance exhibits resonance, we say that none of the individual Lewis structures accurately portrays the bonding in the substance. Why do we draw resonance structures?

59. Benzene (C_6H_6) consists of a six-membered ring of carbon atoms with one hydrogen bonded to each carbon. Draw Lewis structures for benzene, including resonance structures.

60. An important observation supporting the need for resonance in the LE model is that there are only three different structures of dichlorobenzene ($C_6H_4Cl_2$). How does this fact support the need for the concept of resonance?

61. Borazine ($B_3N_3H_6$) has often been called "inorganic" benzene. Draw Lewis structures for borazine. Borazine is a six-membered ring of alternating boron and nitrogen atoms.

62. Draw all the possible Lewis structures for dimethylborazine [$(CH_3)_2B_3N_3H_4$]. (See Exercise 61.) Would there be a different number of structures if there was no resonance?

63. Which of the following statements is(are) true? Correct the false statements.
    a. It is impossible to satisfy the octet rule for all atoms in XeF_2.
    b. Because SF_4 exists, OF_4 should also exist because oxygen is in the same family as sulfur.
    c. The bond in NO^+ should be stronger than the bond in NO^-.
    d. As predicted from the two Lewis structures for ozone, one oxygen–oxygen bond is stronger than the other oxygen–oxygen bond.

64. Lewis structures can be used to understand why some molecules react in certain ways. Write the Lewis structures for the reactants and products in the reactions described below.
    a. Nitrogen dioxide dimerizes to produce dinitrogen tetroxide.
    b. Boron trihydride accepts a pair of electrons from ammonia, forming BH_3NH_3.

Give a possible explanation for why these two reactions occur.

65. The most common type of exception to the octet rule are compounds or ions with central atoms having more than eight electrons around them. PF_5, SF_4, ClF_3, and Br_3^- are examples of this type of exception. Draw the Lewis structures for these compounds or ions. Which elements, when they have to, can have more than eight electrons around them? How is this rationalized?

66. SF_6, ClF_5, and XeF_4 are three compounds whose central atoms do not follow the octet rule. Draw Lewis structures for these compounds.

67. Consider the following bond lengths:

$$\text{C—O } 1.43\ \text{Å} \qquad \text{C=O } 1.23\ \text{Å} \qquad \text{C≡O } 1.09\ \text{Å}$$

In the CO_3^{2-} ion, all three C—O bonds have identical bond lengths of 1.36 Å. Why?

68. Order the following species with respect to the carbon–oxygen bond length (longest to shortest):

$$CO,\ CO_2,\ CO_3^{2-},\ CH_3OH$$

What is the order from the weakest to the strongest carbon–oxygen bond?

69. Place the species below in order of shortest to longest nitrogen–nitrogen bond.

$$N_2,\ N_2F_4,\ N_2F_2$$

(N_2F_4 exists as $F_2N—NF_2$, and N_2F_2 exists as FN—NF.)

Formal Charge

70. Nitrous oxide (N_2O) has three possible Lewis structures:

$$:\ddot{N}=N=\ddot{O}: \longleftrightarrow :N\equiv N-\ddot{\underset{..}{O}}: \longleftrightarrow :\ddot{\underset{..}{N}}-N\equiv O:$$

Given the following bond lengths,

N—N	167 pm	N=O	115 pm
N=N	120 pm	N—O	147 pm
N≡N	110 pm		

rationalize the observations that the N—N bond length in N_2O is 112 pm and that the N—O bond length is 119 pm. Assign formal charges to the resonance structures for N_2O. Can you eliminate any of the resonance structures on the basis of formal charges? Is this consistent with observation?

71. Draw Lewis structures that obey the octet rule for the following species. Assign the formal charge to each central atom.
a. $POCl_3$ c. ClO_4^- e. SO_2Cl_2 g. ClO_3^-
b. SO_4^{2-} d. PO_4^{3-} f. XeO_4 h. NO_4^{3-}

72. Draw the Lewis structures that involve minimum formal charges for the species in Exercise 71.

73. When molten sulfur reacts with chlorine gas, a vile-smelling orange liquid forms that has an empirical formula of SCl. The structure of this compound has a formal charge of zero on all elements in the compound. Draw the Lewis structure for the vile-smelling orange liquid.

74. Oxidation of the cyanide ion produces the stable cyanate ion (OCN^-). The fulminate ion (CNO^-), on the other hand, is very unstable. Fulminate salts explode when struck; $Hg(CNO)_2$ is used in blasting caps. Write the Lewis structures and assign formal charges for the cyanate and fulminate ions. Why is the fulminate ion so unstable? (C is the central atom in OCN^- and N is the central atom in CNO^-.)

Molecular Structure and Polarity

75. Predict the molecular structure and the bond angles for each molecule or ion in Exercises 51, 52, and 54.

76. Predict the molecular structure and the bond angles for each of the following.
a. SeO_3 b. SeO_2 c. PCl_3 d. SCl_2 e. SiF_4

77. There are several molecular structures based on the trigonal bipyramid geometry. Three such structures are

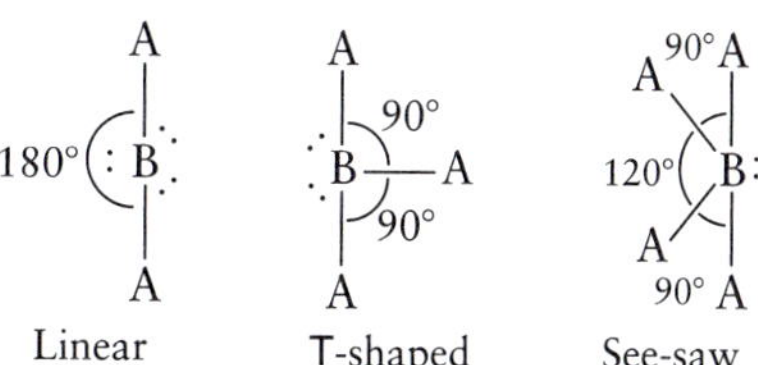

Which of the compounds in Exercises 65 and 66 have these molecular structures?

78. Two variations of the octahedral geometry are illustrated below.

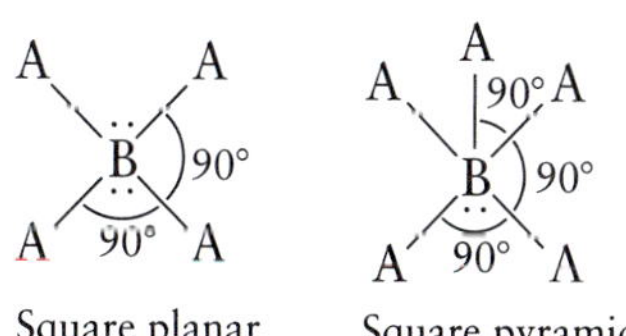

Which of the compounds in Exercises 65 and 66 have these molecular structures?

79. Predict the molecular structure and the bond angles for each of the following. (See Exercises 77 and 78.)
a. $XeCl_2$ b. ICl_3 c. TeF_4 d. PCl_5

80. Predict the molecular structure and the bond angles for each of the following. (See Exercises 77 and 78.)
a. ICl_5 b. $XeCl_4$ c. $SeCl_6$

81. Which of the molecules in Exercise 76 have net dipole moments (are polar)?

82. Which of the molecules in Exercises 79 and 80 have net dipole moments (are polar)?

83. Give two requirements that should be satisfied for a molecule to be polar. Explain why CF_4 and XeF_4 are nonpo-

lar compounds (have no net dipole moments), whereas SF_4 is polar (has a net dipole moment). Is CO_2 polar? What about COS? Explain.

84. What do each of the following sets of compounds/ions have in common with each other? Reference your Lewis structures for Exercises 76, 79, and 80.
   a. $XeCl_4$, $XeCl_2$
   b. ICl_5, TeF_4, ICl_3, PCl_3, SCl_2, SeO_2

85. Which of the following statements is(are) true? Correct the false statements.
   a. The molecules SeS_3, SeS_2, PCl_5, $TeCl_4$, ICl_3, and $XeCl_2$ all exhibit at least one bond angle which is approximately 120 degrees.
   b. The bond angle in SO_2 should be similar to the bond angle in CS_2 or SCl_2.
   c. Of the compounds CF_4, KrF_4, and SeF_4, only SeF_4 exhibits an overall dipole moment (is polar).
   d. Central atoms in a molecule adopt a geometry of the bonded atoms and lone pairs about the central atom in order to maximize electron repulsions.

86. Consider the following Lewis structure, where E is an unknown element:

$$\left[\begin{array}{c} :\ddot{\underset{\cdot\cdot}{O}}-\ddot{E}-\ddot{\underset{\cdot\cdot}{O}}: \\ | \\ :\underset{\cdot\cdot}{\ddot{O}}: \end{array}\right]^-$$

What are some possible identities for element E? Predict the molecular structure (including bond angles) for this ion.

87. Consider the following Lewis structure, where E is an unknown element:

$$\left[\begin{array}{c} \quad\quad\quad\ \ddot{\underset{\cdot\cdot}{O}}: \\ :\ddot{\underset{\cdot\cdot}{F}}-\ddot{E} \quad\ \ \\ \quad\quad\quad\ \ddot{\underset{\cdot\cdot}{F}}: \end{array}\right]^{2-}$$

What are some possible identities for element E? Predict the molecular structure (including bond angles) for this ion. (See Exercises 77 and 78.)

88. Although the VSEPR model is correct in predicting that CH_4 is tetrahedral, NH_3 is pyramidal, and H_2O is bent, the model in its simplest form does not account for the fact that these molecules do not have exactly the same bond angles ($\angle$HCH is 109.5 degrees, as expected for a tetrahedron, but $\angle$HNH is 107.3 degrees and $\angle$HOH is 104.5 degrees). Explain these deviations from the tetrahedral angle.

89. Draw Lewis structures and predict the molecular structures of the following. (See Exercises 77 and 78.)
   a. OCl_2, KrF_2, BeH_2, SO_2
   b. SO_3, NF_3, IF_3
   c. CF_4, SeF_4, KrF_4
   d. IF_5, AsF_5

   Which of the above compounds have net dipole moments (are polar)?

90. Which of the following molecules have net dipole moments? For the molecules that are polar, indicate the polarity of each bond and the direction of the net dipole moment of the molecule.
   a. CH_2Cl_2, $CHCl_3$, CCl_4
   b. CO_2, N_2O
   c. PH_3, NH_3

91. The molecules BF_3, CF_4, CO_2, PF_5, and SF_6 are all nonpolar, even though they contain polar bonds. Why?

Additional Exercises

92. Although both the Br_3^- and I_3^- ions are known, the F_3^- ion does not exist. Explain.

93. Write Lewis structures for CO_3^{2-}, HCO_3^-, and H_2CO_3. When acid is added to an aqueous solution containing carbonate or bicarbonate ions, carbon dioxide gas is formed. We generally say that carbonic acid (H_2CO_3) is unstable. Use bond energies to estimate ΔH for the reaction (in the gas phase):

$$H_2CO_3 \longrightarrow CO_2 + H_2O$$

Specify a possible cause for the instability of carbonic acid.

94. The structure of TeF_5^- is

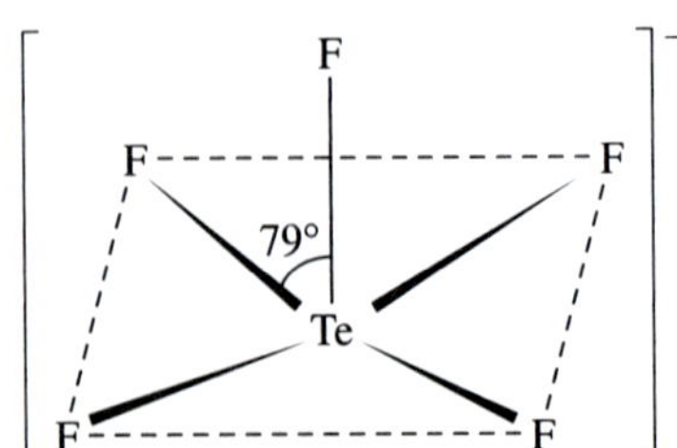

Draw a complete Lewis structure for TeF_5^-, and explain the distortion from the ideal square pyramidal structure. (See Exercise 78.)

95. The compound NF_3 is quite stable, but NCl_3 is very unstable (NCl_3 was first synthesized in 1811 by P. L. Dulong, who lost three fingers and an eye studying its properties). The compounds NBr_3 and NI_3 are unknown, although the explosive compound $NI_3 \cdot NH_3$ is known. Account for the instability of these halides of nitrogen.

96. There are two possible structures of XeF_2Cl_2, where Xe is the central atom. Draw them and describe how measurements of dipole moments might be used to distinguish among them.

97. Which member of the following pairs would you expect to be more energetically stable? Justify each choice.
   a. NaBr or $NaBr_2$
   b. ClO_4 or ClO_4^-
   c. SO_4 or XeO_4
   d. OF_4 or SeF_4

98. Many times, extra stability is characteristic of a molecule or ion in which resonance is possible. How could this feature be used to explain the acidities of the following

compounds? (The acidic hydrogen is marked by an asterisk.) Part c shows resonance in the phenyl ring (C_6H_5).

a. $H{-}\overset{\overset{\displaystyle O}{\|}}{C}{-}OH^*$

b. $CH_3{-}\overset{\overset{\displaystyle O}{\|}}{C}{-}CH{=}\overset{\overset{\displaystyle OH^*}{|}}{C}{-}CH_3$

c. (benzene ring bearing OH^*)

99. Arrange the following in order of increasing radius and increasing ionization energy.
   a. N^+, N, N^-
   b. Se, Se^-, Cl, Cl^+
   c. Br^-, Rb^+, Sr^{2+}

100. Using bond energies, estimate ΔH for the following reaction:

$$CH_3CH_2OH(aq) + HO\overset{\overset{\displaystyle O}{\|}}{C}CH_3(aq) \longrightarrow CH_3CH_2O\overset{\overset{\displaystyle O}{\|}}{C}CH_3(aq) + H_2O(l)$$

101. Give a rationale for the octet rule and the duet rule for H in terms of orbitals.

102. Draw a Lewis structure for the *N*,*N*-dimethylformamide molecule. The skeletal structure is

$$H{-}\overset{\overset{\displaystyle O}{|}}{C}{-}\underset{\underset{\displaystyle CH_3}{|}}{N}{-}CH_3$$

Various types of evidence lead to the conclusion that there is some double-bond character to one of the C—N bonds. Draw one or more resonance structures that support this observation.

103. A compound, XF_5, is 42.81% fluorine by mass. Identify the element X. What is the molecular structure of XF_5?

104. The study of carbon-containing compounds and their properties is called organic chemistry. Besides carbon atoms, *organic compounds* also can contain hydrogen, oxygen, and nitrogen atoms (as well as other types of atoms). A common trait of simple organic compounds is to have Lewis structures in which all atoms have a formal charge of zero. Consider the following incomplete Lewis structure for an organic compound called *histidine* (one of the amino acids, which are the building blocks of proteins found in human bodies):

```
      H—C—N
        |   \
        |    C—H
        |   /
        C—N—H
        |
      H—C—H
   2    |     1
 H—N—C—C—O—H
   |  |  |
   H  H  O
```

Draw a complete Lewis structure for histidine in which all atoms have a formal charge of zero. What are the approximate bond angles about the carbon atom labeled 1 and the nitrogen atom labeled 2?

105. Do the Lewis structures obtained in Exercises 71 and 72 predict the same molecular structure for each case?

106. Predict the molecular structure for each of the following. (See Exercises 77 and 78.)
   a. $BrFI_2$ b. XeO_2F_2 c. $TeF_2Cl_3^-$

   For each formula, there are at least two different structures that can be drawn using the same central atom. Draw the possible structures for each formula.

Challenge Problems

107. Predict the molecular structure of KrF_2. Using hyperconjugation, draw the Lewis structures for KrF_2 that obey the octet rule. Show all resonance forms.

108. Consider the following computer-generated model of caffeine.

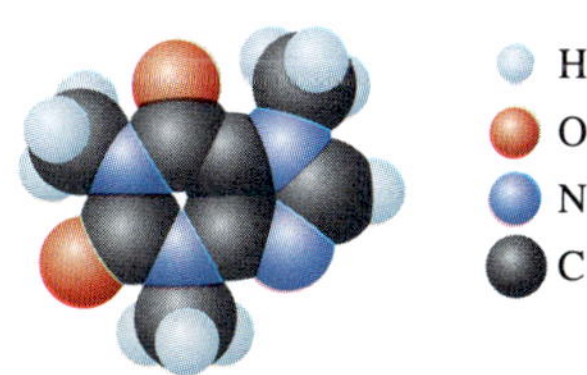

Draw a Lewis structure for caffeine in which all atoms have a formal charge of zero.

109. Given the following information:

Heat of sublimation of Li(s) = 166 kJ/mol
Bond energy of HCl = 427 kJ/mol
Ionization energy of Li(g) = 520. kJ/mol
Electron affinity of Cl(g) = −349 kJ/mol
Lattice energy of LiCl(s) = −829 kJ/mol
Bond energy of H_2 = 432 kJ/mol

Calculate the net change in energy for the following reaction:

$$2Li(s) + 2HCl(g) \longrightarrow 2LiCl(s) + H_2(g)$$

110. Use data in this chapter and Chapter 12 to discuss why MgO is an ionic compound but CO is not an ionic compound.

111. A promising new material with great potential as a fuel in solid rocket motors is ammonium dinitramide [$NH_4N(NO_2)_2$].
   a. Draw Lewis structures (including resonance forms) for the dinitramide ion [$N(NO_2)_2^-$].
   b. Predict the bond angles around each nitrogen in the dinitramide ion.
   c. Ammonium dinitramide can decompose explosively to nitrogen, water, and oxygen. Write a balanced equation for this reaction and use bond energies to estimate ΔH for the explosive decomposition of this compound.
   d. To estimate ΔH from bond energies, you made several assumptions. What are some of your assumptions?

112. Think of forming an ionic compound as three steps (this is a simplification, as with all models): (1) removing an electron from the metal, (2) adding an electron to the nonmetal, and (3) allowing the metal cation and nonmetal anion to come together.
   a. What is the sign of the energy change for each of these three processes?
   b. In general, what is the sign of the sum of the first two processes? Use examples to support your answer.
   c. What must be the sign of the sum of the three processes?
   d. Given your answer to part c, why do ionic bonds occur?
   e. Given your explanations to part d, why is NaCl stable but not Na_2Cl_2 and $NaCl_2$? What about MgO compared to MgO_2 and Mg_2O?

113. The compound hexaazaisowurtzitane is the highest-energy explosive known (*C & E News*, p. 26, Jan. 17, 1994). The compound, also known as CL-20, was first synthesized in 1987. The method of synthesis and detailed performance data are still classified information because of CL-20's potential military application in rocket boosters and in warheads of "smart" weapons. The structure of CL-20 is

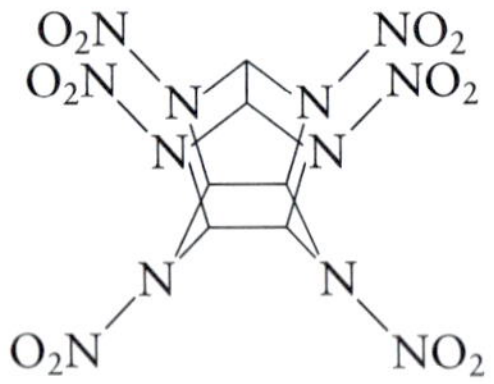

In such shorthand structures, each point where lines meet represents a carbon atom. In addition, the hydrogens attached to the carbon atoms are omitted. Each of the six carbon atoms has one hydrogen atom attached. Three possible reactions for the explosive decomposition of CL-20 are

   i. $C_6H_6N_{12}O_{12}(s) \longrightarrow 6CO(g) + 6N_2(g) + 3H_2O(g) + \frac{3}{2}O_2(g)$
   ii. $C_6H_6N_{12}O_{12}(s) \longrightarrow 3CO(g) + 3CO_2(g) + 6N_2(g) + 3H_2O(g)$
   iii. $C_6H_6N_{12}O_{12}(s) \longrightarrow 6CO_2(g) + 6N_2(g) + 3H_2(g)$

   a. Use bond energies to estimate ΔH for these three reactions.
   b. Which of the above reactions releases the largest amount of energy per kilogram of CL-20?

114. In molecules of the type X—O—H, as the electronegativity of X increases, the acid strength increases. In addition, if the electronegativity of X is a very small value, the molecule acts like a base. Explain these observations and provide examples.

115. Calculate the standard heat of formation of the compound ICl(*g*) at 25°C. (*Hint:* Use Table 13.6 and Appendix 4.)

116. An ionic compound made from the metal M and the diatomic gas X_2 has the formula M_aX_b, in which $a = 1$ or 2 and $b = 1$ or 2. Use the data provided to determine the most likely values for a and b, along with the most likely charges for each of the ions in the ionic compound.

Data (in units of kJ/mol)

Successive ionization energies of M: 480., 4750.

Successive electron affinity values for X: −175, 920.

Enthalpy of sublimation for $M(s) \rightarrow M(g)$: 110.

Bond energy of X_2: 250.

Lattice energy for MX (M^+ and X^-): −1200. kJ/mol

Lattice energy for MX_2 (M^{2+} and X^-): −3500. kJ/mol

Lattice energy for M_2X (M^+ and X^{2-}): −3600. kJ/mol

Lattice energy for MX (M^{2+} and X^{2-}): −4800. kJ/mol

Marathon Problem

117. Identify the following five compounds of H, N, and O. For each compound, write a Lewis structure that is consistent with the information given.
   a. All the compounds are electrolytes, although not all are strong electrolytes. Compounds C and D are ionic and compound B is covalent.

b. Nitrogen occurs in its highest possible oxidation state in compounds A and C; nitrogen occurs in its lowest possible oxidation state in compounds C, D, and E. The formal charge on both nitrogens in compound C is +1; the formal charge on the only nitrogen in compound B is 0.

c. Compounds A and E exist in solution. Both solutions give off gases. Commercially available concentrated solutions of compound A are normally 16 *M*. The commercial, concentrated solution of compound E is 15 *M*.

d. Commercial solutions of compound E are labeled with a misnomer that implies that a binary, gaseous compound of nitrogen and hydrogen has reacted with water to produce ammonium ions and hydroxide ions. Actually, this reaction occurs to only a slight extent.

e. Compound D is 43.7% N and 50.0% O by mass. If compound D were a gas at STP, it would have a density of 2.86 g/L.

f. A formula unit of compound C has one more oxygen than a formula unit of compound D. Compounds C and A have one ion in common when compound A is acting as a strong electrolyte.

g. Solutions of compound C are weakly acidic; solutions of compound A are strongly acidic; solutions of compounds B and E are basic. The titration of 0.726 g of compound B requires 21.98 mL of 1.000 *M* HCl for complete neutralization.

Media Summary

Visit the Student Website at **college.hmco.com/pic/zumdahlCP6e** to help prepare for class, study for quizzes and exams, understand core concepts, and visualize molecular-level interactions. The following media activities are available for this chapter:

Prepare for Class

Video Lessons *Mini-lectures from chemistry experts*

- Lewis Dot Structures for Covalent Bonds
- Valence Electrons and Chemical Bonding
- Ionic Bonds
- Using Bond Dissociation Energies
- Bond Properties
- Resonance Structures
- Predicting Lewis Dot Structures
- Formal Charge
- Electronegativity, Formal Charge, and Resonance
- Valence Shell Electron-Pair Repulsion Theory
- Molecular Shapes for Steric Numbers 2–4
- Molecular Shapes for Steric Numbers 5 & 6
- Predicting Molecular Characteristics Using VSEPR Theory

Improve Your Grade

Visualizations *Molecular-level animations and lab demonstration videos*

- Born–Haber Cycle for NaCl(*s*)
- Determining the Atomic Radius of a Nonmetal (Carbon)
- Formation of C=C Double Bond in Ethylene
- Ionic Radii
- Polar Molecules
- Structure of an Ionic Solid (NaCl)
- VSEPR
- VSEPR: Four-Electron Pair
- VSEPR: Iodine Pentafluoride
- VSEPR: Three-Electron Pair
- VSEPR: Two-Electron Pair

Tutorials *Animated examples and interactive activities*

Lewis Structures
VSEPR Theory

Flashcards *Key terms and definitions*

Online flashcards

ACE the Test

Multiple-choice quizzes
3 ACE Practice Tests

Access these resources using your passkey, available free with new texts or for purchase separately.

Appendixes

Appendix One
Mathematical Procedures

A1.1 Exponential Notation

The numbers characteristic of scientific measurements are often very large or very small; thus it is convenient to express them by using powers of 10. For example, the number 1,300,000 can be expressed as 1.3×10^6, which means multiply 1.3 by 10 six times:

$$1.3 \times 10^6 = 1.3 \times \underbrace{10 \times 10 \times 10 \times 10 \times 10 \times 10}_{10^6 = 1 \text{ million}}$$

Note that each multiplication by 10 moves the decimal point one place to the right, and the easiest way to interpret the notation 1.3×10^6 is that it means move the decimal point in 1.3 to the right six times.

In this notation the number 1985 can be expressed as 1.985×10^3. Note that the usual convention is to write the number that appears before the power of 10 as a number between 1 and 10. Some other examples are given below.

Number	Exponential Notation
5.6	5.6×10^0 or 5.6×1
39	3.9×10^1
943	9.43×10^2
1126	1.126×10^3

To represent a number smaller than 1 in exponential notation, start with a number between 1 and 10 and *divide* by the appropriate power of 10:

$$0.0034 = \frac{3.4}{10 \times 10 \times 10} = \frac{3.4}{10^3} = 3.4 \times 10^{-3}$$

Division by 10 moves the decimal point one place to the *left*. Thus the number 0.00000014 can be written as 1.4×10^{-7}.

To summarize, we can write any number in the form

$$N \times 10^{\pm n}$$

where N is between 1 and 10 and the exponent n is an integer. If the sign preceding n is positive, it means the decimal point in N should be moved n places to the right. If a negative sign precedes n, the decimal point in N should be moved n places to the left.

Multiplication and Division

When two numbers expressed in exponential notation are multiplied, the initial numbers are multiplied and the exponents of 10 are added:

$$(M \times 10^m)(N \times 10^n) = (MN) \times 10^{m+n}$$

For example,

$$(3.2 \times 10^4)(2.8 \times 10^3) = 9.0 \times 10^7$$

When the numbers are multiplied, if a result greater than 10 is obtained for the initial number, the number is adjusted to conventional notation:

$$(5.8 \times 10^2)(4.3 \times 10^8) = 24.9 \times 10^{10} = 2.49 \times 10^{11} = 2.5 \times 10^{11}$$

Division of two numbers expressed in exponential notation involves normal division of the initial numbers and *subtraction* of the exponent of the divisor from that of the dividend. For example,

$$\frac{4.8 \times 10^8}{\underbrace{2.1 \times 10^3}_{\text{Divisor}}} = \frac{4.8}{2.1} \times 10^{(8-3)} = 2.3 \times 10^5$$

Addition and Subtraction

When we add or subtract numbers expressed in exponential notation, *the exponents of the numbers must be the same.* For example, to add 1.31×10^5 and 4.2×10^4, rewrite one number so that the exponents of both are the same:

$$\begin{array}{r} 13.1 \times 10^4 \\ +\ \ 4.2 \times 10^4 \\ \hline 17.3 \times 10^4 \end{array}$$

In correct exponential notation the result is expressed as 1.73×10^5.

Powers and Roots

When a number expressed in exponential notation is taken to some power, the initial number is taken to the appropriate power and the exponent of 10 is multiplied by that power:

$$(N \times 10^n)^m = N^m \times 10^{m \cdot n}$$

For example,*

$$(7.5 \times 10^2)^3 = 7.5^3 \times 10^{3 \cdot 2} = 422 \times 10^6 = 4.22 \times 10^8$$

$$= 4.2 \times 10^8 \text{ (rounded to 2 significant figures)}$$

When a root is taken of a number expressed in exponential notation, the root of the initial number is taken and the exponent of 10 is divided by the number representing the root:

$$\sqrt{N \times 10^n} = (N \times 10^n)^{1/2} = \sqrt{N} \times 10^{n/2}$$

*Refer to the instruction booklet for your calculator for directions concerning how to take roots and powers of numbers.

For example, $(2.9 \times 10^6)^{1/2} = \sqrt{2.9} \times 10^{6/2} = 1.7 \times 10^3$

Because the exponent of the result must be an integer, we may sometimes have to change the form of the number so that the power divided by the root equals an integer; for example,

$$\begin{aligned}\sqrt{1.9 \times 10^3} &= (1.9 \times 10^3)^{1/2} = (0.19 \times 10^4)^{1/2} \\ &= \sqrt{0.19} \times 10^2 = 0.44 \times 10^2 \\ &= 4.4 \times 10^1\end{aligned}$$

The same procedure is followed for roots other than square roots; for example,

$$\begin{aligned}\sqrt[3]{4.6 \times 10^{10}} &= (4.6 \times 10^{10})^{1/3} = (46 \times 10^9)^{1/3} \\ &= \sqrt[3]{46} \times 10^3 = 3.6 \times 10^3\end{aligned}$$

A1.2 Logarithms

A logarithm is an exponent. Any number N can be expressed as follows:

$$N = 10^x$$

For example,

$$\begin{aligned}1000 &= 10^3 \\ 100 &= 10^2 \\ 10 &= 10^1 \\ 1 &= 10^0\end{aligned}$$

The common, or base 10, logarithm of a number is the power to which 10 must be taken to yield that number. Thus, since $1000 = 10^3$,

$$\log 1000 = 3$$

Similarly,

$$\begin{aligned}\log 100 &= 2 \\ \log 10 &= 1 \\ \log 1 &= 0\end{aligned}$$

For a number between 10 and 100, the required exponent of 10 will be between 1 and 2. For example, $65 = 10^{1.8129}$; that is, $\log 65 = 1.8129$. For a number between 100 and 1000, the exponent of 10 will be between 2 and 3. For example, $650 = 10^{2.8129}$ and $\log 650 = 2.8129$.

A number N greater than 0 and less than 1 can be expressed as follows:

$$N = 10^{-x} = \frac{1}{10^x}$$

For example,

$$\begin{aligned}0.001 &= \frac{1}{1000} = \frac{1}{10^3} = 10^{-3} \\ 0.01 &= \frac{1}{100} = \frac{1}{10^2} = 10^{-2} \\ 0.1 &= \frac{1}{10} = \frac{1}{10^1} = 10^{-1}\end{aligned}$$

Thus

$$\log 0.001 = -3$$

$$\log 0.01 = -2$$

$$\log 0.1 = -1$$

Although common logs are often tabulated, the most convenient method for obtaining such logs is to use a calculator.

Since logs are simply exponents, they are manipulated according to the rules for exponents. For example, if $A = 10^x$ and $B = 10^y$, then their product is

$$A \cdot B = 10^x \cdot 10^y = 10^{x+y}$$

and

$$\log AB = x + y = \log A + \log B$$

For division we have

$$\frac{A}{B} = \frac{10^x}{10^y} = 10^{x-y}$$

and

$$\log \frac{A}{B} = x - y = \log A - \log B$$

For a number raised to a power we have

$$A^n = (10^x)^n = 10^{nx}$$

and

$$\log A^n = nx = n \log A$$

It follows that

$$\log \frac{1}{A^n} = \log A^{-n} = -n \log A$$

or for $n = 1$,

$$\log \frac{1}{A} = -\log A$$

When a common log is given, to find the number it represents, we must carry out the process of exponentiation. For example, if the log is 2.673, then $N = 10^{2.673}$. The process of exponentiation is also called taking the antilog, or the inverse logarithm, and is easily carried out by using a calculator.

A second type of logarithm, the natural logarithm, is based on the number 2.7183, which is referred to as e. In this case a number is represented as $N = e^x = 2.7183^x$. For example,

$$N = 7.15 = e^x$$

$$\ln 7.15 = x = 1.967$$

If a natural logarithm is given, to find the number it represents, we must carry out exponentiation to the base e (2.7183) by using a calculator.

A1.3 Graphing Functions

In the interpretation of the results of a scientific experiment, it is often useful to make a graph. It is usually most convenient to graph the function in a form that gives a straight line. The equation for a straight line (a *linear equation*) can be represented by the general form

$$y = mx + b$$

where y is the *dependent variable*, x is the *independent variable*, m is the *slope*, and b is the *intercept* with the y axis.

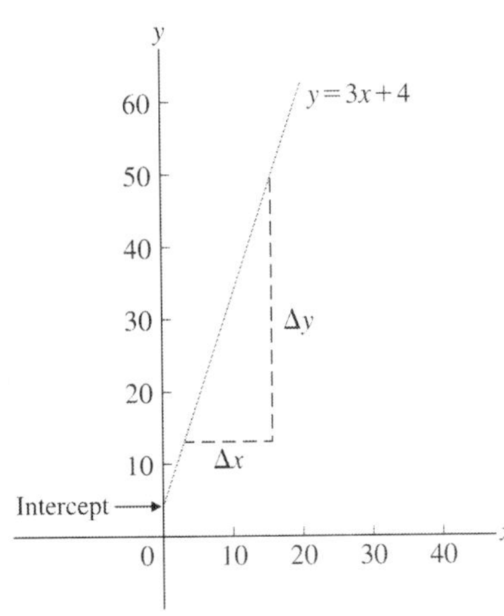

FIGURE A1.1

Graph of the linear equation $y = 3x + 4$.

As an illustration of the characteristics of a linear equation, the function $y = 3x + 4$ is plotted in Fig. A1.1. For this equation, $m = 3$ and $b = 4$. Note that the y intercept occurs when $x = 0$. In this case the intercept is 4, as can be seen from the equation ($b = 4$).

The slope of a straight line is defined as the ratio of the rate of change in y to that in x:

$$m = \text{slope} = \frac{\Delta y}{\Delta x}$$

For the equation $y = 3x + 4$, y changes three times as fast as x (since x has a coefficient of 3). Thus the slope in this case is 3. This can be verified from the graph. For the triangle shown in Fig. A1.1:

$$\text{Slope} = \frac{\Delta y}{\Delta x} = \frac{24}{8} = 3$$

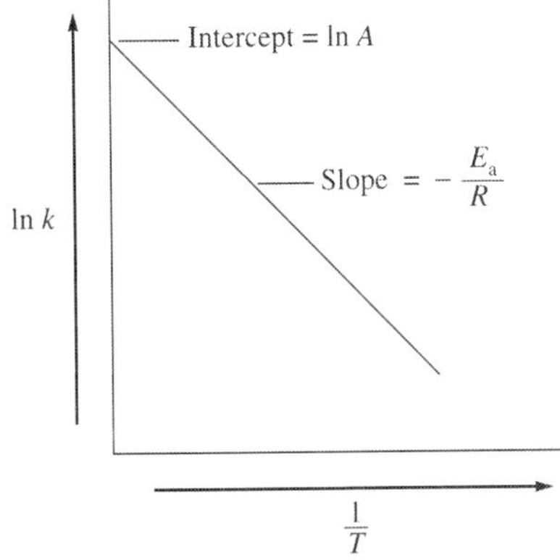

FIGURE A1.2

Graph of ln k versus $1/T$.

Sometimes an equation that is not in standard form can be changed to the form $y = mx + b$ by rearrangement or mathematical manipulation. An example is the equation $k = Ae^{-E_a/RT}$, where A, E_a, and R are constants, k is the dependent variable, and $1/T$ is the independent variable. This equation can be changed to standard form by taking the natural logarithm of both sides,

$$\ln k = \ln Ae^{-E_a/RT} = \ln A + \ln e^{-E_a/RT} = \ln A - \frac{E_a}{RT}$$

noting that the log of a product is equal to the sum of the logs of the individual terms and that the natural log of $e^{-E_a/RT}$ is simply the exponent $-E_a/RT$. Thus in standard form the equation $k = Ae^{-E_a/RT}$ is written

$$\underbrace{\ln k}_{y} = \underbrace{-\frac{E_a}{R}}_{m}\underbrace{\left(\frac{1}{T}\right)}_{x} + \underbrace{\ln A}_{b}$$

A plot of ln k versus $1/T$ (see Fig. A1.2) gives a straight line with slope $-E_a/R$ and intercept ln A.

Of course, many relationships that arise from the description of natural systems are nonlinear, and the "slope" of a curve is continuously changing. In this case the instantaneous slope is given by the tangent to the curve at that point, which is described by a new function obtained by taking the derivative of the original function. For example, for the function in x, $f = ax^2$, the derivative (df/dx) is $2ax$. Thus the slope at each point on the curve defined by the function ax^2 is given by $2ax$.

A1.4 Solving Quadratic Equations

A *quadratic equation,* a polynomial in which the highest power of x is 2, can be written as

$$ax^2 + bx + c = 0$$

One method for finding the two values of x that satisfy a quadratic equation is to use the *quadratic formula:*

$$x = \frac{-b \pm \sqrt{b^2 - 4ac}}{2a}$$

where a, b, and c represent the coefficients of x^2 and x and the constant, respectively. For example, in the determination of $[H^+]$ in a solution of 1.0×10^{-4} M acetic acid, the following expression arises:

$$1.8 \times 10^{-5} = \frac{x^2}{1.0 \times 10^{-4} - x}$$

which yields $\quad x^2 + (1.8 \times 10^{-5})x - 1.8 \times 10^{-9} = 0$

where $a = 1$, $b = 1.8 \times 10^{-5}$, and $c = -1.8 \times 10^{-9}$. Using the quadratic formula, we have

$$x = \frac{-b \pm \sqrt{b^2 - 4ac}}{2a}$$

$$= \frac{-1.8 \times 10^{-5} \pm \sqrt{3.24 \times 10^{-10} - (4)(1)(-1.8 \times 10^{-9})}}{2(1)}$$

and $\quad x = \dfrac{6.9 \times 10^{-5}}{2} = 3.5 \times 10^{-5}$

or $\quad x = \dfrac{-10.5 \times 10^{-5}}{2} = -5.2 \times 10^{-5}$

Note that there are two roots, as there always will be for a polynomial in x^2. In this case x represents a concentration of H^+ (see Section 7.5). Thus the positive root is the one that solves the problem, since a concentration cannot be a negative number.

A second method for solving quadratic equations is by *successive approximations,* a systematic method of trial and error. A value of x is guessed and substituted into the equation everywhere x (or x^2) appears, except for one place. For example, for the equation

$$x^2 + (1.8 \times 10^{-5})x - 1.8 \times 10^{-9} = 0$$

we might guess $x = 2 \times 10^{-5}$. Substituting that value into the equation gives

$$x^2 + (1.8 \times 10^{-5})(2 \times 10^{-5}) - 1.8 \times 10^{-9} = 0$$

or $\quad x^2 = 1.8 \times 10^{-9} - 3.6 \times 10^{-10} = 1.4 \times 10^{-9}$

Thus $\quad x = 3.7 \times 10^{-5}$

Note that the guessed value of x (2×10^{-5}) is not the same as the value of x that is calculated (3.7×10^{-5}) after inserting the estimated value. This means that $x = 2 \times 10^{-5}$ is not the correct solution, and we must try another guess.

We take the calculated value (3.7×10^{-5}) as our next guess:

$$x^2 + (1.8 \times 10^{-5})(3.7 \times 10^{-5}) - 1.8 \times 10^{-9} = 0$$

$$x^2 = 1.8 \times 10^{-9} - 6.7 \times 10^{-10} = 1.1 \times 10^{-9}$$

Thus $\quad x = 3.3 \times 10^{-5}$

Now we compare the two values of x again:

Guessed: $\quad x = 3.7 \times 10^{-5}$

Calculated: $\quad x = 3.3 \times 10^{-5}$

These values are closer but still not identical.

Next, we try 3.3×10^{-5} as our guess:

$$x^2 + (1.8 \times 10^{-5})(3.3 \times 10^{-5}) - 1.8 \times 10^{-9} = 0$$

$$x^2 = 1.8 \times 10^{-9} - 5.9 \times 10^{-10} = 1.2 \times 10^{-9}$$

Thus $\qquad x = 3.5 \times 10^{-5}$

Compare:

Guessed: $\qquad x = 3.3 \times 10^{-5}$

Calculated: $\qquad x = 3.5 \times 10^{-5}$

Next, we guess $x = 3.5 \times 10^{-5}$, which leads to

$$x^2 + (1.8 \times 10^{-5})(3.5 \times 10^{-5}) - 1.8 \times 10^{-9} = 0$$

$$x^2 = 1.8 \times 10^{-9} - 6.3 \times 10^{-10} = 1.2 \times 10^{-9}$$

Thus $\qquad x = 3.5 \times 10^{-5}$

Now the guessed value and the calculated value are the same; we have found the correct solution. Note that this agrees with one of the roots found with the quadratic formula in the first method above.

To further illustrate the method of successive approximations, we will solve Example 7.9 by using this procedure. In solving for $[H^+]$ for 0.010 *M* H_2SO_4, we obtain the following expression:

$$1.2 \times 10^{-2} = \frac{x(0.010 + x)}{0.010 - x}$$

which can be rearranged to give

$$x = (1.2 \times 10^{-2})\left(\frac{0.010 - x}{0.010 + x}\right)$$

We will guess a value for x, substitute it into the right side of the equation, and then calculate a value for x. In guessing a value for x, we know it must be less than 0.010, since a larger value would make the calculated value for x negative and the guessed and calculated values will never match. We start by guessing $x = 0.005$.

The results of the successive approximations are shown in the following table:

Trial	Guessed Value for x	Calculated Value for x
1	0.0050	0.0040
2	0.0040	0.0051
3	0.00450	0.00455
4	0.00452	0.00453

Note that the first guess was close to the actual value and that there was oscillation between 0.004 and 0.005 for the guessed and calculated values. For trial 3, an average of these values was used as the guess, and this led rapidly to the correct value (0.0045 to the correct number of significant figures). Also note that it is useful to carry extra digits until the correct value is obtained, which is then rounded off to the correct number of significant figures.

The method of successive approximations is especially useful for solving polynomials containing x to a power of 3 or higher. The procedure is the same as for quadratic equations: Substitute a guessed value for x into the equation for every x term but one, and then solve for x. Continue this process until the guessed and calculated values agree.

A1.5 Uncertainties in Measurements

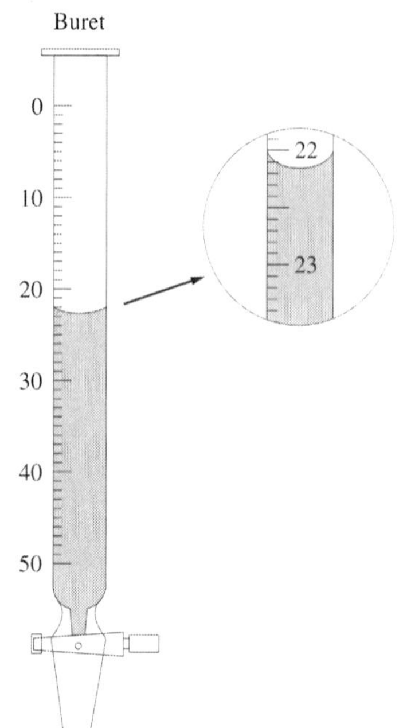

FIGURE A1.3

Measurement of volume using a buret. The volume is read at the bottom of the liquid curve (called the meniscus).

The number associated with a measurement is obtained by using some measuring device. For example, consider the measurement of the volume of a liquid in a buret, as shown in Fig. A1.3, where the scale is greatly magnified. The volume is about 22.15 mL. Note that the last number must be estimated by interpolating between the 0.1-mL marks. Since the last number is estimated, its value may vary depending on who makes the measurement. If five different people read the same volume, the results might be as follows:

Person	Result of Measurement
1	22.15 mL
2	22.14 mL
3	22.16 mL
4	22.17 mL
5	22.16 mL

Note from these results that the first three numbers (22.1) remain the same regardless of who makes the measurement; these are called certain digits. However, the digit to the right of the 1 must be estimated and thus varies; it is called an uncertain digit. We customarily report a measurement by recording all the certain digits plus the *first* uncertain digit. In our example it would not make any sense to try to record the volume to thousandths of a milliliter because the value for hundredths of a milliliter must be estimated when using the buret.

It is very important to realize that a *measurement always has some degree of* ***uncertainty***. The uncertainty of a measurement depends on the precision of the measuring device. For example, using a bathroom scale, you might estimate that the mass of a grapefruit is about 1.5 pounds. Weighing the same grapefruit on a highly precise balance might produce a result of 1.476 pounds. In the first case the uncertainty occurs in the tenths of a pound place; in the second case the uncertainty occurs in the thousandths of a pound place. Suppose we weigh two similar grapefruit on the two devices and obtain the following results:

	Bathroom Scale	Balance
Grapefruit 1	1.5 lb	1.476 lb
Grapefruit 2	1.5 lb	1.518 lb

Do the two grapefruits have the same mass? The answer depends on which set of results you consider. Thus a conclusion based on a series of measurements depends on the certainty of those measurements. For this reason, it is important to indicate the uncertainty in any measurement. This is done by always recording the certain digits and the first uncertain digit (the estimated number). These numbers are called the **significant figures** of a measurement.

The convention of significant figures automatically gives an indication of the uncertainty in a measurement. The uncertainty in the last number (the estimated number) is usually assumed to be ±1 unless otherwise indicated. For example, the measurement 1.86 kilograms can be interpreted to mean 1.86 ± 0.01 kilograms.

Precision and Accuracy

Two terms often used to describe uncertainty in measurements are *precision* and *accuracy*. Although these words are frequently used interchangeably in everyday life, they have different meanings in the scientific context. **Accuracy** refers to the agreement of a particular value with the true value. **Precision** refers to the degree of agreement among several measurements of the same quantity. Precision reflects the *reproducibility* of a given type of measurement. The difference between these terms is illustrated by the results of three different target practices shown in Fig. A1.4.

Two different types of errors are also introduced in Fig. A1.4. A **random error** (also called an indeterminate error) means that a measurement has an equal probability of being high or low. This type of error occurs in estimating the value of the last digit of a measurement. The second type of error is called **systematic error** (or determinate error). This type of error occurs in the same direction each time; it is either always high or always low. Figure A1.4(a) indicates large random errors (poor technique). Figure A1.4(b) indicates small random errors but a large systematic error, and Fig. A1.4(c) indicates small random errors and no systematic error.

In quantitative work precision is often used as an indication of accuracy; we assume that the *average* of a series of precise measurements (which should "average out" the random errors because of their equal probability of being high or low) is accurate, or close to the "true" value. However, this assumption is valid only if systematic errors are absent. Suppose we weigh a piece of brass five times on a very precise balance and obtain the following results:

Weighing	Result
1	2.486 g
2	2.487 g
3	2.485 g
4	2.484 g
5	2.488 g

Normally, we would assume that the true mass of the piece of brass is very close to 2.486 grams, which is the average of the five results. However, if the balance has a defect causing it to give a result that is consistently 1.000 gram too high (a systematic error of +1.000 gram), then 2.486 grams would be seriously in error. The point here is that high precision among several measurements is an indication of accuracy *only* if you can be sure that systematic errors are absent.

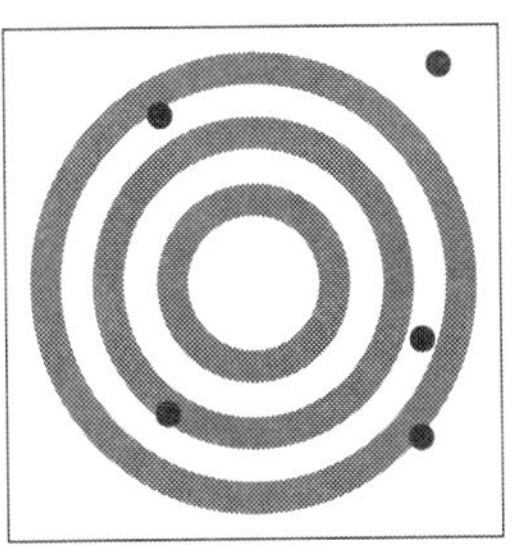

(a)

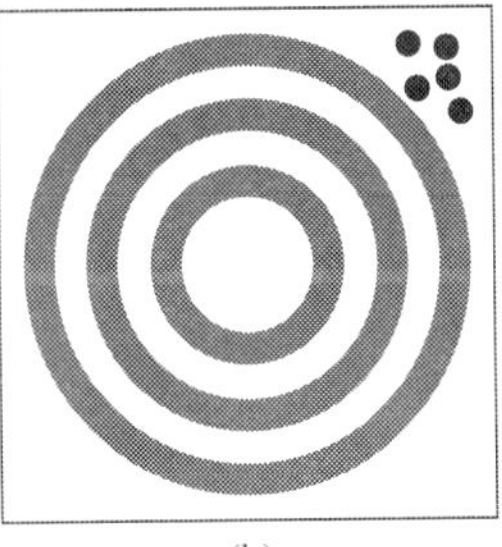

(b)

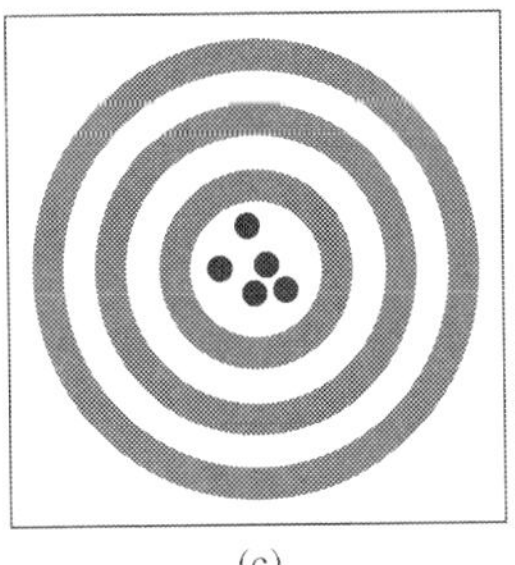

(c)

FIGURE A1.4

Shooting targets show the difference between *precise* and *accurate*.
(a) Neither accurate nor precise (large random errors). (b) Precise but not accurate (small random errors, large systematic error). (c) Bull's-eye! Both precise and accurate (small random errors, no systematic error).

Expression of Experimental Results

The accuracy of a measurement refers to how close it is to the true value. An inaccurate result occurs as a result of some flaw (systematic error) in the measurement: the presence of an interfering substance, incorrect calibration of an instrument, operator error, and so on. The goal of chemical analysis is to eliminate systematic error, but random errors can only be minimized. In practice, an experiment is almost always done in order to find an unknown value (the true value is not known—someone is trying to obtain that value by doing the experiment). In this case the precision of several replicate determinations is used to assess the accuracy of the result. The results of the replicate experiments are expressed as an average (which we assume is close to the true value) with an error limit that gives some indication of how close the average value may be to the true value. The error limit represents the uncertainty of the experimental result.

To illustrate this procedure, consider a situation that might arise in the pharmaceutical industry. Assume that the specification for a commercial 500-mg acetaminophen (the active painkiller in Tylenol) tablet is that each batch of tablets must contain 450 to 550 mg of acetaminophen per tablet. Suppose that chemical analysis gave the following results for a batch of acetaminophen tablets: 428, 479, 442, and 435 mg. How can these results be used to decide whether the batch of tablets meets the specification? Although the details of how to draw such conclusions from measured data are beyond the scope of this discussion, we will consider some aspects of this process. We will focus here on the types of experimental uncertainty, the expression of experimental results, and a simplified method for estimating experimental uncertainty when several types of measurements contribute to the final result.

There are two common ways of expressing an average: the mean and the median. The mean ($\overline{x}$) is the arithmetic average of the results, or

$$\text{Mean} = \overline{x} = \sum_{i=1}^{n} \frac{x_i}{n} = \frac{x_1 + x_2 + \cdots + x_n}{n}$$

where Σ means take the sum of the values. The mean is equal to the sum of all the measurements divided by the number of measurements. For the acetaminophen results given previously, the mean is

$$\overline{x} = \frac{428 + 479 + 442 + 435}{4} = 446 \text{ mg}$$

The median is the value that lies in the middle among the results. Half of the measurements are above the median and half are below the median. For results of 465, 485, and 492 mg, the median is 485 mg. When there is an even number of results, the median is the average of the two middle results. For the acetaminophen results, the median is

$$\frac{442 + 435}{2} = 439 \text{ mg}$$

There are several advantages to using the median. If a small number of measurements is made, one value can greatly affect the mean. Consider the results for the analysis of acetaminophen: 428, 479, 442, and 435 mg. The mean is 446 mg, which is larger than three of the four results. The median is 439 mg, which lies near the three values that are relatively close to one another.

In addition to expressing an average value for a series of results, we must also express the uncertainty. This usually means expressing either the precision of the measurements or the observed range of the measurements. The range of a series of measurements is defined by the smallest value and the largest value. For the analytical results on the acetaminophen tablets, the range is from 428 to 479 mg. Using this range, we can express the results by saying that the true value lies between 428 and 479 mg. That is, we can express the amount of acetaminophen in a typical tablet as 446 ± 33 mg, where the error limit is chosen to give the observed range (approximately).

The most common way to specify precision is by the standard deviation s, which for a small number of measurements is given by the formula

$$s = \left[\frac{\sum_{i=1}^{n} (x_i - \overline{x})^2}{n - 1} \right]^{1/2}$$

where x_i is an individual result, $\overline{x}$ is the average (either mean or median), and n is the total number of measurements. For the acetaminophen example we have

$$s = \left[\frac{(428 - 446)^2 + (479 - 446)^2 + (442 - 446)^2 + (435 - 446)^2}{4 - 1} \right]^{1/2} = 23$$

Thus we can say that the amount of acetaminophen in a typical tablet in the batch of tablets is 446 mg with a sample standard deviation of 23 mg. Statistically, this means that any additional measurement has a 68% probability (68 chances out of 100) of being between 423 mg (446 − 23) and 469 mg (446 + 23). Thus the standard deviation is a measure of the precision of a given type of measurement.

In scientific calculations it is also useful to be able to estimate the precision of a procedure that involves several measurements by combining the precisions of the individual steps. That is, we want to answer the following question: How do the uncertainties propagate when we combine the results of several different types of measurements? There are many ways to deal with the propagation of uncertainty. We will discuss one simple method below.

Worst-Case Method for Estimating Experimental Uncertainty

To illustrate this method, we will consider the determination of the density of an irregularly shaped solid. In this determination we make three measurements. First, we measure the mass of the object on a balance. Next, we must obtain the volume of the solid. The easiest method for doing this is to partially fill a graduated cylinder with a liquid and record the volume. Then we add the solid and record the volume again. The difference in the measured volumes is the volume of the solid. We can then calculate the density of the solid from the equation

$$D = \frac{M}{V_2 - V_1}$$

where M is the mass of the solid, V_1 is the initial volume of liquid in the graduated cylinder, and V_2 is the volume of liquid plus solid. Suppose we get the following results:

$$M = 23.06 \text{ g}$$

$$V_1 = 10.4 \text{ mL}$$

$$V_2 = 13.5 \text{ mL}$$

The calculated density is

$$\frac{23.06\ \text{g}}{13.5\ \text{mL} - 10.4\ \text{mL}} = 7.44\ \text{g/mL}$$

Now suppose that the precision of the balance used is ±0.02 g and that the volume measurements are precise to ±0.05 mL. How do we estimate the uncertainty of the density? We can do this by assuming a worst case. That is, we assume the largest uncertainties in all measurements, and we see what combinations of measurements will give the largest and smallest possible results (the greatest range). Since the density is the mass divided by the volume, the largest value of the density will be that obtained by using the largest possible mass and the smallest possible volume:

Largest possible mass = 23.06 + 0.02

$$D_{\text{max}} = \frac{23.08}{13.45 - 10.45} = 7.69\ \text{g/mL}$$

Smallest possible V_2 Largest possible V_1

The smallest value of the density is

Smallest possible mass

$$D_{\text{min}} = \frac{23.04}{13.55 - 10.35} = 7.20\ \text{g/mL}$$

Largest possible V_2 Smallest possible V_1

Thus the calculated range is from 7.20 to 7.69, and the average of these values is 7.45. The error limit is the number that gives the high and low range values when added and subtracted from the average. Therefore, we can express the density as 7.45 ± 0.25 g/mL, which is the average value plus or minus the quantity that gives the range calculated by assuming the largest uncertainties.

Analysis of the propagation of uncertainties is useful in drawing qualitative conclusions from the analysis of measurements. For example, suppose that we obtained the preceding results for the density of an unknown alloy and we want to know if it is one of the following alloys:

Alloy A: $D = 7.58$ g/mL

Alloy B: $D = 7.42$ g/mL

Alloy C: $D = 8.56$ g/mL

We can safely conclude that the alloy is not C. But the values of the densities for alloys A and B are both within the inherent uncertainty of our method. To distinguish between A and B, we need to improve the precision of our determination. The obvious choice is to improve the precision of the volume measurement.

The worst-case method is useful for estimating the maximum uncertainty expected when the results of several measurements are combined to obtain a result. We assume the maximum uncertainty in each measurement and then calculate the minimum and maximum possible results. These extreme values describe the range and thus the maximum error limit associated with a particular determination.

Confidence Limits

A more sophisticated method for estimating the uncertainty of a particular type of determination involves the use of confidence limits. A confidence limit is defined as

$$\text{Confidence limit} = \pm \frac{ts}{\sqrt{n}}$$

where t = a weighting factor based on statistical analysis

s = the standard deviation

n = the number of experiments carried out

In this context an experiment may refer to a single type of measurement (for example, weighing an object) or to a procedure that requires various types of measurements to obtain a given final result (for example, obtaining the percentage of iron in a particular sample of iron ore). Some representative values of t are listed in Table A1.1.

TABLE A1.1

Values of t for 90% and 95% Confidence Levels

	Values of t for Confidence Intervals	
n	90%	95%
2	6.31	12.7
3	2.92	4.30
4	2.35	3.18
5	2.13	2.78
6	2.02	2.57
7	1.94	2.45
8	1.90	2.36
9	1.86	2.31
10	1.83	2.26

A 95% confidence level means that the true value (the *average* obtained if the experiment were repeated an *infinite* number of times) will lie within $\pm ts/\sqrt{n}$ of the *observed* average (obtained from n experiments) with a 95% probability (95 of 100 times). Thus the factor $\pm ts/\sqrt{n}$ represents an error limit for a given set of results from a particular type of experiment. Thus we might represent the result of n determinations as

$$\bar{x} \pm \frac{ts}{\sqrt{n}}$$

where $\bar{x}$ is the average of the results from the n experiments. This type of error limit is expected to be considerably smaller than that obtained from a worst-case analysis.

A1.6 Significant Figures

Calculating the final result for an experiment usually involves adding, subtracting, multiplying, or dividing the results of various types of measurements. Thus it is important to be able to estimate the uncertainty in the final result. In the previous section we have considered this process in some detail. A closely related matter concerns the number of digits that should be retained in the result of a given calculation. In other words, how many of the digits in the result are significant (meaningful) relative to the uncertainty expected in the result? From statistical analyses of how uncertainties accumulate when arithmetic operations are carried out, rules have been developed for determining the correct number of significant figures in a final result. First, we must consider how to count the number of significant figures (digits) represented in a particular number.

Rules for Counting Significant Figures (Digits)

1. *Nonzero integers.* Nonzero integers always count as significant figures.
2. *Zeros.* There are three classes of zeros:
   a. *Leading zeros* are zeros that *precede* all the nonzero digits. They do not count as significant figures. In the number 0.0025 the three zeros simply

indicate the position of the decimal point. This number has only two significant figures.

b. *Captive zeros* are zeros *between* nonzero digits. They always count as significant figures. The number 1.008 has four significant figures.

c. *Trailing zeros* are zeros at the *right end* of the number. They are significant only if the number contains a decimal point. The number 100 has only one significant figure, whereas the number 1.00×10^2 has three significant figures. The number one hundred written as 100. also has three significant figures.

3. *Exact numbers.* Many times calculations involve numbers that were not obtained by using measuring devices but were determined by counting: 10 experiments, 3 apples, 8 molecules. Such numbers are called *exact numbers.* They can be assumed to have an infinite number of significant figures. Other examples of exact numbers are the 2 in $2\pi r$ (the circumference of a circle) and the 4 and the 3 in $\frac{4}{3}\pi r^3$ (the volume of a sphere). Exact numbers can also arise from definitions. For example, one inch is defined as exactly 2.54 centimeters. Thus, in the statement 1 in = 2.54 cm, neither the 2.54 nor the 1 limits the number of significant figures when used in a calculation.

The following rules apply for determining the number of significant figures in the result of a calculation.

Rules for Significant Figures in Mathematical Operations*

1. For *multiplication or division* the number of significant figures in the result is the same as the number in the least precise measurement used in the calculation. For example, consider this calculation:

$$4.56 \times 1.4 = 6.38 \xrightarrow{\text{Corrected}} 6.4$$

Limiting term has two significant figures (1.4) — Two significant figures (6.4)

The correct product has only two significant figures, since 1.4 has two significant figures.

2. *For addition or subtraction* the result has the same number of decimal places as the least precise measurement used in the calculation. For example, consider the following sum:

$$\begin{array}{r} 12.11 \\ 18.0 \\ 1.013 \\ \hline 31.123 \end{array} \xrightarrow{\text{Corrected}} 31.1$$

18.0 ← Limiting term has one decimal place

31.1 ↑ One decimal place

The correct result is 31.1, since 18.0 has only one decimal place.

*Although these rules work well for most cases, they can give misleading results in certain cases. For a discussion of this, see L. M. Schwartz, "Propagation of Significant Figures," *J. Chem. Ed.* **62** (1985):693.

Note that for multiplication and division significant figures are counted. For addition and subtraction the decimal places are counted.

In most calculations you will need to round off numbers to obtain the correct number of significant figures. The following rules should be applied for rounding.

Rules for Rounding

1. In a series of calculations, carry the extra digits through to the final result, *then* round off.*
2. If the digit to be removed†
   a. is less than 5, the preceding digit stays the same. For example, 1.33 rounds to 1.3.
   b. is equal to or greater than 5, the preceding digit is increased by 1. For example, 1.36 rounds to 1.4.

When rounding, use only the first number to the right of the last significant figure. Do not round off sequentially. For example, the number 4.348 when rounded to two significant figures is 4.3, not 4.4.

Appendix Two
Units of Measurement and Conversions Among Units

A2.1 Measurements

Making observations is fundamental to all science. A quantitative observation, or **measurement,** always consists of two parts: a *number* and a scale (a *unit*). Both parts must be present for the measurement to be meaningful.

The two most widely used systems of units are the *English system* used in the United States and the *metric system* used by most of the rest of the industrialized world. This duality obviously causes a good deal of trouble; for example, parts as simple as bolts are not interchangeable between machines built using the different systems. As a result, the United States has begun to adopt the metric system.

For many years, most scientists worldwide have used the metric system. In 1960 an international agreement established a system of units called the *International System* (*le Système International* in French), abbreviated **SI**. This system is based on the metric system and the units derived from the metric system. The fundamental SI units are listed in Table A2.1.

*This practice will not usually be followed in the examples in this text because we want to show the correct number of significant figures in each step. However, in the answers to the end-of-chapter exercises, only the final answer is rounded.

†This procedure is consistent with the operation of calculators.

TABLE A2.1

The Fundamental SI Units

Physical Quantity	Name of Unit	Abbreviation
Mass	kilogram	kg
Length	meter	m
Time	second	s
Temperature	Kelvin	K
Electric current	ampere	A
Amount of substance	mole	mol
Luminous intensity	candela	cd

TABLE A2.2

The Prefixes Used in the SI System

Prefix	Symbol	Meaning	Exponential Notation*
exa	E	1,000,000,000,000,000,000	10^{18}
peta	P	1,000,000,000,000,000	10^{15}
tera	T	1,000,000,000,000	10^{12}
giga	G	1,000,000,000	10^{9}
mega	**M**	**1,000,000**	$\mathbf{10^{6}}$
kilo	**k**	**1000**	$\mathbf{10^{3}}$
hecto	h	100	10^{2}
deka	da	10	10^{1}
—	—	1	10^{0}
deci	**d**	**0.1**	$\mathbf{10^{-1}}$
centi	**c**	**0.01**	$\mathbf{10^{-2}}$
milli	**m**	**0.001**	$\mathbf{10^{-3}}$
micro	μ	**0.000001**	$\mathbf{10^{-6}}$
nano	**n**	**0.000000001**	$\mathbf{10^{-9}}$
pico	p	0.000000000001	10^{-12}
femto	f	0.000000000000001	10^{-15}
atto	a	0.000000000000000001	10^{-18}

*The most common notations are shown in bold. See Appendix A1.1 if you need a review of exponential notation.

Because the fundamental units are not always convenient (expressing the mass of a pin in kilograms is awkward), the SI system uses prefixes to change the size of the unit. These prefixes are listed in Table A2.2.

One physical quantity that is very important in chemistry is *volume*, which is not a fundamental SI unit; it is derived from length. A cube with dimensions of 1 m on each edge has a volume of $(1\ m)^3 = 1\ m^3$. Then, recognizing that there are 10 decimeters (dm) in a meter, the volume of the cube is $(10\ dm)^3 = 1000\ dm^3$. A cubic decimeter, dm^3, is commonly called a liter (L), which is a unit of volume slightly larger than a quart. Similarly, since 1 dm equals 10 centimeters (cm), the liter $(1\ dm)^3$ contains $1000\ cm^3$, or 1000 milliliters (mL).

A2.2 Unit Conversions

It is often necessary to convert results from one system of units to another. The most common way of converting units is by the *unit factor method*, more commonly called **dimensional analysis.** To illustrate the use of this method, we will look at a simple unit conversion.

Consider a pin measuring 2.85 cm in length. What is its length in inches? To solve this problem, we must use the equivalence statement

$$2.54 \text{ cm} = 1 \text{ in} \qquad \text{(exactly)}$$

If we divide both sides of this equation by 2.54 cm, we get

$$\frac{2.54 \text{ cm}}{2.54 \text{ cm}} = 1 = \frac{1 \text{ in}}{2.54 \text{ cm}}$$

Note that the expression 1 in/2.54 cm equals 1. This expression is called a **unit factor.** Since 1 in and 2.54 cm are exactly equivalent, multiplying any expression by this unit factor will not change its value.

The pin has a length of 2.85 cm. Multiplying this length by the unit factor gives

$$2.85 \not{\text{cm}} \times \frac{1 \text{ in}}{2.54 \not{\text{cm}}} = \frac{2.85}{2.54} \text{ in} = 1.12 \text{ in}$$

Note that the centimeter units cancel to give inches for the result. This is exactly what we wanted to accomplish. Note also that the result has three significant figures, as required by the number 2.85. Recall that the 1 and 2.54 in the conversion factor are exact numbers by definition.

STEPS

Converting from One Unit to Another

1. To convert from one unit to another, use the equivalence statement that relates the two units.
2. Derive the appropriate unit factor by noting the direction of the required change (to cancel the unwanted units).
3. Multiply the quantity to be converted by the unit factor to give the quantity with the desired units.

In dimensional analysis your verification that everything has been done correctly is that the correct units are obtained in the end. *In doing chemistry problems, you should always include the units for the quantities used.* Always check to see that the units cancel to give the correct units for the final result. This provides a very valuable check, especially for complicated problems.

Appendix Three
Spectral Analysis

Although volumetric and gravimetric analyses are still commonly used, spectroscopy is the technique most often used for modern chemical analysis. *Spectroscopy* is the study of electromagnetic radiation emitted or absorbed by a

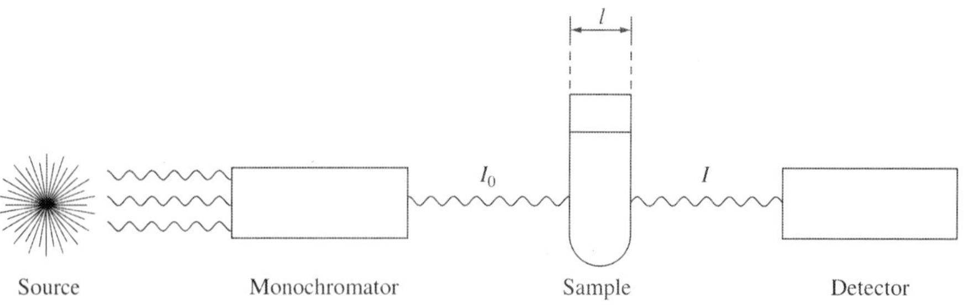

FIGURE A3.1

A schematic diagram of a simple spectrophotometer. The source emits all wavelengths of visible light, which are dispersed by using a prism or grating and then focused, one wavelength at a time, onto the sample. The detector compares the intensity of the incident light (I_0) with the intensity of the light after it has passed through the sample (I).

given chemical species. Since the quantity of radiation absorbed or emitted can be related to the quantity of the absorbing or emitting species present, this technique can be used for quantitative analysis. There are many spectroscopic techniques, since electromagnetic radiation spans a wide range of energies to include microwaves, X rays, and ultraviolet, infrared, and visible light, to name a few of its familiar forms. However, we will consider here only one procedure, which is based on the absorption of visible light.

If a liquid is colored, it is because some component of the liquid absorbs visible light. In a solution the greater the concentration of the light-absorbing substance, the more light is absorbed, and the more intense is the color of the solution.

The quantity of light absorbed by a substance can be measured by a *spectrophotometer,* shown schematically in Fig. A3.1. This instrument consists of a source that emits all wavelengths of light in the visible region (wavelengths of ≈400–700 nm); a monochromator, which selects a given wavelength of light; a sample holder for the solution being measured; and a detector, which compares the intensity of incident light I_0 with the intensity of light after it has passed through the sample I. The ratio I/I_0, called the *transmittance,* is a measure of the fraction of light that passes through the sample. The amount of light absorbed is given by the *absorbance* A, where

$$A = -\log \frac{I}{I_0}$$

The absorbance can be expressed by the *Beer-Lambert law:*

$$A = \epsilon lc$$

where ϵ is the molar absorptivity or the molar extinction coefficient (in L mol^{-1} cm^{-1}), l is the distance the light travels through the solution (in cm), and c is the concentration of the absorbing species (in mol/L). The Beer-Lambert law is the basis for using spectroscopy in quantitative analysis. If ϵ and l are known, determining A for a solution allows us to calculate the concentration of the absorbing species in the solution.

Suppose we have a pink solution containing an unknown concentration of $Co^{2+}(aq)$ ions. A sample of this solution is placed in a spectrophotometer, and the absorbance is measured at a wavelength where ϵ for $Co^{2+}(aq)$ is known to be 12 L mol^{-1} cm^{-1}. The absorbance A is found to be 0.60. The width of the sample tube is 1.0 cm. We want to determine the concentration of $Co^{2+}(aq)$ in the solution. This problem can be solved by a straightforward application of the Beer-Lambert law,

$$A = \epsilon lc$$

where
$$A = 0.60$$
$$\epsilon = \frac{12\ \text{L}}{\text{mol cm}}$$
$$l = \text{light path} = 1.0\ \text{cm}$$

Solving for the concentration gives

$$c = \frac{A}{\epsilon l} = \frac{0.60}{\left(12\ \frac{\text{L}}{\text{mol cm}}\right)(1.0\ \text{cm})} = 5.0 \times 10^{-2}\ \text{mol/L}$$

To obtain the unknown concentration of an absorbing species from the measured absorbance, we must know the product ϵl, since

$$c = \frac{A}{\epsilon l}$$

We can obtain the product ϵl by measuring the absorbance of a solution of *known* concentration, since

Measured using a spectrophotometer

$$\epsilon l = \frac{A}{c}$$

Known from making up the solution

However, a more accurate value of the product ϵl can be obtained by plotting A versus c for a series of solutions. Note that the equation $A = \epsilon lc$ gives a straight line with slope ϵl when A is plotted against c.

For example, consider the following typical spectroscopic analysis. A sample of steel from a bicycle frame is to be analyzed to determine its manganese content. The procedure involves weighing out a sample of the steel, dissolving it in strong acid, treating the resulting solution with a very strong oxidizing agent to convert all the manganese to permanganate ion (MnO_4^-), and then using spectroscopy to determine the concentration of the intensely purple MnO_4^- ions in the solution. To do this, however, the value of ϵl for MnO_4^- must be determined at an appropriate wavelength. The absorbance values for four solutions with known MnO_4^- concentrations were measured to give the following data:

Solution	Concentration of MnO_4^- (mol/L)	Absorbance
1	7.00×10^{-5}	0.175
2	1.00×10^{-4}	0.250
3	2.00×10^{-4}	0.500
4	3.50×10^{-4}	0.875

A plot of absorbance versus concentration for the solutions of known concentration is shown in Fig. A3.2. The slope of this line (change in A/change in c) is 2.48×10^3 L/mol. This quantity represents the product ϵl.

A sample of the steel weighing 0.1523 g was dissolved, and the unknown amount of manganese was converted to MnO_4^- ions. Water was then added to give a solution with a final volume of 100.0 mL. A portion of this solution was placed in a spectrophotometer, and its absorbance was found to be 0.780. We can use these data to calculate the percent manganese in the steel.

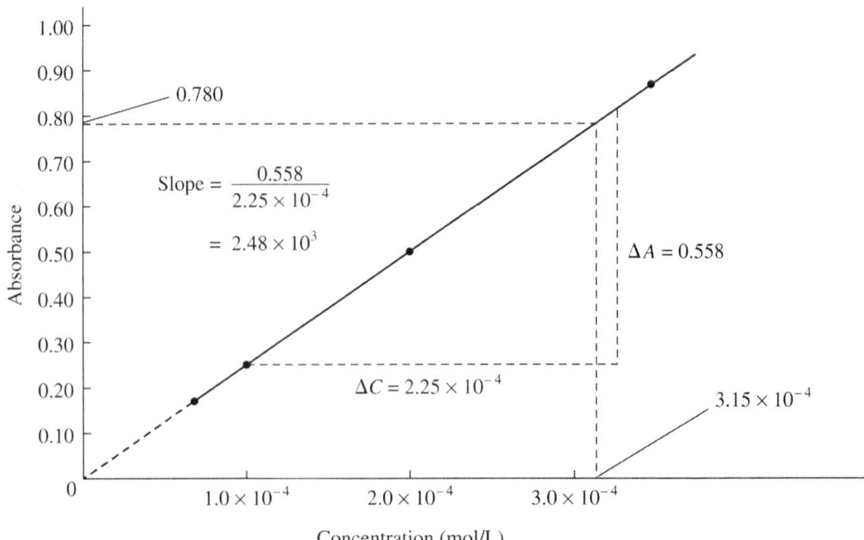

FIGURE A3.2

A plot of absorbance versus concentration of MnO_4^- in a series of solutions of known concentration.

The MnO_4^- ions from the manganese in the dissolved steel sample show an absorbance of 0.780. Using the Beer-Lambert law, we calculate the concentration of MnO_4^- in this solution:

$$c = \frac{A}{\epsilon l} = \frac{0.780}{2.48 \times 10^3 \text{ L/mol}} = 3.15 \times 10^{-4} \text{ mol/L}$$

However, there is a more direct way for finding c. Using a graph such as that in Fig. A3.2 (often called a Beer's law plot), we can read the concentration that corresponds to $A = 0.780$. This interpolation is shown by dashed lines on the graph. By this method, $c = 3.15 \times 10^{-4}$ mol/L, which agrees with the value obtained above.

Recall that the original 0.1523-g steel sample was dissolved, the manganese was converted to permanganate, and the volume was adjusted to 100.0 mL. We now know that the $[MnO_4^-]$ in that solution is 3.15×10^{-4} *M*. Using this concentration, we can calculate the total number of moles of MnO_4^- in that solution:

$$\text{Mol of } MnO_4^- = 100.0 \text{ mL} \times \frac{1 \text{ L}}{1000 \text{ mL}} \times 3.15 \times 10^{-4} \frac{\text{mol}}{\text{L}}$$

$$= 3.15 \times 10^{-5} \text{ mol}$$

Each mole of manganese in the original steel sample yields a mole of MnO_4^-. That is,

$$1 \text{ mol of Mn} \xrightarrow{\text{Oxidation}} 1 \text{ mol of } MnO_4^-$$

so the original steel sample must have contained 3.15×10^{-5} mol of manganese. The mass of manganese present in the sample is

$$3.15 \times 10^{-5} \text{ mol of Mn} \times \frac{54.938 \text{ g of Mn}}{1 \text{ mol of Mn}} = 1.73 \times 10^{-3} \text{ g of Mn}$$

Since the steel sample weighed 0.1523 g, the percent manganese in the steel is

$$\frac{1.73 \times 10^{-3} \text{ g of Mn}}{1.523 \times 10^{-1} \text{ g of sample}} \times 100\% = 1.14\%$$

This example illustrates a typical use of spectroscopy in quantitative analysis. The steps commonly involved are as follows:

1. Preparation of a calibration plot (a Beer's law plot) from the measured absorbance values of a series of solutions with known concentrations.
2. Measurement of the absorbance of the solution of unknown concentration.
3. Use of the calibration plot to determine the unknown concentration.

Appendix Four
Selected Thermodynamic Data*

Substance and State	ΔH_f° (kJ/mol)	ΔG_f° (kJ/mol)	S° (J K^{-1} mol^{-1})
Aluminum			
$Al(s)$	0	0	28
$Al_2O_3(s)$	−1676	−1582	51
$Al(OH)_3(s)$	−1277	—	—
$AlCl_3(s)$	−704	−629	111
Barium			
$Ba(s)$	0	0	67
$BaCO_3(s)$	−1219	−1139	112
$BaO(s)$	−582	−552	70
$Ba(OH)_2(s)$	−946	—	—
$BaSO_4(s)$	−1465	−1353	132
Beryllium			
$Be(s)$	0	0	10
$BeO(s)$	−599	−569	14
$Be(OH)_2(s)$	−904	−815	47
Bromine			
$Br_2(l)$	0	0	152
$Br_2(g)$	31	3	245
$Br_2(aq)$	−3	4	130
$Br^-(aq)$	−121	104	82
$HBr(g)$	−36	−53	199
Cadmium			
$Cd(s)$	0	0	52
$CdO(s)$	−258	−228	55
$Cd(OH)_2(s)$	−561	−474	96
$CdS(s)$	−162	−156	65
$CdSO_4(s)$	−935	−823	123
Calcium			
$Ca(s)$	0	0	41
$CaC_2(s)$	−63	−68	70
$CaCO_3(s)$	−1207	−1129	93
$CaO(s)$	−635	−604	40
$Ca(OH)_2(s)$	−987	−899	83
$Ca_3(PO_4)_2(s)$	−4126	−3890	241
$CaSO_4(s)$	−1433	−1320	107
$CaSiO_3(s)$	−1630	−1550	84
Carbon			
$C(s)$ (graphite)	0	0	6
$C(s)$ (diamond)	2	3	2
$CO(g)$	−110.5	−137	198
$CO_2(g)$	−393.5	−394	214
$CH_4(g)$	−75	−51	186
$CH_3OH(g)$	−201	−163	240
$CH_3OH(l)$	−239	−166	127
$H_2CO(g)$	−116	−110	219
$HCOOH(g)$	−363	−351	249
$HCN(g)$	135.1	125	202
$C_2H_2(g)$	227	209	201
$C_2H_4(g)$	52	68	219
$CH_3CHO(g)$	−166	−129	250
$C_2H_5OH(l)$	−278	−175	161
$C_2H_6(g)$	−84.7	−32.9	229.5
$C_3H_6(g)$	20.9	62.7	266.9
$C_3H_8(g)$	−104	−24	270
$C_2H_4O(g)$ (ethylene oxide)	−53	−13	242
$CH_2{=}CHCN(g)$	185.0	195.4	274
$CH_3COOH(l)$	−484	−389	160
$C_6H_{12}O_6(s)$	−1275	−911	212
$CCl_4(l)$	−135	−65	216
Chlorine			
$Cl_2(g)$	0	0	223
$Cl_2(aq)$	−23	7	121
$Cl^-(aq)$	−167	−131	57
$HCl(g)$	−92	−95	187
Chromium			
$Cr(s)$	0	0	24
$Cr_2O_3(s)$	−1128	−1047	81
$CrO_3(s)$	−579	−502	72
Copper			
$Cu(s)$	0	0	33
$CuCO_3(s)$	−595	−518	88

(continued)

*All values are assumed precise to at least ±1.

Appendix Four (continued)

Substance and State	ΔH_f° (kJ/mol)	ΔG_f° (kJ/mol)	S° (J K^{-1} mol^{-1})
$Cu_2O(s)$	−170	−148	93
$CuO(s)$	−156	−128	43
$Cu(OH)_2(s)$	−450	−372	108
$CuS(s)$	−49	−49	67
Fluorine			
$F_2(g)$	0	0	203
$F^-(aq)$	−333	−279	−14
$HF(g)$	−271	−273	174
Hydrogen			
$H_2(g)$	0	0	131
$H(g)$	217	203	115
$H^+(aq)$	0	0	0
$OH^-(aq)$	−230	−157	−11
$H_2O(l)$	−286	−237	70
$H_2O(g)$	−242	−229	189
Iodine			
$I_2(s)$	0	0	116
$I_2(g)$	62	19	261
$I_2(aq)$	23	16	137
$I^-(aq)$	−55	−52	106
Iron			
$Fe(s)$	0	0	27
$Fe_3C(s)$	21	15	108
$Fe_{0.95}O(s)$ (wustite)	−264	−240	59
$FeO(s)$	−272	−255	61
$Fe_3O_4(s)$ (magnetite)	−1117	−1013	146
$Fe_2O_3(s)$ (hematite)	−826	−740	90
$FeS(s)$	−95	−97	67
$FeS_2(s)$	−178	−166	53
$FeSO_4(s)$	−929	−825	121
Lead			
$Pb(s)$	0	0	65
$PbO_2(s)$	−277	−217	69
$PbS(s)$	−100	−99	91
$PbSO_4(s)$	−920	−813	149
Magnesium			
$Mg(s)$	0	0	33
$MgCO_3(s)$	−1113	−1029	66
$MgO(s)$	−602	−569	27
$Mg(OH)_2(s)$	−925	−834	64
Manganese			
$Mn(s)$	0	0	32
$MnO(s)$	−385	−363	60
$Mn_3O_4(s)$	−1387	−1280	149
$Mn_2O_3(s)$	−971	−893	110
$MnO_2(s)$	−521	−466	53
$MnO_4^-(aq)$	−543	−449	190
Mercury			
$Hg(l)$	0	0	76
$Hg_2Cl_2(s)$	−265	−211	196
$HgCl_2(s)$	−230	−184	144
$HgO(s)$	−90	−59	70
$HgS(s)$	−58	−49	78
Nickel			
$Ni(s)$	0	0	30
$NiCl_2(s)$	−316	−272	107
$NiO(s)$	−241	−213	38
$Ni(OH)_2(s)$	−538	−453	79
$NiS(s)$	−93	−90	53
Nitrogen			
$N_2(g)$	0	0	192
$NH_3(g)$	−46	−17	193
$NH_3(aq)$	−80	−27	111
$NH_4^+(aq)$	−132	−79	113
$NO(g)$	90	87	211
$NO_2(g)$	34	52	240
$N_2O(g)$	82	104	220
$N_2O_4(g)$	10	98	304
$N_2O_4(l)$	−20	97	209
$N_2O_5(s)$	−42	134	178
$N_2H_4(l)$	51	149	121
$N_2H_3CH_3(l)$	54	180	166
$HNO_3(aq)$	−207	−111	146
$HNO_3(l)$	−174	−81	156
$NH_4ClO_4(s)$	−295	−89	186
$NH_4Cl(s)$	−314	−203	96
Oxygen			
$O_2(g)$	0	0	205
$O(g)$	249	232	161
$O_3(g)$	143	163	239
Phosphorus			
$P(s)$ (white)	0	0	41
$P(s)$ (red)	−18	−12	23
$P(s)$ (black)	−39	−33	23
$P_4(g)$	59	24	280
$PF_5(g)$	−1578	−1509	296
$PH_3(g)$	5	13	210
$H_3PO_4(s)$	−1279	−1119	110
$H_3PO_4(l)$	−1267	—	—

Appendix Four (continued)

Substance and State	ΔH_f° (kJ/mol)	ΔG_f° (kJ/mol)	S° (J K^{-1} mol^{-1})
$H_3PO_4(aq)$	−1288	−1143	158
$P_4O_{10}(s)$	−2984	−2698	229
Potassium			
$K(s)$	0	0	64
$KCl(s)$	−436	−408	83
$KClO_3(s)$	−391	−290	143
$KClO_4(s)$	−433	−304	151
$K_2O(s)$	−361	−322	98
$K_2O_2(s)$	−496	−430	113
$KO_2(s)$	−283	−238	117
$KOH(s)$	−425	−379	79
$KOH(aq)$	−481	−440	9.20
Silicon			
$SiO_2(s)$ (quartz)	−911	−856	42
$SiCl_4(l)$	−687	−620	240
Silver			
$Ag(s)$	0	0	43
$Ag^+(aq)$	105	77	73
$AgBr(s)$	−100	−97	107
$AgCN(s)$	146	164	84
$AgCl(s)$	−127	−110	96
$Ag_2CrO_4(s)$	−712	−622	217
$AgI(s)$	−62	−66	115
$Ag_2O(s)$	−31	−11	122
$Ag_2S(s)$	−32	−40	146
Sodium			
$Na(s)$	0	0	51
$Na^+(aq)$	−240	−262	59
$NaBr(s)$	−360	−347	84
$Na_2CO_3(s)$	−1131	−1048	136
$NaHCO_3(s)$	−948	−852	102
$NaCl(s)$	−411	−384	72
$NaH(s)$	−56	−33	40
$NaI(s)$	−288	−282	91
$NaNO_2(s)$	−359	—	—
$NaNO_3(s)$	−467	−366	116
$Na_2O(s)$	−416	−377	73
$Na_2O_2(s)$	−515	−451	95
$NaOH(s)$	−427	−381	64
$NaOH(aq)$	−470	−419	50
Sulfur			
$S(s)$ (rhombic)	0	0	32
$S(s)$ (monoclinic)	0.3	0.1	33
$S^{2-}(aq)$	33	86	−15
$S_8(g)$	102	50	431
$SF_6(g)$	−1209	−1105	292
$H_2S(g)$	−21	−34	206
$SO_2(g)$	−297	−300	248
$SO_3(g)$	−396	−371	257
$SO_4^{2-}(aq)$	−909	−745	20
$H_2SO_4(l)$	−814	−690	157
$H_2SO_4(aq)$	−909	−745	20
Tin			
$Sn(s)$ (white)	0	0	52
$Sn(s)$ (gray)	−2	0.1	44
$SnO(s)$	−285	−257	56
$SnO_2(s)$	−581	−520	52
$Sn(OH)_2(s)$	−561	−492	155
Titanium			
$TiCl_4(g)$	−763	−727	355
$TiO_2(s)$	−945	−890	50
Uranium			
$U(s)$	0	0	50
$UF_6(s)$	−2137	−2008	228
$UF_6(g)$	−2113	−2029	380
$UO_2(s)$	−1084	−1029	78
$U_3O_8(s)$	−3575	−3393	282
$UO_3(s)$	−1230	−1150	99
Xenon			
$Xe(g)$	0	0	170
$XeF_2(g)$	−108	−48	254
$XeF_4(s)$	−251	−121	146
$XeF_6(g)$	−294	—	—
$XeO_3(s)$	402		
Zinc			
$Zn(s)$	0	0	42
$ZnO(s)$	−348	−318	44
$Zn(OH)_2(s)$	−642	—	—
$ZnS(s)$ (wurtzite)	−193	—	—
$ZnS(s)$ (zinc blende)	−206	−201	58
$ZnSO_4(s)$	−983	−874	120

Appendix Five
Equilibrium Constants and Reduction Potentials

TABLE A5.1

Values of K_a for Some Common Monoprotic Acids

Name	Formula	Value of K_a
Hydrogen sulfate ion	HSO_4^-	1.2×10^{-2}
Chlorous acid	$HClO_2$	1.2×10^{-2}
Monochloracetic acid	$HC_2H_2ClO_2$	1.35×10^{-3}
Hydrofluoric acid	HF	7.2×10^{-4}
Nitrous acid	HNO_2	4.0×10^{-4}
Formic acid	HCO_2H	1.8×10^{-4}
Lactic acid	$HC_3H_5O_3$	1.38×10^{-4}
Benzoic acid	$HC_7H_5O_2$	6.4×10^{-5}
Acetic acid	$HC_2H_3O_2$	1.8×10^{-5}
Hydrated aluminum(III) ion	$[Al(H_2O)_6]^{3+}$	1.4×10^{-5}
Propanoic acid	$HC_3H_5O_2$	1.3×10^{-5}
Hypochlorous acid	$HOCl$	3.5×10^{-8}
Hypobromous acid	$HOBr$	2×10^{-9}
Hydrocyanic acid	HCN	6.2×10^{-10}
Boric acid	H_3BO_3	5.8×10^{-10}
Ammonium ion	NH_4^+	5.6×10^{-10}
Phenol	HOC_6H_5	1.6×10^{-10}
Hypoiodous acid	HOI	2×10^{-11}

TABLE A5.2

Stepwise Dissociation Constants for Several Common Polyprotic Acids

Name	Formula	K_{a_1}	K_{a_2}	K_{a_3}
Phosphoric acid	H_3PO_4	7.5×10^{-3}	6.2×10^{-8}	4.8×10^{-13}
Arsenic acid	H_3AsO_4	5×10^{-3}	8×10^{-8}	6×10^{-10}
Carbonic acid	H_2CO_3	4.3×10^{-7}	4.8×10^{-11}	
Sulfuric acid	H_2SO_4	Large	1.2×10^{-2}	
Sulfurous acid	H_2SO_3	1.5×10^{-2}	1.0×10^{-7}	
Hydrosulfuric acid	H_2S	1.0×10^{-7}	$\sim 10^{-19}$	
Oxalic acid	$H_2C_2O_4$	6.5×10^{-2}	6.1×10^{-5}	
Ascorbic acid (vitamin C)	$H_2C_6H_6O_6$	7.9×10^{-5}	1.6×10^{-12}	
Citric acid	$H_3C_6H_5O_7$	8.4×10^{-4}	1.8×10^{-5}	4.0×10^{-6}

TABLE A5.3

Values of K_b for Some Common Weak Bases

Name	Formula	Conjugate Acid	K_b
Ammonia	NH_3	NH_4^+	1.8×10^{-5}
Methylamine	CH_3NH_2	$CH_3NH_3^+$	4.38×10^{-4}
Ethylamine	$C_2H_5NH_2$	$C_2H_5NH_3^+$	5.6×10^{-4}
Diethylamine	$(C_2H_5)_2NH$	$(C_2H_5)_2NH_2^+$	1.3×10^{-3}
Triethylamine	$(C_2H_5)_3N$	$(C_2H_5)_3NH^+$	4.0×10^{-4}
Hydroxylamine	$HONH_2$	$HONH_3^+$	1.1×10^{-8}
Hydrazine	H_2NNH_2	$H_2NNH_3^+$	3.0×10^{-6}
Aniline	$C_6H_5NH_2$	$C_6H_5NH_3^+$	3.8×10^{-10}
Pyridine	C_5H_5N	$C_5H_5NH^+$	1.7×10^{-9}

TABLE A5.4

Values of K_{sp} at 25°C for Common Ionic Solids

Ionic Solid	K_{sp} (at 25°C)
Fluorides	
BaF_2	2.4×10^{-5}
MgF_2	6.4×10^{-9}
PbF_2	4×10^{-8}
SrF_2	7.9×10^{-10}
CaF_2	4.0×10^{-11}
Chlorides	
$PbCl_2$	1.6×10^{-5}
$AgCl$	1.6×10^{-10}
Hg_2Cl_2*	1.1×10^{-18}
Bromides	
$PbBr_2$	4.6×10^{-6}
$AgBr$	5.0×10^{-13}
Hg_2Br_2*	1.3×10^{-22}
Iodides	
PbI_2	1.4×10^{-8}
AgI	1.5×10^{-16}
Hg_2I_2*	4.5×10^{-29}
Sulfates	
$CaSO_4$	6.1×10^{-5}
Ag_2SO_4	1.2×10^{-5}
$SrSO_4$	3.2×10^{-7}
$PbSO_4$	1.3×10^{-8}
$BaSO_4$	1.5×10^{-9}
Chromates	
$SrCrO_4$	3.6×10^{-5}
Chromates (*continued*)	
Hg_2CrO_4*	2×10^{-9}
$BaCrO_4$	8.5×10^{-11}
Ag_2CrO_4	9.0×10^{-12}
$PbCrO_4$	2×10^{-16}
Carbonates	
$NiCO_3$	1.4×10^{-7}
$CaCO_3$	8.7×10^{-9}
$BaCO_3$	1.6×10^{-9}
$SrCO_3$	7×10^{-10}
$CuCO_3$	2.5×10^{-10}
$ZnCO_3$	2×10^{-10}
$MnCO_3$	8.8×10^{-11}
$FeCO_3$	2.1×10^{-11}
Ag_2CO_3	8.1×10^{-12}
$CdCO_3$	5.2×10^{-12}
$PbCO_3$	1.5×10^{-15}
$MgCO_3$	1×10^{-15}
Hg_2CO_3*	9.0×10^{-15}
Hydroxides	
$Ba(OH)_2$	5.0×10^{-3}
$Sr(OH)_2$	3.2×10^{-4}
$Ca(OH)_2$	1.3×10^{-6}
$AgOH$	2.0×10^{-8}
$Mg(OH)_2$	8.9×10^{-12}
$Mn(OH)_2$	2×10^{-13}
$Cd(OH)_2$	5.9×10^{-15}
$Pb(OH)_2$	1.2×10^{-15}
$Fe(OH)_2$	1.8×10^{-15}
Hydroxides (*continued*)	
$Co(OH)_3$	2.5×10^{-16}
$Ni(OH)_2$	1.6×10^{-16}
$Zn(OH)_2$	4.5×10^{-17}
$Cu(OH)_2$	1.6×10^{-19}
$Hg(OH)_2$	3×10^{-26}
$Sn(OH)_2$	3×10^{-27}
$Cr(OH)_3$	6.7×10^{-31}
$Al(OH)_3$	2×10^{-32}
$Fe(OH)_3$	4×10^{-38}
$Co(OH)_3$	2.5×10^{-43}
Sulfides	
MnS	2.3×10^{-13}
FeS	3.7×10^{-19}
NiS	3×10^{-21}
CoS	5×10^{-22}
ZnS	2.5×10^{-22}
SnS	1×10^{-26}
CdS	1.0×10^{-28}
PbS	7×10^{-29}
CuS	8.5×10^{-45}
Ag_2S	1.6×10^{-49}
HgS	1.6×10^{-54}
Phosphates	
Ag_3PO_4	1.8×10^{-18}
$Sr_3(PO_4)_2$	1×10^{-31}
$Ca_3(PO_4)_2$	1.3×10^{-32}
$Ba_3(PO_4)_2$	6×10^{-39}
$Pb_3(PO_4)_2$	1×10^{-54}

*Contains Hg_2^{2+} ions. $K_{sp} = [Hg_2^{2+}][X^-]^2$ for Hg_2X_2 salts.

TABLE A5.5

Standard Reduction Potentials at 25°C (298 K) for Many Common Half-Reactions

Half-Reaction	ℰ° (V)	Half-Reaction	ℰ° (V)
$F_2 + 2e^- \longrightarrow 2F^-$	2.87	$O_2 + 2H_2O + 4e^- \longrightarrow 4OH^-$	0.40
$Ag^{2+} + e^- \longrightarrow Ag^+$	1.99	$Cu^{2+} + 2e^- \longrightarrow Cu$	0.34
$Co^{3+} + e^- \longrightarrow Co^{2+}$	1.82	$Hg_2Cl_2 + 2e^- \longrightarrow 2Hg + 2Cl^-$	0.27
$H_2O_2 + 2H^+ + 2e^- \longrightarrow 2H_2O$	1.78	$AgCl + e^- \longrightarrow Ag + Cl^-$	0.22
$Ce^{4+} + e^- \longrightarrow Ce^{3+}$	1.70	$SO_4^{2-} + 4H^+ + 2e^- \longrightarrow H_2SO_3 + H_2O$	0.20
$PbO_2 + 4H^+ + SO_4^{2-} + 2e^- \longrightarrow PbSO_4 + 2H_2O$	1.69	$Cu^{2+} + e^- \longrightarrow Cu^+$	0.16
$MnO_4^+ + 4H^+ + 3e^- \longrightarrow MnO_2 + 2H_2O$	1.68	$2H^+ + 2e^- \longrightarrow H_2$	0.00
$IO_4^- + 2H^+ + 2e^- \longrightarrow IO_3^- + H_2O$	1.60	$Fe^{3+} + 3e^- \longrightarrow Fe$	−0.036
$MnO_4^+ + 8H^+ + 5e^- \longrightarrow Mn^{2+} + 4H_2O$	1.51	$Pb^{2+} + 2e^- \longrightarrow Pb$	−0.13
$Au^{3+} + 3e^- \longrightarrow Au$	1.50	$Sn^{2+} + 2e^- \longrightarrow Sn$	−0.14
$PbO_2 + 4H^+ + 2e^- \longrightarrow Pb^{2+} + 2H_2O$	1.46	$Ni^{2+} + 2e^- \longrightarrow Ni$	−0.23
$Cl_2 + 2e^- \longrightarrow 2Cl^-$	1.36	$PbSO_4 + 2e^- \longrightarrow Pb + SO_4^{2-}$	−0.35
$Cr_2O_7^{2-} + 14H^+ + 6e^- \longrightarrow 2Cr^{3+} + 7H_2O$	1.33	$Cd^{2+} + 2e^- \longrightarrow Cd$	−0.40
$O_2 + 4H^+ + 4e^- \longrightarrow 2H_2O$	1.23	$Fe^{2+} + 2e^- \longrightarrow Fe$	−0.44
$MnO_2 + 4H^+ + 2e^- \longrightarrow Mn^{2+} + 2H_2O$	1.21	$Cr^{3+} + e^- \longrightarrow Cr^{2+}$	−0.50
$IO_3^- + 6H^+ + 5e^- \longrightarrow \frac{1}{2}I_2 + 3H_2O$	1.20	$Cr^{3+} + 3e^- \longrightarrow Cr$	−0.73
$Br_2 + 2e^- \longrightarrow 2Br^-$	1.09	$Zn^{2+} + 2e^- \longrightarrow Zn$	−0.76
$VO_2^+ + 2H^+ + e^- \longrightarrow VO^{2+} + H_2O$	1.00	$2H_2O + 2e^- \longrightarrow H_2 + 2OH^-$	−0.83
$AuCl_4^- + 3e^- \longrightarrow Au + 4Cl^-$	0.99	$Mn^{2+} + 2e^- \longrightarrow Mn$	−1.18
$NO_3^- + 4H^+ + 3e^- \longrightarrow NO + 2H_2O$	0.96	$Al^{3+} + 3e^- \longrightarrow Al$	−1.66
$ClO_2 + e^- \longrightarrow ClO_2^-$	0.954	$H_2 + 2e^- \longrightarrow 2H^-$	−2.23
$2Hg^{2+} + 2e^- \longrightarrow Hg_2^{2+}$	0.91	$Mg^{2+} + 2e^- \longrightarrow Mg$	−2.37
$Ag^+ + e^- \longrightarrow Ag$	0.80	$La^{3+} + 3e^- \longrightarrow La$	−2.37
$Hg_2^{2+} + 2e^- \longrightarrow 2Hg$	0.80	$Na^+ + e^- \longrightarrow Na$	−2.71
$Fe^{3+} + e^- \longrightarrow Fe^{2+}$	0.77	$Ca^{2+} + 2e^- \longrightarrow Ca$	−2.76
$O_2 + 2H^+ + 2e^- \longrightarrow H_2O_2$	0.68	$Ba^{2+} + 2e^- \longrightarrow Ba$	−2.90
$MnO_4^- + e^- \longrightarrow MnO_4^{2-}$	0.56	$K^+ + e^- \longrightarrow K$	−2.92
$I_2 + 2e^- \longrightarrow 2I^-$	0.54	$Li^+ + e^- \longrightarrow Li$	−3.05
$Cu^+ + e^- \longrightarrow Cu$	0.52		

Glossary

Note to the Student: The Glossary includes brief definitions of some of the fundamental terms used in chemistry. It does not include complex concepts that require detailed explanation for understanding. Please refer to the appropriate sections of the text for complete discussion of particular topics or concepts.

Accuracy: the agreement of a particular value with the true value. (A1.5)

Acid: a substance that produces hydrogen ions in solution; a proton donor. (4.2)

Acid–base indicator: a substance that marks the endpoint of an acid–base titration by changing color. (8.6)

Acid dissociation constant (K_a): the equilibrium constant for a reaction in which a proton is removed from an acid by H_2O to form the conjugate base and H_3O^+. (7.1)

Acid rain: a result of air pollution by sulfur dioxide. (5.11)

Actinide series: a group of 14 elements following actinium in the periodic table, in which the 5*f* orbitals are being filled. (12.13; 18.1)

Activated complex (transition state): the arrangement of atoms found at the top of the potential energy barrier as a reaction proceeds from reactants to products. (15.8)

Activation energy: the threshold energy that must be overcome to produce a chemical reaction. (15.8)

Addition polymerization: a type of polymerization in which the monomers simply add together to form the polymer, with no other products. (21.5)

Addition reaction: a reaction in which atoms add to a carbon–carbon multiple bond. (21.2)

Adiabatic process: a process that occurs without the transfer of energy as heat. (10.14)

Adsorption: the collection of one substance on the surface of another. (15.9)

Air pollution: contamination of the atmosphere, mainly by the gaseous products of transportation and production of electricity. (5.11)

Alcohol: an organic compound in which the hydroxyl group is a substituent on a hydrocarbon. (21.4)

Aldehyde: an organic compound containing the carbonyl group bonded to at least one hydrogen atom. (21.4)

Alkali metal: a Group 1A metal. (2.8; 18.2)

Alkaline earth metal: a Group 2A metal. (2.8; 18.4)

Alkane: a saturated hydrocarbon with the general formula C_nH_{2n+2}. (21.1)

Alkene: an unsaturated hydrocarbon containing a carbon–carbon double bond. The general formula is C_nH_{2n}. (21.2)

Alkyne: an unsaturated hydrocarbon containing a triple carbon–carbon bond. The general formula is C_nH_{2n-2}. (21.2)

Alloy: a substance that contains a mixture of elements and has metallic properties. (16.4)

Alloy steel: a form of steel containing carbon plus other metals such as chromium, cobalt, manganese, and molybdenum. (19.2)

Alpha (α) particle: a helium nucleus. (20.1)

Alpha-particle production: a common mode of decay for radioactive nuclides in which the mass number changes. (20.1)

Amine: an organic base derived from ammonia in which one or more of the hydrogen atoms are replaced by organic groups. (7.6; 21.4)

α-Amino acid: an organic acid in which an amino group and an R group are attached to the carbon atom next to the carboxyl group. (21.6)

Amorphous solid: a solid with considerable disorder in its structure. (16.3)

Ampere: the unit of electric current equal to one coulomb of charge per second. (11.7)

Amphoteric substance: a substance that can behave either as an acid or as a base. (7.2)

Angular momentum quantum number (ℓ): the quantum number relating to the shape of an atomic orbital, which can assume any integral value from 0 to $n - 1$ for each value of n. (12.9)

Anion: a negative ion. (2.7)

Anode: the electrode in a galvanic cell at which oxidation occurs. (11.1)

Antibonding molecular orbital: an orbital higher in energy than the atomic orbitals of which it is composed. (14.2)

Aqueous solution: a solution in which water is the dissolving medium or solvent. (4)

Aromatic hydrocarbon: one of a special class of cyclic unsaturated hydrocarbons, the simplest of which is benzene. (21.3)

Arrhenius concept: a concept postulating that acids produce hydrogen ions in aqueous solution, whereas bases produce hydroxide ions. (7.1)

Arrhenius equation: the equation representing the rate constant as $k = Ae^{-E_a/RT}$ where A represents the product of the collision frequency and the steric factor, and $e^{-E_a/RT}$ is the fraction of collisions with sufficient energy to produce a reaction. (15.8)

Atmosphere: the mixture of gases that surrounds the earth's surface. (5.11)

Atomic mass (average): the weighted average mass of the atoms in a naturally occurring element. (2.3)

Atomic number: the number of protons in the nucleus of an atom. (2.6)

Atomic radius: half the distance between the nuclei in a molecule consisting of identical atoms. (12.15)

Atomic solid: a solid that contains atoms at the lattice points. (16.3)

Aufbau principle: the principle stating that as protons are added one by one to the nucleus to build up the elements, electrons are similarly added to hydrogenlike orbitals. (12.13)

Autoionization: the transfer of a proton from one molecule to another of the same substance. (7.2)

Avogadro's law: equal volumes of gases at the same temperature and pressure contain the same number of particles. (5.2)

Avogadro's number: the number of atoms in exactly 12 grams of pure ^{12}C, equal to 6.022×10^{23}. (3.2)

Ball-and-stick model: a molecular model that distorts the sizes of atoms, but shows bond relationships clearly. (2.7)

Band model: a molecular model for metals in which the electrons are assumed to travel around the metal crystal in molecular orbitals formed from the valence atomic orbitals of the metal atoms. (16.4)

Barometer: a device for measuring atmospheric pressure. (5.1)

Base: a substance that produces hydroxide ions in aqueous solution, a proton acceptor. (7.2)

Base dissociation constant (K_b): the equilibrium constant for the reaction of a base with water to produce the conjugate acid and hydroxide ion. (7.6)

Basic oxide: an ionic oxide that dissolves in water to produce a basic solution. (18.4)

Battery: a group of galvanic cells connected in series. (11.5)

Beta (β) particle: an electron produced in radioactive decay. (20.1)

Beta-particle production: a decay process for radioactive nuclides in which the mass number remains constant and the atomic number changes. The net effect is to change a neutron to a proton. (20.1)

Bidentate ligand: a ligand that can form two bonds to a metal ion. (19.3)

Bimolecular step: a reaction involving the collision of two molecules. (15.6)

Binary compound: a two-element compound. (2.9)

Binding energy (nuclear): the energy required to decompose a nucleus into its component nucleons. (20.5)

Biomolecule: a molecule responsible for maintaining and/or reproducing life. (22)

Bond energy: the energy required to break a given chemical bond. (13.1)

Bond length: the distance between the nuclei of the two atoms connected by a bond; the distance where the total energy of a diatomic molecule is minimal. (13.1)

Bond order: the difference between the number of bonding electrons and the number of antibonding electrons, divided by two. It is an index of bond strength. (14.2)

Bonding molecular orbital: an orbital lower in energy than the atomic orbitals of which it is composed. (14.2)

Bonding pair: an electron pair found in the space between two atoms. (13.9)

Borane: a covalent hydride of boron. (18.5)

Boyle's law: the volume of a given sample of gas at constant temperature varies inversely with the pressure. (5.2)

Breeder reactor: a nuclear reactor in which fissionable fuel is produced while the reactor runs. (20.6)

Brønsted–Lowry definition (model): a model proposing that an acid is a proton donor, and a base is a proton acceptor. (7.1)

Buffer capacity: the ability of a buffered solution to absorb protons or hydroxide ions without a significant change in pH; determined by the magnitudes of [HA] and $[A^-]$ in the solution. (8.4)

Buffered solution: a solution that resists a change in its pH when either hydroxide ions or protons are added. (8.2)

Calorimetry: the science of measuring heat flow. (9.4)

Capillary action: the spontaneous rising of a liquid in a narrow tube. (16.2)

Carbohydrate: a polyhydroxyl ketone or polyhydroxyl aldehyde or a polymer composed of these. (21.6)

Carboxyhemoglobin: a stable complex of hemoglobin and carbon monoxide that prevents normal oxygen uptake in the blood. (19.8)

Carboxyl group: the —COOH group in an organic acid. (7.2; 21.4)

Carboxylic acid: an organic compound containing the carboxyl group; an acid with the general formula RCOOH. (21.4)

Catalyst: a substance that speeds up a reaction without being consumed. (15.9)

Cathode: the electrode in a galvanic cell at which reduction occurs. (11.1)

Cathode rays: the "rays" emanating from the negative electrode (cathode) in a partially evacuated tube; a stream of electrons. (2.5)

Cathodic protection: a method in which an active metal, such as magnesium, is connected to steel to protect it from corrosion. (11.6)

Cation: a positive ion. (2.7)

Cell potential (electromotive force): the driving force in a galvanic cell that pulls electrons from the reducing agent in one compartment to the oxidizing agent in the other. (11.1)

Ceramic: a nonmetallic material made from clay and hardened by firing at high temperature; it contains minute silicate crystals suspended in a glassy cement. (16.5)

Chain reaction (nuclear): a self-sustaining fission process caused by the production of neutrons that proceed to split other nuclei. (20.6)

Charge balance: the positive and negative charges carried by the ions in an aqueous solution must balance. (7.9)

Charles's law: the volume of a given sample of gas at constant pressure is directly proportional to the temperature in kelvins. (5.2)

Chelating ligand (chelate): a ligand having more than one atom with a lone pair that can be used to bond to a metal ion. (19.3)

Chemical bond: the energy that holds two atoms together in a compound. (2.7)

Chemical equation: a representation of a chemical reaction showing the relative numbers of reactant and product molecules. (3.6)

Chemical equilibrium: a dynamic reaction system in which the concentrations of all reactants and products remain constant as a function of time. (6)

Chemical formula: the representation of a molecule in which the symbols for the elements are used to indicate the types of atoms present and subscripts are used to show the relative numbers of atoms. (2.7)

Chemical kinetics: the area of chemistry that concerns reaction rates. (15)

Chemical stoichiometry: the calculation of the quantities of material consumed and produced in chemical reactions. (3)

Chirality: the quality of having nonsuperimposable mirror images. (19.4)

Chlor-alkali process: the process for producing chlorine and sodium hydroxide by electrolyzing brine in a mercury cell. (11.8)

Coagulation: the destruction of a colloid by causing particles to aggregate and settle out. (17.8)

Codons: organic bases in sets of three that form the genetic code. (21.6)

Colligative properties: properties of a solution that depend on the number, and not on the identity, of the solute particles. (17.5)

Collision model: a model based on the idea that molecules must collide to react; used to account for the observed characteristics of reaction rates. (15.8)

Colloid: a suspension of particles in a dispersing medium. (17.8)

Combustion reaction: the vigorous and exothermic reaction that takes place between certain substances, particularly organic compounds, and oxygen. (21.1)

Common ion effect: the shift in an equilibrium position caused by the addition or presence of an ion involved in the equilibrium reaction. (8.1)

Complete ionic equation: an equation that shows all substances that are strong electrolytes as ions. (4.6)

Complex ion: a charged species consisting of a metal ion surrounded by ligands. (8.9; 19.1)

Compound: a substance with constant composition that can be broken down into elements by chemical processes. (2.7)

Concentration cell: a galvanic cell in which both compartments contain the same components, but at different concentrations. (11.4)

Condensation: the process by which vapor molecules re-form a liquid. (16.10)

Condensation polymerization: a type of polymerization in which the formation of a small molecule, such as water, accompanies the extension of the polymer chain. (21.5)

Condensed states of matter: liquids and solids. (16.1)

Conduction bands: the molecular orbitals that can be occupied by mobile electrons, which are free to travel throughout a metal crystal to conduct electricity or heat. (16.4)

Conjugate acid: the species formed when a proton is added to a base. (7.1)

Conjugate acid–base pair: two species related to each other by the donating and accepting of a single proton. (7.1)

Conjugate base: what remains of an acid molecule after a proton is lost. (7.1)

Continuous spectrum: a spectrum that exhibits all the wavelengths of visible light. (12.3)

Control rods: rods in a nuclear reactor composed of substances that absorb neutrons. These rods regulate the power level of the reactor. (20.6)

Coordinate covalent bond: a metal–ligand bond resulting from the interaction of a Lewis base (the ligand) and a Lewis acid (the metal ion). (19.3)

Coordination compound: a compound composed of a complex ion and counter ions sufficient to give no net charge. (19.3)

Coordination isomerism: isomerism in a coordination compound in which the composition of the coordination sphere of the metal ion varies. (19.4)

Coordination number: the number of bonds formed between the metal ion and the ligands in a complex ion. (19.3)

Copolymer: a polymer formed from the polymerization of more than one type of monomer. (21.5)

Core electron: an inner electron in an atom; one not in the outermost (valence) principal quantum level. (12.13)

Corrosion: the process by which metals are oxidized in the atmosphere. (11.6)

Coulomb's law: $E = 2.31 \times 10^{-19} (Q_1Q_2/r)$, where E is the energy of interaction between a pair of ions, expressed in joules; r is the distance between the ion centers in nm; and Q_1 and Q_2 are the numerical ion charges. (13.1)

Counter ions: anions or cations that balance the charge on the complex ion in a coordination compound. (19.3)

Covalent bonding: a type of bonding in which electrons are shared by atoms. (2.7; 13.1)

Critical mass: the mass of fissionable material required to produce a self-sustaining chain reaction. (20.6)

Critical point: the point on a phase diagram at which the temperature and pressure have their critical values; the endpoint of the liquid–vapor line. (16.11)

Critical pressure: the minimum pressure required to produce liquefaction of a substance at the critical temperature. (16.11)

Critical reaction (nuclear): a reaction in which exactly one neutron from each fission event causes another fission event, thus sustaining the chain reaction. (20.6)

Critical temperature: the temperature above which vapor cannot be liquefied, no matter what pressure is applied. (16.11)

Crosslinking: the existence of bonds between adjacent chains in a polymer, thus adding strength to the material. (21.5)

Crystal field model: a model used to explain the magnetism and colors of coordination complexes through the splitting of the *d* orbital energies. (19.6)

Crystalline solid: a solid with a regular arrangement of its components. (16.3)

Cubic closest packed (ccp) structure: a solid modeled by the closest packing of spheres with an *abcabc* arrangement of layers; the unit cell is face-centered cubic. (16.4)

Cyclotron: a type of particle accelerator in which an ion introduced at the center is accelerated in an expanding spiral path by use of alternating electric fields in the presence of a magnetic field. (20.3)

Cytochromes: a series of iron-containing species composed of heme and a protein. Cytochromes are the principal electron-transfer molecules in the respiratory chain. (19.8)

Dalton's law of partial pressures: for a mixture of gases in a container, the total pressure exerted is the sum of the pressures that each gas would exert if it were alone. (5.5)

Degenerate orbitals: a group of orbitals with the same energy. (12.9)

Dehydrogenation reaction: a reaction in which two hydrogen atoms are removed from adjacent carbons of a saturated hydrocarbon, giving an unsaturated hydrocarbon. (21.1)

Delocalization: the condition where the electrons in a molecule are not localized between a pair of atoms but can move throughout the molecule. (13.9)

Denaturation: the breaking down of the three-dimensional structure of a protein resulting in the loss of its function. (21.6)

Denitrification: the return of nitrogen from decomposed matter to the atmosphere by bacteria that change nitrates to nitrogen gas. (18.8)

Deoxyribonucleic acid (DNA): a huge nucleotide polymer having a double-helical structure with complementary bases on the two strands. Its major functions are protein synthesis and the storage and transport of genetic information. (21.6)

Desalination: the removal of dissolved salts from an aqueous solution. (17.6)

Dialysis: a phenomenon in which a semipermeable membrane allows transfer of both solvent molecules and small solute molecules and ions. (17.6)

Diamagnetism: a type of magnetism, associated with paired electrons, that causes a substance to be repelled from the inducing magnetic field. (14.3)

Differential rate law: an expression that gives the rate of a reaction as a function of concentrations; often called the rate law. (15.2)

Diffraction: the scattering of light from a regular array of points or lines, producing constructive and destructive interference. (12.2)

Diffusion: the mixture of gases. (5.7)

Dilution: the process of adding solvent to lower the concentration of solute in a solution. (4.3)

Dimer: a molecule formed by the joining of two identical monomers. (21.5)

Dipole–dipole attraction: the attractive force resulting when polar molecules line up so that the positive and negative ends are close to each other. (16.1)

Dipole moment: a property of a molecule whose charge distribution can be represented by a center of positive charge and a center of negative charge. (13.3)

Disaccharide: a sugar formed from two monosaccharides joined by a glycoside linkage. (21.6)

Disproportionation reaction: a reaction in which a given element is both oxidized and reduced. (18.13)

Disulfide linkage: a S—S bond that stabilizes the tertiary structure of many proteins. (21.6)

Double bond: a bond in which two pairs of electrons are shared by two atoms. (13.8)

Downs cell: a cell used for electrolyzing molten sodium chloride. (11.8)

Dry cell battery: a common battery used in calculators, watches, radios, and portable audio players. (11.5)

Dual nature of light: the statement that light exhibits both wave and particulate properties. (12.2)

$E = mc^2$: Einstein's equation proposing that energy has mass; E is energy, m is mass, and c is the speed of light. (12.2)

Effective nuclear charge: the apparent nuclear charge exerted on a particular electron, equal to the actual nuclear charge minus the effect of electron repulsions. (12.11)

Effusion: the passage of a gas through a tiny orifice into an evacuated chamber. (5.7)

Electrical conductivity: the ability to conduct an electric current. (4.2)

Electrochemistry: the study of the interchange of chemical and electrical energy. (11)

Electrolysis: a process that involves forcing a current through a cell to cause a nonspontaneous chemical reaction to occur. (11.7)

Electrolyte: a material that dissolves in water to give a solution that conducts an electric current. (4.2)

Electrolytic cell: a cell that uses electrical energy to produce a chemical change that would otherwise not occur spontaneously. (11.7)

Electromagnetic radiation: radiant energy that exhibits wavelike behavior and travels through space at the speed of light in a vacuum. (12.1)

Electron: a negatively charged particle that moves around the nucleus of an atom. (2.5)

Electron affinity: the energy change associated with the addition of an electron to a gaseous atom. (12.15)

Electron capture: a process in which one of the inner-orbital electrons in an atom is captured by the nucleus. (20.1)

Electron sea model: a model for metals postulating a regular array of cations in a "sea" of electrons. (16.4)

Electron spin quantum number: a quantum number representing one of the two possible values for the electron spin; either $+\frac{1}{2}$ or $-\frac{1}{2}$. (12.10)

Electronegativity: the tendency of an atom in a molecule to attract shared electrons to itself. (13.2)

Element: a substance that cannot be decomposed into simpler substances by chemical or physical means. (2.1)

Elementary step: a reaction whose rate law can be written from its molecularity. (15.6)

Empirical formula: the simplest whole number ratio of atoms in a compound. (3.5)

Enantiomers: isomers that are nonsuperimposable mirror images of each other. (19.4)

Endpoint: the point in a titration at which the indicator changes color. (4.9)

Endothermic: refers to a reaction where energy (as heat) flows into the system. (9.1)

Energy: the capacity to do work or to cause heat flow. (9.1)

Enthalpy: a property of a system equal to $E + PV$, where E is the internal energy of the system, P is the pressure of the system, and V is the volume of the system. At constant pressure, where only PV work is allowed, the change in enthalpy equals the energy flow as heat. (9.2)

Enthalpy of fusion: the enthalpy change that occurs to melt a solid at its melting point. (16.10)

Entropy: a thermodynamic function that measures randomness or disorder. (10.1)

Enzyme: a large molecule, usually a protein, that catalyzes biological reactions. (15.9)

Equilibrium (thermodynamic definition): the position where the free energy of a reaction system has its lowest possible value. (10.11)

Equilibrium constant: the value obtained when equilibrium concentrations of the chemical species are substituted in the equilibrium expression. (6.2)

Equilibrium expression: the expression (from the law of mass action) obtained by multiplying the product concentrations and dividing by the multiplied reactant concentrations, with each concentration raised to a power represented by the coefficient in the balanced equation. (6.2)

Equilibrium position: a particular set of equilibrium concentrations. (6.2)

Equivalence point (stoichiometric point): the point in a titration when enough titrant has been added to react exactly with the substance in solution being titrated. (4.9; 8.4)

Exothermic: refers to a reaction where energy (as heat) flows out of the system. (9.1)

Exponential notation: expresses a number as $N \times 10^M$, a convenient method for representing a very large or very small number and for easily indicating the number of significant figures. (A1.1)

Faraday: a constant representing the charge on one mole of electrons; 96,485 coulombs. (11.3)

First law of thermodynamics: the energy of the universe is constant; same as the law of conservation of energy. (9.1)

Fission: the process of using a neutron to split a heavy nucleus into two nuclei with smaller mass numbers. (20.6)

Formal charge: the charge assigned to an atom in a molecule or polyatomic ion derived from a specific set of rules. (13.12)

Formation constant (stability constant): the equilibrium constant for each step of the formation of a complex ion by the addition of an individual ligand to a metal ion or complex ion in aqueous solution. (8.9)

Fossil fuel: coal, petroleum, or natural gas; consists of carbon-based molecules derived from decomposition of once-living organisms. (9.7)

Frasch process: the recovery of sulfur from underground deposits by melting it with hot water and forcing it to the surface by air pressure. (18.12)

Free energy: a thermodynamic function equal to the enthalpy (H) minus the product of the entropy (S) and the kelvin temperature (T); $G = H - TS$. Under certain conditions the change in free energy for a process is equal to the maximum useful work. (10.7)

Free radical: a species with an unpaired electron. (21.5)

Frequency: the number of waves (cycles) per second that pass a given point in space. (12.1)

Fuel cell: a galvanic cell for which the reactants are continuously supplied. (11.5)

Functional group: an atom or group of atoms in hydrocarbon derivatives that contains elements in addition to carbon and hydrogen. (21.4)

Fusion: the process of combining two light nuclei to form a heavier, more stable nucleus. (20.6)

Galvanic cell: a device in which chemical energy from a spontaneous redox reaction is changed to electrical energy that can be used to do work. (11.1)

Galvanizing: a process in which steel is coated with zinc to prevent corrosion. (11.6)

Gamma (γ) ray: a high-energy photon. (20.1)

Geiger-Müller counter (Geiger counter): an instrument that measures the rate of radioactive decay based on the ions and electrons produced as a radioactive particle passes through a gas-filled chamber. (20.4)

Gene: a given segment of the DNA molecule that contains the code for a specific protein. (21.6)

Geometrical (*cis-trans*) isomerism: isomerism in which atoms or groups of atoms can assume different positions around a rigid ring or bond. (19.4; 21.2)

Glass: an amorphous solid obtained when silica is mixed with other compounds, heated above its melting point, and then cooled rapidly. (16.5)

Glass electrode: an electrode for measuring pH from the potential difference that develops when it is dipped into an aqueous solution containing H^+ ions. (11.4)

Glycosidic linkage: a C—O—C bond formed between the rings of two cyclic monosaccharides by the elimination of water. (21.6)

Graham's law of effusion: the rate of effusion of a gas is inversely proportional to the square root of the mass of its particles. (5.7)

Gravimetric analysis: a method for determining the amount of a given substance in a solution by precipitation, filtration, drying, and weighing. (4.8)

Greenhouse effect: a warming effect exerted by the earth's atmosphere (particularly CO_2 and H_2O) due to thermal energy retained by absorption of infrared radiation. (9.7)

Ground state: the lowest possible energy state of an atom or molecule. (12.4)

Group (of the periodic table): a vertical column of elements having the same valence electron configuration and showing similar properties. (2.8)

Haber process: the manufacture of ammonia from nitrogen and hydrogen, carried out at high pressure and high temperature with the aid of a catalyst. (3.9; 6.1; 18.8)

Half-life (of a radioactive sample): the time required for the number of nuclides in a radioactive sample to reach half of the original value. (20.2)

Half-life (of a reaction): the time required for a reactant to reach half of its original concentration. (15.4)

Half-reactions: the two parts of an oxidation–reduction reaction, one representing oxidation, the other reduction. (4.11; 11.1)

Halogen: a Group 7A element. (2.8; 18.13)

Halogenation: the addition of halogen atoms to unsaturated hydrocarbons. (21.2)

Hard water: water from natural sources that contains relatively large concentrations of calcium and magnesium ions. (18.4)

Heat: energy transferred between two objects caused by a temperature difference between them. (9.1)

Heat capacity: the amount of energy required to raise the temperature of an object by one degree Celsius. (9.4)

Heat of fusion: the enthalpy change that occurs to melt a solid at its melting point. (16.10)

Heat of hydration: the enthalpy change associated with placing gaseous molecules or ions in water; the sum of the energy needed to expand the solvent and the energy released from the solvent–solute interactions. (17.2)

Heat of solution: the enthalpy change associated with dissolving a solute in a solvent; the sum of the energies needed to expand both solvent and solute in a solution and the energy released from the solvent–solute interactions. (17.2)

Heat of vaporization: the energy required to vaporize one mole of a liquid at a pressure of one atmosphere. (16.10)

Heating curve: a plot of temperature versus time for a substance where energy is added at a constant rate. (16.10)

Heisenberg uncertainty principle: a principle stating that there is a fundamental limitation to how precisely both the position and momentum of a particle can be known at a given time. (12.5)

Heme: an iron complex. (19.8)

Hemoglobin: a biomolecule composed of four myoglobin-like units (proteins plus heme) that can bind and transport four oxygen molecules in the blood. (19.8)

Henderson–Hasselbalch equation: an equation giving the relationship between the pH of an acid–base system and the concentrations of base and acid

$$\mathrm{pH} = \mathrm{p}K_a + \log\left(\frac{[\mathrm{base}]}{[\mathrm{acid}]}\right). \quad (8.2)$$

Henry's law: the amount of a gas dissolved in a solution is directly proportional to the pressure of the gas above the solution. (17.3)

Hess's law: in going from a particular set of reactants to a particular set of products, the enthalpy change is the same whether the reaction takes place in one step or in a series of steps; in summary, enthalpy is a state function. (9.5)

Heterogeneous equilibrium: an equilibrium involving reactants and/or products in more than one phase. (6.5)

Hexagonal closest packed (hcp) structure: a structure composed of closest packed spheres with an *ababab* arrangement of layers; the unit cell is hexagonal. (16.4)

Homogeneous equilibrium: an equilibrium system where all reactants and products are in the same phase. (6.5)

Homopolymer: a polymer formed from the polymerization of only one type of monomer. (21.5)

Hund's rule: the lowest-energy configuration for an atom is the one having the maximum number of unpaired electrons allowed by the Pauli exclusion principle in a particular set of degenerate orbitals, with all unpaired electrons having parallel spins. (12.13)

Hybrid orbitals: a set of atomic orbitals adopted by an atom in a molecule different from those of the atom in the free state. (14.1)

Hybridization: a mixing of the native orbitals on a given atom to form special atomic orbitals for bonding. (14.1)

Hydration: the interaction between solute particles and water molecules. (4.1)

Hydride: a binary compound containing hydrogen. The hydride ion, H^-, exists in ionic hydrides. The three classes of hydrides are covalent, interstitial, and ionic. (18.3)

Hydrocarbon: a compound composed of carbon and hydrogen. (23.1)

Hydrocarbon derivative: an organic molecule that contains one or more elements in addition to carbon and hydrogen. (21.4)

Hydrogen bonding: unusually strong dipole–dipole attractions that occur among molecules in which hydrogen is bonded to a highly electronegative atom. (16.1)

Hydrogenation reaction: a reaction in which hydrogen is added, with a catalyst present, to a carbon–carbon multiple bond. (21.2)

Hydrohalic acid: an aqueous solution of a hydrogen halide. (18.13)

Hydronium ion: the H_3O^+ ion; a hydrated proton. (7.1)

Hypothesis: one or more assumptions put forth to explain the observed behavior of nature. (1.3)

Ideal gas: a gas that obeys the equation, $PV = nRT$. (5.2)

Ideal gas law: an equation of state for a gas, where the state of the gas is its condition at a given time; expressed by $PV = nRT$, where P = pressure, V = volume, n = moles of the gas, R = the universal gas constant, and T = absolute temperature. This equation expresses behavior approached by real gases at high T and low P. (5.3)

Ideal solution: a solution whose vapor pressure is directly proportional to the mole fraction of solvent present. (17.4)

Indicator: a chemical that changes color and is used to mark the endpoint of a titration. (4.9; 8.5)

Inert pair effect: the tendency for the heavier Group 3A elements to exhibit the +1 as well as the expected +3 oxidation states, and Group 4A elements to exhibit the +2 as well as the +4 oxidation states. (18.5)

Integrated rate law: an expression that shows the concentration of a reactant as a function of time. (15.2)

Intermediate: a species that is neither a reactant nor a product but that is formed and consumed in the reaction sequence. (15.6)

Intermolecular forces: relatively weak interactions that occur between molecules. (16.1)

Internal energy: a property of a system that can be changed by a flow of work, heat, or both; $\Delta E = q + w$, where ΔE is the change in the internal energy of the system, q is heat, and w is work. (9.1)

Ion: an atom or a group of atoms that has a net positive or negative charge. (2.7)

Ion exchange (water softening): the process in which an ion-exchange resin removes unwanted ions (for example, Ca^{2+} and Mg^{2+}) and replaces them with Na^{+} ions, which do not interfere with soap and detergent action. (18.4)

Ion pairing: a phenomenon occurring in solution when oppositely charged ions aggregate and behave as a single particle. (17.7)

Ion-product constant (K_w): the equilibrium constant for the autoionization of water; $K_w = [H^+][OH^-]$. At 25°C, K_w equals 1.0×10^{-14}. (7.2)

Ion-selective electrode: an electrode sensitive to the concentration of a particular ion in solution. (11.4)

Ionic bonding: the electrostatic attraction between oppositely charged ions. (2.7; 13.1)

Ionic compound (binary): a compound that results when a metal reacts with a nonmetal to form a cation and an anion. (13.1)

Ionic solid: a solid containing cations and anions that dissolves in water to give a solution containing the separated ions, which are mobile and thus free to conduct electric current. (16.3)

Ionization energy: the quantity of energy required to remove an electron from a gaseous atom or ion. (12.15)

Irreversible process: any real process. When a system undergoes the changes State 1 → State 2 → State 1 by any real pathway, the universe is different than before the cyclic process took place in the system. (10.2)

Isoelectronic ions: ions containing the same number of electrons. (13.4)

Isomers: species with the same formula but different properties. (19.4)

Isothermal process: a process in which the temperature remains constant. (10.2)

Isotonic solutions: solutions having identical osmotic pressures. (17.6)

Isotopes: atoms of the same element (the same number of protons) with different numbers of neutrons. They have identical atomic numbers but different mass numbers. (2.6)

Ketone: an organic compound containing the carbonyl group

bonded to two carbon atoms. (21.4)

Kinetic energy ($\frac{1}{2}mv^2$): energy resulting from the motion of an object; dependent on the mass of the object and the square of its velocity. (9.1)

Kinetic molecular theory: a model that assumes that an ideal gas is composed of tiny particles (molecules) in constant motion. (5.6)

Lanthanide contraction: the decrease in the atomic radii of the lanthanide series elements, going from left to right in the periodic table. (19.1)

Lanthanide series: a group of 14 elements following lanthanum in the periodic table, in which the $4f$ orbitals are being filled. (12.13; 18.1; 19.1)

Lattice: a three-dimensional system of points designating the positions of the centers of the components of a solid (atoms, ions, or molecules). (16.3)

Lattice energy: the energy change occurring when separated gaseous ions are packed together to form an ionic solid. (13.5)

Law of conservation of energy: energy can be converted from one form to another but can be neither created nor destroyed. (9.1)

Law of conservation of mass: mass is neither created nor destroyed. (2.2)

Law of definite proportion: a given compound always contains exactly the same proportion of elements by mass. (2.2)

Law of mass action: a general description of the equilibrium condition; it defines the equilibrium constant expression. (6.2)

Law of multiple proportions: when two elements form a series of compounds, the ratios of the masses of the second element that combine with one gram of the first element can always be reduced to small whole numbers. (2.2)

Lead storage battery: a battery (used in cars) in which the anode is lead, the cathode is lead coated with lead dioxide, and the electrolyte is a sulfuric acid solution. (11.5)

Le Châtelier's principle: if a change is imposed on a system at equilibrium, the position of the equilibrium will shift in a direction that tends to reduce the effect of that change. (6.8)

Lewis acid: an electron-pair acceptor. (19.3)

Lewis base: an electron-pair donor. (19.3)

Lewis structure: a diagram of a molecule showing how the valence electrons are arranged among the atoms in the molecule. (13.10)

Ligand: a neutral molecule or ion having a lone pair of electrons that can be used to form a bond to a metal ion; a Lewis base. (19.3)

Lime-soda process: a water-softening method in which lime and soda ash are added to water to remove calcium and magnesium ions by precipitation. (7.6)

Limiting reactant (limiting reagent): the reactant that is completely consumed when a reaction is run to completion. (3.9)

Line spectrum: a spectrum showing only certain discrete wavelengths. (12.3)

Linear accelerator: a type of particle accelerator in which a changing electric field is used to accelerate a positive ion along a linear path. (20.3)

Linkage isomerism: isomerism involving a complex ion where the ligands are all the same but the point of attachment of at least one of the ligands differs. (19.4)

Liquefaction: the transformation of a gas into a liquid. (18.1)

Localized electron (LE) model: a model that assumes that a molecule is composed of atoms that are bound together by sharing pairs of electrons using the atomic orbitals of the bound atoms. (13.9)

London dispersion forces: the forces, existing among noble gas atoms and nonpolar molecules, that involve an accidental dipole that induces a momentary dipole in a neighbor. (16.1)

Lone pair: an electron pair that is localized on a given atom; an electron pair not involved in bonding. (13.9)

Magnetic quantum number (m_ℓ): the quantum number relating to the orientation of an orbital in space relative to the other orbitals with the same ℓ quantum number. It can have integral values between ℓ and $-\ell$, including zero. (12.9)

Main-group (representative) elements: elements in the groups labeled 1A, 2A, 3A, 4A, 5A, 6A, 7A, and 8A in the periodic table. The group number gives the sum of valence *s* and *p* electrons. (12.13; 18.1)

Major species: the components present in relatively large amounts in a solution. (7.4)

Manometer: a device for measuring the pressure of a gas in a container. (5.1)

Mass defect: the change in mass occurring when a nucleus is formed from its component nucleons. (20.5)

Mass number: the total number of protons and neutrons in the atomic nucleus of an atom. (2.6)

Mass percent: the percent by mass of a component of a mixture (17.1) or of a given element in a compound. (3.4)

Mass spectrometer: an instrument used to determine the relative masses of atoms by the deflection of their ions in a magnetic field. (3.1)

Matter: the material of the universe.

Mean free path: the average distance a molecule in a given gas sample travels between collisions with other molecules. (5.6; 5.9)

Measurement: a quantitative observation. (A1.5)

Messenger RNA (mRNA): a special RNA molecule built in the cell nucleus that migrates into the cytoplasm and participates in protein synthesis. (21.6)

Metal: an element that gives up electrons relatively easily and is lustrous, malleable, and a good conductor of heat and electricity. (2.8)

Metalloids (semimetals): elements along the division line in the periodic table between metals and nonmetals. These elements exhibit both metallic and nonmetallic properties. (12.16; 18.1)

Metallurgy: the process of separating a metal from its ore and preparing it for use. (18.1)

Millimeters of mercury (mm Hg): a unit of pressure, also called a torr; 760 mm Hg = 760 torr = 101,325 Pa = 1 standard atmosphere. (5.1)

Mixture: a material of variable composition that contains two or more substances.

Model (theory): a set of assumptions put forth to explain the observed behavior of matter. The models of chemistry usually involve assumptions about the behavior of individual atoms or molecules. (1.3)

Moderator: a substance used in a nuclear reactor to slow down the neutrons. (20.6)

Molal boiling-point elevation constant: a constant characteristic of a particular solvent that gives the change in boiling point as a function of solution molality; used in molecular weight determinations. (17.5)

Molal freezing-point depression constant: a constant characteristic of a particular solvent that gives the change in freezing point as a function of the solution molality; used in molecular weight determinations. (17.5)

Molality: the number of moles of solute per kilogram of solvent in a solution. (17.1)

Molar heat capacity: the energy required to raise the temperature of one mole of a substance by one degree Celsius. (9.3; 9.4)

Molar mass: the mass in grams of one mole of molecules or formula units of a substance; also called molecular weight. (3.3)

Molar volume: the volume of one mole of an ideal gas; equal to 22.42 liters at STP. (5.4)

Molarity: moles of solute per volume of solution in liters. (4.3; 17.1)

Mole (mol): the number equal to the number of carbon atoms in exactly 12 grams of pure ^{12}C; Avogadro's number. One mole represents 6.022×10^{23} units. (3.2)

Mole fraction: the ratio of the number of moles of a given component in a mixture to the total number of moles in the mixture. (5.5; 17.1)

Mole ratio (stoichiometry): the ratio of moles of one substance to moles of another substance in a balanced chemical equation. (3.8)

Molecular equation: an equation representing a reaction in solution showing the reactants and products in undissociated form, whether they are strong or weak electrolytes. (4.6)

Molecular formula: the exact formula of a molecule, giving the types of atoms and the number of each type. (3.5)

Molecular orbital (MO) model: a model that regards a molecule as a collection of nuclei and electrons, where the electrons are assumed to occupy orbitals much as they do in atoms, but having the orbitals extend over the entire molecule. In this model the electrons are assumed to be delocalized rather than always located between a given pair of atoms. (14.2)

Molecular orientations (kinetics): orientations of molecules during collisions, some of which can lead to a reaction and some of which cannot. (15.8)

Molecular solid: a solid composed of neutral molecules at the lattice points. (16.3)

Molecular structure: the three-dimensional arrangement of atoms in a molecule. (13.13)

Molecular weight: the mass in grams of one mole of molecules or formula units of a substance; also called molar mass. (3.3)

Molecularity: the number of species that must collide to produce the reaction represented by an elementary step in a reaction mechanism. (15.6)

Molecule: a bonded collection of two or more atoms of the same or different elements. (2.7)

Monodentate (unidentate) ligand: a ligand that can form one bond to a metal ion. (19.3)

Monoprotic acid: an acid with one acidic proton. (7.2)

Monosaccharide (simple sugar): a polyhydroxy ketone or aldehyde containing from three to nine carbon atoms. (21.6)

Myoglobin: an oxygen-storing biomolecule consisting of a heme complex and a protein. (19.8)

Natural law: a statement that expresses generally observed behavior. (1.3)

Nernst equation: an equation relating the potential of an electrochemical cell to the concentrations of the cell components

$$\mathscr{E} = \mathscr{E}° - \frac{0.0591}{n}\log(Q) \text{ at } 25°\text{C}. \quad (11.4)$$

Net ionic equation: an equation for a reaction in solution, where strong electrolytes are written as ions, showing only those components that are directly involved in the chemical change. (4.6)

Network solid: an atomic solid containing strong directional covalent bonds. (16.5)

Neutralization reaction: an acid–base reaction. (4.9)

Neutron: a particle in the atomic nucleus with mass virtually equal to the proton's but with no charge. (2.6)

Nitrogen cycle: the conversion of N_2 to nitrogen-containing compounds, followed by the return of nitrogen gas to the atmosphere by natural decay processes. (18.8)

Nitrogen fixation: the process of transforming N_2 to nitrogen-containing compounds useful to plants. (18.8)

Nitrogen-fixing bacteria: bacteria in the root nodules of plants that can convert atmospheric nitrogen to ammonia and other nitrogen-containing compounds useful to plants. (18.8)

Noble gas: a Group 8A element. (2.8; 18.14)

Node: an area of an orbital having zero electron probability. (12.9)

Nonelectrolyte: a substance that, when dissolved in water, gives a nonconducting solution. (4.2)

Nonmetal: an element not exhibiting metallic characteristics. Chemically, a typical nonmetal accepts electrons from a metal. (2.8)

Normal boiling point: the temperature at which the vapor pressure of a liquid is exactly one atmosphere. (16.10)

Normal melting point: the temperature at which the solid and liquid states have the same vapor pressure under conditions where the total pressure on the system is one atmosphere. (16.10)

Normality: the number of equivalents of a substance dissolved in a liter of solution. (17.1)

Nuclear atom: an atom having a dense center of positive charge (the nucleus) with electrons moving around the outside. (2.5)

Nuclear transformation: the change of one element into another. (20.3)

Nucleon: a particle in an atomic nucleus, either a neutron or a proton. (2.6)

Nucleotide: a monomer of the nucleic acids composed of a five-carbon sugar, a nitrogen-containing base, and phosphoric acid. (21.6)

Nucleus: the small, dense center of positive charge in an atom. (2.5)

Nuclide: the general term applied to each unique atom; represented by ${}^{A}_{Z}X$, where X is the symbol for a particular element. (20.2)

Octet rule: the observation that atoms of nonmetals tend to form the most stable molecules when they are surrounded by eight electrons (to fill their valence orbitals). (13.10)

Optical isomerism: isomerism in which the isomers have opposite effects on plane-polarized light. (19.4)

Orbital: a specific wave function for an electron in an atom. The square of this function gives the probability distribution for the electron. (12.5)

***d*-Orbital splitting:** a splitting of the *d* orbitals of the metal ion in a complex such that the orbitals pointing at the ligands have higher energies than those pointing between the ligands. (19.6)

Order (of reactant): the positive or negative exponent, determined by experiment, of the reactant concentration in a rate law. (15.2)

Organic acid: an acid with a carbon-atom backbone; often contains the carboxyl group. (7.2)

Organic chemistry: the study of carbon-containing compounds (typically chains of carbon atoms) and their properties. (21)

Osmosis: the flow of solvent into a solution through a semipermeable membrane. (17.6)

Osmotic pressure (π): the pressure that must be applied to a solution to stop osmosis; $= MRT$. (17.6)

Ostwald process: a commercial process for producing nitric acid by the oxidation of ammonia. (18.8)

Oxidation: an increase in oxidation state (a loss of electrons). (4.10; 11.1)

Oxidation–reduction (redox) reaction: a reaction in which one or more electrons are transferred. (4.4; 4.10; 11.1)

Oxidation states: a concept that provides a way to keep track of electrons in oxidation–reduction reactions according to certain rules. (4.10)

Oxidizing agent (electron acceptor): a reactant that accepts electrons from another reactant. (4.10; 11.1)

Oxyacid: an acid in which the acidic proton is attached to an oxygen atom. (7.2)

Ozone: O_3, the form of elemental oxygen in addition to the much more common O_2. (18.11)

Paramagnetism: a type of induced magnetism, associated with unpaired electrons, that causes a substance to be attracted into the inducing magnetic field. (14.3)

Partial pressures: the independent pressures exerted by different gases in a mixture. (5.5)

Particle accelerator: a device used to accelerate nuclear particles to very high speeds. (20.3)

Pascal: the SI unit of pressure; equal to newtons per meter squared. (5.1)

Pauli exclusion principle: in a given atom no two electrons can have the same set of four quantum numbers. (12.10)

Penetration effect: the effect whereby a valence electron penetrates the core electrons, thus reducing the shielding effect and increasing the effective nuclear charge. (12.14)

Peptide linkage: the bond resulting from the condensation reaction between amino acids; represented by

$$\begin{array}{cc} \text{O} & \text{H} \\ \| & | \\ -\text{C}- & \text{N}- \end{array} \qquad (22.6)$$

Percent dissociation: the ratio of the amount of a substance that is dissociated at equilibrium to the initial concentration of the substance in a solution, multiplied by 100. (7.5)

Percent yield: the actual yield of a product as a percentage of the theoretical yield. (3.9)

Periodic table: a chart showing all the elements arranged in columns with similar chemical properties. (2.8)

pH curve (titration curve): a plot showing the pH of a solution being analyzed as a function of the amount of titrant added. (8.5)

pH scale: a log scale based on 10 and equal to $-\log[H^+]$; a convenient way to represent solution acidity. (7.3)

Phase diagram: a convenient way of representing the phases of a substance in a closed system as a function of temperature and pressure. (16.11)

Phenyl group: the benzene molecule minus one hydrogen atom. (21.3)

Photochemical smog: air pollution produced by the action of light on oxygen, nitrogen oxides, and unburned fuel from auto exhaust to form ozone and other pollutants. (5.11)

Photon: a quantum of electromagnetic radiation. (12.2)

Physical change: a change in the form of a substance, but not in its chemical composition; chemical bonds are not broken in a physical change.

Pi (π) bond: a covalent bond in which parallel p orbitals share an electron pair occupying the space above and below the line joining the atoms. (14.1)

Planck's constant: the constant relating the change in energy for a system to the frequency of the electromagnetic radiation absorbed or emitted; equal to 6.626×10^{-34} J s. (12.2)

Polar covalent bond: a covalent bond in which the electrons are not shared equally because one atom attracts them more strongly than the other. (13.1)

Polar molecule: a molecule that has a permanent dipole moment. (4.1)

Polyatomic ion: an ion containing a number of atoms. (2.7)

Polyelectronic atom: an atom with more than one electron. (12.11)

Polymer: a large, usually chainlike molecule built from many small molecules (monomers). (21.5)

Polymerization: a process in which many small molecules (monomers) are joined together to form a large molecule. (21.2)

Polypeptide: a polymer formed from amino acids joined together by peptide linkages. (21.6)

Polyprotic acid: an acid with more than one acidic proton. It dissociates in a stepwise manner, one proton at a time. (7.7)

Porous disk: a disk in a tube connecting two different solutions in a galvanic cell that allows ion flow without extensive mixing of the solutions. (11.1)

Porphyrin: a planar ligand with a central ring structure and various substituent groups at the edges of the ring. (19.8)

Positional probability: a type of probability that depends on the number of arrangements in space that yield a particular state. (10.1)

Positron production: a mode of nuclear decay in which a particle is formed having the same mass as an electron but opposite charge. The net effect is to change a proton to a neutron. (20.1)

Potential energy: energy resulting from position or composition. (9.1)

Precipitation reaction: a reaction in which an insoluble substance forms and separates from the solution. (4.5)

Precision: the degree of agreement among several measurements of the same quantity; the reproducibility of a measurement. (A1.5)

Primary structure (of a protein): the order (sequence) of amino acids in the protein chain. (21.6)

Principal quantum number: the quantum number relating to the size and energy of an orbital; it can have any positive integer value. (12.9)

Probability distribution: the square of the wave function indicating the probability of finding an electron at a particular point in space. (12.8)

Product: a substance resulting from a chemical reaction. It is shown to the right of the arrow in a chemical equation. (3.6)

Protein: a natural high-molecular-weight polymer formed by condensation reactions between amino acids. (21.6)

Proton: a positively charged particle in an atomic nucleus. (2.6; 20)

Qualitative analysis: the separation and identification of individual ions from a mixture. (4.7; 8.9)

Quantization: the concept that energy can occur only in discrete units called quanta. (12.2)

Rad: a unit of radiation dosage corresponding to 10^{-2} J of energy deposited per kilogram of tissue (from *r*adiation *a*bsorbed *d*ose). (20.7)

Radioactive decay (radioactivity): the spontaneous decomposition of a nucleus to form a different nucleus. (20.1)

Radiocarbon dating (carbon-14 dating): a method for dating ancient wood or cloth based on the rate of radioactive decay of the nuclide $^{14}_{6}C$. (20.4)

Radiotracer: a radioactive nuclide, introduced into an organism for diagnostic purposes, whose pathway can be traced by monitoring its radioactivity. (20.4)

Random error: an error that has an equal probability of being high or low. (A1.5)

Raoult's law: the vapor pressure of a solution is directly proportional to the mole fraction of solvent present. (17.4)

Rate constant: the proportionality constant in the relationship between reaction rate and reactant concentrations. (15.2)

Rate of decay: the change in the number of radioactive nuclides in a sample per unit time. (20.2)

Rate-determining step: the slowest step in a reaction mechanism, the one determining the overall rate. (15.6)

Rate law (differential rate law): an expression that shows how the rate of reaction depends on the concentration of reactants. (15.2)

Reactant: a starting substance in a chemical reaction. It appears to the left of the arrow in a chemical equation. (3.6)

Reaction mechanism: the series of elementary steps involved in a chemical reaction. (15.6)

Reaction quotient: a quotient obtained by applying the law of mass action to initial concentrations rather than to equilibrium concentrations. (6.6)

Reaction rate: the change in concentration of a reactant or product per unit time. (15.1)

Reactor core: the part of a nuclear reactor where the fission reaction takes place. (20.6)

Reducing agent (electron donor): a reactant that donates electrons to another substance to reduce the oxidation state of one of its atoms. (4.10; 11.1)

Reduction: a decrease in oxidation state (a gain of electrons). (4.10; 11.1)

Rem: a unit of radiation dosage that accounts for both the energy of the dose and its effectiveness in causing biological damage (from *r*oentgen *e*quivalent for *m*an). (20.7)

Resonance: a condition occurring when more than one valid Lewis structure can be written for a particular molecule. The actual electronic structure is not represented by any one of the Lewis structures but by the average of all of them. (13.11)

Reverse osmosis: the process occurring when the external pressure on a solution causes a net flow of solvent through a semipermeable membrane from the solution to the solvent. (17.6)

Reversible process: a cyclic process carried out by a hypothetical pathway, which leaves the universe exactly the same as it was before the process. No real process is reversible. (10.2)

Ribonucleic acid (RNA): a nucleotide polymer that transmits the genetic information stored in DNA to the ribosomes for protein synthesis. (21.6)

Root mean square velocity: the square root of the average of the squares of the individual velocities of gas particles. (5.6)

Salt: an ionic compound. (7.8)

Salt bridge: a U-tube containing an electrolyte that connects the two compartments of a galvanic cell, allowing ion flow without extensive mixing of the different solutions. (11.1)

Scientific method: the process of studying natural phenomena, involving observations, forming laws and theories, and testing of theories by experimentation. (1.3)

Scintillation counter: an instrument that measures radioactive decay by sensing the flashes of light produced in a substance by the radiation. (20.4)

Second law of thermodynamics: in any spontaneous process, there is always an increase in the entropy of the universe. (10.5)

Secondary structure (of a protein): the three-dimensional structure of the protein chain (for example, α-helix, random coil, or pleated sheet). (21.6)

Selective precipitation: a method of separating metal ions from an aqueous mixture by using a reagent whose anion forms a precipitate with only one or a few of the ions in the mixture. (4.7; 8.8)

Semiconductor: a substance conducting only a slight electric current at room temperature, but showing increased conductivity at higher temperatures. (16.5)

Semipermeable membrane: a membrane that allows solvent but not solute molecules to pass through. (17.6)

Shielding: the effect by which the other electrons screen, or shield, a given electron from some of the nuclear charge. (12.14)

SI units: International System of units based on the metric system and units derived from the metric system. (A2.1)

Side chain (of amino acid): the hydrocarbon group on an amino acid represented by H, CH_3, or a more complex substituent. (21.6)

Sigma (σ) bond: a covalent bond in which the electron pair is shared in an area centered on a line running between the atoms. (14.1)

Significant figures: the certain digits and the first uncertain digit of a measurement. (A1.5)

Silica: the fundamental silicon–oxygen compound, which has the empirical formula SiO_2, and forms the basis of quartz and certain types of sand. (16.5)

Silicates: salts that contain metal cations and polyatomic silicon–oxygen anions that are usually polymeric. (16.5)

Single bond: a bond in which one pair of electrons is shared by two atoms. (13.8)

Solubility: the amount of a substance that dissolves in a given volume of solvent at a given temperature. (4.2)

Solubility product constant: the constant for the equilibrium expression representing the dissolving of an ionic solid in water. (8.8)

Solute: a substance dissolved in a liquid to form a solution. (4.2; 17.1)

Solution: a homogeneous mixture. (17)

Solvent: the dissolving medium in a solution. (4.2)

Somatic damage: radioactive damage to an organism resulting in its sickness or death. (20.7)

Space-filling model: a model of a molecule showing the relative sizes of the atoms and their relative orientations. (2.7)

Specific heat capacity: the energy required to raise the temperature of one gram of a substance by one degree Celsius. (9.4)

Spectator ions: ions present in solution that do not participate directly in a reaction. (4.6)

Spectrochemical series: a listing of ligands in order based on their ability to produce *d*-orbital splitting. (19.6)

Spectroscopy: the study of the interaction of electromagnetic radiation within matter. (14.7)

Spontaneous fission: the spontaneous splitting of a heavy nuclide into two lighter nuclides. (20.1)

Spontaneous process: a process that occurs without outside intervention. (10.1)

Standard atmosphere: a unit of pressure equal to 760 mm Hg. (5.1)

Standard enthalpy of formation: the enthalpy change that accompanies the formation of one mole of a compound at 25°C from its elements, with all substances in their standard states at that temperature. (9.6)

Standard free energy change: the change in free energy that will occur for one unit of reaction if the reactants in their standard states are converted to products in their standard states. (10.9)

Standard free energy of formation: the change in free energy that accompanies the formation of one mole of a substance from its constituent elements with all reactants and products in their standard states. (10.9)

Standard hydrogen electrode: a platinum conductor in contact with 1 *M* H^+ ions and bathed by hydrogen gas at one atmosphere. (11.2)

Standard reduction potential: the potential of a half-reaction under standard state conditions, as measured against the potential of the standard hydrogen electrode. (11.2)

Standard solution: a solution whose concentration is accurately known. (4.3)

Standard state: a reference state for a specific substance defined according to a set of conventional definitions. (9.6)

Standard temperature and pressure (STP): the condition 0°C and 1 atm of pressure. (5.4)

Standing wave: a stationary wave as on a string of a musical instrument; in the wave mechanical model, the electron in the hydrogen atom is considered to be a standing wave. (12.5)

State function: a property that is independent of the pathway. (9.1)

States of matter: the three different forms in which matter can exist: solid, liquid, and gas. (5)

Stereoisomerism: isomerism in which all the bonds in the isomers are the same but the spatial arrangements of the atoms are different. (19.4)

Steric factor: the factor (always less than one) that reflects the fraction of collisions with orientations that can produce a chemical reaction. (15.8)

Stoichiometric quantities: quantities of reactants mixed in exactly the correct amounts so that all are used up at the same time. (3.9)

Strong acid: an acid that completely dissociates to produce a H^+ ion and the conjugate base. (4.2; 7.2)

Strong base: a metal hydroxide salt that completely dissociates into its ions in water. (4.2; 7.6)

Strong electrolyte: a material that, when dissolved in water, gives a solution that conducts an electric current very efficiently. (4.2)

Structural formula: the representation of a molecule in which the relative positions of the atoms are shown and the bonds are indicated by lines. (2.7)

Structural isomerism: isomerism in which the isomers contain the same atoms but one or more bonds differ. (19.4; 21.1)

Subcritical reaction (nuclear): a reaction in which less than one neutron causes another fission event and the process dies out. (20.6)

Sublimation: the process by which a substance goes directly from the solid to the gaseous state without passing through the liquid state. (16.10)

Subshell: a set of orbitals with a given angular momentum quantum number. (12.9)

Substitution reaction (hydrocarbons): a reaction in which an atom, usually a halogen, replaces a hydrogen atom in a hydrocarbon. (21.1)

Supercooling: the process of cooling a liquid below its freezing point without its changing to a solid. (16.10)

Supercritical reaction (nuclear): a reaction in which more than one neutron from each fission event causes another fission event. The process rapidly escalates to a violent explosion. (20.6)

Superheating: the process of heating a liquid above its boiling point without its boiling. (16.10)

Superoxide: a compound containing the O_2^- anion. (18.2)

Surface tension: the resistance of a liquid to an increase in its surface area. (16.2)

Surroundings: everything in the universe surrounding a thermodynamic system. (9.1)

Syngas: synthetic gas, a mixture of carbon monoxide and hydrogen, obtained by coal gasification. (9.8)

System (thermodynamic): that part of the universe on which attention is to be focused. (9.1)

Systematic error: an error that always occurs in the same direction. (A1.5)

Termolecular step: a reaction involving the simultaneous collision of three molecules. (15.6)

Tertiary structure (of a protein): the overall shape of a protein, long and narrow or globular, maintained by different types of intramolecular interactions. (21.6)

Theoretical yield: the maximum amount of a given product that can be formed when the limiting reactant is completely consumed. (3.9)

Theory: a set of assumptions put forth to explain some aspect of the observed behavior of matter. (1.3)

Thermal pollution: the oxygen-depleting effect on lakes and rivers of using water for industrial cooling and returning it to its natural source at a higher temperature. (17.3)

Thermodynamic stability (nuclear): the potential energy of a particular nucleus as compared with the sum of the potential energies of its component protons and neutrons. (20.1)

Thermodynamics: the study of energy and its interconversions. (9.1)

Third law of thermodynamics: the entropy of a perfect crystal at 0 K is zero. (10.8)

Titration: a technique in which one solution is used to analyze another. (4.9)

Torr: another name for millimeter of mercury (mm Hg). (5.1)

Transfer RNA (tRNA): a small RNA fragment that finds specific amino acids and attaches them to the protein chain as dictated by the codons in mRNA. (21.6)

Transition metals: several series of elements in which inner orbitals (*d* or *f* orbitals) are being filled. (12.13; 18.1)

Transuranium elements: the elements beyond uranium that are made artificially by particle bombardment. (20.3)

Triple bond: a bond in which three pairs of electrons are shared by two atoms. (13.8)

Triple point: the point on a phase diagram at which all three states of a substance are present. (16.11)

Tyndall effect: the scattering of light by particles in a suspension. (17.8)

Uncertainty (in measurement): the characteristics that any measurement involves estimates and cannot be exactly reproduced. (A1.5)

Unimolecular step: a reaction step involving only one molecule. (15.6)

Unit cell: the smallest repeating unit of a lattice. (16.3)

Unit factor: an equivalence statement between units used for converting from one unit to another. (A2.2)

Universal gas constant: the combined proportionality constant in the ideal gas law; 0.08206 L atm/K mol or 8.3145 J/K mol. (5.3)

Valence electrons: the electrons in the outermost principal quantum level of an atom. (12.13)

Valence shell electron-pair repulsion (VSEPR) model: a model whose main postulate is that the structure around a given atom in a molecule is determined principally by minimizing electron-pair repulsions. (13.13)

van der Waals's equation: a mathematical expression for describing the behavior of real gases. (5.10)

van't Hoff factor: the ratio of moles of particles in solution to moles of solute dissolved. (17.7)

Vapor pressure: the pressure of the vapor over a liquid at equilibrium. (16.10)

Vaporization: the change in state that occurs when a liquid evaporates to form a gas. (16.10)

Viscosity: the resistance of a liquid to flow. (16.2)

Volt: the unit of electrical potential defined as one joule of work per coulomb of charge transferred. (11.1)

Voltmeter: an instrument that measures cell potential by drawing electric current through a known resistance. (11.1)

Volumetric analysis: a process involving titration of one solution with another. (4.9)

Wave function: a function of the coordinates of an electron's position in three-dimensional space that describes the properties of the electron. (12.5)

Wave mechanical model: a model for the hydrogen atom in which the electron is assumed to behave as a standing wave. (12.7)

Wavelength: the distance between two consecutive peaks or troughs in a wave. (12.1)

Weak acid: an acid that dissociates only slightly in aqueous solution. (4.2; 7.2)

Weak base: a base that reacts with water to produce hydroxide ions to only a slight extent in aqueous solution. (4.2; 7.6)

Weak electrolyte: a material that, when dissolved in water, gives a solution that conducts only a small electric current. (4.2)

Weight: the force exerted on an object by gravity. (2.3)

Work: force acting over a distance. (9.1)

X-ray diffraction: a technique for establishing the structures of crystalline solids by directing X rays of a single wavelength at a crystal and obtaining a diffraction pattern from which interatomic spaces can be determined. (16.3)

Zone of nuclear stability: the area encompassing the stable nuclides on a plot of their positions as a function of the number of protons and the number of neutrons in the nucleus. (20.1)

Answers to Selected Exercises

The answers listed here are from the *Complete Solutions Guide*, in which rounding is carried out at each intermediate step in a calculation in order to show the correct number of significant figures for that step. Therefore, an answer given here may differ in the last digit from the result obtained by carrying extra digits throughout the entire calculation and rounding at the end (the procedure you should follow).

Chapter 2

19. ClF_3 **21.** All the masses of hydrogen in these three compounds can be expressed as simple whole-number ratios. The g H/g N in hydrazine, ammonia, and hydrogen azide are in the ratios 6:9:1. **23.** O, 7.94; Na, 22.8; Mg, 11.9; O and Mg are incorrect by a factor of ≈2; correct formulas are H_2O, Na_2O, and MgO. **25.** d(nucleus) = 3×10^{15} g/cm^3; d(atom) = 0.4 g/cm^3 **27.** Since all charges are whole-number multiples of 6.40×10^{-13} zirkombs, then the charge on one electron could be 6.40×10^{-13} zirkombs. However, 6.40×10^{-13} zirkombs could be the charge of two electrons (or three electrons, etc.). All one can conclude is that the charge of an electron is 6.40×10^{-13} zirkombs or an integer fraction of 6.40×10^{-13}. **29.** If the plum pudding model were correct (a diffuse positive charge with electrons scattered throughout), then α particles should have traveled through the thin foil with very minor deflections in their path. This was not the case because a few of the α particles were deflected at very large angles. Rutherford reasoned that the large deflections of these α particles could be caused only by a center of concentrated positive charge that contains most of the atom's mass (the nuclear model of the atom). **31.** The atomic number of an element is equal to the number of protons in the nucleus of an atom of that element. The mass number is the sum of the number of protons plus neutrons in the nucleus. The atomic mass is the actual mass of a particular isotope (including electrons). As we will see in Chapter 3, the average mass of an atom is taken from a measurement made on a large number of atoms. The average atomic mass value is listed in the periodic table. **33.** a. The noble gases are He, Ne, Ar, Kr, Xe, and Rn (helium, neon, argon, krypton, xenon, and radon). Radon has only radioactive isotopes. In the periodic table, the whole number enclosed in parentheses is the mass number of the longest-lived isotope of the element. b. promethium (Pm) and technetium (Tc) **35.** a. Five; F, Cl, Br, I, and At; b. Six; Li, Na, K, Rb, Cs, Fr (H is not considered an alkali metal.) c. 14; Ce, Pr, Nd, Pm, Sm, Eu, Gd, Tb, Dy, Ho, Er, Tm, Yb, and Lu; d. 40; all elements in the block defined by Sc, Zn, Uub, and Ac are transition metals. **37.** a. 12 p, 12 n, 12 e; b. 12 p, 12 n, 10 e; c. 27 p, 32 n, 25 e; d. 27 p, 32 n, 24 e; e. 27 p, 32 n, 27 e; f. 34 p, 45 n, 34 e; g. 34 p, 45 n, 36 e; h. 28 p, 35 n, 28 e; i. 28 p, 31 n, 26 e **39.** ${}^{151}_{63}Eu^{3+}$; ${}^{118}_{50}Sn^{2+}$ **41.** a. Lose 2 e^- to form Ra^{2+}; b. Lose 3 e^- to form In^{3+}; c. Gain 3 e^- to form P^{3-}; d. Gain 2 e^- to form Te^{2-}; e. Gain 1 e^- to form Br^-; f. Lose 1 e^- to form Rb^+ **43.** $AlCl_3$, aluminum chloride; $CrCl_3$, chromium(III) chloride; ICl_3, iodine trichloride; $AlCl_3$ and $CrCl_3$ are ionic compounds following the rules for naming ionic compounds. The major difference is that $CrCl_3$ contains a transition metal (Cr) that generally exhibits two or more stable charges when in ionic compounds. We need to indicate which charged ion we have in the compound. This is generally true whenever the metal in the ionic compound is a transition metal. ICl_3 is made from only nonmetals and is a covalent compound. Predicting formulas for covalent compounds is extremely difficult. Because of this, we need to indicate the number of each nonmetal in the binary covalent compound. The exception is when there is only one of the first species present in the formula; when this is the case, *mono-* is not used (it is assumed). **45.** a. sulfur difluoride; b. dinitrogen tetroxide; c. iodine trichloride; d. tetraphosphorus hexoxide **47.** a. copper(I) iodide; b. copper(II) iodide; c. cobalt(II) iodide; d. sodium carbonate; e. sodium hydrogen carbonate or sodium bicarbonate; f. tetrasulfur tetranitride; g. sulfur hexafluoride; h. sodium hypochlorite; i. barium chromate; j. ammonium nitrate **49.** a. SO_2; b. SO_3; c. Na_2SO_3; d. $KHSO_3$; e. Li_3N; f. $Cr_2(CO_3)_3$; g. $Cr(C_2H_3O_2)_2$; h. SnF_4; i. NH_4HSO_4 (composed of NH_4^+ and HSO_4^- ions); j. $(NH_4)_2HPO_4$; k. $KClO_4$; l. NaH; m. HBrO; n. HBr **51.** a. lead(II) acetate; b. copper(II) sulfate; c. calcium oxide; d. magnesium sulfate; e. magnesium hydroxide; f. calcium sulfate; g. dinitrogen monoxide or nitrous oxide (common) **53.** a. nitric acid, HNO_3; b. perchloric acid, $HClO_4$; c. acetic acid, $HC_2H_3O_2$; d. sulfuric acid, H_2SO_4; e. phosphoric acid, H_3PO_4 **55.** 116 g S; 230. g O **57.** a. True; b. False. The isotope has 34 protons; c. False. The isotope has 45 neutrons; d. False. The identity is selenium, Se. **59.** Ra; 142 n **61.** SeO_4^{2-}: selenate; SeO_3^{2-}: selenite; TeO_4^{2-}: tellurate; TeO_3^{2-}: tellurite **63.** InO, atomic mass of In = 76.54; In_2O_3, atomic mass of In = 114.8 **65.** $SbCl_3$; antimony(III) chloride **67.** chlorine; 18 electrons **69.** a. The compounds have the same number and types of atoms (same formula), but the atoms in the molecules are bonded together differently. Therefore, the two compounds are different compounds with different properties. The compounds are called isomers of each other. b. When wood burns, most of the solid material in wood is converted to gases, which escape. The gases produced are most likely CO_2 and H_2O. c. The atom is not an indivisible particle, but is instead composed of other smaller particles—electrons, neutrons, and protons. d. The two hydride samples contain different isotopes of either hydrogen or lithium. Although the compounds are composed of different isotopes, their properties are similar because different isotopes of the same element have similar properties (except, of course, their mass). **71.** The ratio of the masses of R that combine with 1.00 g Q is 3:1, as expected by the law of multiple proportions. R_3Q **73.** C:H ratio = 8:18 or 4:9

Chapter 3

23. 47.88 amu; Ti **25.** 185 amu **27.** There are three peaks in the mass spectrum, each two mass units apart. This is consistent with two isotopes, differing in mass by two mass units. The peak at 157.84 corresponds to a Br_2 molecule composed of two atoms of the lighter isotope. This isotope has mass equal to 157.84/2 or 78.92, which corresponds to ^{79}Br. The second isotope is ^{81}Br with mass equal to 161.84/2 = 80.92. The peaks in the mass spectrum correspond to $^{79}Br_2$, $^{79}Br^{81}Br$, and $^{81}Br_2$, in order of increasing mass. The intensities of the highest and lowest masses tell us the two isotopes are present at about equal abundance. The actual abundance is 50.68% ^{79}Br and 49.32% ^{81}Br. **29.** GaAs can be either $^{69}GaAs$ or $^{71}GaAs$. The mass spectrum for GaAs will have 2 peaks at 144 (69 + 75) and 146 (71 + 75) with intensities in the ratio of 60:40 or 3:2. Ga_2As_2 can be $^{69}Ga_2As_2$, $^{69}Ga^{71}GaAs_2$, or $^{71}Ga_2As_2$. The mass spectrum will have 3 peaks at 288, 290, and 292 with intensities in the ratio of 36:48:16 or 9:12:4. **31.** a. 1.03×10^{-4} mol; b. 4.52×10^{-3} mol; c. 3.41×10^{-2} mol **33.** 4.0 g He < 1.0 mol F_2 < 44.0 g CO_2 < 4.0 g H_2 < 146 g SF_6 **35.** a. 165.39 g/mol; b. 3.023 mol; c. 3.3 g; d. 5.5×10^{22} atoms; e. 1.6 g; f. 1.373×10^{-19} g **37.** 71.40% C; 8.689% H; 5.648% F; 14.26% O **39.** a. 46.681%; b. 30.447%; c. 30.447%; d. 63.649% **41.** 6.54×10^4 g/mol **43.** a. 40.002% C, 6.7135% H, 53.285% O; b. 40.002% C, 6.7136% H, 53.284% O; c. 40.002% C, 6.7135% H, 53.285% O; All three compounds have the same empirical formula, CH_2O, but different molecular formulas. The composition of all three in mass percent is also the same (within rounding differences). Therefore, elemental analysis will give us only the empirical formula. **45.** HgO and Hg_2O **47.** $C_7H_5N_3O_6$ **49.** $C_3H_5O_2$; $C_6H_{10}O_4$ **51.** $C_3H_4O_3$; $C_6H_8O_6$ **53.** a. $C_6H_{12}O_6(s) + 6O_2(g) \rightarrow 6CO_2(g) + 6H_2O(g)$; b. $Fe_2S_3(s) + 6HCl(g) \rightarrow 2FeCl_3(s) + 3H_2S(g)$; c. $CS_2(l) + 2NH_3(g) \rightarrow H_2S(g) + NH_4SCN(s)$ **55.** a. $16Cr(s) + 3S_8(s) \rightarrow 8Cr_2S_3(s)$; b. $2NaHCO_3(s) \rightarrow Na_2CO_3(s) + CO_2(g) + H_2O(g)$; c. $2KClO_3(s) \rightarrow 2KCl(s) + 3O_2(g)$; d. $2Eu(s) + 6HF(g) \rightarrow 2EuF_3(s) + 3H_2(g)$; e. $2C_6H_6(l) + 15O_2(g) \rightarrow 12CO_2(g) + 6H_2O(g)$ **57.** 4355 g **59.** 21.5 g Fe_2O_3; 7.26 g Al; 13.7 g Al_2O_3 **61.** 2.8 days **63.** 150 g **65.** $2NO(g) + O_2(g) \rightarrow 2NO_2(g)$; NO is limiting. **67.** 0.301 g H_2O_2; 3.6×10^{-2} g HCl **69.** 1.20 metric tons **71.** 2.81×10^6 g HCN; 5.63×10^6 g H_2O **73.** 99.8 g F_2 **75.** 4.30×10^{-2} mol; 2.50 g **77.** M is yttrium, and X is chlorine. Yttrium(III) chloride; 1.84 g **79.** 5 **81.** Al_2Se_3 **83.** 42.8% **85.** 86.2% **87.** 86.92 amu **89.** I. NH_3; II. N_2H_4; III. HN_3; If we set the atomic mass of H equal to 1.008, then the atomic mass of N is 14.01. **91.** 87.8 amu **93.** $C_{20}H_{30}O$ **95.** The gas mixture consists of $^{16}O^{16}O$, $^{16}O^{18}O$, and ^{40}Ar. The isotope composition is 42.82% ^{16}O, 8.6×10^{-2}% ^{18}O, and 57.094% ^{40}Ar. **97.** 207 amu; Pb **99.** Ge **101.** 184 amu **103.** 1.05 mol **105.** 10.% La^{2+}, 90.% La^{3+} **107.** 32.9% **109.** 0.48 mol

Chapter 4

11. a. *Polarity* is a term applied to covalent compounds. Polar covalent compounds have an unequal sharing of electrons in bonds that results in unequal charge distribution in the overall molecule. Polar molecules have a partial negative end and a partial positive end. These are not full charges as in ionic compounds but are charges much smaller in magnitude. Water is a polar molecule and dissolves other polar solutes readily. The oxygen end of water (the partial negative end of the polar water molecule) aligns with the partial positive end of the polar solute, whereas the hydrogens of water (the partial positive end of the polar water molecule) align with the partial negative end of the solute. These opposite-charge attractions stabilize polar solutes in water. This process is called *hydration*. Nonpolar solutes do not have permanent partial negative and partial positive ends; nonpolar solutes are not stabilized in water and do not dissolve. b. KF is a soluble ionic compound, so it is a strong electrolyte. KF(*aq*) actually exists as separate hydrated K^+ ions and hydrated F^- ions in solution: $C_6H_{12}O_6$ is a polar covalent molecule that is a nonelectrolyte. $C_6H_{12}O_6$ is hydrated as described in part a. c. RbCl is a soluble ionic compound, so it exists as separate hydrated Rb^+ ions and hydrated Cl^- ions in solution. AgCl is an insoluble ionic compound, so the ions stay together in solution and fall to the bottom of the container as a precipitate. d. HNO_3 is a strong acid and exists as separate hydrated H^+ ions and hydrated NO_3^- ions in solution. CO is a polar covalent molecule and is hydrated as explained in part a. **13.** a. picture iv; b. picture ii; c. picture iii; d. picture i; **15.** a. Place 20.0 g NaOH in a 2-L volumetric flask; add water to dissolve the NaOH, and fill to the mark with water, mixing several times along the way. b. Add 500. mL of 1.00 *M* NaOH stock solution to a 2-L volumetric flask; fill to the mark with water, mixing several times along the way. c. Similar to the solution made in part a, instead using 38.8 g K_2CrO_4. d. Similar to the solution made in part b, instead using 114 mL of the 1.75 *M* K_2CrO_4 stock solution. **17.** 4.5 *M* **19.** 5.95×10^{-8} *M* **21.** a. 2.5×10^{-8} *M*; b. 8.4×10^{-9} *M*; c. 1.33×10^{-4} *M*; d. 2.8×10^{-7} *M* **23.** Bromides: NaBr, KBr, and NH_4Br (and others) would be soluble, and AgBr, $PbBr_2$, and Hg_2Br_2 would be insoluble. Sulfates: Na_2SO_4, K_2SO_4, and $(NH_4)_2SO_4$ (and others) would be soluble, and $BaSO_4$, $CaSO_4$, and $PbSO_4$ (or Hg_2SO_4) would be insoluble. Hydroxides: NaOH, KOH, $Ca(OH)_2$ (and others) would be soluble, and $Al(OH)_3$, $Fe(OH)_3$, and $Cu(OH)_2$ (and others) would be insoluble. Phosphates: Na_3PO_4, K_3PO_4, $(NH_4)_3PO_4$ (and others) would be soluble, and Ag_3PO_4, $Ca_3(PO_4)_2$, and $FePO_4$ (and others) would be insoluble. Lead: $PbCl_2$, $PbBr_2$, PbI_2, $Pb(OH)_2$, $PbSO_4$, and PbS (and others) would be insoluble. $Pb(NO_3)_2$ would be a soluble Pb^{2+} salt. **25.** a. $(NH_4)_2SO_4(aq) + Ba(NO_3)_2(aq) \rightarrow 2NH_4NO_3(aq) + BaSO_4(s)$; $2NH_4^+(aq) + SO_4^{2-}(aq) + Ba^{2+}(aq) + 2NO_3^-(aq) \rightarrow 2NH_4^+(aq) + 2NO_3^-(aq) + BaSO_4(s)$; $Ba^{2+}(aq) + SO_4^{2-}(aq) \rightarrow BaSO_4(s)$; b. $Pb(NO_3)_2(aq) + 2NaCl(aq) \rightarrow PbCl_2(s) + 2NaNO_3(aq)$; $Pb^{2+}(aq) + 2NO_3^-(aq) + 2Na^+(aq) + 2Cl^-(aq) \rightarrow PbCl_2(s) + 2Na^+(aq) + 2NO_3^-(aq)$; $Pb^{2+}(aq) + 2Cl^-(aq) \rightarrow PbCl_2(s)$; c. No reaction occurs since all possible products are soluble. d. No reaction occurs since all possible products are soluble. e. $CuCl_2(aq) + 2NaOH(aq) \rightarrow Cu(OH)_2(s) + 2NaCl(aq)$; $Cu^{2+}(aq) + 2Cl^-(aq) + 2Na^+(aq) + 2OH^-(aq) \rightarrow Cu(OH)_2(s) + 2Na^+(aq) + 2Cl^-(aq)$; $Cu^{2+}(aq) + 2OH^-(aq) \rightarrow Cu(OH)_2(s)$ **27.** a. When $CuSO_4(aq)$ is added to $Na_2S(aq)$, the precipitate that forms is CuS(*s*). Therefore, Na^+ (the gray spheres) and SO_4^{2-} (the bluish green spheres) are the spectator ions. $CuSO_4(aq) + Na_2S(aq) \rightarrow CuS(s) + Na_2SO_4(aq)$; $Cu^{2+}(aq) + S^{2-}(aq) \rightarrow CuS(s)$ b. When $CoCl_2(aq)$ is added to NaOH(*aq*), the precipitate that forms is $Co(OH)_2(s)$. Therefore, Na^+ (the gray spheres) and Cl^- (the green spheres) are the spectator ions. $CoCl_2(aq) + 2NaOH(aq) \rightarrow Co(OH)_2(s) + 2NaCl(aq)$; $Co^{2+}(aq) + 2OH^-(aq) \rightarrow Co(OH)_2(s)$ c. When $AgNO_3(aq)$ is added to KI(*aq*), the precipitate that forms is AgI(*s*). Therefore, K^+ (the red spheres) and NO_3^- (the blue spheres) are the spectator ions. $AgNO_3(aq) + KI(aq) \rightarrow AgI(s) + KNO_3(aq)$; $Ag^+(aq) + I^-(aq) \rightarrow AgI(s)$ **29.** From the solubility rules in Table 4.1, the possible cations could be Ba^{2+} and Ca^{2+}. **31.** 2.9 g AgCl; 0.050 *M* Cl^-; 0.10 *M* NO_3^-; 0.075 *M* Ca^{2+} **33.** 0.607 g **35.** 16.2% **37.** 39.49 mg/tablet; 67.00% **39.** 23 amu; Na **41.** a. Perchloric acid reacted with potassium hydroxide is a possibility. $HClO_4(aq) + KOH(aq) \rightarrow H_2O(l) + KClO_4(aq)$; b. Nitric acid reacted with cesium hydroxide is a possibility. $HNO_3(aq) + CsOH(aq) \rightarrow H_2O(l) + CsNO_3(aq)$; c. Hydroiodic acid reacted with calcium hydroxide is a possibility. $2HI(aq) + Ca(OH)_2(aq) \rightarrow 2H_2O(l) + CaI_2(aq)$ **43.** a. 50.0 mL; b. 25.0 mL; c. 8.33 mL; d. 33.3 mL; e. 25.0 mL; f. 8.33 mL **45.** The acid is a diprotic acid (H_2A), meaning that it has two H^+ ions in the formula to donate to a base. The reaction is $H_2A(aq) + 2NaOH(aq) \rightarrow 2H_2O(l) + Na_2A(aq)$, where A^{2-} is what is left over from the acid formula when the two protons (H^+ ions) are reacted. For the HCl reaction, the base has the ability to accept two protons. The most common examples are $Ca(OH)_2$, $Sr(OH)_2$, and $Ba(OH)_2$. A possible reaction would be $2HCl(aq) + Ca(OH)_2(aq) \rightarrow 2H_2O(l) + CaCl_2(aq)$. **47.** 4.7×10^{-2} *M* **49.** a. 0.8393 *M*; b. 5.010% **51.** The resulting solution is not neutral. 5.9×10^{-3} *M* OH^- **53.** 2.0×10^{-2} *M* OH^- **55.** a. K, +1; O, −2; Mn, +7; b. Ni, +4; O, −2; c. Fe, +2; d. H, +1; O, −2; N, −3; P, +5; e. P, +3; O, −2; f. O, −2; Fe, $+\frac{8}{3}$; g. O, −2; F, −1; Xe, +6; h. S, +4; F, −1; i. C, +2; O, −2; j. C, 0; H, +1; O, −2 **57.** a. Sr, +2; O, −2; Cr, +6; b. Cu, +2; Cl, −1; c. O, 0; d. H, +1; O, −1; e. Mg, +2; O, −2; C, +4; f. Ag, 0; g. Pb, +2; O, −2; S, +4; h. O, −2; Pb, +4; i. Na, +1; O, −2; C, +3; j. O, −2; C, +4; k. H, +1; N, −3; O, −2; S, +6; Ce, +4; l. O, −2; Cr, +3 **59.** a. $2Al(s) + 6HCl(aq) \rightarrow 2AlCl_3(aq) + 3H_2(g)$; H is reduced, and Al is oxidized. b. $CH_4(g) + 4S(s) \rightarrow CS_2(l) + 2H_2S(g)$; S is reduced, and C is oxidized. c. $C_3H_8(g) + 5O_2(g) \rightarrow 3CO_2(g) + 4H_2O(l)$; O is reduced, and C is oxidized. d. $Cu(s) + 2Ag^+(aq) \rightarrow 2Ag(s) + Cu^{2+}(aq)$; Ag is reduced, and Cu is oxidized. **61.** a. $3Cu(s) + 8H^+(aq) + 2NO_3^-(aq) \rightarrow 3Cu^{2+}(aq) + 2NO(g) + 4H_2O(l)$; b. $14H^+(aq) + Cr_2O_7^{2-}(aq) + 6Cl^-(aq) \rightarrow 3Cl_2(g) + 2Cr^{3+}(aq) + 7H_2O(l)$; c. $Pb(s) + 2H_2SO_4(aq) + PbO_2(s) \rightarrow 2PbSO_4(s) + 2H_2O(l)$; d. $14H^+(aq) + 2Mn^{2+}(aq) + 5NaBiO_3(s) \rightarrow 2MnO_4^-(aq) + 5Bi^{3+}(aq) + 5Na^+(aq) + 7H_2O(l)$; e. $8H^+(aq) + H_3AsO_4(aq) + 4Zn(s) \rightarrow 4Zn^{2+}(aq) + AsH_3(g) + 4H_2O(l)$; f. $7H_2O(l) + 4H^+(aq) + 3As_2O_3(s) + 4NO_3^-(aq) \rightarrow 4NO(g) + 6H_3AsO_4(aq)$; g. $16H^+(aq) + 2MnO_4^-(aq) + 10Br^-(aq) \rightarrow 5Br_2(l) + 2Mn^{2+}(aq) + 8H_2O(l)$; h. $8H^+(aq) + 3CH_3OH(aq) + Cr_2O_7^{2-}(aq) \rightarrow 2Cr^{3+}(aq) + 3CH_2O(aq) + 7H_2O(l)$ **63.** a. $8HCl(aq) + 2Fe(s) \rightarrow 2HFeCl_4(aq) + 3H_2(g)$; b. $6H^+(aq) + 8I^-(aq) + IO_3^-(aq) \rightarrow 3I_3^-(aq) + 3H_2O(l)$; c. $97Ce^{4+}(aq) + 54H_2O(l) + Cr(NCS)_6^{4-}(aq) \rightarrow 97Ce^{3+}(aq) + Cr^{3+}(aq) + 6NO_3^-(aq) + 6CO_2(g) + 6SO_4^{2-}(aq) + 108H^+(aq)$; d. $64OH^-(aq) + 2CrI_3(s) + 27Cl_2(g) \rightarrow 54Cl^-(aq) + 2CrO_4^{2-}(aq) + 6IO_4^-(aq) + 32H_2O(l)$; e. $258OH^-(aq) + Fe(CN)_6^{4-}(aq) + 61Ce^{4+}(aq) \rightarrow Fe(OH)_3(s) + 61Ce(OH)_3(s) + 6CO_3^{2-}(aq) + 6NO_3^-(aq) + 36H_2O(l)$ **65.** 1.622×10^{-2} *M* **67.** 34.6% **69.** 49.4 mL **71.** 173 mL **73.** a. 14.2%; b. 8.95 mL **75.** a. 24.8% Co, 29.7% Cl, 5.09% H, 40.4% O; b. $CoCl_2 \cdot 6H_2O$; c. $CoCl_2 \cdot 6H_2O(aq) + 2AgNO_3(aq) \rightarrow 2AgCl(s) + Co(NO_3)_2(aq) + 6H_2O(l)$, $CoCl_2 \cdot 6H_2O(aq) + 2NaOH(aq) \rightarrow Co(OH)_2(s) + 2NaCl(aq) + 6H_2O(l)$, $4Co(OH)_2(s) + O_2(g) \rightarrow 2Co_2O_3(s) + 4H_2O(l)$ **77.** 72.4% KCl; 27.6% NaCl **79.** 2.00 *M* **81.** 0.0785 ± 0.0002 *M* **83.** three acidic hydrogens **85.** a. 31.3%; b. 6.00 *M* **87.** 14.6 g Zn, 14.4 g Ag **89.** 77.1% KCl; 22.9% KBr **91.** a. $\frac{\text{mass of AgCl}}{\text{mass of PCB}} = \frac{143.4\,n}{154.20 + 34.44\,n}$ or $\text{mass}_{AgCl}\,(154.20 + 34.44\,n) = \text{mass}_{PCB}\,(143.4\,n)$; b. 7.097 **93.** a. 5.35×10^4 L/s; b. 4.25 ppm; c. 1.69×10^6 g; d. 10.3 ppm **95.** a. $YBa_2Cu_3O_{6.5}$: +2; only Cu^{2+} present; $YBa_2Cu_3O_7$: +2.33; two Cu^{2+} and one Cu^{3+} present; $YBa_2Cu_3O_8$: +3; only Cu^{3+} present; b. $2Cu^{2+}(aq) + 5I^-(aq) \rightarrow 2CuI(s) + I_3^-(aq)$; $Cu^{3+}(aq) + 4I^-(aq) \rightarrow CuI(s) + I_3^-(aq)$; $2S_2O_3^{2-}(aq) + I_3^-(aq) \rightarrow 3I^-(aq) + S_4O_6^{2-}(aq)$; c. $YBa_2Cu_3O_{7.25}$; +2.50 **97.** 24.99% $AgNO_3$; 40.07% $CuCl_2$; 34.94% $FeCl_3$

Chapter 5

21. 47.5 torr; 6.33×10^3 Pa; 6.25×10^{-2} atm **23.** 1.01×10^5 Pa; 10.3 m **25.** a. 3.6×10^3 mm Hg; b. 3.6×10^3 torr; c. 4.9×10^5 Pa; d. 71 psi **27.** $P_{H_2} = 317$ torr; $P_{N_2} = 50.7$ torr; $P_{Total} = 368$ torr **29.** 309 g **31.** 3.08 atm; $P_{CO_2} = 3.08$ atm, $P_{Total} = 4.05$ atm **33.** 4.44×10^3 g He; 2.24×10^3 g H_2 **35.** 7.0×10^{2}°C **37.** 12.5 mL **39.** $n_2/n_1 = 0.921$ **41.** BrF_3 **43.** 3.69 L **45.** a. $\chi_{CH_4} = 0.412$, $\chi_{O_2} = 0.588$; b. 0.161 mol; c. 1.06 g CH_4, 3.03 g O_2 **47.** $P_{methane} = 1.32$ atm; $P_{ethane} = 0.12$ atm; 33.7 g **49.** N_2H_4 **51.** The calculated molar masses are 209 g/mol from data set I and 202 g/mol from data set II. These values are close to the expected molar mass (207 g/mol) for the divalent metal compound, $Be(C_5H_7O_2)_2$. **53.** 1.16% **55.** 1.5×10^7 g Fe; 2.6×10^7 g 98% H_2SO_4 **57.** 46.5% **59.** $P_{N_2} = 0.74$ atm; $P_{Total} = 2.2$ atm **61.** 18.0% **63.** 13.3% **65.** 0.333 atm **67.** Rigid container (constant volume): As reactants are converted to products, the moles of gas particles present decrease by one-half. As *n* decreases, the pressure will decrease (by one-half). Density is the mass per unit volume. Mass is conserved in a chemical reaction, so the density of the gas will not change since mass and volume do not change. Flexible container (constant pressure): Pressure is constant since the container changes volume to keep a constant pressure. As the moles of gas particles decrease by a factor of 2, the volume of the container will decrease (by one-half). We have the same mass of gas in a smaller volume, so the gas density will increase (doubles). **69.** $2NH_3(g) \rightarrow N_2(g) + 3H_2(g)$: As reactants are converted into products, we go from 2 mol of gaseous reactants to 4 mol of gaseous products (1 mol N_2 + 3 mol H_2). Because the moles of gas double as reactants are converted into products, the volume of the gases will double (at constant *P* and *T*). Pressure is directly related to *n* at constant *T* and *V*. As the reaction occurs, the moles of gas will double, so the pressure will double. Because 1 mol of N_2 is produced for every 2 mol of NH_3 reacted, $P_{N_2} = \frac{1}{2}P^{\circ}_{NH_3}$. Due to the 3:2 mole ratio in the balanced equation, $P_{H_2} = \frac{3}{2}P^{\circ}_{NH_3}$. Note: $P_{Total} = P_{H_2} + P_{N_2} = \frac{3}{2}P^{\circ}_{NH_3} + \frac{1}{2}P^{\circ}_{NH_3} = 2P^{\circ}_{NH_3}$. As said earlier, the total pressure will double from the initial pressure of NH_3 as the reactants are completely converted into products. **71.** a. Both gas samples have the same number of molecules present (*n* is constant); b. Since *T* is constant, $(KE)_{avg}$ must be the same for both gases $[(KE)_{avg} = \frac{3}{2}RT]$; c. The lighter gas A molecules will have the faster average velocity; d. The heavier gas B molecules do collide more forcefully, but gas A molecules, with the faster average velocity, collide more frequently. The end result is that *P* is constant between the two containers. **73.** 3.40×10^3 J/mol $= 5.65 \times 10^{-21}$ J/molecule (for each gas at 273 K); 6.81×10^3 J/mol $= 1.13 \times 10^{-20}$ J/molecule (for each gas at 546 K) **75.** No; there is a distribution of energies with only the average kinetic energy equal to $\frac{3}{2}RT$. Similarly, there is always a distribution of velocities for a gas sample at some temperature. **77.** a. All the same; b. Flask C; c. Flask A **79.** CF_2Cl_2 **81.** 63.7 g/mol **83.** a. 12.24 atm; b. 12.13 atm; c. The ideal gas law is high by 0.91%. **85.** The kinetic molecular theory assumes that gas particles do not exert forces on each other and that gas particles are volumeless. Real gas particles do exert attractive forces on each other, and real gas particles do have volumes. A gas behaves most ideally at low pressures and high temperatures. The effect of attractive forces is minimized at high temperatures since the gas particles are moving very rapidly. At low pressure, the container volume is relatively large (*P* and *V* are inversely related), so the volume of the container taken up by the gas particles is negligible. **87.** The pressure measured for real gases is too low compared to ideal gases. This is due to the attractions gas particles do have for each other; these attractions "hold" them back from hitting the container walls as forcefully. To make up for this slight decrease in pressure for real gases, a factor is added to the measured pressure. The measured volume is too large. A fraction of the space of the container volume is taken up by the volume of gas of the molecules themselves. Therefore, the actual volume available to real gas molecules is slightly less than the container volume. A term is subtracted from the container volume to correct for the volume taken up by real gas molecules. **89.** CO_2 since it has the largest *a* value. **91.** u_{rms} = 667 m/s; u_{mp} = 545 m/s; u_{avg} = 615 m/s **93.** Impact force (H_2)/impact force (He) = 0.7097 **95.** The change in momentum per impact is 2.827 times larger for O_2 molecules than for He atoms. There are 2.827 times as many impacts per second for He as compared with those for O_2. **97.** 1.0×10^9 collisions/s; 1.3×10^{-6} m **99.** a. 0.19 torr; b. 6.6×10^{21} molecules/m^3; c. 6.6×10^{15} molecules/cm^3 **101.** Benzene: 9.47×10^{-3} ppmv; 2.31×10^{11} molecules/cm^3; toluene: 1.37×10^{-2} ppmv; 3.33×10^{11} molecules/cm^3 **103.** 46 mL **105.** $MnCl_4$ **107.** 1.61×10^3 g **109.** 0.990 atm; 0.625 g Zn **111.** P_{He} = 50.0 torr; P_{Ne} = 76.0 torr; P_{Ar} = 90.0 torr; P_{Total} = 216.0 torr **113.** a. 78.0%; b. 0.907 L **115.** 1490 **117.** 60.6 kJ **119.** 7.00 mL **121.** a. $2CH_4(g) + 2NH_3(g) + 3O_2(g) \rightarrow 2HCN(g) + 6H_2O(g)$ b. 15.6 g/s **123.** 30.% **125.** 29.0% **127.** $dT = \frac{P(\text{molar mass})}{R}$ = constant so $d = \text{constant}\left(\frac{1}{T}\right)$; $-272.6°C$ **129.** 16.03 g/mol **131.** From Figure 5.16 of the text, as temperature increases, the probability that a gas particle has the most probable velocity decreases. Since the probability of the gas particle with the most probable velocity decreased by one-half, the temperature must be higher than 300. K. The temperature is 1.20×10^3 K. **133.** 1.3 L **135.** χ_{CO} = 0.291; χ_{CO_2} = 0.564; χ_{O_2} = 0.145 **137.** a. 8.7×10^3 L air/min; b. χ_{CO} = 0.0017, χ_{CO_2} = 0.032, χ_{O_2} = 0.13, χ_{N_2} = 0.77, χ_{H_2O} = 0.067; c. P_{CO} = 0.0017 atm, P_{CO_2} = 0.032 atm, P_{O_2} = 0.13 atm, P_{N_2} = 0.77 atm, P_{H_2O} = 0.067 atm **139.** 2.1×10^2 stages **141.** a. A given volume of air at a given set of conditions has a larger density than helium at those conditions. We need to heat the air to greater than 25°C to lower the air density (by driving air out of the hot-air balloon) until the density is the same as that for helium (at 25°C and 1.00 atm). b. 2150 K **143.** C_3H_8 is possible.

Chapter 6

11. $2NOCl(g) \rightleftharpoons 2NO(g) + Cl_2(g)$; $K = 1.6 \times 10^{-5}$ mol/L
The expression for *K* is the product concentrations divided by the reactant concentrations. When *K* has a value much less than 1, the product concentrations are relatively small and the reactant concentrations are relatively large.

$$2NO(g) \rightleftharpoons N_2(g) + O_2(g);\ K = 1 \times 10^{31}$$

When *K* has a value much greater than 1, the product concentrations are relatively large and the reactant concentrations are relatively small. In both cases, however, the rate of the forward reaction equals the rate of the reverse reaction at equilibrium (this is a definition of equilibrium). **13.** No, it doesn't matter in which direction the equilibrium position is reached. Both experiments will give the same equilibrium position since both experiments started with stoichiometric amounts of reactants or products. **15.** When equilibrium is reached, there is no net change in the amount of reactants and products present since the rates of the forward and reverse reactions are equal. The first diagram has 4 A_2B molecules, 2 A_2 molecules, and 1 B_2 molecule present. The second diagram has 2 A_2B molecules, 4 A_2 molecules, and 2 B_2 molecules. The first diagram cannot represent equilibrium because there was a net change in reactants and products. Is the second diagram the equilibrium mixture? That depends on whether there is a net change between reactants and products when going from the second diagram to the third diagram. The third diagram contains the same number and type of molecules as the second diagram, so the second diagram is the first illustration that represents equilibrium. The reaction container initially contained only A_2B. From the first diagram, 2 A_2 molecules and 1 B_2 molecule are present (along with 4 A_2B molecules). From the balanced reaction, these 2 A_2 molecules and 1 B_2 molecule were formed when 2 A_2B molecules decomposed. Therefore, the initial number of A_2B molecules present equals 4 + 2 = 6 A_2B molecules. **17.** *K* and K_p are equilibrium constants as determined by the law of mass action. For *K*, concentration units of mol/L are used, and for K_p, partial pressures in units of atm are used (generally). *Q* is called the reaction quotient. *Q* has the exact same form as *K* or K_p, but instead of equilibrium concentrations, initial concentrations are used to calculate the *Q* value. The use of *Q* is when it is compared to the *K* value. When $Q = K$ (or when $Q_p = K_p$), the reaction is at equilibrium. When $Q \neq K$, the reaction is not at equilibrium, and one can deduce the net change that must occur for the system to get to equilibrium.
19. a. $K = \frac{[H_2O]}{[NH_3]^2[CO_2]}$; $K_p = \frac{P_{H_2O}}{P_{NH_3}{}^2 \times P_{CO_2}}$; b. $K = [N_2][Br_2]^3$; $K_p = P_{N_2} \times P_{Br_2}{}^3$; c. $K = [O_2]^3$; $K_p = P_{O_2}{}^3$; d. $K = \frac{[H_2O]}{[H_2]}$; $K_p = \frac{P_{H_2O}}{P_{H_2}}$ **21.** 4.6×10^3 atm^3 **23.** 4.08×10^8 L/mol; yes, this set of concentrations represents a system at equilibrium because the calculated value of *K* using these concentrations gives 4.08×10^8. **25.** 4.07 **27.** 0.72 atm; 0.017 mol/L **29.** 0.056 mol/L **31.** a. decrease; b. will not change; c. will not change; d. increase **33.** a. [HOCl] = 9.2×10^{-3} *M*, [Cl_2O] = 1.8×10^{-2} *M*, [H_2O] = 5.1×10^{-2} *M*; b. [HOCl] = 0.07 *M*, [Cl_2O] = [H_2O] = 0.22 *M* **35.** P_{SO_2} = 0.38 atm, P_{O_2} = 0.44 atm, P_{SO_3} = 0.12 atm **37.** The assumption comes from the value of *K* being much less than 1. For these reactions, the equilibrium mixture will not have a lot of products present; mostly reactants are present at equilibrium. If we define the change that must occur in terms of *x* as the amount (molarity or partial pressure) of a reactant that must react to reach equilibrium, then *x* must be a small number because *K* is a very small number. We want to know the value of *x* in order to solve the problem, so we don't assume *x* = 0. Instead, we concentrate on the equilibrium row in the ICE table. Those reactants (or products) that have equilibrium concentrations in the form of $0.10 - x$ or $0.25 + x$ or $3.5 - 3x$, etc., is where an important assumption can be made. The assumption is that because $K \ll 1$, *x* will be small ($x \ll 1$) and when we add *x* or subtract *x* from some initial concentration, it will make little or no difference. That is, we assume that $0.10 - x \approx 0.10$ or $0.25 + x \approx 0.25$ or $3.5 - 3x \approx 3.5$; we assume that the initial concentration of a substance is equal to the equilibrium concentration. This assumption makes the math much easier and usually gives a value of *x* that is well within 5% of the true value of *x* (we get about the same answer with a lot less work). When the 5% rule fails, the equation must be solved exactly or by using the method of successive approximations (see Appendix A1.4). **39.** [CO_2] = 0.39 *M*; [CO] = 8.6×10^{-3} *M*; [O_2] = 4.3×10^{-3} *M* **41.** 66.0% **43.** a. 1.5×10^8 atm^{-1}; b. $P_{CO} = P_{Cl_2} = 1.8 \times 10^{-4}$ atm; P_{COCl_2} = 5.0 atm **45.** Only statement d is correct. Addition of a catalyst has no effect on the equilibrium position; the reaction just reaches equilibrium more quickly. Statement a is false for reactants that are either solids or liquids (adding more of these has no effect on the equilibrium). Statement b is false always. If temperature remains constant, then the value of *K* is constant. Statement c is false for exothermic reactions where an increase in temperature decreases the value of *K*. **47.** a. no effect; b. shifts left; c. shifts right **49.** $H^+ + OH^- \rightarrow H_2O$; sodium hydroxide (NaOH) will react with the H^+ on the product side of the reaction. This effectively removes H^+ from the equilibrium, which will shift the reaction

to the right to produce more H^+ and CrO_4^{2-}. Since more CrO_4^{2-} is produced, the solution turns yellow. **51.** a. right; b. right; c. no effect; d. left; e. no effect **53.** a. left; b. right; c. left; d. no effect; e. no effect; f. right **55.** An endothermic reaction, where heat is a reactant, will shift right to products with an increase in temperature. The amount of $NH_3(g)$ will increase as the reaction shifts right, so the smell of ammonia will increase. **57.** a. 2×10^3 molecules/cm^3 b. There is more NO in the atmosphere than expected from the value of *K*. The reason is the slow rate (kinetics) of the reaction at low temperatures. Nitric oxide is produced in high-energy or high-temperature environments. In nature some NO is produced by lightning, and the primary man-made source is automobiles. The production of NO is endothermic. At high temperatures *K* will increase and the rates of the reaction will also increase, resulting in a higher production of NO. Once the NO gets into a more normal temperature environment, it doesn't go back to N_2 and O_2 because of the slow rate of the reaction. **59.** 2.6×10^{81} L/mol **61.** $[H_2]_0 = 11.0\ M$, $[N_2]_0 = 10.0\ M$ **63.** a. $P_{PCl_3} = P_{Cl_2} = 0.2230$ atm; $P_{PCl_5} = 0.0259$ atm; $K_p = 1.92$ atm; b. $P_{PCl_3} = 0.0650$ atm; $P_{Cl_2} = 5.44$ atm; $P_{PCl_5} = 0.1839$ atm **65.** $P_{NO_2} = 0.704$ atm; $P_{N_2O_4} = 0.12$ atm **67.** a. $P_{CO_2} = P_{H_2O} = 0.50$ atm; b. 7.5 g $NaHCO_3$, 1.6 g Na_2CO_3; c. 3.9 L **69.** 192 g NH_4HS; 1.3 atm **71.** 6.74×10^{-6} mol^3/L^3 **73.** a. 134 atm^{-1}; b. $P_{NO} = 0.052$ atm; $P_{Br_2} = 0.18$ atm; $P_{NOBr} = 0.25$ atm **75.** $P_{P_4} = 0.73$ atm, $P_{P_2} = 0.270$ atm; 0.16 dissociated **77.** 1.5 atm^{-1} **79.** 0.23 mol^2/L^2 **81.** a. $P_{PCl_5} = 0.137$ atm; $P_{PCl_3} = P_{Cl_2} = 0.191$ atm; b. 39.4 g **83.** 71 atm **85.** 0.63 atm$^{1/2}$ **87.** See the *Solutions Guide* for the plot. The data for the plot follow. At $P_{Total} = 1.0$ atm, $P_{NH_3} = 0.024$ atm; at $P_{Total} = 10.0$ atm, $P_{NH_3} = 1.4$ atm; at $P_{Total} = 100.$ atm, $P_{NH_3} = 32$ atm; at $P_{Total} = 1000.$ atm, $P_{NH_3} = 440$ atm; notice that as P_{Total} increases larger fractions of both N_2 and H_2 are converted to NH_3, i.e., as P_{Total} increases (*V* decreases), the reaction shifts further to the right, as predicted by Le Châtelier's principle. **89.** 4.81 g/L; 5.5×10^{-3} mol/L

Chapter 7

17.

	Acid	Base	Conjugate Base of Acid	Conjugate Acid of Base
a.	H_2CO_3	H_2O	HCO_3^-	H_3O^+
b.	$C_5H_5NH^+$	H_2O	C_5H_5N	H_3O^+
c.	$C_5H_5NH^+$	HCO_3^-	C_5H_5N	H_2CO_3

19. a. $HC_2H_3O_2(aq) \rightleftharpoons H^+(aq) + C_2H_3O_2^-(aq)$; $K_a = \dfrac{[H^+][C_2H_3O_2^-]}{[HC_2H_3O_2]}$; b. $Co(H_2O)_6^{3+}(aq) \rightleftharpoons H^+(aq) + Co(H_2O)_5(OH)^{2+}(aq)$; $K_a = \dfrac{[H^+][Co(H_2O)_5(OH)^{2+}]}{[Co(H_2O)_6^{3+}]}$;

c. $CH_3NH_3^+(aq) \rightleftharpoons H^+(aq) + CH_3NH_2(aq)$; $K_a = \dfrac{[H^+][CH_3NH_2]}{[CH_3NH_3^+]}$

21. The beaker on the left represents a strong acid in solution; the acid HA is 100% dissociated into the H^+ and A^- ions. The beaker on the right represents a weak acid in solution; only a little bit of the acid HB dissociates into ions, so the acid exists mostly as undissociated HB molecules in water.
a. HNO_2: weak acid beaker; b. HNO_3: strong acid beaker; c. HCl: strong acid beaker; d. HF: weak acid beaker; e. $HC_2H_3O_2$: weak acid beaker **23.** $HClO_4 > HClO_2 > NH_4^+ > H_2O$ **25.** a. H_2SO_4; b. HOCl; c. $HC_2H_2ClO_2$ **27.** a. H_2O and $CH_3CO_2^-$; b. An acid–base reaction can be thought of as a competition between two opposing bases. Since this equilibrium lies far to the left ($K_a < 1$), then $CH_3CO_2^-$ is a stronger base than H_2O. c. The acetate ion is a better base than water and produces basic solutions in water. When we put acetate ion into solution as the only major basic species, the reaction is

$$CH_3CO_2^- + H_2O \rightleftharpoons CH_3CO_2H + OH^-$$

Now the competition is between $CH_3CO_2^-$ and OH^- for the proton. Hydroxide ion is the strongest base possible in water. The above equilibrium lies far to the left, resulting in a K_b value less than 1. Those species we specifically call weak bases ($10^{-14} < K_b < 1$) lie between H_2O and OH^- in base strength. Weak bases are stronger than water but are weaker bases than OH^-.
29. a. weak acid; b. strong acid; c. weak base; d. strong base; e. weak base; f. weak acid; g. weak acid; h. strong base; i. strong acid **31.** a. $[H^+] = [OH^-] = 1.71 \times 10^{-7}\ M$; b. 6.767; c. 12.54 **33.** a. $[H^+] = 4.0 \times 10^{-8}\ M$; $[OH^-] = 2.5 \times 10^{-7}\ M$; basic; b. $[H^+] = 5 \times 10^{-16}\ M$; $[OH^-] = 20\ M$; basic; c. $[H^+] = 10\ M$; $[OH^-] = 1 \times 10^{-15}\ M$; acidic; d. $[H^+] = 6.3 \times 10^{-4}\ M$; $[OH^-] = 1.6 \times 10^{-11}\ M$; acidic; e. $[H^+] = 1 \times 10^{-9}\ M$; $[OH^-] = 1 \times 10^{-5}\ M$; basic; f. $[H^+] = 4.0 \times 10^{-5}\ M$; $[OH^-] = 2.5 \times 10^{-10}\ M$; acidic **35.** a. 1.00; b. −0.70; c. 7.00 **37.** Use 4.2 mL of 12 *M* HCl with enough water added to make 1600 mL of solution. **39.** a. H^+, Br^-, H_2O; 0.602; b. H^+, ClO_4^-, H_2O; 0.602; c. H^+, NO_3^-, H_2O; 0.602; d. HNO_2, H_2O; 2.00; e. CH_3CO_2H, H_2O; 2.68; f. HCN, H_2O; 4.92 **41.** a. $[H^+] = [OCl^-] = 8.4 \times 10^{-5}\ M$; $[OH^-] = 1.2 \times 10^{-10}\ M$; $[HOCl] = 0.20\ M$; 4.08; b. $[H^+] = [OC_6H_5^-] = 1.5 \times 10^{-5}\ M$; $[OH^-] = 6.7 \times 10^{-10}\ M$; $[HOC_6H_5] = 1.5\ M$; 4.82; c. $[H^+] = [F^-] = 3.5 \times 10^{-3}\ M$; $[OH^-] = 2.9 \times 10^{-12}\ M$; $[HF] = 0.017\ M$; 2.46 **43.** 2.02 **45.** $[C_6H_5CO_2H] = 4.1 \times 10^{-3}\ M$; $[C_6H_5CO_2^-] = [H^+] = 5.1 \times 10^{-4}\ M$; $[OH^-] = 1.9 \times 10^{-11}\ M$; 3.29 **47.** 2.68 **49.** a. 1.00; b. 1.30 **51.** 0.033 **53.** 3.5×10^{-4} **55.** 0.024 *M* **57.** $NH_3 > C_5H_5N > H_2O > NO_3^-$ **59.** a. $C_6H_5NH_2$; b. $C_6H_5NH_2$; c. OH^-; d. CH_3NH_2 **61.** a. 13.00; b. 7.00; c. 14.30 **63.** $1.6 \times 10^{-4}\ M$ **65.** Neutrally charged organic compounds containing at least one nitrogen atom generally behave as weak bases. The nitrogen atom has an unshared pair of electrons around it. This lone pair of electrons is used to form a bond to H^+. **67.** 12.00 **69.** 9.59 **71.** a. 1.3%; b. 4.2%; c. 6.4% **73.** 1.0×10^{-9}
75. $H_3C_6H_5O_7(aq) \rightleftharpoons H_2C_6H_5O_7^-(aq) + H^+(aq)$; $K_{a_1} = \dfrac{[H_2C_6H_5O_7^-][H^+]}{[H_3C_6H_5O_7]}$; $H_2C_6H_5O_7^-(aq) \rightleftharpoons$

$HC_6H_5O_7^{2-}(aq) + H^+(aq)$; $K_{a_2} = \dfrac{[HC_6H_5O_7^{2-}][H^+]}{[H_2C_6H_5O_7^-]}$;

$HC_6H_5O_7^{2-}(aq) \rightleftharpoons C_6H_5O_7^{3-}(aq) + H^+(aq)$;

$K_{a_3} = \dfrac{[C_6H_5O_7^{3-}][H^+]}{[HC_6H_5O_7^{2-}]}$ **77.** $[H^+] = 3 \times 10^{-2}\ M$, $[OH^-] = 3 \times 10^{-13}\ M$, $[H_3AsO_4] = 0.17\ M$, $[H_2AsO_4^-] = 3 \times 10^{-2}\ M$; $[HAsO_4^{2-}] = 8 \times 10^{-8}\ M$, $[AsO_4^{3-}] = 2 \times 10^{-15}\ M$ **79.** 3.00 **81.** −0.30 **83.** a. These are strong acids such as HCl, HBr, HI, HNO_3, H_2SO_4, and $HClO_4$. b. These are salts of the conjugate acids of the bases in Table 7.3. These conjugate acids are all weak acids. NH_4Cl, $CH_3NH_3NO_3$, and $C_2H_5NH_3Br$ are three examples. Note that the anions used to form these salts (Cl^-, NO_3^-, and Br^-) are conjugate bases of strong acids; this is so because they have no acidic or basic properties in water (with the exception of HSO_4^-, which has weak acid properties). c. These are strong bases such as LiOH, NaOH, KOH, RbOH, CsOH, $Ca(OH)_2$, $Sr(OH)_2$, and $Ba(OH)_2$. d. These are salts of the conjugate bases of the neutrally charged weak acids in Table 7.2. The conjugate bases of weak acids are weak bases themselves. Three examples are $NaClO_2$, $KC_2H_3O_2$, and CaF_2. The cations used to form these salts are Li^+, Na^+, K^+, Rb^+, Cs^+, Ca^{2+}, Sr^{2+}, and Ba^{2+} since these cations have no acidic or basic properties in water. Notice that these are the cations of the strong bases you should memorize. e. There are two ways to make a neutral salt. The easiest way is to combine a conjugate base of a strong acid (except for HSO_4^-) with one of the cations from a strong base. These ions have no acidic/basic properties in water so salts of these ions are neutral. Three examples are NaCl, KNO_3, and SrI_2. Another type of strong electrolyte that can produce neutral solutions are salts that contain an ion with weak acid properties combined with an ion of opposite charge having weak base properties. If the K_a for the weak acid ion is equal to the K_b for the weak base ion, then the salt will produce a neutral solution. The most common example of this type of salt is ammonium acetate ($NH_4C_2H_3O_2$). For this salt, K_a for NH_4^+ = K_b for $C_2H_3O_2^-$ = 5.6×10^{-10}. This salt at any concentration produces a neutral solution. **85.** a. HI < HF < NaI < NaF; b. HBr < NH_4Br < KBr < NH_3; c. HNO_3 < $C_6H_5NH_3NO_3$ < HOC_6H_5 < $NaNO_3$ < $C_6H_5NH_2$ < KOC_6H_5 < NaOH **87.** a. Neutral; K^+ and Cl^- have no acidic or basic properties; b. Basic; K_b for CN^- > K_a for $C_2H_5NH_3^+$ (CN^- is a better base than $C_2H_5NH_3^+$ is as an acid); c. Acidic; K_a for $C_5H_5NH^+$ > K_b for F^-; d. Neutral; K_a for NH_4^+ = K_b for $C_2H_3O_2^-$; e. Acidic; HSO_3^- is a stronger acid than a base because K_a for HSO_3^- > K_b for HSO_3^-; f. Basic; HCO_3^- is a stronger base than an acid because K_b for HCO_3^- > K_a for HCO_3^-. **89.** a. 8.23; b. 10.56; c. 4.82 **91.** $[HN_3] = [OH^-] = 2.3 \times 10^{-6}$ *M*; $[N_3^-] = 0.010$ *M*; $[Na^+] = 0.010$ *M*; $[H^+] = 4.3 \times 10^{-9}$ *M* **93.** NaF; this was determined by calculating K_b for F^-. **95.** 3.00 **97.** 8.37 **99.** 7.4; when an acid is added to water, the pH cannot be basic. Must account for the autoionization of water. **101.** 6.24 **103.** 6.15 **105.** a. 1.8×10^9; b. 2.5×10^3; c. 3.1×10^{-5}; d. 1.0×10^{14}; e. 5.6×10^4; f. 4.0×10^{10} **107.** NH_4Cl **109.** 11.77 **111.** 3.36 **113.** a. 1.66; b. −0.78; c. Because of the lower charge, $Fe^{2+}(aq)$ will not be as strong an acid as $Fe^{3+}(aq)$. A solution of iron(II) nitrate will be less acidic (have a higher pH) than a solution with the same concentration of iron(III) nitrate. **115.** 0.022 *M* **117.** 0.022 *M* **119.** a. 2.80; b. 1.1×10^{-3} *M* **121.** 1.96 **123.** a. 1.1×10^{-4}; b. $[H_2CO_3] = [CO_3^{2-}]$; c. $pH = \frac{pK_{a_1} + pK_{a_2}}{2}$ (see *Solutions Guide* for derivation); d. 8.35 **125.** 20.0 g **127.** K_a for HX = 1.0×10^{-5} **129.** 2.492 **131.** 6.18 L **133.** 10.00 **135.** 7.20 **137.** 6.72

Chapter 8

15. A buffer solution is one that resists a change in its pH when either hydroxide ions or protons (H^+) are added. Any solution that contains a weak acid and its conjugate base or a weak base and its conjugate acid is classified as a buffer. The pH of a buffer depends on the [base]/[acid] ratio. When H^+ is added to a buffer, the weak base component of the buffer reacts with the H^+ and forms the acid component of the buffer. Even though the concentrations of the acid and base components of the buffer change some, the ratio of [base]/[acid] does not change that much. This translates into a pH that doesn't change much. When OH^- is added to a buffer, the weak acid component is converted into the base component of the buffer. Again, the ratio of [base]/[acid] does not change a lot (unless a large quantity of OH^- is added), so the pH does not change much. $H^+(aq) + CO_3^{2-}(aq) \rightarrow HCO_3^-(aq)$; $OH^-(aq) + HCO_3^-(aq) \rightarrow CO_3^{2-}(aq) + H_2O(l)$ **17.** When [weak acid] > [conjugate base], pH < pK_a. When [conjugate base] > [weak acid], pH > pK_a. **19.** a. 2.96; b. 8.94; c. 7.00; d. 4.89 **21.** a. 4.29; b. 12.30; c. 12.30; d. 5.07 **23.** 3.37 **25.** 3.48; 3.14 **27.** 4.37 **29.** a. 0.19; b. 0.59; c. 1.0; d. 1.9 **31.** a. 0.50 mol; b. 0.78 mol; c. 0.36 mol **33.** 15 g **35.** a. 1.2 *M*; b. 4.32 **37.** a. 0.091; b. 1.1 ≈ 1; c. A best buffer has approximately equal concentrations of weak acid and conjugate base so that pH ≈ pK_a for a best buffer. The pK_a value for a $H_3PO_4/H_2PO_4^-$ buffer is $-\log(7.5 \times 10^{-3}) = 2.12$. A pH of 7.1 is too high for a $H_3PO_4/H_2PO_4^-$ buffer to be effective. At this high pH, there would be so little H_3PO_4 present that we could hardly consider it a buffer; this solution would not be effective in resisting pH changes, especially when a strong base is added. **39.** Only mixture c results in a buffered solution. **41.** HOCl; there are many possibilities. One possibility is a solution with [HOCl] = 1.0 *M* and [NaOCl] = 0.35 *M*. **43.** 7.0×10^{-7} *M* **45.** 6.89 **47.**

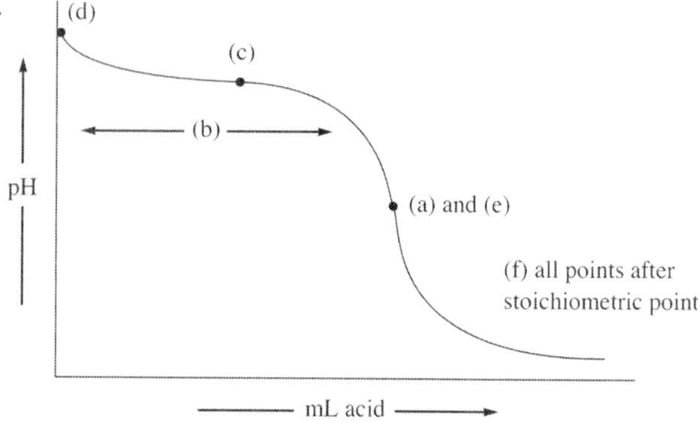

$B + H^+ \rightarrow BH^+$; added H^+ from the strong acid converts the weak base, B, into its conjugate acid, BH^+. Initially, before any H^+ is added (point d), B is the dominant species present. After H^+ is added, both B and BH^+ are present and a buffered solution results (region b). At the equivalence point (points a and e), exactly enough H^+ has been added to convert all of the weak base present initially into its conjugate acid, BH^+. Past the equivalence point (region f), excess H^+ is present. For the answer to part b, we included almost the entire buffer region. The maximum buffer region is around the halfway point to equivalence (point c) where [B] = $[BH^+]$. Here, pH = pK_a, which is a characteristic of a best buffer. **49.** a. all the same; b. i < iv < iii < ii; c. i < iv < iii < ii; d. iii < ii = i < iv; the only different answer would be part c. The ordering would be i < iii < iv < ii. **51.** a. f; b. a; the best point to look at to differentiate a strong acid from a weak acid titration (if initial concentrations are not known) is the equivalence point pH. If the pH = 7.00, the acid titrated is a strong acid; if the pH is greater than 7.00, the acid titrated is a weak acid. c. d **53.** a. 0.699; b. 0.854; c. 1.301; d. 7.00; e. 12.15 **55.** a. 2.72; b. 4.26; c. 4.74; d. 5.22; e. 8.79; f. 12.15

57.

Volume (mL)	pH
0.0	2.43
4.0	3.14
8.0	3.53
12.5	3.86
20.0	4.46
24.0	5.24
24.5	5.6
24.9	6.3
25.0	8.28
25.1	10.3
26.0	11.29
28.0	11.75
30.0	11.96

See *Solutions Guide* for pH plot.

59.

Volume (mL)	pH
0.0	11.11
4.0	9.97
8.0	9.58
12.5	9.25
20.0	8.65
24.0	7.87
24.5	7.6
24.9	6.9
25.0	5.28
25.1	3.7
26.0	2.71
28.0	2.24
30.0	2.04

See *Solutions Guide* for pH plot.

61. a. 4.19, 8.45; b. 10.74, 5.96; c. 0.89, 7.00 **63.** a. 100. g/mol; b. 3.02 **65.** a. yellow; b. 8.0; c. blue **67.** The pH is between 5 and 8. **69.** Bromthymol blue or phenol red are possible indicators for Exercise 53, and *o*-cresolphthalein or phenolphthalein are possible indicators for Exercise 55. **71.** Phenolphthalein is a possible indicator for Exercise 57, and bromcresol green is a possible indicator for Exercise 59. **73.** Methyl red changes color in a pH range of about pH = $pK_a \pm 1 = 5.3 \pm 1$. Therefore, methyl red is a useful indicator at pH values approximately between 4.3 and 6.3. In titrating a weak acid with a base, we start off with an acidic solution with pH < 4.3 so the color would change from red to reddish-orange at pH ~ 4.3. In titrating a weak base with an acid, the color change would be from yellow to yellowish-orange at pH ~ 6.3. Only a weak base–strong acid titration would have an acidic pH at the equivalence point, so only in this type of titration would the color change of methyl red indicate the approximate endpoint. **75.** a. 2.18; b. 2.55; c. 3.00; d. 5.00; e. 6.40; f. 7.00; g. 9.50; h. 11.63; i. 12.00; j. 12.34 **77.** a. $K_{a_1} = 1.5 \times 10^{-4}$; $K_{a_2} = 2.8 \times 10^{-7}$; $K_{a_3} = 3.6 \times 10^{-10}$; b. The pH at the third halfway point to equivalence (60.5 mL of NaOH added) will be equal to pK_{a_3} = 9.44. The pH at 59.0 mL NaOH added should be slightly less than 9.44. c. 9.34 **79.** a. Na^+ is present in all solutions. A. CO_3^{2-}, H_2O; B. CO_3^{2-}, HCO_3^-, H_2O, Cl^-; C. HCO_3^-, H_2O, Cl^-; D. HCO_3^-, $CO_2(H_2CO_3)$, H_2O, Cl^-; E. $CO_2(H_2CO_3)$, H_2O, Cl^-; F. H^+ (excess), $CO_2(H_2CO_3)$, H_2O, Cl^-; b. A, 11.66; B, 10.32; C, 8.35; D, 6.37; E, 3.92 **81.** a. 1.6×10^{-5} mol/L; 6.7×10^{-3} g/L; b. 9.3×10^{-5} mol/L; 9.3×10^{-3} g/L; c. 6.5×10^{-7} mol/L; 3.1×10^{-4} g/L **83.** a. 2.3×10^{-9}; b. 8.20×10^{-19} **85.** 5.3×10^{-12} **87.** a. CaF_2 has the smallest molar solubility since it has the smaller K_{sp} value. b. $FePO_4$ has the smallest molar solubility (must be calculated). **89.** a. 4×10^{-17} mol/L; b. 4×10^{-11} mol/L; c. 4×10^{-29} mol/L **91.** 1.5×10^{-19} g **93.** If the anion in the salt can act as a base in water, then the solubility of the salt will increase as the solution becomes more acidic. Added H^+ will react with the base, forming the conjugate acid. As the basic anion is removed, more of the salt will dissolve to replenish the basic anion. The salts with basic anions are Ag_3PO_4, $CaCO_3$, $CdCO_3$, and $Sr_3(PO_4)_2$. Hg_2Cl_2 and PbI_2 do not have any pH dependence since Cl^- and I^- are terrible bases (the conjugate bases of strong acids). $Ag_3PO_4(s) + H^+(aq) \rightarrow 3Ag^+(aq) + HPO_4^{2-}(aq) \xrightarrow{\text{Excess } H^+} 3Ag^+(aq) + H_3PO_4(aq)$; $CaCO_3(s) + H^+(aq) \rightarrow Ca^{2+}(aq) + HCO_3^-(aq) \xrightarrow{\text{Excess } H^+} Ca^{2+}(aq) + H_2CO_3(aq)\ [H_2O(l) + CO_2(g)]$; $CdCO_3(s) + H^+(aq) \rightarrow Cd^{2+}(aq) + HCO_3^-(aq) \rightarrow Cd^{2+}(aq) + H_2CO_3(aq)\ [H_2O(l) + CO_2(g)]$; $Sr_3(PO_4)_2(s) + 2H^+(aq) \rightarrow 3Sr^{2+}(aq) + 2HPO_4^{2-} \xrightarrow{\text{Excess } H^+} 3Sr^{2+}(aq) + 2H_3PO_4(aq)$ **95.** a. AgF; b. $Pb(OH)_2$; c. $Sr(NO_2)_2$; d. $Ni(CN)_2$ **97.** $[Ba^{2+}] = 6.0 \times 10^{-5}$ *M*; $[Br^-] = 1.2 \times 10^{-4}$ *M*; $[K^+] = 4.8 \times 10^{-4}$ *M*; $[C_2O_4^{2-}] = 2.4 \times 10^{-4}$ *M* **99.** When $[AgNO_3]_0$ is greater than 5.6×10^{-5} *M*, then $Ag_3PO_4(s)$ will precipitate. **101.** See the *Solutions Guide* for the flowchart for each separation. A possible order of chemicals necessary to separate the ions follows. a. $NaCl(aq)$ followed by $NH_3(aq)$ followed by $H_2S(aq)$; b. $NaCl(aq)$ followed by $Na_2SO_4(aq)$ followed by the basic addition of $H_2S(aq)$; c. $AgNO_3(aq)$ followed by $NH_3(aq)$ followed by $Na_2S_2O_3(aq)$; d. Na_2SO_4 followed by the basic addition of $H_2S(aq)$ **103.** $Hg^{2+}(aq) + 2I^-(aq) \rightarrow HgI_2(s)$, orange ppt.; $HgI_2(s) + 2I^-(aq) \rightarrow HgI_4^{2-}(aq)$, soluble complex ion **105.** 3.3×10^{-32} *M* **107.** $[Ag^+] = 4.6 \times 10^{-9}$ *M*; $[NH_3] = 1.6$ *M*; $[Ag(NH_3)_2^+] = 0.20$ *M*; $[AgNH_3^+] = 1.5 \times 10^{-5}$ *M* **109.** a. 1.6×10^{-6}; b. 0.056 mol/L **111.** 42 g **113.** Test tube 1: added Cl^- reacts with Ag^+ to form a silver chloride precipitate. The net ionic equation is $Ag^+(aq) + Cl^-(aq) \rightarrow AgCl(s)$. Test tube 2: added NH_3 reacts with Ag^+ ions to form a soluble complex ion, $Ag(NH_3)_2^+$. As this complex ion forms, Ag^+ is removed from the solution, which causes the $AgCl(s)$ to dissolve. When enough NH_3 is added, all of the silver chloride precipitate will dissolve. The equation is $AgCl(s) + 2NH_3(aq) \rightarrow Ag(NH_3)_2^+(aq) + Cl^-(aq)$. Test tube 3: added H^+ reacts with the weak base, NH_3, to form NH_4^+. As NH_3 is removed from the $Ag(NH_3)_2^+$ complex ion, Ag^+ ions are released to solution and can then react with Cl^- to re-form $AgCl(s)$. The equations are $Ag(NH_3)_2^+(aq) + 2H^+(aq) \rightarrow Ag^+(aq) + 2NH_4^+(aq)$ and $Ag^+(aq) + Cl^-(aq) \rightarrow AgCl(s)$. **115.** a. 8.1; b. pH = 7.00: 0.083; pH = 9.00: 8.3; c. 8.08; 7.95 **117.** a. potassium fluoride + HCl; b. benzoic acid + NaOH; c. sodium acetate + acetic acid; d. HOCl + NaOH or ammonium chloride + sodium acetate; e. ammonium chloride + NaOH **119.** 99.5%; for a strong base–strong acid titration, the equivalence point occurs at pH = 7.0. Bromthymol blue is a good indicator choice since it changes color at pH ~ 7 (from base color to acid color). **121.** Since the equivalence point occurs at pH = 8.9, phenolphthalein would be a good indicator choice because it changes color at pH ~ 9 (from acid color to base color). **123.** a. 6.7×10^{-6} *M*; b. 1.2×10^{-13} *M*; c. 2.3×10^{-19} *M*; no, since *Q* (= 2.3×10^{-21}) is less than the K_{sp} value. **125.** 65 mL **127.** 4.92 **129.** 180. g/mol; 3.3×10^{-4} **131.** K_{a_3} is so small (4.8×10^{-13}) that a break is not seen at the third stoichiometric point. **133.** 49 mL **135.** a. See the *Solutions Guide* for the derivation. b. ammonium formate, 6.50; ammonium acetate, 7.00; ammonium bicarbonate, 7.81; c. $NH_4^+(aq) + OH^-(aq) \rightarrow NH_3(aq) + H_2O(l)$, $C_2H_3O_2^-(aq) + H^+(aq) \rightarrow HC_2H_3O_2(aq)$ **137.** a. 0.33 mol/L; b. 0.33 *M*; c. 4.8×10^{-3} *M* **139.** a. 5.8×10^{-4} mol/L; b. Greater; since F^- is a weak base, the F^- concentration is lowered by reaction with water, which causes more $SrF_2(s)$ to dissolve. c. 3.5×10^{-3} mol/L **141.** a. See the *Solutions Guide* for derivation of the equation; b. The data for the plot follow. See the *Solutions Guide* for the plot. As can be seen from the data, the solubility of $Al(OH)_3(s)$ is increased by very acidic conditions or by very basic conditions.

pH	Solubility (*S*, mol/L)	log *S*
4.0	2×10^{-2}	−1.7
5.0	2×10^{-5}	−4.7
6.0	4.2×10^{-7}	−6.38
7.0	4.0×10^{-6}	−5.40
8.0	4.0×10^{-5}	−4.40
9.0	4.0×10^{-4}	−3.40
10.0	4.0×10^{-3}	−2.40
11.0	4.0×10^{-2}	−1.40
12.0	4.0×10^{-1}	−0.40

143. 3.9 L **145.** a. 2.21; b. 2.30×10^{-2} *M* **147.** 12.34

149. pH = 9.50: 10.0 mL; pH = 4.00: 4.55 mL **151.** a. Since there are two sources of HCO_3^-, $V_2 > V_1$. b. Since OH^- will be titrated first, $V_1 > V_2$. c. 57.1% Na_2CO_3; 42.9% $NaHCO_3$

Chapter 9

15. KE = 78 J; PE = 118 J **17.** Path-dependent functions for a trip from Chicago to Denver are those quantities that depend on the route taken. One can fly directly from Chicago to Denver or one could fly from Chicago to Atlanta to Los Angeles and then to Denver. Some path-dependent quantities are miles traveled, fuel consumption of the airplane, time traveling, airplane snacks eaten, etc. State functions are path-independent; they depend only on the initial and final states. Some state functions for an airplane trip from Chicago to Denver would be longitude change, latitude change, elevation change, and overall time zone change. **19.** 70. J **21.** $q = 30.9$ kJ; $w = -12.4$ kJ; $\Delta E = 18.5$ kJ **23.** −37.56 kJ **25.** $\Delta H = \Delta E + P\Delta V$ at constant P; from the strict definition of enthalpy, the difference between ΔH and ΔE is the quantity $P\Delta V$. Thus, when a system at constant P can do pressure-volume work, then $\Delta H \neq \Delta E$. When the system cannot do PV work, then $\Delta H = \Delta E$ at constant pressure. An important way to differentiate ΔH from ΔE is to concentrate on q, the heat flow; the heat flow by a system at constant pressure equals ΔH, and the heat flow by a system at constant volume equals ΔE. **27.** Since the reaction is exothermic (heat is a product), one should provide cooling for the reaction mixture to prevent $H_2SO_4(aq)$ from boiling. **29.** a. 1650 kJ of heat released; b. 826 kJ of heat released; c. 7.39 kJ of heat released; d. 34.4 kJ of heat released **31.** When a liquid is converted into gas, there is an increase in volume. The 2.5 kJ/mol quantity is the work done by the vaporization process in pushing back the atmosphere. **33.** Constant V: $\Delta E = q = 74.3$ kJ, $w = 0$, $\Delta H = 88.1$ kJ; constant P: $\Delta H = q = 88.1$ kJ, $w = -13.8$ kJ, $\Delta E = 74.3$ kJ **35.** Pathway one: step 1: $q = 30.4$ kJ, $w = -12.2$ kJ, $\Delta E = 18.2$ kJ, $\Delta H = 30.4$ kJ; step 2: $q = -28.1$ kJ, $w = 21.3$ kJ, $\Delta E = -6.8$ kJ, $\Delta H = -11$ kJ; total: $q = 2.3$ kJ, $w = 9.1$ kJ, $\Delta E = 11.4$ kJ, $\Delta H = 19$ kJ; pathway two: step 3: $q = 6.84$ kJ, $w = 0$, $\Delta E = 6.84$ kJ, $\Delta H = 11.40$ kJ; step 4: $q = 7.6$ kJ, $w = -3.0$ kJ, $\Delta E = 4.6$ kJ, $\Delta H = 7.6$ kJ; total: $q = 14.4$ kJ, $w = -3.0$ kJ, $\Delta E = 11.4$ kJ, $\Delta H = 19.0$ kJ; state functions are independent of the particular pathway taken between two states; path functions are dependent on the particular pathway. In this problem, the overall values of ΔH and ΔE for the two pathways are the same; hence, ΔH and ΔE are state functions. The overall values of q and w for the two pathways are different; hence, q and w are path functions. **37.** In calorimetry, heat flow is determined into or out of the surroundings. Because $\Delta E_{univ} = 0$ by the first law of thermodynamics, $\Delta E_{sys} = -\Delta E_{surr}$; what happens to the surroundings is the exact opposite of what happens to the system. To determine heat flow, we need to know the heat capacity of the surroundings, the mass of the surroundings that accepts/donates the heat, and the change in temperature. If we know these quantities, q_{surr} can be calculated and then equated to q_{sys} ($-q_{surr} = q_{sys}$). For an endothermic reaction, the surroundings (the calorimeter contents) donate heat to the system. This is accompanied by a decrease in temperature of the surroundings. For an exothermic reaction, the system donates heat to the surroundings (the calorimeter) so temperature increases. $q_p = \Delta H$; $q_v = \Delta E$; a coffee cup calorimeter is at constant (atmospheric) pressure. The heat released or gained at constant pressure is ΔH. A bomb calorimeter is at constant volume. The heat released or gained at constant volume is ΔE. **39.** $H_2O(l)$, 2.30×10^3 J; $Hg(l)$, 140°C **41.** 75.0 g **43.** −66 kJ/mol **45.** 39.2°C **47.** −25 kJ/g; −2700 kJ/mol **49.** a. $C_{12}H_{22}O_{11}(s) + 12O_2(g) \rightarrow 12CO_2(g) + 11H_2O(l)$; b. −5630 kJ/mol; c. −5630 kJ/mol **51.** 226 kJ **53.** −713 kJ **55.** −158 kJ **57.** −202.6 kJ **59.** 1268 kJ; since the reaction is very endothermic (requires a lot of heat), high energy costs would make it an impractical way of making ammonia. **61.** a. −940. kJ; b. −265 kJ; c. −176 kJ; d. −1235 kJ; e. −320. kJ; f. −37 kJ **63.** −832 kJ; −368 kJ; −133 kJ; in both cases sodium metal reacts with the "extinguishing agent." Both reactions are exothermic and each reaction produces a flammable gas—H_2 and CO, respectively. **65.** −4594 kJ **67.** a. 632 kJ; b. Since $3C_2H_2(g)$ is higher in energy than $C_6H_6(l)$, acetylene will release more energy per gram when burned in air. **69.** −169 kJ/mol **71.** 3.97 g **73.** −22.7 kJ/g versus −29.67 kJ/g for ethanol. Ethanol has a higher fuel value than methanol. **75.** 25 J **77.** 282.1 kJ **79.** $C_{H_2O} = 4.18$ kJ/°C; $C_{cal} = 6.66$ kJ/°C **81.** −306 kJ/mol **83.** 4.2 kJ of heat released **85.** When $\Delta V > 0$ ($\Delta n > 0$), then $w < 0$ and the system does work on the surroundings (c and e). When $\Delta V < 0$ ($\Delta n < 0$), then $w > 0$ and the surroundings do work on the system (a and d). When $\Delta V = 0$ ($\Delta n = 0$), then $w = 0$ (b). **87.** a. $2Al(s) + \frac{3}{2}O_2(g) \rightarrow Al_2O_3(s)$; b. $C_2H_5OH(l) + 3O_2(g) \rightarrow 2CO_2(g) + 3H_2O(l)$; c. $Ba(OH)_2(aq) + 2HCl(aq) \rightarrow 2H_2O(l) + BaCl_2(aq)$ d. $2C(\text{graphite}) + \frac{3}{2}H_2(g) + \frac{1}{2}Cl_2(g) \rightarrow C_2H_3Cl(g)$; e. $C_6H_6(l) + \frac{15}{2}O_2(g) \rightarrow 6CO_2(g) + 3H_2O(l)$ (Note: ΔH_{comb} values generally assume 1 mol of compound combusted); f. $NH_4Br(s) \rightarrow NH_4^+(aq) + Br^-(aq)$ **89.** 6.02 kJ/mol **91.** 43.58 kJ/mol; from Appendix 4 data, $\Delta H° = 44$ kJ/mol. The ΔH values agree to two significant figures (as they should). **93.** 37 m² **95.** For an isothermal expansion of an ideal gas, $\Delta T = 0$, so $\Delta E = 0$ (and $\Delta H = 0$); therefore, $q = -w = P\Delta V$. As long as the gas expands against a nonzero external pressure, $q \neq 0$ because $w \neq 0$.

Chapter 10

13.

2 kJ	___	___	x	___	x	xx
1 kJ	___	x	___	xx	x	___
0 kJ	xx	x	x	___	___	___
Total E =	0 kJ	1 kJ	2 kJ	2 kJ	3 kJ	4 kJ

The most likely total energy is 2 kJ.

15. c, e, f **17.** a. Positional probability increases; there is a greater volume accessible to the randomly moving gas molecules, which increases positional probability. b. The positional probability does not change. There is no change in volume and thus no change in the numbers of positions of the molecules. c. Positional probability decreases; volume decreases. **19.** There are six ways to get a seven, more than any other number. The seven is not favored by energy; rather it is favored by probability. To change the probability we would have to expend energy (do work). **21.** $q_v = \Delta E = 71.9$ kJ; $q_p = \Delta H = 85.3$ kJ; $\Delta E = 71.9$ kJ for both constant-volume and constant-pressure processes. **23.** 77.0°C **25.** $w = q = 0$; $q_{rev} = 350$ J **27.** a. constant V, 1.51 kJ; constant P, 1.94 kJ; b. 219.63 J K^{-1} mol^{-1}; c. 218.30 J K^{-1} mol^{-1} **29.** 2.50×10^4 J **31.** 2.9 J/K **33.** Living organisms need an external source of energy to carry out these processes. Green plants use the energy from sunlight to produce glucose from carbon dioxide and water by photosynthesis. In the human body, the energy released from the metabolism of glucose helps drive the synthesis of proteins. For all processes combined, ΔS_{univ} must be greater than zero (second law). **35.** ΔS_{surr} is primarily determined by heat flow. This heat flow into or out of the surroundings comes from the heat flow out of or into the system. In an exothermic process ($\Delta H < 0$), heat flows into the surroundings from the system. The heat flow into the surroundings increases the random motions in the surroundings and increases the entropy of the surroundings ($\Delta S_{surr} > 0$). This is a favorable driving force for spontaneity. In an endothermic reaction ($\Delta H > 0$), heat is transferred from the surroundings into the system. This heat flow out of the surroundings decreases the random motions in the surroundings and decreases the entropy of the surroundings ($\Delta S_{surr} < 0$). This is unfavorable. The magnitude of ΔS_{surr} also depends on the temperature. The relationship is inverse; at low temperatures, a specific amount of heat exchange makes a larger percent change in the surroundings than the same amount of heat flow at a higher temperature. The negative sign in the $\Delta S_{surr} = -\Delta H/T$ equation is necessary to get the signs correct. For an exothermic reaction where ΔH is negative, this increases ΔS_{surr}, so the negative sign converts the negative ΔH value into a positive quantity. For an endothermic process where ΔH is positive, the sign of ΔS_{surr} is negative and the negative sign converts the positive ΔH value into a negative quantity. **37.** a. 7.45×10^3 J/K; b. -376 J/K **39.** a. negative; b. positive; c. negative; d. negative; e. negative; f. positive **41.** 262 J K^{-1} mol^{-1} **43.** 184 J K^{-1} mol^{-1} **45.** 629.7 K **47.** a. Yes, NH_3 will melt since $\Delta G < 0$. b. 196 K **49.** 43.7 K **51.** -16 kJ/mol **53.** -5.40 kJ; 328.6 K; $\Delta G°$ is negative below 328.6 K where the favorable $\Delta H°$ term dominates. **55.** a. 464 kJ; b. Since $\Delta G°$ is positive, this reaction is not spontaneous at standard conditions at 298 K. c. This reaction will be spontaneous at standard conditions ($\Delta G° < 0$) at $T > 2890$ K where the favorable entropy term will dominate. **57.** $CH_4(g) + CO_2(g) \rightarrow CH_3CO_2H(l)$, $\Delta H° = -16$ kJ, $\Delta S° = -240.$ J/K, $\Delta G° = 56$ kJ; $CH_3OH(g) + CO(g) \rightarrow CH_3CO_2H(l)$, $\Delta H° = -173$ kJ, $\Delta S° = -278$ J/K, $\Delta G° = -90.$ kJ; the second reaction is preferred at standard conditions since it will be spontaneous at high enough temperatures so that the rate of the reaction should be reasonable. It should be run at temperatures below 622 K. **59.** Enthalpy is not favorable, so ΔS must provide the driving force for the change. Thus ΔS is positive. There is an increase in positional probability, so the original enzyme has the more ordered structure. **61.** Since there are more product gas molecules than reactant gas molecules ($\Delta n > 0$), ΔS will be positive. From the signs of ΔH and ΔS, this reaction is spontaneous at all temperatures. It will cost money to heat the reaction mixture. Since there is no thermodynamic reason to do this, the purpose of the elevated temperature must be to increase the rate of the reaction, i.e., kinetic reasons. **63.** The sign of ΔG (positive or negative) tells us which reaction is spontaneous (the forward or reverse reaction). If $\Delta G < 0$, then the forward reaction is spontaneous and if $\Delta G > 0$, then the reverse reaction is spontaneous. If $\Delta G = 0$, then the reaction is at equilibrium (neither the forward reaction nor the reverse reaction is spontaneous). $\Delta G°$ gives the equilibrium position by determining K for a reaction utilizing the equation $\Delta G° = -RT \ln K$. $\Delta G°$ can be used to predict spontaneity only when all reactants and products are present at standard pressures of 1 atm and/or standard concentrations of 1 M. **65.** a. shifts right; b. no shift (at equilibrium); c. shifts left **67.** $\Delta H° = -92$ kJ, $\Delta S° = -199$ J/K, $\Delta G° = -34$ kJ, $K = 9.1 \times 10^5$; a. $\Delta G = -67$ kJ; b. $\Delta G = -68$ kJ; c. $\Delta G = -85$ kJ; d. $\Delta G = -41$ kJ **69.** To determine K at a temperature other than 25°C, one needs to know $\Delta G°$ at that temperature. We assume $\Delta H°$ and $\Delta S°$ are temperature-independent and use the equation $\Delta G° = \Delta H° - T\Delta S°$ to estimate $\Delta G°$ at the different temperature. For $K = 1$, we want $\Delta G° = 0$, which occurs when $\Delta H° = T\Delta S°$. Again, assume $\Delta H°$ and $\Delta S°$ are temperature-independent, and then solve for T ($= \Delta H°/\Delta S°$). At this temperature, $K = 1$ because $\Delta G° = 0$. This works only for reactions where the signs of $\Delta H°$ and $\Delta S°$ are the same (either both positive or both negative). When the signs are opposite, K will always be greater than one (when $\Delta H°$ is negative and $\Delta S°$ is positive) or K will always be less than one (when $\Delta H°$ is positive and $\Delta S°$ is negative). When the signs of $\Delta H°$ and $\Delta S°$ are opposite, K can never equal one. **71.** -71 kJ/mol **73.** -4.1 kJ/mol **75.** 60 **77.** At 25.0°C, $K = 8.72$; at 100.0°C, $K = 0.0789$ **79.** a. $\Delta H° = 57.5$ kJ, $\Delta S° = -75.6$ J/K; b. 106.4 kJ **81.** 310 K; -310 J K^{-1} mol^{-1} **83.** 12.8 L; 3.84×10^3 J **85.** a. $q = 4.01 \times 10^3$ J; $w = -4.01 \times 10^3$ J; $\Delta E = 0$; b. $q = 1.99 \times 10^3$ J; $w = -1.99 \times 10^3$ J; $\Delta E = 0$; c. $q = 0$; $w = \Delta E = -2.76 \times 10^3$ J **87.** All are positive. **89.** $w_{max} = \Delta G$; when ΔG is negative, the magnitude of ΔG is equal to the maximum possible useful work obtainable from the process (at constant T and P). When ΔG is positive, the magnitude of ΔG is equal to the minimum amount of work that must be expended to make the process spontaneous. Due to waste energy (heat) in any real process, the amount of useful work obtainable from a spontaneous process is always less than w_{max}, and for a nonspontaneous reaction, an amount of work greater than w_{max} must be applied to make the process spontaneous. **91.** a. $\Delta G = \Delta G° = 1.8 \times 10^4$ J/mol, shifts left; b. $\Delta G = 0$, at equilibrium; c. $\Delta G = -1.1 \times 10^4$ J/mol, shifts right; d. $\Delta G = 0$, at equilibrium; e. $\Delta G = 2 \times 10^3$ J/mol, shifts left **93.** $\Delta S°$ will be negative because 2 mol of gaseous reactants forms 1 mol of gaseous product. For $\Delta G°$ to be negative, $\Delta H°$ must be negative (exothermic). For this sign combination, K decreases as T increases. Therefore, the ratio of the partial pressure of PCl_5 to the partial pressure of PCl_3 will decrease when T is raised. **95.** Using Le Châtelier's principle: A decrease in pressure (volume increases) will favor the side with the greater number of particles. Thus $2I(g)$ will be favored at low pressure. Looking at ΔG: $\Delta G = \Delta G° + RT \ln(P_I^2/P_{I_2})$; $\ln(P_I^2/P_{I_2}) > 0$ for $P_I = P_{I_2} = 10$ atm and ΔG is positive (not spontaneous). But at $P_I = P_{I_2} = 0.10$ atm, the logarithm term is negative. If $|RT \ln(Q)| > \Delta G°$, then ΔG becomes negative, and the reaction is spontaneous. **97.** 6 M **99.** ΔS is more favorable for reaction two than for reaction one. In reaction one, seven particles in solution are forming one particle in solution. In reaction two, four particles are forming one particle, which results in a smaller decrease in positional probability than for reaction one and a larger equilibrium constant value for reaction two. **101.** Note that these substances are not in the solid state but are in the aqueous state; water molecules are also present. There is an apparent increase in ordering (decrease in positional probability) when these ions are placed in water as compared to the separated state. The hydrating water molecules must be in a highly ordered arrangement when surrounding these anions. **103.** 1.6×10^6 **105.** $\Delta G = 11.5$ kJ; $\Delta H = 0$; $\Delta S = -38.3$ J/K **107.** a. $\Delta E = 0$, $\Delta H = 0$, $\Delta S = 7.62$ J/K, $\Delta G = -2320$ J, $w = -1500$ J, $q = 1500$ J; b. Since ΔS_{univ} (2.7 J/K) is positive, the process is spontaneous. **109.** a. 226 K; b. 239 K **111.** a. -2.87 J/K; b. 2.87 J/K **113.** 88 J/K **115.** greater than 7.5 torr **117.** $\Delta H° = 286$ kJ, $\Delta G° = 326$ kJ, $K = 7.22 \times 10^{-58}$; $P_{O_3} = 3.3 \times 10^{-41}$ atm; this partial pressure represents one molecule of ozone per 9.5×10^{17} L of air. Equilibrium is probably not maintained under these conditions since the concentration of ozone is not large enough to maintain equilibrium. **119.** a. 0.333; b. $P_A = 1.50$ atm, $P_B = 0.50$ atm; c. $\Delta G = 2722$ J + (8.3145)(298) ln(0.50/1.50) = 2722 J − 2722 J = 0 (carrying extra sig. figs.)

121.

	q	w	ΔE	ΔS	ΔH	ΔG
Step 1	1480 J	−1480 J	0	7.56 J/K	0	−2250 J
Step 2	1240 J	−1240 J	0	5.77 J/K	0	−1720 J
Step 3	1250 J	−1250 J	0	5.81 J/K	0	−1730 J
Total	3970 J	−3970 J	0	19.14 J/K	0	-5.70×10^3 J

123. a. 1.24×10^{90}; b. −154 J/K **125.** $\Delta S = 29$ J K^{-1} mol^{-1} **127.** $\Delta S_{sys} = -34.3$ J/K, $\Delta S_{surr} = 33.7$ J/K

Chapter 11

15. Electrochemistry is the study of the interchange of chemical and electrical energy. A redox (oxidation–reduction) reaction is a reaction in which one or more electrons are transferred. In a galvanic cell, a spontaneous redox reaction occurs that produces an electric current. In an electrolytic cell, electricity is used to force a nonspontaneous redox reaction to occur. **17.** See Fig. 11.2 for a typical galvanic cell. The anode compartment contains the oxidation half-reaction compounds/ions, and the cathode compartment contains the reduction half-reaction compounds/ions. The electrons flow from the anode to the cathode. In the salt bridge, cations flow to the cathode and anions flow to the anode. For each of the following answers, all solutes are 1.0 *M* and all gases are at 1.0 atm. a. $7H_2O(l) + 2Cr^{3+}(aq) + 3Cl_2(g) \rightarrow Cr_2O_7^{2-}(aq) + 6Cl^-(aq) + 14H^+(aq)$; $\mathscr{E}^\circ_{cell} = 0.03$ V; cathode: Pt electrode; Cl_2 bubbled into solution, Cl^- in solution; anode: Pt electrode; Cr^{3+}, H^+, and $Cr_2O_7^{2-}$ in solution; b. $Cu^{2+}(aq) + Mg(s) \rightarrow Cu(s) + Mg^{2+}(aq)$; $\mathscr{E}^\circ_{cell} = 2.71$ V; cathode: Cu electrode; Cu^{2+} in solution; anode: Mg electrode; Mg^{2+} in solution; c. $12H^+(aq) + 2IO_3^-(aq) + 10Fe^{2+}(aq) \rightarrow 10Fe^{3+}(aq) + I_2(s) + 6H_2O(l)$; $\mathscr{E}^\circ_{cell} = 0.43$ V; cathode: Pt electrode; IO_3^-, I_2, and H_2SO_4 (H^+ source) in solution; anode: Pt electrode; Fe^{2+} and Fe^{3+} in solution; d. $Zn(s) + 2Ag^+(aq) \rightarrow 2Ag(s) + Zn^{2+}(aq)$; $\mathscr{E}^\circ = 1.56$ V; cathode: Ag electrode; Ag^+ in solution; anode: Zn electrode; Zn^{2+} in solution **19.** See Exercise 17 for a description of a galvanic cell. For each of the following answers, all solutes are 1.0 *M* and all gases are at 1.0 atm. a. $Cl_2(g) + 2Br^-(aq) \rightarrow Br_2(aq) + 2Cl^-(aq)$; $\mathscr{E}^\circ = 0.27$ V; cathode: Pt electrode; $Cl_2(g)$ bubbled in, Cl^- in solution; anode: Pt electrode; Br_2 and Br^- in solution; b. $3H_2O(l) + 5IO_4^-(aq) + 2Mn^{2+}(aq) \rightarrow 5IO_3^-(aq) + 2MnO_4^-(aq) + 6H^+(aq)$; $\mathscr{E}^\circ = 0.09$ V; cathode: Pt electrode; IO_4^-, IO_3^-, and H_2SO_4 (as a source of H^+) in solution; anode: Pt electrode; Mn^{2+}, MnO_4^-, and H_2SO_4 in solution; c. $2H_2O_2(aq) \rightarrow 2H_2O(l) + O_2(g)$; $\mathscr{E}^\circ_{cell} = 1.10$ V; cathode: Pt electrode; H_2O_2 and H^+ in solution; anode: Pt electrode; $O_2(g)$ bubbled in, H_2O_2 and H^+ in solution; d. $2Fe^{3+}(aq) + 3Mn(s) \rightarrow 2Fe(s) + 3Mn^{2+}(aq)$; $\mathscr{E}^\circ_{cell} = 1.14$ V; cathode: Fe electrode; Fe^{3+} in solution; anode: Mn electrode; Mn^{2+} in solution **21.** a. Pt | Br^- (1.0 *M*), Br_2 (1.0 *M*) ‖ Cl_2 (1.0 atm) | Cl^- (1.0 *M*) | Pt; b. Pt | Mn^{2+} (1.0 *M*), MnO_4^- (1.0 *M*), H^+ (1.0 *M*) ‖ IO_4^- (1.0 *M*), IO_3^- (1.0 *M*), H^+ (1.0 *M*) | Pt; c. Pt | H_2O_2 (1.0 *M*), H^+ (1.0 *M*) | O_2 (1.0 atm) ‖ H_2O_2 (1.0 *M*), H^+ (1.0 *M*) | Pt; d. Mn | Mn^{2+} (1.0 *M*) ‖ Fe^{3+} (1.0 *M*) | Fe **23.** a. no; b. yes; c. yes; d. no **25.** $F^- < H_2O < I_2 < Cu^+ < H^- < K$ **27.** a. $Cr_2O_7^{2-}$, O_2, MnO_2, IO_3^-; b. $PbSO_4$, Cd^{2+}, Fe^{2+}, Cr^{3+}, Zn^{2+}, H_2O **29.** Since $\mathscr{E}^\circ_{cell}$ is positive for this reaction, at standard conditions ClO^- can spontaneously oxidize NH_3 to the somewhat toxic N_2H_4. **31.** An extensive property is one that depends directly on the amount of substance. The free energy change for a reaction depends on whether 1 mole of product is produced or 2 moles of product are produced or 1 million moles of product are produced. This is not the case for cell potentials, which do not depend on the amount of substance. The equation that relates ΔG to $\mathscr{E}$ is $\Delta G = -nF\mathscr{E}$. It is the *n* term that converts the intensive property $\mathscr{E}$ into the extensive property ΔG. *n* is the number of moles of electrons transferred in the balanced reaction that ΔG is associated with. **33.** a. −388 kJ; b. −270. kJ **35.** a. $\Delta G^\circ = -52$ kJ, $K = 1.4 \times 10^9$; b. $\Delta G^\circ = -90$ kJ, $K = 2 \times 10^{15}$; c. $\Delta G^\circ = -212$ kJ, $K = 1.68 \times 10^{37}$; d. $\Delta G^\circ = -660.$ kJ, $K = 5.45 \times 10^{115}$ **37.** a. $3Mn(s) + 8H^+(aq) + 2NO_3^-(aq) \rightarrow 2NO(g) + 4H_2O(l) + 3Mn^{2+}(aq)$; $5IO_4^-(aq) + 2Mn^{2+}(aq) + 3H_2O(l) \rightarrow 5IO_3^-(aq) + 2MnO_4^-(aq) + 6H^+(aq)$; b. reaction one: $\mathscr{E}^\circ_{cell} = 2.14$ V, $\Delta G^\circ = -1240$ kJ, $K \approx 10^{217}$; reaction two: $\mathscr{E}^\circ_{cell} = 0.09$ V, $\Delta G^\circ = -90$ kJ, $K = 2 \times 10^{15}$ **39.** $\mathscr{E}^\circ = \frac{T\Delta S^\circ}{nF} - \frac{\Delta H^\circ}{nF}$; if we graph $\mathscr{E}^\circ$ vs. *T*, we should get a straight line ($y = mx + b$). The slope of the line is equal to $\Delta S^\circ/nF$ and the *y*-intercept is equal to $-\Delta H^\circ/nF$. $\mathscr{E}^\circ$ will have a small temperature dependence when ΔS° is close to zero. **41.** a. spontaneous; −35 kJ; 1.2×10^6; b. not spontaneous; c. spontaneous; −85 kJ; 7.8×10^{14} **43.** 6.5×10^{20} **45.** −0.19 V **47.** a. $Au^{3+}(aq) + 3Tl(s) \rightarrow Au(s) + 3Tl^+(aq)$; $\mathscr{E}^\circ_{cell} = 1.84$ V; b. $\Delta G^\circ = -533$ kJ; $K = 2.52 \times 10^{93}$; c. 2.04 V **49.** 1.50 V **51.** a. 2.12 V; b. 1.98 V; c. 2.05 V; d. Cell potential decreases with decreasing *T* (also, oil becomes more viscous). **53.** Concentration cell: a galvanic cell in which both compartments contain the same components but at different concentrations. All concentration cells have $\mathscr{E}^\circ_{cell} = 0$ because both compartments contain the same contents. The driving force for the cell is the different ion concentrations at the anode and cathode. The cell produces a voltage as long as the ion concentrations are different. The lower ion concentration is always at the anode. The magnitude of the cell potential depends on the magnitude of the differences in ion concentrations between the anode and cathode. The larger the difference in ion concentrations, the more negative is the log *Q* term, and the more positive is the cell potential. Thus, as the difference in ion concentrations between the anode and cathode compartments increase, the cell potential increases. This can be accomplished by decreasing the ion concentration at the anode and/or by increasing the ion concentration at the cathode. **55.** Electron flow is always from the anode to the cathode. For the cells with a nonzero cell potential, we will identify the cathode, which means the other compartment is the anode. a. 0.0 V; b. 8.9×10^{-3} V; compartment with $[Ni^{2+}] = 2.0$ *M* is the cathode; c. 0.030 V; compartment with $[Ni^{2+}] = 1.0$ *M* is the cathode; d. 0.13 V; compartment with $[Ni^{2+}] = 1.0$ *M* is the cathode; e. 0.0 V **57.** a. 0.23 V; b. 0.16 V (the reverse reaction from part a occurs); c. 1.2×10^{-5} *M*; d. A graph of $\mathscr{E}$ versus $\log[Cu^{2+}]$ will yield a straight line with slope equal to 0.0296 V or 29.6 mV. **59.** a. 1.4×10^{-14} *M*; b. 2.9×10^{13} **61.** a. 16 g; b. 25 g; c. 71 g; d. 4.9 g **63.** 554 s **65.** a. cathode: $Ni^{2+} + 2e^- \rightarrow Ni$, anode: $2Br^- \rightarrow Br_2 + 2e^-$; b. cathode: $Al^{3+} + 3e^- \rightarrow Al$, anode: $2F^- \rightarrow F_2 + 2e^-$; c. cathode: $Mn^{2+} + 2e^- \rightarrow Mn$, anode: $2I^- \rightarrow I_2 + 2e^-$; d. cathode: $Ni^{2+} + 2e^- \rightarrow Ni$, anode: $2Br^- \rightarrow Br_2 + 2e^-$; e. cathode: $2H_2O + 2e^- \rightarrow H_2 + 2OH^-$, anode: $2H_2O \rightarrow O_2 + 4H^+ + 4e^-$; f. cathode: $2H_2O + 2e^- \rightarrow H_2 + 2OH^-$, anode: $2I^- \rightarrow I_2 + 2e^-$ **67.** Au(*s*) will plate out first since it has the most positive reduction potential, followed by Ag(*s*), which is followed by Ni(*s*), and finally Cd(*s*) will plate out last since it has the most negative reduction potential of the metals listed. **69.** Yes, since the range of potentials for plating out each metal do not overlap, we should be able to separate the three metals. The order of plating will be Ir(*s*) first, followed by Pt(*s*), and finally Pd(*s*) as the potential is gradually increased. **71.** $MgCl_2$ **73.** 9.12 L F_2 at the anode; 29.2 g K at the cathode **75.** 6.00 L **77.** 0.98 V **79.** 1.14 V **81.** −212 kJ **83.** a. Paint: covers the metal surface so no contact occurs between the metal and air. This works only as long as the painted surface is not scratched. b. Durable oxide coatings: covers the metal surface so no contact occurs between the metal and air. c. Galvanizing: coating steel with zinc; Zn forms an effective oxide coating over steel; also, zinc is more easily oxidized than the iron in the steel. d. Sacrificial metal: a more easily oxidized metal attached to an iron surface; the more active metal is preferentially oxidized instead of iron. e. Alloying: adding chromium and nickel to steel; the added Cr and Ni form oxide coatings on the steel surface. f. Cathodic protection: a more easily oxidized metal is placed in electrical contact with the metal we are trying to protect. It is oxidized in preference to the protected metal. The protected metal becomes the cathode electrode, thus, "cathodic protection." **85.** It is easier to oxidize Zn than Fe, so the Zn will be oxidized, protecting the iron of the *Monitor*'s hull. **87.** The potential oxidizing agents are NO_3^- and H^+. Hydrogen ion cannot oxidize Pt under either condition. Nitrate cannot oxidize Pt unless there is Cl^- in the solution. Aqua regia has both Cl^- and NO_3^-. The overall reaction is $12Cl^-(aq) + 3Pt(s) + 2NO_3^-(aq) + 8H^+(aq) \rightarrow 3PtCl_4^{2-}(aq) + 2NO(g) + 4H_2O(l)$; $\mathscr{E}^\circ_{cell} = 0.21$ V **89.** a. ± 0.02 pH units; $\pm 6 \times 10^{-6}$ M H^+; b. ± 0.001 V **91.** a. decrease; b. increase; c. decrease; d. decrease; e. no effect **93.** osmium(IV) nitrate; copper, Cu **95.** a. 1.14 V; b. 1.09 V; c. 78 g Zn, 319 g Cu; d. 13.4 h **97.** a. 5.77×10^{-10}; b. 1.9 M **99.** 2.39×10^{-7} **101.** a. 0.16 V; b. 8.6 mol **103.** 0.64 V **105.** a. yes; b. $3H_2O(l) + Zr(s) \rightarrow 2H_2(g) + ZrO_2 \cdot H_2O(s)$; c. $\mathscr{E}^\circ = 1.53$ V; $\Delta G^\circ = -590.$ kJ; $K \approx 10^{104}$; d. 4.42×10^4 g H_2; 2.3×10^6 L; e. Probably yes; less radioactivity overall was released by venting the H_2 than what would have been released if the H_2 had exploded inside the reactor (as happened at Chernobyl). Neither alternative is pleasant, but venting the radioactive hydrogen is the less unpleasant of the two alternatives.

Chapter 12

21. 3.0×10^{10} s^{-1}; 2.0×10^{-23} J/photon; 12 J/mol **23.** a. 5.0×10^{-6} m; b. infrared; c. 4.0×10^{-20} J/photon; 2.4×10^4 J/mol; d. less **25.** 427.7 nm **27.** 4.36×10^{-19} J **29.** The photoelectric effect refers to the phenomenon in which electrons are emitted from the surface of a metal when light strikes it. The light must have a certain minimum frequency (energy) in order to remove electrons from the surface of a metal. Light having a frequency below the minimum results in no electrons being emitted, whereas light at or higher than the minimum frequency does cause electrons to be emitted. For light having a frequency higher than the minimum frequency, the excess energy is transferred into kinetic energy for the emitted electron. Albert Einstein explained the photoelectric effect by applying quantum theory. **31.** a. 1.32×10^{-13} m; b. 5.3×10^3 m/s **33.** Ca **35.** a. 656.7 nm, visible electromagnetic radiation (red light); b. 486.4 nm, visible electromagnetic radiation (green-blue light); c. 121.6 nm, ultraviolet electromagnetic radiation; d. 1876 nm, infrared electromagnetic radiation **37.** a. False; It takes less energy to ionize an electron from $n = 3$ than from the ground state. b. True; c. False; The energy difference between $n = 3$ and $n = 2$ is smaller than the energy difference between $n = 3$ and $n = 1$. Thus the wavelength of light emitted is longer for the $n = 3$ to $n = 2$ electronic transition than for the $n = 3$ to $n = 1$ transition. d. True; e. False; $n = 2$ is the first excited state and $n = 3$ is the second excited state. **39.** $n = 7$ **41.** $n = 4$ **43.** $n = 4$ **45.** a. 5.79×10^{-4} m; b. 3.64×10^{-33} m; c. The diameter of an H atom is roughly 1.0×10^{-8} cm. The uncertainty in position is much larger than the size of the atom. d. The uncertainty is insignificant compared to the size of a baseball. **47.** 3.50 nm **49.** As L increases, E_n will decrease and the spacing between energy levels will also decrease. **51.** Since E_n is inversely proportional to L^2, the electron in the larger box (10^{-6} m) has the lowest ground-state energy. **53.** The 2*p* orbitals differ from each other in the direction in which they point in space. The 2*p* and 3*p* orbitals differ from each other in their size, energy, and number of nodes. A nodal surface in an atomic orbital is a surface in which the probability of finding an electron is zero. The 1*p*, 1*d*, 2*d*, 1*f*, 2*f*, and 3*f* orbitals are not allowed solutions to the Schrödinger equation. For $n = 1$, $\ell \neq 1, 2, 3$, etc., so 1*p*, 1*d*, and 1*f* orbitals are forbidden. For $n = 2$, $\ell \neq 2, 3, 4$, etc., so 2*d* and 2*f* orbitals are forbidden. For $n = 3$, $\ell \neq 3, 4, 5$, etc., so 3*f* orbitals are forbidden. The penetrating term refers to the fact that there is a higher probability of finding a 4*s* electron closer to the nucleus than a 3*d* electron. This leads to a lower energy for the 4*s* orbital relative to the 3*d* orbitals in polyelectronic atoms and ions. **55.** a. For $n = 3$, $\ell = 3$ is not possible. d. m_s cannot equal -1. e. ℓ cannot be a negative number. f. For $\ell = 1$, m_ℓ cannot equal 2. **57.** 1*p*, 0 electrons ($\ell \neq 1$ when $n = 1$); $6d_{x^2-y^2}$, 2 electrons (specifies one atomic orbital); 4*f*, 14 electrons (7 orbitals have 4*f* designation); $7p_y$, 2 electrons (specifies one atomic orbital); 2*s*, 2 electrons (specifies one atomic orbital); $n = 3$, 18 electrons (3*s*, 3*p*, and 3*d* orbitals are possible; there are one 3*s* orbital, three 3*p* orbitals, and five 3*d* orbitals) **59.** The diagrams of the orbitals in the text give only 90% probabilities of where the electron may reside. We can never be 100% certain of the location of the electrons due to Heisenberg's uncertainty principle. **61.** $\theta = 0 : 2.46 \times 10^{28}$; $\theta = 90^\circ : 0$ **63.** Valence electrons are the electrons in the outermost principal quantum level of an atom (those electrons in the highest n value orbitals). The electrons in the lower n value orbitals are all inner core or just core electrons. The key is that the outermost electrons are the valence electrons. When atoms interact with each other, it will be the outermost electrons that are involved in these interactions. In addition, how tightly the nucleus holds these outermost electrons determines atomic size, ionization energy, and other properties of atoms. Elements in the same group have similar valence electron configurations and, as a result, have similar chemical properties. **65.** a. 32; b. 8; c. 25; d. 10; e. 6; f. 0; g. 1; h. 9; i. 0; j. 2 **67.** Sc: $1s^22s^22p^63s^23p^64s^23d^1$; Fe: $1s^22s^22p^63s^23p^64s^23d^6$; P: $1s^22s^22p^63s^23p^3$; Cs: $1s^22s^22p^63s^23p^64s^23d^{10}4p^65s^24d^{10}5p^66s^1$; Eu: $1s^22s^22p^63s^23p^64s^23d^{10}4p^65s^24d^{10}5p^66s^24f^65d^1$ (actual: $[Xe]6s^24f^7$); Pt: $1s^22s^22p^63s^23p^64s^23d^{10}4p^65s^24d^{10}5p^66s^24f^{14}5d^8$ (actual: $[Xe]6s^14f^{14}5d^9$); Xe: $1s^22s^22p^63s^23p^64s^23d^{10}4p^65s^24d^{10}5p^6$; Br: $1s^22s^22p^63s^23p^64s^23d^{10}4p^5$ **69.** Cr, Cu, Nb, Mo, Tc, Ru, Rh, Pd, Ag, Pt, Au **71.** Cr: $1s^22s^22p^63s^23p^64s^13d^5$; Cr has 6 unpaired electrons.

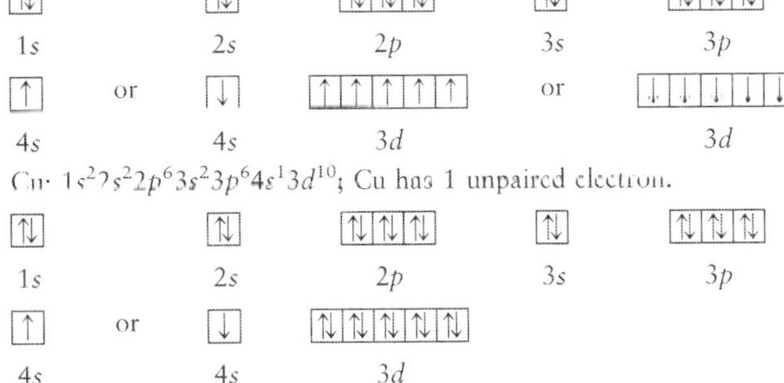

73. a. 32; b. 28; c. 23; d. 56 or 59 **75.** Li (1 unpaired electron), N (3 unpaired electrons), Ni (2 unpaired electrons), and Te (2 unpaired electrons) are paramagnetic. **77.** H, Li, Na, K, B, Al, Ga, F, Cl, Br, Sc, and Cu all have one unpaired electron in the ground state. **79.** O, 2; O^+, 3; O^-, 1; Os, 4; Zr, 2; S, 2; F, 1; Ar, 0 **81.** Ionization energy: $P(g) \rightarrow P^+(g) + e^-$; electron affinity: $P(g) + e^- \rightarrow P^-(g)$ **83.** As successive electrons are removed, the net positive charge on the resultant ion increases. This increase in positive charge binds the remaining electrons more firmly, and the ionization energy increases. The electron configuration for Si is $1s^22s^22p^63s^23p^2$. There is a large jump in ionization energy when going from the removal of valence electrons to the removal of core electrons. For silicon, this occurs when the fifth electron is removed since we go from the valence electrons in $n = 3$ to the core electrons in $n = 2$. There should be another big jump when the thirteenth electron is removed (i.e., when a 1s electron is removed). **85.** a. S < Se < Te; b. Br < Ni < K; c. F < Si < Ba; d. Be < Na < Rb; e. Ne < Se < Sr; f. O < P < Fe **87.** a. Ba; b. K; c. O; d. S^{2-}; e. Cs **89.** As: $[Ar]4s^23d^{10}4p^3$; Se: $[Ar]4s^23d^{10}4p^4$; The general ionization energy trend predicts that Se should have a higher ionization energy than As. Se is an exception to the general ionization energy trend. There are extra electron–electron repulsions in Se because two electrons are in the same 4*p* orbital, resulting in a lower ionization energy for Se than predicted. **91.** Size also decreases going across a period. Sc and Ti plus Y and Zr are adjacent elements. There are 14 elements (the lanthanides) between La and Hf, making Hf considerably smaller. **93.** a. $[Rn]7s^25f^{14}6d^{10}7p^5$; b. At; c. NaUus, $Mg(Uus)_2$, $C(Uus)_4$, $O(Uus)_2$; d. $UusO^-$, $UusO_2^-$, $UusO_3^-$, $UusO_4^-$ **95.** Electron–electron repulsions must be considered when adding an electron to an atom. Smaller atoms attract electrons more strongly yet have larger electron–electron repulsions. The two factors oppose one another, leading to relatively small variations in electron affinities. **97.** Electron–electron repulsions are much greater in O^- than in S^- because the electron goes into a smaller 2*p* orbital versus the larger 3*p* orbital in sulfur. This results in a more favorable (more exothermic) electron affinity for sulfur. **99.** a. −1445 kJ/mol; b. −580 kJ/mol; c. 348.7 kJ/mol; d. 1255 kJ/mol; e. −1255 kJ/mol **101.** Yes; ionization energy generally decreases down a group, and atomic radius generally increases down a group. The data in Table 12.9 confirm both of these general trends. **103.** a. $6Li(s) + N_2(g) \rightarrow 2Li_3N(s)$; b. $2Rb(s) + S(s) \rightarrow Rb_2S(s)$; c. $2Cs(s) + 2H_2O(l) \rightarrow 2CsOH(aq) + H_2(g)$; d. $2Na(s) + Cl_2(g) \rightarrow 2NaCl(s)$ **105.** For 589.0 nm: $5.090 \times 10^{14}\ s^{-1}$, 3.373×10^{-19} J/photon, 203.1 kJ/mol; for 589.6 nm: $5.085 \times 10^{14}\ s^{-1}$, 3.369×10^{-19} J/photon, 202.9 kJ/mol **107.** 119: $[Rn]7s^25f^{14}6d^{10}7p^68s^1$ **109.** a. *n*; b. *n* and ℓ **111.** Size decreases from left to right and increases going down the periodic table. So going one element right and one element down would result in a similar size for the two elements diagonal to each other. The ionization energies will be similar for the diagonal elements since the periodic trends also oppose each other. Electron affinities are harder to predict, but atoms with similar size and ionization energy should also have similar electron affinities. **113.** 200 s (about 3 minutes) **115.** greenish yellow light **117.** When the *p* and *d* orbital functions are evaluated at various points in space, the results sometimes have positive values and sometimes have negative values. The term "phase" is often associated with the + and − signs. For example, a sine wave has alternating positive and negative phases. This is analogous to the positive and negative values (phases) in the *p* and *d* orbitals. **119.** a. 146 kJ; b. 407 kJ; c. 1117 kJ; d. 1524 kJ **121.** alkaline earth metal family **123.** a. 24; b. 6; c. 12; d. 2; e. 26; f. 24.9 g; g. $1s^22s^22p^63s^23p^64s^13d^5$ **125.** a. 4; b. four elements in the first period and 16 elements in the second period; c. 20; d. 28 **127.** At $x = 0$, the value of the square of the wave function must be zero. The particle must be inside the box. For $\psi = A\cos(Lx)$, at $x = 0$, $\cos(0) = 1$ and $\psi^2 = A^2$. This violates the boundary condition. **129.** a. Line A: 6 → 3; Line B: 5 → 3; b. 121.6 nm **131.** X = carbon; $m = 5(C^{5+})$ **133.** a. The quantum numbers are:

ground state (E_{11}) → $n_x = 1, n_y = 1$
first excited state (E_{21}) → $n_x = 2, n_y = 1$
second excited state (E_{12}) → $n_x = 1, n_y = 2$

b. 4.5×10^{-5} m **135.** a. 2.2×10^{-9}; b. 1.5×10^{-9}; c. 2.9×10^{-10}; d. 1.9×10^{-4}; e. 1×10^{-3} **137.** 2470 nm **139.** a. 43.33; b. Z_{eff} is slightly less than *Z*. Electrons in other orbitals can penetrate the 1*s* orbital. Thus a 1*s* electron can be slightly shielded from the nucleus, giving a Z_{eff} close to but less than *Z*.

Chapter 13

11. Electronegativity is the ability of an atom in a molecule to attract electrons to itself. Electronegativity is a bonding term. Electron affinity is the energy change when an electron is added to a substance. Electron affinity deals with isolated atoms in the gas phase. A covalent bond is a sharing of an electron pair in a bond between two atoms. An ionic bond is a complete transfer of electrons from one atom to another to form ions. The electrostatic attraction of the oppositely charged ions is the ionic bond. A pure covalent bond is an equal sharing of a shared electron pair in a bond. A polar covalent bond is an unequal sharing. Ionic bonds form when there is a large difference in electronegativity between the two atoms bonding together. This usually occurs when a metal with a small electronegativity is bonded to a nonmetal having a large electronegativity. A pure covalent bond forms between atoms having identical or nearly identical electronegativities. A polar covalent bond forms when there is an intermediate electronegativity difference. In general, nonmetals bond together by forming covalent bonds, either pure covalent or polar covalent. Ionic bonds form due to the strong electrostatic attraction between two oppositely charged ions. Covalent bonds form because the shared electrons in the bond are attracted to two different nuclei, unlike the isolated atoms, in which electrons are attracted to only one nuclei. The attraction to another nuclei overrides the added electron–electron repulsions. **13.** a. C < N < O; b. Se < S < Cl; c. Sn < Ge < Si; d. Tl < Ge < S; e. Rb < K < Na; f. Ga < B < O **15.** a. C (2.6) < N (3.0) < O (3.4), same as predicted; b. Se (2.6) = S (2.6) < Cl (3.2), different; c. Si (1.9) < Ge (2.0) = Sn (2.0), different; d. Tl (2.0) = Ge (2.0) < S (2.6), different; e. Rb = K < Na, different; f. Ga < B < O, same; Most polar bonds in Exercise 14 using actual EN values: a. Si—F (Ge—F predicted); b. P—Cl (same as predicted); c. S—F (same as predicted); d. Ti—Cl (same as predicted); e. C—H (Sn—H predicted); f. Al—Br (Tl—Br predicted) **17.** Br—Br < N—O < C—F < Ca—O < K—F **19.** Anions are larger than the neutral atom, and cations are smaller than the neutral atom. For anions, the added electrons increase the electron–electron repulsions. To counteract this, the size of the electron cloud increases, placing the electrons further apart from one another. For cations, as electrons are removed, there are fewer electron–electron repulsions, and the electron cloud can be pulled closer to the nucleus. Isoelectronic: same number of electrons. Two variables, the number of protons and the number of electrons, determine the size of an ion. Keeping the number of electrons constant, we have to consider only the number of protons to predict trends in size. The ion with the most protons attracts the same number of electrons most strongly, resulting in a smaller size. **21.** a. $Cu > Cu^+ > Cu^{2+}$; b. $Pt^{2+} > Pd^{2+} > Ni^{2+}$; c. $O^{2-} > O^- > O$; d. $La^{3+} > Eu^{3+} > Gd^{3+} > Yb^{3+}$; e. $Te^{2-} > I^- > Cs^+ > Ba^{2+} > La^{3+}$ **23.** a. Cs_2S is composed of Cs^+ and S^{2-}. Cs^+ has the same electron configuration as Xe, and S^{2-} has the same configuration as Ar. b. SrF_2; Sr^{2+} has the Kr electron configuration and F^- has the Ne configuration. c. Ca_3N_2; Ca^{2+} has the Ar electron configuration and N^{3-} has the Ne configuration. d. $AlBr_3$; Al^{3+} has the Ne electron configuration and Br^- has the Kr configuration. **25.** Se^{2-}, Br^-, Rb^+, Sr^{2+}, Y^{3+}, and Zr^{4+} are some ions that are isoelectronic with Kr (36 electrons). In terms of size, the ion with the most protons will hold the electrons tightest and will be the smallest. The size trend is

$$Zr^{4+} < Y^{3+} < Sr^{2+} < Rb^+ < Br^- < Se^{2-}$$

smallest — largest

27. a. Al_2S_3, aluminum sulfide; b. K_3N, potassium nitride; c. $MgCl_2$, magnesium chloride; d. CsBr, cesium bromide **29.** −437 kJ/mol **31.** 181 kJ/mol **33.** a. From the data given, less energy is required to produce $Mg^+(g) + O^-(g)$ than to produce $Mg^{2+}(g) + O^{2-}(g)$. However, the lattice energy for $Mg^{2+}O^{2-}$ will be much more exothermic than for Mg^+O^- (because of the greater charges in $Mg^{2+}O^{2-}$). The favorable lattice energy term will dominate and $Mg^{2+}O^{2-}$ forms. b. Mg^+ and O^- both have unpaired electrons. In Mg^{2+} and O^{2-}, there are no unpaired electrons. Hence, Mg^+O^- would be paramagnetic; $Mg^{2+}O^{2-}$ would be diamagnetic. Paramagnetism can be detected by measuring the mass of a sample in the presence and absence of a magnetic field. The apparent mass of a paramagnetic substance will be larger in a magnetic field because of the force between the unpaired electrons and the field. **35.** Ca^{2+} has greater charge than Na^+, and Se^{2-} is smaller than Te^{2-}. The effect of charge on the lattice energy is greater than the effect of size. We expect the trend from most exothermic to least exothermic to be

$$\underset{(-2862)}{CaSe} > \underset{(-2721)}{CaTe} > \underset{(-2130)}{Na_2Se} > \underset{(-2095\ kJ/mol)}{Na_2Te}$$

37. a. −183 kJ; b. −109 kJ; c. −158 kJ; d. −1169 kJ **39.** −42 kJ **41.** −1228 kJ **43.** −5681 kJ **45.** Since both reactions are highly exothermic, the high temperature is not needed to provide energy. It must be necessary for some other reason. This will be discussed in Chapter 15 on kinetics. **47.** a. 1549 kJ; b. 1390. kJ; c. 1312 kJ; d. 1599 kJ **49.** D_{calc} = 389 kJ/mol as compared with 391 kJ/mol in the table.

51. 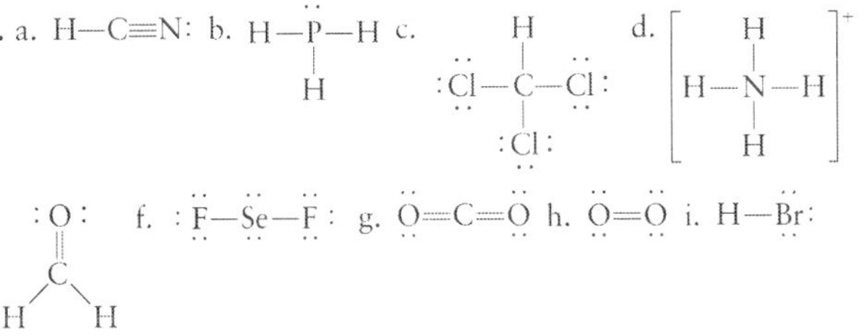 a. H—C≡N: b. H—P—H (with H below P) c. :Cl—C—Cl: (with H above C and :Cl: below C) d. [H—N—H with H above and H below N]$^+$

e. :O: double bonded to C, C bonded to H and H f. :F—Se—F: g. O=C=O h. O=O i. H—Br:

53. Molecules or ions that have the same number of valence electrons and the same number of atoms will have similar Lewis structures.

55. 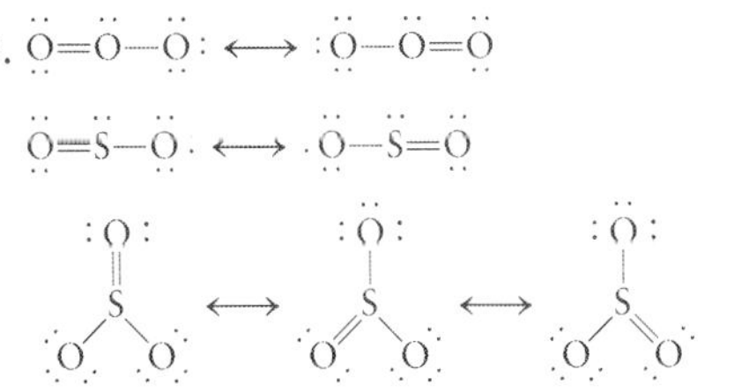 O=O—O: ⟷ :O—O=O

O=S—O· ⟷ ·O—S=O

(:O: double bonded to S, S bonded to O and O) ⟷ (:O: single bonded to S, S double bonded to left O) ⟷ (:O: single bonded to S, S double bonded to right O)

57. 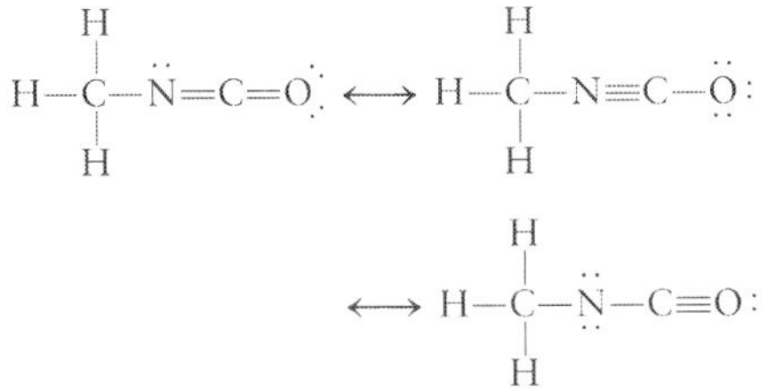 H—C(H)(H)—N=C=O ⟷ H—C(H)(H)—N≡C—O: ⟷ H—C(H)(H)—N—C≡O:

59. 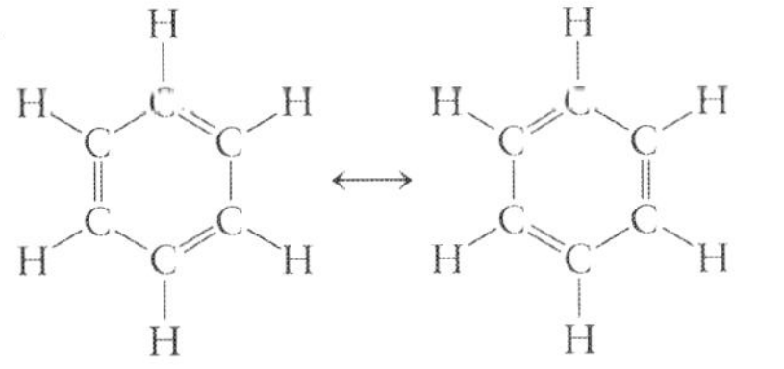 (two benzene-type resonance structures, C_6H_6 ring with alternating double bonds) ⟷

61. 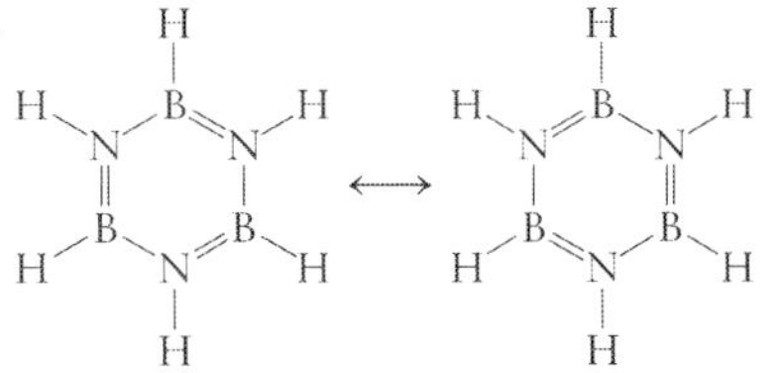 (two resonance structures of $B_3N_3H_6$ ring, with alternating B=N double bonds, each B and N bearing one H) ⟷

63. Statements a and c are true. For statement b, SF_4 has 5 electron pairs around the sulfur in the best Lewis structure; it is an exception to the octet rule. Because OF_4 has the same number of valence electrons as SF_4, OF_4 would also have to be an exception to the octet rule. However, Row 2 elements like O never have more than 8 electrons around them, so OF_4 does not exist. For statement d, two resonance structures can be drawn for ozone:

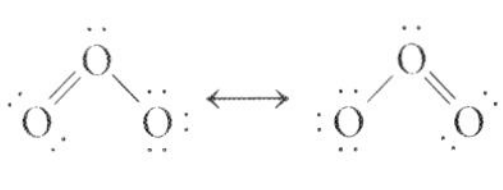

O=O—O: ⟷ :O—O=O (bent)

When resonance structures can be drawn, the actual bond lengths and strengths are all equal to each other. Even though each Lewis structure implies the two O—O bonds are different, this is not the case in real life. In real life, both of the O—O bonds are equivalent. When resonance structures can be drawn, you can think of the bonding as an average of all of the resonance structures.

65. 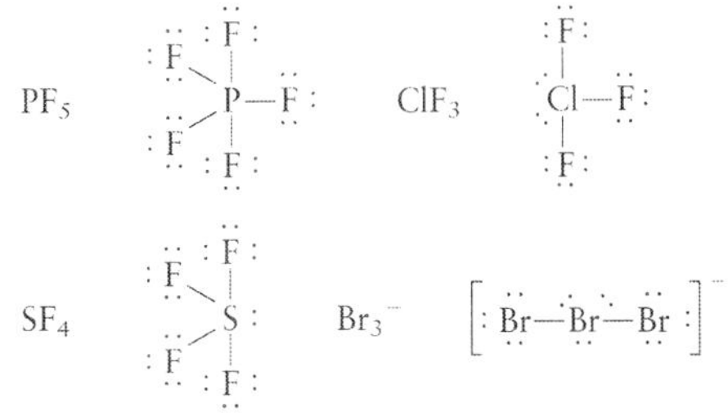 PF_5 (P bonded to five :F:) ClF_3 (Cl bonded to three :F:, with two lone pairs on Cl)

SF_4 (S bonded to four :F:, with one lone pair on S) Br_3^- [:Br—Br—Br:]$^-$

Row 3 and heavier nonmetals can have more than 8 electrons around them when they have to. Row 3 and heavier elements have empty *d* orbitals which are close in energy to the valence *s* and *p* orbitals. These empty *d* orbitals can accept extra electrons.
67. Three resonance structures can be drawn for CO_3^{2-}. The actual structure for CO_3^{2-} is an average of these three resonance structures. That is, the three C—O bond lengths are all equivalent, with a length somewhere between a single bond and a double bond. The actual bond length of 136 pm is consistent with this resonance view of CO_3^{2-}. **69.** $N_2 < N_2F_2 < N_2F_4$ **71.** a–f and h all have similar Lewis structures:

:Y—X—Y: (X bonded to four :Y:) g. ClO_3^- [:O—Cl—O: with :O: below Cl]$^-$

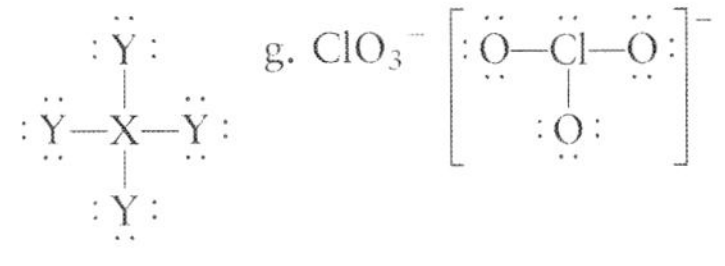

Formal charges: a. +1; b. +2; c. +3; d. +1; e. +2; f. +4; g. +2; h. +1

73.  :Cl—S—S—Cl:

75. [51] a. linear, 180°, b. trigonal pyramid, < 109.5°; c. tetrahedral, 109.5°; d. tetrahedral, 109.5°; e. trigonal planar, 120°; f. V-shaped, < 109.5°; g. linear, 180°; h. and i. linear, no bond angle in diatomic molecules; [52] a. All are tetrahedral, 109.5°; b. All are trigonal pyramid, < 109.5°; c. All are V-shaped, < 109.5°; [54] a. NO_2^-: V-shaped, ≈ 120°; NO_3^-: trigonal planar, 120°;

N_2O_4: trigonal planar about both nitrogens, 120°; b. All are linear; 180° **77.** Br_3^- is linear; ClF_3 is T-shaped; SF_4 is see-saw. **79.** a. linear, 180°; b. T-shaped, ≈ 90°; c. see-saw, ≈ 120° and ≈ 90°; d. trigonal bipyramid, 120° and 90° **81.** SeO_2, PCl_3, and SCl_2 have net dipole moments (are polar). **83.** The two general requirements for a polar molecule are (1) polar bonds and (2) a structure such that the bond dipoles of the polar bonds do not cancel. In CF_4, the fluorines are symmetrically arranged about the central carbon atom. The net result is for all of the individual C—F bond dipoles to cancel each other out, giving a nonpolar molecule. In XeF_4 the 4 Xe—F bond dipoles are also symmetrically arranged and XeF_4 is also nonpolar. The individual bond dipoles cancel out when summed together. In SF_4 we also have 4 polar bonds. But in SF_4 the bond dipoles are not symmetrically arranged and they do not cancel each other out. SF_4 is polar. It is the positioning of the lone pair that disrupts the symmetry in SF_4. CO_2 is nonpolar because the individual bond dipoles cancel each other out, but COS is polar. By replacing an O with a less electronegative S atom, the molecule is not symmetric any more. The individual bond dipoles do not cancel since the C—S bond dipole is smaller than the C—O bond dipole, resulting in a polar molecule. **85.** Only statement c is true. The bond dipoles in CF_4 and KrF_4 are arranged in a manner that they all cancel each other out, making them nonpolar molecules (CF_4 has a tetrahedral molecular structure, while KrF_4 has a square planar molecular structure). In SeF_4 the bond dipoles in this see-saw molecule do not cancel each other out, so SeF_4 is polar. For statement a, all the molecules have either a trigonal planar geometry or a trigonal bipyramidal geometry, both of which have 120-degree bond angles. However, $XeCl_2$ has three lone pairs and two bonded chlorine atoms around it. $XeCl_2$ has a linear molecular structure with a 180-degree bond angle. With three lone pairs, we no longer have a 120-degree bond angle in $XeCl_2$. For statement b, SO_2 has a V-shaped molecular structure with a bond angle of about 120 degrees. CS_2 is linear with a 180-degree bond angle, and SCl_2 is V-shaped but with an approximately 109.5-degree bond angle. The three compounds do not have the same bond angle. For statement d, central atoms adopt a geometry to minimize electron repulsions, not maximize them. **87.** Element E must belong to the Group 6A elements because it has 6 valence electrons. E must also be a Period 3 or heavier element since this ion has more than 8 electrons around the central E atom (Period 2 elements never have more than 8 electrons around them). Some possible identities for E are S, Se, and Te. The ion has a T-shaped molecular structure with bond angles of ≈ 90°.

89. a. OCl_2 — V-shaped, polar

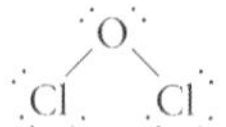

KrF_2 — :F—Kr—F: — Linear, nonpolar

BeH_2 — H—Be—H — Linear, nonpolar

SO_2 — V-shaped, polar

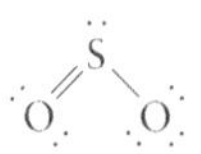

(One other resonance structure possible)

b. SO_3 — Trigonal planar, nonpolar

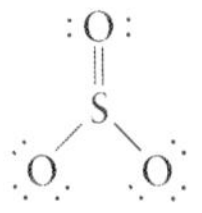

(Two other resonance structures possible)

NF_3 — Trigonal pyramid, polar

IF_3 — T-shaped, polar

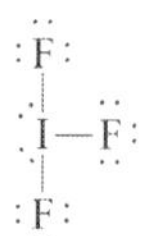

c. CF_4 — Tetrahedral, nonpolar

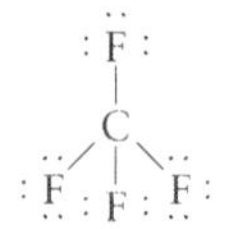

SeF_4 — See-saw, polar

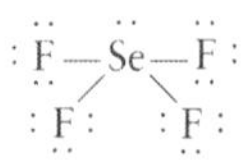

KrF_4 — Square planar, nonpolar

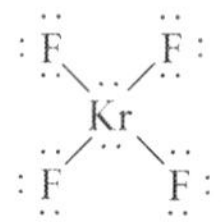

d. IF_5 — Square pyramid, polar

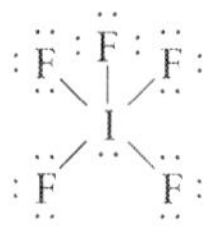

AsF_5 — Trigonal bipyramid, nonpolar

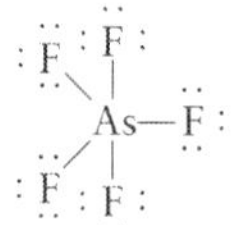

91. All these molecules have polar bonds that are symmetrically arranged around the central atom. In each molecule, the individual bond dipoles cancel to give no net overall dipole moment.

93. $[CO_3]^{2-}$ ⟷ $[CO_3]^{2-}$ ⟷ $[CO_3]^{2-}$ (three resonance structures)

$[HCO_3]^-$ ⟷ $[HCO_3]^-$ (two resonance structures); H_2CO_3

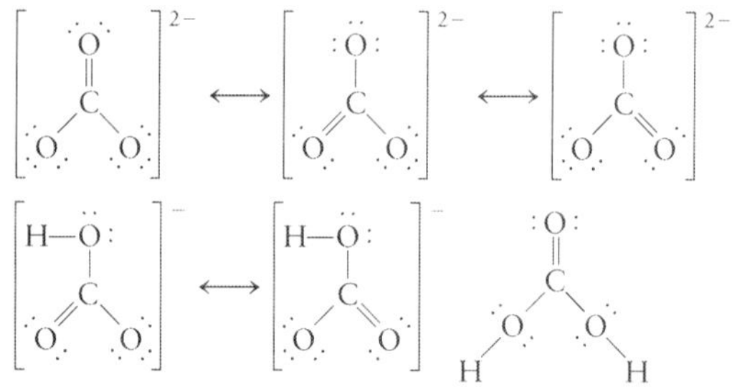

$\Delta H = -83$ kJ; the carbon–oxygen double bond is stronger than two carbon–oxygen single bonds; hence CO_2 and H_2O are more stable than H_2CO_3. **95.** As the halogen atoms get larger, it becomes more difficult to fit three halogen atoms around the small nitrogen atom, and the NX_3 molecule becomes less stable. **97.** a. NaBr: In $NaBr_2$ the sodium ion would have a +2 charge assuming each bromine has a −1 charge. Sodium doesn't form stable Na^{2+} compounds. b. ClO_4^-: ClO_4 has 31 valence electrons so it is impossible to satisfy the octet rule for all atoms in ClO_4. The extra electron from the −1 charge in ClO_4^- allows for complete octets for all atoms. c. XeO_4: We can't draw a Lewis structure that obeys the octet rule for SO_4 (30 electrons), unlike with XeO_4 (32 electrons). d. SeF_4: Both compounds require the central atom to expand its octet. O is too small and doesn't have low-energy *d* orbitals to expand its octet (which is true for all Period 2 elements). **99.** a. radius: $N^+ < N < N^-$; IE: $N^- < N < N^+$; b. radius: $Cl^+ < Cl < Se < Se^-$; IE: $Se^- < Se < Cl < Cl^+$; c. radius: $Sr^{2+} < Rb^+ < Br^-$; IE: $Br^- < Rb^+ < Sr^{2+}$ **101.** Nonmetals, which form covalent bonds to each other, have valence electrons in the *s* and *p* orbitals. Since there are

4 total s and p orbitals, there is room for only 8 valence electrons (the octet rule). The valence shell for hydrogen is just the 1s orbital. This orbital can hold 2 electrons, so hydrogen follows the duet rule. **103.** I; square pyramid **105.** Yes; each structure has the same number of effective pairs around the central atom, giving the same predicted molecular structure for each compound/ion. (A multiple bond is counted as a single group of electrons.) **107.** From VSEPR, the molecular structure would be linear. Using hyperconjugation, the resonance structures are

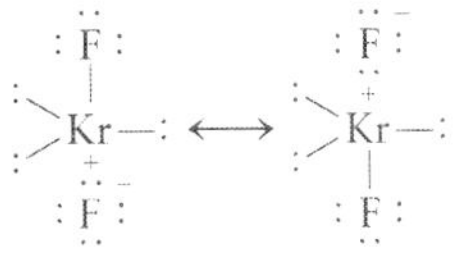

:F: — Kr — : ⟷ :F: — Kr — :

109. −562 kJ **111.** a. The most likely structures are

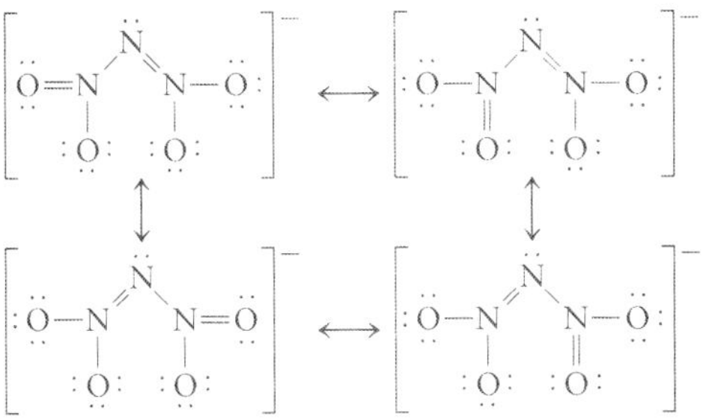

[O=N—N(—O)—N—O]$^-$ ⟷ [O—N—N—N—O]$^-$ ⟷ [O—N—N—N=O]$^-$ ⟷ [O—N—N—N—O]$^-$

There are other possible resonance structures, but these are most likely. b. The NNN, ONN, and ONO bond angles should be about 120°. c. $NH_4N(NO_2)_2(s) \rightarrow 2N_2(g) + 2H_2O(g) + O_2(g)$, $\Delta H =$ −893 kJ; d. To estimate ΔH, we ignored the ionic interactions between NH_4^+ and $N(NO_2)_2^-$. In addition, we assumed the bond energies in Table 13.6 applied to the $N(NO_2)^-$ bonds in any one of the resonance structures above. This is a bad assumption since molecules that exhibit resonance generally have stronger overall bonds than predicted. All of these assumptions give an estimated ΔH value that is too negative. **113.** a. i. −2636 kJ; ii. −3471 kJ; iii. −3543 kJ; b. reaction iii, −8085 kJ/kg **115.** 17 kJ/mol

Chapter 14

9. The valence orbitals of the nonmetals are the s and p orbitals. The lobes of the p orbitals are 90 degrees and 180 degrees apart from each other. If the p orbitals were used to form bonds, then all bond angels should be 90 degrees or 180 degrees. This is not the case. In order to explain the observed geometry (bond angles) that molecules exhibit, we need to make up (hybridize) orbitals that point to where the bonded atoms and lone pairs are located. We know the geometry; we hybridize orbitals to explain the geometry. Sigma bonds have shared electrons in the area centered on a line joining the atoms. The orbitals that overlap to form the sigma bonds must overlap head to head or end to end. The hybrid orbitals about a central atom always are directed at the bonded atoms. Hybrid orbitals will always overlap head to head to form sigma bonds. **11.** We use d orbitals when we have to; that is, we use d orbitals when the central atom on a molecule has more than eight electrons around it. The d orbitals are necessary to accommodate more than eight electrons. Row 2 elements never have more than eight electrons around them so they never hybridize d orbitals. We rationalize this by saying there are no d orbitals close in energy to the valence 2s and 2p orbitals (2d orbitals are forbidden energy levels). However, for Row 3 and heavier elements, there are 3d, 4d, 5d, etc. orbitals, which will be close in energy to the valence s and p orbitals. It is Row 3 and heavier nonmetals that hybridize d orbitals when they have to. For phosphorus, the valence electrons are in 3s and 3p orbitals. Therefore, 3d orbitals are closest in energy and are available for hybridization. Arsenic would hybridize 4d orbitals to go with the valence 4s and 4p orbitals, whereas iodine would hybridize 5d orbitals since the valence electrons are in $n = 5$.

13. H_2O: H—O—H

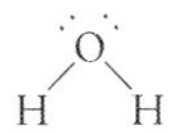

H_2O has a tetrahedral arrangement of the electron pairs around the O atom that requires sp^3 hybridization. Two of the sp^3 hybrid orbitals are used to form bonds to the two hydrogen atoms, and the other two sp^3 hybrid orbitals hold the two lone pairs on oxygen. The two O—H bonds are formed from overlap of the sp^3 hybrid orbitals from oxygen with the 1s atomic orbitals from the hydrogen atoms. **15.** For ethane, the carbon atoms are sp^3 hybridized. The six C—H sigma bonds are formed from overlap of the sp^3 hybrid orbitals on C with the 1s atomic orbitals from the hydrogen atoms. The carbon–carbon sigma bond is formed from overlap of an sp^3 hybrid orbital on each C atom. For ethanol, the two C atoms and the O atom are sp^3 hybridized. All bonds are formed from overlap with these sp^3 hybrid orbitals. The C—H and O—H sigma bonds are formed from overlap of sp^3 hybrid orbitals with hydrogen 1s atomic orbitals. The C—C and C—O sigma bonds are formed from overlap of the sp^3 hybrid orbitals on each atom. **17.** [51] a. sp; b. sp^3; c. sp^3; d. sp^3; e. sp^2; f. sp^3; g. sp; h. each O is sp^2 hybridized; i. Br is sp^3 hybridized [52] a. All are sp^3 hybridized. b. All are sp^3 hybridized. c. All are sp^3 hybridized. [54] a. NO_2^-, sp^2; NO_3^-, sp^2; N_2O_4, sp^2 for both nitrogens; b. All are sp hybridized.

19. a. Tetrahedral, 109.5°, sp^3, nonpolar (CF_4)

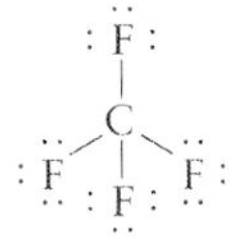

b. Trigonal pyramid, <109.5°, sp^3, polar (NF_3)

c. V-shaped, <109.5°, sp^3, polar (OF_2)

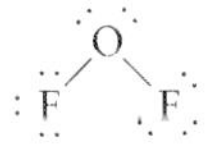

d. Trigonal planar, 120°, sp^2, nonpolar (BF_3)

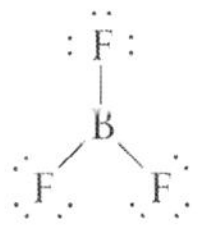

e. Linear, 180°, sp, nonpolar

H—Be—H

f. (TeF_4) $a \approx 120°$, see-saw, dsp^3, polar; $b \approx 90°$

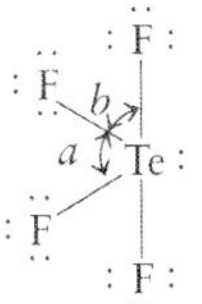

g. (AsF_5) $a = 90°$, trigonal bipyramid, dsp^3, nonpolar; $b = 120°$

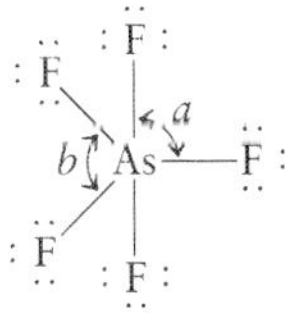

h. :F—Kr—F: Linear, 180°, dsp^3, nonpolar

i. F–Kr(–F)(–F)–F Square planar, 90°, d^2sp^3, nonpolar

j. F–Se(–F)(–F)(–F)(–F)–F Octahedral, 90°, d^2sp^3, nonpolar

k. F–I(–F)(–F)(–F)–F Square pyramid, ≈90°; d^2sp^3, polar

l. F–I(–F)–F T-shaped, ≈90°, dsp^3, polar

21. For the p orbitals to properly line up to form the π bond, all six atoms are forced into the same plane. If the atoms are not in the same plane, then the π bond could not form since the p orbitals would no longer be parallel to each other.

23. Biacetyl

H / H–C–C(=O)–C(=O)–C–H / H, H (labels: sp^2, sp^3)

All CCO angles are 120°. The six atoms are not forced to lie in the same plane because of free rotation around the carbon–carbon single (σ) bonds. 11 σ and 2 π bonds.

Acetoin

H–C(H)(H)–C(H)(O–H)–C(=O)–C(H)(H)–H (labels: sp^2, a, b, sp^3, sp^3)

Angle a = 120°, angle b = 109.5°, 13 σ bonds and 1 π bond

25. To complete the Lewis structure, add lone pairs to complete octets for each atom. a. 6; b. 4; c. The center N in —N=N=N group; d. 33 σ; e. 5 π; f. 180°; g. ≈109.5°; h. sp^3

27. CO :C≡O: CO_2 O=C=O C_3O_2 O=C=C=C=O

There is no molecular structure for the diatomic CO molecule. The carbon in CO is sp hybridized. CO_2 is a linear molecule, and the central carbon atom is sp hybridized. C_3O_2 is a linear molecule with all of the central carbon atoms exhibiting sp hybridization. **29.** Bonding and antibonding molecular orbitals are both solutions to the quantum mechanical treatment of the molecule. Bonding orbitals form when in-phase orbitals combine to give constructive interference. This results in enhanced electron probability located between the two nuclei. The end result is that a bonding MO is lower in energy than the atomic orbitals from which it is composed. Antibonding orbitals form when out-of-phase orbitals combine. The mismatched phases produce destructive interference leading to a node of electron probability between the two nuclei. With electron distribution pushed to the outside, the energy of an antibonding orbital is higher than the energy of the atomic orbitals from which it is composed. **31.** a. H_2 has two valence electrons to put in the MO diagram whereas He_2 has 4 valence electrons.

H_2: $(\sigma_{1s})^2$ Bond order = B.O. = (2 − 0)/2 = 1
He_2: $(\sigma_{1s})^2(\sigma_{1s}{}^*)^2$ B.O. = (2 − 2)/2 = 0

H_2 has a nonzero bond order, so MO theory predicts it will exist. The H_2 molecule is stable with respect to the two free H atoms. He_2 has a bond order of zero, so it should not form. The He_2 molecule is not more stable than the two free He atoms. b. See Fig. 14.41 for the MO energy-level diagrams of B_2, C_2, N_2, O_2, and F_2. B_2 and O_2 have unpaired electrons in their electron configurations, so they are predicted to be paramagnetic. C_2, N_2, and F_2 have no unpaired electrons in their MO diagrams; they are all diamagnetic. c. From the MO energy diagram in Fig. 14.41, N_2 maximizes the number of electrons in the lower-energy bonding orbitals and has no electrons in the antibonding $2p$ molecular orbitals. N_2 has the highest possible bond order of 3, so it should be a very strong (stable) bond. d. NO^+ has 5 + 6 − 1 = 10 valence electrons to place in the MO diagram and NO^- has 5 + 6 + 1 = 12 valence electrons. The MO diagram for these two ions is assumed to be the same as that used for N_2.

NO^+: $(\sigma_{2s})^2(\sigma_{2s}{}^*)^2(\pi_{2p})^4(\sigma_{2p})^2$ B.O. = (8 − 2)/2 = 3
NO^-: $(\sigma_{2s})^2(\sigma_{2s}{}^*)^2(\pi_{2p})^4(\sigma_{2p})^2(\pi_{2p}{}^*)^2$ B.O. = (8 − 4)/2 = 2

NO^+ has a larger bond order than NO^-, so NO^+ should be more stable than NO^-.

33.

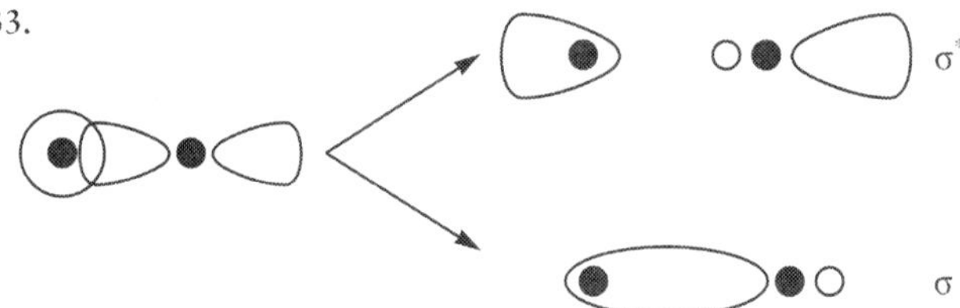

These MOs are σ MOs since the electron density is cylindrically symmetric around the internuclear axis. **35.** $N_2{}^+$ and $N_2{}^-$ both have a bond order of 2.5.
37. CN: $(\sigma_{2s})^2(\sigma_{2s}{}^*)^2(\pi_{2p})^4(\sigma_{2p})^1$;
NO: $(\sigma_{2s})^2(\sigma_{2s}{}^*)^2(\pi_{2p})^4(\sigma_{2p})^2(\pi_{2p}{}^*)^1$;
$O_2{}^{2+}$: $(\sigma_{2s})^2(\sigma_{2s}{}^*)^2(\sigma_{2p})^2(\pi_{2p})^4$;
$N_2{}^{2+}$: $(\sigma_{2s})^2(\sigma_{2s}{}^*)^2(\pi_{2p})^4$; If the added electron goes into a bonding orbital, the bond order would increase, making the species more stable and more likely to form. Between CN and NO, CN would most likely form CN^- since the bond order increases (unlike NO^-, in which the added electron goes into an antibonding orbital). Between $O_2{}^{2+}$ and $N_2{}^{2+}$, $N_2{}^+$ would most likely form since the bond order increases (unlike $O_2{}^+$). **39.** The π bonds between S atoms and between C and S atoms are not as strong. The atomic orbitals do not overlap with each other as well as the smaller atomic orbitals of C and O overlap. **41.** π molecular orbital **43.** a. The electron density would be closer to F on the average. The F atom is more electronegative than the H atom, and the $2p$ orbital of F is lower in energy than the 1s orbital of H; b. The bonding MO would have more fluorine $2p$ character because it is closer in energy to the fluorine $2p$ atomic orbital; c. The antibonding MO would place more electron density closer to H and would have a greater contribution from the higher-energy hydrogen 1s atomic orbital. **45.** Molecules that exhibit resonance have delocalized π bonding. This is a fancy way of saying that the π electrons are not permanently stationed between two specific atoms but instead can roam about over the surface of a molecule. We use the concept of delocalized π electrons to explain why molecules that exhibit resonance have equal bonds in terms of strength. Because the π electrons can roam about over the entire surface of the molecule, the π electrons are shared by all of the

atoms in the molecule, giving rise to equal bond strengths. The classic example of delocalized π electrons is benzene (C_6H_6). Figure 14.50 shows the π molecular orbital system for benzene. Each carbon in benzene is sp^2 hybridized, leaving one unhybridized p atomic orbital. All six of the carbon atoms in benzene have an unhybridized p orbital pointing above and below the planar surface of the molecule. Instead of just two unhybridized p orbitals overlapping, we say all six of the unhybridized p orbitals overlap, resulting in delocalized π electrons roaming about above and below the entire surface of the benzene molecule.

S O O ⟷ S O O

In SO_2 the central sulfur atom is sp^2 hybridized. The unhybridized p atomic orbital on the central sulfur atom will overlap with parallel p orbitals on each adjacent O atom. All three of these p orbitals overlap together, resulting in the π electrons moving about above and below the surface of the SO_2 molecule. With the delocalized π electrons, the S—O bond lengths in SO_2 are equal (and not different as each individual Lewis structure indicates). **47.** The Lewis structures for CO_3^{2-} (24 e^-) are:

$[CO_3]^{2-}$ ⟷ $[CO_3]^{2-}$ ⟷ $[CO_3]^{2-}$

In the localized electron view, the central carbon atom is sp^2 hybridized; the sp^2 hybrid orbitals are used to form the three sigma bonds in CO_3^{2-}. The central C atom also has one unhybridized p atomic orbital that overlaps with another p atomic orbital from one of the oxygen atoms to form the π bond in each resonance structure. This localized π bond moves (resonates) from one position to another. In the molecular orbital model for CO_3^{2-}, all four atoms in CO_3^{2-} have a p atomic orbital that is perpendicular to the plane of the ion. All four of these p orbitals overlap at the same time to form a delocalized π bonding system where the π electrons can roam above and below the entire surface of the ion. The π molecular orbital system for CO_3^{2-} is analogous to that for NO_3^-, which is shown in Fig. 14.51 of the text. **49.** 411.7 N m^{-1} **51.** a. 113 pm; b. 3.45×10^{11} s^{-1} **53.** a. quadruplet in iv; b. singlet in i; c. triplet in iii; d. doublet in ii

55. a. Trigonal pyramid; sp^3

b. Tetrahedral; sp^3

c. Square pyramid; d^2sp^3

d. T-shaped; dsp^3

e. Trigonal bipyramid; dsp^3

57. a. No, some atoms are in different places. Thus these are not resonance structures; they are different compounds. b. For the first Lewis structure, all nitrogens are sp^3 hybridized and all carbons are sp^2 hybridized. In the second Lewis structure, all nitrogens and carbons are sp^2 hybridized. c. The first structure with the carbon–oxygen double bonds is slightly more stable.

59.

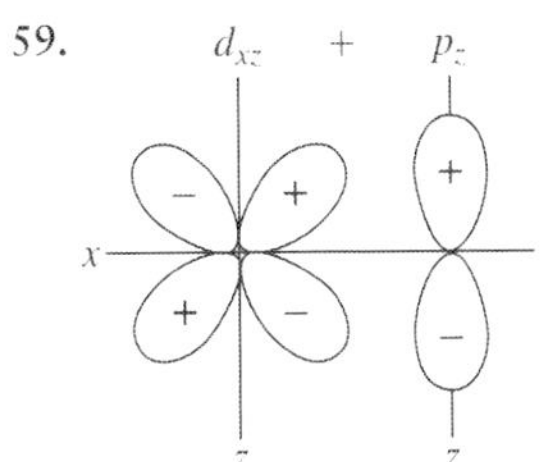

The two orbitals overlap side to side, so when the orbitals are in phase, a π bonding molecular orbital would form.

61.

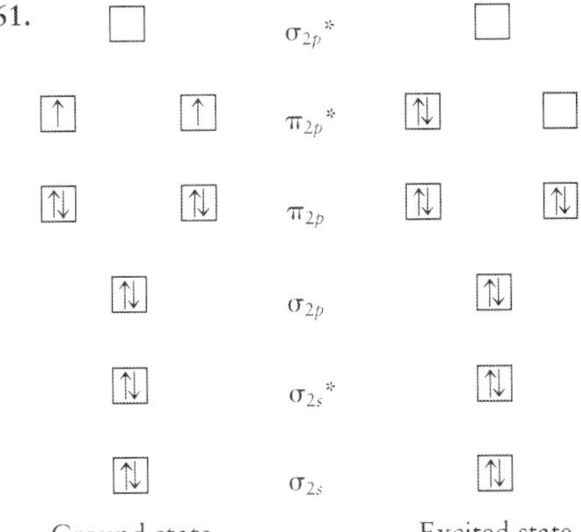

Lewis structure shows no unpaired electrons. The ground-state MO diagram for O_2 has two unpaired electrons. It takes energy to pair electrons in the same orbital. The MO diagram with no unpaired electrons is at a higher energy and is an excited state.

63. a. Trigonal planar; nonpolar; 120°; sp^2

b. Can also be:

Polar — Nonpolar

V-shaped about both N atoms; ≈120°; sp^2.
These are distinctly different molecules.

c. Trigonal planar about each carbon; nonpolar; 120°; sp^2 for all carbons

d. T-shaped; polar; ≈90°; dsp^3

65. 6 C and N atoms are sp^2 hybridized; 6 C and N atoms are sp^3 hybridized; 0 C and N atoms are sp hybridized (linear geometry); 25 σ bonds and 4 π bonds **67.** a. The NNO structure is correct.

From the Lewis structures we would predict both NNO and NON to be linear, but NON would be nonpolar. NNO is polar.

b. $:\ddot{N}=N=\ddot{O}: \longleftrightarrow :N\equiv N-\ddot{O}: \longleftrightarrow :\ddot{N}-N\equiv O:$

−1 +1 0	0 +1 −1	−2 +1 +1 Formal charges

The central N is *sp* hybridized. We can probably ignore the third resonance structure on the basis of formal charge. c. *sp* hybrid orbitals from the center N overlap with atomic orbitals (or hybrid orbitals) from the other two atoms to form the two sigma bonds. The remaining *p* orbitals from the center N overlap with *p* orbitals from the other N to form the two π bonds. **69.** a. 116 kJ/mol;

b. F_2: $(\sigma_{2s})^2(\sigma_{2s}^*)^2(\sigma_{2p})^2(\pi_{2p})^4(\pi_{2p}^*)^4$ B.O. = (8 − 6)/2 = 1
F_2^-: $(\sigma_{2s})^2(\sigma_{2s}^*)^2(\sigma_{2p})^2(\pi_{2p})^4(\pi_{2p}^*)^4(\sigma_{2p}^*)^1$ B.O. = (8 − 7)/2 = 0.5

MO theory predicts that F_2 should have a stronger bond than F_2^- because F_2 has the larger bond order. The calculated F_2^- bond energy is indeed smaller than the F_2 bond energy (154 kJ/mol). **71.** a. N_2, N_2^+, N, and N^+ will all be present, assuming excess N_2. b. Light with wavelengths in the range of 85.33 nm $< \lambda \le$ 127 nm will produce N but no ions. c. Consider the MO energy-level diagram for N_2 (see Figs. 14.40 and 14.41). The electron removed from N_2 is in the σ_{2p} MO, which is lower in energy than the 2*p* atomic orbital from which the electron in atomic nitrogen is removed. Since the electron removed from N_2 is lower in energy than the electron in N, the ionization energy of N_2 is greater than for N. **73.** The complete Lewis structure follows. All but two of the carbon atoms are sp^3 hybridized. The two carbon atoms that contain the double bond are sp^2 hybridized (see*).

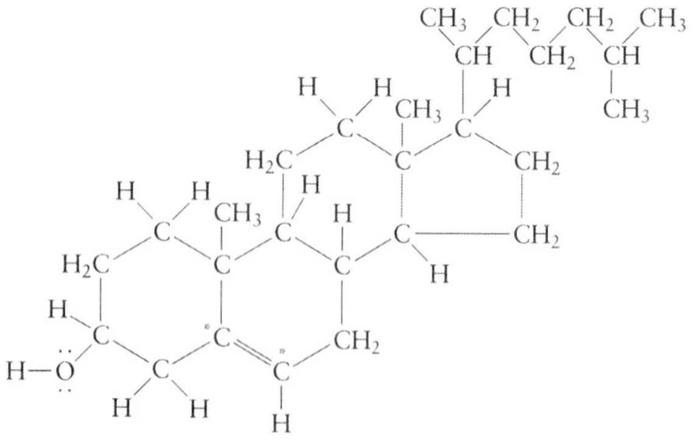

```
                                  CH3  CH2  CH2  CH3
                                    \  /  \ /  \ /
                                     CH    CH2  CH
                    H     H          |    H     |
                     \   /  CH3      |   /      CH3
                       C     |      C
                     /   \   |    /   \
                 H2C       C          CH2
       H    H     |  H     |            |
        \  /  CH3 | /  H   |            |
         C     |  C    |   C --------- CH2
       /   \   | /  \  |  /  \
   H2C       C        C       H
    |        |        |
H    |        |        |
  \  C       *C ===== *C      CH2
..  /  \    /          \
H—O      C              C
..     /   \            |
      H     H           H
```

No; most of the carbons are not in the same plane since a majority of carbon atoms exhibit a tetrahedral structure. *Note:* CH, CH_2, and CH_3 are shorthand for carbon atoms singly bonded to hydrogen atoms. **75.** a. The CO bond is polar with the negative end around the more electronegative oxygen atom. We would expect metal cations to be attracted to and bond to the oxygen end of CO on the basis of electronegativity. b. $:C\equiv O:$ FC (carbon) = −1; FC (oxygen) = +1. From formal charge, we would expect metal cations to bond to the carbon (with the negative formal charge). c. In MO theory, only orbitals with proper symmetry overlap to form bonding orbitals. The metals that form bonds to CO are usually transition metals, all of which have outer electrons in the *d* orbitals. The only MOs of CO that have proper symmetry to overlap with *d* orbitals are the π_{2p}^* orbitals, whose shape is similar to the *d* orbitals (see Fig. 14.36). Since the antibonding MOs have more carbon character, one would expect the bond to form through carbon. **77.** ΔH_f° = 237 kJ/mol (calculated); there is a 154 kJ discrepancy. Benzene is more stable (lower in energy) by 154 kJ than we expect from bond energies. Two equivalent Lewis structures can be drawn for benzene. The π bonding system implied by each Lewis structure consists of three localized π bonds. This, however, does not adequately explain the six equivalent C—C bonds in benzene. To explain the equivalent C—C bonds, the π electrons are said to be delocalized over the entire surface of the molecule (see Section 14.5). The large discrepancy between the ΔH_f° values is due to the effect of a delocalized π bonding system that was not considered when calculating ΔH_f°. The extra stability associated with benzene can be called resonance stabilization. **79.** The two isomers having the formula C_2H_6O are

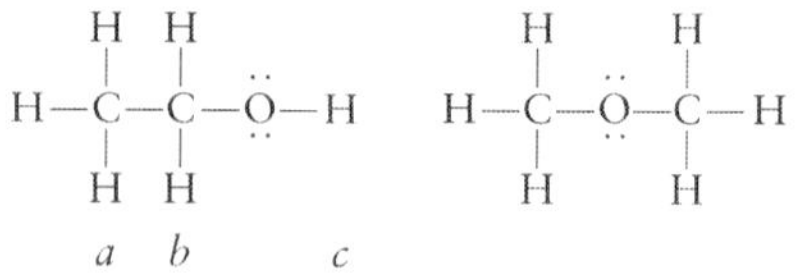

```
     H  H                 H       H
     |  |    ..           |  ..   |
  H—C—C—O—H           H—C—O—C—H
     |  |    ..           |  ..   |
     H  H                 H       H
     a  b     c
```

The first structure has three types of hydrogens (*a*, *b*, and *c*), so the signals should be seen in three different regions of the NMR spectrum. The overall relative areas of the three signals should be in a 3:2:1 ratio due to the number of *a*, *b*, and *c* hydrogens. The signal for the *a* hydrogens will be split into a triplet signal due to the two neighboring *b* hydrogens. The signal for the *b* hydrogens will be split into a quartet signal due to the three neighboring *a* hydrogens. In theory, the *c* hydrogen should be split into a triplet signal due to the two neighboring *b* hydrogens. In practice, the *c* hydrogen signal is usually not split (which you could not predict). In the second structure, all 6 hydrogens are equivalent, so only one signal will appear in the NMR spectrum. This signal will be a singlet signal because the hydrogens in the two $-CH_3$ groups are more than three sigma bonds apart from each other (no splitting occurs).

Chapter 15

11. A possible balanced equation is 2A + 3B → 4C.
13. $\frac{d[NH_3]}{dt} = -\frac{2}{3}\frac{d[H_2]}{dt}$ **15.** $L^{1/2}\ mol^{-1/2}\ s^{-1}$
17. a. Rate = $k[NO]^2[Cl_2]$; b. $k_{mean} = 1.8 \times 10^2\ L^2\ mol^{-2}\ min^{-1}$ **19.** a. rate = $k[NOCl]^2$; b. $k_{mean} = 6.6 \times 10^{-29}\ cm^3\ molecules^{-1}\ s^{-1}$; c. $4.0 \times 10^{-8}\ L\ mol^{-1}\ s^{-1}$ **21.** a. Rate = $k[ClO_2]^2[OH^-]$; $k_{mean} = 2.30 \times 10^2\ L^2\ mol^{-2}\ s^{-1}$; b. 0.594 mol $L^{-1}\ s^{-1}$ **23.** Rate = $k[N_2O_5]$; $k_{mean} = 1.19 \times 10^{-2}\ s^{-1}$ **25.** Rate = $\frac{k[I^-][OCl^-]}{[OH^-]}$; k_{mean} = 60. s^{-1} **27.** The first-order half-life is independent of concentration, the zero-order half-life is directly related to the concentration, and the second-order half-life is inversely related to concentration. For a first-order reaction, if the first half-life equals 20. s, the second half-life will also be 20. s because the half-life for a first-order reaction is concentration-independent. The second half-life for a zero-order reaction will be 1/2(20.) = 10. s. This is so because the half-life for a zero-order reaction has a direct relationship with concentration (as the concentration decreases by a factor of 2, the half-life decreases by a factor of 2). Because a second-order reaction has an inverse relationship between $t_{1/2}$ and $[A]_0$, the second half-life will be 40. s (twice the first half-life value).
29. a. Rate = $k[A]$; $\ln[A] = -kt + \ln[A]_0$; $k = 2.97 \times 10^{-2}\ min^{-1}$; b. 23.3 min; c. 69.9 min **31.** $\ln[H_2O_2] = -kt + \ln[H_2O_2]_0$; Rate = $k[H_2O_2]$; $k = 8.3 \times 10^{-4}\ s^{-1}$; 0.037 *M* **33.** $\frac{1}{[NO_2]} = kt + \frac{1}{[NO_2]_0}$; Rate = $k[NO_2]^2$; $k = 2.08 \times 10^{-4}\ L\ mol^{-1}\ s^{-1}$; 0.131 *M* **35.** The rate law and integrated rate law are Rate = $k = 1.7 \times 10^{-4}$ atm/s;

$$P_{C_2H_5OH} = -kt + 250.\ \text{torr}\left(\frac{1\ \text{atm}}{760\ \text{torr}}\right) = -kt + 0.329\ \text{atm};$$

At 900. s, $P_{C_2H_5OH}$ = 130 torr **37.** a. The reaction is first order with respect to NO and first order with respect to O_3. b. Rate = $k[NO][O_3]$; c. $k' = 1.8\ s^{-1}$; $k'' = 3.6\ s^{-1}$; d. $1.8 \times 10^{-14}\ cm^3\ molecules^{-1}\ s^{-1}$ **39.** 150. s **41.** 12.5 s **43.** a. 160. s for both; b. 532 s **45.** a. $[A] = -kt + [A]_0$ where $k = 5.0 \times 10^{-2}\ mol\ L^{-1}\ s^{-1}$ and $[A]_0 = 1.00 \times 10^{-3}$ *M*; b. 1.0×10^{-2} s; c. 2.5×10^{-4} *M* **47.** a. $1.15 \times 10^2\ L^3\ mol^{-3}\ s^{-1}$; b. 87.0 s; c. [B] = 1.00 *M*; [A] = 1.27×10^{-5} *M* **49.** In a unimolecular reaction, a single-reactant molecule decomposes to products. In a bimolecular reaction, two molecules collide to give products. The probability of the simultaneous collision of three molecules with enough energy and proper orientation is very small, making termolecular steps very unlikely. **51.** a. Rate = $k[CH_3NC]$; b. Rate = $k[O_3][NO]$; c. Rate = $k[O_3]$; d. Rate = $k[O_3][O]$; e. Rate = $k[^{14}_{6}C]$ or rate = kN where N = the number of $^{14}_{6}C$ atoms **53.** Rate = $k[C_4H_9Br]$; $C_4H_9Br + 2H_2O \rightarrow C_4H_9OH + Br^- + H_3O^+$; the intermediates are $C_4H_9^+$ and $C_4H_9OH_2^+$ **55.** mechanism b **57.** Rate = $\frac{k_3k_2k_1}{k_{-2}k_{-1}}[Br^-][BrO_3^-][H^+]^2 = k[Br^-][BrO_3^-][H^+]^2$ **59.** Rate = $\frac{k_2k_1[I^-][OCl^-]}{k_{-1}[OH^-]} = \frac{k[I^-][OCl^-]}{[OH^-]}$ **61.** a. $MoCl_5^-$; b. Rate = $\frac{d[NO_2^-]}{dt} = \frac{k_1k_2[NO_3^-][MoCl_6^{2-}]}{k_{-1}[Cl^-] + k_2[NO_3^-]}$ **63.** a. Rate = $\frac{d[E]}{dt} = \frac{k_1k_2[B]^2}{k_{-1}[B] + k_2}$; b. When $k_2 << k_{-1}[B]$, then the reaction is first order in B. c. Collisions between B molecules only transfer energy from one B to another. This occurs at a much faster rate than the decomposition of an energetic B molecule (B*). **65.** Two reasons are: a. The collision must involve enough energy to produce the reaction; that is, the collision energy must be equal to or exceed the activation energy. b. The relative orientation of the reactants when they collide must allow formation of any new bonds necessary to produce products. **67.** $9.5 \times 10^{-5}\ L\ mol^{-1}\ s^{-1}$ **69.** 51°C **71.** a. 91.5 kJ/mol; b. $3.54 \times 10^{14}\ s^{-1}$; c. $3.24 \times 10^{-2}\ s^{-1}$ **73.** A plot of ln(k) versus $\frac{1}{T}$ gives a straight line with negative slope equal to $-E_a/R$ (see *Solutions Guide* for plot). E_a = 11.2 kJ/mol **75.**

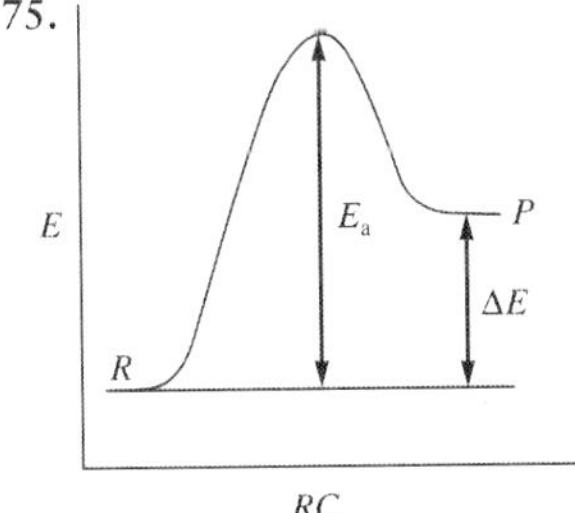

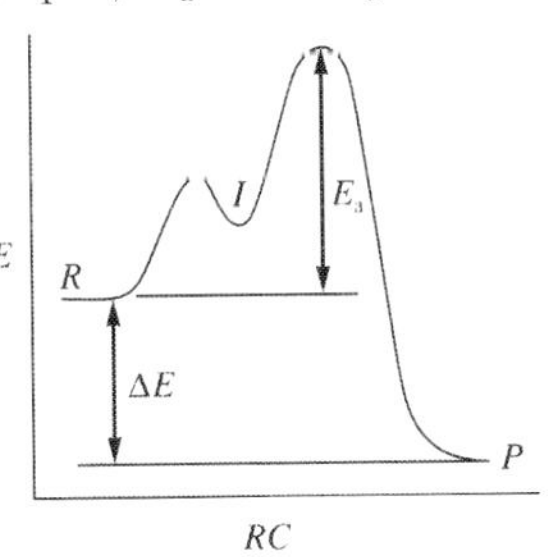

d. See plot for the intermediate plateau, which represents the energy of the intermediate. The two steps in the mechanism are $R \rightarrow I$ and $I \rightarrow P$. In a mechanism, the rate of the slowest step determines the rate of the reaction. The activation energy for the slowest step will be the largest energy barrier that the reaction must overcome. Since the second hump in the diagram is at the highest energy, then the second step ($I \rightarrow P$) has the largest activation energy and will be the rate-determining step (the slow step). **77.** positive ΔE value **79.** A metal-catalyzed reaction is dependent on the number of adsorption sites on the metal surface. Once the metal surface is saturated with reactant, the rate of reaction becomes independent of concentration. **81.** a. W, because it has a lower activation energy than the Os catalyst. b. The W-catalyzed reaction is approximately 10^{30} times faster than the uncatalyzed reaction. c. Because $[H_2]$ is in the denominator of the rate law, the presence of H_2 decreases the rate of the reaction. For the decomposition to occur, NH_3 molecules must be adsorbed on the surface of the catalyst. If H_2 is also adsorbed on the catalyst surface, then there are fewer sites for NH_3 molecules to be adsorbed, and the rate decreases. **83.** Since the chlorine atom–catalyzed reaction has a lower activation energy, then the Cl-catalyzed rate is faster. Hence, Cl is a more effective catalyst. **85.** At high [S], the enzyme is completely saturated with substrate. Once the enzyme is completely saturated, the rate of decomposition of ES can no longer increase and the overall rate remains constant. **87.** ~1 second **89.** The most common method to experimentally determine the differential rate law is the method of initial rates. Once the differential rate law is determined experimentally, the integrated rate law can be derived. However, sometimes it is more convenient and more accurate to collect concentration versus time data for a reactant. When this is the case, then we do "proof" plots to determine the integrated rate law. Once the integrated rate law is determined, the differential rate law can be determined. Either experimental procedure allows determination of both the integrated and the differential rate; which rate law is determined by experiment and which is derived is usually decided by which data is easiest and most accurately collected. **91.** a. untreated: $k = 0.465\ day^{-1}$; deacidifying agent: $k = 0.659\ day^{-1}$; antioxidant: $k = 0.779\ day^{-1}$; b. No, the silk degrades more rapidly with the additives; c. untreated: $t_{1/2}$ = 1.49 day; deacidifying agent: $t_{1/2}$ = 1.05 day; antioxidant: $t_{1/2}$ = 0.890 day **93.** Carbon cannot form the fifth bond necessary for the transition state because of the small atomic size of carbon and because carbon doesn't have low-energy *d* orbitals available to expand the octet. **95.** a. Second order; b. 17.1 s ≈ 20 s; c. 34.6 s ≈ 30 s; d. 847 s; e. 31 kJ/mol **97.** 29.8 kJ/mol **99.** a. 53.9 kJ/mol; b. 83 chirps per minute per insect; c.

T (°C)	T (°F)	k (min^{-1})	42 + 0.80 (k/4)
25.0	77.0	178	78°F
20.3	68.5	126	67°F
17.3	63.1	100.	62°F
15.0	59.0	83	59°F

The rule of thumb appears to be fairly accurate, about ±1°F.

101. Rate = $\frac{d[Cl_2]}{dt} = \frac{k_1k_2[NO_2Cl]^2}{k_{-1}[NO_2] + k_2[NO_2Cl]}$ **103.** a. Rate = $k[CH_3X]$, $k = 0.93\ h^{-1}$; b. 8.80×10^{-10} h; c. 3.0×10^2 kJ/mol; d. The activation energy is close to the C—X bond energy. A plausible mechanism is

$$CH_3X \longrightarrow CH_3 + X \quad \text{(slow)}$$
$$CH_3 + Y \longrightarrow CH_3Y \quad \text{(fast)}$$

105. 10. s **107.** Rate = $k[A][B]^2$, $k = 1.4 \times 10^{-2}\ L^2\ mol^{-2}\ s^{-1}$ **109.** Rate = $\frac{-d[N_2O_5]}{dt} = \frac{2k_1k_2[M][N_2O_5]}{k_{-1}[M] + 2k_2}$ **111.** a. Both are first order. b. $k_1 = 0.82\ L\ mol^{-1}\ min^{-1}$; $k_2 = 9.5\ L^2\ mol^{-2}\ min^{-1}$ c. There are two pathways, one involving H^+ with rate = $k_2[H^+][I^-][H_2O_2]$ and another pathway not involving H^+ with rate = $k_1[I^-][H_2O_2]$. The overall rate of reaction depends on which of these two pathways dominates, and this depends on the H^+ concentration.

Chapter 16

11. Intermolecular forces are the relatively weak forces between molecules that hold the molecules together in the solid and liquid phases. Intramolecular forces are the forces within a molecule. These are the covalent bonds in a molecule. Intramolecular forces (covalent bonds) are much stronger than intermolecular forces. Dipole forces are the forces that act between polar molecules. The electrostatic attraction between the partial positive end of one polar molecule and the partial negative end of another is the dipole force. Dipole forces are generally weaker than hydrogen bonding. Both of these forces are due to dipole moments in molecules. Hydrogen bonding is given a separate name from dipole forces because hydrogen bonding is a particularly strong dipole force. Any neutral molecule that has a hydrogen covalently bonded to N, O, or F exhibits the relatively strong hydrogen-bonding intermolecular forces. London dispersion forces are accidental or induced dipole forces. Like dipole forces, London dispersion forces are electrostatic in nature. Dipole forces are the electrostatic forces between molecules having a permanent dipole. London dispersion forces are the electrostatic forces between molecules having an accidental or induced dipole. All covalent molecules (polar and nonpolar) have London dispersion forces, but only polar molecules (those with permanent dipoles) exhibit dipole forces. **13.** *Fusion* refers to a solid converting to a liquid and *vaporization* refers to a liquid converting to a gas. Only a fraction of the hydrogen bonds are broken in going from the solid phase to the liquid phase. Most of the hydrogen bonds are still present in the liquid phase and must be broken during the liquid-to-gas transition. The enthalpy of vaporization is much larger than the enthalpy of fusion because more intermolecular forces are broken during the vaporization process. **15.** a. LD (London dispersion); b. dipole, LD; c. hydrogen bonding, LD; d. ionic; e. LD; f. dipole, LD; g. ionic; h. ionic; i. LD mostly; C—F bonds are polar, but polymers like Teflon are so large that LD forces are the predominant intermolecular forces. j. LD; k. dipole, LD; l. hydrogen bonding, LD; m. dipole, LD; n. LD **17.** a. $H_2NCH_2CH_2NH_2$; More extensive hydrogen bonding is possible since two NH_2 groups are present. b. H_2CO; H_2CO has dipole forces in addition to LD forces. CH_3CH_3 has only LD forces; c. CH_3OH; CH_3OH can form hydrogen-bonding interactions; H_2CO does not. d. HF; HF is capable of forming hydrogen-bonding interactions; HBr is not. **19.** a. Neopentane is more compact than *n*-pentane. There is less surface area contact between neopentane molecules. This leads to weaker LD forces and a lower boiling point. b. HF is capable of hydrogen bonding; HCl is not. c. LiCl is ionic and HCl is a molecular solid with only dipole forces and LD forces. Ionic forces are much stronger than the forces for molecular solids. d. *n*-Hexane is a larger molecule so it has stronger LD forces. **21.** Ar exists as individual atoms that are held together in the condensed phases by London dispersion forces. The molecule that will have a boiling point closest to Ar will be a nonpolar substance with about the same molar mass as Ar (39.95 g/mol); this same size nonpolar substance will have about the equivalent strength of London dispersion forces. Of the choices, only Cl_2 (70.90 g/mol) and F_2 (38.00 g/mol) are nonpolar. Because F_2 has a molar mass closest to that of Ar, one would expect the boiling point of F_2 to be close to that of Ar. **23.** ethanol, 78.5°C; dimethyl ether, −23°C; propane, −42.1°C **25.** 46.7 kJ/mol; 90.% **27.** a. Surface tension: the resistance of a liquid to an increase in its surface area. b. Viscosity: the resistance of a liquid to flow. c. Melting point: the temperature (at constant pressure) where a solid converts entirely to a liquid as long as heat is applied. A more detailed definition is the temperature at which the solid and liquid states have the same vapor pressure under conditions where the total pressure is constant. d. Boiling point: the temperature (at constant pressure) where a liquid converts entirely to a gas as long as heat is applied. The detailed definition is the temperature at which the vapor pressure of the liquid is exactly equal to the external pressure. e. Vapor pressure: the pressure of the vapor over a liquid at equilibrium. As the strengths of intermolecular forces increase, surface tension, viscosity, melting point, and boiling point increase, while vapor pressure decreases. **29.** Water is a polar substance and wax is a nonpolar substance. A molecule at the surface of a drop of water is subject to attractions only by molecules below it and to each side. The effect of this uneven pull on the surface molecules tends to draw them into the body of the liquid and causes the droplet to assume the shape that has the minimum surface area, a sphere. **31.** The structure of H_2O_2 is H—O—O—H, which produces greater hydrogen bonding than in water. **33.** a. crystalline solid: regular, repeating structure; amorphous solid: irregular arrangement of atoms or molecules; b. ionic solid: made up of ions held together by ionic bonding; molecular solid: Made up of discrete covalently bonded molecules held together in the solid phase by weaker forces (LD, dipole, or hydrogen bonds); c. molecular solid: discrete, individual molecules; covalent network solid: no discrete molecules; A covalent network solid is one large molecule. The interparticle forces are the covalent bonds between atoms. d. metallic solid: completely delocalized electrons, conductor of electricity (ions in a sea of electrons); covalent network solid: localized electrons; insulator or semiconductor **35.** a. Both forms of carbon are network solids. In diamond, each carbon atom is surrounded by a tetrahedral arrangement of other carbon atoms to form a huge molecule. Each carbon atom is covalently bonded to four other carbon atoms. The structure of graphite is based on layers of carbon atoms arranged in fused six-membered rings. Each carbon atom in a particular layer of graphite is surrounded by three other carbons in a trigonal planar arrangement. This requires sp^2 hybridization. Each carbon has an unhybridized p atomic orbital; all of these p orbitals in each six-membered ring overlap with each other to form a delocalized π electron system. b. Silica is a network solid having an empirical formula of SiO_2. The silicon atoms are singly bonded to four oxygens. Each silicon atom is at the center of a tetrahedral arrangement of oxygen atoms that are shared with other silicon atoms. The structure of silica is based on a network of SiO_4 tetrahedra with shared oxygen atoms rather than discrete SiO_2 molecules. Silicates closely resemble silica. The structure is based on interconnected SiO_4 tetrahedra. However, in contrast to silica, where the O/Si ratio is 2:1, silicates have O/Si ratios greater than 2:1 and contain silicon–oxygen anions. To form a neutral solid silicate, metal cations are needed to balance the charge. In other words, silicates are salts containing metal cations and polyatomic silicon–oxygen anions. When silica is heated above its melting point and cooled rapidly, an amorphous (disordered) solid called glass results. Glass more closely resembles a very viscous solution than it does a crystalline solid. To affect the properties of glass, several different additives are thrown into the mixture. Some of these additives are Na_2CO_3, B_2O_3, and K_2O, with each compound serving a specific purpose relating to the properties of the glass. **37.** A crystalline solid will have the simpler diffraction pattern because a regular, repeating arrangement is necessary to produce planes of atoms that will diffract the X rays in regular patterns. An amorphous solid does not have a regular repeating arrangement and will produce a complicated diffraction pattern. **39.** 0.704 Å **41.** 3.13 Å = 313 pm **43.** 1.54 g/cm^3 **45.** 136 pm **47.** edge length of unit cell = 328 pm; radius = 142 pm **49.** The measured density is consistent with the calculated density, assuming a face-centered cubic structure. **51.** Conductor: The energy difference between the filled and unfilled molecular orbitals is minimal. We call this energy difference the band gap. Because the band gap is minimal, electrons can easily move into the conduction bands (the unfilled molecular orbitals). Insulator: Large band gap; electrons do not move from the filled molecular orbitals to the conduction bands since the energy difference is large. Semiconductor: Small band gap; the energy difference between the filled and unfilled molecular orbitals is smaller than in insulators, so some electrons can jump into the conduction bands. The band gap, however, is not as small as with conductors, so semiconductors have intermediate conductivity. a. As the temperature is increased, more electrons in the filled molecular orbitals have sufficient kinetic energy to jump into the conduction bands (the unfilled molecular orbitals). b. A photon of light is absorbed by an electron, which then has sufficient energy to jump into the conduction bands. c. An impurity either adds electrons at an energy near that of the conduction bands (n-type) or creates holes (unfilled energy levels) at energies in the previously filled molecular orbitals (p-type). Both n-type and p-type semiconductors increase conductivity by creating an easier path for electrons to jump from filled to unfilled energy levels. In conductors, electrical conductivity is inversely proportional to temperature. Increases in temperature increase the motions of the atoms, which gives rise to increased resistance (decreased conductivity). In a semiconductor, electrical conductivity is directly proportional to temperature. An increase in temperature provides more electrons with enough kinetic energy to jump from the filled molecular orbitals to the conduction bands, increasing conductivity. **53.** A rectifier is a device that produces a current that flows in one direction from an alternating current, which flows in both directions. In a p–n junction, a p-type and an n-type semiconductor are connected. The natural flow of electrons in a p–n junction is for the excess electrons in the n-type semiconductor to move to the empty energy levels (holes) of the p-type semiconductor. Only when an external electric potential is connected so that electrons flow in this natural direction will the current flow easily (forward bias). If the external electric potential is connected in reverse of the natural flow of electrons, no current flows through the system (reverse bias). A p–n junction transmits a current only under forward bias, thus converting the alternating current to direct current. **55.** In has fewer valence electrons than Se; thus Se doped with In would be a p-type semiconductor. **57.** 5.0×10^2 nm **59.** The structures of most binary ionic solids can be explained by the closest packing of spheres. Typically, the larger ions, usually the anions, are packed in one of the closest packing arrangements and the smaller cations fit into holes among the closest packed anions. There are different types of holes within the closest packed anions that are determined by the number of spheres that form them. Which of the three types of holes are filled usually depends on the relative size of the cation to the anion. Ionic solids will always try to maximize electrostatic attractions among oppositely charged ions and minimize the repulsions among ions with like charges. The structure of sodium chloride can be described in terms of a cubic closest packed array of Cl^- ions with Na^+ ions in all of the octahedral holes. An octahedral hole is formed between 6 Cl^- anions. The number of octahedral holes is the same as the number of packed ions. So, in the face-centered unit cell of sodium chloride, there are 4 net Cl^- ions and 4 net octahedral holes. Because the stoichiometry dictates a 1:1 ratio between the number of Cl^- anions and Na^+ cations, all of the octahedral holes must be filled with Na^+ ions. In zinc sulfide, the sulfide anions also occupy the lattice points of a cubic closest packing arrangement. But instead of having the cations in octahedral holes, the Zn^{2+} cations occupy tetrahedral holes. A tetrahedral hole is the empty space created when four spheres are packed together. There are twice as many tetrahedral holes as packed anions in the closest packed structure. Therefore, each face-centered unit cell of sulfide anions contains 4 net S^{2-} ions and 8 net tetrahedral holes. For the 1:1 stoichiometry to work out, only half the tetrahedral holes are filled with Zn^{2+} ions. This gives 4 S^{2-} ions and 4 Zn^{2+} ions per unit cell for an empirical formula of ZnS. **61.** CoF_2 **63.** +3 **65.** From density data, 358 pm; from ionic radii data, 350. pm. The two values differ by less than 2.5%. **67.** CsBr: $r_+/r_- = 0.867$; From the radius ratio, Cs^+ should occupy cubic holes. The structure should be the CsCl structure. The actual structure is the CsCl structure. KF: $r_+/r_- = 0.978$; Again, we would predict a structure similar to CsCl, i.e., cations in the middle of a simple cubic array of anions. The actual structure is the NaCl structure. The rule fails for KF. Exceptions are common for crystal structures. **69.** a. XeF_2; b. 4.29 g/cm^3 **71.** The formula is Al_2MgO_4. Half of the octahedral holes are filled with Al^{3+} ions, and one-eighth of the tetrahedral holes are filled with Mg^{2+} ions. **73.** a. $YBa_2Cu_3O_9$; b. The structure of this superconductor material follows the alternative perovskite structure described in Exercise 72b. The $YBa_2Cu_3O_9$ structure is three of these cubic perovskite unit cells stacked on top of each other. The atoms are in the same places, Cu takes the place of Ti, two of the Ca atoms are replaced by two Ba atoms, and one Ca is replaced by Y. c. $YBa_2Cu_3O_7$ **75.** A volatile liquid is one that evaporates relatively easily. Volatile liquids have large vapor pressures because the intermolecular forces that prevent evaporation are relatively weak. **77.** a. As the strength of the intermolecular forces increases, the rate of evaporation decreases. b. As temperature increases, the rate of evaporation increases. c. As surface area increases, the rate of evaporation increases. **79.** $C_2H_5OH(l) \rightarrow C_2H_5OH(g)$ is an endothermic process. Heat is absorbed when liquid ethanol vaporizes; the internal heat from the body provides this heat, which results in the cooling of the body. **81.** If one plots $\ln(P_{vap})$ versus $1/T$ with temperature in kelvin, the slope of the straight line is $-\Delta H_{vap}/R$. Because ΔH_{vap} is always positive, the slope of the straight line will be negative. **83.** 77°C **85.** The plots of $\ln(P_{vap})$ versus $1/T$ are linear with negative slope equal to $-\Delta H_{vap}/R$ (see *Solutions Guide* for plot). For Li, ΔH_{vap} = 158 kJ/mol; for Mg, ΔH_{vap} = 139 kJ/mol. The bonding is stronger in Li since ΔH_{vap} is larger for Li.

87.

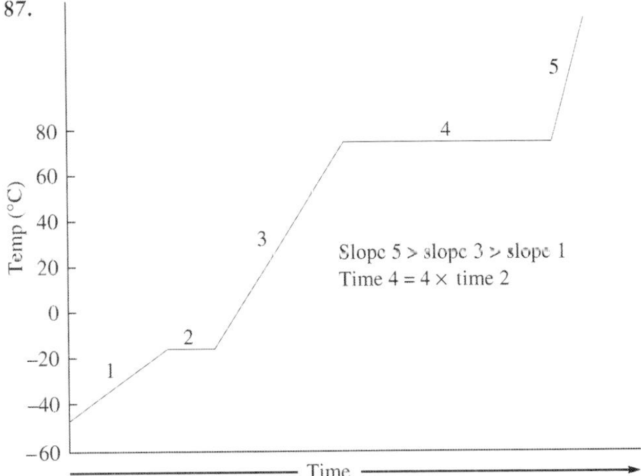

89. 1680 kJ **91.** The reaction doesn't release enough heat to melt all the ice. The temperature will remain at 0°C. **93.** 1490 g **95.** See Figs. 16.55 and 16.58 for the phase diagrams of H_2O and CO_2. Most substances exhibit only three different phases: solid, liquid, and gas. This is true for H_2O and CO_2. Also typical of phase diagrams is the positive slopes for both the liquid/gas equilibrium line and the solid/gas equilibrium line. This is also true for both H_2O and CO_2. The solid/liquid equilibrium line also generally has a positive slope. This is true for CO_2 but not for H_2O. In the H_2O phase diagram, the slope of the solid/liquid line is negative. The determining factor for the slope of the solid/liquid line is the relative densities of the solid and liquid phases. The solid phase is denser than the liquid phase in most substances; for these substances, the slope of the solid/liquid equilibrium line is positive. For water, the liquid phase is denser than the solid phase, which corresponds to a negative-sloping solid/liquid equilibrium line. Another difference between H_2O and CO_2 is the normal melting points and normal boiling points. The term "normal" just dictates a pressure of 1 atm. H_2O has a normal melting point (0°C) and a normal boiling point (100°C), but CO_2 does not. At 1 atm pressure, CO_2 only sublimes (goes from the solid phase directly to the gas phase). There are no temperatures at 1 atm for CO_2 where the solid and liquid phases are in equilibrium or where the liquid and gas phases are in equilibrium. There are other differences, but those discussed above are the major ones. The relationship between melting point and pressure is determined by the slope of the solid/liquid equilibrium line. For most substances (CO_2 included), the positive slope of the solid/liquid line shows a direct relationship between the melting point and pressure. As pressure increases, the melting point increases. Water is just the opposite since the slope of the solid/liquid line in water is negative. Here the melting point of water is inversely related to the pressure. For boiling points, the positive slope of the liquid/gas equilibrium line indicates a direct relationship between the boiling point and pressure. This direct relationship is true for all substances including H_2O and CO_2. The critical temperature for a substance is defined as the temperature above which the vapor cannot be liquefied no matter what pressure is applied. The critical temperature, like the boiling-point temperature, is directly related to the strength of the intermolecular forces. Since H_2O exhibits relatively strong hydrogen-bonding interactions and CO_2 exhibits only London dispersion forces, one would expect a higher critical temperature for H_2O than for CO_2. **97.** The critical temperature is the temperature above which the vapor cannot be liquefied no matter what pressure is applied. Since N_2 has a critical temperature below room temperature (~22°C), it cannot be liquefied at room temperature. NH_3, with a critical temperature above room temperature, can be liquefied at room temperature. **99.** A: solid; B: liquid; C: vapor; D: solid + vapor; E: solid + liquid + vapor (triple point); F: liquid + vapor; G: liquid + vapor (critical point); H: vapor; the first dashed line (at the lower temperature) is the normal melting point, and the second dashed line is the normal boiling point. The solid phase is denser because of the positive slope of the solid/liquid equilibrium line. **101.** a. two; b. Higher-pressure triple point: graphite, diamond, and liquid; lower-pressure triple point: graphite, liquid, and vapor; c. It is converted to diamond (the more dense solid form). d. Diamond is more dense, which is why graphite can be converted to diamond by applying pressure. **103.** Because the density of the liquid phase is greater than the density of the solid phase, the slope of the solid/liquid boundary line is negative (as in H_2O). With a negative slope, the melting points increase with a decrease in pressure, so the normal melting point of X should be greater than 225°C.

105.

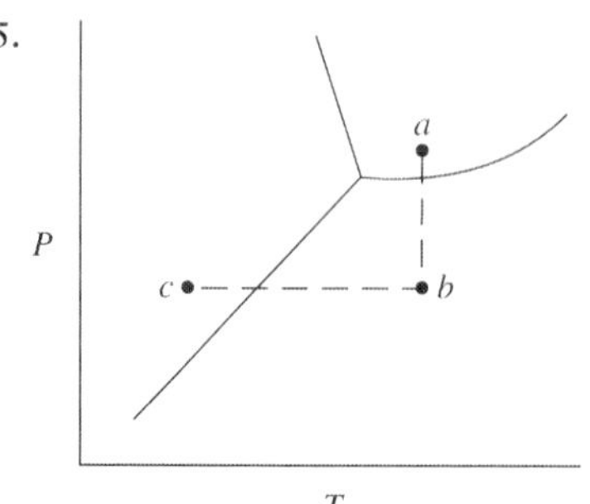

As P is lowered, we go from a to b on the phase diagram. The water boils. The evaporation of the water is endothermic and the water is cooled ($b \rightarrow c$), forming some ice. If the pump is left on, the ice will sublime until none is left. This is the basis of freeze drying. **107.** If TiO_2 conducts electricity as a liquid, then it is an ionic solid; if not, then TiO_2 is a network solid. **109.** B_2H_6, molecular solid; SiO_2, network solid; CsI, ionic solid; W, metallic solid **111.** The cation must have a radius that is 0.155 times the radius of the spheres to just fit into the trigonal hole. **113.** 1.71 g/cm^3 **115.** The formula is $TiO_{1.182}$ or $Ti_{0.8462}O$. 63.7% of the titanium is Ti^{2+} and 36.3% is Ti^{3+}. **117.** 57.8 torr **119.** 92.47% of the energy goes to increase the internal energy of the water. The remainder of the energy (7.53%) goes to do work against the atmosphere. **121.** The empirical formula is AB_2. Each A atom is in a cubic hole of B atoms so 8 B atoms surround each A atom. This will also be true in the extended lattice. The structure of B atoms in the unit cell is a cubic arrangement with B atoms at every face, edge, corner, and center of the cube. **123.** 46.5 pm **125.** Rh; 12.42 g/cm^3 **127.** distance (liquid)/distance (vapor) = 0.03123 **129.** a. structure (a), $TlBa_2CuO_5$; structure (b), $TlBa_2CaCu_2O_7$; structure (c), $TlBa_2Ca_2Cu_3O_9$; structure (d), $TlBa_2Ca_3Cu_4O_{11}$; b. (a) $<$ (b) $<$ (c) $<$ (d); c. structure (a), only Cu^{3+} is present; structure (b), each formula unit contains 1 Cu^{2+} ion and 1 Cu^{3+} ion; structure (c), each formula unit contains 2 Cu^{2+} ions and 1 Cu^{3+} ion; structure (d), each formula unit contains 3 Cu^{2+} ions and 1 Cu^{3+} ion; d. This superconductor material achieves variable copper oxidation states by varying the numbers of Ca, Cu, and O in each unit cell. The mixtures of copper oxidation states are discussed in part c. The superconductor material in Exercise 73 achieves variable copper oxidation states by omitting oxygen at various sites in the lattice. **131.** Calculated Li^+ radius = 75 pm; calculated Cl^- radius = 182 pm; From Fig. 13.8, the Li^+ radius is 60 pm and the Cl^- radius is 181 pm. The Li^+ ion is much smaller than calculated. This probably means that the ions are not actually in contact with each other. The octahedral holes are larger than the Li^+ ions.

Chapter 17

13. a. $HNO_3(l) \rightarrow H^+(aq) + NO_3^-(aq)$; b. $Na_2SO_4(s) \rightarrow 2Na^+(aq) + SO_4^{2-}(aq)$; c. $Al(NO_3)_3(s) \rightarrow Al^{3+}(aq) + 3NO_3^-(aq)$; d. $SrBr_2(s) \rightarrow Sr^{2+}(aq) + 2Br^-(aq)$; e. $KClO_4(s) \rightarrow K^+(aq) + ClO_4^-(aq)$; f. $NH_4Br(s) \rightarrow NH_4^+(aq) + Br^-(aq)$; g. $NH_4NO_3(s) \rightarrow NH_4^+(aq) + NO_3^-(aq)$; h. $CuSO_4(s) \rightarrow Cu^{2+}(aq) + SO_4^{2-}(aq)$; i. $NaOH(s) \rightarrow Na^+(aq) + OH^-(aq)$ **15.** 10.7 mol/kg; 6.77 mol/L; 0.162 **17.** 35%; 0.39; 7.3 mol/kg; 3.1 mol/L **19.** 23.9%; 1.6 mol/kg; 0.028 **21.** 1.06 g/cm^3; 0.0180; 0.981 mol/L; 1.02 mol/kg **23.** "Like dissolves like" refers to the nature of the intermolecular forces. Polar solutes and ionic solutes dissolve in polar solvents because the types of intermolecular forces present in solute and solvent are similar. When they dissolve, the strengths of the intermolecular forces in solution are about the same as in pure solute and pure solvent. The same is true for nonpolar solutes in nonpolar solvents. The strengths of the intermolecular forces (London dispersion forces) are about the same in solution as in pure solute and pure solvent. In all cases of like dissolves like, the magnitude of ΔH_{soln} is either a small positive number (endothermic) or a small negative number (exothermic). For polar solutes in nonpolar solvents and vice versa, ΔH_{soln} is a very large, unfavorable value (very endothermic). Because the energetics are so unfavorable, polar solutes do not dissolve in nonpolar solvents and vice versa. **25.** $NaI(s) \rightarrow Na^+(aq) + I^-(aq)$, $\Delta H_{soln} = -8$ kJ/mol **27.** Both $Al(OH)_3$ and NaOH are ionic compounds. Since the lattice energy is proportional to the charge of the ions, the lattice energy of aluminum hydroxide is greater than that of sodium hydroxide. The attraction of water molecules for Al^{3+} and OH^- cannot overcome the larger lattice energy and $Al(OH)_3$ is insoluble. For NaOH, the favorable hydration energy is large enough to overcome the smaller lattice energy and NaOH is soluble. **29.** Water is a polar molecule capable of hydrogen bonding. Polar molecules, especially molecules capable of hydrogen bonding, and ions are all attracted to water. For covalent compounds, as polarity increases, the attraction to water increases. For ionic compounds, as the charge of the ions increases and/or the size of the ions decreases, the attraction to water increases. a. CH_3CH_2OH; CH_3CH_2OH is polar, whereas $CH_3CH_2CH_3$ is nonpolar. b. $CHCl_3$; $CHCl_3$ is polar, whereas CCl_4 is nonpolar. c. CH_3CH_2OH; CH_3CH_2OH is much more polar than $CH_3(CH_2)_{14}CH_2OH$. **31.** As the length of the hydrocarbon chain increases, the solubility decreases. The —OH end of the alcohols can hydrogen-bond with water. The hydrocarbon chain, however, is basically nonpolar and interacts poorly with water. As the hydrocarbon chain gets longer, a greater portion of the molecule cannot interact with the water molecules and the solubility decreases; i.e., the effect of the —OH group decreases as the alcohols get larger. **33.** Structure effects refer to solute and solvent having similar polarities in order for solution formation to occur. Hydrophobic solutes are mostly nonpolar substances that are "water-fearing." Hydrophilic solutes are mostly polar or ionic substances that are "water-loving." Pressure has little effect on the solubilities of solids or liquids; it does significantly affect the solubility of a gas. Henry's law states that the amount of a gas dissolved in a solution is directly proportional to the pressure of the gas above the solution ($C = kP$). The equation for Henry's law works best for dilute solutions of gases that do not dissociate in or react with the solvent. $HCl(g)$ does not follow Henry's law because it dissociates into $H^+(aq)$ and $Cl^-(aq)$ in solution (HCl is a strong acid). For O_2 and N_2, Henry's law works well since these gases do not react with the water solvent. An increase in temperature can either increase or decrease the solubility of a solid solute in water. It is true that a solute dissolves more rapidly with an increase in temperature, but the amount of solid solute that dissolves to form a saturated solution can either decrease or increase with temperature. The temperature effect is difficult to predict for solid solutes. However, the temperature effect for gas solutes is easier to predict as the solubility of a gas typically decreases with increasing temperature. **35.** 962 L atm/mol; 1.14×10^{-3} mol/L **37.** As the temperature increases, the gas molecules will have a greater average kinetic energy. A greater fraction of the gas molecules in solution will have kinetic energy greater than the attractive forces between the gas molecules and the solvent molecules. More gas molecules escape to the vapor phase and the solubility of the gas decreases. **39.** 136 torr **41.** solution d **43.** 3.0×10^2 g/mol **45.** a. 290 torr; b. 0.69 **47.** $\chi_{meth} = \chi_{prop} = 0.500$ **49.** $P_{ideal} = 188.6$ torr; $\chi_{acetone} = 0.512$, $\chi_{methanol} = 0.488$; since the actual vapor pressure of the solution is smaller than the ideal vapor pressure, this solution exhibits a negative deviation from Raoult's law. This occurs when solute–solvent attractions are stronger than for the pure substances. **51.** No, the solution is not ideal. For an ideal solution, the strength of the intermolecular forces in the solution is the same as in pure solute and pure solvent. This results in $\Delta H_{soln} = 0$ for an ideal solution. ΔH_{soln} for methanol/water is not zero. Since $\Delta H_{soln} < 0$, this solution exhibits a negative deviation from Raoult's law. **53.** Solutions of A and B have vapor pressures less than ideal (see Fig. 17.11), so this plot shows negative deviations from Raoult's law. Negative deviations occur when the intermolecular forces are stronger in solution than in pure solvent and solute. This results in an exothermic enthalpy of solution. The only statement that is false is e. A substance boils when the vapor pressure equals the external pressure. Since $\chi_B = 0.6$ has a lower vapor pressure at the temperature of the plot than either pure A or pure B, one would expect this solution to require the highest temperature for the vapor pressure to reach the external pressure. Therefore, the solution with $\chi_B = 0.6$ will have a higher boiling point than either pure A or pure B. (Note that since $P_B^\circ > P_A^\circ$, then B is more volatile than A.) **55.** Osmotic pressure: the pressure that must be applied to a solution to stop osmosis; osmosis is the flow of solvent into the solution through a semipermeable membrane. The equation to calculate osmotic pressure π is

$$\pi = MRT$$

where M is the molarity of the solution, R is the gas constant, and T is the Kelvin temperature. The molarity of a solution approximately equals the molality of the solution when 1 kg solvent $\approx$ 1 L solution. This occurs for dilute solutions of water since $d_{H_2O} = 1.00$ g/cm^3. **57.** 101.5°C **59.** −29.9°C; 108.2°C **61.** 100.08°C **63.** 498 g/mol **65.** 776 g/mol **67.** 2.0×10^{-5} °C; 0.20 torr **69.** ~30 m **71.** With addition of salt or sugar, the osmotic pressure inside the fruit cells (and bacteria) is less than outside the cell. Water will leave the cells, which will dehydrate any bacteria present, causing them to die. **73.** 0.0880; 59.2 torr **75.** a. 0.010 m Na_3PO_4 and 0.020 m KCl; b. 0.020 m HF; c. 0.020 m $CaBr_2$ **77.** The van't Hoff factor i is the number of moles of particles (ions) produced for every mole of solute dissolved. For NaCl, $i = 2$ since Na^+ and Cl^- are produced in water; for $Al(NO_3)_3$, $i = 4$ since Al^{3+} and 3 NO_3^- ions are produced when $Al(NO_3)_3$ dissolves in water. In real life, the van't Hoff factor is rarely the value predicted by the number of ions a salt dissolves into; i is generally something less than the predicted number of ions. This is due to a phenomenon called ion pairing, where at any instant a small percentage of oppositely charged ions pair up and act like a single solute particle. Ion pairing occurs most when the concentration of ions is large. Therefore, dilute solutions behave most ideally; here i is close to that determined by the number of ions in a salt. **79.** A pressure greater than 4.8 atm should be applied. **81.** a. −0.25°C; 100.069°C; b. −0.32°C; 100.087°C **83.** 97.8 g/mol **85.** Both solutions and colloids have suspended particles in some medium. The major difference between the two is the size of the particles. A colloid is a suspension of relatively large particles as compared with a solution. Because of this, colloids will scatter light, whereas solutions will not. The scattering of light by a colloidal suspension is called the Tyndall effect. **87.** Coagulation is the destruction of a colloid by the aggregation of many suspended particles to form a large particle that settles out of solution. **89.** 7.9 M **91.** a. 100.77°C; b. 23.1 mm Hg; c. We assumed ideal behavior in solution formation and assumed $i = 1$ (no ions form). **93.** Benzoic acid is capable of hydrogen bonding, but a significant part of benzoic acid is the nonpolar benzene ring, which is composed of only carbon and hydrogen. In benzene, a hydrogen-bonded dimer forms:

O---H—O
C C
O—H---O

The dimer is relatively nonpolar and thus more soluble in benzene than in water. Since benzoic acid forms dimers in benzene, the effective solute particle concentration will be less than 1.0 molal. Therefore, the freezing-point depression would be less than 5.12°C ($\Delta T_f = K_f m$). **95.** a. 26.6 kJ/mol; b. −657 kJ/mol **97.** 0.050 **99.** C_7H_4O; $C_{14}H_8O_2$ **101.** a. 303 ± 9 g/mol; b. No, codeine could not be eliminated since its molar mass is in the possible range including the uncertainty; c. We would like the uncertainty to be ±1 g/mol. We need the freezing-point depression to be about 10 times what it was in this problem. Two possibilities are (1) make the solution ten times more concentrated (may be solubility problem) or (2) use a solvent with a larger K_f value, e.g., camphor. **103.** 5.08×10^{-17} **105.** 1.0×10^{-3} **107.** 72% $MgCl_2$ **109.** $\chi_{CCl_4} = 0.554$, $\chi_{C_6H_6} = 0.446$ **111.** 0.286 **113.** a. If we assume $MgCO_3$ does not dissociate, 46 L of water must be processed. If $MgCO_3$ does dissociate, then 47 L of water must be processed. b. No; A reverse osmosis system that applies 8.0 atm can only purify water with a solute concentration less than 0.32 mol/L. Salt water has a solute concentration of 2(0.60 *M*) = 1.2 mol/L ions. The solute concentration of salt water is much too high for this reverse osmosis unit to work. **115.** a. 6.11 atm; b. 8.3×10^{-3} **117.** a. 0.25 or 25%; b. −0.562°C

Chapter 18

1. The gravity of the earth is not strong enough to keep H_2 in the atmosphere. **3.** a. $\Delta H° = -92$ kJ; $\Delta S° = -199$ J/K; $\Delta G° = -33$ kJ; b. yes; c. $T < 460$ K **5.** Ionic, covalent, and metallic (or interstitial); The ionic and covalent hydrides are true compounds obeying the law of definite proportions and differ from each other in the type of bonding. The interstitial hydrides are more like solid solutions of hydrogen with a transition metal, and do not obey the law of definite proportions. **7.** Alkali metals have an ns^1 valence shell electron configuration. Alkali metals lose this valence electron with relative ease to form M^+ cations when in ionic compounds. They all are easily oxidized. Therefore, in order to prepare the pure metals, alkali metals must be produced in the absence of materials (H_2O, O_2) that are capable of oxidizing them. The method of preparation is electrochemical processes, specifically, electrolysis of molten chloride salts and reduction of alkali salts with Mg and H_2. In all production methods, H_2O and O_2 must be absent. **9.** Hydrogen forms many compounds in which the oxidation state is +1, as do the Group 1A elements. Consider, for example, H_2SO_4 and HCl compared with Na_2SO_4 and NaCl. On the other hand, hydrogen forms diatomic H_2 molecules and is a nonmetal, while the Group 1A elements are metals. Hydrogen also forms compounds with a −1 oxidation state, which is not characteristic of Group 1A metals, e.g., NaH. **11.** $4Li(s) + O_2(g) \rightarrow 2Li_2O(s)$; $16Li(s) + S_8(s) \rightarrow 8Li_2S(s)$; $2Li(s) + Cl_2(g) \rightarrow 2LiCl(s)$; $12Li(s) + P_4(s) \rightarrow 4Li_3P(s)$; $2Li(s) + H_2(g) \rightarrow 2LiH(s)$; $2Li(s) + 2H_2O(l) \rightarrow 2LiOH(aq) + H_2(g)$; $2Li(s) + 2HCl(aq) \rightarrow 2LiCl(aq) + H_2(g)$ **13.** $MgCl_2$ **15.** The alkaline earth ions that give water the hard designation are Ca^{2+} and Mg^{2+}. These ions interfere with the action of detergents and form unwanted precipitates with soaps. Large-scale water softeners remove Ca^{2+} by precipitating out the calcium ions as $CaCO_3$. In homes, Ca^{2+} and Mg^{2+} (plus other cations) are removed by ion exchange. See Fig. 18.6 for a schematic of a typical cation exchange resin. **17.** $CaCO_3(s) + H_2SO_4(aq) \rightarrow CaSO_4(aq) + H_2O(l) + CO_2(g)$ **19.** $CaF_2(s)$ will precipitate when $[Ca^{2+}]_0 > 2 \times 10^{-2}$ *M*. Therefore, hard water should have a calcium ion concentration of less than 2×10^{-2} *M* to avoid precipitate formation. **21.** The valence electron configuration of Group 3A elements is ns^2np^1. The lightest Group 3A element, boron, is a nonmetal because most of its compounds are covalent. Aluminum, although commonly thought of as a metal, does have some nonmetallic properties because its bonds to other nonmetals have significant covalent character. The other Group 3A elements have typical metal characteristics; their compounds formed with nonmetals are ionic. From this discussion, metallic character increases as the Group 3A elements get larger. As mentioned previously, boron is a nonmetal in both properties and compounds formed. However, aluminum has physical properties of metals such as high thermal and electrical conductivities and a lustrous appearance. The compounds of aluminum with other nonmetals, however, do have some nonmetallic properties because the bonds have significant covalent character. **23.** $B_2H_6(g) + 3O_2(g) \rightarrow 2B(OH)_3(s)$ **25.** $2Ga(s) + 3F_2(g) \rightarrow 2GaF_3(s)$; $4Ga(s) + 3O_2(g) \rightarrow 2Ga_2O_3(s)$; $16Ga(s) + 3S_8(s) \rightarrow 8Ga_2S_3(s)$; $2Ga(s) + 6HCl(aq) \rightarrow 2GaCl_3(aq) + 3H_2(g)$ **27.** The valence electron configuration of Group 4A elements is ns^2np^2. The two most important elements on earth are both Group 4A elements. They are carbon, found in all biologically important molecules, and silicon, found in most of the compounds that make up the earth's crust. They are important because they are so prevalent in compounds necessary for life and the geologic world. As with Group 3A, Group 4A shows an increase in metallic character as the elements get heavier. Carbon is a typical nonmetal, silicon and germanium have properties of both metals and nonmetals, so they are classified as semimetals, whereas tin and lead have typical metallic characteristics. **29.** White tin is stable at normal temperatures. Gray tin is stable at temperatures below 13.2°C. Thus for the phase change Sn(gray) → Sn(white), ΔG is negative at $T > 13.2$°C and ΔG is positive at $T < 13.2$°C. This is possible only if ΔH is positive and ΔS is positive. Thus gray tin has the more ordered structure (has the smaller positional probability). **31.** $Sn(s) + 2F_2(g) \rightarrow SnF_4(s)$, tin(IV) fluoride; $Sn(s) + F_2(g) \rightarrow SnF_2(s)$, tin(II) fluoride. **33.** In each formula unit of Pb_3O_4, two atoms are Pb(II) and one atom is Pb(IV) (2:1 ratio). **35.** Both NO_4^{3-} and PO_4^{3-} have 32 valence electrons, so both have similar Lewis structures. From the Lewis structure for NO_4^{3-}, the central N atom has a tetrahedral arrangement of electron pairs. N is small. There is probably not enough room for all 4 oxygen atoms around N. P is larger; thus PO_4^{3-} is stable. PO_3^- and NO_3^- both have 24 valence electrons, so both have similar Lewis structures. From the Lewis structure for PO_3^-, PO_3^- has a trigonal arrangement of electron pairs around the central P atom (two single bonds and one double bond). P=O bonds are not particularly stable, whereas N=O bonds are stable. Thus NO_3^- is stable. **37.** N: $1s^22s^22p^3$; the extremes of the oxidation states for N can be rationalized by examining the electron configuration of N. Nitrogen is three electrons short of the stable Ne electron configuration of $1s^22s^22p^6$. Having an oxidation state of −3 makes sense. The +5 oxidation state corresponds to N "losing" its 5 valence electrons. In compounds with oxygen, the N—O bonds are polar covalent, with N having the partial positive end of the bond dipole. In the world of oxidation states, electrons in polar covalent bonds are assigned to the more electronegative atom; this is oxygen in N—O bonds. N can form enough bonds to oxygen to give it a +5 oxidation state. This loosely corresponds to losing all of the valence electrons. **39.** This is due to nitrogen's ability to form strong π bonds, whereas heavier Group 5A elements do not form strong π bonds. Therefore, P_2, As_2, and Sb_2 do not form since two π bonds are required to form these diatomic substances.

41. a. [Lewis structure of NO_2] plus other resonance structures; [Lewis structure of N_2O_4] plus other resonance structures;

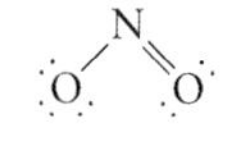

b. [Lewis structures of BF_3, NH_3, and $F_3B{-}NH_3$]

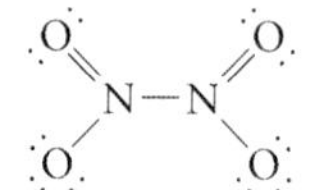

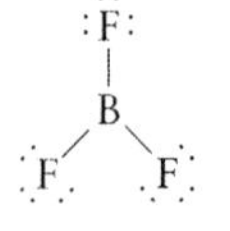

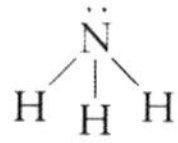

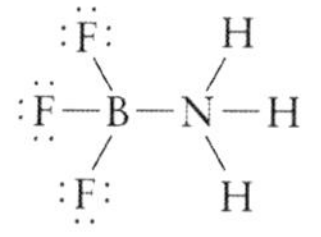

In reaction a, NO_2 has an odd number of electrons so it is impossible to satisfy the octet rule. By dimerizing to form N_2O_4, the odd electrons on two NO_2 molecules can pair up, giving a species whose Lewis structure can satisfy the octet rule. In general odd-electron species are very reactive. In reaction b, BF_3 can be considered electron-deficient. Boron has only six electrons around it. By forming BF_3NH_3, the boron atom satisfies the octet rule by accepting a lone pair of electrons from NH_3 to form a fourth bond. **43.** exothermic **45.** $N_2H_4(l) + 2F_2(g) \longrightarrow 4HF(g) + N_2(g); \Delta H° = -1169$ kJ

47. $2Bi_2S_3(s) + 9O_2(g) \longrightarrow 2Bi_2O_3(s) + 6SO_2(g);$
$2Bi_2O_3(s) + 3C(s) \longrightarrow 4Bi(s) + 3CO_2(g)$

$2Sb_2S_3(s) + 9O_2(g) \longrightarrow 2Sb_2O_3(s) + 6SO_2(g);$
$2Sb_2O_3(s) + 3C(s) \longrightarrow 4Sb(s) + 3CO_2(g)$

49. TSP = Na_3PO_4; PO_4^{3-} is the conjugate base of the weak acid HPO_4^{2-} ($K_a = 4.8 \times 10^{-13}$). All conjugate bases of weak acids are effective bases ($K_b = K_w/K_a = 1.0 \times 10^{-14}/4.8 \times 10^{-13} = 2.1 \times 10^{-2}$). The weak-base reaction of PO_4^{3-} with H_2O is $PO_4^{3-}(aq) + H_2O(l) \rightleftharpoons HPO_4^{2-}(aq) + OH^-(aq)$ $K_b = 2.1 \times 10^{-2}$.

51.
$4As(s) + 3O_2(g) \longrightarrow As_4O_6(s); 4As(s) + 5O_2(g) \longrightarrow As_4O_{10}(s);$
$As_4O_6(s) + 6H_2O(l) \longrightarrow 4H_3AsO_3(aq); As_4O_{10}(s) + 6H_2O(l) \longrightarrow 4H_3AsO_4(aq)$

53. $\frac{1}{2}N_2(g) + \frac{1}{2}O_2(g) \rightarrow NO(g)$, $\Delta G° = \Delta G°_{f,\,NO} = 87$ kJ/mol; By definition, $\Delta G°_f$ for a compound equals the free energy change that would accompany the formation of 1 mol of that compound from its elements in their standard states. NO (and some other oxides of nitrogen) have weaker bonds as compared to the triple bond of N_2 and the double bond of O_2. Because of this, NO (and some other oxides of nitrogen) have higher (positive) free energies of formation as compared to the relatively stable N_2 and O_2 molecules. **55.** The pollution provides nitrogen and phosphorus nutrients so the algae can grow. The algae consume oxygen, causing fish to die. **57.** −4594 kJ **59.** Hydrazine also can hydrogen bond because it has covalent N—H bonds as well as a lone pair of electrons on each N. The high boiling point for hydrazine's relatively small size supports this. **61.** The two allotropic forms of oxygen are O_2 and O_3.

O_2: O=O O_3: O–O=O ⟷ O=O–O (bent resonance structures)

The MO electron configuration of O_2 has two unpaired electrons in the degenerate pi antibonding (π_{2p}^*) orbitals. A substance with unpaired electrons is paramagnetic (see Fig. 14.41). Ozone has a V-shape molecular structure with a bond angle of 117°, slightly less than the predicted 120° trigonal planar bond angle. **63.** In the upper atmosphere, O_3 acts as a filter for UV radiation: $O_3 \xrightarrow{h\nu} O_2 + O$; O_3 is also a powerful oxidizing agent. It irritates the lungs and eyes, and at high concentration it is toxic. The smell of a "fresh spring day" is O_3 formed during lightning discharges. Toxic materials don't necessarily smell bad. For example, HCN smells like almonds. **65.** +6 oxidation state: SO_4^{2-}, SO_3, SF_6; +4 oxidation state: SO_3^{2-}, SO_2, SF_4; +2 oxidation state: SCl_2; 0 oxidation state: S_8 and all other elemental forms of sulfur; −2 oxidation state: H_2S, Na_2S **67.** $H_2SeO_4(aq) + 3SO_2(g) \rightarrow Se(s) + 3SO_3(g) + H_2O(l)$

69. F—O—O—F

	F	O	O	F
Formal Charge	0	0	0	0
Oxidation State	−1	+1	+1	−1

Oxidation states are more useful. We are forced to assign +1 as the oxidation state for oxygen. Oxygen is very electronegative and +1 is not a stable oxidation state for this element. **71.** Fluorine is the most reactive of the halogens because it is the most electronegative atom and the bond in the F_2 molecule is very weak. **73.** $ClO^-(aq) + 2NH_3(aq) \rightarrow Cl^-(aq) + N_2H_4(aq) + H_2O(l)$, $\mathscr{E}°_{cell} = 1.00$ V; Since $\mathscr{E}°_{cell}$ is positive for this reaction, then at standard conditions ClO^- can spontaneously oxidize NH_3 to the somewhat toxic N_2H_4. **75.** a. $AgCl(s) \xrightarrow{h\nu} Ag(s) + Cl$; the reactive chlorine atom is trapped in the crystal. When light is removed, Cl reacts with silver atoms to re-form AgCl; that is, the reverse reaction occurs. In pure AgCl, the Cl atoms escape, making the reverse reaction impossible. b. Over time, chlorine is lost, and the dark silver metal is permanent. **77.** Helium is unreactive and doesn't combine with any other elements. It is a very light gas and would easily escape the earth's gravitational pull as the planet was formed **79.** 2×10^4 g; 2×10^{26} atoms in room; 5×10^{20} atoms in one breath; Since Ar and Rn are both noble gases, both species will be relatively unreactive. However, all nuclei of Rn are radioactive, unlike most nuclei of Ar. The radioactive decay products of Rn can cause biological damage when inhaled. **81.** One would expect RnF_2 and RnF_4 to form in fashion similar to XeF_2 and XeF_4. The chemistry of radon is difficult to study because all radon isotopes are radioactive. The hazards of dealing with radioactive materials are immense. **83.** Solids have stronger intermolecular forces than liquids. In order to maximize the hydrogen bonding in the solid phase, ice is forced into an open structure. This open structure is why $H_2O(s)$ is less dense than $H_2O(l)$. **85.** Strontium and calcium are both alkaline earth metals, so both have similar chemical properties. Since milk is a good source of calcium, strontium could replace some calcium in milk without much difficulty. **87.** In solution Tl^{3+} can oxidize I^- to I_3^-. Thus we expect TlI_3 to be thallium(I) triiodide. **89.** The inert pair effect refers to the difficulty of removing the pair of *s* electrons from some of the elements in the fifth and sixth periods of the periodic table. As a result, multiple oxidation states are exhibited for the heavier elements of Groups 3A and 4A. In^+, In^{3+}, Tl^+, and Tl^{3+} oxidation states are all important to the chemistry of In and Tl. **91.** 5.4×10^4 kJ (Hall–Heroult process) versus 4.0×10^2 kJ; it is feasible to recycle Al by melting the metal because, in theory, it takes less than 1% of the energy required to produce the same amount of Al by the Hall–Heroult process. **93.** Carbon cannot form the fifth bond necessary for the transition state since carbon doesn't have low-energy *d* orbitals available to expand the octet. **95.** a. The NNO structure is correct. From the Lewis structures we would predict both NNO and NON to be linear. However, we would predict NNO to be polar and NON to be nonpolar. Since experiments show N_2O to be polar, then NNO is the correct structure.

b. N=N=O ⟷ N≡N—O ⟷ N—N≡O

Formal charges: (−1, +1, 0); (0, +1, −1); (−2, +1, +1)

The formal charges for the atoms in the various resonance structures appear below each atom. The central N is *sp* hybridized in all of the resonance structures. We can probably ignore the third resonance structure on the basis of the relatively large formal charges as compared to the first two resonance structures. c. The *sp* hybrid orbitals on the center N overlap with atomic orbitals (or hybrid orbitals) on the other two atoms to form the two σ bonds. The remaining two unhybridized *p* orbitals on the center N overlap with two *p* orbitals on the peripheral N to form the two π bonds. **97.** $\frac{[H_2PO_4^-]}{[HPO_4^{2-}]} = \frac{1}{0.9} = 1.1 \approx 1$; a best buffer has approximately equal concentrations of weak acid and conjugate base so that pH $\approx$ pK_a for a best buffer. The pK_a value for a $H_3PO_4/H_2PO_4^-$ buffer is $-\log(7.5 \times 10^{-3}) = 2.12$. A pH of 7.1 is too high for a $H_3PO_4/H_2PO_4^-$ buffer to be effective. At this high a pH, there would be so little H_3PO_4 present that we could hardly consider it a buffer. This solution would not be effective in resisting pH changes, especially when a strong base is added. **99.** 4.8×10^{-11} *M* **101.** $\Delta H° = 286$ kJ; $\Delta G° = 326$ kJ; $K_p = 7.22 \times 10^{-58}$; $P_{O_3} = 3.3 \times 10^{-41}$ atm; The volume occupied by one molecule of ozone at 230. K is 9.5×10^{17} L. Equilibrium is probably not maintained under these conditions. When only two ozone molecules are in a volume of 9.5×10^{17} L, the reaction is not at equilibrium. Under these conditions, $Q > K$ and the reaction shifts to the left. But with only two ozone molecules in this huge volume, it is extremely unlikely that they will collide with each other. Under these conditions, the concentration of ozone is not large enough to maintain equilibrium. **103.** a. NO is the catalyst; b. NO_2 is an intermediate; c. $k_{cat}/k_{uncat} = 2.3$; d. The proposed mechanism for the chlorine-catalyzed destruction of ozone is:

$$O_3 + Cl \longrightarrow O_2 + ClO \quad \text{Slow}$$
$$ClO + O \longrightarrow O_2 + Cl \quad \text{Fast}$$
$$O_3 + O \longrightarrow 2O_2$$

e. $k_{Cl}/k_{NO} = 52$; At 25°C, the Cl-catalyzed reaction is roughly 52 times faster (more efficient) than the NO-catalyzed reaction, assuming the frequency factor *A* is the same for each reaction and assuming similar rate laws. **105.** 2.0×10^{-37} *M* **107.** a. 8.8 kJ; b. Other ions will have to be transported to maintain electroneutrality. Either anions must be transported into the cells or cations (Na^+) in the cell must be transported to the blood. The latter is what happens: $[Na^+]$ in blood is greater than $[Na^+]$ in cells as a result of this pumping. c. 0.28 mol.

Chapter 19

7. a. Cr: $[Ar]4s^13d^5$, Cr^{2+}: $[Ar]3d^4$, Cr^{3+}: $[Ar]3d^3$; b. Cu: $[Ar]4s^13d^{10}$, Cu^+: $[Ar]3d^{10}$, Cu^{2+}: $[Ar]3d^9$; c. V: $[Ar]4s^23d^3$, V^{2+}: $[Ar]3d^3$, V^{3+}: $[Ar]3d^2$ **9.** The lanthanide elements are located just before the 5*d* transition metals. The lanthanide contraction is the steady decrease in the atomic radii of the lanthanide elements when going from left to right across the periodic table. As a result of the lanthanide contraction, the sizes of the 4*d* and 5*d* elements are very similar (see the following exercise). This leads to a greater similarity in the chemistry of the 4*d* and 5*d* elements in a given vertical group. **11.** a. molybdenum(IV) sulfide; molybdenum(VI) oxide; b. MoS_2, +4; MoO_3, +6; $(NH_4)_2Mo_2O_7$, +6; $(NH_4)_6Mo_7O_{24} \cdot 4H_2O$, +6 **13.** TiF_4: Ionic compound containing Ti^{4+} ions and F^- ions. $TiCl_4$, $TiBr_4$, and TiI_4: Covalent compounds containing discrete, tetrahedral TiX_4 molecules. As these molecules get larger, the bp and mp increase because the London dispersion forces increase. TiF_4 has the highest bp since the interparticle forces are stronger in ionic compounds as compared with those in covalent compounds. **15.** $H^+ + OH^- \rightarrow H_2O$; Sodium hydroxide (NaOH) will react with the H^+ on the product side of the reaction. This effectively removes H^+ from the equilibrium, which will shift the reaction to the right to produce more H^+ and CrO_4^{2-}. Since more CrO_4^{2-} is produced, the solution turns yellow. **17.** Because transition metals form bonds to species that donate lone pairs of electrons, transition metals are Lewis acids (electron-pair acceptors). The Lewis bases in coordination compounds are the ligands, all of which have an unshared pair of electrons to donate. The coordinate covalent bond between the ligand and the transition metal just indicates that both electrons in the bond originally came from one of the atoms in the bond. Here, the electrons in the bond come from the ligand.

19. $Fe_2O_3(s) + 6H_2C_2O_4(aq) \rightarrow 2Fe(C_2O_4)_3^{3-}(aq) + 3H_2O(l) + 6H^+(aq)$; the oxalate anion forms a soluble complex ion with iron in rust (Fe_2O_3), which allows rust stains to be removed. **21.** a. The correct name is tetraamminecopper(II) chloride. The complex ion is named incorrectly in several ways. b. The correct name is bis(ethylenediamine)nickel(II) sulfate. The ethylenediamine ligands are neutral and sulfate has a 2– charge. Therefore, Ni^{2+} is present, not Ni^{4+}. c. The correct name is potassium diaquatetrachlorochromate(III). Because the complex ion is an anion, the *-ate* suffix ending is added to the name of the metal. Also, the ligands were not in alphabetical order (*a* in aqua comes before *c* in chloro). d. The correct name is sodium tetracyanooxalatocobaltate(II). The only error is that *tetra* should be omitted in front of sodium. That four sodium ions are needed to balance charge is deduced from the name of the complex ion. **23.** a. pentaamminechlororuthenium(III) ion; b. hexacyanoferrate(II) ion; c. tris(ethylenediamine)manganese(II) ion; d. pentaamminenitrocobalt(III) ion **25.** a. $K_2[CoCl_4]$; b. $[Pt(H_2O)(CO)_3]Br_2$; c. $Na_3[Fe(CN)_2(C_2O_4)_2]$; d. $[Cr(NH_3)_3Cl(H_2NCH_2CH_2NH_2)]I_2$

27.

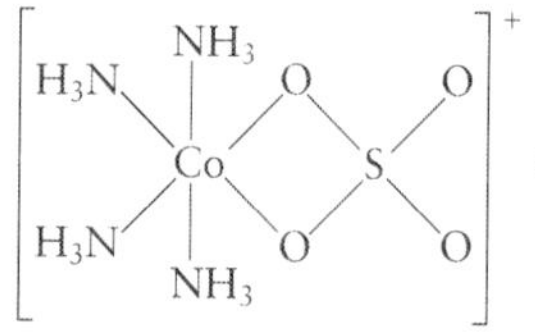

29. Three moles of AgI will precipitate per mole of $[Co(NH_3)_6]I_3$, 2 moles of AgI will precipitate per mole of $[Pt(NH_3)_4I_2]I_2$, 0 moles of AgI will precipitate per mole of $Na_2[PtI_6]$, and 1 mole of AgI will precipitate per mole of $[Cr(NH_3)_4I_2]I$. **31.** a. Isomers: Species with the same formulas but different properties. See text for examples of the following types of isomers. b. Structural isomers: Isomers that have one or more bonds that are different. c. Stereoisomers: Isomers that contain the same bonds but differ in how the atoms are arranged in space. d. Coordination isomers: Structural isomers that differ in the atoms that make up the complex ion. e. Linkage isomers: Structural isomers that differ in how one or more ligands are attached to the transition metal. f. Geometric isomers: (*cis–trans* isomerism); stereoisomers that differ in the positions of atoms with respect to a rigid ring, bond, or each other. g. Optical isomers: Stereoisomers that are nonsuperimposable mirror images of each other; that is, they are different in the same way that our left and right hands are different.

33. a.

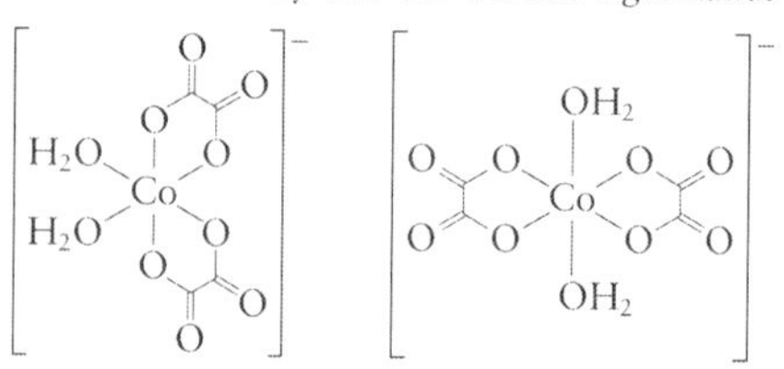

cis *trans*

b.

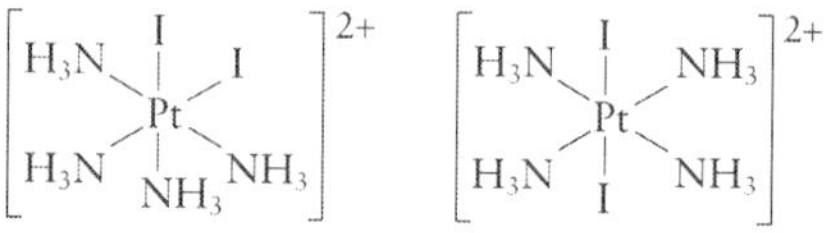

cis *trans*

c.

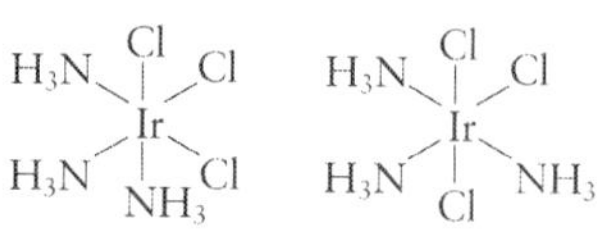

cis *trans*

d. 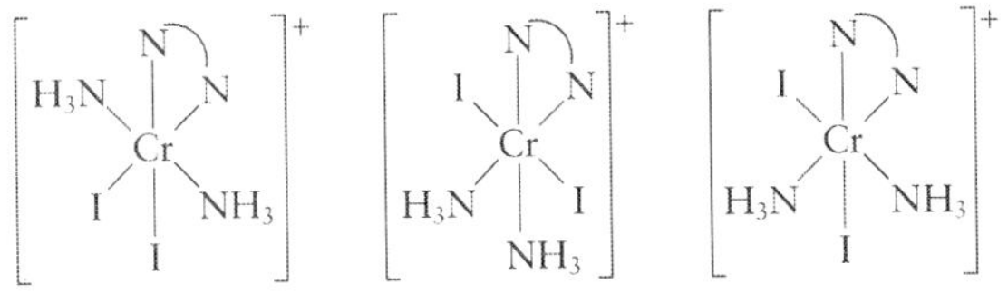 $[Cr(en)(NH_3)I_2]^+$ isomers (structures): H_3N, N, N, I, NH_3, I around Cr; I, N, N, H_3N, I, NH_3 around Cr; I, N, N, H_3N, NH_3, I around Cr

en = N⌒N = $NH_2CH_2CH_2NH_2$

35. No; both the *trans* and the *cis* forms of $Co(NH_3)_4Cl_2^+$ have mirror images that are superimposable. For the *cis* form, the mirror image needs only a 90° rotation to produce the original structure. Hence neither the *trans* nor *cis* form is optically active.

37. (structure: M bonded to O and H_2N; ring O–C(=O)–CH_2–H_2N)

(structures: two $Cu(H_2NCH_2CO_2)_2$ isomers: O=C–O, NH_2–CH_2, Cu, H_2C–H_2N, O–C=O) and (O=C–O, O–C=O, Cu, H_2C–H_2N, NH_2–CH_2)

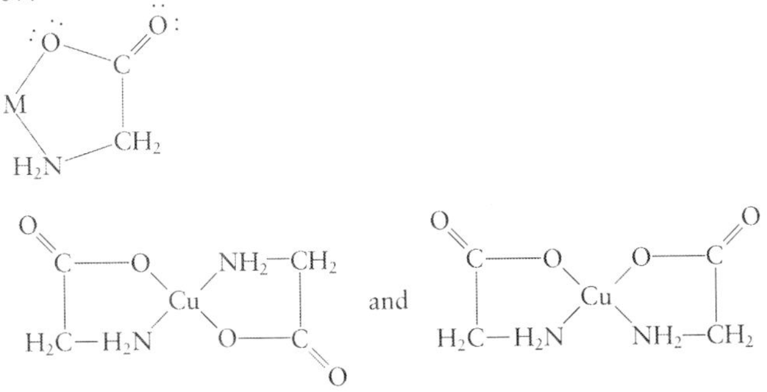

39. Linkage isomers differ in the way that the ligand bonds to the metal. SCN^- can bond through the sulfur or through the nitrogen atom. NO_2^- can bond through the nitrogen or through the oxygen atom. OCN^- can bond through the oxygen or through the nitrogen atom. N_3^-, $NH_2CH_2CH_2NH_2$, and I^- are not capable of linkage isomerism. 41. $Cr(acac)_3$ and *cis*-$Cr(acac)_2(H_2O)_2$ are optically active. 43. a. ligand that will give complex ions with the maximum number of unpaired electrons; b. ligand that will give complex ions with the minimum number of unpaired electrons; c. complex with a minimum number of unpaired electrons (low spin = strong field); d. complex with a maximum number of unpaired electrons (high spin = weak field) 45. Sc^{3+} has no electrons in *d* orbitals. Ti^{3+} and V^{3+} have *d* electrons present. The color of transition metal complexes results from electron transfer between split *d* orbitals. If no *d* electrons are present, no electron transfer can occur and the compounds are not colored.

47. 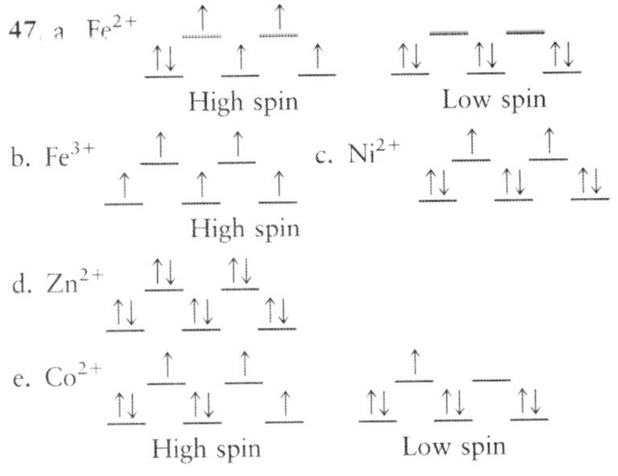 a. Fe^{2+} High spin: ↑ ↑ (upper); ↑↓ ↑ ↑ (lower) — Low spin: __ __ (upper); ↑↓ ↑↓ ↑↓ (lower)

b. Fe^{3+} High spin: ↑ ↑ (upper); ↑ ↑ ↑ (lower)

c. Ni^{2+} ↑ ↑ (upper); ↑↓ ↑↓ ↑↓ (lower)

d. Zn^{2+} ↑↓ ↑↓ (upper); ↑↓ ↑↓ ↑↓ (lower)

e. Co^{2+} High spin: ↑ ↑ (upper); ↑↓ ↑↓ ↑ (lower) — Low spin: ↑ __ (upper); ↑↓ ↑↓ ↑↓ (lower)

49. a. 0; b. 2; c. 2 51. Replacement of water ligands by ammonia ligands resulted in shorter wavelengths of light being absorbed. Energy and wavelength are inversely related, so the presence of the NH_3 ligands resulted in a larger *d*-orbital splitting (larger Δ). Therefore, NH_3 is a stronger field ligand than H_2O. 53. The violet complex ion absorbs yellow-green light ($\lambda \approx 570$ nm) and is $Cr(H_2O)_6^{3+}$. The yellow complex ion absorbs blue light ($\lambda \approx 450$ nm) and is $Cr(NH_3)_6^{3+}$. The green complex ion absorbs red light ($\lambda \approx 650$ nm) and is $Cr(H_2O)_4Cl_2^+$.

55. a. optical isomerism; b.  __ __ (upper); ↑↓ ↑↓ ↑↓ (lower)

57. The crystal field diagrams are different because the geometries of where the ligands point is different. The tetrahedrally oriented ligands point differently in relationship to the *d*-orbitals than do the octahedrally oriented ligands. Plus, there are more ligands in an octahedral complex. The *d*-orbital splitting in tetrahedral complexes is less than one-half the *d*-orbital splitting in octahedral complexes. There are no known ligands powerful enough to produce the strong-field case; hence all tetrahedral complexes are weak-field or high-spin. 59. $CoBr_6^{4-}$ has an octahedral structure and $CoBr_4^{2-}$ has a tetrahedral structure (as do most Co^{2+} complexes with four ligands). Coordination complexes absorb electromagnetic radiation (EMR) of energy equal to the energy difference between the split *d* orbitals. Since the tetrahedral *d*-orbital splitting is less than one-half of the octahedral *d*-orbital splitting, tetrahedral complexes will absorb lower-energy EMR, which corresponds to longer-wavelength EMR ($E = hc/\lambda$). Therefore, $CoBr_6^{4-}$ will absorb EMR having a wavelength shorter than 3.4×10^{-6} m. 61. $Hg^{2+}(aq) + 2I^-(aq) \rightarrow HgI_2(s)$, orange ppt.; $HgI_2(s) + 2I^-(aq) \rightarrow HgI_4^{2-}(aq)$, soluble complex ion; Hg^{2+} is a d^{10} ion. Color is the result of electron transfer between split *d* orbitals. This cannot occur for the filled *d* orbitals in Hg^{2+}. Therefore, we would not expect Hg^{2+} complex ions to form colored solutions. 63. zero (carbon monoxide is a neutrally charged ligand) 65. $Cr(NH_3)_5I_3$ is the empirical formula. Cr(III) forms octahedral complexes. So compound A is made of the octahedral $[Cr(NH_3)_5I]^{2+}$ complex ion and two I^- ions; that is, $[Cr(NH_3)_5I]I_2$.

67. 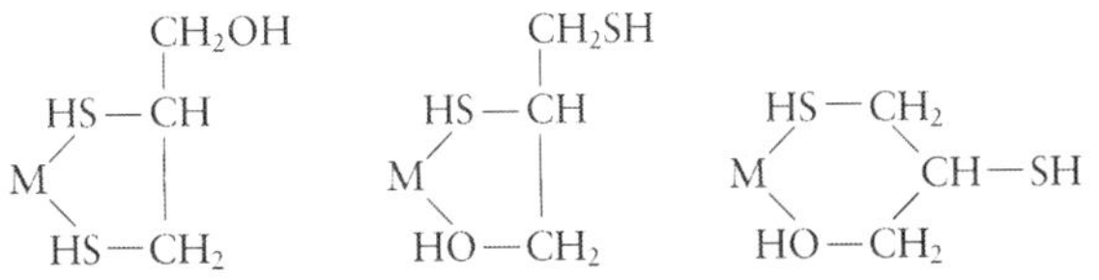 (structures: M bonded to HS–CH(CH_2OH) and HS–CH_2; M bonded to HS–CH(CH_2SH) and HO–CH_2; M bonded to HS–CH_2–CH(SH)... HO–CH_2)

CH_2OH / HS—CH / M / HS—CH_2 ; CH_2SH / HS—CH / M / HO—CH_2 ; HS—CH_2 / M / CH—SH / HO—CH_2

where M = metal ion 69. No; in all three cases, six bonds are formed between Ni^{2+} and nitrogen, so ΔH values should be similar. $\Delta S°$ for formation of the complex ion is most negative for six NH_3 molecules reacting with a metal ion (seven independent species become one). For penten reacting with a metal ion, two independent species become one, so $\Delta S°$ is least negative of all three of the reactions. Thus the chelate effect occurs because the more bonds a chelating agent can form to the metal, the more favorable $\Delta S°$ is for the formation of the complex ion and the larger the formation constant.

$$71.\ \overset{II}{(H_2O)_5Cr} - Cl - \overset{III}{Co(NH_3)_5} \rightarrow \overset{III}{(H_2O)_5Cr} - Cl - \overset{II}{Co(NH_3)_5} \rightarrow Cr(H_2O)_5Cl^{2+} + \text{Co(II) complex}$$

Yes, this is consistent. After the oxidation, the ligands on Cr(III) won't exchange. Since Cl^- is in the coordination sphere, then it must have formed a bond to Cr(II) before the electron transfer occurred (as proposed through the formation of the intermediate). 73. a. 1.66; b. Because of the lower charge, $Fe^{2+}(aq)$ will not be as strong an acid as $Fe^{3+}(aq)$. A solution of iron(II) nitrate will be less acidic (have a higher pH) than a solution with the same concentration of iron(III) nitrate. 75. a. In the lungs, there is a lot of O_2, and the equilibrium favors $Hb(O_2)_4$. In the cells, there is a deficiency of O_2, and the equilibrium favors HbH_4^{4+}. b. CO_2 is a weak acid in water, $CO_2 + H_2O \rightleftharpoons HCO_3^- + H^+$. Removing CO_2 essentially decreases H^+. $Hb(O_2)_4$ is then favored, and O_2 is not released by hemoglobin in the cells. Breathing into a paper bag increases $[CO_2]$ in the blood, thus increasing $[H^+]$ and shifting the reaction to the left. c. CO_2 builds up in the blood, and it becomes too acidic, driving the equilibrium to the left. Hemoglobin can't bind O_2 as strongly in the lungs. Bicarbonate ion acts as a base in water and neutralizes the excess acidity.

77.

— — $d_{x^2-y^2}$, d_{xy}

— d_{z^2}

— — d_{xz}, d_{yz}

The $d_{x^2-y^2}$ and d_{xy} orbitals are in the plane of the three ligands and should be destabilized the most. The amount of destabilization should be about equal when all the possible interactions are considered. The d_{z^2} orbital has some electron density in the xy plane (the doughnut) and should be destabilized a lesser amount as compared to the $d_{x^2-y^2}$ and d_{xy} orbitals. The d_{xz} and d_{yz} orbitals have no electron density in the plane and should be lowest in energy.

79.

— d_{z^2}

— — $d_{x^2-y^2}$, d_{xy}

— — d_{xz}, d_{yz}

The d_{z^2} orbital will be destabilized much more than in the trigonal planar case (see Exercise 19.77). The d_{z^2} orbital has electron density on the z-axis directed at the two axial ligands. The $d_{x^2-y^2}$ and d_{xy} orbitals are in the plane of the three trigonal planar ligands and should be destabilized a lesser amount as compared to the d_{z^2} orbital; only a portion of the electron density in the $d_{x^2-y^2}$ and d_{xy} orbitals is directed at the ligands. The d_{xz} and d_{yz} orbitals will be destabilized the least since the electron density is directed between the ligands. **81.** a. -0.26 V; b. The stronger oxidizing agent is the species more easily reduced. From the reduction potentials, Co^{3+} ($\mathscr{E}° = 1.82$ V) is a much stronger oxidizing agent than $Co(en)_3^{3+}$ ($\mathscr{E}° = -0.26$ V). c. In aqueous solution, Co^{3+} forms the hydrated transition metal complex $Co(H_2O)_6^{3+}$. In both complexes, $Co(H_2O)_6^{3+}$ and $Co(en)_3^{3+}$, cobalt exists as Co^{3+}, which has 6 d electrons. Assuming a strong-field case, the d-orbital splitting diagram for each is

— — e_g

↑↓ ↑↓ ↑↓ t_{2g}

When each complex gains an electron, the electron enters the higher-energy e_g orbitals. Since en is a stronger field ligand than H_2O, the d-orbital splitting is larger for $Co(en)_3^{3+}$ and it takes more energy to add an electron to $Co(en)_3^{3+}$ than to $Co(H_2O)_6^{3+}$. Therefore, it is more favorable for $Co(H_2O)_6^{3+}$ to gain an electron than for $Co(en)_3^{3+}$ to gain an electron. **83.** a. 7.1×10^{-7} mol/L; b. 8.7×10^{-3} mol/L; c. The presence of NH_3 increases the solubility of AgBr. Added NH_3 removes Ag^+ from solution by forming the complex ion $Ag(NH_3)_2^+$. As Ag^+ is removed, more AgBr(s) will dissolve to replenish the Ag^+ concentration. d. 0.41g; e. Added HNO_3 will have no effect on the AgBr(s) solubility in pure water. Neither H^+ nor NO_3^- reacts with Ag^+ or Br^- ions. Br^- is the conjugate base of the strong acid HBr, so it is a terrible base. Added H^+ will not react with Br^- to any great extent. However, added HNO_3 will reduce the solubility of AgBr(s) in the ammonia solution. NH_3 is a weak base ($K_b = 1.8 \times 10^{-5}$). Added H^+ will react with NH_3 to form NH_4^+. As NH_3 is removed, a smaller amount of the $Ag(NH_3)_2^+$ complex ion will form, resulting in a smaller amount of AgBr(s) which will dissolve.

Chapter 20

1. a. Thermodynamic stability: the potential energy of a particular nucleus as compared to the sum of the potential energies of its component protons and neutrons. b. Kinetic stability: the probability that a nucleus will undergo decomposition to form a different nucleus. c. Radioactive decay: a spontaneous decomposition of a nucleus to form a different nucleus. d. β-particle production: a decay process for radioactive nuclides where an electron is produced; the mass number remains constant and the atomic number changes. e. α-particle production: a common mode of decay for heavy radioactive nuclides where a helium nucleus is produced causing the atomic number and the mass number to change. f. Positron production: a mode of nuclear decay in which a particle is formed having the same mass as an electron but opposite in charge. g. Electron capture: a process in which one of the inner-orbital electrons in an atom is captured by the nucleus. h. γ-ray emissions: the production of high-energy photons called γ rays that frequently accompany nuclear decays and particle reactions. **3.** a. $^{51}_{24}Cr + ^{0}_{-1}e \rightarrow ^{51}_{23}V$; b. $^{131}_{53}I \rightarrow ^{0}_{-1}e + ^{131}_{54}Xe$; c. $^{32}_{15}P \rightarrow ^{0}_{-1}e + ^{32}_{16}S$; d. $^{235}_{92}U \rightarrow ^{4}_{2}He + ^{231}_{90}Th$ **5.** a. $^{68}_{31}Ga + ^{0}_{-1}e \rightarrow ^{68}_{30}Zn$; b. $^{62}_{29}Cu \rightarrow ^{0}_{+1}e + ^{62}_{28}Ni$; c. $^{212}_{87}Fr \rightarrow ^{4}_{2}He + ^{208}_{85}At$; d. $^{129}_{51}Sb \rightarrow ^{0}_{-1}e + ^{129}_{52}Te$ **7.** 10 α particles and 5 β particles **9.** Refer to Table 20.2 for potential radioactive decay processes. ^{17}F and ^{18}F contain too many protons or too few neutrons. Electron capture or positron production are both possible decay mechanisms that increase the neutron-to-proton ratio. α-particle production also increases the neutron-to-proton ratio, but it is not likely for these light nuclei. ^{21}F contains too many neutrons or too few protons. β-particle production lowers the neutron-to-proton ratio, so we expect ^{21}F to be a β-emitter. **11.** a. $^{249}_{98}Cf + ^{18}_{8}O \rightarrow ^{263}_{106}Sg + 4^{1}_{0}n$; b. $^{259}_{104}Rf$ **13.** 6.35×10^{11} **15.** a. 6.27×10^{-7} s^{-1}; b. 1.65×10^{14} decays/s; c. 25.6 days **17.** 0.219 or 21.9% of the ^{90}Sr remains. **19.** 6.22 mg ^{32}P **21.** Plants take in CO_2 in the photosynthesis process, which incorporates carbon, including ^{14}C, into its molecules. As long as the plant is alive, the $^{14}C/^{12}C$ ratio in the plant will equal the ratio in the atmosphere. When the plant dies, ^{14}C is not replenished, and ^{14}C decays by β-particle production. By measuring the ^{14}C activity today in the artifact and comparing this to the assumed ^{14}C activity when the plant died to make the artifact, an age can be determined for the artifact. The assumptions are that the ^{14}C level in the atmosphere is constant or that the ^{14}C level at the time the plant died can be calculated. A constant ^{14}C level is a pure assumption, and accounting for variation is complicated. Another problem is that some of the material must be destroyed to determine the ^{14}C level. **23.** No; from ^{14}C dating, the painting was produced (at the earliest) during the late 1800s. **25.** 1975 **27.** 3.8×10^9 yr **29.** 4.3×10^6 kg/s **31.** 1.408×10^{-12} J/nucleon **33.** ^{12}C, 1.230×10^{-12} J/nucleon; ^{235}U, 1.2154×10^{-12} J/nucleon; Since ^{26}Fe is the most stable known nucleus, the binding energy per nucleon for ^{56}Fe (1.41×10^{-12} J/nucleon) will be larger than that for ^{12}C or ^{235}U (see Fig. 20.9). **35.** 26.9830 amu **37.** -2.820×10^{-12} J/nucleus; -1.698×10^{12} J/mol **39.** The Geiger-Müller tube has a certain response time. After the gas in the tube ionizes to produce a "count," some time must elapse for the gas to return to an electrically neutral state. The response of the tube levels because at high activities, radioactive particles are entering the tube faster than the tube can respond to them. **41.** Fission: Splitting of a heavy nucleus into two (or more) lighter nuclei. Fusion: Combining two light nuclei to form a heavier nucleus. The maximum binding energy per nucleon occurs at Fe. Nuclei smaller than Fe become more stable by fusing to form heavier nuclei closer in mass to Fe. Nuclei larger than Fe form more stable nuclei by splitting to form lighter nuclei closer in mass to Fe. **43.** The moderator slows the neutrons to increase the efficiency of the fission reaction. The control rods absorb neutrons to slow or halt the fission reaction. **45.** In order to sustain a nuclear chain reaction, the neutrons produced by the fission must be contained within the fissionable material so that they can go on to cause other fissions. The fissionable material must be closely packed together to ensure that neutrons are not lost to the outside. The critical mass is the mass of material in which exactly one neutron from each fission event causes another fission event so that the process sustains itself. A supercritical situation occurs when more than one neutron from each fission event causes another fission event. In this case the process rapidly escalates and the heat build-up causes a violent explosion. **47.** A nonradioactive substance can be put in equilibrium with a radioactive substance. The two materials can then be checked to see whether all the radioactivity remains in the original material or if it has been scrambled by the equilibrium. **49.** All evolved oxygen in O_2 comes from water and not from carbon dioxide. **51.** Some factors for the biological effects of radiation exposure are as follows: a. The energy of the radiation. The higher the energy, the more damage it can cause. b. The penetrating ability of radiation. The ability of specific radiation to penetrate human tissue and cause damage must be considered. c. The ionizing ability of the radiation. When biomolecules are ionized, their function is usually disturbed. d. The chemical properties of the radiation source. Specifically, the radioactive substance either is readily incorporated into the body or is inert chemically and passes through the body relatively quickly. ^{90}Sr will be incorporated into the body by replacing calcium in the bones. Once incorporated, ^{90}Sr can cause leukemia and bone cancer. Krypton is chemically inert so it will not be incorporated into the body. **53.** (i) and (ii) mean that Pu is not a significant threat outside the body. Our skin is sufficient to keep out the α particles. If Pu gets inside the body, it is easily oxidized to Pu^{4+} (iv), which is chemically similar to Fe^{3+} (iii). Thus Pu^{4+} will concentrate in tissues where Fe^{3+} is found, including the bone marrow, where red blood cells are produced. Once inside the body, α particles cause considerable damage. **55.** ~900 g ^{235}U (assuming that all ^{235}U present undergoes fission) **57.** $\frac{1}{9}$ of the NO_2 is $N^{16}O_2$, $\frac{4}{9}$ of the NO_2 is $N^{18}O_2$, and $\frac{4}{9}$ of the NO_2 is $N^{16}O^{18}O$. **59.** Assuming that (1) the radionuclide is long-lived enough that no significant decay occurs during the time of the experiment, and (2) the total activity is uniformly distributed only in the rat's blood; $V = 10.$ mL. **61.** 5×10^9 K **63.** 4.3×10^{-29} **65.** a. $^{235}UF_6$; At constant temperature, average velocity is proportional to $(1/M)^{1/2}$. Therefore, the lighter the molecule, the faster the average velocity. b. In theory, 345 stages (steps) are required. c. $^{235}U/^{238}U = 1.01 \times 10^{-2}$

Chapter 21

1. A hydrocarbon is a compound composed of only carbon and hydrogen. A saturated hydrocarbon has only carbon–carbon single bonds in the molecule. An unsaturated hydrocarbon has one or more carbon–carbon multiple bonds but may also contain carbon–carbon single bonds. A normal hydrocarbon has one chain of consecutively bonded carbon atoms. A branched hydrocarbon has at least one carbon atom not bonded to the end carbon of a chain of consecutively bonded carbon atoms. Instead, at least one carbon atom forms a bond to an inner carbon atom in the chain of consecutively bonded carbon atoms. **3.** In order to form, cyclopropane and cyclobutane are forced to form bond angles much smaller than the preferred 109.5° bond angles. Cyclopropane and cyclobutane easily react in order to obtain the preferred 109.5° bond angles.

5. a.

CH_3
$CH_3\underset{|}{C}HCH_2CH_2CH_2CH_2CH_3$;
2-Methylheptane

CH_3
$CH_3CH_2CHCH_2CH_2CH_2CH_3$;
3-Methylheptane

CH_3
$CH_3CH_2CH_2CHCH_2CH_2CH_3$;
4-Methylheptane

b.

CH_3
$CH_3CCH_2CH_2CH_2CH_3$;
CH_3
2,2-Dimethylhexane

CH_3
$CH_3CHCHCH_2CH_2CH_3$;
CH_3
2,3-Dimethylhexane

CH_3
$CH_3CHCH_2CHCH_2CH_3$;
CH_3
2,4-Dimethylhexane

CH_3
$CH_3CHCH_2CH_2CHCH_3$;
CH_3
2,5-Dimethylhexane

CH_3
$CH_3CH_2CCH_2CH_2CH_3$;
CH_3
3,3-Dimethylhexane

CH_3
$CH_3CH_2CHCHCH_2CH_3$;
CH_3
3,4-Dimethylhexane

CH_2CH_3
$CH_3CH_2CHCH_2CH_2CH_3$;
3-Ethylhexane

c.

$H_3C\ \ CH_3$
$CH_3-C-CH-CH_2-CH_3$;
CH_3
2,2,3-Trimethylpentane

$CH_3\ \ \ \ CH_3$
$CH_3-C-CH_2-CH-CH_3$;
CH_3
2,2,4-Trimethylpentane

$CH_3\ \ CH_3$
$CH_3-CH-C-CH_2-CH_3$;
CH_3
2,3,3-Trimethylpentane

$CH_3\ \ CH_3\ \ CH_3$
$CH_3-CH-CH-CH-CH_3$;
2,3,4-Trimethylpentane

$CH_3\ \ CH_2CH_3$
$CH_3-CH-CH-CH_2-CH_3$;
3-Ethyl-2-methylpentane

CH_2CH_3
$CH_3-CH_2-C-CH_2-CH_3$;
CH_3
3-Ethyl-3-methylpentane

d.

$CH_3\ CH_3$
$CH_3-C-C-CH_3$
$CH_3\ CH_3$
2,2,3,3-Tetramethylbutane

7. London dispersion (LD) forces are the primary intermolecular forces exhibited by hydrocarbons. The strength of the LD forces depends on the surface area contact among neighboring molecules. As branching increases, there is less surface area contact among neighboring molecules, leading to weaker LD forces and lower boiling points.

9. a. CH_3CH_2—CH(CH—CH_3 with CH_3 on top carbon)

$$\begin{array}{l} \quad\;\; CH_3 \\ \quad\;\; | \\ CH_3-CH-CH_2 \\ \qquad\qquad\;\; | \\ CH_3CH_2-CH-CH_2CH_2CH_3 \end{array}$$

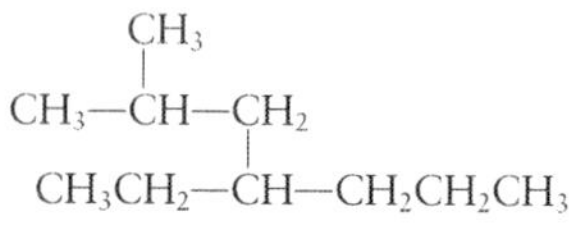

b.

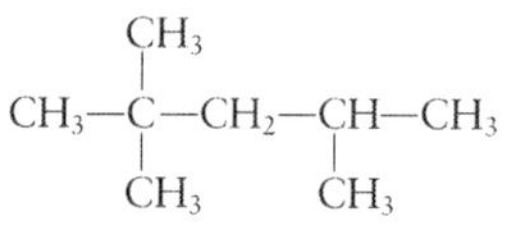

$$\begin{array}{l} \qquad\quad CH_3 \\ \qquad\quad | \\ CH_3-C-CH_2-CH-CH_3 \\ \qquad\quad | \qquad\quad\;\; | \\ \qquad\; CH_3 \qquad CH_3 \end{array}$$

c.

$$\begin{array}{l} CH_3-CH-CH_2CH_2CH_3 \\ \qquad\;\; | \\ CH_3-C-CH_3 \\ \qquad\;\; | \\ \qquad CH_3 \end{array}$$

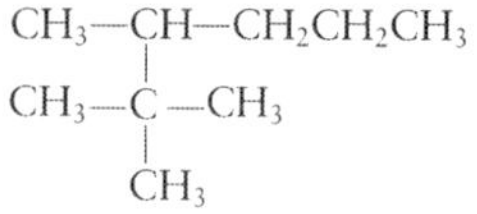

d. 4-ethyl-2-methylheptane, 2,2,3-trimethylhexane **11.** a. 2,2,4-trimethylhexane; b. 5-methylnonane; c. 2,2,4,4-tetramethylpentane; d. 3-ethyl-3-methyloctane **13.** a. 1-butene; b. 2-methyl-2-butene; c. 2,5-dimethyl-3-heptene; d. 2,3-dimethyl-1-pentene; e. 1-ethyl-3-methylcyclopentene; f. 4-ethyl-3-methylcyclopentene; g. 4-methyl-2-pentyne **15.** a. 1,3-dichlorobutane; b. 1,1,1-trichlorobutane; c. 2,3-dichloro-2,4-dimethylhexane; d. 1,2-difluoroethane; e. 3-iodo-1-butene; f. 2-bromotoluene (or 1-bromo-2-methylbenzene); g. 1-bromo-2-methylcyclohexane; h. 4-bromo-3-methylcyclohexene **17.** isopropylbenzene or 2-phenylpropane **19.** All six of these compounds are the same. They differ from each other only by rotations about one or more carbon–carbon single bonds. Only one isomer of C_7H_{16} is present in all of these names, 3-methylhexane. **21.** Compounds c and f exhibit *cis–trans* isomerism. See Exercise 25 for an example of *cis–trans* isomerism in ring compounds. **23.** a. Compounds ii and iii are identical compounds, so they would have the same physical properties. b. Compound i is a *trans* isomer because the bulkiest groups bonded to the carbon atoms in the $C_3{=}C_4$ double bond are as far apart as possible. c. Compound iv does not have carbon atoms in a double bond that each have two different groups attached. Compound iv does not exhibit *cis–trans* isomerism.

25. (chair cyclohexane structures: H, H up / CH_3, CH_3 down)

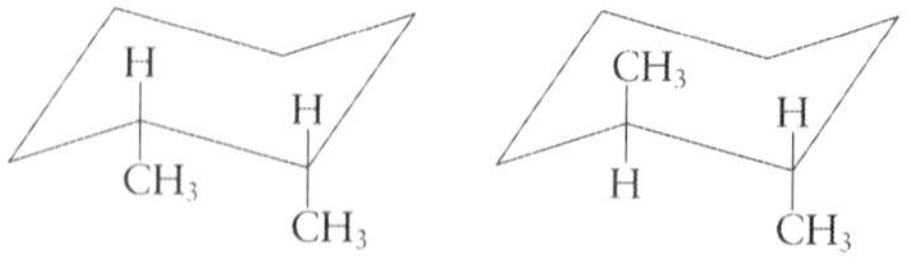

cis (H, CH_3 down; H, CH_3 down) — *trans* (CH_3 up, H down; H up, CH_3 down)

27. Cl(H)C=C(CH_3)(H) (Cl and CH_3 on same side); H(Cl)C=C(CH_3)(H) (H and CH_3 on same side); chlorocyclopropane (Cl on cyclopropane ring)

$CH_2{=}C(Cl){-}CH_3$ $\quad$ $CH_2{=}CH{-}CH_2Cl$

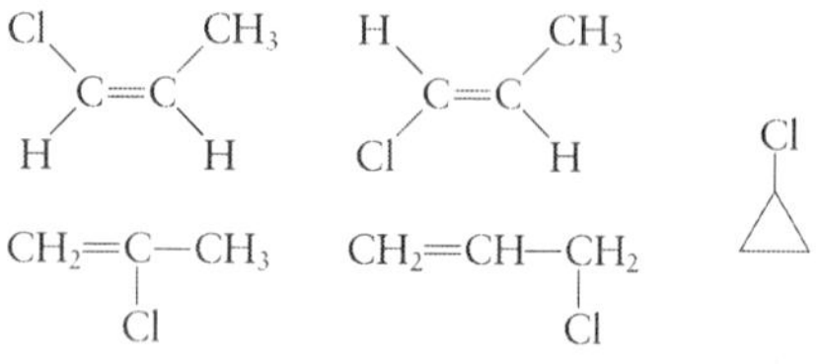

29. $CH_2{=}CHCH_2CH_2CH_3$ $\quad$ $CH_3CH{=}CHCH_2CH_3$

$CH_2{=}C(CH_3)CH_2CH_3$ $\quad$ $CH_3C(CH_3){=}CHCH_3$ $\quad$ $CH_3CH(CH_3)CH{=}CH_2$

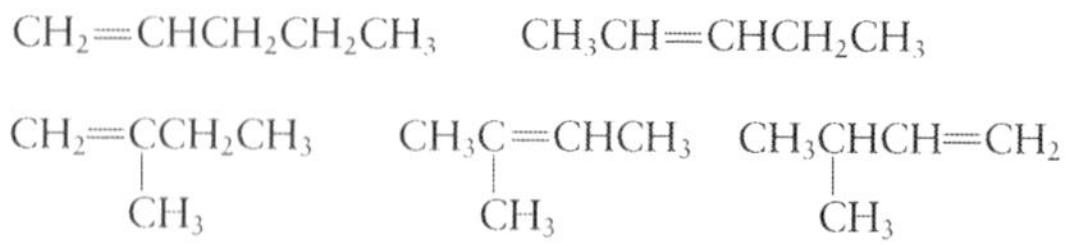

Only 2-pentene exhibits *cis–trans* isomerism. The isomers are

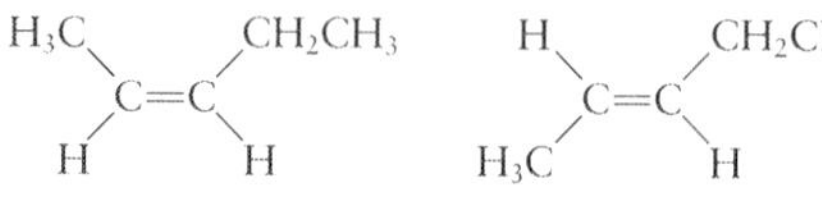

H_3C(H)C=C(CH_2CH_3)(H) — *cis*

H(H_3C)C=C(CH_2CH_3)(H) — *trans*

31. a. *cis*-1-bromo-1-propene; b. *cis*-4-ethyl-3-methyl-3-heptene; c. *trans*-1,4-diiodo-2-propyl-1-pentene

33. a. (dichlorobenzenes)

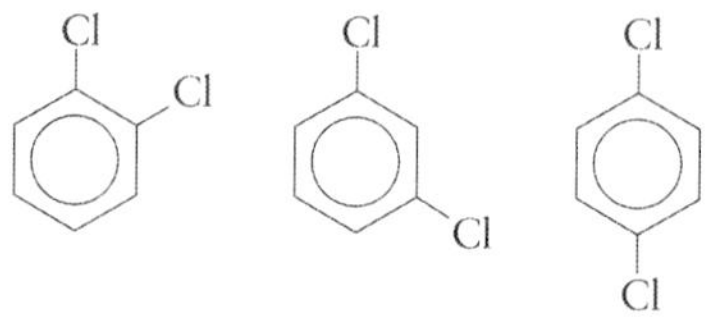

ortho $\quad$ *meta* $\quad$ *para*

b. There are three trichlorobenzenes (1,2,3-trichlorobenzene, 1,2,4-trichlorobenzene, and 1,3,5-trichlorobenzene). c. The *meta* isomer will be very difficult to synthesize. d. 1,3,5-trichlorobenzene will be the most difficult to synthesize since all Cl groups are *meta* to each other in this compound.

35. Carboxylic acid: R—C(=Ö)—Ö—H, RCOOH $\quad$ Aldehyde: R—C(=Ö)—H, RCHO

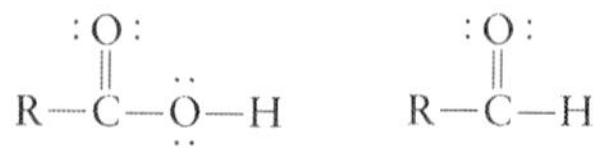

The R designation refers to the rest of the organic molecule beyond the specific functional group indicated in the formula. The R group may sometimes be a hydrogen, but is usually a hydrocarbon fragment. The major point in the R group designation is that if the R group is an organic fragment, then the first atom in the R group is a carbon atom. What the R group has after the first carbon is not important to the functional group designation. **37.** a. ketone; b. aldehyde; c. carboxylic acid; d. amine

39. a. (structure with labels: ketone, amine, amine, carboxylic acid, alcohol)

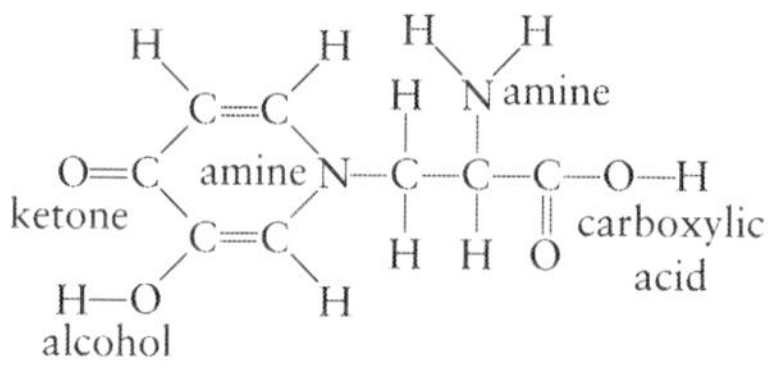

b. 5 carbons in ring and the carbon in —CO_2H: sp^2, the other two carbons: sp^3; c. 24 σ bonds, 4 π bonds **41.** a. 3-chloro-1-butanol, primary alcohol; b. 3-methyl-3-hexanol, tertiary alcohol; c. 2-methylcyclopentanol, secondary alcohol

43.

$CH_3CH_2CH_2CH_2CH_2OH$ — 1-Pentanol

$CH_3CH_2CH_2CH(OH)CH_3$ — 2-Pentanol

$CH_3CH_2CH(OH)CH_2CH_3$ — 3-Pentanol

$CH_3CH_2CH(CH_3)CH_2OH$ — 2-Methyl-1-butanol

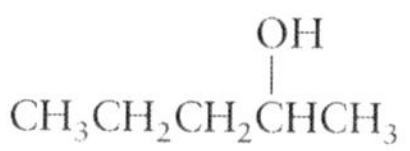

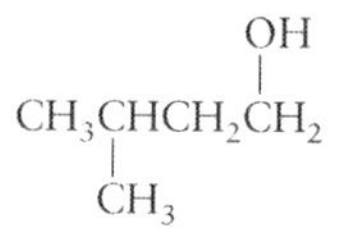

$CH_3CH(CH_3)CH_2CH_2OH$ — 3-Methyl-1-butanol

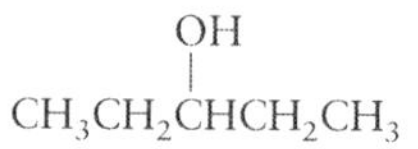

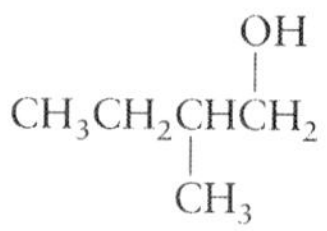

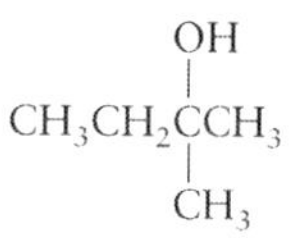

$CH_3CH_2C(OH)(CH_3)CH_3$ — 2-Methyl-2-butanol

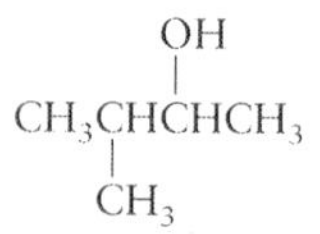

$CH_3CH(CH_3)CH(OH)CH_3$ — 3-Methyl-2-butanol

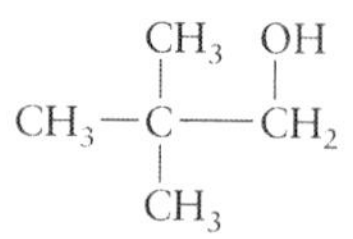

$CH_3{-}C(CH_3)_2{-}CH_2OH$ — 2,2-Dimethyl-1-propanol

There are six isomeric ethers with the formula $C_5H_{12}O$. The structures follow.

$CH_3-O-CH_2CH_2CH_2CH_3$ $\quad$ $CH_3-O-CH(CH_3)CH_2CH_3$

$CH_3-O-CH_2CH(CH_3)CH_3$ $\quad$ $CH_3-O-C(CH_3)_2-CH_3$

$CH_3CH_2-O-CH_2CH_2CH_3$ $\quad$ $CH_3CH_2-O-CH(CH_3)-CH_3$

45. a. 4,5-dichloro-3-hexanone; b. 2,3-dimethylpentanal; c. 3-methylbenzaldehyde or *m*-methylbenzaldehyde

47. a. $H-C(=O)-H$ $\quad$ b. $CH_3CH_2CH_2C(=O)CH_2CH_2CH_3$

c. $H-C(=O)CH_2CH(Cl)CH_3$ $\quad$ d. $CH_3C(=O)CH_2CH_2C(CH_3)_2CH_3$

49. a. H H H / H / H H H H (cyclobutane) or H H / H CH_3 / H H (cyclopropane ring)

b. $CH_3C(=O)-O-CH_3$ or $HC(=O)-O-CH_2CH_3$

c. $CH_3CH_2C(=O)CH_3$

d. $CH_3-N(CH_2CH_2CH_3)-H$ or $CH_3-N(CH(CH_3)CH_3)-H$ or $CH_3CH_2-N(CH_2CH_3)-H$

e. $CH_3-N(CH_2CH_3)-CH_3$

f. $CH_3-O-CH_2CH_2CH_3$ or $CH_3-O-CH(CH_3)-CH_3$ or $CH_3CH_2-O-CH_2CH_3$

g. $CH_3CH(OH)CH_2CH_3$

51. a. 2-Chloro-2-butyne would have 5 bonds to the second carbon. Carbon never expands its octet. b. 2-Methyl-2-propanone would have 5 bonds to the second carbon. c. Carbon-1 in 1,1-dimethylbenzene would have 5 bonds. d. You cannot have an aldehyde functional group bonded to a middle carbon in a chain. Aldehyde groups can only be at the beginning and/or the end of a chain of carbon atoms. e. You cannot have a carboxylic acid group bonded to a middle carbon in a chain. Carboxylic groups must be at the beginning and/or the end of a chain of carbon atoms. f. In cyclobutanol, the 1 and 5 positions refer to the same carbon atom. 5,5-Dibromo-1-cyclobutanol would have five bonds to carbon-1. This is impossible; carbon never expands its octet. **53.** Substitution: An atom or group is replaced by another atom or group; e.g., H in benzene is replaced by Cl. $C_6H_6 + Cl_2 \xrightarrow{\text{Catalyst}} C_6H_5Cl + HCl$ Addition: Atoms or groups are added to a molecule; e.g., Cl_2 adds to ethene. $CH_2{=}CH_2 + Cl_2 \longrightarrow CH_2Cl-CH_2Cl$

55. a. $CH_2{=}CH_2 + H_2O \xrightarrow{H^+} CH_2(OH)-CH_2(H)$

b. $CH_3CH_2(OH) \xrightarrow{\text{Oxidation}} CH_3C(=O)H \xrightarrow{\text{Oxidation}} CH_3C(=O)-OH$

c. $CH_3CH(OH)CH_3 \xrightarrow{\text{Oxidation}} CH_3C(=O)CH_3$

d. $CH_3-O-H + HO-C(=O)CH_3 \xrightarrow{H^+} CH_3-O-C(=O)CH_3 + H_2O$

57. a. $CH_3CH(H)-CH(H)CH_3$ $\quad$ b. $CH_2(Cl)-CH(Cl)CH(CH_3)CH(Cl)-CH(Cl)(CH_3)$

c. $C_6H_5-Cl + HCl$

d. $C_4H_8(g) + 6O_2(g) \longrightarrow 4CO_2(g) + 4H_2O(g)$

59. When $CH_2{=}CH_2$ reacts with HCl, there is only one possible product, chloroethane. When Cl_2 is reacted with CH_3CH_3 (in the presence of light), there are six possible products because any number of the six hydrogens in ethane can be substituted for by Cl. The light-catalyzed substitution reaction is very difficult to control; hence it is not a very efficient method of producing monochlorinated alkanes.

61. a. $H-C(=O)-CH_2CH(CH_3)CH_3 + HO-C(=O)-CH_2CH(CH_3)CH_3$

b. $CH_3-C(=O)-CH(CH_3)CH_3$ $\quad$ c. no reaction

d. $C_6H_5-C(=O)-H + C_6H_5-C(=O)-OH$

e. (cyclohexane ring with O) $-CH_3$

f. (cyclohexane ring with O) OH, $-CH_3$, $-C(=O)-H$ + (cyclohexane ring with O) OH, $-CH_3$, $-C(=O)-OH$

63. The products of the reactions with excess $KMnO_4$ are 2-propanone and propanoic acid. 65. a. $CH_3CH{=}CH_2 + Br_2 \longrightarrow CH_3CHBrCH_2Br$ (Addition reaction of Br_2 with propene)

b.  $CH_3-CH(OH)-CH_3 \xrightarrow{\text{Oxidation}} CH_3-C(=O)-CH_3$

Oxidation of 2-propanol yields acetone (2-propanone).

c. 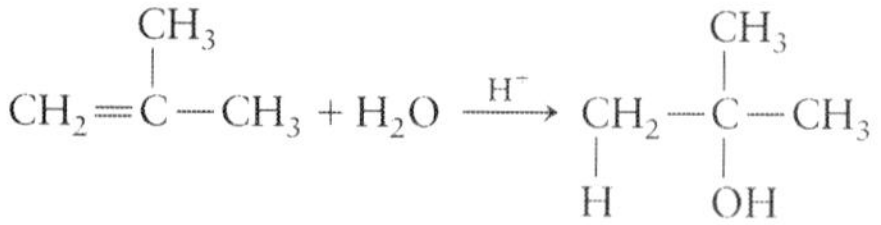 $CH_2{=}C(CH_3)-CH_3 + H_2O \xrightarrow{H^+} CH_2(H)-C(CH_3)(OH)-CH_3$

Addition of H_2O to 2-methylpropene would yield *tert*-butyl alcohol (2-methyl-2-propanol) as the major product.

d. 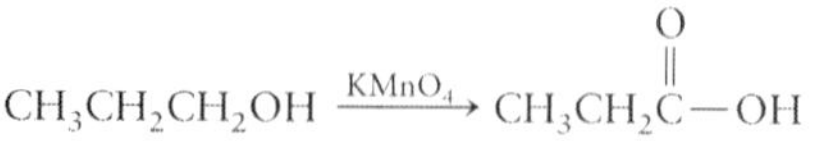 $CH_3CH_2CH_2OH \xrightarrow{KMnO_4} CH_3CH_2C(=O)-OH$

Oxidation of 1-propanol would eventually yield propanoic acid. Propanal is produced first in this reaction and is then oxidized to propanoic acid.

67. 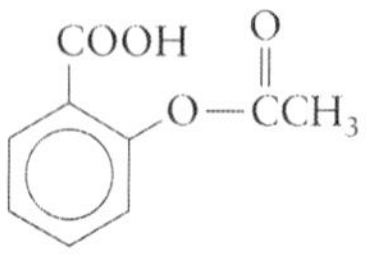 Acetylsalicylic acid (aspirin): benzene ring with COOH and $O-C(=O)CH_3$ (ortho)

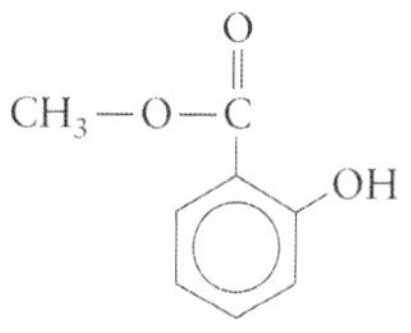

Methyl salicylate: $CH_3-O-C(=O)-$ benzene ring with OH (ortho)

69. a. Addition polymer: a polymer that forms by adding monomer units together (usually by reacting double bonds). Teflon, polyvinyl chloride, and polyethylene are examples of addition polymers. b. Condensation polymer: a polymer that forms when two monomers combine by eliminating a small molecule (usually H_2O or HCl). Nylon and Dacron are examples of condensation polymers. c. Copolymer: a polymer formed from more than one type of monomer. Nylon and Dacron are copolymers. d. Homopolymer: a polymer formed from the polymerization of only one type of monomer. Polyethylene, Teflon, and polystyrene are examples of homopolymers. e. Polyester: a condensation polymer whose monomers link together by formation of the ester functional group. Dacron is a polyester. f. Polyamide: a condensation polymer whose monomers link together by formation of the amide functional group. Nylon is a polyamide as are proteins in the human body. 71. a. A polyester forms when an alcohol functional group reacts with a carboxylic acid functional group. The monomer for a homopolymer polyester must have an alcohol functional group and a carboxylic acid functional group present in the structure. b. A polyamide forms when an amine functional group reacts with a carboxylic acid functional group. For a copolymer polyamide, one monomer would have at least two amine functional groups present, and the other monomer would have at least two carboxylic acid functional groups present. For polymerization to occur, each monomer must have two reactive functional groups present. c. To form a typical addition polymer, a carbon–carbon double bond must be present. To form a polyester, the monomer would need the alcohol and carboxylic acid functional groups present. To form a polyamide, the monomer would need the amine and carboxylic acid functional groups present. The two possibilities are for the monomer to have a carbon–carbon double bond, an alcohol functional group, and a carboxylic acid functional group present or to have a carbon–carbon double bond, an amine functional group, and a carboxylic acid functional group present.

73.  $F(Cl)C{=}CF_2$

75. a. 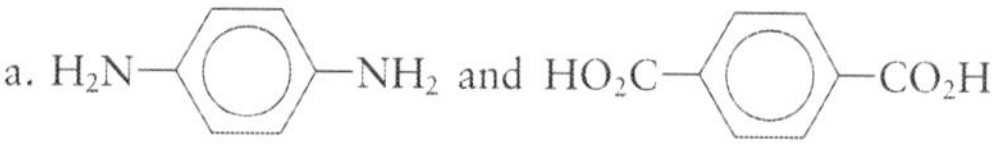 $H_2N-C_6H_4-NH_2$ and $HO_2C-C_6H_4-CO_2H$

b. Repeating unit: $\left(\!-HN-C_6H_4-NHC(=O)-C_6H_4-C(=O)-\right)_n$ (meta-substituted rings)

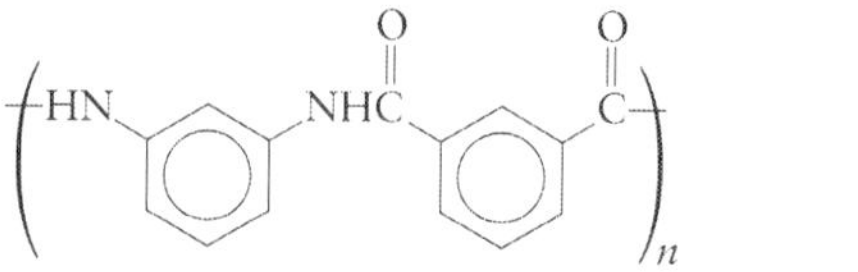

The two polymers differ in the substitution pattern on the benzene rings. The Kevlar chain is straighter, and there is more efficient hydrogen bonding between Kevlar chains than between Nomex chains.

77. 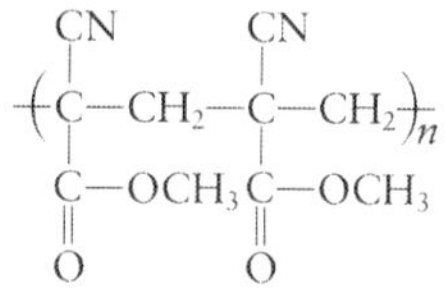 $\left(\!-C(CN)(C(=O)-OCH_3)-CH_2-C(CN)(C(=O)-OCH_3)-CH_2-\right)_n$

79. Divinylbenzene (a crosslinking agent) has two reactive double bonds, which are both reacted when divinylbenzene inserts itself into two adjacent polymer chains. The chains cannot move past each other because the crosslinks bond adjacent polymer chains together making the polymer more rigid.

81. a. $\left(\!-O-C_6H_4-C(CH_3)_2-C_6H_4-O-C(=O)-O-C_6H_4-C(CH_3)_2-C_6H_4-O-C(=O)-\right)$

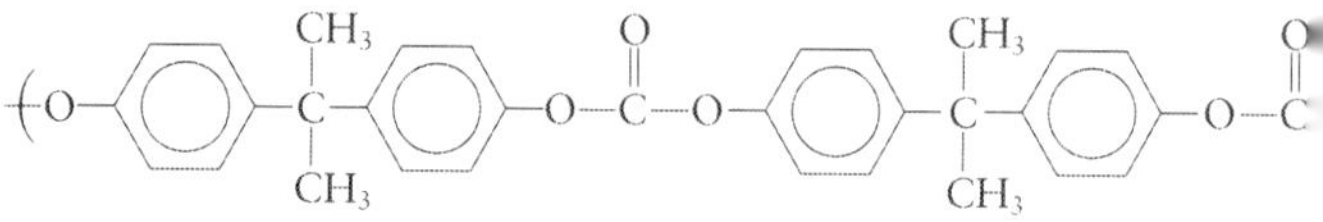

b. Condensation; HCl is eliminated when the polymer bonds form.

83. Polyacrylonitrile: $\left(\!-CH_2-CH(C{\equiv}N)-\right)_n$

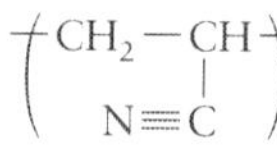

The CN triple bond is very strong and will not easily break in the combustion process. A likely combustion product is the toxic gas hydrogen cyanide [HCN(g)].

85. 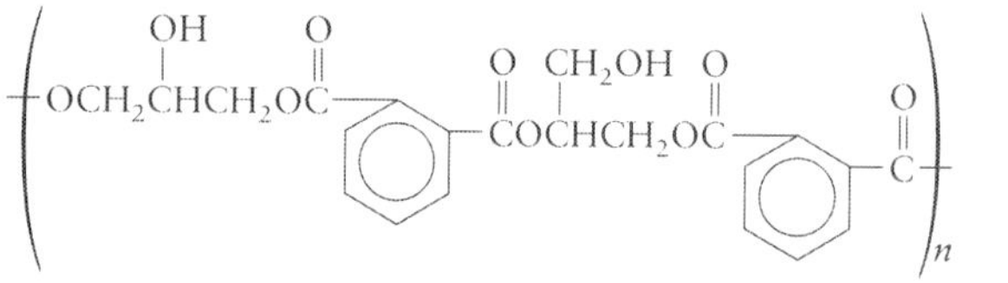 $\left(\!-OCH_2CH(OH)CH_2OC(=O)-C_6H_4-C(=O)OCH(CH_2OH)CH_2OC(=O)-C_6H_4-C(=O)-\right)_n$

Two linkages are possible with glycerol. A possible repeating unit with both types of linkages is shown above. With either linkage, there are free OH groups on the polymer chains. These unreacted OH groups on adjacent polymer chains can react with the acid groups of phthalic acid to form crosslinks between various polymer chains.

87. 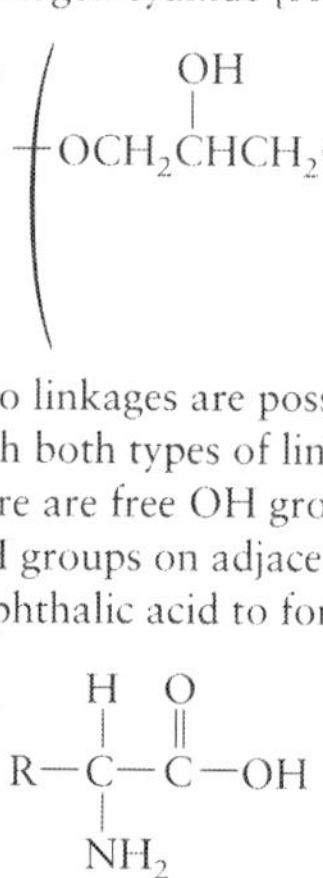 $R-C(H)(NH_2)-C(=O)-OH$

Hydrophilic (water-loving) and hydrophobic (water-fearing) refer to the polarity of the R groups. When the R group consists of a polar group, then the amino acid is hydrophilic. When the R group consists of a nonpolar group, then the amino acid is hydrophobic.

89. Denaturation changes the three-dimensional structure of a protein. Once the structure is affected, the function of the protein will also be affected. **91.** a. serine, tyrosine, and threonine; b. aspartic acid and glutamic acid; c. histidine, lysine, arginine, and tryptophan; d. glutamine and asparagine **93.** a. aspartic acid and phenylalanine; b. Aspartame contains the methyl ester of phenylalanine. This ester can hydrolyze to form methanol, $R—CO_2CH_3 + H_2O \rightleftharpoons RCO_2H + CH_3OH$.

95.

```
         O                                  O
         ||                                 ||
  H2NCHC—NHCHCO2H              H2NCHC—NHCHCO2H
      |        |                   |        |
     CH2      CH3                 CH3      CH2
      |                                     |
     OH                                    OH
        ser-ala                        ala-ser
```

97. a. Six tetrapeptides are possible. From NH_2 to CO_2H end: phe-phe-gly-gly, gly-gly-phe-phe, gly-phe-phe-gly, phe-gly-gly-phe, phe-gly-phe-gly, gly-phe-gly-phe; b. Twelve tetrapeptides are possible. From NH_2 to CO_2H end: phe-phe-gly-ala, phe-phe-ala-gly, phe-gly-phe-ala, phe-gly-ala-phe, phe-ala-phe-gly, phe-ala-gly-phe, gly-phe-phe-ala, gly-phe-ala-phe, gly-ala-phe-phe, ala-phe-phe-gly, ala-phe-gly-phe, ala-gly-phe-phe **99.** a. covalent; b. hydrogen bonding; c. ionic; d. London dispersion **101.** Glutamic acid: $R = —CH_2CH_2CO_2H$; Valine: $R = —CH(CH_3)_2$; A polar side chain is replaced by a nonpolar side chain. This could affect the tertiary structure of hemoglobin and the ability of hemoglobin to bind oxygen.

103.

```
H2N—CH—CO2H              or   H2N—CH—CO2⁻Na⁺
     |                             |
     CH2CH2CO2⁻Na⁺                 CH2CH2CO2H
```

The first structure is MSG, which is impossible for you to predict.

105.

```
  CH2OH     H                    CH2OH
      \ O /                      |----O
       /  \                 H   /H      \  H
  H   H   H  OH            HO   OH HO      OH
      |   |                      |    |
     OH   OH                     H    H
    D-Ribose                    D-Mannose
```

107. The aldohexoses contain 6 carbons and the aldehyde functional group. Glucose, mannose, and galactose are aldohexoses. Ribose and arabinose are aldopentoses since they contain 5 carbons with the aldehyde functional group. The ketohexose (6 carbons + ketone functional group) is fructose, and the ketopentose (5 carbons + ketone functional group) is ribulose. **109.** A disaccharide is a carbohydrate formed by bonding two monosaccharides (simple sugars) together. In sucrose, the simple sugars are glucose and fructose, and the bond formed between these two monosaccharides is called a glycoside linkage. **111.** The α and β forms of glucose differ in the orientation of a hydroxyl group on one specific carbon in the cyclic forms (see Fig. 21.31). Starch is a polymer composed of only α-D-glucose, and cellulose is a polymer composed of only β-D-glucose.

113.

```
        H                          H
        |                          |
  H3C—C*—CH2CH3              H3C—C*—OH
        |                          |
  H2N—C*—CO2H                H2N—C*—CO2H
        |                          |
        H                          H
     Isoleucine                 Threonine
```

The chiral carbons are marked with asterisks.

115.

```
        Cl
        |
   Br—C*—CH=CH2
        |
        H
```

is optically active. The chiral carbon is marked with an asterisk. **117.** They all contain nitrogen atoms with lone pairs of electrons. **119.** C-C-A-G-A-T-A-T-G **121.** Uracil will hydrogen-bond to adenine.

```
                                H
Uracil                          |
                   O------H—N        N=
                                       
                    N—H-----N           N
                  N                  =N        Sugar
            Sugar       O          Adenine
```

123. a. glu: CTT, CTC; val: CAA, CAG, CAT, CAC; met: TAC; trp: ACC; phe: AAA, AAG; asp: CTA, CTG; b. ACC-CTT-AAA-TAC;

```
                         or    or
                        CTC   AAG
```

c. Due to glu and phe, there is a possibility of four different DNA sequences; d. met-asp-phe; e. TAC-CTA-AAG; TAC-CTA-AAA; TAC-CTG-AAA **125.** A deletion may change the entire code for a protein, thus giving an entirely different sequence of amino acids. A substitution will change only one single amino acid in a protein. **127.** $CH_3CH_2CH_2CH_2CH_2CH_2CH_2COOH + OH^- \rightarrow CH_3{-}(CH_2)_6{-}COO^- + H_2O$; Octanoic acid is more soluble in 1 *M* NaOH. Added OH^- will remove the acidic proton from octanoic acid, creating a charged species. As is the case with any substance with an overall charge, solubility in water increases. When morphine is reacted with H^+, the amine group is protonated, creating a positive charge on morphine ($R_3N + H^+ \rightarrow R_3\overset{+}{N}H$). By treating morphine with HCl, an ionic compound results that is more soluble in water and in the bloodstream than is the neutral covalent form of morphine. **129.** To substitute for the benzene ring hydrogens, an iron(III) catalyst must be present. Without this special iron catalyst, the benzene ring hydrogens are unreactive. To substitute for an alkane hydrogen, light must be present. For toluene, the light-catalyzed reaction substitutes a chlorine for a hydrogen in the methyl group attached to the benzene ring. **131.** 1-butene **133.** a. The bond angles in the ring are about 60°. VSEPR predicts bond angles close to 109°. The bonding electrons are closer together than they prefer, resulting in strong electron–electron repulsions. Thus ethylene oxide is unstable (reactive). b. The ring opens up during polymerization; monomers link together through the formation of O—C bonds.

$$\text{+O—CH}_2\text{CH}_2\text{—O—CH}_2\text{CH}_2\text{—O—CH}_2\text{CH}_2\text{+}_n$$

135.

```
 (    O                    O               O                O  )
 (    ||                   ||              ||               || )
 (-OCH2CH2OCN—(C6H4)—NCOCH2CH2OCN—(C6H4)—NC- )
 (           |             |               |                |  )
 (           H             H               H                H  )n
```

137. *n*-hexane, 69°C; pentanal, 103°C; 1-pentanol, 137°C; butanoic acid, 164°C; the strength of the intermolecular forces increases when going from *n*-hexane to pentanal to 1-pentanol to butanoic acid. Hence the boiling points will increase in the same order.

139.

```
   O
   ||
  HC—O—CH3
```

141. a. The longest chain is 4 carbons long. The correct name is 2-methylbutane. b. The longest chain is 7 carbons long, and we would start the numbering system at the other end for lowest possible numbers. The correct name is 3-iodo-3-methylheptane. c. This compound cannot exhibit *cis–trans* isomerism because one of the double-bonded carbons has the same two groups (CH_3) attached. The numbering system should also start at the other end to give the double bond the lowest possible number. 2-Methyl-2-pentene is correct. d. The OH functional group gets the lowest number. 3-Bromo-2-butanol is correct. **143.** For the reaction, we break a P—O and O—H bond and form a P—O and O—H bond, so $\Delta H \approx 0$. ΔS for this process should be negative since two molecules are reacting to form one molecule (positional probability decreases). Thus ΔG should be positive and the reaction is not expected to be spontaneous. **145.** a. No; the mirror image is superimposable.

b.

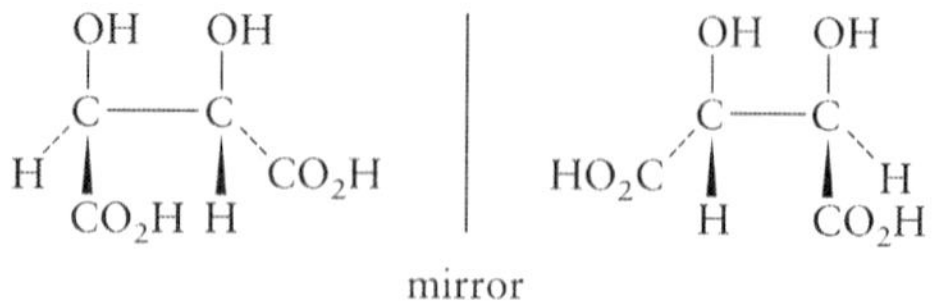

```
     OH   OH          |          OH   OH
     |    |           |          |    |
     C————C           |          C————C
   H/ |    | \CO2H    |   HO2C/ |    | \H
  CO2H H              |          H    CO2H
                    mirror
```

These two forms of tartaric acid are nonsuperimposable.

147. pH = 6.07 = isoelectric point **149.** 0.11% **151.** C_2H_6; ethane

153. a. $CH_3CHCH_2CH_3$ (with CH_3 on C-2)

b. $CH_2{=}CCH_3$ (with CH_3 on C-2)

c. Methylenecyclohexane-type structures: cyclohexene ring bearing CH_3 (1-methylcyclohexene, all ring H shown) and cyclohexane ring bearing $={CH_2}$ (methylenecyclohexane, all ring H shown)

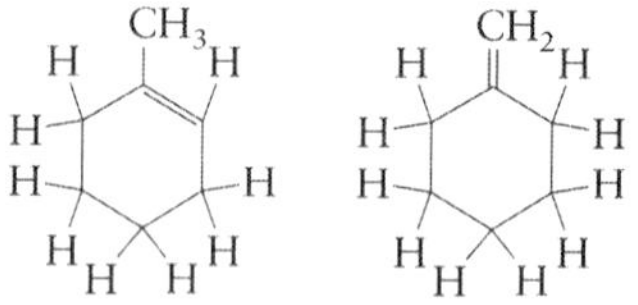

d. $CH_2{=}CHCH_3$

e. $CH_2CH_2CH_2CH_2CH_3$ with OH on C-1; $CH_2CHCH_2CH_3$ with CH_3 on C-2 and OH on C-1

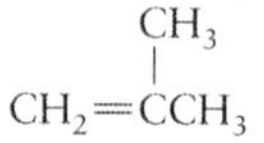

$CH_3CHCH_2CH_2$ with CH_3 on C-2 and OH on C-4; $CH_2{-}C{-}CH_3$ with OH on CH_2 and two CH_3 on the central C

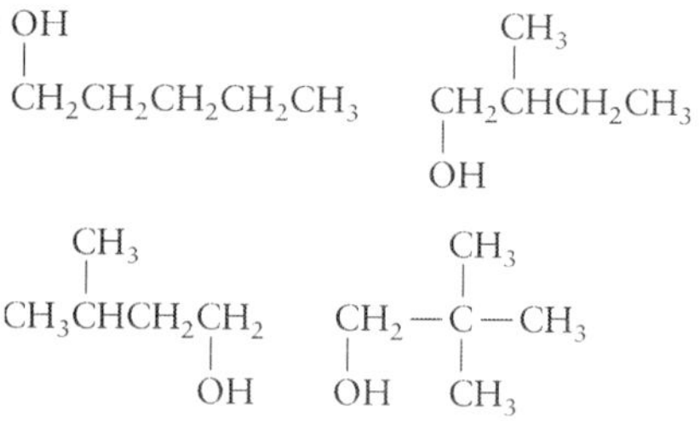

155. a. 25.0 mL, 2.37; 50.0 mL, 6.07; 75.0 mL, 9.77

b.

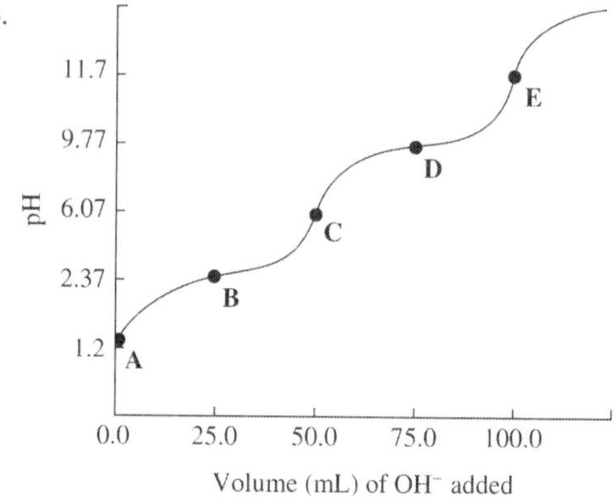

The major amino acid species present are:

point A (0.0 mL OH^-):	$H_3\overset{+}{N}CH_2COOH$
point B (25.0 mL OH^-):	$H_3\overset{+}{N}CH_2COOH$, $H_3\overset{+}{N}CH_2COO^-$
point C (50.0 mL OH^-):	$H_3\overset{+}{N}CH_2COO^-$
point D (75.0 mL OH^-):	$H_3\overset{+}{N}CH_2COO^-$, $H_2NCH_2COO^-$
point E (100.0 mL OH^-):	$H_2NCH_2COO^-$

c. 6.07; d. $+\frac{1}{2}$, 2.37; $-\frac{1}{2}$, 9.77

Photo Credits

Chapter 1: p. 1, © Beathan/Corbis Royalty Free; p. 2, Will & Deni McIntyre/Photo Researchers, Inc.; p. 3, Kristen Brochmann, Fundamental Photographs; p. 4, Denise Applewhite/Office of Communications/Princeton University; p. 5, Courtesy, Dow Corning Corporation; p. 9, NASA; p. 10, Simon Fraser/Science Photo Library/Photo Researchers, Inc., p. 13, Brownie Harris/ Stock Market/Corbis.

Chapter 2: p. 15, Dr. Dennis Marinus Hansen; p. 16, Roald Hoffman/Cornell University; p. 17 (Detail), *Antoine Laurent Lavoisier & His Wife* by Jacques Louis David, The Metropolitan Museum of Art; p. 18, Manchester Literary & Philosophical Society; p. 20, The Granger Collection; p. 22 (both), Courtesy, IBM Corporation, Research Division, Almaden Research Center; p. 24 (top), Bettmann/Corbis; p. 24 (bottom), Richard Megna/Fundamental Photographs; p. 26: Bettmann/ Corbis, p. 28, *Portrait of Ernest Rutherford*, 1932 by Oswald Hornby Birley, Royal Society, London, UK/Bridgeman Art Library; p. 31 (left, center), Frank Cox/Photograph © Houghton Mifflin Company. All rights reserved; p. 31 (right), Ken O'Donoghue/Photograph © Houghton Mifflin Company. All rights reserved; p. 33 (left), Ken O'Donoghue/Photograph © Houghton Mifflin Company. All rights reserved; p. 33 (center, left, and right), Photograph © Houghton Mifflin Company. All rights reserved; p. 33 (right), Charles D. Winters/Photo Researchers; p. 33 (bottom), Ken O'Donoghue/Photograph © Houghton Mifflin Company. All rights reserved; p. 35 (top), Tom Pantages; p. 35 (bottom), Photograph © Houghton Mifflin Company. All rights reserved; p. 39, Richard Megna/ Fundamental Photographs/Photograph © Houghton Mifflin Company. All rights reserved; p. 43 (both), Microtrace, LLC.

Chapter 3: p. 52, Jim Sugar/Science Faction; p. 53, Agricultural Research Service/USDA Photo by Scott Baur; p. 54, Kennan Ward/Stock Market/Corbis; p. 56, Ken O'Donoghue/ Photograph © Houghton Mifflin Company. All rights reserved; p. 61 (top), Phil Degginger/Stone/Getty Images; p. 61 (bottom), The Nobel Foundation; p. 65 (top), Ken O'Donoghue/ Photograph © Houghton Mifflin Company. All rights reserved; p. 65 (bottom), Frank Cox/Photograph © Houghton Mifflin Company. All rights reserved; p. 67 (top), Owen Franken/Stock Boston; p. 67 (bottom), Photograph © Houghton Mifflin Company. All rights reserved; p. 70 (both), Photograph © Houghton Mifflin Company. All rights reserved; p. 73, Larry Larimer/Artville/PictureQuest.

Chapter 4: p. 90, Photograph © Houghton Mifflin Company. All rights reserved; p. 94 (all), Ken O'Donoghue/Photograph © Houghton Mifflin Company. All rights reserved; p. 101, Richard Megna/Fundamental Photographs/Photograph © Houghton Mifflin Company. All rights reserved; p. 102, Photograph © Houghton Mifflin Company. All rights reserved; p. 103, Photograph © Houghton Mifflin Company. All rights reserved; p. 104 (all), Photograph © Houghton Mifflin Company. All rights reserved; p. 105, Ken O'Donoghue/Photograph © Houghton Mifflin Company. All rights reserved; p. 106, Photograph © Houghton Mifflin Company. All rights reserved; p. 110, Photograph © Houghton Mifflin Company. All rights reserved; p. 112, Jeff Daly/Visuals Unlimited; p. 114 (all), Richard Megna/Fundamental Photographs/Photograph © Houghton Mifflin Company. All rights reserved; p. 115, AFP Photo/Toru Yamanaka/Corbis; p. 118 (left), Photograph © Houghton Mifflin Company. All rights reserved; p. 118 (center two), Ken O'Donoghue/Photograph © Houghton Mifflin Company. All rights reserved; p. 118 (right), Photograph © Houghton Mifflin Company. All rights reserved; p. 119, Tammy Peluso/Tom Stack and Associates; p. 120, Richard Megna/Fundamental Photographs/Photograph © Houghton Mifflin Company. All rights reserved; p. 130, Photograph © Houghton Mifflin Company. All rights reserved.

Chapter 5: p. 141, AP Photo/Keystone/Martial Trezzini/AP/ Wide World Photos; p. 153, Ford Motor Company; p. 160 (all), Ken O'Donoghue; p. 165, Ken O'Donoghue; p. 167, U.S. Department of Energy; p. 174, Jim Gray/Courtesy of Kenneth S. Suslick; p. 179 (top two), Michael Barnes; p. 179 (bottom), John Lawlor; p. 180, M. Edwards/Peter Arnold, Inc.; p. 181 (both), Field Museum, Chicago, #CSGN40263 and #GN83213_6c.

Chapter 6: p. 196, Lee Foster/Taxi/Getty Images; p. 198, Photograph © Houghton Mifflin Company. All rights reserved; p. 203, Danny Eilers/Alamy; p. 206, Larry Andre Maslennikov/Peter Arnold, Inc.; p. 216, Science Photo Library/ Photo Researchers; p. 217, Photograph © Houghton Mifflin Company. All rights reserved; p. 220 (all), Ken O'Donoghue; p. 221 (both), Photograph © Houghton Mifflin Company. All rights reserved; p. 227 (all), Photograph © Houghton Mifflin Company. All rights reserved.

Chapter 7: p. 233, Sinclair Stammers/Photo Researchers, Inc.; p. 240 (both), Ken O'Donoghue/Photograph © Houghton Mifflin Company. All rights reserved; p. 248, Patrick Ward/Corbis; p. 249, Rick Poley/Visuals Unlimited; p. 253, R. Konig/Jacana/Photo Researchers; p. 261, Dan Loh/AP/Wide World Photos; p. 266, Photograph © Houghton Mifflin Company. All rights reserved; p. 267, Photograph © Houghton Mifflin Company. All rights reserved.

Chapter 8: p. 286, David Scharf/Science Faction; p. 289, Barry Slaven/Visuals Unlimited; p. 290, Ken O'Donoghue/Photograph © Houghton Mifflin Company. All rights reserved; p. 292 (both), Ken O'Donoghue/Photograph © Houghton Mifflin Company. All rights reserved; p. 304, American Color; p. 319 (both), Photograph © Houghton Mifflin Company. All rights reserved; p. 320 (all), Ken O'Donoghue/Photograph © Houghton Mifflin Company. All rights reserved; p. 328, CNRI/Science Photo Library/Photo Researchers; p. 329, Ken O'Donoghue/Photograph © Houghton Mifflin Company. All rights reserved; p. 332, Ken O'Donoghue/Photograph © Houghton Mifflin Company. All rights reserved; p. 335, Bruce Roberts/Photo Researchers; p. 338 (both), Photograph © Houghton Mifflin Company. All rights reserved; p. 339, Courtesy of Yi Lu/University of Illinois at Urbana; p. 340 (bottom, right), Photograph © Houghton Mifflin Company. All rights reserved; p. 340 (top; bottom left), Ken O'Donoghue/Photograph © Houghton Mifflin Company. All rights reserved; p. 344 (both), Photograph © Houghton Mifflin Company. All rights reserved.

Chapter 9: p. 358, Jeremy Holden; p. 376 (both), Photograph © Houghton Mifflin Company. All rights reserved; p. 379, Phil Degginger/Color-Pic, Inc.; p. 381, Argonne National Laboratory; p. 383, Itsuo Inouye/AP/Wide World Photos; p. 391, Klaus Andrews/Argus Fotoarchiv/Peter Arnold, Inc.; p. 396, Courtesy FPL Energy, LLC; p. 400, The Image Bank/Getty.

Chapter 10: p. 410, Matt Meadows/Science Photo Library/Photo Researchers; p. 417, Stephen Frisch/Stock Boston; p. 425, Photograph © Houghton Mifflin Company. All rights reserved; p. 431, Inga Spence/Visuals Unlimited; p. 432, Photograph © Houghton Mifflin Company. All rights reserved; p. 443, Phil A. Harrington/Peter Arnold, Inc.; p. 451, Michael S. Yamashita/Corbis.

Chapter 11: p. 472, Fog Stock/Tips Images; p. 478, Richard Megna/Fundamental Photographs/Photograph © Houghton Mifflin Company. All rights reserved; p. 483, Royal Institution/London; p. 485, Photograph © Houghton Mifflin Company. All rights reserved; p. 486, Bettmann/Corbis; p. 488, Photograph © Houghton Mifflin Company. All rights reserved; p. 489, Ken O'Donoghue/Photograph © Houghton Mifflin Company. All rights reserved; p. 496, © Car Culture/Corbis; p. 499, Ron Watts/Corbis; p. 504, Yoav Levy/Corbis; p. 506, Oberlin College Archives/Oberlin College; p. 508, Tom Hollyman/Photo Researchers.

Chapter 12: p. 521, Tetra Images/Alamy; p. 526, Photograph © Houghton Mifflin Company. All rights reserved; p. 527, The Granger Collection; p. 529, Lester V. Bergman/Corbis; p. 532, The Granger Collection; p. 534, James Bergquist and D. Wineland/National Institute of Standards and Technology; p. 537, Photodisc/Getty Images; p. 540 (both), Research Division, Almaden Research Center/IBM Corporation; p. 543 (left), Stephen Jensen/Tufts University/Charles Sykes; p. 543 (right), M. El Kouedi/Charles Sykes; p. 559: Bettmann/Corbis; p. 560, *Annalen der Chemie und Pharmacia*, VIII, Supplementary Volume for 1872; p. 574, Charles D. Winters/Photo Researchers; p. 582, E.R. Degginger/Color-Pic, Inc.

Chapter 13: p. 592, Kenneth Eward/Science Photo Library/Photo Researchers; p. 593 (top), Photograph © Houghton Mifflin Company. All rights reserved; p. 593 (bottom), Photodisc Red/Getty Images; p. 602 (all), Ken O'Donoghue/Photograph © Houghton Mifflin Company. All rights reserved; p. 612, Ken O'Donoghue/Photograph © Houghton Mifflin Company. All rights reserved; p. 614, Will & Deni McIntyre/Photo Researchers; p. 615, Ken Eward/Science Source/Photo Researchers; p. 620 (left), The Bancroft Library; p. 620 (right), From G.N. Lewis, *Valence*, Dover Publications, Inc. New York, 1966; p. 627, Tom Pantages; p. 638: Steve Borick/American Color; p. 641 (all), Ken O'Donoghue/Photograph © Houghton Mifflin Company. All rights reserved; p. 644, Argonne National Laboratory; p. 647, Scott Camazine/Photo Researchers; p. 648: Courtesy, Cyrano Science, Inc. p. 650, Ken O'Donoghue/Photograph © Houghton Mifflin Company. All rights reserved.

Chapter 14: p. 660, Russell Kightley/Science Photo Library/Photo Researchers, Inc.; p. 664, Comstock Images/Royalty Free/Comstock Mountainside, NJ; p. 682 (both), Donald Clegg; p. 690, Courtesy of Tim Kenny/Jonathan E. Kenny; p. 692, Joseph Nettis/Stock Boston/PictureQuest; p. 694, Naturelle/Alamy Images; p. 704, Joshua Lutz/Redux Pictures; p. 705 (left), SIU/Visuals Unlimited; p. 705 (right), Alfred Pasieka/Science Photo Library/Photo Researchers.

Chapter 15: p. 714, Mike Powell/Allsports Concepts/Getty Images; p. 718, Ahmed Zewail/California Institute of Technology; p. 725 (both), Photograph © Houghton Mifflin Company. All rights reserved; p. 727, Photograph © Houghton Mifflin Company. All rights reserved; p. 738, Courtesy of Gabriela Petruck/Christopher Rose-Petruck; p. 739, Photograph © Houghton Mifflin Company. All rights reserved; p. 742, Dr. Wilson Ho/University of California, Irvine; p. 752, Phil Degginger/Color-Pic, Inc.; p. 754, Graham MacIndoe/Wellesley College; p. 755, Delphi Automotive Systems; p. 760, Alfred Pasieka/Science Photo Library/Photo Researchers, Inc.

Chapter 16: p. 777, Artform No. 1 by Mark Ho; p. 782, Paul D. Stewart/Dr. Kellar Autumn/Lewis and Clark College; p. 783 (top two), Photograph © Houghton Mifflin Company. All rights reserved; p. 783 (bottom), Ray Massey/Stone/Getty Images; p. 784 (both), Courtesy, Lord Corporation; p. 786 (top left, top right), Photograph © Houghton Mifflin Company. All rights reserved; p. 786 (bottom left), Jose Manuel Sanchis Calvete/Corbis; p. 786 (bottom right), Brian Parker/Tom Stack and Associates; p. 789 (bottom left): Joanna Aizenberg; p. 789 (top right), Kevin Schafer/Vireo; p. 789 (bottom right), Courtesy, Dr. Marc A. Meyers/University of California, San Diego; p. 792, Denise Applewhite/Princeton University; p. 793, Photograph © Houghton Mifflin Company. All rights reserved; p. 798, Chip Clark; p. 800, Ken Eward/Photo Researchers; p. 803, Yoav Levy/Phototake; p. 806, Dr. Thomas Thundat/Oak Ridge National Laboratory; p. 812 (top), IBM Corporation; p. 812 (bottom left), Ken O'Donoghue/Photograph © Houghton Mifflin Company. All rights reserved; p. 812 (bottom right),

Richard Megna/Fundamental Photographs/Photograph © Houghton Mifflin Company. All rights reserved; p. 823, Stephen Frisch/Stock Boston; p. 826, Photograph © Houghton Mifflin Company. All rights reserved; p. 832, Nanophase Technologies Corporation; p. 833, National Geographic; p. 834, Zina Deretsky/National Science Foundation.

Chapter 17: p. 846, Charles D. Winters/Photo Researchers, Inc.; p. 848, Richard Nowitz; p. 851, AP Photo/David Adema/POOL/AP/Wide World Photos; p. 852 (both), USDA; p. 854 (both), Frank Cox/Photograph © Houghton Mifflin Company. All rights reserved; p. 855, Photograph © Houghton Mifflin Company. All rights reserved; p. 858 (bottom left and right), Betz/Visuals Unlimited; p. 859, T. Orban/Sygma/Corbis; p. 871, Courtesy, PUR/Recovery Engineering, Inc., Minneapolis, MN; p. 873, Stephen Frisch/Stock Boston.

Chapter 18: p. 885, Science Photo Library/Photo Researchers, Inc.; p. 890, John Chard/Stone/Getty Images; p. 891, Larry Stepanowicz/Fundamental Photographs; p. 892, Ken O'Donoghue; p. 893 (both), Richard Megna/Fundamental Photographs/Photograph © Houghton Mifflin Company. All rights reserved; p. 895, Photograph © Houghton Mifflin Company. All rights reserved; p. 898 (top), The Advertizing Archive, London; p. 898 (bottom), Photograph © Houghton Mifflin Company. All rights reserved; p. 900, Photodisc Green/Getty Images, Royalty Free; p. 901, The Granger Collection; p. 903, Roger Ressmeyer/Corbis; p. 906, Hugh Spencer/Photo Researchers; p. 907, Photograph © Houghton Mifflin Company. All rights reserved; p. 908, Richard Megna/Fundamental Photographs; p. 911, Stephen Frisch/Stock Boston; p. 914, Eyebyte/Royalty Free Image/Alamy Images; p. 915, Fred J. Maroon/Photo Researchers; p. 916, E.R. Degginger/Color-Pic, Inc.; p. 917, Farrell Grehan/Photo Researchers; p. 918 (top left), Ken O'Donoghue/Photograph © Houghton Mifflin Company. All rights reserved; p. 918 (top right), E.R. Degginger/Color-Pic, Inc.; p. 918 (bottom), Ken O'Donoghue/Photograph © Houghton Mifflin Company. All rights reserved; p. 919, Photograph © Houghton Mifflin Company. All rights reserved; p. 920, Yoav Levy/Phototake; p. 934, AP Photo/Donna McWilliam/AP/Wide World Photos.

Chapter 19: p. 933, Walter Bibikow/Getty Images; p. 935 (top left), Paul Silverman/Fundamental Photographs; p. 935 (top right and bottom left), Photograph © Houghton Mifflin Company. All rights reserved; p. 935 (bottom right), Ken O'Donoghue/Photograph © Houghton Mifflin Company. All rights reserved; p. 936, Photograph © Houghton Mifflin Company. All rights reserved; p. 940, James King-Holmes/Science Photo Library/Photo Researchers; p. 941, Ken O'Donoghue/Photograph © Houghton Mifflin Company. All rights reserved; p. 942, Jodi Jacobson/Peter Arnold, Inc.; p. 943, John Cunningham/Visuals Unlimited; p. 945, Photograph © Houghton Mifflin Company. All rights reserved; p. 946, Keith Weller/USDA Photo; p. 950 (both), Photograph © Houghton Mifflin Company. All rights reserved; p. 954, Martin Bough/Fundamental Photographs/Photograph © Houghton Mifflin Company. All rights reserved; p. 965 (both), Courtesy, International Colored Gemstone Association; p. 972, Stanley Flegler/Visuals Unlimited; p. 974 (all), Photograph © Houghton Mifflin Company. All rights reserved.

Chapter 20: p. 981, NASA; p. 986, Courtesy, CERN/Geneva, Switzerland; p. 988, Jeff Hester/Arizona State University/NASA; p. 990, U.S. Department of Energy/Science Photo Library/Photo Researchers; p. 991, James A. Sugar/Corbis; p. 994, University Museum/University of Pennsylvania Photo Archives; p. 996 (both), SIU/Visuals Unlimited; p. 1003, Peter Arnold, Inc.; p. 1005, Courtesy, Fermilab Visual Media/Batavia, IL.

Chapter 21: p. 1013, Food Image Source/Getty Images; p. 1015, Iris Kappers/Koppert Biological Systems; p. 1028, Digital Vision/Royalty Free/Getty Images; p. 1030, AP Photo/Nati Harnik/AP/Wide World Photos; p. 1032, Inga Spence/Visuals Unlimited; p. 1034, Laguna Design/Science Photo Library/Photo Researchers; p. 1035 (top), AP Photo/Indianapolis Star/Karen Ducey/AP/Wide World Photos; p. 1035 (bottom), Phil Nelson, Woodinville, WA, p. 1036, Ron Boardman/Frank Lane Picture Agency/Corbis; p. 1037, AP Photo/Elise Amendola/AP/Wide World Photos; p. 1039, Photograph © Houghton Mifflin Company. All rights reserved; p. 1041, Courtesy Dupont/Dupont, Wilmington, Delaware; p. 1043, Scott White/University of Illinois, Urbana-Champaign; p. 1047, Ken Eward/Biografx/Science Source/Photo Researchers; p. 1052 (both), Photograph © Houghton Mifflin Company. All rights reserved; p. 1058, Alfred Pasieka/Science Source/Photo Researchers.

Index

Student Solutions

Student Solutions

CHAPTER 9

ENERGY, ENTHALPY, AND THERMOCHEMISTRY

The Nature of Energy

15. Ball A: $PE = mgz = 2.00\ kg \times \frac{9.81\ m}{s^2} \times 10.0\ m = \frac{196\ kg\ m^2}{s^2} = 196\ J$

    At point I: All this energy is transferred to ball B. All of B's energy is kinetic energy at this point. $E_{total} = KE = 196\ J$. At point II, the sum of the total energy will equal 196 J.

    At point II: $PE = mgz = 4.00\ kg \times \frac{9.81\ m}{s^2} \times 3.00\ m = 118\ J$

    $KE = E_{total} - PE = 196\ J - 118\ J = 78\ J$

17. Path-dependent functions for a trip from Chicago to Denver are those quantities that depend on the route taken. One can fly directly from Chicago to Denver or one could fly from Chicago to Atlanta to Los Angeles and then to Denver. Some path-dependent quantities are miles traveled, fuel consumption of the airplane, time traveling, airplane snacks eaten, etc. State functions are path independent; they only depend on the initial and final states. Some state functions for an airplane trip from Chicago to Denver would be longitude change, latitude change, elevation change, and overall time zone change.

19. Step 1: $\Delta E_1 = q + w = 72\ J + 35\ J = 107\ J$; step 2: $\Delta E_2 = 35\ J - 72\ J = -37\ J$

    $\Delta E_{overall} = \Delta E_1 + \Delta E_2 = 107\ J - 37\ J = 70.\ J$

21. $q = \text{molar heat capacity} \times \text{mol} \times \Delta T = \frac{20.8\ J}{^\circ C\ mol} \times 39.1\ mol \times (38.0 - 0.0)^\circ C = 30{,}900\ J$

    $= 30.9\ kJ$

    $w = -P\Delta V = -1.00\ atm \times (998\ L - 876\ L) = -122\ L\ atm \times \frac{101.3\ J}{L\ atm} = -12{,}400\ J = -12.4\ kJ$

    $\Delta E = q + w = 30.9\ kJ + (-12.4\ kJ) = 18.5\ kJ$

23. $H_2O(g) \rightarrow H_2O(l)$; $\Delta E = q + w$; $q = -40.66$ kJ; $w = -P\Delta V$

$$\text{Volume of 1 mol } H_2O(l) = 1.000 \text{ mol } H_2O(l) \times \frac{18.02 \text{ g}}{\text{mol}} \times \frac{1 \text{ cm}^3}{0.996 \text{ g}} = 18.1 \text{ cm}^3 = 18.1 \text{ mL}$$

$$w = -P\Delta V = -1.00 \text{ atm} \times (0.0181 \text{ L} - 30.6 \text{ L}) = 30.6 \text{ L atm} \times \frac{101.3 \text{ J}}{\text{L atm}} = 3.10 \times 10^3 \text{ J} = 3.10 \text{ kJ}$$

$$\Delta E = q + w = -40.66 \text{ kJ} + 3.10 \text{ kJ} = -37.56 \text{ kJ}$$

Properties of Enthalpy

25. $\Delta H = \Delta E + P\Delta V$ at constant P; from the definition of enthalpy, the difference between ΔH and ΔE at constant P is the quantity $P\Delta V$. Thus when a system at constant P can do pressure-volume work, then $\Delta H \neq \Delta E$. When the system cannot do PV work, then $\Delta H = \Delta E$ at constant pressure. An important way to differentiate ΔH from ΔE is to concentrate on q, the heat flow; the heat flow by a system at constant pressure equals ΔH, and the heat flow by a system at constant volume equals ΔE.

27. One should try to cool the reaction mixture or provide some means of removing heat since the reaction is very exothermic (heat is released). The $H_2SO_4(aq)$ will get very hot and possibly boil unless cooling is provided.

29. $4\ Fe(s) + 3\ O_2(g) \rightarrow 2\ Fe_2O_3(s)$ $\Delta H = -1652$ kJ; note that 1652 kJ of heat is released when 4 mol Fe reacts with 3 mol O_2 to produce 2 mol Fe_2O_3.

a. $4.00 \text{ mol Fe} \times \dfrac{-1652 \text{ kJ}}{4 \text{ mol Fe}} = -1650 \text{ kJ}$; 1650 kJ of heat released

b. $1.00 \text{ mol } Fe_2O_3 \times \dfrac{-1652 \text{ kJ}}{2 \text{ mol } Fe_2O_3} = -826 \text{ kJ}$; 826 kJ of heat released

c. $1.00 \text{ g Fe} \times \dfrac{1 \text{ mol Fe}}{55.85 \text{ g}} \times \dfrac{-1652 \text{ kJ}}{4 \text{ mol Fe}} = -7.39 \text{ kJ}$; 7.39 kJ of heat released

d. $10.0 \text{ g Fe} \times \dfrac{1 \text{ mol Fe}}{55.85 \text{ g}} = 0.179 \text{ mol Fe}$; $2.00 \text{ g } O_2 \times \dfrac{1 \text{ mol } O_2}{32.00 \text{ g}} = 0.0625 \text{ mol } O_2$

0.179 mol Fe/0.0625 mol O_2 = 2.86; the balanced equation requires a 4 mol Fe/3 mol O_2 = 1.33 mole ratio. O_2 is limiting since the actual mole Fe/mole O_2 ratio is greater than the required mole ratio.

$0.0625 \text{ mol } O_2 \times \dfrac{-1652 \text{ kJ}}{3 \text{ mol } O_2} = -34.4 \text{ kJ}$; 34.4 kJ of heat released

31. When a liquid is converted into gas, there is an increase in volume. The 2.5 kJ/mol quantity is the work done by the vaporization process in pushing back the atmosphere.

The Thermodynamics of Ideal Gases

33. Consider the constant volume process first.

$$n = 1.00 \times 10^3 \text{ g} \times \frac{1 \text{ mol}}{30.07 \text{ g}} = 33.3 \text{ mol } C_2H_6; \; C_v = \frac{44.60 \text{ J}}{\text{K mol}} = \frac{44.60 \text{ J}}{°\text{C mol}}$$

$\Delta E = nC_v\Delta T = (33.3 \text{ mol})(44.60 \text{ J °C}^{-1} \text{ mol}^{-1})(75.0 - 25.0°\text{C}) = 74{,}300 \text{ J} = 74.3 \text{ kJ}$

$\Delta E = q + w$; since $\Delta V = 0$, $w = 0$; $\Delta E = q_v = 74.3 \text{ kJ}$

$\Delta H = \Delta E + \Delta PV = \Delta E + nR\Delta T$

$\Delta H = 74.3 \text{ kJ} + (33.3 \text{ mol})(8.3145 \text{ J K}^{-1} \text{ mol}^{-1})(50.0 \text{ K})(1 \text{ kJ}/1000 \text{ J})$

$\Delta H = 74.3 \text{ kJ} + 13.8 \text{ kJ} = 88.1 \text{ kJ}$

Now consider the constant pressure process.

$q_p = \Delta H = nC_p\Delta T = (33.3 \text{ mol})(52.92 \text{ J K}^{-1} \text{ mol}^{-1})(50.0 \text{ K})$

$q_p = 88{,}100 \text{ J} = 88.1 \text{ kJ} = \Delta H$

$w = -P\Delta V = -nR\Delta T = -(33.3 \text{ mol})(8.3145 \text{ J K}^{-1} \text{ mol}^{-1})(50.0 \text{ K}) = -13{,}800 \text{ J} = -13.8 \text{ kJ}$

$\Delta E = q + w = 88.1 \text{ kJ} - 13.8 \text{ kJ} = 74.3 \text{ kJ}$

Summary	**Constant V**	**Constant P**
q	74.3 kJ	88.1 kJ
ΔE	74.3 kJ	74.3 kJ
ΔH	88.1 kJ	88.1 kJ
w	0	−13.8 kJ

35. Pathway I:

Step 1: (5.00 mol, 3.00 atm, 15.0 L) → (5.00 mol, 3.00 atm, 55.0 L)

$w = -P\Delta V = -(3.00 \text{ atm})(55.0 - 15.0 \text{ L}) = -120. \text{ L atm}$

$$w = -120. \text{ L atm} \times \frac{101.3 \text{ J}}{\text{L atm}} \times \frac{1 \text{ kJ}}{1000 \text{ J}} = -12.2 \text{ kJ}$$

$$\Delta H = q_p = nC_p\Delta T = nC_p \times \frac{\Delta(PV)}{nR} = \frac{C_p\Delta(PV)}{R}; \; \Delta(PV) = (P_2V_2 - P_1V_1)$$

For an ideal monatomic gas: $C_p = \frac{5}{2}R$; $\Delta H = \left(\frac{5}{2}R\right)\frac{\Delta(PV)}{R} = \frac{5}{2}\Delta(PV)$

$$\Delta H = q_p = \frac{5}{2}\Delta(PV) = \frac{5}{2}(165 - 45.0)\text{ L atm} = 300.\text{ L atm}$$

$$\Delta H = q_p = 300.\text{ L atm} \times \frac{101.3\text{ J}}{\text{L atm}} \times \frac{1\text{ kJ}}{1000\text{ J}} = 30.4\text{ kJ}$$

$$\Delta E = q + w = 30.4\text{ kJ} - 12.2\text{ kJ} = 18.2\text{ kJ}$$

Step 2: (5.00 mol, 3.00 atm, 55.0 L) → (5.00 mol, 6.00 atm, 20.0 L)

$$\Delta E = nC_v\Delta T = n\left(\frac{3}{2}R\right)\left(\frac{\Delta(PV)}{nR}\right) = \frac{3}{2}\Delta PV$$

$$\Delta E = \frac{3}{2}(120. - 165)\text{ L atm} = -67.5\text{ L atm}$$ (Carry an extra significant figure.)

$$\Delta E = -67.5\text{ L atm} \times \frac{101.3\text{ J}}{\text{L atm}} \times \frac{1\text{ kJ}}{1000\text{ J}} = -6.8\text{ kJ}$$

$$\Delta H = nC_p\Delta T = n\left(\frac{5}{2}R\right)\left(\frac{\Delta(PV)}{nR}\right) = \frac{5}{2}\Delta PV$$

$$\Delta H = \frac{5}{2}(-45\text{ L atm}) = -113\text{ L atm}$$ (Carry an extra significant figure.)

$$\Delta H = -113\text{ L atm} \times \frac{101.3\text{ J}}{\text{L atm}} \times \frac{1\text{ kJ}}{1000\text{ J}} = -11.4 = -11\text{ kJ}$$

$$w = -P_{ext}\Delta V = -(6.00\text{ atm})(20.0 - 55.0)\text{ L} = 210.\text{ L atm}$$

$$w = 210.\text{ L atm} \times \frac{101.3\text{ J}}{\text{L atm}} \times \frac{1\text{ kJ}}{1000\text{ J}} = 21.3\text{ kJ}$$

$$\Delta E = q + w,\ -6.8\text{ kJ} = q + 21.3\text{ kJ},\ q = -28.1\text{ kJ}$$

Summary:

Path I	Step 1	Step 2	Total
q	30.4 kJ	−28.1 kJ	2.3 kJ
w	−12.2 kJ	21.3 kJ	9.1 kJ
ΔE	18.2 kJ	−6.8 kJ	11.4 kJ
ΔH	30.4 kJ	−11 kJ	19 kJ

Pathway II:

Step 3: (5.00 mol, 3.00 atm, 15.0 L) $\rightarrow$ (5.00 mol, 6.00 atm, 15.0 L)

$$\Delta E = q_v = \frac{3}{2}\Delta(PV) = \frac{5}{2}(90.0 - 45.0)\text{ L atm} = 67.5\text{ L atm}$$

$$\Delta E = q_v = 67.5\text{ L atm} \times \frac{101.3\text{ J}}{\text{L atm}} \times \frac{1\text{ kJ}}{1000\text{ J}} = 6.84\text{ kJ}$$

$w = -P\Delta V = 0$ because $\Delta V = 0$

$\Delta H = \Delta E + \Delta(PV) = 67.5\text{ L atm} + 45.0\text{ L atm} = 112.5\text{ L atm} = 11.40\text{ kJ}$

Step 4: (5.00 mol, 6.00 atm, 15.0 L) $\rightarrow$ (5.00 mol, 6.00 atm, 20.0 L)

$$\Delta H = q_p = nC_p\Delta T = \left(\frac{5}{2}R\right)\left(\frac{\Delta(PV)}{nR}\right) = \frac{5}{2}\Delta PV$$

$$\Delta H = \frac{5}{2}(120. - 90.0)\text{ L atm} = 75\text{ L atm}$$

$$\Delta H = q_p = 75\text{ L atm} \times \frac{101.3\text{ J}}{\text{L atm}} \times \frac{1\text{ kJ}}{1000\text{ J}} = 7.6\text{ kJ}$$

$w = -P\Delta V = -(6.00\text{ atm})(20.0 - 15.0)\text{ L} = -30.\text{ L atm}$

$$w = -30.\text{ L atm} \times \frac{101.3\text{ J}}{\text{L atm}} \times \frac{1\text{ kJ}}{1000\text{ J}} = -3.0\text{ kJ}$$

$\Delta E = q + w = 7.6\text{ kJ} - 3.0\text{ kJ} = 4.6\text{ kJ}$

Summary:

Path II	**Step 3**	**Step 4**	**Total**
q	6.84 kJ	7.6 kJ	14.4 kJ
w	0	−3.0 kJ	−3.0 kJ
ΔE	6.84 kJ	4.6 kJ	11.4 kJ
ΔH	11.40 kJ	7.6 kJ	19.0 kJ

State functions are independent of the particular pathway taken between two states; path functions are dependent on the particular pathway. In this problem, the overall values of ΔH and ΔE for the two pathways are the same; hence ΔH and ΔE are state functions. The overall values of q and w for the two pathways are different; hence q and w are path functions.

Calorimetry and Heat Capacity

37. In calorimetry, heat flow is determined into or out of the surroundings. Because $\Delta E_{univ} = 0$ by the first law of thermodynamics, $\Delta E_{sys} = -\Delta E_{surr}$; what happens to the surroundings is the exact opposite of what happens to the system. To determine heat flow, we need to know the heat capacity of the surroundings, the mass of the surroundings that accepts/donates the heat, and the change in temperature. If we know these quantities, q_{surr} can be calculated and then equated to q_{sys} ($-q_{surr} = q_{sys}$). For an endothermic reaction, the surroundings (the calorimeter contents) donates heat to the system. This is accompanied by a decrease in temperature of the surroundings. For an exothermic reaction, the system donates heat to the surroundings (the calorimeter), so temperature increases.

    $q_P = \Delta H$; $q_V = \Delta E$; a coffee-cup calorimeter is at constant (atmospheric) pressure. The heat released or gained at constant pressure is ΔH. A bomb calorimeter is at constant volume. The heat released or gained at constant volume is ΔE.

39. Specific heat capacity is defined as the amount of heat necessary to raise the temperature of one gram of substance by one degree Celsius. Therefore, $H_2O(l)$ with the largest heat capacity value requires the largest amount of heat for this process. The amount of heat for $H_2O(l)$ is:

$$\text{energy} = s \times m \times \Delta T = \frac{4.18\text{ J}}{^\circ\text{C g}} \times 25.0\text{ g} \times (37.0^\circ\text{C} - 15.0^\circ\text{C}) = 2.30 \times 10^3\text{ J}$$

    The largest temperature change when a certain amount of energy is added to a certain mass of substance will occur for the substance with the smallest specific heat capacity. This is Hg(l), and the temperature change for this process is:

$$\Delta T = \frac{\text{energy}}{s \times m} = \frac{10.7\text{ kJ} \times \frac{1000\text{ J}}{\text{kJ}}}{\frac{0.14\text{ J}}{^\circ\text{C g}} \times 550.\text{ g}} = 140^\circ\text{C}$$

41. Heat loss by hot water = heat gain by cold water; keeping all quantities positive to avoid sign errors:

$$\frac{4.18\text{ J}}{^\circ\text{C g}} \times m_{hot} \times (55.0^\circ\text{C} - 37.0^\circ\text{C}) = \frac{4.18\text{ J}}{^\circ\text{C g}} \times 90.0\text{ g} \times (37.0\ ^\circ\text{C} - 22.0^\circ\text{C})$$

$$m_{hot} = \frac{90.0\text{ g} \times 15.0^\circ\text{C}}{18.0^\circ\text{C}} = 75.0\text{ g hot water needed}$$

43. 50.0×10^{-3} L × 0.100 mol/L = 5.00×10^{-3} mol of both $AgNO_3$ and HCl are reacted. Thus 5.00×10^{-3} mol of AgCl will be produced because there is a 1 : 1 mole ratio between reactants.

    Heat lost by chemicals = heat gained by solution

$$\text{Heat gain} = \frac{4.18\ \text{J}}{^\circ\text{C g}} \times 100.0\ \text{g} \times (23.40 - 22.60)^\circ\text{C} = 330\ \text{J}$$

Heat loss = 330 J; this is the heat evolved (exothermic reaction) when 5.00×10^{-3} mol of AgCl is produced. So q = −330 J and ΔH (heat per mol AgCl formed) is negative with a value of:

$$\Delta H = \frac{-330\ \text{J}}{5.00 \times 10^{-3}\ \text{mol}} \times \frac{1\ \text{kJ}}{1000\ \text{J}} = -66\ \text{kJ/mol}$$

Note: Sign errors are common with calorimetry problems. However, the correct sign for ΔH can be determined easily from the ΔT data; i.e., if ΔT of the solution increases, then the reaction is exothermic because heat was released, and if ΔT of the solution decreases, then the reaction is endothermic because the reaction absorbed heat from the water. For calorimetry problems, keep all quantities positive until the end of the calculation and then decide the sign for ΔH. This will help eliminate sign errors.

45. Because ΔH is exothermic, the temperature of the solution will increase as $CaCl_2(s)$ dissolves. Keeping all quantities positive:

$$\text{heat loss as } CaCl_2 \text{ dissolves} = 11.0\ \text{g } CaCl_2 \times \frac{1\ \text{mol } CaCl_2}{110.98\ \text{g } CaCl_2} \times \frac{81.5\ \text{kJ}}{\text{mol } CaCl_2} = 8.08\ \text{kJ}$$

$$\text{heat gained by solution} = 8.08 \times 10^3\ \text{J} = \frac{4.18\ \text{J}}{^\circ\text{C g}} \times (125 + 11.0)\ \text{g} \times (T_f - 25.0^\circ\text{C})$$

$$T_f - 25.0^\circ\text{C} = \frac{8.08 \times 10^3}{4.18 \times 136} = 14.2^\circ\text{C},\ \ T_f = 14.2^\circ\text{C} + 25.0^\circ\text{C} = 39.2^\circ\text{C}$$

47. $$\text{Heat gain by calorimeter} = \frac{1.56\ \text{kJ}}{^\circ\text{C}} \times 3.2^\circ\text{C} = 5.0\ \text{kJ} = \text{heat loss by quinone}$$

Heat loss = 5.0 kJ, which is the heat evolved (exothermic reaction) by the combustion of 0.1964 g of quinone. Because we are at constant volume, $q_v = \Delta E$.

$$\Delta E_{comb} = \frac{-5.0\ \text{kJ}}{0.1964\ \text{g}} = -25\ \text{kJ/g}; \qquad \Delta E_{comb} = \frac{-25\ \text{kJ}}{\text{g}} \times \frac{108.09\ \text{g}}{\text{mol}} = -2700\ \text{kJ/mol}$$

49. a. $C_{12}H_{22}O_{11}(s) + 12\ O_2(g) \rightarrow 12\ CO_2(g) + 11\ H_2O(l)$

b. A bomb calorimeter is at constant volume, so heat released = $q_v = \Delta E$:

$$\Delta E = \frac{-24.00\ \text{kJ}}{1.46\ \text{g}} \times \frac{342.30\ \text{g}}{\text{mol}} = -5630\ \text{kJ/mol } C_{12}H_{22}O_{11}$$

c. ΔH = ΔE + Δ(PV) = ΔE + Δ(nRT) = ΔE + ΔnRT, where Δn = moles of gaseous products − moles of gaseous reactants.

For this reaction, Δn = 12 −12 = 0, so ΔH = ΔE = −5630 kJ/mol.

Hess's Law

51.

$2\ C + 2\ O_2 \rightarrow 2\ CO_2$	$\Delta H = 2(-394\ kJ)$
$H_2 + 1/2\ O_2 \rightarrow H_2O$	$\Delta H = -286\ kJ$
$2\ CO_2 + H_2O \rightarrow C_2H_2 + 5/2\ O_2$	$\Delta H = -(-1300.kJ)$
$2\ C(s) + H_2(g) \rightarrow C_2H_2(g)$	$\Delta H = 226\ kJ$

Note: The enthalpy change for a reaction that is reversed is the negative quantity of the enthalpy change for the original reaction. If the coefficients in a balanced reaction are multiplied by an integer, then the value of ΔH is multiplied by the same integer.

53.

$CaC_2 \rightarrow Ca + 2\ C$	$\Delta H = -(-62.8\ kJ)$
$CaO + H_2O \rightarrow Ca(OH)_2$	$\Delta H = -653.1\ kJ$
$2\ CO_2 + H_2O \rightarrow C_2H_2 + 5/2\ O_2$	$\Delta H = -(-1300.\ kJ)$
$Ca + 1/2\ O_2 \rightarrow CaO$	$\Delta H = -635.5\ kJ$
$2\ C + 2\ O_2 \rightarrow 2\ CO_2$	$\Delta H = 2(-393.5\ kJ)$
$CaC_2(s) + 2\ H_2O(l) \rightarrow Ca(OH)_2(aq) + C_2H_2(g)$	$\Delta H = -713\ kJ$

55.

$C_4H_4(g) + 5\ O_2(g) \rightarrow 4\ CO_2(g) + 2\ H_2O(l)$	$\Delta H_{comb} = -2341\ kJ$
$C_4H_8(g) + 6\ O_2(g) \rightarrow 4\ CO_2(g) + 4\ H_2O(l)$	$\Delta H_{comb} = -2755\ kJ$
$H_2(g) + 1/2\ O_2(g) \rightarrow H_2O(l)$	$\Delta H_{comb} = -286\ kJ$

By convention, $H_2O(l)$ is produced when enthalpies of combustion are given, and because per mole quantities are given, the combustion reaction refers to 1 mole of that quantity reacting with $O_2(g)$.

Using Hess's Law to solve:

$C_4H_4(g) + 5\ O_2(g) \rightarrow 4\ CO_2(g) + 2\ H_2O(l)$	$\Delta H_1 = -2341\ kJ$
$4\ CO_2(g) + 4\ H_2O(l) \rightarrow C_4H_8(g) + 6\ O_2(g)$	$\Delta H_2 = -(-2755\ kJ)$
$2\ H_2(g) + O_2(g) \rightarrow 2\ H_2O(l)$	$\Delta H_3 = 2(-286\ kJ)$
$C_4H_4(g) + 2\ H_2(g) \rightarrow C_4H_8(g)$	$\Delta H = \Delta H_1 + \Delta H_2 + \Delta H_3 = -158\ kJ$

57.

$C_6H_4(OH)_2 \rightarrow C_6H_4O_2 + H_2$	$\Delta H = 177.4\ kJ$
$H_2O_2 \rightarrow H_2 + O_2$	$\Delta H = -(-191.2\ kJ)$
$2\ H_2 + O_2 \rightarrow 2\ H_2O(g)$	$\Delta H = 2(-241.8\ kJ)$
$2\ H_2O(g) \rightarrow 2\ H_2O(l)$	$\Delta H = 2(-43.8\ kJ)$
$C_6H_4(OH)_2(aq) + H_2O_2(aq) \rightarrow C_6H_4O_2(aq) + 2\ H_2O(l)$	$\Delta H = -202.6\ kJ$

59.

$2\ N_2(g) + 6\ H_2(g) \rightarrow 4\ NH_3(g)$	$\Delta H = -4(46\ kJ)$
$6\ H_2O(g) \rightarrow 6\ H_2(g) + 3\ O_2(g)$	$\Delta H = -3(-484\ kJ)$
$2\ N_2(g) + 6\ H_2O(g) \rightarrow 3\ O_2(g) + 4\ NH_3(g)$	$\Delta H = 1268\ kJ$

No, because the reaction is very endothermic (requires a lot of heat to react), it would not be a practical way of making ammonia because of the high energy costs required.

Standard Enthalpies of Formation

61. In general: $\Delta H° = \sum n_p \Delta H^o_{f,\ products} - \sum n_r \Delta H^o_{f,\ reactants}$, and all elements in their standard state have $\Delta H^o_f = 0$ by definition.

a. The balanced equation is: $2\ NH_3(g) + 3\ O_2(g) + 2\ CH_4(g) \rightarrow 2\ HCN(g) + 6\ H_2O(g)$

$$\Delta H° = (2\text{ mol HCN} \times \Delta H^o_{f,\ HCN} + 6\text{ mol } H_2O(g) \times \Delta H^o_{f,\ H_2O}) - (2\text{ mol } NH_3 \times \Delta H^o_{f,\ NH_3} + 2\text{ mol } CH_4 \times \Delta H^o_{f,\ CH_4})$$

$$\Delta H° = [2(135.1) + 6(-242)] - [2(-46) + 2(-75)] = -940.\text{ kJ}$$

b. $Ca_3(PO_4)_2(s) + 3\ H_2SO_4(l) \rightarrow 3\ CaSO_4(s) + 2\ H_3PO_4(l)$

$$\Delta H° = \left[3\text{ mol } CaSO_4(s)\left(\frac{-1433\text{ kJ}}{\text{mol}}\right) + 2\text{ mol } H_3PO_4(l)\left(\frac{-1267\text{ kJ}}{\text{mol}}\right)\right] - \left[1\text{ mol } Ca_3(PO_4)_2(s)\left(\frac{-4126\text{ kJ}}{\text{mol}}\right) + 3\text{ mol } H_2SO_4(l)\left(\frac{-814\text{ kJ}}{\text{mol}}\right)\right]$$

$$\Delta H° = -6833\text{ kJ} - (-6568\text{ kJ}) = -265\text{ kJ}$$

c. $NH_3(g) + HCl(g) \rightarrow NH_4Cl(s)$

$$\Delta H° = (1\text{ mol } NH_4Cl \times \Delta H^o_{f,\ NH_4Cl}) - (1\text{ mol } NH_3 \times \Delta H^o_{f,\ NH_3} + 1\text{ mol HCl} \times \Delta H^o_{f,\ HCl})$$

$$\Delta H° = \left[1\text{ mol}\left(\frac{-314\text{ kJ}}{\text{mol}}\right)\right] - \left[1\text{ mol}\left(\frac{-46\text{ kJ}}{\text{mol}}\right) + 1\text{ mol}\left(\frac{-92\text{ kJ}}{\text{mol}}\right)\right]$$

$$\Delta H° = -314\text{ kJ} + 138\text{ kJ} = -176\text{ kJ}$$

d. The balanced equation is: $C_2H_5OH(l) + 3\ O_2(g) \rightarrow 2\ CO_2(g) + 3\ H_2O(g)$

$$\Delta H° = \left[2\text{ mol}\left(\frac{-393.5\text{ kJ}}{\text{mol}}\right) + 3\text{ mol}\left(\frac{-242\text{ kJ}}{\text{mol}}\right)\right] - \left[1\text{ mol}\left(\frac{-278\text{ kJ}}{\text{mol}}\right)\right]$$

$$\Delta H° = -1513\text{ kJ} - (-278\text{ kJ}) = -1235\text{ kJ}$$

e. $SiCl_4(l) + 2\ H_2O(l) \rightarrow SiO_2(s) + 4\ HCl(aq)$

Because HCl(aq) is $H^+(aq) + Cl^-(aq)$, $\Delta H_f^\circ = 0 - 167 = -167$ kJ/mol.

$$\Delta H^\circ = \left[4\text{ mol}\left(\frac{-167\text{ kJ}}{\text{mol}}\right) + 1\text{ mol}\left(\frac{-911\text{ kJ}}{\text{mol}}\right)\right] - \left[1\text{ mol}\left(\frac{-687\text{ kJ}}{\text{mol}}\right) + 2\text{ mol}\left(\frac{-286\text{ kJ}}{\text{mol}}\right)\right]$$

$\Delta H^\circ = -1579\text{ kJ} - (-1259\text{ kJ}) = -320.\text{ kJ}$

f. $MgO(s) + H_2O(l) \rightarrow Mg(OH)_2(s)$

$$\Delta H^\circ = \left[1\text{ mol}\left(\frac{-925\text{ kJ}}{\text{mol}}\right)\right] - \left[1\text{ mol}\left(\frac{-602\text{ kJ}}{\text{mol}}\right) + 1\text{ mol}\left(\frac{-286\text{ kJ}}{\text{mol}}\right)\right]$$

$\Delta H^\circ = -925\text{ kJ} - (-888\text{ kJ}) = -37\text{ kJ}$

63. $4\ Na(s) + O_2(g) \rightarrow 2\ Na_2O(s)$, $\Delta H^\circ = 2\text{ mol}\left(\frac{-416\text{ kJ}}{\text{mol}}\right) = -832\text{ kJ}$

$2\ Na(s) + 2\ H_2O(l) \rightarrow 2\ NaOH(aq) + H_2(g)$

$$\Delta H^\circ = \left[2\text{ mol}\left(\frac{-470.\text{ kJ}}{\text{mol}}\right)\right] - \left[2\text{ mol}\left(\frac{-286\text{ kJ}}{\text{mol}}\right)\right] = -368\text{ kJ}$$

$2Na(s) + CO_2(g) \rightarrow Na_2O(s) + CO(g)$

$$\Delta H^\circ = \left[1\text{ mol}\left(\frac{-416\text{ kJ}}{\text{mol}}\right) + 1\text{ mol}\left(\frac{-110.5\text{ kJ}}{\text{mol}}\right)\right] - \left[1\text{ mol}\left(\frac{-393.5\text{ kJ}}{\text{mol}}\right)\right] - 133\text{ kJ}$$

In reactions 2 and 3, sodium metal reacts with the "extinguishing agent." Both reactions are exothermic and each reaction produces a flammable gas, H_2 and CO, respectively.

65. $5\ N_2O_4(l) + 4\ N_2H_3CH_3(l) \rightarrow 12\ H_2O(g) + 9\ N_2(g) + 4\ CO_2(g)$

$$\Delta H^\circ = \left[12\text{ mol}\left(\frac{-242\text{ kJ}}{\text{mol}}\right) + 4\text{ mol}\left(\frac{-393.5\text{ kJ}}{\text{mol}}\right)\right] - \left[5\text{ mol}\left(\frac{-20.\text{ kJ}}{\text{mol}}\right) + 4\text{ mol}\left(\frac{54\text{ kJ}}{\text{mol}}\right)\right] = -4594\text{ kJ}$$

67. a. $\Delta H° = 3\ mol(227\ kJ/mol) - 1\ mol(49\ kJ/mol) = 632\ kJ$

b. Because 3 $C_2H_2(g)$ is higher in energy than $C_6H_6(l)$, acetylene will release more energy per gram when burned in air.

69. $2\ ClF_3(g) + 2\ NH_3(g) \rightarrow N_2(g) + 6\ HF(g) + Cl_2(g) \quad \Delta H° = -1196\ kJ$

$$\Delta H° = (6\,\Delta H^\circ_{f,\ HF}) - (2\,\Delta H^\circ_{f,\ ClF_3} + 2\Delta H^\circ_{f,\ NH_3})$$

$$-1196\ kJ = 6\ mol\left(\frac{-271\ kJ}{mol}\right) - 2\,\Delta H^\circ_{f,\ ClF_3} - 2\ mol\left(\frac{-46\ kJ}{mol}\right)$$

$$-1196\ kJ = -1626\ kJ - 2\,\Delta H^\circ_{f,\ ClF_3} + 92\ kJ,\ \ \Delta H^\circ_{f,\ ClF_3} = \frac{(-1626 + 92 + 1196)\ kJ}{2\ mol} = \frac{-169\ kJ}{mol}$$

Energy Consumption and Sources

71. Mass of $H_2O = 1.00\ gal \times \frac{3.785\ L}{gal} \times \frac{1000\ mL}{L} \times \frac{1.00\ g}{mL} = 3790\ g\ H_2O$

Energy required (theoretical) $= s \times m \times \Delta T = \frac{4.18\ J}{°C\ g} \times 3790\ g \times 10.0\ °C = 1.58 \times 10^5\ J$

For an actual (80.0% efficient) process, more than this quantity of energy is needed since heat is always lost in any transfer of energy. The energy required is:

$$1.58 \times 10^5\ J \times \frac{100.\ J}{80.0\ J} = 1.98 \times 10^5\ J$$

Mass of $C_2H_2 = 1.98 \times 10^5\ J \times \frac{1\ mol\ C_2H_2}{1300. \times 10^3\ J} \times \frac{26.04\ g\ C_2H_2}{mol\ C_2H_2} = 3.97\ g\ C_2H_2$

73. $CH_3OH(l) + 3/2\ O_2(g) \rightarrow CO_2(g) + 2\ H_2O(l)$

$\Delta H° = [-393.5\ kJ + 2(-286\ kJ)] - (-239\ kJ) = -727\ kJ/mol\ CH_3OH$

$$\frac{-727\ kJ}{mol} \times \frac{1\ mol}{32.04\ g} = -22.7\ kJ/g \text{ versus } -29.67\ kJ/g \text{ for ethanol}$$

Ethanol has a higher fuel value than methanol.

Additional Exercises

75. $\Delta E_{overall} = \Delta E_{step\ 1} + \Delta E_{step\ 2}$; this is a cyclic process, which means that the overall initial state and final state are the same. Because ΔE is a state function, $\Delta E_{overall} = 0$ and $\Delta E_{step\ 1} = -\Delta E_{step\ 2}$.

$\Delta E_{step\ 1} = q + w = 45\ J + (-10.\ J) = 35\ J$

$\Delta E_{step\ 2} = -\Delta E_{step\ 1} = -35\ J = q + w,\ \ -35\ J = -60.\ J +\ w,\ \ w = 25\ J$

77. $H_2(g) + 1/2\ O_2(g) \rightarrow H_2O(l)\quad \Delta H° = \Delta H^{o}_{f,\ H_2O(l)} = -285.8\ kJ$

$H_2O(l) \rightarrow\ H_2(g) + 1/2\ O_2(g)\quad \Delta H° = 285.8\ kJ$

$\Delta E° = \Delta H° - P\Delta V = \Delta H° - \Delta nRT$

$$\Delta E° = 285.8\ kJ - (1.50 - 0\ mol)(8.3145\ J\ K^{-1}\ mol^{-1})(298\ K)\left(\frac{1\ kJ}{1000\ J}\right)$$

$\Delta E° = 285.8\ kJ - 3.72\ kJ = 282.1\ kJ$

79. The specific heat of water is 4.18 J °C^{-1} g^{-1}, which is equal to 4.18 kJ °C^{-1} kg^{-1}

We have 1.00 kg of H_2O, so: $1.00\ kg \times \dfrac{4.18\ kJ}{^{o}C\ kg} = 4.18\ kJ/°C$

This is the portion of the heat capacity that can be attributed to H_2O.

Total heat capacity = $C_{cal} + C_{H_2O}$, $\quad C_{cal} = 10.84 - 4.18 = 6.66\ kJ/°C$

81. $Na_2SO_4(aq) + Ba(NO_3)_2(aq) \rightarrow\ BaSO_4(s) + 2\ NaNO_3(aq)\qquad \Delta H = ?$

$$1.00\ L \times \frac{2.00\ mol}{L} = 2.00\ mol\ Na_2SO_4;\ \ 2.00\ L \times \frac{0.750\ mol}{L} = 1.50\ mol\ Ba(NO_3)_2$$

The balanced equation requires a 1 : 1 mole ratio between Na_2SO_4 and $Ba(NO_3)_2$. Because we have fewer moles of $Ba(NO_3)_2$ present, it is limiting and 1.50 mol $BaSO_4$ will be produced [there is a 1 : 1 mole ratio between $Ba(NO_3)_2$ and $BaSO_4$].

Heat gain by solution = heat loss by reaction

$$\text{Mass of solution} = 3.00\ L\ \times \frac{1000\ mL}{1\ L} \times \frac{2.00\ g}{mL} = 6.00 \times 10^{3}\ g$$

$$\text{Heat gain by solution} = \frac{6.37\ J}{^{o}C\ g} \times 6.00 \times 10^{3}\ g \times (42.0 - 30.0)°C = 4.59 \times 10^{5}\ J$$

Because the solution gained heat, the reaction is exothermic; $q = -4.59 \times 10^{5}$ J for the reaction.

$$\Delta H = \frac{-4.59 \times 10^{5}\ J}{1.50\ mol\ BaSO_4} = -3.06 \times 10^{5}\ J/mol = -306\ kJ/mol$$

83. $HCl(aq) + NaOH(aq) \rightarrow H_2O(l) + NaCl(aq) \quad \Delta H = -56\ kJ$

$$0.2000\ L \times \frac{0.400\ mol\ HCl}{L} = 8.00 \times 10^{-2}\ mol\ HCl$$

$$0.1500\ L \times \frac{0.500\ mol\ NaOH}{L} = 7.50 \times 10^{-2}\ mol\ NaOH$$

Because the balanced reaction requires a 1 : 1 mole ratio between HCl and NaOH, and because fewer moles of NaOH are actually present as compared with HCl, NaOH is the limiting reagent.

$$7.50 \times 10^{-2}\ mol\ NaOH \times \frac{-56\ kJ}{mol\ NaOH} = -4.2\ kJ;$$ 4.2 kJ of heat is released.

85. $w = -P\Delta V$; Δn = moles of gaseous products – moles of gaseous reactants. Only gases can do PV work (we ignore solids and liquids). When a balanced reaction has more moles of product gases than moles of reactant gases (Δn positive), the reaction will expand in volume (ΔV positive), and the system will do work on the surroundings. For example, in reaction c, $\Delta n = 2 - 0 = 2$ moles, and this reaction would do expansion work against the surroundings. When a balanced reaction has a decrease in the moles of gas from reactants to products (Δn negative), the reaction will contract in volume (ΔV negative), and the surroundings will do compression work on the system, e.g., reaction a, where $\Delta n = 0 - 1 = -1$. When there is no change in the moles of gas from reactants to products, $\Delta V = 0$ and $w = 0$, e.g., reaction b, where $\Delta n = 2 - 2 = 0$.

When $\Delta V > 0$ ($\Delta n > 0$), then $w < 0$, and the system does work on the surroundings (c and e).

When $\Delta V < 0$ ($\Delta n < 0$), then $w > 0$, and the surroundings do work on the system (a and d).

When $\Delta V = 0$ ($\Delta n = 0$), then $w = 0$ (b).

87. a. aluminum oxide = Al_2O_3; $2\ Al(s) + 3/2\ O_2(g) \rightarrow Al_2O_3(s)$

b. $C_2H_5OH(l) + 3\ O_2(g) \rightarrow 2\ CO_2(g) + 3\ H_2O(l)$

c. $Ba(OH)_2(aq) + 2\ HCl(aq) \rightarrow 2\ H_2O(l) + BaCl_2(aq)$

d. $2\ C(graphite, s) + 3/2\ H_2(g) + 1/2\ Cl_2(g) \rightarrow C_2H_3Cl(g)$

e. $C_6H_6(l) + 15/2\ O_2(g) \rightarrow 6\ CO_2(g) + 3\ H_2O(l)$

Note: ΔH_{comb} values assume 1 mole of compound combusted.

f. $NH_4Br(s) \rightarrow NH_4^+(aq) + Br^-(aq)$

Challenge Problems

89. $H_2O(s) \rightarrow H_2O(l)$ $\Delta H = \Delta H_{fus}$; for 1 mol of supercooled water at −15.0°C (or 258.2 K), $\Delta H_{fus,\,258.2\,K} = 10.9$ kJ/2.00 mol = 5.45 kJ/mol. Using Hess's law and the equation $\Delta H = nC_p\Delta T$:

$H_2O(s, 273.2\ K) \rightarrow H_2O(s, 258.2\ K)$ $\quad \Delta H_1 = 1\ mol(37.5\ J\ K^{-1}\ mol^{-1})(-15.0\ K) = -563\ J = -0.563\ kJ$

$H_2O(s, 258.2\ K) \rightarrow H_2O(l, 258.2\ K)$ $\quad \Delta H_2 = 1\ mol(5.45\ kJ/mol) = 5.45\ kJ$

$H_2O(l, 258.2\ K) \rightarrow H_2O(l, 273.2\ K)$ $\quad \Delta H_3 = 1\ mol(75.3\ J\ K^{-1}\ mol^{-1})(15.0\ K) = 1130\ J = 1.13\ kJ$

$H_2O(s, 273.\ 2\ K) \rightarrow H_2O(l, 273.2\ K)$ $\quad \Delta H_{fus,\,273.2} = \Delta H_1 + \Delta H_2 + \Delta H_3$

$\Delta H_{fus,\,273.2} = -0.563\ kJ + 5.45\ kJ + 1.13\ kJ$, $\quad \Delta H_{fus,\,273.2} = 6.02\ kJ/mol$

91. Molar heat capacity of $H_2O(l) = 4.184\ J\ K^{-1}\ g^{-1}(18.015\ g/mol) = 75.37\ J\ K^{-1}\ mol^{-1}$

Molar heat capacity of $H_2O(g) = 2.02\ J\ K^{-1}\ g^{-1}(18.015\ g/mol) = 36.4\ J\ K^{-1}\ mol^{-1}$

Using Hess's law and the equation $\Delta H = nC_p\Delta T$:

$H_2O(l, 298.2\ K) \rightarrow H_2O(l, 373.2\ K)$ $\quad \Delta H_1 = 1\ mol(75.37\ J\ K^{-1}\ mol^{-1})(75.0\ K)(1\ kJ/1000\ J) = 5.65\ kJ$

$H_2O(l, 373.2\ K) \rightarrow H_2O(g, 373.2\ K)$ $\quad \Delta H_2 = 1\ mol(40.66\ kJ/mol) = 40.66\ kJ$

$H_2O(g, 373.2\ K) \rightarrow H_2O(g, 298.2\ K)$ $\quad \Delta H_3 = 1\ mol(36.4\ J\ K^{-1}\ mol^{-1})(-75.0\ K)(1\ kJ/1000\ J) = -2.73\ kJ$

$H_2O(l, 298.2\ K) \rightarrow H_2O(g, 298.2\ K)$ $\quad \Delta H_{vap,\,298.2\,K} = \Delta H_1 + \Delta H_2 + \Delta H_3 = 43.58\ kJ/mol$

Using ΔH_f° values in Appendix 4 (which are determined at 25° C):

$\Delta H^\circ_{vap} = -242\ kJ - (-286\ kJ) = 44\ kJ$

To two significant figures, the two calculated ΔH_{vap} values agree (as they should).

93. Energy used in 8.0 hours = 40. kWh = $\frac{40.\ kJ\ h}{s} \times \frac{3600\ s}{h} = 1.4 \times 10^5\ kJ$

Energy from the sun in 8.0 hours = $\frac{10.\ kJ}{s\ m^2} \times \frac{60\ s}{min} \times \frac{60\ min}{h} \times 8.0\ h = 2.9 \times 10^4\ kJ/m^2$

Only 13% of the sunlight is converted into electricity:

$0.13 \times (2.9 \times 10^4\ kJ/m^2) \times area = 1.4 \times 10^5\ kJ$, $\quad area = 37\ m^2$

95. For an isothermal (constant T) process involving the expansion or compression of a gas, $\Delta E = nC_v\Delta T = 0$ (ΔH is also zero). Because $\Delta E = q + w = 0$, $q = -w = -(-P_{ex}\Delta V)$. So if the expansion or compression occurs against some nonzero external pressure, $w \neq 0$ and $q \neq 0$. Instead, $q = -w = P_{ex}\Delta V$ (if the gas expands or contracts against some constant pressure).

CHAPTER 10

SPONTANEITY, ENTROPY, AND FREE ENERGY

Spontaneity and Entropy

13. We draw all the possible arrangements of the two particles in the three levels.

2 kJ	__	__	x_	__	x_	xx_
1 kJ	__	x_	__	xx_	x_	__
0 kJ	xx_	x_	x_	__	__	__
Total E =	0 kJ	1 kJ	2 kJ	2 kJ	3 kJ	4 kJ

The most likely total energy is 2 kJ.

15. Processes a, b, d, and g are spontaneous. Processes c, e, and f require an external source of energy in order to occur since they are nonspontaneous.

17. a. Positional probability increases; there is a greater volume accessible to the randomly moving gas molecules, which increases disorder.

    b. The positional probability doesn't change. There is no change in volume and thus no change in the numbers of positions of the molecules.

    c. Positional probability decreases because the volume decreases (P and V are inversely related).

19. There are more ways to roll a seven. We can consider all the possible throws by constructing a table.

One die	1	2	3	4	5	6
1	2	3	4	5	6	7
2	3	4	5	6	7	8
3	4	5	6	7	8	9
4	5	6	7	8	9	10
5	6	7	8	9	10	11
6	7	8	9	10	11	12

Sum of the two dice

There are six ways to get a seven, more than any other number. The seven is not favored by energy; rather, it is favored by probability. To change the probability, we would have to expend energy (do work).

Energy, Enthalpy, and Entropy Changes Involving Ideal Gases and Physical Changes

21. $1.00 \times 10^3 \text{ g } C_2H_6 \times \dfrac{1 \text{ mol}}{30.07 \text{ g}} = 33.3 \text{ mol}$

$q_v = \Delta E = nC_v\Delta T = 33.3 \text{ mol}(44.60 \text{ J K}^{-1} \text{ mol}^{-1})(48.4 \text{ K}) = 7.19 \times 10^4 \text{ J} = 71.9 \text{ kJ}$

At constant volume, 71.9 kJ of energy is required, and $\Delta E = 71.9$ kJ.

At constant pressure (assuming ethane acts as an ideal gas):

$C_p = C_v + R = 44.60 + 8.31 = 52.91 \text{ J K}^{-1} \text{ mol}^{-1}$

Energy required $= q_p = \Delta H = nC_p\Delta T = 33.3 \text{ mol}(52.91 \text{ J K}^{-1} \text{ mol}^{-1})(48.4 \text{ K}) = 8.53 \times 10^4 \text{ J} = 85.3 \text{ kJ}$

For the constant pressure process, $\Delta E = 71.9$ kJ, as calculated previously (ΔE is unchanged).

23. Heat gain by He = heat loss by N_2; because ΔT in °C = ΔT in K, the units on the heat capacities also could be J °C^{-1} mol^{-1}.

$(0.400 \text{ mol})(12.5 \text{ J °C}^{-1} \text{ mol}^{-1})(T_f - 20.0 \text{ °C}) = (0.600 \text{ mol})(20.7 \text{ J °C}^{-1} \text{ mol}^{-1})(100.0 \text{ °C} - T_f)$

$(5.00)T_f - 100. = 1240 - (12.4)T_f, \quad T_f = \dfrac{1340}{17.4} = 77.0 \text{ °C}$

25. $P_1V_1 = P_2V_2$ since n and T are constant; $P_2 = \dfrac{P_1V_1}{V_2} = \dfrac{(5.0)(1.0)}{(2.0)} = 2.5 \text{ atm}$

Gas expands isothermally against no pressure, so $\Delta E = 0$, $w = 0$, and $q = 0$.

$\Delta E = 0$, so $q_{rev} = -w_{rev} = nRT \ln(V_2/V_1)$; $T = \dfrac{PV}{nR} = 61 \text{ K}$

$q_{rev} = (1.0 \text{ mol})(8.3145 \text{ J K}^{-1} \text{ mol}^{-1})(61 \text{ K}) \ln(2.0/1.0) = 350 \text{ J}$

27. a. $q_v = \Delta E = nC_v\Delta T = (1.000 \text{ mol})(28.95 \text{ J K}^{-1} \text{ mol}^{-1})(350.0 - 298.0 \text{ K})$

$q_v = 1.51 \times 10^3 \text{ J} = 1.51 \text{ kJ}$

$q_p = \Delta H = nC_p\Delta T = (1.000)(37.27)(350.0 - 298.0) = 1.94 \times 10^3 \text{ J} = 1.94 \text{ kJ}$

b. $\Delta S = S_{350} - S_{298} = nC_p \ln(T_2/T_1)$

$S_{350} - 213.64\ J/K = (1.000\ mol)(37.27\ J\ K^{-1}\ mol^{-1}) \ln(350.0/298.0)$

$S_{350} = 213.64\ J/K + 5.994\ J/K = 219.63\ J/K$ = molar entropy at 350.0 K and 1.000 atm

c. $\Delta S = nR \ln(V_2/V_1)$, $V = nRT/P$, $\Delta S = nR \ln(P_1/P_2) = S_{(350,\ 1.174)} - S_{(350,\ 1.000)}$

$\Delta S = S_{(350,\ 1.174)} - 219.63\ J/K = (1.000\ mol)(8.3145\ J\ K^{-1}\ mol^{-1}) \ln(1.000\ atm/1.174\ atm)$

$\Delta S = -1.334\ J/K = S_{(350,\ 1.174)} - 219.63$, $S_{(350,\ 1.174)} = 218.30\ J\ K^{-1}\ mol^{-1}$

29. For A(l, 125°C) → A(l, 75°C):

$\Delta S = nC_p \ln(T_2/T_1) = 1.00\ mol(75.0\ J\ K^{-1}\ mol^{-1}) \ln(348\ K/398\ K) = -10.1\ J/K$

For A(l, 75°C) → A(g, 155°C): $\Delta S = 75.0\ J\ K^{-1}\ mol^{-1}$

For A(g, 155°C) → A(g,125°C):

$\Delta S = nC_p \ln(T_2/T_1) = 1.00\ mol(29.0\ J\ K^{-1}\ mol^{-1}) \ln(398\ K/428\ K) = -2.11\ J/K$

The sum of the three step gives A(l, 125°C) → A(g, 125°C). ΔS for this process is the sum of ΔS for each of the three steps.

$\Delta S = -10.1 + 75.0 - 2.11 = 62.8\ J/K$

For a phase change, $\Delta S = \Delta H/T$. At 125°C: $\Delta H_{vap} = T\Delta S = 398\ K(62.8\ J/K) = 2.50 \times 10^4\ J$

31. Calculate the final temperature by equating heat loss to heat gain. Keep all quantities positive to avoid sign errors.

$(3.00\ mol)(75.3\ J\ °C^{-1}\ mol^{-1})(T_f - 0°C) = (1.00\ mol)(75.3\ J\ °C^{-1}\ mol^{-1})(100.°C - T_f)$

Solving: $T_f = 25°C = 298\ K$

Now we can calculate ΔS for the various changes using $\Delta S = nC_p \ln (T_2/T_1)$.

Heat 3 mol H_2O: $\Delta S_1 = (3.00\ mol)(75.3\ J\ K^{-1}\ mol^{-1}) \ln(298\ K/273\ K) = 19.8\ J/K$

Cool 1 mol H_2O: $\Delta S_2 = (1.00\ mol)(75.3\ J\ K^{-1}\ mol^{-1}) \ln(298/373) = -16.9\ J/K$

$\Delta S_{total} = \Delta S_{heat} + \Delta S_{cool} = 19.8 - 16.9 = 2.9\ J/K$

Entropy and the Second Law of Thermodynamics: Free Energy

33. Living organisms need an external source of energy to carry out these processes. Green plants use the energy from sunlight to produce glucose from carbon dioxide and water by

photosynthesis. In the human body the energy released from the metabolism of glucose helps drive the synthesis of proteins. For all processes combined, ΔS_{univ} must be greater than zero (second law).

35. ΔS_{surr} is primarily determined by heat flow. This heat flow into or out of the surroundings comes from the heat flow out of or into the system. In an exothermic process ($\Delta H < 0$), heat flows into the surroundings from the system. The heat flow into the surroundings increases the random motions in the surroundings and increases the entropy of the surroundings ($\Delta S_{surr} > 0$). This is a favorable driving force for spontaneity. In an endothermic reaction ($\Delta H > 0$), heat is transferred from the surroundings into the system. This heat flow out of the surroundings decreases the random motions in the surroundings and decreases the entropy of the surroundings ($\Delta S_{surr} < 0$). This is unfavorable. The magnitude of ΔS_{surr} also depends on the temperature. The relationship is inverse; at low temperatures, a specific amount of heat exchange makes a larger percent change in the surroundings than the same amount of heat flow at a higher temperature. The negative sign in the $\Delta S_{surr} = -\Delta H/T$ equation is necessary to get the signs correct. For an exothermic reaction where ΔH is negative, this increases ΔS_{surr}, so the negative sign converts the negative ΔH value into a positive quantity. For an endothermic process where ΔH is positive, the sign of ΔS_{surr} is negative, and the negative sign converts the positive ΔH value into a negative quantity.

37. a. $\Delta S_{surr} = \dfrac{-\Delta H}{T} = \dfrac{-(-2221\text{ kJ})}{298\text{ K}} = 7.45\text{ kJ/K} = 7.45 \times 10^3\text{ J/K}$

b. $\Delta S_{surr} = \dfrac{-\Delta H}{T} = \dfrac{-112\text{ kJ}}{298\text{ K}} = -0.376\text{ kJ/K} = -376\text{ J/K}$

39. a. Decrease in positional probability; $\Delta S°(-)$

b. Increase in positional probability; $\Delta S°(+)$

c. Decrease in positional probability ($\Delta n < 0$); $\Delta S°(-)$

d. Decrease in positional probability ($\Delta n < 0$); $\Delta S°(-)$

e. HCl(g) has a greater positional probability due to the huge volume of gas, $\Delta S°(-)$.

f. Increase in positional probability; $\Delta S°(+)$

For c, d, and e, concentrate on the gaseous products and reactants. When there are more gaseous product molecules than gaseous reactant molecules ($\Delta n > 0$), then $\Delta S°$ will be positive. When Δn is negative, then $\Delta S°$ is negative.

41. $C_2H_2(g) + 4\,F_2(g) \rightarrow 2\,CF_4(g) + H_2(g)$; $\Delta S° = 2S^o_{CF_4} + S^o_{H_2} - [S^o_{C_2H_2} + 4S^o_{F_2}]$

$-358\text{ J/K} = (2\text{ mol})S^o_{CF_4} + 131\text{ J/K} - [201\text{ J/K} + 4(203\text{ J/K})]$, $S^o_{CF_4} = 262\text{ J K}^{-1}\text{ mol}^{-1}$

43. $-144\text{ J/K} = (2\text{ mol})\,S^o_{AlBr_3} - [2(28\text{ J/K}) + 3(152\text{ J/K})]$, $S^o_{AlBr_3} = 184\text{ J K}^{-1}\text{ mol}^{-1}$

45. At the boiling point, $\Delta G = 0$ so $\Delta H - T\Delta S$. $T = \dfrac{\Delta H}{\Delta S} = \dfrac{58.51 \times 10^3\text{ J/mol}}{92.92\text{ J K}^{-1}\text{ mol}^{-1}} = 629.7\text{ K}$

47. a. $NH_3(s) \rightarrow NH_3(l)$; $\Delta G = \Delta H - T\Delta S = 5650 \text{ J/mol} - 200.\text{ K}(28.9 \text{ J K}^{-1} \text{ mol}^{-1})$

$\Delta G = 5650 \text{ J/mol} - 5780 \text{ J/mol} = -130 \text{ J/mol}$

Yes, NH_3 will melt because $\Delta G < 0$ at this temperature.

b. At the melting point, $\Delta G = 0$, so $T = \dfrac{\Delta H}{\Delta S} = \dfrac{5650 \text{ J/mol}}{28.9 \text{ J K}^{-1} \text{ mol}^{-1}} = 196 \text{ K}$.

49. Solid I $\rightarrow$ solid II; equilibrium occurs when $\Delta G = 0$.

$$\Delta G = \Delta H - T\Delta S,\ \Delta H = T\Delta S,\ T = \Delta H/\Delta S = \frac{-743.1 \text{ J/mol}}{-17.0 \text{ J K}^{-1} \text{ mol}^{-1}} = 43.7 \text{ K} = -229.5°\text{C}$$

Free Energy and Chemical Reactions

51. $-5490.\text{ kJ} = 8(-394 \text{ kJ}) + 10(-237 \text{ kJ}) - 2\,\Delta G^\circ_{f,C_4H_{10}}$, $\Delta G^\circ_{f,C_4H_{10}} = -16 \text{ kJ/mol}$

53. $\Delta G° = -58.03 \text{ kJ} - (298 \text{ K})(-0.1766 \text{ kJ/K}) = -5.40 \text{ kJ}$

$$\Delta G° = 0 = \Delta H° - T\Delta S°,\ T = \frac{\Delta H^\circ}{\Delta S^\circ} = \frac{-58.03 \text{ kJ}}{-0.1766 \text{ kJ/K}} = 328.6 \text{ K}$$

$\Delta G°$ is negative below 328.6 K, where the favorable $\Delta H°$ term dominates.

55. a. $\Delta G° = 2(-270.\text{ kJ}) - 2(-502 \text{ kJ}) = 464 \text{ kJ}$

b. Because $\Delta G°$ is positive, this reaction is not spontaneous at standard conditions at 298 K.

c. $\Delta G° = \Delta H° - T\Delta S°$, $\Delta H° = \Delta G° + T\Delta S° = 464 \text{ kJ} + 298 \text{ K}(0.179 \text{ kJ/K}) = 517 \text{ kJ}$

We need to solve for the temperature when $\Delta G° = 0$:

$$\Delta G° = 0 = \Delta H° - T\Delta S°,\ T = \frac{\Delta H^\circ}{\Delta S^\circ} = \frac{517 \text{ kJ}}{0.179 \text{ kJ/K}} = 2890 \text{ K}$$

This reaction will be spontaneous at standard conditions ($\Delta G° < 0$) when $T > 2890$ K. At these temperatures, the favorable entropy term will dominate.

57. $CH_4(g) + CO_2(g) \rightarrow CH_3CO_2H(l)$

$\Delta H° = -484 - [-75 + (-393.5)] = -16 \text{ kJ}$; $\Delta S° = 160. - (186 + 214) = -240.\text{ J/K}$

$\Delta G° = \Delta H° - T\Delta S° = -16 \text{ kJ} - (298 \text{ K})(-0.240 \text{ kJ/K}) = 56 \text{ kJ}$

At standard concentrations, where $\Delta G = \Delta G°$, this reaction is spontaneous only at temperatures below $T = \Delta H°/\Delta S° = 67$ K (where the favorable $\Delta H°$ term will dominate, giving a negative $\Delta G°$ value). This is not practical. Substances will be in condensed phases and rates will be very slow at this extremely low temperature.

$CH_3OH(g) + CO(g) \rightarrow CH_3CO_2H(l)$

$\Delta H° = -484 - [-110.5 + (-201)] = -173$ kJ; $\Delta S° = 160. - (198 + 240.) = -278$ J/K

$\Delta G° = -173 \text{ kJ} - (298 \text{ K})(-0.278 \text{ kJ/K}) = -90. \text{ kJ}$

This reaction also has a favorable enthalpy and an unfavorable entropy term. This reaction is spontaneous at temperatures below $T = \Delta H°/\Delta S° = 622$ K (assuming standard concentrations). The reaction of CH_3OH and CO will be preferred at standard conditions. It is spontaneous at high enough temperatures that the rates of reaction should be reasonable.

59. Enthalpy is not favorable, so ΔS must provide the driving force for the change. Thus ΔS is positive. There is an increase in positional probability, so the original enzyme has the more ordered structure (has the smaller positional probability).

61. Because there are more product gas molecules than reactant gas molecules ($\Delta n > 0$), ΔS will be positive. From the signs of ΔH and ΔS, this reaction is spontaneous at all temperatures. It will cost money to heat the reaction mixture. Because there is no thermodynamic reason to do this, the purpose of the elevated temperature must be to increase the rate of the reaction, i.e., kinetic reasons.

Free Energy: Pressure Dependence and Equilibrium

63. At constant temperature and pressure, the sign of ΔG (positive or negative) tells us which reaction is spontaneous (the forward or reverse reaction). If $\Delta G < 0$, then the forward reaction is spontaneous and if $\Delta G > 0$, then the reverse reaction is spontaneous. If $\Delta G = 0$, then the reaction is at equilibrium (neither the forward nor reverse reaction is spontaneous). $\Delta G°$ gives the equilibrium position by deter-mining K for a reaction utilizing the equation $\Delta G° = -RT \ln K$. $\Delta G°$ can only be used to predict spontaneity when all reactants and products are present at standard pressures of 1 atm and/or standard concentrations of 1 *M*.

65. $\Delta G = \Delta G° + RT \ln Q = \Delta G° + RT \ln \dfrac{P_{N_2O_4}}{P_{NO_2}^2}$

$\Delta G° = 1 \text{ mol}(98 \text{ kJ/mol}) - 2 \text{ mol}(52 \text{ kJ/mol}) = -6 \text{ kJ}$

a. These are standard conditions, so $\Delta G = \Delta G°$ since $Q = 1$ and $\ln Q = 0$. Because $\Delta G°$ is negative, the forward reaction is spontaneous. The reaction shifts right to reach equilibrium.

b. $\Delta G = -6 \times 10^3 \text{ J} + 8.3145 \text{ J K}^{-1} \text{ mol}^{-1}(298 \text{ K}) \ln \dfrac{0.50}{(0.21)^2}$

$\Delta G = -6 \times 10^3 \text{ J} + 6.0 \times 10^3 \text{ J} = 0$

Since $\Delta G = 0$, this reaction is at equilibrium (no shift).

c. $\Delta G = -6 \times 10^3 \text{ J} + 8.3145 \text{ J K}^{-1} \text{ mol}^{-1}(298 \text{ K}) \ln \dfrac{1.6}{(0.29)^2}$

$\Delta G = -6 \times 10^3 \text{ J} + 7.3 \times 10^3 \text{ J} = 1.3 \times 10^3 \text{ J} = 1 \times 10^3 \text{ J}$

Since ΔG is positive, the reverse reaction is spontaneous, so the reaction shifts to the left to reach equilibrium.

67. $\Delta H° = 2\,\Delta H^\circ_{f,\,NH_3} = 2(-46) = -92$ kJ; $\Delta G° = 2\Delta G^\circ_{f,\,NH_3} = 2(-17) = -34$ kJ

$\Delta S° = 2(193 \text{ J/K}) - [192 \text{ J/K} + 3(131 \text{ J/K})] = -199$ J/K; $\Delta G° = -RT \ln K$

$$K = \frac{-\Delta G^\circ}{RT} = \exp\left(\frac{-(-34{,}000 \text{ J})}{(8.3145 \text{ J K}^{-1} \text{ mol}^{-1})(298 \text{ K})}\right) = e^{13.72} = 9.1 \times 10^5$$

Note: When determining exponents, we will round off after the calculation is complete. This helps eliminate excessive round off error.

a. $\Delta G = \Delta G° + RT \ln \dfrac{P^2_{NH_3}}{P_{N_2} \times P^3_{H_2}} = -34 \text{ kJ} + \dfrac{(8.3145 \text{ J K}^{-1} \text{ mol}^{-1})(298 \text{ K})}{1000 \text{ J/kJ}} \ln \dfrac{(50.)^2}{(200.)(200.)^3}$

$\Delta G = -34 \text{ kJ} - 33 \text{ kJ} = -67 \text{ kJ}$

b. $\Delta G = -34 \text{ kJ} + \dfrac{(8.3145 \text{ J K}^{-1} \text{ mol}^{-1})(298 \text{ K})}{1000 \text{ J/kJ}} \ln \dfrac{(200.)^2}{(200.)(600.)^3}$

$\Delta G = -34 \text{ kJ} - 34.4 \text{ kJ} = -68 \text{ kJ}$

c. Assume $\Delta H°$ and $\Delta S°$ are temperature independent.

$\Delta G^\circ_{100} = \Delta H° - T\Delta S°$, $\Delta G^\circ_{100} = -92 \text{ kJ} - (100. \text{ K})(-0.199 \text{ kJ/K}) = -72 \text{ kJ}$

$\Delta G_{100} = \Delta G^\circ_{100} + RT \ln Q = -72 \text{ kJ} + \dfrac{(8.3145 \text{ J K}^{-1} \text{ mol}^{-1})(100. \text{ K})}{1000 \text{ J/kJ}} \ln \dfrac{(10.)^2}{(50.)(200.)^3}$

$\Delta G_{100} = -72 \text{ kJ} - 13 \text{ kJ} = -85 \text{ kJ}$

d. $\Delta G^\circ_{700} = -92 \text{ kJ} - (700. \text{ K})(-0.199 \text{ kJ/K}) = 47 \text{ kJ}$

$\Delta G_{700} = 47 \text{ kJ} + \dfrac{(8.3145 \text{ J K}^{-1} \text{ mol}^{-1})(700. \text{ K})}{1000 \text{ J/kJ}} \ln \dfrac{(10.)^2}{(50.)(200.)^3}$

$\Delta G_{700} = 47 \text{ kJ} - 88 \text{ kJ} = -41 \text{ kJ}$

69. $\Delta G° = -RT \ln K$. To determine K at a temperature other than 25°C, one needs to know $\Delta G°$ at that temperature. We assume $\Delta H°$ and $\Delta S°$ are temperature-independent and use the equation $\Delta G° = \Delta H° - T\Delta S°$ to estimate $\Delta G°$ at the different temperature. For K = 1, we want $\Delta G° = 0$, which occurs when $\Delta H° = T\Delta S°$. Again, assume $\Delta H°$ and $\Delta S°$ are temperature independent, and then solve for T (= $\Delta H°/\Delta S°$). At this temperature, K = 1 because $\Delta G° = 0$. This only works for reactions where the signs of $\Delta H°$ and $\Delta S°$ are the same (either both positive or both negative). When the signs are opposite, K will always be greater than one (when $\Delta H°$ is negative and $\Delta S°$ is positive) or K will always be less than one (when $\Delta H°$ is positive and $\Delta S°$ is negative). When the signs of $\Delta H°$ and $\Delta S°$ are opposite, K can never equal one.

71. $$K = \frac{P_{NF_3}^2}{P_{N_2}^2 \times P_{F_2}^3} = \frac{(0.48)^2}{0.021(0.063)^3} = 4.4 \times 10^4$$

$$\Delta G^\circ_{800} = -RT \ln K = -8.3145 \text{ J K}^{-1} \text{ mol}^{-1}(800.\text{ K}) \ln(4.4 \times 10^4) = -7.1 \times 10^4 \text{ J/mol} = -71 \text{ kJ/mol}$$

73. Because the partial pressure of C(g) decreased, the net change that occurs for this reaction to reach equilibrium is for products to convert to reactants.

	A(g)	+ 2 B(g)	⇌	C(g)
Initial	0.100 atm	0.100 atm		0.100 atm
Change	$+x$	$+2x$	←	$-x$
Equil.	$0.100 + x$	$0.100 + 2x$		$0.100 - x$

From the problem, $P_C = 0.040$ atm $= 0.100 - x$, $x = 0.060$ atm.

The equilibrium partial pressures are: $P_A = 0.100 + x = 0.100 + 0.060 = 0.160$ atm,

$P_B = 0.100 + 2(0.60) = 0.220$ atm, and $P_C = 0.040$ atm.

$$K = \frac{0.040}{0.160(0.220)^2} = 5.2$$

$$\Delta G° = -RT \ln K = -8.3145 \text{ J K}^{-1} \text{ mol}^{-1}(298 \text{ K}) \ln(5.2) = -4.1 \times 10^3 \text{ J/mol} = -4.1 \text{ kJ/mol}$$

75.
$$HgbO_2 \rightarrow Hgb + O_2 \quad \Delta G° = -(-70 \text{ kJ})$$
$$Hgb + CO \rightarrow HgbCO \quad \Delta G° = -80 \text{ kJ}$$
$$HgbO_2 + CO \rightarrow HgbCO + O_2 \quad \Delta G° = -10 \text{ kJ}$$

$$\Delta G° = -RT \ln K, \quad K = \exp\left(\frac{-\Delta G°}{RT}\right) = \exp\left(\frac{-(-10 \times 10^3 \text{ J})}{(8.3145 \text{ J K}^{-1} \text{ mol}^{-1})(298 \text{ K})}\right) = 60$$

77. At 25.0°C: $\Delta G° = \Delta H° - T\Delta S° = -58.03 \times 10^3 \text{ J/mol} - (298.2 \text{ K})(-176.6 \text{ J K}^{-1} \text{ mol}^{-1})$
$= -5.37 \times 10^3$ J/mol

$$\Delta G° = -RT \ln K, \ln K = \frac{-\Delta G°}{RT} = \frac{-(-5.37 \times 10^3 \text{ J/mol})}{(8.3145 \text{ J K}^{-1} \text{ mol}^{-1})(298.2 \text{ K})} = 2.166; \quad K = e^{2.166} = 8.72$$

At 100.0°C: $\Delta G° = -58.03 \times 10^3$ J/mol $-$ (373.2 K)(-176.6 J K^{-1} mol^{-1}) $= 7.88 \times 10^3$ J/mol

$$\ln K = \frac{-(7.88 \times 10^3 \text{ J/mol})}{(8.3145 \text{ J K}^{-1} \text{ mol}^{-1})(373.2 \text{ K})} = -2.540, \quad K = e^{-2.540} = 0.0789$$

79. A graph of ln K vs. 1/T will yield a straight line with slope equal to $-\Delta H°/R$ and y intercept equal to $\Delta S°/R$.

a.

Temp (°C)	T(K)	1000/T (K^{-1})	K_w	ln K_w
0	273	3.66	1.14×10^{-15}	−34.408
25	298	3.36	1.00×10^{-14}	−32.236
35	308	3.25	2.09×10^{-14}	−31.499
40.	313	3.19	2.92×10^{-14}	−31.165
50.	323	3.10	5.47×10^{-14}	−30.537

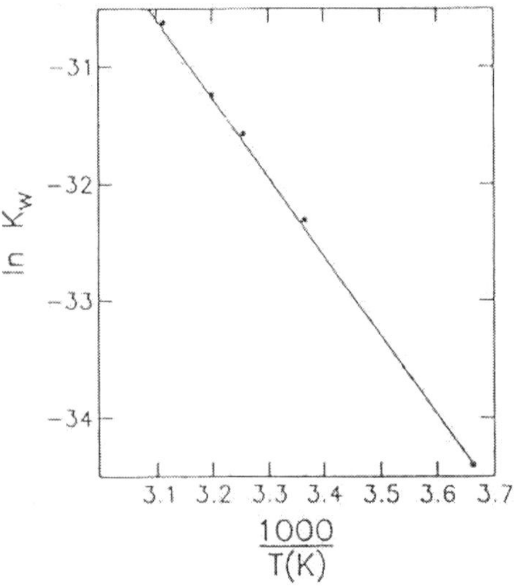

The straight-line equation (from a calculator) is: $\ln K = -6.91 \times 10^3 \left(\frac{1}{T}\right) - 9.09$

$$\text{Slope} = -6.91 \times 10^3 \text{ K} = \frac{-\Delta H°}{R}$$

$\Delta H° = -(-6.91 \times 10^3 \text{ K} \times 8.3145 \text{ J K}^{-1} \text{ mol}^{-1}) = 5.75 \times 10^4$ J/mol $= 57.5$ kJ/mol

y intercept $= -9.09 = \frac{\Delta S°}{R}$, $\Delta S° = -9.09 \times 8.3145$ J K^{-1} mol^{-1} $= -75.6$ J K^{-1} mol^{-1}

b. From part a, $\Delta H° = 57.5$ kJ/mol and $\Delta S° = -75.6$ J K^{-1} mol^{-1}. Assuming that $\Delta H°$ and $\Delta S°$ are temperature independent:

$$\Delta G° = 57{,}500 \text{ J/mol} - 647 \text{ K}(-75.6 \text{ J K}^{-1} \text{ mol}^{-1}) = 106{,}400 \text{ J/mol} = 106.4 \text{ kJ/mol}$$

81. From Exercise 10.78, $\ln K = \frac{-\Delta H°}{RT} + \frac{\Delta S°}{R}$, $R = 8.3145$ J K^{-1} mol^{-1}

For two sets of K and T:

$$\ln K_1 = \frac{-\Delta H°}{R}\left(\frac{1}{T_1}\right) + \frac{\Delta S°}{R}; \quad \ln K_2 = \frac{-\Delta H°}{R}\left(\frac{1}{T_2}\right) + \frac{\Delta S°}{R}$$

Subtracting the first expression from the second:

$$\ln K_2 - \ln K_1 = \frac{\Delta H°}{R}\left(\frac{1}{T_1} - \frac{1}{T_2}\right) \text{ or } \ln\frac{K_2}{K_1} = \frac{\Delta H°}{R}\left(\frac{1}{T_1} - \frac{1}{T_2}\right)$$

$$\ln\left(\frac{3.25 \times 10^{-2}}{8.84}\right) = \frac{\Delta H°}{8.3145 \text{ J K}^{-1} \text{ mol}^{-1}}\left(\frac{1}{298 \text{ K}} - \frac{1}{348 \text{ K}}\right)$$

$$-5.61 = (5.8 \times 10^{-5} \text{ mol/J})(\Delta H°), \quad \Delta H° = -9.7 \times 10^4 \text{ J/mol}$$

For $K = 8.84$ at $T = 25°C$:

$$\ln(8.84) = \frac{-(-9.7 \times 10^4 \text{ J/mol})}{(8.3145 \text{ J K}^{-1} \text{ mol}^{-1})(298 \text{ K})} + \frac{\Delta S°}{8.3145 \text{ J K}^{-1} \text{ mol}^{-1}}$$

$$\frac{\Delta S°}{8.3145} = -37, \quad \Delta S° = -310 \text{ J K}^{-1} \text{ mol}^{-1}$$

We get the same value for $\Delta S°$ using $K = 3.25 \times 10^{-2}$ at $T = 348$ K data. $\Delta G° = -RT \ln K$. When $K = 1.00$ then $\Delta G° = 0$ since $\ln(1.00) = 0$. $\Delta G° = 0 = \Delta H° - T\Delta S°$. Assuming that $\Delta H°$ and $\Delta S°$ do not depend on temperature:

$$\Delta H° = T\Delta S°, \quad T = \frac{\Delta H°}{\Delta S°} = \frac{-9.7 \times 10^4 \text{ J/mol}}{-310 \text{ J K}^{-1} \text{ mol}^{-1}} = 310 \text{ K}$$

Adiabatic Processes

83. For reversible adaiabatic process, $P_1V_1^{\gamma} = P_2V_2^{\gamma}$, where $\gamma = C_p/C_v$. For an ideal gas, $C_p = C_v + R$.

$$C_p = C_v + R = 20.5 \text{ J K}^{-1} \text{ mol}^{-1} + 8.3145 \text{ J K}^{-1} \text{ mol}^{-1} = 28.8 \text{ J K}^{-1} \text{ mol}^{-1}$$

$$V_2{}^{\gamma} = \frac{P_1V_1^{\gamma}}{P_2}, \text{ where } \gamma = \frac{C_p}{C_v} = \frac{28.8}{20.5} = 1.40$$

$$V_1 = \frac{nRT_1}{P_1} = \frac{1.75\text{ mol}(0.08206\text{ L atm K}^{-1}\text{ mol}^{-1})(294\text{ K})}{1.50\text{ atm}} = 28.1\text{ L}$$

$$V_2{}^{1.40} = \frac{1.50\text{ atm}(28.1\text{ L})^{1.40}}{4.50\text{ atm}} = 35.6,\ \ V_2 = (35.6)^{1/1.40} = 12.8\text{ L}$$

To calculate $w = \Delta E = nC_V\Delta T$, we need T_2 in order to calculate ΔT.

$$T_2 = \frac{P_2V_2}{nR} = \frac{4.50\text{ atm}(12.8\text{ L})}{1.75\text{ mol}(0.08206\text{ L atm K}^{-1}\text{ mol}^{-1})} = 401\text{ K}$$

Note: we also could have used the formula $T_2/T_1 = (V_1/V_2)^{\gamma-1}$ to calculate T_2.

$w = \Delta E = nC_V\Delta T = 1.75\text{ mol}(20.5\text{ J K}^{-1}\text{ mol}^{-1})(401\text{ K} - 294\text{ K}) = 3.84 \times 10^3\text{ J}$

Note that for the adiabatic reversible compression, T increases, so ΔE increases. As work is added during the compression, the internal energy of the system increases ($\Delta E = w$).

85. a. Isothermal reversible expansion: $\Delta T = 0$ so $\Delta E = 0$

$w = -nRT\ln(V_2/V_1) = -nRT\ln(P_1/P_2)$

$w = -1.00\text{ mol}(8.3145\text{ J K}^{-1}\text{ mol}^{-1})(300.\text{ K})\ln(5.00/1.00) = -4.01 \times 10^3\text{ J}$

Since $\Delta E = 0$, $q = -w = 4.01 \times 10^3$ J.

b. Isothermal irreversible expansion against a constant external pressure of 1.00 atm, so $\Delta E = 0$, and $q = -w$.

$$w = -P_{ex}\Delta V;\ \ V_1 = \frac{1.00\text{ mol}(0.08206\text{ L atm K}^{-1}\text{ mol}^{-1})300.\text{ K}}{5.00\text{ atm}} = 4.92\text{ L}$$

$$V_f = \frac{1.00\text{ mol}(0.08206)(300.\text{ K})}{1.00\text{ atm}} = 24.6\text{ L}$$

$w = -(1.00\text{ atm})(24.6\text{ L} - 4.92\text{ L}) \times 101.3\text{ J L}^{-1}\text{ atm}^{-1} = -1.99 \times 10^3\text{ J}$

$q = -w = 1.99 \times 10^3$ J

As expected, we get more work from the reversible process.

c. Adiabatic reversible expansion: $q = 0$ so $\Delta E = w$, $\Delta E = nC_v\Delta T$; we need C_v and ΔT to calculate ΔE and, in turn, w. In order to calculate T_2, we need to determine V_2.

$P_1V_1^\gamma = P_2V_2^\gamma$, where $\gamma = C_p/C_v$. From part b, $V_1 = 4.92$ L.

For a gas behaving ideally:

$$C_p = C_v + R,\ C_v = C_p - R = 37.1\ \text{J K}^{-1}\ \text{mol}^{-1} - 8.3145\ \text{J K}^{-1}\ \text{mol}^{-1} = 28.8\ \text{J K}^{-1}\ \text{mol}^{-1}$$

$$\frac{C_P}{C_V} = \frac{37.1}{28.8} = 1.29;\ V_2^{1.29} = \frac{P_1V_1^{1.29}}{P_2}$$

$$V_2^{1.29} = \frac{5.00\ \text{atm}(4.92\ \text{L})^{1.29}}{1.00\ \text{atm}} = 39.0,\ V_2 = 17.1\ \text{L}$$

$$T_2 = \frac{P_2V_2}{nR} = \frac{1.00\ \text{atm}(17.1\ \text{L})}{1.00\ \text{mol}(0.08206\ \text{L atm K}^{-1}\ \text{mol}^{-1})} = 208\ \text{K}$$

$$\Delta E = w = nC_v\Delta T = 1.00\ \text{mol}(28.8\ \text{J K}^{-1}\ \text{mol}^{-1})(208\ \text{K} - 300.\ \text{K}) = -2.65 \times 10^3\ \text{J}$$

As expected, we do not get as much work in the adiabatic reversible process as when the gas expands isothermally and reversibly.

Additional Exercises

87. When an ionic solid dissolves, positional probability increases, so ΔS_{sys} is positive. Since temperature increased as the solid dissolved, this is an exothermic process, and ΔS_{surr} is positive ($\Delta S_{surr} = -\Delta H/T$). Since the solid did dissolve, the dissolving process is spontaneous, so ΔS_{univ} is positive (as it must be when ΔS_{sys} and ΔS_{surr} are both positive).

89. $w_{max} = \Delta G$; when ΔG is negative, the magnitude of ΔG is equal to the maximum possible useful work obtainable from the process (at constant T and P). When ΔG is positive, the magnitude of ΔG is equal to the minimum amount of work that must be expended to make the process spontaneous. Due to waste energy (heat) in any real process, the amount of useful work obtainable from a spontaneous process is always less than w_{max}, and for a nonspontaneous reaction, an amount of work greater than w_{max} must be applied to make the process spontaneous.

91. $HF(aq) \rightleftharpoons H^+(aq) + F^-(aq)$; $\Delta G = \Delta G° + RT \ln\frac{[H^+][F^-]}{[HF]}$

$$\Delta G° = -RT \ln K = -(8.3145\ \text{J K}^{-1}\ \text{mol}^{-1})(298\ \text{K}) \ln(7.2 \times 10^{-4}) = 1.8 \times 10^4\ \text{J/mol}$$

a. The concentrations are all at standard conditions, so $\Delta G = \Delta G = 1.8 \times 10^4$ J/mol (since $Q = 1.0$ and $\ln Q = 0$). Because $\Delta G°$ is positive, the reaction shifts left to reach equilibrium.

b. $\Delta G = 1.8 \times 10^4 \text{ J/mol} + (8.3145 \text{ J K}^{-1} \text{ mol}^{-1})(298 \text{ K}) \ln \dfrac{(2.7 \times 10^{-2})^2}{0.98}$

$\Delta G = 1.8 \times 10^4 \text{ J/mol} - 1.8 \times 10^4 \text{ J/mol} = 0$

Because $\Delta G = 0$, the reaction is at equilibrium (no shift).

c. $\Delta G = 1.8 \times 10^4 + 8.3145(298) \ln \dfrac{(1.0 \times 10^{-5})^2}{1.0 \times 10^{-5}} = -1.1 \times 10^4$ J/mol; shifts right

d. $\Delta G = 1.8 \times 10^4 + 8.3145(298) \ln \dfrac{7.2 \times 10^{-4}(0.27)}{0.27} = 1.8 \times 10^4 - 1.8 \times 10^4 = 0$; at equilibrium

e. $\Delta G = 1.8 \times 10^4 + 8.3145(298) \ln \dfrac{1.0 \times 10^{-3}(0.67)}{0.52} = 2 \times 10^3$ J/mol; shifts left

93. $\Delta S°$ will be negative because 2 moles of gaseous reactants forms 1 mole of gaseous product. For $\Delta G°$ to be negative, ΔH must be negative (exothermic). For this sign combination of $\Delta H°$ and $\Delta S°$, K decreases as T increases because $\Delta G°$ becomes more positive ($\Delta G° = -RT \ln K$). Therefore, the ratio of the partial pressure of PCl_5 (a product) to the partial pressure of PCl_3 (a reactant) will decrease when T is raised.

95. Using Le Chatelier's principle: A decrease in pressure (volume increases) will favor the side with the greater number of particles. Thus 2 I(g) will be favored at low pressure.

Looking at ΔG: $\Delta G = \Delta G° + RT \ln(P_I^2 / P_{I_2})$; $\ln(P_I^2 / P_{I_2}) > 0$ when $P_I = P_{I_2} = 10$ atm, and ΔG is positive (not spontaneous). But at $P_I = P_{I_2} = 0.10$ atm, the logarithm term is negative. If $|RT \ln Q| > \Delta G°$, then ΔG becomes negative, and the reaction is spontaneous.

97. $NaCl(s) \rightleftharpoons Na^+(aq) + Cl^-(aq) \qquad K = K_{sp} = [Na^+][Cl^-]$

$\Delta G° = [(-262 \text{ kJ}) + (-131 \text{ kJ})] - (-384 \text{ kJ}) = -9 \text{ kJ} = -9000 \text{ J}$

$\Delta G° = -RT \ln K_{sp}, \; K_{sp} = \exp\left[\dfrac{-(-9000 \text{ J})}{8.3145 \text{ J K}^{-1} \text{ mol}^{-1} \times 298 \text{ K}}\right] = 38 = 40$

	$NaCl(s) \rightleftharpoons$	$Na^+(aq)$	$+ \; Cl^-(aq)$	$K_{sp} = 40$
Initial	s = solubility (mol/L)	0	0	
Equil.		s	s	

$K_{sp} = 40 = s(s), \; s = (40)^{1/2} = 6.3 = 6 \; M = [Cl^-]$

99. ΔS is more favorable (less negative) for reaction 2 than for reaction 1, resulting in $K_2 > K_1$. In reaction 1, seven particles in solution are forming one particle in solution. In reaction 2, four particles are forming one, which results in a smaller decrease in positional probability than for reaction 1.

101. Note that these substances are not in the solid state but are in the aqueous state; water molecules are also present. There is an apparent increase in ordering (decrease in positional probability) when these ions are placed in water as compared to the separated state. The hydrating water molecules must be in a highly ordered arrangement when surrounding these anions.

103. $S = k \ln \Omega$; S has units of $J\ K^{-1}\ mol^{-1}$, and k has units of J/K ($k = 1.38 \times 10^{-23}$ J/K).

To make units match: $S\ (J\ K^{-1}\ mol^{-1}) = N_A k \ln \Omega$ when N_A = Avogadro's number

$$189\ J\ K^{-1}\ mol^{-1} = 8.31\ J\ K^{-1}\ mol^{-1} \ln \Omega_g$$
$$-(70.\ J\ K^{-1}\ mol^{-1} = 8.31\ J\ K^{-1}\ mol^{-1} \ln \Omega_l)$$

Subtracting: $119\ J\ K^{-1}\ mol^{-1} = 8.31\ J\ K^{-1}\ mol^{-1}(\ln \Omega_g - \ln \Omega_l)$

$$14.3 = \ln(\Omega_g/\Omega_l), \quad \frac{\Omega_g}{\Omega_l} = e^{14.3} = 1.6 \times 10^6$$

105. Isothermal: $\Delta H = 0$ (assume ideal gas)

$$\Delta S = nR \ln\left(\frac{V_2}{V_1}\right) = (1.00\ mol)(8.3145\ J\ K^{-1}\ mol^{-1}) \ln\left(\frac{1.00\ L}{100.0\ L}\right) = -38.3\ J/K$$

$$\Delta G = \Delta H - T\Delta S = 0 - (300.\ K)(-38.3\ J/K) = +11{,}500\ J = 11.5\ kJ$$

107. a. Isothermal: $\Delta E = 0$ and $\Delta H = 0$ if gas is ideal.

$$\Delta S = nR \ln(P_1/P_2) = (1.00\ mol)(8.3145\ J\ K^{-1}\ mol^{-1}) \ln(5.00\ atm/2.00\ atm) = 7.62\ J/K$$

$$T = \frac{PV}{nR} = \frac{5.00\ atm \times 5.00\ L}{1.00\ mol \times 0.08206\ L\ atm\ K^{-1}\ mol^{-1}} = 305\ K$$

$$\Delta G = \Delta H - T\Delta S = 0 - (305\ K)(7.62\ J/K) = -2320\ J$$

$$w = -P\Delta V = -(2.00\ atm)\Delta V, \text{ where } V_f = \frac{nRT}{2.00\ atm} = 12.5\ L$$

$$w = -2.00\ atm\ (12.5 - 5.00\ L) \times (101.3\ J\ L^{-1}\ atm^{-1}) = -1500\ J$$

$$\Delta E = 0 = q + w, \quad q = 1500\ J$$

b. Second law, $\Delta S_{univ} > 0$ for spontaneous processes:

$$\Delta S_{univ} = \Delta S_{sys} + \Delta S_{surr} = \Delta S_{sys} - \frac{q_{actual}}{T}$$

$$\Delta S_{univ} = 7.62 \text{ J/K} - \frac{1500 \text{ J}}{305 \text{ K}} = 7.62 - 4.9 = 2.7 \text{ J/K};$$ thus the process is spontaneous.

Challenge Problems

109. a. $$V_1 = \frac{nRT_1}{P_1} = \frac{2.00 \text{ mol} \times 0.08206 \text{ L atm K}^{-1} \text{ mol}^{-1} \times 298 \text{ K}}{2.00 \text{ atm}} = 24.5 \text{ L}$$

For an adiabatic reversible process, $P_1V_1^{\gamma} = P_2V_2^{\gamma}$ and $T_1V_1^{\gamma-1} = T_2V_2^{\gamma-1}$, where $\gamma = C_p/C_v$. Because argon is a monoatomic gas, $C_p = (5/2)R$ and $C_v = (3/2)R$, so $\gamma = 5/3$.

$$V_2^{\gamma} = \frac{P_1V_1^{\gamma}}{P_2} = \frac{2.00 \text{ atm}(24.5 \text{ L})^{5/3}}{1.00 \text{ atm}} = 413, \; V_2 = (413)^{3/5} = 37.1 \text{ L}$$

We can either use the ideal gas law or the $T_1V^{\gamma-1} = T_2V_2^{\gamma-1}$ equation to calculate the final temperature. Using the ideal gas law:

$$T_2 = \frac{P_2V_2}{nR} = \frac{1.00 \text{ atm} \times 37.1 \text{ L}}{2.00 \text{ mol} \times 0.08206 \text{ L atm K}^{-1} \text{ mol}^{-1}} = 226 \text{ K}$$

b. For an adiabatic process ($q = 0$), $\Delta E = w = nC_v\Delta T$. For an expansion against a fixed external pressure, $w = -P\Delta V$. From the ideal gas equation (see part a), $V_1 = 24.5$ L.

$$w = -P\Delta V = nC_v\Delta T$$

$$1.00 \text{ atm}(V_2 - 24.5 \text{ L})\left(\frac{101.3 \text{ J}}{\text{L atm}}\right) = 2.00 \text{ mol}(3/2)\left(\frac{8.3145 \text{ J}}{\text{mol K}}\right)(T_2 - 298 \text{ K})$$

We will ignore units from here. Note that both sides of the equation are in units of J.

$$(-101)V_2 + 2480 = (24.9)T_2 - 7430$$

To solve for T_2, we need to find an expression for V_2. Using the ideal gas equation:

$$V_2 = \frac{nRT_2}{P_2} = \frac{2.00 \text{ mol}(0.08206)T_2}{1.00} = (0.164)T_2;$$ substituting:

$$-101(0.164\, T_2) + 2480 = (24.9)T_2 - 7430, \; (41.5)T_2 = 9910, \; T_2 = 239 \text{ K}$$

111. a. $\Delta S = \frac{q_{rev}}{T}$; isothermal: $\Delta T = 0$ so $\Delta E = 0$ and $q = -w$.

Reversible expansion: $q_{rev} = 855$ J; $\Delta S = \frac{q_{rev}}{T} = \frac{855 \text{ J}}{298 \text{ K}} = 2.87$ J/K

For the compression, we go back to the initial state for the overall process (expansion then compression). Because the initial and overall final state are the same, all state functions like ΔS must equal zero.

$\Delta S_{overall} = 0 = \Delta S_{exp} + \Delta S_{comp}$, $\Delta S_{comp} = -\Delta S_{exp} = -2.87$ J/K

Note: Even though the compression step is not a reversible process, it still must have $q_{rev} = -855$ J.

$$\Delta S_{comp} = \frac{q_{rev}}{T} = \frac{-855 \text{ J}}{298 \text{ K}} = -2.87 \text{ J/K}$$

b. $\Delta S_{univ,\ overall} = \Delta S_{sys,\ overall} + \Delta S_{surr,\ overall}$

$\Delta S_{sys,\ overall} = 2.87 \text{ J/K} - 2.87 \text{ J/K} = 0$

$$\Delta S_{surr,\ exp} = \frac{-q_{actual}}{T} = \frac{-855 \text{ J}}{298 \text{ K}} = -2.87 \text{ J/K}$$

From the problem, the compression step is isothermal, so $q = -w$. The work done on the system is 2(855) J, so $q = -2(855)$ J.

$$\Delta S_{surr,\ comp} = \frac{-q_{actual}}{T} = \frac{-[-2(855) \text{ J}]}{298 \text{ K}} = 5.74 \text{ J/K}$$

$\Delta S_{surr,\ overall} = -2.87 \text{ J/K} + 5.74 \text{ J/K} = 2.87 \text{ J/K}$

$\Delta S_{univ,\ overall} = 0 + 2.87 \text{ J/K} = 2.87 \text{ J/K}$

113. The system is the 1.00-L sample of water. The process is the cooling of the water from 90.°C to 25°C. The surroundings are the room (at 25°C) and everything else.

$$1.00 \times 10^3 \text{ mL} \times \frac{1.00 \text{ g}}{\text{mL}} \times \frac{1 \text{ mol}}{18.02 \text{ g}} = 55.5 \text{ mol } H_2O$$

$\Delta S = nC_p \ln(T_2/T_1) = 55.5 \text{ mol}(75.3 \text{ J K}^{-1} \text{ mol}^{-1}) \ln(298/363) = -825$ J/K

$\Delta S_{surr} = \frac{-q_{actual}}{T}$; $q_{actual} = nC_p\Delta T = 55.5 \text{ mol}(75.3 \text{ J K}^{-1} \text{ mol}^{-1})(-65 \text{ K}) = -2.72 \times 10^5$ J

$$\Delta S_{surr} = \frac{-q_{actual}}{T} = \frac{-(-2.72 \times 10^5\ J)}{298\ K} = 913\ J/K$$

$\Delta S_{univ} = \Delta S + \Delta S_{surr} = -825\ J/K + 913\ J/K = 88\ J/K$

Not too surprising, this is a spontaneous process due to the favorable ΔS_{surr} term.

115. $K_p = P_{CO_2}$; to prevent Ag_2CO_3 from decomposing, P_{CO_2} should be greater than K_p.

From Exercise 10.78, $\ln K = \frac{-\Delta H^\circ}{RT} + \frac{\Delta S^\circ}{R}$. For two conditions of K and T, the equation is:

$$\ln\frac{K_2}{K_1} = \frac{\Delta H^\circ}{R}\left(\frac{1}{T_1} - \frac{1}{T_2}\right)$$

Let $T_1 = 25°C = 298\ K$, $K_1 = 6.23 \times 10^{-3}$ torr; $T_2 = 110.°C = 383\ K$, $K_2 = ?$

$$\ln\frac{K_2}{6.23 \times 10^{-3}\ torr} = \frac{79.14 \times 10^3\ J/mol}{8.3145\ J\ K^{-1}\ mol^{-1}}\left(\frac{1}{298\ K} - \frac{1}{383\ K}\right)$$

$$\ln\frac{K_2}{6.23 \times 10^{-3}} = 7.1,\quad \frac{K_2}{6.23 \times 10^{-3}} = e^{7.1} = 1.2 \times 10^3,\ K_2 = 7.5\ torr$$

To prevent decomposition of Ag_2CO_3, the partial pressure of CO_2 should be greater than 7.5 torr.

117. $3\ O_2(g) \rightleftharpoons 2\ O_3(g)$; $\Delta H° = 2(143) = 286\ kJ$; $\Delta G° = 2(163) = 326\ kJ$

$$\ln K = \frac{-\Delta G^\circ}{RT} = \frac{-326 \times 10^3\ J}{(8.3145\ J\ K^{-1}\ mol^{-1})(298\ K)} = -131.573,\ K = e^{-131.573} = 7.22 \times 10^{-58}$$

We need the value of K at 230. K. From Exercise 10.78: $\ln K = \frac{-\Delta H^\circ}{RT} + \frac{\Delta S^\circ}{R}$

For two sets of K and T:

$$\ln\frac{K_2}{K_1} = \frac{\Delta H^\circ}{R}\left(\frac{1}{T_1} - \frac{1}{T_2}\right)$$

Let $K_2 = 7.22 \times 10^{-58}$, $T_2 = 298$; $K_1 = K_{230}$, $T_1 = 230.\ K$; $\Delta H° = 286 \times 10^3\ J$

$$\ln\frac{7.22 \times 10^{-58}}{K_{230}} = \frac{286 \times 10^3}{8.3145}\left(\frac{1}{230.} - \frac{1}{298}\right) = 34.13$$

$$\frac{7.22 \times 10^{-58}}{K_{230}} = e^{34.13} = 6.6 \times 10^{14}, \ K_{230} = 1.1 \times 10^{-72}$$

$$K_{230} = 1.1 \times 10^{-72} = \frac{P_{O_3}^2}{P_{O_2}^3} = \frac{P_{O_3}^2}{(1.0 \times 10^{-3} \text{ atm})^3}, \quad P_{O_3} = 3.3 \times 10^{-41} \text{ atm}$$

The volume occupied by one molecule of ozone is:

$$V = \frac{nRT}{P} = \frac{(1/6.022 \times 10^{23} \text{ mol})(0.8206 \text{ L atm K}^{-1} \text{ mol}^{-1})(230.\text{ K})}{(3.3 \times 10^{-41} \text{ atm})}, \ V = 9.5 \times 10^{17} \text{ L}$$

Equilibrium is probably not maintained under these conditions. When only two ozone molecules are in a volume of 9.5×10^{17} L, the reaction is not at equilibrium. Under these conditions, Q > K and the reaction shifts left. But with only two ozone molecules in this huge volume, it is extremely unlikely that they will collide with each other. At these conditions, the concentration of ozone is not large enough to maintain equilibrium.

119. a. $\Delta G° = G_B^o - G_A^o = 11{,}718 - 8996 = 2722$ J

$$K = \exp\left(\frac{-\Delta G°}{RT}\right) = \exp\left[\frac{-2722 \text{ J}}{(8.3145 \text{ J K}^{-1} \text{ mol}^{-1})(298 \text{ K})}\right] = 0.333$$

b. Since Q = 1.00 > K, reaction shifts left. Let x = atm of B(g) that reacts to reach equilibrium.

	A(g)	⇌	B(g)	$K = P_B/P_A$
Initial	1.00 atm		1.00 atm	
Equil.	$1.00 + x$		$1.00 - x$	

$$K = \frac{1.00 - x}{1.00 + x} = 0.333, \ 1.00 - x = 0.333 + (0.333)x, \ x = 0.50 \text{ atm}$$

$P_B = 1.00 - 0.50 = 0.50$ atm; $P_A = 1.00 + 0.50 = 1.50$ atm

c. $\Delta G = \Delta G° + RT \ln Q = \Delta G° + RT \ln(P_B/P_A)$

$\Delta G = 2722 \text{ J} + (8.3145)(298) \ln(0.50/1.50) = 2722 \text{ J} - 2722 \text{ J} = 0$

(carrying extra sig. figs.)

121. Step 1: $\Delta E = 0$ and $\Delta H = 0$ since $\Delta T = 0$

$$w = -P\Delta V = -(9.87 \times 10^{-3} \text{ atm})\Delta V; \ V = \frac{nRT}{P}, \ R = 0.08206 \text{ L atm K}^{-1} \text{ mol}^{-1}$$

$$\Delta V = V_f - V_i = nRT\left(\frac{1}{P_f} - \frac{1}{P_i}\right)$$

$$\Delta V = 1.00 \text{ mol}(0.08206)(298 \text{ K})\left(\frac{1}{9.87 \times 10^{-3} \text{ atm}} - \frac{1}{2.45 \times 10^{-2} \text{ atm}}\right)$$

$\Delta V = 1480$ L (we will carry all values to three sig. figs.)

$$w = -(9.87 \times 10^{-3} \text{ atm})(1480 \text{ L}) = -14.6 \text{ L atm}(101.3 \text{ J L}^{-1} \text{ atm}^{-1}) = -1480 \text{ J}$$

$$\Delta E = q + w = 0,\ \ q = -w = +1480 \text{ J};\ \ \Delta S = nR \ln\left(\frac{P_1}{P_2}\right)$$

$$\Delta S = 1.00 \text{ mol}(8.3145 \text{ J K}^{-1} \text{ mol}^{-1}) \ln\left(\frac{2.45 \times 10^{-2} \text{ atm}}{9.87 \times 10^{-3} \text{ atm}}\right),\ \ \Delta S = 7.56 \text{ J/K}$$

$$\Delta G = \Delta H - T\Delta S = 0 - 298 \text{ K}(7.56 \text{ J/K}) = -2250 \text{ J}$$

Step 2: $\Delta E = 0,\ \ \Delta H = 0$

$$w = -(4.93 \times 10^{-3} \text{ atm})\left(\frac{nRT}{4.93 \times 10^{-3}} - \frac{nRT}{9.87 \times 10^{-3} \text{ atm}}\right) \text{L} \times \frac{101.3 \text{ J}}{\text{L atm}} = -1240 \text{ J}$$

$$q = -w = 1240 \text{ J};\ \ \Delta S = nR \ln\left(\frac{9.87 \times 10^{-3} \text{ atm}}{4.93 \times 10^{-3} \text{ atm}}\right) = 5.77 \text{ J/K}$$

$$\Delta G = 0 - 298\text{K}(5.77 \text{ J/K}) = -1720 \text{ J}$$

Step 3: $\Delta E = 0,\ \ \Delta H = 0$

$$w = -(2.45 \times 10^{-3} \text{ atm})\left(\frac{nRT}{2.45 \times 10^{-3}} - \frac{nRT}{4.93 \times 10^{-3}}\right) \text{L} \times \frac{101.3 \text{ J}}{\text{L atm}} = -1250 \text{ J}$$

$$q = -w = 1250 \text{ J};\ \ \Delta S = nR \ln\left(\frac{4.93 \times 10^{-3} \text{ atm}}{2.45 \times 10^{-3} \text{ atm}}\right) = 5.81 \text{ J/K}$$

$$\Delta G = 0 - 298 \text{ K}(5.81 \text{ J/K}) = -1730 \text{ J}$$

	q	w	ΔE	ΔS	ΔH	ΔG
Step 1	1480 J	−1480 J	0	7.56 J/K	0	−2250 J
Step 2	1240 J	−1240 J	0	5.77 J/K	0	−1720 J
Step 3	1250 J	−1250 J	0	5.81 J/K	0	−1730 J
Total	3970 J	−3970 J	0	19.14 J/K	0	-5.70×10^3 J

123. a. $\Delta G° = 2\text{ mol}(-394\text{ kJ/mol}) - 2\text{ mol}(-137\text{ kJ/mol}) = -514\text{ kJ}$

$$K = \exp\left(\frac{-\Delta G°}{RT}\right) = \exp\left(\frac{-(-514{,}000\text{ J})}{(8.3145\text{ J K}^{-1}\text{ mol}^{-1})(298\text{ K})}\right) = 1.24 \times 10^{90}$$

b. $\Delta S° = 2(214\text{ J/K}) - [2(198\text{ J/K}) + 205\text{ J/K}] = -173\text{ J/K}$

$2\,CO(1.00\text{ atm}) + O_2(1.00\text{ atm}) \rightarrow 2\,CO_2(1.00\text{ atm})$	$\Delta S° = -173$ J/K
$2\,CO_2(1.00\text{ atm}) \rightarrow 2\,CO_2(10.0\text{ atm})$	$\Delta S = nR\ln(P_1/P_2) = -38.3$ J/K
$2\,CO(10.0\text{ atm}) \rightarrow 2\,CO(1.00\text{ atm})$	$\Delta S = nR\ln(P_1/P_2) = 38.3$ J/K
$O_2(10.0\text{ atm}) \rightarrow O_2(1.00\text{ atm})$	$\Delta S = nR\ln(P_1/P_2) = 19.1$ J/K
$2\,CO(10.0\text{ atm}) + O_2(10.0\text{ atm}) \rightarrow CO_2(10.0\text{ atm})$	$\Delta S = -173 + 19.1 = -154$ J/K

125.

T(°C)	T(K)	C_p ($J\,K^{-1}\,mol^{-1}$)	C_p/T ($J\,K^{-2}\,mol^{-1}$)
−200.	73	12	0.16
-180.	93	15	0.16
-160.	113	17	0.15
-140.	133	19	0.14
-100.	173	24	0.14
-60.	213	29	0.14
-30.	243	33	0.14
-10.	263	36	0.14
0	273	37	0.14

Total area of C_p/T versus T plot = ΔS = I + II + III (See following plot.)

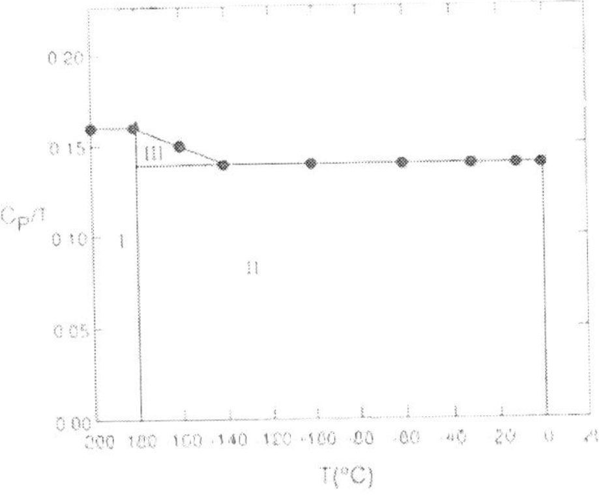

$\Delta S = (0.16 \text{ J K}^{-2} \text{ mol}^{-1})(20.\text{ K}) + (0.14 \text{ J K}^{-2} \text{ mol}^{-1})(180.\text{ K}) + 1/2\ (0.02 \text{ J K}^{-2} \text{ mol}^{-1})(40.\text{ K})$

$\Delta S = 3.2 + 25 + 0.4 = 29 \text{ J K}^{-1} \text{ mol}^{-1}$

127. To calculate ΔS_{sys} at 10.0°C, we need a place to start. From the data in the problem, we can calculate ΔS_{sys} at the melting point (5.5°C). For a phase change, $\Delta S_{sys} = q_{rev}/T = \Delta H/T$, where ΔH is determined at the melting point (5.5°C). Solving for ΔH at 5.5°C (using a thermochemical cycle):

$C_6H_6(l, 25.0°C) \rightarrow C_6H_6(s, 25.0°C)$ $\Delta H = -10.04$ kJ
$C_6H_6(l, 5.5°C) \rightarrow C_6H_6(l, 25.0°C)$ $\Delta H = nC_p\Delta T/1000 = 2.59$ kJ
$C_6H_6(s, 25.0°C) \rightarrow C_6H_6(s, 5.5°C)$ $\Delta H = nC_p\Delta T/1000 = -1.96$ kJ

$C_6H_6(l, 5.5°C) \rightarrow C_6H_6(s, 5.5°C)$ $\Delta H = -9.41$ kJ

At the melting point, $\Delta S_{sys} = \dfrac{\Delta H}{T} = \dfrac{-9.41 \times 10^3 \text{ J}}{278.7 \text{ K}} = -33.8$ J/K.

For the phase change at 10.0°C (283.2 K):

$C_6H_6(l, 278.7\text{ K}) \rightarrow C_6H_6(s, 278.7\text{ K})$ $\Delta S = -33.8$ J/K
$C_6H_6(l, 283.2\text{ K}) \rightarrow C_6H_6(l, 278.7\text{ K})$ $\Delta S = nC_p \ln(T_2/T_1) = -2.130$ J/K
$C_6H_6(s, 278.7\text{ K}) \rightarrow C_6H_6(s, 283.2\text{ K})$ $\Delta S = nC_p \ln(T_2/T_1) = 1.608$ J/K

$C_6H_6(l, 283.2\text{ K}) \rightarrow C_6H_6(s, 283.2\text{ K})$ $\Delta S_{sys} = -34.3$ J/K

To calculate ΔS_{surr}, we need ΔH at 10.0°C ($\Delta S_{surr} = \dfrac{-\Delta H}{T}$).

$C_6H_6(l, 25.0°C) \rightarrow C_6H_6(s, 25.0°C)$ $\Delta H = -10.04$ kJ
$C_6H_6(l, 10.0°C) \rightarrow C_6H_6(l, 25.0°C)$ $\Delta H = nC_p\Delta T/1000 = 2.00$ kJ
$C_6H_6(s, 25.0°C) \rightarrow C_6H_6(s, 10.0°C)$ $\Delta H = nC_p\Delta T/1000 = -1.51$ kJ

$C_6H_6(l, 10.0°C) \rightarrow C_6H_6(s, 10.0°C)$ $\Delta H = -9.55$ kJ

$$\Delta S_{surr} = \frac{-\Delta H}{T} = \frac{-(-9.55 \times 10^3 \text{ J})}{283.2 \text{ K}} = 33.7 \text{ J/K}$$

CHAPTER 11

ELECTROCHEMISTRY

Galvanic Cells, Cell Potentials, and Standard Reduction Potentials

15. Electrochemistry is the study of the interchange of chemical and electrical energy. A redox (oxidation-reduction) reaction is a reaction in which one or more electrons are transferred. In a galvanic cell, a spontaneous redox reaction occurs that produces an electric current. In an electrolytic cell, electricity is used to force a nonspontaneous redox reaction to occur.

17. A typical galvanic cell diagram is:

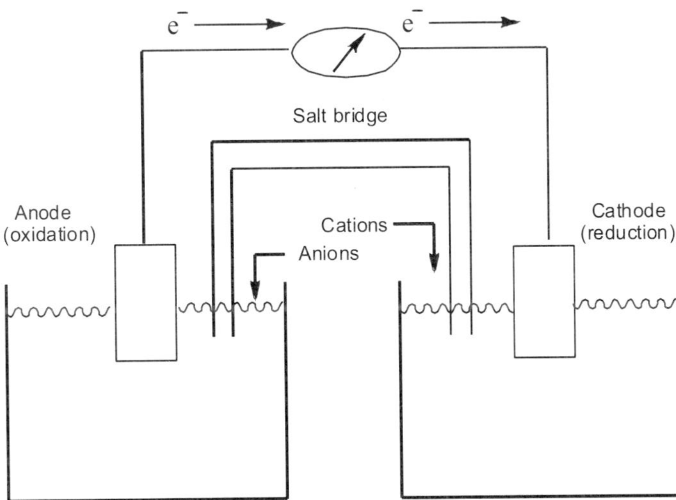

The diagram for all cells will look like this. The contents of each half-cell will be identified for each reaction, with all concentrations at 1.0 *M* and partial pressures at 1.0 atm. Note that cations always flow into the cathode compartment and anions always flow into the anode compartment. This is required to keep each compartment electrically neutral.

a. Reference Table 11.1 for standard reduction potentials. Remember that E°_{cell} = E°(cathode) − E°(anode); in the *Solutions Guide*, we will represent E°(cathode) as E°_c and represent −E°(anode) as $-E^\circ_a$. Also remember that standard potentials are *not* multiplied by the integer used to obtain the overall balanced equation.

$$(Cl_2 + 2\,e^- \rightarrow 2\,Cl^-) \times 3 \qquad E^o_c = 1.36\text{ V}$$
$$7\,H_2O + 2\,Cr^{3+} \rightarrow Cr_2O_7^{2-} + 14\,H^+ + 6\,e^- \qquad -E^o_a = -1.33\text{ V}$$

$$7\,H_2O(l) + 2\,Cr^{3+}(aq) + 3\,Cl_2(g) \rightarrow Cr_2O_7^{2-}(aq) + 6\,Cl^-(aq) + 14\,H^+(aq) \qquad E^o_{cell} = 0.03\text{ V}$$

The contents of each compartment are:

Cathode: Pt electrode; Cl_2 bubbled into solution, Cl^- in solution

Anode: Pt electrode; Cr^{3+}, H^+, and $Cr_2O_7^{2-}$ in solution

We need a nonreactive metal to use as the electrode in each case since all the reactants and products are in solution. Pt is the most common choice. Another possibility is graphite.

b.

$$Cu^{2+} + 2\,e^- \rightarrow Cu \qquad E^o_c = 0.34\text{ V}$$
$$Mg \rightarrow Mg^{2+} + 2\,e^- \qquad -E^o_a = 2.37\text{ V}$$

$$Cu^{2+}(aq) + Mg(s) \rightarrow Cu(s) + Mg^{2+}(aq) \qquad E^o_{cell} = 2.71\text{ V}$$

Cathode: Cu electrode; Cu^{2+} in solution

Anode: Mg electrode; Mg^{2+} in solution

c.

$$5\,e^- + 6\,H^+ + IO_3^- \rightarrow 1/2\,I_2 + 3\,H_2O \qquad E^o_c = 1.20\text{ V}$$
$$(Fe^{2+} \rightarrow Fe^{3+} + e^-) \times 5 \qquad -E^o_a = -0.77\text{ V}$$

$$6\,H^+ + IO_3^- + 5\,Fe^{2+} \rightarrow 5\,Fe^{3+} + 1/2\,I_2 + 3\,H_2O \qquad E^o_{cell} = 0.43\text{ V}$$

or $12\,H^+(aq) + 2\,IO_3^-(aq) + 10\,Fe^{2+}(aq) \rightarrow 10\,Fe^{3+}(aq) + I_2(s) + 6\,H_2O(l) \qquad E^o_{cell} = 0.43\text{ V}$

Cathode: Pt electrode; IO_3^-, I_2, and H_2SO_4 (H^+ source) in solution

Anode: Pt electrode; Fe^{2+} and Fe^{3+} in solution

Note: $I_2(s)$ would make a poor electrode since it sublimes.

d.

$$(Ag^+ + e^- \rightarrow Ag) \times 2 \qquad E^o_c = 0.80\text{ V}$$
$$Zn \rightarrow Zn^{2+} + 2\,e^- \qquad -E^o_a = 0.76\text{ V}$$

$$Zn(s) + 2\,Ag^+(aq) \rightarrow 2\,Ag(s) + Zn^{2+}(aq) \qquad E^o_{cell} = 1.56\text{ V}$$

Cathode: Ag electrode; Ag^+ in solution

Anode: Zn electrode; Zn^{2+} in solution

19. Reference Exercise 11.17 for a typical galvanic cell design. The contents of each half-cell compartment are identified below, with all solute concentrations at 1.0 *M* and all gases at 1.0 atm. For each pair of half-reactions, the half-reaction with the largest standard reduction potential will be the cathode reaction, and the half-reaction with the smallest reduction potential will be reversed to become the anode reaction. Only this combination gives a spontaneous overall reaction, i.e., a reaction with a positive overall standard cell potential.

a.

$$Cl_2 + 2\,e^- \rightarrow 2\,Cl^- \qquad E^\circ_c = 1.36\text{ V}$$
$$2\,Br^- \rightarrow Br_2 + 2\,e^- \qquad -E^\circ_a = -1.09\text{ V}$$

$$Cl_2(g) + 2\,Br^-(aq) \rightarrow Br_2(aq) + 2\,Cl^-(aq) \qquad E^\circ_{cell} = 0.27\text{ V}$$

The contents of each compartment are:

Cathode: Pt electrode; $Cl_2(g)$ bubbled in, Cl^- in solution

Anode: Pt electrode; Br_2 and Br^- in solution

b.

$$(2\,e^- + 2\,H^+ + IO_4^- \rightarrow IO_3^- + H_2O) \times 5 \qquad E^\circ_c = 1.60\text{ V}$$
$$(4\,H_2O + Mn^{2+} \rightarrow MnO_4^- + 8\,H^+ + 5\,e^-) \times 2 \qquad -E^\circ_a = -1.51\text{ V}$$

$$10\,H^+ + 5\,IO_4^- + 8\,H_2O + 2\,Mn^{2+} \rightarrow 5\,IO_3^- + 5\,H_2O + 2\,MnO_4^- + 16\,H^+ \qquad E^\circ_{cell} = 0.09\text{ V}$$

This simplifies to:

$$3\,H_2O(l) + 5\,IO_4^-(aq) + 2\,Mn^{2+}(aq) \rightarrow 5\,IO_3^-(aq) + 2\,MnO_4^-(aq) + 6\,H^+(aq)$$
$$E^\circ_{cell} = 0.09\text{ V}$$

Cathode: Pt electrode; IO_4^-, IO_3^-, and H_2SO_4 (as a source of H^+) in solution

Anode: Pt electrode; Mn^{2+}, MnO_4^-, and H_2SO_4 in solution

c.

$$H_2O_2 + 2\,H^+ + 2\,e^- \rightarrow 2\,H_2O \qquad E^\circ_c = 1.78\text{ V}$$
$$H_2O_2 \rightarrow O_2 + 2\,H^+ + 2\,e^- \qquad -E^\circ_a = -0.68\text{ V}$$

$$2\,H_2O_2(aq) \rightarrow 2\,H_2O(l) + O_2(g) \qquad E^\circ_{cell} = 1.10\text{ V}$$

Cathode: Pt electrode; H_2O_2 and H^+ in solution

Anode: Pt electrode; $O_2(g)$ bubbled in, H_2O_2 and H^+ in solution

d.

$$(Fe^{3+} + 3\,e^- \rightarrow Fe) \times 2 \qquad E^\circ_c = -0.036\text{ V}$$
$$(Mn \rightarrow Mn^{2+} + 2\,e^-) \times 3 \qquad -E^\circ_a = 1.18\text{ V}$$

$$2\,Fe^{3+}(aq) + 3\,Mn(s) \rightarrow 2\,Fe(s) + 3\,Mn^{2+}(aq) \qquad E^\circ_{cell} = 1.14\text{ V}$$

Cathode: Fe electrode; Fe^{3+} in solution; Anode: Mn electrode; Mn^{2+} in solution

21. In standard line notation, the anode is listed first and the cathode is listed last. A double line separates the two compartments. By convention, the electrodes are on the ends, with all solutes and gases toward the middle. A single line is used to indicate a phase change. We also included all concentrations.

19a. $Pt \mid Br^-$ (1.0 *M*), Br_2 (1.0 *M*) $\|$ Cl_2 (1.0 atm) $\mid Cl^-$ (1.0 *M*) $\mid Pt$

19b. $Pt \mid Mn^{2+}$ (1.0 *M*), MnO_4^- (1.0 *M*), H^+ (1.0 *M*) $\mid\mid$ IO_4^- (1.0 *M*), IO_3^- (1.0 *M*), H^+ (1.0 *M*) $\mid Pt$

19c. $Pt \mid H_2O_2$ (1.0 *M*), H^+ (1.0 *M*) $\mid O_2$ (1.0 atm) $\mid\mid$ H_2O_2 (1.0 *M*), H^+ (1.0 *M*) $\mid Pt$

19d. $Mn \mid Mn^{2+}$ (1.0 *M*) $\mid\mid$ Fe^{3+} (1.0 *M*) $\mid Fe$

23. a. $2\,H^+ + 2\,e^- \rightarrow H_2$ $E° = 0.00$ V; $Cu \rightarrow Cu^{2+} + 2\,e^-$ $-E° = -0.34$ V

$E°_{cell} = -0.34$ V; no, H^+ cannot oxidize Cu to Cu^{2+} at standard conditions ($E°_{cell} < 0$).

b. $Fe^{3+} + e^- \rightarrow Fe^{2+}$ $E° = 0.77$ V; $2\,I^- \rightarrow I_2 + 2\,e^-$ $-E° = -0.54$ V

$E°_{cell} = 0.77 - 0.54 = 0.23$ V; yes, Fe^{3+} can oxidize I^- to I_2.

c. $H_2 \rightarrow 2\,H^+ + 2\,e^-$ $-E° = 0.00$ V; $Ag^+ + e^- \rightarrow Ag$ $E° = 0.80$ V

$E°_{cell} = 0.80$ V; yes, H_2 can reduce Ag^+ to Ag at standard conditions ($E°_{cell} > 0$).

d. $Fe^{2+} \rightarrow Fe^{3+} + e^-$ $-E° = -0.77$ V; $Cr^{3+} + e^- \rightarrow Cr^{2+}$ $E° = -0.50$ V

$E°_{cell} = -0.50 - 0.77 = -1.27$ V; no, Fe^{2+} cannot reduce Cr^{3+} to Cr^{2+} at standard conditions.

25. Good reducing agents are easily oxidized. The reducing agents are on the right side of the reduction half-reactions listed in Table 11.1. The best reducing agents have the most negative standard reduction potentials (E°); i.e., the best reducing agents have the most positive -E° value. The ordering from worst to best reducing agents is:

	F^-	<	H_2O	<	I_2	<	Cu^+	<	H^-	<	K
-E°(V)	-2.92		-1.23		-1.20		-0.16		2.23		2.92

27. a. $2\,Br^- \rightarrow Br_2 + 2\,e^-$ $-E°_a = -1.09$ V; $2\,Cl^- \rightarrow Cl_2 + 2\,e^-$ $-E°_a = -1.36$ V; $E°_c > 1.09$ V to oxidize Br^-; $E°_c < 1.36$ V to not oxidize Cl^-; $Cr_2O_7^{2-}$, O_2, MnO_2, and IO_3^- are all possible because when all these oxidizing agents are coupled with Br^-, they give $E°_{cell} > 0$, and when coupled with Cl^-, they give $E°_{cell} < 0$ (assuming standard conditions).

b. $Mn \rightarrow Mn^{2+} + 2\,e^- \quad -E^\circ_a = 1.18$ V; $Ni \rightarrow Ni^{2+} + 2\,e^- \quad -E^\circ_a = 0.23$ V; any oxidizing agent with $-0.23\text{ V} > E^\circ_c > -1.18$ V will work. $PbSO_4$, Cd^{2+}, Fe^{2+}, Cr^{3+}, Zn^{2+}, and H_2O will be able to oxidize Mn but not oxidize Ni (assuming standard conditions).

29.

$$\begin{array}{lr} ClO^- + H_2O + 2\,e^- \rightarrow 2\,OH^- + Cl^- & E^\circ_c = 0.90\text{ V} \\ 2\,NH_3 + 2\,OH^- \rightarrow N_2H_4 + 2\,H_2O + 2\,e^- & -E^\circ_a = 0.10\text{ V} \\ \hline ClO^-(aq) + 2\,NH_3(aq) \rightarrow Cl^-(aq) + N_2H_4(aq) + H_2O(l) & E^\circ_{cell} = 1.00\text{ V} \end{array}$$

Because E°_{cell} is positive for this reaction, at standard conditions ClO^- can spontaneously oxidize NH_3 to the somewhat toxic N_2H_4.

Cell Potential, Free Energy, and Equilibrium

31. An extensive property is one that depends directly on the amount of substance. The free energy change for a reaction depends on whether 1 mole of product is produced or 2 moles of product are produced or 1 million moles of product are produced. This is not the case for cell potentials, which do not depend on the amount of substance. The equation that relates ΔG to E is ΔG = −nFE. It is the n term that converts the intensive property E into the extensive property ΔG. n is the number of moles of electrons transferred in the balanced reaction that ΔG is associated with.

33. Because the cells are at standard conditions, $w_{max} = \Delta G = \Delta G^\circ = -nFE^\circ_{cell}$. See Exercise 11.20 for the balanced overall equations and E°_{cell}.

20a. $w_{max} = -(3\text{ mol }e^-)(96{,}485\text{ C/mol }e^-)(1.34\text{ J/C}) = -3.88 \times 10^5\text{ J} = -388\text{ kJ}$

20b. $w_{max} = -(2\text{ mol }e^-)(96{,}485\text{ C/mol }e^-)(1.40\text{ J/C}) = -2.70 \times 10^5\text{ J} = -270.\text{ kJ}$

35. Reference Exercise 11.19 for the balanced reactions and standard cell potentials. The balanced reactions are necessary to determine n, the moles of electrons transferred.

19a. $Cl_2(aq) + 2\,Br^-(aq) \rightarrow Br_2(aq) + 2\,Cl^-(aq) \quad E^\circ_{cell} = 0.27\text{ V} = 0.27\text{ J/C},\ n = 2\text{ mol }e^-$

$$\Delta G^\circ = -nFE^\circ_{cell} = -(2\text{ mol }e^-)(96{,}485\text{ C/mol }e^-)(0.27\text{ J/C}) = -5.2 \times 10^4\text{ J} = -52\text{ kJ}$$

$$E^\circ_{cell} = \frac{0.0591}{n}\log K,\ \log K = \frac{nE^\circ}{0.0591} = \frac{2(0.27)}{0.0591} = 9.14,\ K = 10^{9.14} = 1.4 \times 10^9$$

Note: When determining exponents, we will round off to the correct number of significant figures after the calculation is complete in order to help eliminate excessive round-off errors.

19b. $\Delta G° = -(10 \text{ mol e}^-)(96{,}485 \text{ C/mol e}^-)(0.09 \text{ J/C}) = -9 \times 10^4 \text{ J} = -90 \text{ kJ}$

$$\log K = \frac{10(0.09)}{0.0591} = 15.2,\ \ K = 10^{15.2} = 2 \times 10^{15}$$

19c. $\Delta G° = -(2 \text{ mol e}^-)(96{,}485 \text{ C/mol e}^-)(1.10 \text{ J/C}) = -2.12 \times 10^5 \text{ J} = -212 \text{ kJ}$

$$\log K = \frac{2(1.10)}{0.0591} = 37.225,\ \ K = 1.68 \times 10^{37}$$

19d. $\Delta G° = -(6 \text{ mol e}^-)(96{,}485 \text{ C/mol e}^-)(1.14 \text{ J/C}) = -6.60 \times 10^5 \text{ J} = -660. \text{ kJ}$

$$\log K = \frac{6(1.14)}{0.0591} = 115.736,\ \ K = 5.45 \times 10^{115}$$

37. a.

$$(4\ H^+ + NO_3^- + 3\ e^- \rightarrow NO + 2\ H_2O) \times 2 \qquad E_c^o = 0.96\ V$$
$$(Mn \rightarrow Mn^{2+} + 2\ e^-) \times 3 \qquad -E_a^o = 1.18\ V$$

$$3\ Mn(s) + 8\ H^+(aq) + 2\ NO_3^-(aq) \rightarrow 2\ NO(g) + 4\ H_2O(l) + 3\ Mn^{2+}(aq) \qquad E_{cell}^o = 2.14\ V$$

$$5 \times (2\ e^- + 2\ H^+ + IO_4^- \rightarrow IO_3^- + H_2O) \qquad E_c^o = 1.60\ V$$
$$2 \times (Mn^{2+} + 4\ H_2O \rightarrow MnO_4^- + 8\ H^+ + 5\ e^-) \qquad -E_a^o = -1.51\ V$$

$$5\ IO_4^-(aq) + 2\ Mn^{2+}(aq) + 3\ H_2O(l) \rightarrow 5\ IO_3^-(aq) + 2\ MnO_4^-(aq) + 6\ H^+(aq)$$
$$E_{cell}^o = 0.09\ V$$

b. Nitric acid oxidation (see part a for E_{cell}^o):

$$\Delta G° = -\ nFE_{cell}^o = -(6 \text{ mol e}^-)(96{,}485 \text{ C/mol e}^-)(2.14 \text{ J/C}) = -1.24 \times 10^6 \text{ J} = -1240 \text{ kJ}$$

$$\log K = \frac{nE^o}{0.0591} = \frac{6(2.14)}{0.0591} = 217,\ \ K \approx 10^{217}$$

Periodate oxidation (see part a for E_{cell}^o):

$$\Delta G° = -(10 \text{ mol e}^-)(96{,}485 \text{ C/mol e}^-)(0.09 \text{ J/C})(1 \text{ kJ/1000 J}) = -90 \text{ kJ}$$

$$\log K = \frac{10(0.09)}{0.0591} = 15.2,\ \ K = 10^{15.2} = 2 \times 10^{15}$$

39. $\Delta G° = -nFE° = \Delta H° - T\Delta S°,\ \ E° = \dfrac{T\Delta S^o}{nF} - \dfrac{\Delta H^o}{nF}$

If we graph E° versus T, we should get a straight line ($y = mx + b$). The slope (m) of the line is equal to ΔS°/nF, and the y intercept is equal to -ΔH°/nF. From the equation above, E° will have a small temperature dependence when ΔS° is close to zero.

41. a.

$$Cu^+ + e^- \rightarrow Cu \qquad E^\circ_c = 0.52\ V$$
$$Cu^+ \rightarrow Cu^{2+} + e^- \qquad -E^\circ_a = -0.16\ V$$

$$2\ Cu^+(aq) \rightarrow Cu^{2+}(aq) + Cu(s) \qquad E^\circ_{cell} = 0.36\ V;\ \text{spontaneous}$$

$$\Delta G^\circ = -nFE^\circ_{cell} = -(1\ mol\ e^-)(96{,}485\ C/mol\ e^-)(0.36\ J/C) = -34{,}700\ J = -35\ kJ$$

$$E^\circ_{cell} = \frac{0.0591}{n}\log K,\ \log K = \frac{nE^\circ}{0.0591} = \frac{1(0.36)}{0.0591} = 6.09,\ K = 10^{6.09} = 1.2 \times 10^6$$

b.

$$Fe^{2+} + 2\ e^- \rightarrow Fe \qquad E^\circ_c = -0.44\ V$$
$$(Fe^{2+} \rightarrow Fe^{3+} + e^-) \times 2 \qquad -E^\circ_a = -0.77\ V$$

$$3\ Fe^{2+}(aq) \rightarrow 2\ Fe^{3+}(aq) + Fe(s) \qquad E^\circ_{cell} = -1.21\ V;\ \text{not spontaneous}$$

c.

$$HClO_2 + 2\ H^+ + 2\ e^- \rightarrow HClO + H_2O \qquad E^\circ_c = 1.65\ V$$
$$HClO_2 + H_2O \rightarrow ClO_3^- + 3\ H^+ + 2\ e^- \qquad -E^\circ_a = -1.21\ V$$

$$2\ HClO_2(aq) \rightarrow ClO_3^-(aq) + H^+(aq) + HClO(aq) \qquad E^\circ_{cell} = 0.44\ V;\ \text{spontaneous}$$

$$\Delta G^\circ = -nFE^\circ_{cell} = -(2\ mol\ e^-)(96{,}485\ C/mol\ e^-)(0.44\ J/C) = -84{,}900\ J = -85\ kJ$$

$$\log K = \frac{nE^\circ}{0.0591} = \frac{2(0.44)}{0.0591} = 14.89,\ K = 7.8 \times 10^{14}$$

43.

$$Al^{3+} + 3\ e^- \rightarrow Al \qquad E^\circ_c = -1.66\ V$$
$$Al + 6\ F^- \rightarrow AlF_6^{3-} + 3\ e^- \qquad -E^\circ_a = 2.07\ V$$

$$Al^{3+}(aq) + 6\ F^-(aq) \rightarrow AlF_6^{3-}(aq) \qquad E^\circ_{cell} = 0.41\ V \qquad K = ?$$

$$\log K = \frac{nE^\circ}{0.0591} = \frac{3(0.41)}{0.0591} = 20.81,\ K = 10^{20.81} = 6.5 \times 10^{20}$$

45.

$$CuI + e^- \rightarrow Cu + I^- \qquad E^\circ_{CuI} = ?$$
$$Cu \rightarrow Cu^+ + e^- \qquad -E^\circ_a = -0.52\ V$$

$$CuI(s) \rightarrow Cu^+(aq) + I^-(aq) \qquad E^\circ_{cell} = E^\circ_{CuI} - 0.52\ V$$

For this overall reaction, $K = K_{sp} = 1.1 \times 10^{-12}$:

$$E^\circ_{cell} = \frac{0.0591}{n}\log K_{sp} = \frac{0.0591}{1}\log(1.1 \times 10^{-12}) = -0.71\ V$$

$$E^\circ_{cell} = -0.71\ V = E^\circ_{CuI} - 0.52,\quad E^\circ_{CuI} = -0.19\ V$$

Galvanic Cells: Concentration Dependence

47. a.

$$\begin{array}{ll} Au^{3+} + 3\ e^- \rightarrow Au & E^\circ_c = 1.50\ V \\ (Tl \rightarrow Tl^+ + e^-) \times 3 & -E^\circ_a = 0.34\ V \\ \hline Au^{3+}(aq) + 3\ Tl(s) \rightarrow Au(s) + 3\ Tl^+(aq) & E^\circ_{cell} = 1.84\ V \end{array}$$

b. $\Delta G^\circ = -nFE^\circ_{cell} = -(3\ mol\ e^-)(96{,}485\ C/mol\ e^-)(1.84\ J/C) = -5.33 \times 10^5\ J = -533\ kJ$

$$\log K = \frac{nE^\circ}{0.0591} = \frac{3(1.84)}{0.0591} = 93.401,\quad K = 10^{93.401} = 2.52 \times 10^{93}$$

c. At 25°C, $E_{cell} = E^\circ_{cell} - \frac{0.0591}{n}\log Q$, where $Q = \frac{[Tl^+]^3}{[Au^{3+}]}$.

$$E_{cell} = 1.84\ V - \frac{0.0591}{3}\log\frac{[Tl^+]^3}{[Au^{3+}]} = 1.84 - \frac{0.0591}{3}\log\frac{(1.0 \times 10^{-4})^3}{1.0 \times 10^{-2}}$$

$$E_{cell} = 1.84 - (-0.20) = 2.04\ V$$

49.

$$\begin{array}{ll} (Pb^{2+} + 2\ e^- \rightarrow Pb) \times 3 & E^\circ_c = -0.13\ V \\ (Al \rightarrow Al^{3+} + 3\ e^-) \times 2 & -E^\circ_a = 1.66\ V \\ \hline 3\ Pb^{2+}(aq) + 2\ Al(s) \rightarrow 3\ Pb(s) + 2\ Al^{3+}(aq) & E^\circ_{cell} = 1.53\ V \end{array}$$

From the balanced reaction, when the $[Al^{3+}]$ has increased by 0.60 mol/L (Al^{3+} is a product in the spontaneous reaction), then the Pb^{2+} concentration has decreased by 3/2 (0.60 mol/L) = 0.90 *M*.

$$E_{cell} = 1.53\ V - \frac{0.0591}{6}\log\frac{[Al^{3+}]^2}{[Pb^{2+}]^3} = 1.53 - \frac{0.0591}{6}\log\frac{(1.60)^2}{(0.10)^3}$$

$$E_{cell} = 1.53\ V - 0.034\ V = 1.50\ V$$

51. a. $n = 2$ for this reaction (lead goes from $Pb \rightarrow Pb^{2+}$ in $PbSO_4$).

$$E = E^\circ_{cell} - \frac{0.0591}{2}\log\left(\frac{1}{[H^+]^2[HSO_4^-]^2}\right) = 2.04\ V - \frac{0.0591}{2}\log\frac{1}{(4.5)^2(4.5)^2}$$

$$2.04\ V - (-0.077\ V) = 2.12\ V$$

b. We can calculate $\Delta G°$ from $\Delta G° = \Delta H° - T\Delta S°$ and then $E°$ from $\Delta G° = -nFE°$, or we can use the equation derived in Exercise 11.39.

$$E^{\circ}_{-20} = \frac{T\Delta S^{\circ} - \Delta H^{\circ}}{nF} = \frac{(253\ K)(263.5\ J/K) + 315.9 \times 10^3\ J}{(2\ mol\ e^-)(96{,}485\ C/mol\ e^-)} = 1.98\ J/C = 1.98\ V$$

c. $$E_{-20} = E^{\circ}_{-20} - \frac{RT}{nF}\ln Q = 1.98\ V - \frac{RT}{nF}\ln\frac{1}{[H^+]^2[HSO_4^-]^2}$$

$$E_{-20} = 1.98\ V - \frac{(8.3145\ J\ K^{-1}\ mol^{-1})(253\ K)}{(2\ mol\ e^-)(96{,}485\ C/mol\ e^-)}\ln\frac{1}{(4.5)^2(4.5)^2} = 1.98\ V - (-0.066\ V)$$

$$= 2.05\ V$$

d. As the temperature decreases, the cell potential decreases. Also, oil becomes more viscous at lower temperatures, which adds to the difficulty of starting an engine on a cold day. The combination of these two factors results in batteries failing more often on cold days than on warm days.

53. Concentration cell: a galvanic cell in which both compartments contain the same components but at different concentrations. All concentration cells have $E^{\circ}_{cell} = 0$ because both compartments contain the same contents. The driving force for the cell is the different ion concentrations at the anode and cathode. The cell produces a voltage as long as the ion concentrations are different. Equilibrium for a concentration cell is reached ($E = 0$) when the ion concentrations in the two compartments are equal.

The net reaction in a concentration cell is:

$$M^{a+}(\text{cathode}, x\ M) \rightarrow M^{a+}(\text{anode}, y\ M) \quad E^{\circ}_{cell} = 0$$

and the Nernst equation is:

$$E = E^{\circ} - \frac{0.0591}{n}\log Q = -\frac{0.0591}{a}\log\frac{[M^{a+}(\text{anode})]}{[M^{a+}(\text{cathode})]},$$ where a is the number of electrons transferred.

To register a potential ($E > 0$), the log Q term must be a negative value. This occurs when M^{a+}(cathode) > M^{a+}(anode). The higher ion concentration is always at the cathode, and the lower ion concentration is always at the anode. The magnitude of the cell potential depends on the magnitude of the differences in ion concentrations between the anode and cathode. The larger the difference in ion concentrations, the more negative is the log Q term, and the more positive is the cell potential. Thus, as the difference in ion concentrations between the anode and cathode compartments increases, the cell potential increases. This can be accomplished by decreasing the ion concentration at the anode and/or by increasing the ion concentration at the cathode.

55. As is the case for all concentration cells, $E^o_{cell} = 0$, and the smaller ion concentration is always in the anode compartment. The general Nernst equation for the Ni | Ni^{2+} (x *M*) | | Ni^{2+}(y *M*) | Ni concentration cell is:

$$E_{cell} = E^o_{cell} - \frac{0.0591}{n}\log Q = \frac{0.0591}{2}\log\frac{[Ni^{2+}]_{anode}}{[Ni^{2+}]_{cathode}}$$

a. Because both compartments are at standard conditions ($[Ni^{2+}] = 1.0$ *M*), $E_{cell} = E^o_{cell} = 0$ V. No electron flow occurs.

b. Cathode = 2.0 *M* Ni^{2+}; anode = 1.0 *M* Ni^{2+}; electron flow is always from the anode to the cathode, so electrons flow to the right in the diagram.

$$E_{cell} = \frac{-0.0591}{2}\log\frac{[Ni^{2+}]_{anode}}{[Ni^{2+}]_{cathode}} = \frac{-0.0591}{2}\log\frac{1.0}{2.0} = 8.9 \times 10^{-3}\text{ V}$$

c. Cathode = 1.0 *M* Ni^{2+}; anode = 0.10 *M* Ni^{2+}; electrons flow to the left in the diagram.

$$E_{cell} = \frac{-0.0591}{2}\log\frac{0.10}{1.0} = 0.030\text{ V}$$

d. Cathode = 1.0 *M* Ni^{2+}; anode = 4.0×10^{-5} *M* Ni^{2+}; electrons flow to the left in the diagram.

$$E_{cell} = \frac{-0.0591}{2}\log\frac{4.0 \times 10^{-5}}{1.0} = 0.13\text{ V}$$

e. Since both concentrations are equal, log(2.5/2.5) = log(1.0) = 0, and $E_{cell} = 0$. No electron flow occurs.

57. $Cu^{2+}(aq) + H_2(g) \rightarrow 2\,H^+(aq) + Cu(s)$ $E^o_{cell} = 0.34\text{ V} - 0.00\text{V} = 0.34\text{ V}$ and n = 2

Since $P_{H_2} = 1.0$ atm and $[H^+] = 1.0$ *M*: $E_{cell} = E^o_{cell} - \frac{0.0591}{2}\log\frac{1}{[Cu^{2+}]}$

a. $E_{cell} = 0.34\text{ V} - \frac{0.0591}{2}\log\frac{1}{2.5 \times 10^{-4}} = 0.34\text{ V} - 0.11\text{ V} = 0.23\text{ V}$

b. Use the K_{sp} expression to calculate the Cu^{2+} concentration in the cell.

$Cu(OH)_2(s) \rightleftharpoons Cu^{2+}(aq) + 2\,OH^-(aq)$ $K_{sp} = 1.6 \times 10^{-19} = [Cu^{2+}][OH^-]^2$

From the problem, $[OH^-] = 0.10$ *M*, so: $[Cu^{2+}] = \frac{1.6 \times 10^{-19}}{(0.10)^2} = 1.6 \times 10^{-17}$ *M*

$$E_{cell} = E^{o}_{cell} - \frac{0.0591}{2}\log\frac{1}{[Cu^{2+}]} = 0.34 - \frac{0.0591}{2}\log\frac{1}{1.6 \times 10^{-17}} = 0.34 - 0.50 = -0.16\ V$$

Because $E_{cell} < 0$, the forward reaction is not spontaneous, but the reverse reaction is spontaneous. The Cu electrode becomes the anode, and $E_{cell} = 0.16$ V for the reverse reaction. The cell reaction is $2\ H^+(aq) + Cu(s) \rightarrow Cu^{2+}(aq) + H_2(g)$.

c. $0.195\ V = 0.34\ V - \frac{0.0591}{2}\log\frac{1}{[Cu^{2+}]}$, $\log\frac{1}{[Cu^{2+}]} = 4.91$, $[Cu^{2+}] = 10^{-4.91} = 1.2 \times 10^{-5}\ M$

Note: When determining exponents, we will carry extra significant figures.

d. $E_{cell} = E^{o}_{cell} - (0.0591/2)\log(1/[Cu^{2+}]) = E^{o}_{cell} + 0.0296\log[Cu^{2+}]$; this equation is in the form of a straight-line equation, $y = mx + b$. A graph of E_{cell} versus $\log[Cu^{2+}]$ will yield a straight line with slope equal to 0.0296 V or 29.6 mV.

59. a. $Ag^+(x\ M$, anode) $\rightarrow Ag^+(0.10\ M$, cathode); for the silver concentration cell, $E° = 0.00$ (as is always the case for concentration cells) and n = 1.

$$E = 0.76\ V = 0.00 - \frac{0.0591}{1}\log\frac{[Ag^+]_{anode}}{[Ag^+]_{cathode}}$$

$$0.76 = -0.0591\log\frac{[Ag^+]_{anode}}{0.10},\quad \frac{[Ag^+]_{anode}}{0.10} = 10^{-12.86},\quad [Ag^+]_{anode} = 1.4 \times 10^{-14}\ M$$

b. $Ag^+(aq) + 2\ S_2O_3^{2-}(aq) \rightleftharpoons Ag(S_2O_3)_2^{3-}(aq)$

$$K = \frac{[Ag(S_2O_3)_2^{3-}]}{[Ag^+][S_2O_3^{2-}]^2} = \frac{1.0 \times 10^{-3}}{1.4 \times 10^{-14}(0.050)^2} = 2.9 \times 10^{13}$$

Electrolysis

61. $15\ A = \frac{15\ C}{s} \times \frac{60\ s}{min} \times \frac{60\ min}{mol} = 5.4 \times 10^4$ C of charge passed in 1 hour

a. $5.4 \times 10^4\ C \times \frac{1\ mol\ e^-}{96,485\ C} \times \frac{1\ mol\ Co}{2\ mol\ e^-} \times \frac{58.9\ g}{mol} = 16\ g\ Co$

b. $5.4 \times 10^4\ C \times \frac{1\ mol\ e^-}{96,485\ C} \times \frac{1\ mol\ Hf}{4\ mol\ e^-} \times \frac{178.5\ g}{mol} = 25\ g\ Hf$

c. $2\ I^- \rightarrow I_2 + 2e^-$; $5.4 \times 10^4\ C \times \frac{1\ mol\ e^-}{96,485\ C} \times \frac{1\ mol\ I_2}{2\ mol\ e^-} \times \frac{253.8\ g\ I_2}{mol\ I_2} = 71\ g\ I_2$

d. Cr is in the +6 oxidation state in CrO_3. Six moles of e^- are needed to produce 1 mol Cr from molten CrO_3.

$$5.4 \times 10^4 \text{ C} \times \frac{1 \text{ mol e}^-}{96{,}485 \text{ C}} \times \frac{1 \text{ mol Cr}}{6 \text{ mol e}^-} \times \frac{52.0 \text{ g Cr}}{\text{mol Cr}} = 4.9 \text{ g Cr}$$

63. The oxidation state of bismuth in BiO^+ is +3 because oxygen has a −2 oxidation state in this ion. Therefore, 3 moles of electrons are required to reduce the bismuth in BiO^+ to Bi(s).

$$10.0 \text{ g Bi} \times \frac{1 \text{ mol Bi}}{209.0 \text{ g Bi}} \times \frac{3 \text{ mol e}^-}{\text{mol Bi}} \times \frac{96{,}485 \text{ C}}{\text{mol e}^-} \times \frac{1 \text{ s}}{25.0 \text{ C}} = 554 \text{ s} = 9.23 \text{ min}$$

65. First determine the species present, and then reference Table 11.1 to help you identify each species as a possible oxidizing agent (species reduced) or as a possible reducing agent (species oxidized). Of all the possible oxidizing agents, the species that will be reduced at the cathode will have the most positive E_c° value; the species that will be oxidized at the anode will be the reducing agent with the most positive $-E_a^\circ$ value.

a. Species present: Ni^{2+} and Br^-; Ni^{2+} can be reduced to Ni, and Br^- can be oxidized to Br_2 (from Table 11.1). The reactions are:

Cathode: $Ni^{2+} + 2e^- \rightarrow Ni$ $\quad E_c^\circ = -0.23$ V

Anode: $2\ Br^- \rightarrow Br_2 + 2\ e^-$ $\quad -E_a^\circ = -1.09$ V

b. Species present: Al^{3+} and F^-; Al^{3+} can be reduced, and F^- can be oxidized. The reactions are:

Cathode: $Al^{3+} + 3\ e^- \rightarrow Al$ $\quad E_c^\circ = -1.66$ V

Anode: $2\ F^- \rightarrow F_2 + 2\ e^-$ $\quad -E_a^\circ = -2.87$ V

c. Species present: Mn^{2+} and I^-; Mn^{2+} can be reduced, and I^- can be oxidized. The reactions are:

Cathode: $Mn^{2+} + 2\ e^- \rightarrow Mn$ $\quad E_c^\circ = -1.18$ V

Anode: $2\ I^- \rightarrow I_2 + 2\ e^-$ $\quad -E_a^\circ = -0.54$ V

d. For aqueous solutions, we must now consider H_2O as a possible oxidizing agent and a possible reducing agent. Species present: Ni^{2+}, Br^-, and H_2O. Possible cathode reactions are:

$Ni^{2+} + 2e^- \rightarrow Ni$ $\quad E_c^\circ = -0.23$ V

$2\ H_2O + 2\ e^- \rightarrow H_2 + 2\ OH^-$ $\quad E_c^\circ = -0.83$ V

Because it is easier to reduce Ni^{2+} than H_2O (assuming standard conditions), Ni^{2+} will be reduced by the above cathode reaction.

Possible anode reactions are:

$2\,Br^- \rightarrow Br_2 + 2\,e^-$ $\quad -E^\circ_a = -1.09$ V

$2\,H_2O \rightarrow O_2 + 4\,H^+ + 4\,e^-$ $\quad -E^\circ_a = -1.23$ V

Because Br^- is easier to oxidize than H_2O (assuming standard conditions), Br^- will be oxidized by the above anode reaction.

e. Species present: Al^{3+}, F^-, and H_2O; Al^{3+} and H_2O can be reduced. The reduction potentials are $E^\circ_c = -1.66$ V for Al^{3+} and $E^\circ_c = -0.83$ V for H_2O (assuming standard conditions). H_2O should be reduced at the cathode ($2\,H_2O + 2\,e^- \rightarrow H_2 + 2\,OH^-$).

F^- and H_2O can be oxidized. The oxidation potentials are $-E^\circ_a = -2.87$ V for F^- and $-E^\circ_a = -1.23$ V for H_2O (assuming standard conditions). From the potentials, we would predict H_2O to be oxidized at the anode ($2\,H_2O \rightarrow O_2 + 4\,H^+ + 4\,e^-$).

f. Species present: Mn^{2+}, I^-, and H_2O; Mn^{2+} and H_2O can be reduced. The possible cathode reactions are:

$Mn^{2+} + 2\,e^- \rightarrow Mn$ $\quad E^\circ_c = -1.18$ V

$2\,H_2O + 2\,e^- \rightarrow H_2 + 2\,OH^-$ $\quad E^\circ_c = -0.83$ V

Reduction of H_2O should occur at the cathode because E°_c for H_2O is most positive.

I^- and H_2O can be oxidized. The possible anode reactions are:

$2\,I^- \rightarrow I_2 + 2\,e^-$ $\quad -E^\circ_a = -0.54$ V

$2\,H_2O \rightarrow O_2 + 4\,H^+ + 4\,e^-$ $\quad -E^\circ_a = -1.23$ V

Oxidation of I^- will occur at the anode because $-E^\circ_a$ for I^- is most positive.

67 $Au^{3+} + 3\,e^- \rightarrow Au$ $\quad E^\circ = 1.50$ V $\qquad Ni^{2+} + 2\,e^- \rightarrow Ni$ $\quad E^\circ = -0.23$ V

$Ag^+ + e^- \rightarrow Ag$ $\quad E^\circ = 0.80$ V $\qquad Cd^{2+} + 2\,e^- \rightarrow Cd$ $\quad E^\circ = -0.40$ V

$2\,H_2O + 2e^- \rightarrow H_2 + 2\,OH^-$ $\quad E^\circ = -0.83$ V

Au(s) will plate out first since it has the most positive reduction potential, followed by Ag(s), which is followed by Ni(s), and finally, Cd(s) will plate out last since it has the most negative reduction potential of the metals listed.

69. To begin plating out Pd: $E_c = 0.62\text{ V} - \dfrac{0.0591}{2}\log\dfrac{[Cl^-]^4}{[PdCl_4^{\,2-}]} = 0.62 - \dfrac{0.0591}{2}\log\dfrac{(1.0)^4}{0.020}$

$E_c = 0.62\text{ V} - 0.050\text{ V} = 0.57\text{ V}$

When 99% of Pd has plated out, $[PdCl_4^-] = \frac{1}{100}(0.020) = 0.00020$ *M*.

$$E_c = 0.62 - \frac{0.0591}{2}\log\frac{(1.0)^4}{2.0\times10^{-4}} = 0.62\text{ V} - 0.11\text{V} = 0.51\text{ V}$$

To begin Pt plating: $E_c = 0.73\text{ V} - \frac{0.0591}{2}\log\frac{(1.0)^4}{0.020} = 0.73 - 0.050 = 0.68\text{ V}$

When 99% of Pt is plated: $E_c = 0.73 - \frac{0.0591}{2}\log\frac{(1.0)^4}{2.0\times10^{-4}} = 0.73 - 0.11 = 0.62\text{ V}$

To begin Ir plating: $E_c = 0.77\text{ V} - \frac{0.0591}{3}\log\frac{(1.0)^4}{0.020} = 0.77 - 0.033 = 0.74\text{ V}$

When 99% of Ir is plated: $E_c = 0.77 - \frac{0.0591}{3}\log\frac{(1.0)^4}{2.0\times10^{-4}} = 0.77 - 0.073 = 0.70\text{ V}$

Yes, since the range of potentials for plating out each metal do not overlap, we should be able to separate the three metals. The exact potential to apply depends on the oxidation reaction. The order of plating will be Ir(s) first, followed by Pt(s), and finally, Pd(s) as the potential is gradually increased.

71. Alkaline earth metals form +2 ions, so 2 mol of e^- are transferred to form the metal M.

$$\text{Mol M} = 748\text{ s}\times\frac{5.00\text{ C}}{\text{s}}\times\frac{1\text{ mol e}^-}{96{,}485\text{ C}}\times\frac{1\text{ mol M}}{2\text{ mol e}^-} = 1.94\times10^{-2}\text{ mol M}$$

Molar mass of $M = \frac{0.471\text{ g M}}{1.94\times10^{-2}\text{ mol M}} = 24.3$ g/mol; $MgCl_2$ was electrolyzed.

73. F_2 is produced at the anode: $2\,F^- \rightarrow F_2 + 2\,e^-$

$$2.00\text{ h}\times\frac{60\text{ min}}{\text{h}}\times\frac{60\text{ s}}{\text{min}}\times\frac{10.0\text{ C}}{\text{s}}\times\frac{1\text{ mol e}^-}{96{,}485\text{ C}} = 0.746\text{ mol e}^-$$

$0.746\text{ mol e}^-\times\frac{1\text{ mol F}_2}{2\text{ mol e}^-} = 0.373\text{ mol F}_2$; $PV = nRT$, $V = \frac{nRT}{P}$

$$V = \frac{(0.373\text{ mol})(0.08206\text{ L atm K}^{-1}\text{ mol}^{-1})(298\text{ K})}{1.00\text{ atm}} = 9.12\text{ L F}_2$$

K is produced at the cathode: $K^+ + e^- \rightarrow K$

$$0.746\text{ mol e}^-\times\frac{1\text{ mol K}}{\text{mol e}^-}\times\frac{39.10\text{ g K}}{\text{mol K}} = 29.2\text{ g K}$$

75. In the electrolysis of aqueous sodium chloride, H_2O is reduced in preference to Na^+, and Cl^- is oxidized in preference to H_2O. The anode reaction is $2\ Cl^- \rightarrow Cl_2 + 2\ e^-$, and the cathode reaction is $2\ H_2O + 2\ e^- \rightarrow H_2 + 2\ OH^-$. The overall reaction is $2\ H_2O(l) + 2\ Cl^-\ (aq) \rightarrow Cl_2(g) + H_2(g) + 2\ OH^-\ (aq)$.

From the 1 : 1 mole ratio between Cl_2 and H_2 in the overall balanced reaction, if 6.00 L of $H_2(g)$ is produced, then 6.00 L of $Cl_2(g)$ also will be produced since moles and volume of gas are directly proportional at constant T and P (see Chapter 5 of text).

Additional Exercises

77.

$$(CO + O^{2-} \rightarrow CO_2 + 2\ e^-) \times 2$$
$$O_2 + 4\ e^- \rightarrow 2\ O^{2-}$$
$$\overline{2\ CO + O_2 \rightarrow 2\ CO_2}$$

$$\Delta G = -nFE, \quad E = \frac{-\Delta G}{nF} = \frac{-(-380 \times 10^{-3}\ J)}{(4\ mol\ e^-)(96{,}485\ C/mol\ e^-)} = 0.98\ V$$

79. For C_2H_5OH, H has a +1 oxidation state, and O has a −2 oxidiation state. This dictates a −2 oxidation state for C. For CO_2, O has a −2 oxidiation state, so carbon has a +4 oxidiation state. Six moles of electrons are transferred per mole of carbon oxidized (C goes from −2 → +4). Two moles of carbon are in the balanced reaction, so n = 12.

$$w_{max} = -1320\ kJ = \Delta G = -nFE, \quad -1320 \times 10^3\ J = -nFE = -(12\ mol\ e^-)(96{,}485\ C/mol\ e^-)E$$

$$E = 1.14\ J/C = 1.14\ V$$

81. Cadmium goes from the zero oxidation state to the +2 oxidation state in $Cd(OH)_2$. Because one mole of Cd appears in the balanced reaction, n = 2 mol electrons transferred. At standard conditions:

$$w_{max} = \Delta G° = -nFE°, \quad w_{max} = -(2\ mol\ e^-)(96{,}485\ C/mol\ e^-)(1.10\ J/C) = -2.12 \times 10^5\ J = -212\ kJ$$

83.

a. Paint: covers the metal surface so that no contact occurs between the metal and air. This only works as long as the painted surface is not scratched.

b. Durable oxide coatings: covers the metal surface so that no contact occurs between the metal and air.

c. Galvanizing: coating steel with zinc; Zn forms an effective oxide coating over steel; also, zinc is more easily oxidized than the iron in the steel.

d. Sacrificial metal: attaching a more easily oxidized metal to an iron surface; the more active metal is preferentially oxidized instead of iron.

e. Alloying: adding chromium and nickel to steel; the added Cr and Ni form oxide coatings on the steel surface.

f. Cathodic protection: a more easily oxidized metal is placed in electrical contact with the metal we are trying to protect. It is oxidized in preference to the protected metal. The protected metal becomes the cathode electrode, thus cathodic protection.

85. $Zn \rightarrow Zn^{2+} + 2\,e^- \quad -E^\circ_a = 0.76$ V; $Fe \rightarrow Fe^{2+} + 2\,e^- \quad -E^\circ_a = 0.44$ V

It is easier to oxidize Zn than Fe, so the Zn will be oxidized, protecting the iron of the *Monitor*'s hull.

87. The potential oxidizing agents are NO_3^- and H^+. Hydrogen ion cannot oxidize Pt under either condition. Nitrate cannot oxidize Pt unless there is Cl^- in the solution. Aqua regia has both Cl^- and NO_3^-. The overall reaction is:

$$(NO_3^- + 4\,H^+ + 3\,e^- \rightarrow NO + 2\,H_2O) \times 2 \qquad E^\circ_c = 0.96 \text{ V}$$

$$(4\,Cl^- + Pt \rightarrow PtCl_4^{2-} + 2\,e^-) \times 3 \qquad -E^\circ_a = -0.755 \text{ V}$$

$$12\,Cl^-(aq) + 3\,Pt(s) + 2\,NO_3^-(aq) + 8\,H^+(aq) \rightarrow 3\,PtCl_4^{2-}(aq) + 2\,NO(g) + 4\,H_2O(l) \quad E^\circ_{cell} = 0.21 \text{ V}$$

89. a. $E_{cell} = E_{ref} + 0.05916$ pH, $0.480 \text{ V} = 0.250 \text{ V} + 0.05916$ pH

$$pH = \frac{0.480 - 0.250}{0.05916} = 3.888; \text{ uncertainty} = \pm 1 \text{ mV} = \pm 0.001 \text{ V}$$

$$pH_{max} = \frac{0.481 - 0.250}{0.05916} = 3.905; \quad pH_{min} = \frac{0.479 - 0.250}{0.05916} = 3.871$$

Thus if the uncertainty in potential is ± 0.001 V, then the uncertainty in pH is ± 0.017, or about ± 0.02 pH units. For this measurement, $[H^+] = 10^{-3.888} = 1.29 \times 10^{-4}$ *M*. For an error of +1 mV, $[H^+] = 10^{-3.905} = 1.24 \times 10^{-4}$ *M*. For an error of −1 mV, $[H^+] = 10^{-3.871} = 1.35 \times 10^{-4}$ *M*. So the uncertainty in $[H^+]$ is $\pm 0.06 \times 10^{-4}$ *M* = $\pm 6 \times 10^{-6}$ *M*.

b. From part a, we will be within ± 0.02 pH units if we measure the potential to the nearest ± 0.001 V (± 1 mV).

91. $2\,Ag^+(aq) + Cu(s) \rightarrow Cu^{2+}(aq) + 2\,Ag(s) \quad E^\circ_{cell} = 0.80 - 0.34 \text{ V} = 0.46$ V; A galvanic cell produces a voltage as the forward reaction occurs. Any stress that increases the tendency of the forward reaction to occur will increase the cell potential, whereas a stress that decreases the tendency of the forward reaction to occur will decrease the cell potential.

a. Added Cu^{2+} (a product ion) will decrease the tendency of the forward reaction to occur, which will decrease the cell potential.

b. Added NH_3 removes Cu^{2+} in the form of $Cu(NH_3)_4^{2+}$. Because a product ion is removed, this will increase the tendency of the forward reaction to occur, which will increase the cell potential.

c. Added Cl^- removes Ag^+ in the form of AgCl(s). Because a reactant ion is removed, this will decrease the tendency of the forward reaction to occur, which will decrease the cell potential.

d. $Q_1 = \frac{[Cu^{2+}]_0}{[Ag^+]_0^2}$; as the volume of solution is doubled, each concentration is halved.

$$Q_2 = \frac{1/2\ [Cu^{2+}]_0}{(1/2\ [Ag^+]_0)^2} = \frac{2[Cu^{2+}]_0}{[Ag^+]_0^2} = 2Q_1$$

The reaction quotient is doubled because the concentrations are halved. Because reactions are spontaneous when $Q < K$, and because Q increases when the solution volume doubles, the reaction is closer to equilibrium, which will decrease the cell potential.

e. Because Ag(s) is not a reactant in this spontaneous reaction, and because solids do not appear in the reaction quotient expressions, replacing the silver electrode with a platinum electrode will have no effect on the cell potential.

Challenge Problems

93. Chromium(III) nitrate $[Cr(NO_3)_3]$ has chromium in the +3 oxidation state.

$$1.15\text{ g Cr} \times \frac{1\text{ mol Cr}}{52.00\text{ g}} \times \frac{3\text{ mol e}^-}{\text{mol Cr}} \times \frac{96{,}485\text{ C}}{\text{mol e}^-} = 6.40 \times 10^3\text{ C of charge}$$

For the Os cell, 6.40×10^3 C of charge also was passed.

$$3.15\text{ g Os} \times \frac{1\text{ mol Os}}{190.2\text{ g}} = 0.0166\text{ mol Os};\ 6.40 \times 10^3\text{ C} \times \frac{1\text{ mol e}^-}{96{,}485\text{ C}} = 0.0663\text{ mol e}^-$$

$$\frac{\text{Mol e}^-}{\text{Mol Os}} = \frac{0.0663}{0.0166} = 3.99 \approx 4$$

This salt is composed of Os^{4+} and NO_3^- ions. The compound is $Os(NO_3)_4$, osmium(IV) nitrate.

For the third cell, identify X by determining its molar mass. Two moles of electrons are transferred when X^{2+} is reduced to X.

$$\text{Molar mass} = \frac{2.11\text{ g X}}{6.40 \times 10^3\text{ C} \times \frac{1\text{ mol e}^-}{96{,}485\text{ C}} \times \frac{1\text{ mol X}}{2\text{ mol e}^-}} = 63.6\text{ g/mol}$$

This is copper (Cu).

95. a. $Zn(s) + Cu^{2+}(aq) \rightarrow Zn^{2+}(aq) + Cu(s) \quad E^{\circ}_{cell} = 1.10 \text{ V}$

$$E_{cell} = 1.10 \text{ V} - \frac{0.0591}{2} \log \frac{[Zn^{2+}]}{[Cu^{2+}]}$$

$$E_{cell} = 1.10 \text{ V} - \frac{0.0591}{2} \log \frac{0.10}{2.50} = 1.10 \text{ V} - (-0.041 \text{ V}) = 1.14 \text{ V}$$

b. $10.0 \text{ h} \times \frac{60 \text{ min}}{\text{h}} \times \frac{60 \text{ s}}{\text{min}} \times \frac{10.0 \text{ C}}{\text{s}} \times \frac{1 \text{ mol e}^-}{96{,}485 \text{ C}} \times \frac{1 \text{ mol Cu}}{2 \text{ mol e}^-} = 1.87 \text{ mol Cu produced}$

The Cu^{2+} concentration decreases by 1.87 mol/L, and the Zn^{2+} concentration will increase by 1.87 mol/L.

$[Cu^{2+}] = 2.50 - 1.87 = 0.63$ *M*; $[Zn^{2+}] = 0.10 + 1.87 = 1.97$ *M*

$$E_{cell} = 1.10 \text{ V} - \frac{0.0591}{2} \log \frac{1.97}{0.63} = 1.10 \text{ V} - 0.015 \text{ V} = 1.09 \text{ V}$$

c. $1.87 \text{ mol Zn consumed} \times \frac{65.38 \text{ g Zn}}{\text{mol Zn}} = 122 \text{ g Zn}$

Mass of electrode = 200. − 122 = 78 g Zn

$$1.87 \text{ mol Cu formed} \times \frac{63.55 \text{ g Cu}}{\text{mol Cu}} = 119 \text{ g Cu}$$

Mass of electrode = 200. + 119 = 319 g Cu

d. Three things could possibly cause this battery to go dead:

1. All the Zn is consumed.
2. All the Cu^{2+} is consumed.
3. Equilibrium is reached ($E_{cell} = 0$).

We began with 2.50 mol Cu^{2+} and 200. g Zn × 1 mol Zn/65.38 g Zn = 3.06 mol Zn. Cu^{2+} is the limiting reagent and will run out first. To react all the Cu^{2+} requires:

$$2.50 \text{ mol } Cu^{2+} \times \frac{2 \text{ mol e}^-}{\text{mol } Cu^{2+}} \times \frac{96{,}485 \text{ C}}{\text{mol e}^-} \times \frac{1 \text{ s}}{10.0 \text{ C}} \times \frac{1 \text{ h}}{3600 \text{ s}} = 13.4 \text{ h}$$

For equilibrium to be reached: $E = 0 = 1.10 \text{ V} - \frac{0.0591}{2} \log \frac{[Zn^{2+}]}{[Cu^{2+}]}$

$$\frac{[Zn^{2+}]}{[Cu^{2+}]} = K = 10^{2(1.10)/0.0591} = 1.68 \times 10^{37}$$

This is such a large equilibrium constant that virtually all the Cu^{2+} must react to reach equilibrium. So the battery will go dead in 13.4 hours.

97. a.

$$3 \times (e^- + 2\ H^+ + NO_3^- \rightarrow NO_2 + H_2O) \qquad E_c^o = 0.775\ V$$

$$2\ H_2O + NO \rightarrow NO_3^- + 4\ H^+ + 3\ e^- \qquad -E_a^o = -0.957\ V$$

$$2\ H^+(aq) + 2\ NO_3^-(aq) + NO(g) \rightarrow 3\ NO_2(g) + H_2O(l) \qquad E_{cell}^o = -0.182\ V \quad K = ?$$

$$\log K = \frac{E^o}{0.0591} = \frac{3(-0.182)}{0.0591} = -9.239,\ K = 10^{-9.239} = 5.77 \times 10^{-10}$$

b. Let C = concentration of $HNO_3 = [H^+] = [NO_3^-]$.

$$5.77 \times 10^{-10} = \frac{P_{NO_2}^3}{P_{NO} \times [H^+]^2 \times [NO_3^-]^2} = \frac{P_{NO_2}^3}{P_{NO} \times C^4}$$

If 0.20% NO_2 by moles and P_{total} = 1.00 atm:

$$P_{NO_2} = \frac{0.20\ \text{mol}\ NO_2}{100.\ \text{mol total}} \times 1.00\ \text{atm} = 2.0 \times 10^{-3}\ \text{atm};\ P_{NO} = 1.00 - 0.0020 = 1.00\ \text{atm}$$

$$5.77 \times 10^{-10} = \frac{(2.0 \times 10^{-3})^3}{(1.00)C^4},\ C = 1.9\ M\ HNO_3$$

99.

$$2\ H^+ + 2\ e^- \rightarrow H_2 \qquad E_c^o = 0.000\ V$$

$$Fe \rightarrow Fe^{2+} + 2e^- \qquad -E_a^o = -(-0.440V)$$

$$2\ H^+(aq) + Fe(s) \rightarrow H_2(g) + Fe^{3+}(aq) \qquad E_{cell}^o = 0.440\ V$$

$$E_{cell} = E_{cell}^o - \frac{0.0591}{n} \log Q,\ \text{where n = 2 and } Q = \frac{P_{H_2} \times [Fe^{3+}]}{[H^+]^2}$$

To determine K_a for the weak acid, first use the electrochemical data to determine the H^+ concentration in the half-cell containing the weak acid.

$$0.333\ V = 0.440\ V - \frac{0.0591}{2} \log \frac{1.00\ \text{atm}(1.00 \times 10^{-3}\ M)}{[H^+]^2}$$

$$\frac{0.107(2)}{0.0591} = \log \frac{1.0 \times 10^{-3}}{[H^+]^2},\ \frac{1.0 \times 10^{-3}}{[H^+]^2} = 10^{3.621} = 4.18 \times 10^3,\ [H^+] = 4.89 \times 10^{-4}\ M$$

Now we can solve for the K_a value of the weak acid HA through the normal setup for a weak acid problem.

	$HA(aq)$	$\rightleftharpoons$	$H^+(aq)$	+	$A^-(aq)$	$K_a = \frac{[H^+][A^-]}{[HA]}$
Initial	1.00 *M*		~0		0	
Equil.	$1.00 - x$		x		x	

$$K_a = \frac{x^2}{1.00 - x}, \text{ where } x = [H^+] = 4.89 \times 10^{-4}\ M,\ K_a = \frac{(4.89 \times 10^{-4})^2}{1.00 - 4.89 \times 10^{-4}} = 2.39 \times 10^{-7}$$

101. a.

$$(Ag^+ + e^- \rightarrow Ag) \times 2 \qquad E^\circ_c = 0.80\text{ V}$$
$$Cu \rightarrow Cu^{2+} + 2e^- \qquad -E^\circ_a = -0.34\text{ V}$$

$$2\,Ag^+(aq) + Cu(s) \rightarrow 2\,Ag(s) + Cu^{2+}(aq) \qquad E^\circ_{cell} = 0.46\text{ V}$$

$$E_{cell} = E^\circ_{cell} - \frac{0.0591}{n}\log Q, \text{ where n = 2 and } Q = \frac{[Cu^{2+}]}{[Ag^+]^2}.$$

To calculate E_{cell}, we need to use the K_{sp} data to determine $[Ag^+]$.

	$AgCl(s)$	$\rightleftharpoons$	$Ag^+(aq)$	+	$Cl^-(aq)$	$K_{sp} = 1.6 \times 10^{-10} = [Ag^+][Cl^-]$
Initial	s = solubility (mol/L)		0		0	
Equil.			s		s	

$K_{sp} = 1.6 \times 10^{-10} = s^2$, $s = [Ag^+] = 1.3 \times 10^{-5}$ mol/L

$$E_{cell} = 0.46\text{ V} - \frac{0.0591}{2}\log\frac{2.0}{(1.3 \times 10^{-5})^2} = 0.46\text{ V} - 0.30 = 0.16\text{ V}$$

b. $Cu^{2+}(aq) + 4\,NH_3(aq) \rightleftharpoons Cu(NH_4)_4^{2+}(aq) \quad K = 1.0 \times 10^{13} = \frac{[Cu(NH_3)_4^{2+}]}{[Cu^{2+}][NH_3]^4}$

Because K is very large for the formation of $Cu(NH_3)_4^{2+}$, the forward reaction is dominant. At equilibrium, essentially all the 2.0 *M* Cu^{2+} will react to form 2.0 *M* $Cu(NH_3)_4^{2+}$. This reaction requires 8.0 *M* NH_3 to react with all the Cu^{2+} in the balanced equation. Therefore, the moles of NH_3 added to 1.0-L solution will be larger than 8.0 mol since some NH_3 must be present at equilibrium. In order to calculate how much NH_3 is present at equilibrium, we need to use the electrochemical data to determine the Cu^{2+} concentration.

$$E_{cell} = E^\circ_{cell} - \frac{0.0591}{n}\log Q,\ 0.52\text{ V} = 0.46\text{ V} - \frac{0.0591}{2}\log\frac{[Cu^{2+}]}{(1.3 \times 10^{-5})^2}$$

$$\log\frac{[Cu^{2+}]}{(1.3 \times 10^{-5})^2} = \frac{-0.06(2)}{0.0591} = -2.03,\ \frac{[Cu^{2+}]}{(1.3 \times 10^{-5})^2} = 10^{-2.03} = 9.3 \times 10^{-3}$$

$[Cu^{2+}] = 1.6 \times 10^{-12} = 2 \times 10^{-12}\ M$

(We carried extra significant figures in the calculation.)

Note: Our assumption that the 2.0 M Cu^{2+} essentially reacts to completion is excellent as only $2 \times 10^{-12}\ M$ Cu^{2+} remains after this reaction. Now we can solve for the equilibrium $[NH_3]$.

$$K = 1.0 \times 10^{13} = \frac{[Cu(NH_3)_4^{2+}]}{[Cu^{2+}][NH_3]^4} = \frac{(2.0)}{(2 \times 10^{-12})[NH_3]^4},\quad [NH_3] = 0.6\ M$$

Because 1.0 L of solution is present, 0.6 mol NH_3 remains at equilibrium. The total moles of NH_3 added is 0.6 mol plus the 8.0 mol NH_3 necessary to form 2.0 M $Cu(NH_3)_4^{2+}$. Therefore, 8.0 + 0.6 = 8.6 mol NH_3 was added.

103.

$$\begin{array}{lr} (Ag^+ + e^- \rightarrow Ag) \times 2 & E^\circ_c = 0.80\ V \\ Cd \rightarrow Cd^{2+} + 2\,e^- & -E^\circ_a = 0.40\ V \\ \hline 2\,Ag^+(aq) + Cd(s) \rightarrow Cd^{2+}(aq) + 2\,Ag(s) & E^\circ_{cell} = 1.20\ V \end{array}$$

Overall complex ion reaction:

$$Ag^+(aq) + 2\,NH_3(aq) \rightarrow Ag(NH_3)_2^+(aq) \quad K = K_1K_2 = 2.1 \times 10^3(8.2 \times 10^3) = 1.7 \times 10^7$$

Because K is large, we will let the reaction go to completion, and then solve the back-equilibrium problem.

	Ag^+	+	$2\,NH_3$	$\rightleftharpoons$	$Ag(NH_3)_2^+$	$K = 1.7 \times 10^7$
Before	1.00 M		15.0 M		0	
After	0		13.0		1.00	New initial
Change	x		$+2x$	$\leftarrow$	$-x$	
Equil.	x		$13.0 + 2x$		$1.00 - x$	

$$K = \frac{[Ag(NH_3)_2^+]}{[Ag^+][NH_3]^2};\quad 1.7 \times 10^7 = \frac{1.00 - x}{x(13.0 + 2x)^2} \approx \frac{1.00}{x(13.0)^2}$$

Solving: $x = 3.5 \times 10^{-10}\ M = [Ag^+]$; assumptions good.

$$E = E^\circ - \frac{0.0591}{2}\log\frac{[Cd^{2+}]}{[Ag^+]^2} = 1.20\ V - \frac{0.0591}{2}\log\left[\frac{1.0}{(3.5 \times 10^{-10})^2}\right]$$

$E = 1.20 - 0.56 = 0.64\ V$

105. a. From Table 11.1: $2\ H_2O + 2\ e^- \rightarrow H_2 + 2\ OH^-$ $E° = -0.83$ V

$$E^o_{cell} = E^o_{H_2O} - E^o_{Zr} = -0.83\ V + 2.36\ V = 1.53\ V$$

Yes, the reduction of H_2O to H_2 by Zr is spontaneous at standard conditions since $E^o_{cell} > 0$.

b.

$$(2\ H_2O + 2\ e^- \rightarrow H_2 + 2\ OH^-) \times 2$$

$$Zr + 4\ OH^- \rightarrow ZrO_2{\bullet}H_2O + H_2O + 4\ e^-$$

$$3\ H_2O(l) + Zr(s) \rightarrow 2\ H_2(g) + ZrO_2{\bullet}H_2O(s)$$

c. $\Delta G° = -nFE° = -(4\ \text{mol}\ e^-)(96{,}485\ \text{C/mol}\ e^-)(1.53\ \text{J/C}) = -5.90 \times 10^5\ \text{J} = -590.\ \text{kJ}$

$$E = E° - \frac{0.0591}{n}\log Q;\ \text{at equilibrium, } E = 0 \text{ and } Q = K.$$

$$E° = \frac{0.0591}{n}\log K,\ \log K = \frac{4(1.53)}{0.0591} = 104,\ K \approx 10^{104}$$

d. $$1.00 \times 10^3\ \text{kg Zr} \times \frac{1000\ \text{g}}{\text{kg}} \times \frac{1\ \text{mol Zr}}{91.22\ \text{g Zr}} \times \frac{2\ \text{mol}\ H_2}{\text{mol Zr}} = 2.19 \times 10^4\ \text{mol}\ H_2$$

$$2.19 \times 10^4\ \text{mol}\ H_2 \times \frac{2.016\ \text{g}\ H_2}{\text{mol}\ H_2} = 4.42 \times 10^4\ \text{g}\ H_2$$

$$V = \frac{nRT}{P} = \frac{(2.19 \times 10^4\ \text{mol})(0.08206\ \text{L atm K}^{-1}\ \text{mol}^{-1})(1273\ \text{K})}{1.0\ \text{atm}} = 2.3 \times 10^6\ \text{L}\ H_2$$

e. Probably yes; less radioactivity overall was released by venting the H_2 than what would have been released if the H_2 had exploded inside the reactor (as happened at Chernobyl). Neither alternative is pleasant, but venting the radioactive hydrogen is the less unpleasant of the two alternatives.

CHAPTER 12

QUANTUM MECHANICS AND ATOMIC THEORY

Light and Matter

21. $$\nu = \frac{c}{\lambda} = \frac{3.00 \times 10^8 \text{ m/s}}{1.0 \times 10^{-2} \text{ m}} = 3.0 \times 10^{10} \text{ s}^{-1}$$

$$E = h\nu = 6.63 \times 10^{-34} \text{ J s} \times 3.0 \times 10^{10} \text{ s}^{-1} = 2.0 \times 10^{-23} \text{ J/photon}$$

$$\frac{2.0 \times 10^{-23} \text{ J}}{\text{photon}} = \frac{6.02 \times 10^{23} \text{ photons}}{\text{mol}} = 12 \text{ J/mol}$$

23. a. $$\lambda = \frac{c}{\nu} = \frac{3.00 \times 10^8 \text{ m/s}}{6.0 \times 10^{13} \text{ s}^{-1}} = 5.0 \times 10^{-6} \text{ m}$$

b. From Figure 12.3, this is infrared EMR.

c. $$E = h\nu = 6.63 \times 10^{-34} \text{ J s} \times 6.0 \times 10^{13} \text{ s}^{-1} = 4.0 \times 10^{-20} \text{ J/photon}$$

$$\frac{4.0 \times 10^{-20} \text{ J}}{\text{photon}} \times \frac{6.022 \times 10^{23} \text{ photons}}{\text{mol}} = 2.4 \times 10^4 \text{ J/mol}$$

d. Frequency and photon energy are directly related ($E = h\nu$). Because $5.4 \times 10^{13} \text{ s}^{-1}$ EMR has a lower frequency than $6.0 \times 10^{13} \text{ s}^{-1}$ EMR, the $5.4 \times 10^{13} \text{ s}^{-1}$ EMR will have less energetic photons.

25. The energy needed to remove a single electron is:

$$\frac{279.7 \text{ kJ}}{\text{mol}} \times \frac{1 \text{ mol}}{6.0221 \times 10^{23}} = 4.645 \times 10^{-22} \text{ kJ} = 4.645 \times 10^{-19} \text{ J}$$

$$E = \frac{hc}{\lambda},\ \lambda = \frac{hc}{E} = \frac{6.6261 \times 10^{-34} \text{ J s} \times 2.9979 \times 10^8 \text{ m/s}}{4.645 \times 10^{-19} \text{ J}} = 4.277 \times 10^{-7} \text{ m} = 427.7 \text{ nm}$$

27. The energy to remove a single electron is:

$$\frac{208.4 \text{ kJ}}{\text{mol}} \times \frac{1 \text{ mol}}{6.022 \times 10^{23}} = 3.461 \times 10^{-22} \text{ kJ} = 3.461 \times 10^{-19} \text{ J} = E_w$$

Energy of 254-nm light is:

$$E = \frac{hc}{\lambda} = \frac{(6.626 \times 10^{-34} \text{ J s})(2.998 \times 10^8 \text{ m/s})}{254 \times 10^{-9} \text{ m}} = 7.82 \times 10^{-19} \text{ J}$$

$E_{photon} = E_K + E_w$, $E_K = 7.82 \times 10^{-19}$ J – 3.461×10^{-19} J = 4.36×10^{-19} J = maximum KE

29. The photoelectric effect refers to the phenomenon in which electrons are emitted from the surface of a metal when light strikes it. The light must have a certain minimum frequency (energy) in order to remove electrons from the surface of a metal. Light having a frequency below the minimum results in no electrons being emitted, whereas light at or higher than the minimum frequency does cause electrons to be emitted. For light having a frequency higher than the minimum frequency, the excess energy is transferred into kinetic energy for the emitted electron. Albert Einstein explained the photoelectric effect by applying quantum theory.

31. a. $$\lambda = \frac{h}{mv} = \frac{6.626 \times 10^{-34} \text{ J s}}{1.675 \times 10^{-27} \text{ kg} \times (0.0100 \times 2.998 \times 10^8 \text{ m/s})} = 1.32 \times 10^{-13} \text{ m}$$

b. $$\lambda = \frac{h}{mv},\ v = \frac{h}{\lambda m} = \frac{6.626 \times 10^{-34} \text{ J s}}{75 \times 10^{-12} \text{ m} \times 1.675 \times 10^{-27} \text{ kg}} = 5.3 \times 10^3 \text{ m/s}$$

33. $$m = \frac{h}{\lambda v} = \frac{6.626 \times 10^{-34} \text{ kg m}^2\text{/s}}{3.31 \times 10^{-15} \text{ m} \times (0.0100 \times 2.998 \times 10^8 \text{ m/s})} = 6.68 \times 10^{-26} \text{ kg/atom}$$

$$\frac{6.68 \times 10^{-26} \text{ kg}}{\text{atom}} \times \frac{6.022 \times 10^{23} \text{ atoms}}{\text{mol}} \times \frac{1000 \text{ g}}{\text{kg}} = 40.2 \text{ g/mol}$$

The element is calcium (Ca).

Hydrogen Atom: The Bohr Model

35. For the H atom (Z = 1): $E_n = -2.178 \times 10^{-18}$ J/n^2; for a spectral transition, $\Delta E = E_f - E_i$:

$$\Delta E = -2.178 \times 10^{-18} \text{ J} \left(\frac{1}{n_f^2} - \frac{1}{n_i^2} \right)$$

where n_i and n_f are the levels of the initial and final states, respectively. A positive value of ΔE always corresponds to an absorption of light, and a negative value of ΔE always corresponds to an emission of light.

a. $$\Delta E = -2.178 \times 10^{-18}\text{ J}\left(\frac{1}{2^2} - \frac{1}{3^2}\right) = -2.178 \times 10^{-18}\text{ J}\left(\frac{1}{4} - \frac{1}{9}\right)$$

$$\Delta E = -2.178 \times 10^{-18}\text{ J} \times (0.2500 - 0.1111) = -3.025 \times 10^{-19}\text{ J}$$

The photon of light must have precisely this energy (3.025×10^{-19} J).

$$|\Delta E| = E_{photon} = h\nu = \frac{hc}{\lambda},\ \lambda = \frac{hc}{|\Delta E|} = \frac{6.6261 \times 10^{-34}\text{ J s} \times 2.9979 \times 10^{8}\text{ m/s}}{3.025 \times 10^{-19}\text{ J}}$$

$$= 6.567 \times 10^{-7}\text{ m} = 656.7\text{ nm}$$

From Figure 12.3, this is visible electromagnetic radiation (red light).

b. $$\Delta E = -2.178 \times 10^{-18}\text{ J}\left(\frac{1}{2^2} - \frac{1}{4^2}\right) = -4.084 \times 10^{-19}\text{ J}$$

$$\lambda = \frac{hc}{|\Delta E|} = \frac{6.6261 \times 10^{-34}\text{ J s} \times 2.9979 \times 10^{8}\text{ m/s}}{4.084 \times 10^{-19}\text{ J}} = 4.864 \times 10^{-7}\text{ m} = 486.4\text{ nm}$$

This is visible electromagnetic radiation (green-blue light).

c. $$\Delta E = -2.178 \times 10^{-18}\text{ J}\left(\frac{1}{1^2} - \frac{1}{2^2}\right) = -1.634 \times 10^{-18}\text{ J}$$

$$\lambda = \frac{6.6261 \times 10^{-34}\text{ J s} \times 2.9979 \times 10^{8}\text{ m/s}}{1.634 \times 10^{-18}\text{ J}} = 1.216 \times 10^{-7}\text{ m} = 121.6\text{ nm}$$

This is ultraviolet electromagnetic radiation.

d. $$\Delta E = -2.178 \times 10^{-18}\text{ J}\left(\frac{1}{3^2} - \frac{1}{4^2}\right) = -1.059 \times 10^{-19}\text{ J}$$

$$\lambda = \frac{hc}{|\Delta E|} = \frac{6.6261 \times 10^{-34}\text{ J s} \times 2.9979 \times 10^{8}\text{ m/s}}{1.059 \times 10^{-19}\text{ J}} = 1.876 \times 10^{-6}\text{ m or 1876 nm}$$

This is infrared electromagnetic radiation.

37. a. False; it takes less energy to ionize an electron from $n = 3$ than from the ground state.

b. True

c. False; the energy difference between $n = 3$ and $n = 2$ is smaller than the energy difference to $n = 2$ electronic transition than for the $n = 3$ to $n = 1$ transition. E and λ are inversely proportional to each other (E = hc/λ).

d. True

e. False; the ground state in hydrogen is $n = 1$ and all other allowed energy states are called excited states; $n = 2$ is the first excited state, and $n = 3$ is the second excited state.

39. $$|\Delta E| = E_{photon} = \frac{hc}{\lambda} = \frac{6.6261 \times 10^{-34}\ \text{J s} \times 2.9979 \times 10^{8}\ \text{m/s}}{397.2 \times 10^{-9}\ \text{m}} = 5.001 \times 10^{-19}\ \text{J}$$

$\Delta E = -5.001 \times 10^{-19}$ J because we have an emission.

$$-5.001 \times 10^{-19}\ \text{J} = E_2 - E_n = -2.178 \times 10^{-18}\ \text{J}\left(\frac{1}{2^2} - \frac{1}{n^2}\right)$$

$$0.2296 = \frac{1}{4} - \frac{1}{n^2},\ \frac{1}{n^2} = 0.0204,\ n = 7$$

41. $$\Delta E = E_4 ! E_n = -E_n = 2.178 \times 10^{-18}\ \text{J}\left(\frac{1}{n^2}\right)$$

$$E_{photon} = \frac{hc}{\lambda} = \frac{6.626 \times 10^{-34}\ \text{J s} \times 2.9979 \times 10^{8}\ \text{m/s}}{1460 \times 10^{-9}\ \text{m}} = 1.36 \times 10^{-19}\ \text{J}$$

$$E_{photon} = \Delta E = 1.36 \times 10^{-19}\ \text{J} = 2.178 \times 10^{-18}\left(\frac{1}{n^2}\right),\quad n^2 = 16.0,\ n = 4$$

43. $$E_{photon} = \frac{hc}{\lambda} = \frac{6.6261 \times 10^{-34}\ \text{J s} \times 2.9979 \times 10^{8}\ \text{m/s}}{253.4 \times 10^{-9}\ \text{m}} = 7.839 \times 10^{-19}\ \text{J}$$

$\Delta E = 7.839 \times 10^{-19}$ J

The general energy equation for one-electron ions is $E_n = !2.178 \times 10^{-18}\ \text{J}\ (Z^2)/n^2$, where Z = atomic number.

$$\Delta E = !2.178 \times 10^{-18}\ \text{J}\ (Z)^2\left(\frac{1}{n_f^2} - \frac{1}{n_i^2}\right),\ Z = 4 \text{ for } Be^{3+}$$

$$\Delta E = !7.839 \times 10^{-19}\ \text{J} = !2.178 \times 10^{-18}\ (4)^2\left(\frac{1}{n_f^2} - \frac{1}{5^2}\right)$$

$$\frac{7.839 \times 10^{-19}}{2.178 \times 10^{-18} \times 16} + \frac{1}{25} = \frac{1}{n_f^2}, \quad \frac{1}{n_f^2} = 0.06249, \; n_f = 4$$

This emission line corresponds to the $n = 5 \rightarrow n = 4$ electronic transition.

Wave Mechanics and Particle in a Box

45. a. $\Delta p = m\Delta v = 9.11 \times 10^{-31}\text{ kg} \times 0.100\text{ m/s} = \dfrac{9.11 \times 10^{-32}\text{ kg m}}{\text{s}}$

$$\Delta p \Delta x \geq \frac{h}{4\pi}, \; \Delta x = \frac{h}{4\pi \Delta p} = \frac{6.626 \times 10^{-34}\text{ J s}}{4 \times 3.142 \times (9.11 \times 10^{-32}\text{ kg m/s})} = 5.79 \times 10^{-4}\text{ m}$$

b. $\Delta x = \dfrac{h}{4\pi \Delta p} = \dfrac{6.626 \times 10^{-34}\text{ J s}}{4 \times 3.142 \times 0.145\text{ kg} \times 0.100\text{ m/s}} = 3.64 \times 10^{-33}\text{ m}$

c. The diameter of an H atom is roughly 1.0×10^{-8} cm. The uncertainty in position is much larger than the size of the atom.

d. The uncertainty is insignificant compared to the size of a baseball.

47. $E_n = \dfrac{n^2h^2}{8mL^2}$; $\Delta E = E_3 - E_2 = \dfrac{9h^2}{8mL^2} - \dfrac{4h^2}{8mL^2} = \dfrac{5h^2}{8mL^2}$

$$\Delta E = \frac{hc}{\lambda} = \frac{(6.626 \times 10^{-34}\text{ J s})(2.998 \times 10^{8}\text{ m/s})}{8080 \times 10^{-9}\text{ m}} = 2.46 \times 10^{-20}\text{ J}$$

$$\Delta E - 2.46 \times 10^{-20}\text{ J} = \frac{5h^2}{8mL^2} = \frac{5(6.626 \times 10^{-34}\text{ J s})^2}{8(9.109 \times 10^{-31}\text{ kg})\,L^2}, \; L = 3.50 \times 10^{-9}\text{ m} = 3.50\text{ nm}$$

49. $E_n = \dfrac{n^2h^2}{8mL^2}$; as L increases, E_n will decrease, and the spacing between energy levels will also decrease.

51. $E_n = \dfrac{n^2h^2}{8mL^2}$, $n = 1$ for ground state; from equation, as L increases, E_n decreases.

Using numbers: 10^{-6} m box: $E_1 = \dfrac{h^2}{8m}(1 \times 10^{12}\text{ m}^{-2})$; 10^{-10} m box: $E_1 = \dfrac{h^2}{8m}(1 \times 10^{20}\text{ m}^{-2})$

As expected, the electron in the 1×10^{-6} m box has the lowest ground state energy.

Orbitals and Quantum Numbers

53. The 2p orbitals differ from each other in the direction in which they point in space. The 2p and 3p orbitals differ from each other in their size, energy, and number of nodes. A nodal surface in an atomic orbital is a surface in which the probability of finding an electron is zero.

    The 1p, 1d, 2d, 1f, 2f, and 3f orbitals are not allowed solutions to the Schrödinger equation. For $n = 1$, $\ell \neq 1, 2, 3$, etc., so 1p, 1d, and 1f orbitals are forbidden. For $n = 2$, $\ell \neq 2, 3, 4$, etc., so 2d and 2f orbitals are forbidden. For $n = 3$, $\ell \neq 3, 4, 5$, etc., so 3f orbitals are forbidden.

    The penetrating term refers to the fact that there is a higher probability of finding a 4s electron closer to the nucleus than a 3d electron. This leads to a lower energy for the 4s orbital relative to the 3d orbitals in polyelectronic atoms and ions.

55. a. For $n = 3$, $\ell = 3$ is not possible.

    d. m_s cannot equal !1.

    e. ℓ cannot be a negative number.

    f. For $\ell = 1$, m_ℓ cannot equal 2.

    The quantum numbers in parts b and c are allowed.

57. 1p, 0 electrons ($\ell \neq 1$ when $n = 1$); $6d_{x^2-y^2}$, 2 electrons (specifies one atomic orbital); 4f, 14 electrons (7 orbitals have 4f designation); $7p_y$, 2 electrons (specifies one atomic orbital); 2s, 2 electrons (specifies one atomic orbital); $n = 3$, 18 electrons (3s, 3p, and 3d orbitals are possible; there are one 3s orbital, three 3p orbitals, and five 3d orbitals).

59. The diagrams of the orbitals in the text give only 90% probabilities of where the electron may reside. We can never be 100% certain of the location of the electrons due to Heisenburg's uncertainty principle.

61. For $r = a_o$ and $\theta = 0°$ (Z = 1 for H):

$$\psi_{2p_z} = \frac{1}{4(2\pi)^{1/2}}\left(\frac{1}{5.29 \times 10^{-11}}\right)^{3/2} (1)\ e^{-1/2} \cos 0 = 1.57 \times 10^{14};\ \psi^2 = 2.46 \times 10^{28}$$

    For $r = a_o$ and $\theta = 90°$: $\psi_{2p_z} = 0$ because $\cos 90° = 0$; $\psi^2 = 0$; the xy plane is a node for the $2p_z$ atomic orbital.

Polyelectronic Atoms

63. Valence electrons are the electrons in the outermost principal quantum level of an atom (those electrons in the highest n value orbitals). The electrons in the lower n value orbitals are all inner core or just core electrons. The key is that the outermost electrons are the valence electrons. When atoms interact with each other, it will be the outermost electrons that are involved in these interactions. In addition, how tightly the nucleus holds these outermost

electrons determines atomic size, ionization energy, and other properties of atoms. Elements in the same group have similar valence electron configurations and, as a result, have similar chemical properties.

65. a. $n = 4$: ℓ can be 0, 1, 2, or 3. Thus we have s (2 e^-), p (6 e^-), d (10 e^-) and f (14 e^-) orbitals present. Total number of electrons to fill these orbitals is 32.

b. $n = 5$, $m_\ell = +1$: for $n = 5$, $\ell = 0, 1, 2, 3, 4$; for $\ell = 1, 2, 3, 4$, all can have $m_\ell = +1$. Four distinct orbitals, thus 8 electrons.

c. $n = 5$, $m_s = +1/2$: for $n = 5$, $\ell = 0, 1, 2, 3, 4$. Number of orbitals = 1, 3, 5, 7, 9 for each value of ℓ, respectively. There are 25 orbitals with $n = 5$. They can hold 50 electrons, and 25 of these electrons can have $m_s = +1/2$.

d. $n = 3$, $\ell = 2$: these quantum numbers define a set of 3d orbitals. There are 5 degenerate 3d orbitals that can hold a total of 10 electrons.

e. $n = 2$, $\ell = 1$: these define a set of 2p orbitals. There are 3 degenerate 2p orbitals that can hold a total of 6 electrons.

f. It is impossible for $n = 0$. Thus no electrons can have this set of quantum numbers.

g. The four quantum numbers completely specify a single electron.

h. $n = 3$: 3s, 3p, and 3d orbitals all have $n = 3$. These orbitals can hold 18 electrons, and 9 of these electrons can have $m_s = +1/2$.

i. $n = 2$, $\ell = 2$: this combination is not possible ($\ell \neq 2$ for $n = 2$). Zero electrons in an atom can have these quantum numbers.

j. $n = 1$, $\ell = 0$, $m_\ell = 0$: these define a 1s orbital that can hold 2 electrons.

67. The following are complete electron configurations. Noble gas shorthand notation could also be used.

Sc: $1s^2 2s^2 2p^6 3s^2 3p^6 4s^2 3d^1$; Fe: $1s^2 2s^2 2p^6 3s^2 3p^6 4s^2 3d^6$

P: $1s^2 2s^2 2p^6 3s^2 3p^3$; Cs: $1s^2 2s^2 2p^6 3s^2 3p^6 4s^2 3d^{10} 4p^6 5s^2 4d^{10} 5p^6 6s^1$

Eu: $1s^2 2s^2 2p^6 3s^2 3p^6 4s^2 3d^{10} 4p^6 5s^2 4d^{10} 5p^6 6s^2 4f^6 5d^1$*

Pt: $1s^2 2s^2 2p^6 3s^2 3p^6 4s^2 3d^{10} 4p^6 5s^2 4d^{10} 5p^6 6s^2 4f^{14} 5d^8$*

Xe: $1s^2 2s^2 2p^6 3s^2 3p^6 4s^2 3d^{10} 4p^6 5s^2 4d^{10} 5p^6$; Br: $1s^2 2s^2 2p^6 3s^2 3p^6 4s^2 3d^{10} 4p^5$

**Note*: These electron configurations were written down using only the periodic table.

Actual electron configurations are: Eu: $[Xe]6s^2 4f^7$ and Pt: $[Xe]6s^1 4f^{14} 5d^9$

69. Exceptions: Cr, Cu, Nb, Mo, Tc, Ru, Rh, Pd, Ag, Pt, and Au; Tc, Ru, Rh, Pd, and Pt do not correspond to the supposed extra stability of half-filled and filled subshells.

71. The two exceptions are Cr and Cu.

Cr: $1s^22s^22p^63s^23p^64s^13p^5$; Cr has 6 unpaired electrons.

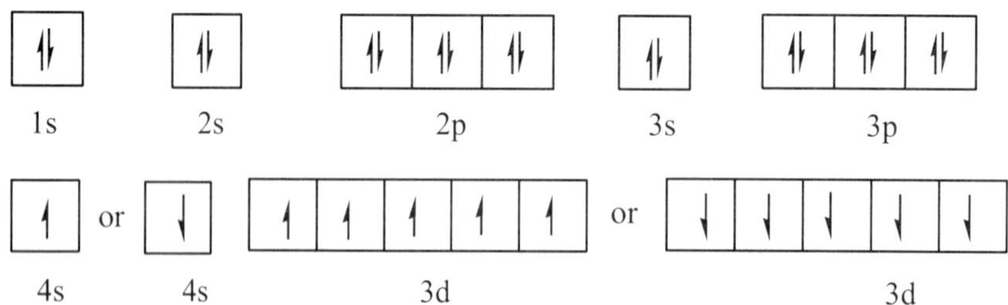

Cu: $1s^22s^22p^63s^23p^64s^13d^{10}$; Cu has 1 unpaired electron.

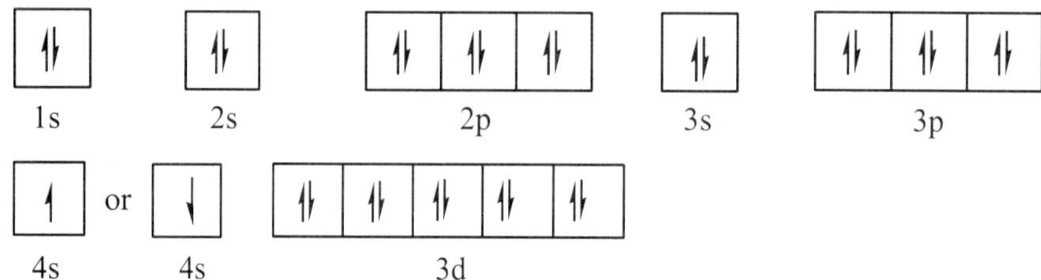

73. Element 115, Uup, is in Group 5A under Bi (bismuth):

Uup: $1s^22s^22p^63s^23p^64s^23d^{10}4p^65s^24d^{10}5p^66s^24f^{14}5d^{10}6p^67s^25f^{14}6d^{10}7p^3$

a. $5s^2$, $5p^6$, $5d^{10}$, and $5f^{14}$; 32 electrons have $n = 5$ as one of their quantum numbers

b. $\ell = 3$ are f orbitals. $4f^{14}$ and $5f^{14}$ are the f orbitals used. They are all filled so 28 electrons have $\ell = 3$.

c. p, d, and f orbitals all have one of the degenerate orbitals with $m_\ell = 1$. There are 6 orbitals with $m_\ell = 1$ for the various p orbitals used; there are 4 orbitals with $m_\ell = 1$ for the various d orbitals used; and there are 2 orbitals with $m_\ell = 1$ for the various f orbitals used. We have a total of $6 + 4 + 2 = 12$ orbitals with $m_\ell = 1$. Eleven of these orbitals are filled with 2 electrons, and the 7p orbitals are only half-filled. The number of electrons with $m_\ell = 1$ is $11 \times (2\ e^-) + 1 \times (1\ e^-) = 23$ electrons.

d. The first 112 electrons are all paired; one-half of these electrons ($56\ e^-$) will have m_s = !1/2. The 3 electrons in the 7p orbitals singly occupy each of the three degenerate 7p orbitals; the three electrons are spin parallel, so the 7p electrons either have $m_s = +1/2$ or m_s = !1/2. Therefore, either 56 electrons have m_s = !1/2 or 59 electrons have m_s = !1/2.

75. We get the number of unpaired electrons by examining the incompletely filled subshells. The paramagnetic substances have unpaired electrons, and the ones with no unpaired electrons are not paramagnetic (they are called diamagnetic).

Li: $1s^22s^1$ $\underset{2s}{\uparrow}$; paramagnetic with 1 unpaired electron.

N: $1s^22s^22p^3$ $\underset{2p}{\uparrow\ \uparrow\ \uparrow}$; paramagnetic with 3 unpaired electrons.

Ni: $[Ar]4s^23d^8$ $\underset{3d}{\uparrow\downarrow\ \uparrow\downarrow\ \uparrow\downarrow\ \uparrow\ \uparrow}$; paramagnetic with 2 unpaired electrons.

Te: $[Kr]5s^24d^{10}5p^4$ $\underset{5p}{\uparrow\downarrow\ \uparrow\ \uparrow}$; paramagnetic with 2 unpaired electrons.

Ba: $[Xe]6s^2$ $\underset{6s}{\uparrow\downarrow}$; not paramagnetic because no unpaired electrons are present.

Hg: $[Xe]6s^24f^{14}5d^{10}$ $\underset{5d}{\uparrow\downarrow\ \uparrow\downarrow\ \uparrow\downarrow\ \uparrow\downarrow\ \uparrow\downarrow}$; not paramagnetic because no unpaired electrons.

77. The s block elements with ns^1 for a valence electron configuration have one unpaired electrons. These are elements H, Li, Na, and K for the first 36 elements. The p block elements with ns^2np^1 or ns^2np^5 valence electron configurations have one unpaired electron. These are elements B, Al, and Ga (ns^2np^1) and elements F, Cl, and Br (ns^2np^5) for the first 36 elements. In the d block, Sc ($[Ar]4s^23d^1$) and Cu ($[Ar]4s^13d^{10}$) each have one unpaired electron. A total of 12 elements from the first 36 elements have one unpaired electron in the ground state.

79. We get the number of unpaired electrons by examining the incompletely filled subshells.

O: $[He]2s^22p^4$	$2p^4$: $\uparrow\downarrow\ \uparrow\ \uparrow$	Two unpaired e^-
O^+: $[He]2s^22p^3$	$2p^3$: $\uparrow\ \uparrow\ \uparrow$	Three unpaired e^-
O^-: $[He]2s^22p^5$	$2p^5$: $\uparrow\downarrow\ \uparrow\downarrow\ \uparrow$	One unpaired e^-
Os: $[Xe]6s^24f^{14}5d^6$	$5d^6$: $\uparrow\downarrow\ \uparrow\ \uparrow\ \uparrow\ \uparrow$	Four unpaired e^-
Zr: $[Kr]5s^24d^2$	$4d^2$: $\uparrow\ \uparrow$ __ __ __	Two unpaired e^-
S: $[Ne]3s^23p^4$	$3p^4$: $\uparrow\downarrow\ \uparrow\ \uparrow$	Two unpaired e^-
F: $[He]2s^22p^5$	$2p^5$: $\uparrow\downarrow\ \uparrow\downarrow\ \uparrow$	One unpaired e^-
Ar: $[Ne]3s^23p^6$	$3p^6$: $\uparrow\downarrow\ \uparrow\downarrow\ \uparrow\downarrow$	Zero unpaired e^-

The Periodic Table and Periodic Properties

81. Ionization energy: $P(g) \rightarrow P^+(g) + e^-$; electron affinity: $P(g) + e^- \rightarrow P^-(g)$

83. As successive electrons are removed, the net positive charge on the resulting ion increases. This increase in positive charge binds the remaining electrons more firmly, and the ionization energy increases.

    The electron configuration for Si is $1s^22s^22p^63s^23p^2$. There is a large jump in ionization energy when going from the removal of valence electrons to the removal of core electrons. For silicon, this occurs when the fifth electron is removed since we go from the valence electrons in $n = 3$ to the core electrons in $n = 2$. There should be another big jump when the thirteenth electron is removed, i.e., when a 1s electron is removed.

85. Size (radius) decreases left to right across the periodic table, and size increases from top to bottom of the periodic table.

    a. S < Se < Te b. Br < Ni < K c. F < Si < Ba

    d. Be < Na < Rb e. Ne < Se < Sr f. O < P < Fe

    All follow the general radius trend.

87. a. Ba b. K

    c. O; in general, Group 6A elements have a lower ionization energy than neighboring Group 5A elements. This is an exception to the general ionization energy trend across the periodic table.

    d. S^{2-}; this ion has the most electrons compared to the other sulfur species present. S^{2-} has the largest number of electron-electron repulsions, which leads to S^{2-} having the largest size and smallest ionization energy.

    e. Cs; this follows the general ionization energy trend.

89. As: $[Ar]4s^23d^{10}4p^3$; Se: $[Ar]4s^23d^{10}4p^4$; the general ionization energy trend predicts that Se should have a higher ionization energy than As. Se is an exception to the general ionization energy trend. There are extra electron-electron repulsions in Se because two electrons are in the same 4p orbital, resulting in a lower ionization energy for Se than predicted.

91. Size also decreases going across a period. Sc and Ti along with Y and Zr are adjacent elements. There are 14 elements (the lanthanides) between La and Hf, making Hf considerably smaller.

93. a. Uus will have 117 electrons. $[Rn]7s^25f^{14}6d^{10}7p^5$

    b. It will be in the halogen family and will be most similar to astatine (At).

    c. Like the other halogens: NaUus, $Mg(Uus)_2$, $C(Uus)_4$, $O(Uus)_2$

d. Like the other halogens: $UusO^-$, $UusO_2^-$, $UusO_3^-$, $UusO_4^-$

95. Electron-electron repulsions become important when we try to add electrons to an atom. From the standpoint of electron-electron repulsions, larger atoms would have more favorable (more exothermic) electron affinities. Considering only electron-nucleus attractions, smaller atoms would be expected to have the more favorable (more exothermic) EA values. These two factors are the opposite of each other. Thus the overall variation in EA is not as great as ionization energy, in which attractions to the nucleus dominate.

97. Electron-electron repulsions are much greater in O^- than in S^- because the electron goes into a smaller 2p orbital versus the larger 3p orbital in sulfur. This results in a more favorable (more exothermic) EA for sulfur.

99. a. The electron affinity of Mg^{2+} is ΔH for $Mg^{2+}(g) + e^- \rightarrow Mg^+(g)$; this is just the reverse of the second ionization energy for Mg. $EA(Mg^{2+}) = -IE_2(Mg) = -1445$ kJ/mol (Table 12.6)

b. EA of Al^+ is ΔH for $Al^+(g) + e^- \rightarrow Al(g)$; $EA(Al^+) = -IE_1(Al) =$ -580 kJ/mol (Table 12.6)

c. IE of Cl^- is ΔH for $Cl^-(g) \rightarrow Cl(g) + e^-$; $IE(Cl^-) = -EA(Cl) = +348.7$ kJ/mol (Table 12.8)

d. $Cl(g) \rightarrow Cl^+(g) + e^-$ $\Delta H = IE_1(Cl) = 1255$ kJ/mol (Table 12.6)

e. $Cl^+(g) + e^- \rightarrow Cl(g)$ $\Delta H = -IE_1(Cl) = -1255$ kJ/mol $= EA(Cl^+)$

The Alkali Metals

101. Yes; the ionization energy general trend is to decrease down a group, and the atomic radius trend is to increase down a group. The data in Table 12.9 confirm both of these general trends.

103. a. $6\ Li(s) + N_2(g) \rightarrow 2\ Li_3N(s)$

b. $2\ Rb(s) + S(s) \rightarrow Rb_2S(s)$

c. $2\ Cs(s) + 2\ H_2O(l) \rightarrow 2\ CsOH(aq) + H_2(g)$

d. $2\ Na(s) + Cl_2(g) \rightarrow 2\ NaCl(s)$

105. For 589.0 nm: $\nu = \frac{c}{\lambda} = \frac{2.9979 \times 10^8\ m/s}{589.0 \times 10^{-9}\ m} = 5.090 \times 10^{14}\ s^{-1}$

$$E = h\nu = 6.6261 \times 10^{-34}\ J\ s \times 5.090 \times 10^{14}\ s^{-1} = 3.373 \times 10^{-19}\ J$$

For 589.6 nm: $\nu = c/\lambda = 5.085 \times 10^{14}\ s^{-1}$; $E = h\nu = 3.369 \times 10^{-19}\ J$

The energies in kJ/mol are:

$$3.373 \times 10^{-19}\ \text{J} \times \frac{1\ \text{kJ}}{1000\ \text{J}} \times \frac{6.0221 \times 10^{23}}{\text{mol}} = 203.1\ \text{kJ/mol}$$

$$3.369 \times 10^{-19}\ \text{J} \times \frac{1\ \text{kJ}}{1000\ \text{J}} \times \frac{6.0221 \times 10^{23}}{\text{mol}} = 202.9\ \text{kJ/mol}$$

107. It should be element 119 with ground state electron configuration: $[Rn]7s^2 5f^{14} 6d^{10} 7p^6 8s^1$

Additional Exercises

109. a. n b. n and ℓ

111. Size decreases from left to right and increases going down the periodic table. Thus going one element right and one element down would result in a similar size for the two elements diagonal to each other. The ionization energies will be similar for the diagonal elements since the periodic trends also oppose each other. Electron affinities are harder to predict, but atoms with similar sizes and ionization energies should also have similar electron affinities.

113. $$60 \times 10^6\ \text{km} \times \frac{1000\ \text{m}}{\text{km}} \times \frac{1\ \text{s}}{3.00 \times 10^8\ \text{m}} = 200\ \text{s}\ \ \text{(about 3 minutes)}$$

115. $$\lambda = \frac{hc}{E} = \frac{6.626 \times 10^{-34}\ \text{J s} \times 2.998 \times 10^8\ \text{m/s}}{3.59 \times 10^{-19}\ \text{J}} = 5.53 \times 10^{-7}\ \text{m} \times \frac{100\ \text{cm}}{\text{m}}$$

$$= 5.53 \times 10^{-5}\ \text{cm}$$

From the spectrum, $\lambda = 5.53 \times 10^{-5}$ cm is greenish yellow light.

117. When the p and d orbital functions are evaluated at various points in space, the results sometimes have positive values and sometimes have negative values. The term phase is often associated with the + and ! signs. For example, a sine wave has alternating positive and negative phases. This is analogous to the positive and negative values (phases) in the p and d orbitals.

119. a.

$Na(g) \rightarrow Na^+(g) + e^-$	$IE_1 = 495$ kJ
$Cl(g) + e^- \rightarrow Cl^-(g)$	EA = !348.7 kJ
$Na(g) + Cl(g) \rightarrow Na^+(g) + Cl^-(g)$	$\Delta H = 146$ kJ

b.

$Mg(g) \rightarrow Mg^+(g) + e^-$	$IE_1 = 735$ kJ
$F(g) + e^- \rightarrow F^-(g)$	EA = !327.8 kJ
$Mg(g) + F(g) \rightarrow Mg^+(g) + F^-(g)$	$\Delta H = 407$ kJ

c.

$Mg^+(g) \rightarrow Mg^{2+}(g) + e^-$	$IE_2 = 1445$ kJ
$F(g) + e^- \rightarrow F^-(g)$	EA = !327.8 kJ
$Mg^+(g) + F(g) \rightarrow Mg^{2+}(g) + F^-(g)$	$\Delta H = 1117$ kJ

d. From parts b and c, we get:

$Mg(g) + F(g) \rightarrow Mg^+(g) + F^-(g)$	$\Delta H = 407$ kJ
$Mg^+(g) + F(g) \rightarrow Mg^{2+}(g) + F^-(g)$	$\Delta H = 1117$ kJ
$Mg(g) + 2\ F(g) \rightarrow Mg^{2+}(g) + 2\ F^-(g)$	$\Delta H = 1524$ kJ

121. Valence electrons are easier to remove than inner core electrons. The large difference in energy between I_2 and I_3 indicates that this element has two valence electrons. This element is most likely an alkaline earth metal since alkaline earth metal elements all have two valence electrons.

123. a. The 4+ ion contains 20 electrons. Thus the electrically neutral atom will contain 24 electrons. The atomic number is 24 which identifies it as chromium.

b. The ground state electron configuration of the ion must be: $1s^2 2s^2 2p^6 3s^2 3p^6 4s^0 3d^2$; there are 6 electrons in s orbitals.

c. 12

d. 2

e. This is the isotope $^{50}_{24}Cr$. There are 26 neutrons in the nucleus.

f. $3.01 \times 10^{23} \text{ atoms} \times \frac{1 \text{ mol}}{6.022 \times 10^{23} \text{ atoms}} \times \frac{49.9 \text{ g}}{\text{mol}} = 24.9 \text{ g}$

g. $1s^2 2s^2 2p^6 3s^2 3p^6 4s^1 3d^5$ is the ground state electron configuration for Cr. Cr is an exception to the normal filling order.

125. a. Each orbital could hold 4 electrons.

b. The first period corresponds to $n = 1$, which can only have 1s orbitals. The 1s orbital could hold 4 electrons; hence the first period would have four elements. The second period corresponds to $n = 2$, which has 2s and 2p orbitals. These four orbitals can each hold four electrons. A total of 16 elements would be in the second period.

c. 20

d. 28

127. At $x = 0$, the value of the square of the wave function must be zero. The particle must be inside the box. For $\psi = A \cos(Lx)$, at $x = 0$, $\cos(0) = 1$ and $\psi^2 = A^2$. This violates the boundary condition.

129. a. Because wavelength is inversely proportional to energy, the spectral line to the right of B (at a larger wavelength) represents the lowest possible energy transition; this is $n = 4$ to $n = 3$. The B line represents the next lowest energy transition, which is $n = 5$ to $n = 3$, and the A line corresponds to the $n = 6$ to $n = 3$ electronic transition.

b. Because this spectrum is for a one-electron ion, $E_n = -2.178 \times 10^{-18}$ J (Z^2/n^2). To determine ΔE and, in turn, the wavelength of spectral line A, we must determine Z, the atomic number of the one electron species. Use spectral line B data to determine Z.

$$\Delta E_{5\to3} = -2.178 \times 10^{-18}\ \text{J}\left(\frac{Z^2}{3^2} - \frac{Z^2}{5^2}\right) = -2.178 \times 10^{-18}\left(\frac{16Z^2}{9 \times 25}\right)$$

$$E = \frac{hc}{\lambda} = \frac{6.6261 \times 10^{-34}\ \text{J s}(2.9979 \times 10^8\ \text{m/s})}{142.5 \times 10^{-9}\ \text{m}} = 1.394 \times 10^{-18}\ \text{J}$$

Because an emission occurs, $\Delta E_{5\to3} = -1.394 \times 10^{-18}$ J.

$$\Delta E = -1.394 \times 10^{-18}\ \text{J} = -2.178 \times 10^{-18}\ \text{J}\left(\frac{16\,Z^2}{9 \times 25}\right),\quad Z^2 = 9.001,\ Z = 3;\ \text{the ion is } Li^{2+}.$$

Solving for the wavelength of line A:

$$\Delta E_{6\to3} = -2.178 \times 10^{-18}(3)^2\left(\frac{1}{3^2} - \frac{1}{6^2}\right) = -1.634 \times 10^{-18}\ \text{J}$$

$$\lambda = \frac{hc}{|\Delta E|} = \frac{6.6261 \times 10^{-34}\ \text{J s}(2.9979 \times 10^8\ \text{m/s})}{1.634 \times 10^{-18}\ \text{J}} = 1.216 \times 10^{-7}\ \text{m} = 121.6\ \text{nm}$$

Challenge Problems

131. For one-electron species, $E_n = -R_H Z^2/n^2$. IE is for the $n = 1 \rightarrow n = \infty$ transition. So:

$$IE = E_\infty - E_1 = -E_1 = R_H Z^2/n^2 = R_H Z^2$$

$$\frac{4.72 \times 10^4\ \text{kJ}}{\text{mol}} \times \frac{1\ \text{mol}}{6.022 \times 10^{23}} \times \frac{1000\ \text{J}}{\text{kJ}} = 2.178 \times 10^{-18}\ \text{J}\ (Z^2);\ \text{solving: } Z = 6$$

Element 6 is carbon (X = carbon), and the charge for a one-electron carbon ion is 5+ ($m = 5$). The one-electron ion is C^{5+}.

133. a. Because the energy levels E_{xy} are inversely proportional to L^2, the $n_x = 2$, $n_y = 1$ energy level will be lower in energy than the $n_x = 1$, $n_y = 2$ energy level since $L_x > L_y$. The first three energy levels E_{xy} in order of increasing energy are:

$$E_{11} < E_{21} < E_{12}$$

The quantum numbers are:

Ground state (E_{11}) $\rightarrow$ $n_x = 1, n_y = 1$
First excited state (E_{21}) $\rightarrow$ $n_x = 2, n_y = 1$
Second excited state (E_{12}) $\rightarrow$ $n_x = 1, n_y = 2$

b. $E_{21} \rightarrow E_{12}$ is the transition. $E_{xy} = \dfrac{h^2}{8m}\left(\dfrac{n_x^2}{L_x^2} + \dfrac{n_y^2}{L_y^2}\right)$

$$E_{12} = \frac{h^2}{8m}\left[\frac{1^2}{(8.00 \times 10^{-9}\ m)^2} + \frac{2^2}{(5.00 \times 10^{-9}\ m)^2}\right] = \frac{1.76 \times 10^{17}\ h^2}{8m}$$

$$E_{21} = \frac{h^2}{8m}\left[\frac{2^2}{(8.00 \times 10^{-9}\ m)^2} + \frac{1^2}{(5.00 \times 10^{-9}\ m)^2}\right] = \frac{1.03 \times 10^{17}\ h^2}{8m}$$

$$\Delta E = E_{12} - E_{21} = \frac{1.76 \times 10^{17}\ h^2}{8m} - \frac{1.03 \times 10^{17}\ h^2}{8m} = \frac{7.3 \times 10^{16}\ h^2}{8m}$$

$$\Delta E = \frac{(7.3 \times 10^{16}\ m^{-2})(6.626 \times 10^{-34}\ J\ s)^2}{8(9.11 \times 10^{-31}\ kg)} = 4.4 \times 10^{-21}\ J$$

$$\lambda = \frac{hc}{\Delta E} = \frac{6.626 \times 10^{-34}\ J\ s(2.998 \times 10^8\ m/s)}{4.4 \times 10^{-21}\ J} = 4.5 \times 10^{-5}\ m$$

135. $\psi_{1s} = \dfrac{1}{\sqrt{\pi}}\left(\dfrac{Z}{a_0}\right)^{3/2} e^{-\sigma}$; $Z = 1$ for H, $\sigma = \dfrac{Zr}{a_0} = \dfrac{-r}{a_0}$, $a_0 = 5.29 \times 10^{-11}$ m

$$\psi_{1s} = \frac{1}{\sqrt{\pi}}\left(\frac{1}{a_0}\right)^{3/2} \exp\left(\frac{-r}{a_0}\right)$$

Probability is proportional to ψ^2: $\psi_{1s}^2 = \dfrac{1}{\pi}\left(\dfrac{1}{a_0}\right)^3 \exp\left(\dfrac{-2r}{a_0}\right)$ (units of $\psi^2 = m^{-3}$)

a. ψ_{1s}^2 (at nucleus) $= \dfrac{1}{\pi}\left(\dfrac{1}{a_0}\right)^3 \exp\left[\dfrac{-2\,(0)}{a_0}\right] = 2.15 \times 10^{30}\ m^{-3}$

If we assume this probability is constant throughout the 1×10^{-3} pm^3 volume, then the total probability p is $\psi_{1s}^2 \times V$.

$1.0 \times 10^{-3}\ pm^3 = (1.0 \times 10^{-3}\ pm) \times (1 \times 10^{-12}\ m/pm)^3 = 1.0 \times 10^{-39}\ m^3$

Total probability = p = $(2.15 \times 10^{30}\ m^{-3}) \times (1.0 \times 10^{-39}\ m^3) = 2.2 \times 10^{-9}$

b. For an electron that is 1.0×10^{-11} m from the nucleus:

$$\psi_{1s}^2 = \frac{1}{\pi}\left(\frac{1}{5.29 \times 10^{-11}}\right)^3 \exp\left[\frac{-2(1.0 \times 10^{-11})}{(5.29 \times 10^{-11})}\right] = 1.5 \times 10^{30}\ \text{m}^{-3}$$

$V = 1.0 \times 10^{-39}\ \text{m}^3;\ p = \psi_{1s}^2 \times V = 1.5 \times 10^{-9}$

c. $\psi_{1s}^2 = 2.15 \times 10^{30}\ \text{m}^{-3} \exp\left[\frac{-2(53 \times 10^{-12})}{(5.29 \times 10^{-11})}\right] = 2.9 \times 10^{29};\ V = 1.0 \times 10^{-39}\ \text{m}^3$

$p = \psi_{1s}^2 \times V = 2.9 \times 10^{-10}$

d. $V = \frac{4}{3}\pi[(10.05 \times 10^{-12}\ \text{m})^3 - (9.95 \times 10^{-12}\ \text{m})^3] = 1.3 \times 10^{-34}\ \text{m}^3$

We shall evaluate ψ_{1s}^2 at the middle of the shell, r = 10.00 pm, and assume ψ_{1s}^2 is constant from r = 9.95 to 10.05 pm. The concentric spheres are assumed centered about the nucleus.

$$\psi_{1s}^2 = 2.15 \times 10^{30}\ \text{m}^{-3} \exp\left[\frac{-2(10.0 \times 10^{-12}\ \text{m})}{(5.29 \times 10^{-11}\ \text{m})}\right] = 1.47 \times 10^{30}\ \text{m}^{-3}$$

$p = (1.47 \times 10^{30}\ \text{m}^{-3})(1.3 \times 10^{-34}\ \text{m}^3) = 1.9 \times 10^{-4}$

e. $V = \frac{4}{3}\pi[(52.95 \times 10^{-12}\ \text{m})^3 - (52.85 \times 10^{-12}\ \text{m})^3] = 4 \times 10^{-33}\ \text{m}^3$

Evaluate ψ_{1s}^2 at r = 52.90 pm: $\psi_{1s}^2 = 2.15 \times 10^{30}\ \text{m}^{-3}(e^{-2}) = 2.91 \times 10^{29}\ \text{m}^{-3};\ p = 1 \times 10^{-3}$

137. $E_{xyz} = \frac{h^2(n_x^2 + n_y^2 + n_z^2)}{8mL^2}$, where $L = L_x = L_y = L_z$.

The first four energy levels will be filled with the 8 electrons. The first four energy levels are:

$$E_{111} = \frac{h^2 + 1^2 + 1^2}{8mL^2} = \frac{3h^2}{8mL^2}$$

$E_{211} = E_{121} = E_{112} = \frac{6h^2}{8mL^2}$ (These three energy levels are degenerate.)

The next energy levels correspond to the first excited state. The energy for these levels are:

$E_{221} = E_{212} = E_{122} = \frac{9h^2}{8mL^2}$ (These three energy levels are degenerate.)

The electronic transition in question is from one of the degenerate E_{211}, E_{121}, or E_{112} levels to one of the degenerate E_{221}, E_{212}, or E_{122} levels.

$$\Delta E = \frac{9h^2}{8mL^2} - \frac{6h^2}{8mL^2} = \frac{3h^2}{8mL^2}$$

$$\Delta E = \frac{3(6.626 \times 10^{-34}\ \text{J s})^2}{8(9.109 \times 10^{-31}\ \text{kg})(1.50 \times 10^{-9}\ \text{m})^2} = 8.03 \times 10^{-20}\ \text{J}$$

$$\lambda = \frac{hc}{\Delta E} = \frac{(6.626 \times 10^{-34}\ \text{J s})(2.998 \times 10^{8}\ \text{m/s})}{8.03 \times 10^{-20}\ \text{J}} = 2.47 \times 10^{-6}\ \text{m} = 2470\ \text{nm}$$

139. a. Assuming the Bohr model applies to the 1s electron, $E_{1s} = -R_H Z^2/n^2 = -R_H Z^2_{eff}$, where $n = 1$.

$IE = E_\infty - E_{1s} = 0 - E_{1s} = R_H Z^2_{eff}$

$$\frac{2.462 \times 10^6\ \text{kJ}}{\text{mol}} \times \frac{1\ \text{mol}}{6.0221 \times 10^{23}} \times \frac{1000\ \text{J}}{\text{kJ}} = 2.178 \times 10^{-18}\ \text{J}\ (Z_{eff})^2$$

Solving: $Z_{eff} = 43.33$

b. Silver is element 47, so Z = 47 for silver. Our calculated Z_{eff} value is slightly less than 47. Electrons in other orbitals can penetrate the 1s orbital. Thus a 1s electron can be slightly shielded from the nucleus, giving a Z_{eff} close to but less than Z.

CHAPTER 13

BONDING: GENERAL CONCEPTS

Chemical Bonds and Electronegativity

11. Electronegativity is the ability of an atom in a molecule to attract electrons to itself. Electronegativity is a bonding term. Electron affinity is the energy change when an electron is added to a substance. Electron affinity deals with isolated atoms in the gas phase.

    A covalent bond is a sharing of electron pair(s) in a bond between two atoms. An ionic bond is a complete transfer of electrons from one atom to another to form ions. The electrostatic attraction of the oppositely charged ions is the ionic bond.

    A pure covalent bond is an equal sharing of shared electron pair(s) in a bond. A polar covalent bond is an unequal sharing.

    Ionic bonds form when there is a large difference in electronegativity between the two atoms bonding together. This usually occurs when a metal with a small electronegativity is bonded to a nonmetal having a large electronegativity. A pure covalent bond forms between atoms having identical or nearly identical eletronegativities. A polar covalent bond forms when there is an intermediate electronegativity difference. In general, nonmetals bond together by forming covalent bonds, either pure covalent or polar covalent.

    Ionic bonds form due to the strong electrostatic attraction between two oppositely charged ions. Covalent bonds form because the shared electrons in the bond are attracted to two different nuclei, unlike the isolated atoms where electrons are only attracted to one nuclei. The attraction to another nuclei overrides the added electron-electron repulsions.

13. Using the periodic table, we expect the general trend for electronegativity to be:

    1. Increase as we go from left to right across a period

    2. Decrease as we go down a group

    a. C < N < O b. Se < S < Cl c. Sn < Ge < Si

    d. Tl < Ge < S e. Rb < K < Na f. Ga < B < O

15. The general trends in electronegativity used in Exercises 13.13 and 13.14 are only rules of thumb. In this exercise we use experimental values of electronegativities and can begin to see several exceptions. The order of EN using Figure 13.3 is:

a. C (2.6) < N (3.0) < O (3.4) same as predicted

b. Se (2.6) = S (2.6) < Cl (3.2) different

c. Si (1.9) < Ge (2.0) = Sn (2.0) different

d. Tl (2.0) = Ge (2.0) < S (2.6) different

e. Rb (0.8) = K (0.8) < Na (0.9) different

f. Ga (1.8) < B (2.0) < O (3.4) same

Most polar bonds using actual EN values:

a. Si–F (Ge–F predicted)

b. P–Cl (same as predicted)

c. S–F (same as predicted)

d. Ti–Cl (same as predicted)

e. C–H (Sn–H predicted)

f. Al–Br (Tl–Br predicted)

17. Ionic character is proportional to the difference in electronegativity values between the two elements forming the bond. Using the trend in electronegativity, the order will be:

Br–Br < N–O < C–F < Ca–O < K–F

least ionic character ... most ionic character

Note that Br–Br, N–O and C–F bonds are all covalent bonds since the elements are all nonmetals. The Ca–O and K–F bonds are ionic, as is generally the case when a metal forms a bond with a nonmetal.

Ionic Compounds

19. Anions are larger than the neutral atom, and cations are smaller than the neutral atom. For anions, the added electrons increase the electron-electron repulsions. To counteract this, the size of the electron cloud increases, placing the electrons further apart from one another. For cations, as electrons are removed, there are fewer electron-electron repulsions, and the electron cloud can be pulled closer to the nucleus.

Isoelectronic: same number of electrons. Two variables, the number of protons and the number of electrons, determine the size of an ion. Keeping the number of electrons constant, we only have to consider the number of protons to predict trends in size. The ion with the most protons attracts the same number of electrons most strongly, resulting in a smaller size.

21. a. $Cu > Cu^{+} > Cu^{2+}$

b. $Pt^{2+} > Pd^{2+} > Ni^{2+}$

c. $O^{2-} > O^{-} > O$

d. $La^{3+} > Eu^{3+} > Gd^{3+} > Yb^{3+}$

e. $Te^{2-} > I^{-} > Cs^{+} > Ba^{2+} > La^{3+}$

For answer a, as electrons are removed from an atom, size decreases. Answers b and d follow the radius trend. For answer c, as electrons are added to an atom, size increases. Answer e follows the trend for an isoelectronic series, i.e., the smallest ion has the most protons.

23. a. Cs_2S is composed of Cs^+ and S^{2-}. Cs^+ has the same electron configuration as Xe, and S^{2-} has the same configuration as Ar.

b. SrF_2; Sr^{2+} has the Kr electron configuration, and F^- has the Ne configuration.

c. Ca_3N_2; Ca^{2+} has the Ar electron configuration, and N^{3-} has the Ne configuration.

d. $AlBr_3$; Al^{3+} has the Ne electron configuration, and Br^- has the Kr configuration.

25. Se^{2-}, Br^-, Rb^+, Sr^{2+}, Y^{3+}, and Zr^{4+} are some ions that are isoelectronic with Kr (36 electrons). In terms of size, the ion with the most protons will hold the electrons tightest and will be the smallest. The size trend is:

$$Zr^{4+} < Y^{3+} < Sr^{2+} < Rb^+ < Br^- < Se^{2-}$$

smallest largest

27. a. Al^{3+} and S^{2-} are the expected ions. The formula of the compound would be Al_2S_3 (aluminum sulfide).

b. K^+ and N^{3-}; K_3N, potassium nitride

c. Mg^{2+} and Cl^-; $MgCl_2$, magnesium chloride

d. Cs^+ and Br^-; CsBr, cesium bromide

29.

$K(s) \rightarrow K(g)$	$\Delta H =$ 64 kJ (sublimation)
$K(g) \rightarrow K^+(g) + e^-$	$\Delta H =$ 419 kJ (ionization energy)
$1/2\ Cl_2(g) \rightarrow Cl(g)$	$\Delta H =$ 239/2 kJ (bond energy)
$Cl(g) + e^- \rightarrow Cl^-(g)$	$\Delta H = -349$ kJ (electron affinity)
$K^+(g) + Cl^-(g) \rightarrow KCl(s)$	$\Delta H = -690.$ kJ (lattice energy)
$K(s) + 1/2\ Cl_2(g) \rightarrow KCl(s)$	$\Delta H_f^\circ = -437$ kJ/mol

31. Use Figure 13.11 as a template for this problem.

$Li(s) \rightarrow Li(g)$	$\Delta H_{sub} = ?$
$Li(g) \rightarrow Li^+(g) + e^-$	$\Delta H = 520.$ kJ
$1/2\ I_2(g) \rightarrow I(g)$	$\Delta H = 151/2$ kJ
$I(g) + e^- \rightarrow I^-(g)$	$\Delta H = -295$ kJ
$Li^+(g) + I^-(g) \rightarrow LiI(s)$	$\Delta H = -753$ kJ
$Li(s) + 1/2\ I_2(g) \rightarrow LiI(s)$	$\Delta H = -272$ kJ

$\Delta H_{sub} + 520. + 151/2 - 295 - 753 = -272$, $\Delta H_{sub} = 181$ kJ

33. a. From the data given, less energy is required to produce $Mg^+(g) + O^-(g)$ than to produce $Mg^{2+}(g) + O^{2-}(g)$. However, the lattice energy for $Mg^{2+}O^{2-}$ will be much more exothermic than for Mg^+O^- (due to the greater charges in $Mg^{2+}O^{2-}$). The favorable lattice energy term will dominate and $Mg^{2+}O^{2-}$ forms.

b. Mg^+ and O^- both have unpaired electrons. In Mg^{2+} and O^{2-} there are no unpaired electrons. Hence Mg^+O^- would be paramagnetic; $Mg^{2+}O^{2-}$ would be diamagnetic. Paramagnetism can be detected by measuring the mass of a sample in the presence and absence of a magnetic field. The apparent mass of a paramagnetic substance will be larger in a magnetic field because of the force between the unpaired electrons and the field.

35. Ca^{2+} has a greater charge than Na^+, and Se^{2-} is smaller than Te^{2-}. The effect of charge on the lattice energy is greater than the effect of size. We expect the trend from most exothermic to least exothermic to be:

$$CaSe > CaTe > Na_2Se > Na_2Te$$

(−2862) (−2721) (−2130) (−2095 kJ/mol) This is what we observe.

Bond Energies

37. a. H—H + Cl—Cl → 2 H—Cl

Bonds broken:

1 H–H (432 kJ/mol)
1 Cl–Cl (239 kJ/mol)

Bonds formed:

2 H–Cl (427 kJ/mol)

$\Delta H = \Sigma D_{broken} - \Sigma D_{formed}$, $\Delta H = 432\text{ kJ} + 239\text{ kJ} - 2(427)\text{ kJ} = -183\text{ kJ}$

b. N≡N + 3 H—H → 2 H—N(—H)—H

Bonds broken:

1 N ≡ N (941 kJ/mol)
3 H–H (432 kJ/mol)

Bonds formed:

6 N–H (391 kJ/mol)

$\Delta H = 941\text{ kJ} + 3(432)\text{ kJ} - 6(391)\text{ kJ} = -109\text{ kJ}$

c. Sometimes some of the bonds remain the same between reactants and products. To save time, only break and form bonds that are involved in the reaction.

$$H{-}C{\equiv}N + 2\ H{-}H \longrightarrow H{-}CH_2{-}NH_2$$

Bonds broken:

1 C≡N (891 kJ/mol)
2 H–H (432 kJ/mol)

Bonds formed:

1 C–N (305 kJ/mol)
2 C–H (413 kJ/mol)
2 N–H (391 kJ/mol)

ΔH = 891 kJ + 2(432 kJ) – [305 kJ + 2(413 kJ) + 2(391 kJ)] = –158 kJ

d. $$H_2N{-}NH_2 + 2\ F{-}F \longrightarrow 4\ H{-}F + N{\equiv}N$$

Bonds broken:

1 N–N (160. kJ/mol)
4 N–H (391 kJ/mol)
2 F–F (154 kJ/mol)

Bonds formed:

4 H–F (565 kJ/mol)
1 N≡N (941 kJ/mol)

ΔH = 160. kJ + 4(391 kJ) + 2(154 kJ) – [4(565 kJ) + 941 kJ] = –1169 kJ

39. $$H_3C{-}N{\equiv}C \longrightarrow H_3C{-}C{\equiv}N$$

Bonds broken: 1 C–N (305 kJ/mol) Bonds formed: 1 C–C (347 kJ/mol)

$\Delta H = \Sigma D_{broken} - \Sigma D_{formed}$, ΔH = 305 – 347 = –42 kJ

Note: Sometimes some of the bonds remain the same between reactants and products. To save time, only break and form bonds that are involved in the reaction.

41. H–C≡C–H + 5/2 O=O → 2 O=C=O + H–O–H

Bonds broken:

2 C–H (413 kJ/mol)
1 C≡C (839 kJ/mol)
5/2 O = O (495 kJ/mol)

Bonds formed:

2 × 2 C=O (799 kJ/mol)
2 O–H (467 kJ/mol)

ΔH = 2(413 kJ) + 839 kJ + 5/2 (495 kJ) – [4(799 kJ) + 2(467 kJ)] = –1228 kJ

43.

$$4\ CH_3NHNH_2 + 5\ O_2N{-}NO_2 \longrightarrow 12\ H{-}O{-}H + 9\ N{\equiv}N + 4\ O{=}C{=}O$$

Bonds broken:

9 N–N (160. kJ/mol)
4 N–C (305 kJ/mol)
12 C–H (413 kJ/mol)
12 N–H (391 kJ/mol)
10 N=O (607 kJ/mol)
10 N–O (201 kJ/mol)

Bonds formed:

24 O–H (467 kJ/mol)
9 N≡N (941 kJ/mol)
8 C=O (799 kJ/mol)

$$\Delta H = 9(160.) + 4(305) + 12(413) + 12(391) + 10(607) + 10(201) - [24(467) + 9(941) + 8(799)]$$

$$\Delta H = 20{,}388\text{ kJ} - 26{,}069\text{ kJ} = -5681\text{ kJ}$$

45. Because both reactions are highly exothermic, the high temperature is not needed to provide energy. It must be necessary for some other reason. The reason is to increase the speed of the reaction. This will be discussed in Chapter 15 on kinetics.

47. a.

$HF(g) \rightarrow H(g) + F(g)$	ΔH = 565 kJ
$H(g) \rightarrow H^+(g) + e^-$	ΔH = 1312 kJ
$F(g) + e^- \rightarrow F^-(g)$	ΔH = –327.8 kJ
$HF(g) \rightarrow H^+(g) + F^-(g)$	ΔH = 1549 kJ

b.

Reaction	ΔH
$HCl(g) \rightarrow H(g) + Cl(g)$	ΔH = 427 kJ
$H(g) \rightarrow H^+(g) + e^-$	ΔH = 1312 kJ
$Cl(g) + e^- \rightarrow Cl^-(g)$	ΔH = –348.7 kJ
$HCl(g) \rightarrow H^+(g) + Cl^-(g)$	ΔH = 1390. kJ

c.

Reaction	ΔH
$HI(g) \rightarrow H(g) + I(g)$	ΔH = 295 kJ
$H(g) \rightarrow H^+(g) + e^-$	ΔH = 1312 kJ
$I(g) + e^- \rightarrow I^-(g)$	ΔH = –295.2 kJ
$HI(g) \rightarrow H^+(g) + I^-(g)$	ΔH = 1312 kJ

d.

Reaction	ΔH
$H_2O(g) \rightarrow OH(g) + H(g)$	ΔH = 467 kJ
$H(g) \rightarrow H^+(g) + e^-$	ΔH = 1312 kJ
$OH(g) + e^- \rightarrow OH^-(g)$	ΔH = –180. kJ
$H_2O(g) \rightarrow H^+(g) + OH^-(g)$	ΔH = 1599 kJ

49. $NH_3(g) \rightarrow N(g) + 3\ H(g)$

$$\Delta H° = 3D_{NH} = 472.7\ kJ + 3(216.0\ kJ) - (-46.1\ kJ) = 1166.8\ kJ$$

$$D_{NH} = \frac{1166.8\ kJ}{3\ mol\ NH\ bonds} = 388.93\ kJ/mol$$

D_{calc} = 389 kJ/mol compared with 391 kJ/mol in the table. There is good agreement.

Lewis Structures and Resonance

51. Drawing Lewis structures is mostly trial and error. However, the first two steps are always the same. These steps are (1) count the valence electrons available in the molecule/ion, and (2) attach all atoms to each other with single bonds (called the skeletal structure). Unless noted otherwise, the atom listed first is assumed to be the atom in the middle, called the central atom, and all other atoms in the formula are attached to this atom. The most notable exceptions to the rule are formulas that begin with H, e.g., H_2O, H_2CO, etc. Hydrogen can never be a central atom since this would require H to have more than two electrons. In these compounds, the atom listed second is assumed to be the central atom.

After counting valence electrons and drawing the skeletal structure, the rest is trial and error. We place the remaining electrons around the various atoms in an attempt to satisfy the octet rule (or duet rule for H). Keep in mind that practice makes perfect. After practicing, you can (and will) become very adept at drawing Lewis structures.

a. HCN has 1 + 4 + 5 = 10 valence electrons.

H—C—N

Skeletal structure

H—C≡N:

Lewis structure

Skeletal structure uses 4 e^-; 6 e^- remain

b. PH_3 has 5 + 3(1) = 8 valence electrons.

H—P—H with a third H bonded below P

Skeletal structure

H—P̈—H with a third H bonded below P

Lewis structure

Skeletal structures uses 6 e^-; 2 e^- remain

c. $CHCl_3$ has 4 + 1 + 3(7) = 26 valence electrons.

d. NH_4^+ has 5 + 4(1) − 1 = 8 valence electrons.

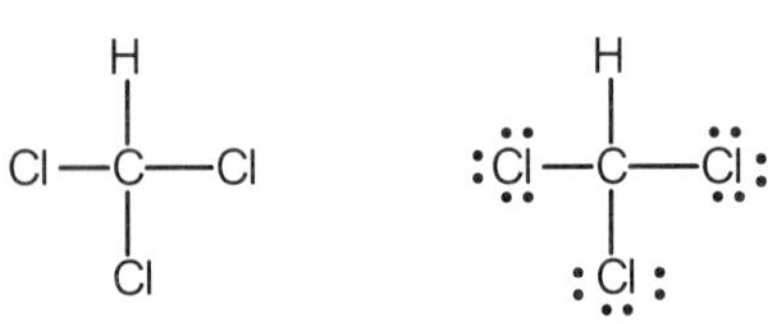

Skeletal structure

Lewis structure

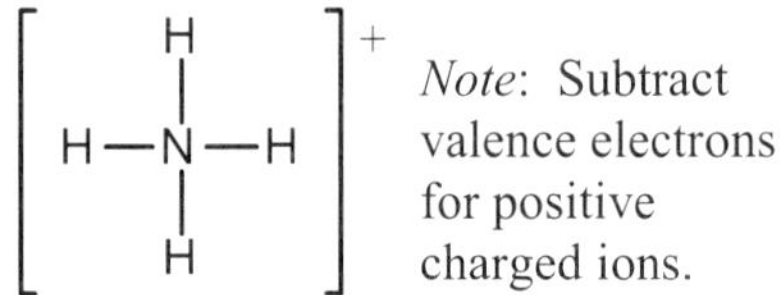

Lewis structure

e. H_2CO has 2(1) + 4 + 6 = 12 valence electrons.

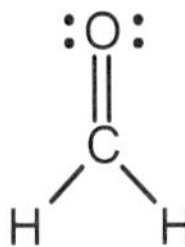

f. SeF_2 has 6 + 2(7) = 20 valence electrons.

g. CO_2 has 4 + 2(6) = 16 valence electrons.

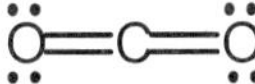

h. O_2 has 2(6) = 12 valence electrons.

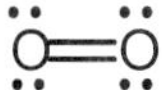

i. HBr has 1 + 7 = 8 valence electrons.

53. Molecules/ions that have the same number of valence electrons and the same number of atoms will have similar Lewis structures.

55. Ozone: O_3 has 3(6) = 18 valence electrons. Two resonance structures can be drawn.

Sulfur dioxide: SO_2 has 6 + 2(6) = 18 valence electrons. Two resonance structures are possible.

Sulfur trioxide: SO_3 has 6 + 3(6) = 24 valence electrons. Three resonance structures are possible.

57. CH_3NCO has 4 + 3(1) + 5 + 4 + 6 = 22 valence electrons. The order of the elements in the formula give the skeletal structure.

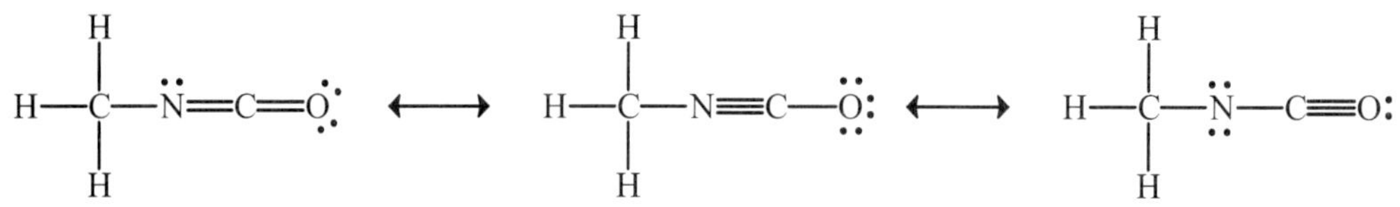

59. Benzene has 6(4) + 6(1) = 30 valence electrons. Two resonance structures can be drawn for benzene. The actual structure of benzene is an average of these two resonance structures; i.e., all carbon-carbon bonds are equivalent with a bond length and bond strength somewhere between a single and a double bond.

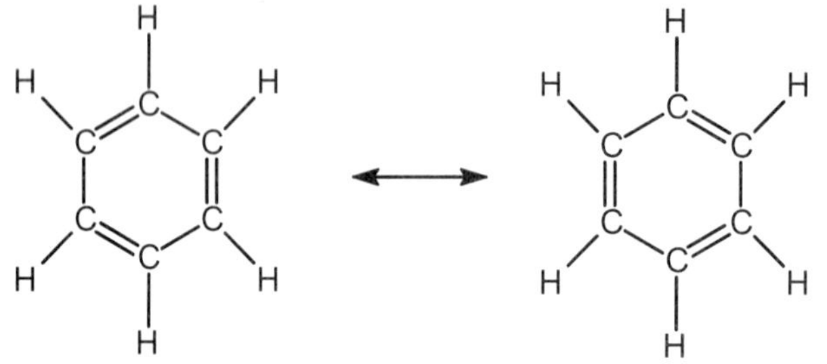

61. Borazine ($B_3N_3H_6$) has 3(3) + 3(5) + 6(1) = 30 valence electrons. The possible resonance structures are similar to those of benzene in Exercise 13.59.

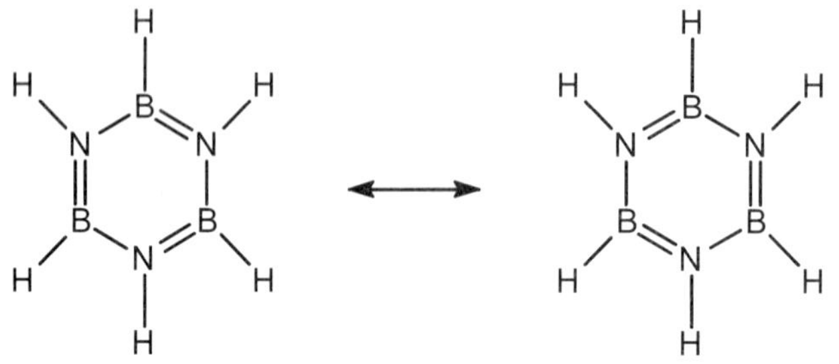

63. Statements a and c are true. For statement a, XeF_2 has 22 valence electrons and it is impossible to satisfy the octet rule for all atoms with this number of electrons. The best Lewis structure is:

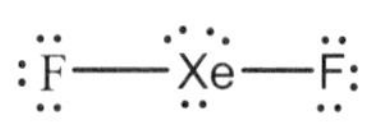

For statement c, NO^+ has 10 valence electrons, whereas NO^- has 12 valence electrons. The Lewis structures are:

$$[:N\equiv O:]^+ \qquad [N=O]^-$$

Because a triple bond is stronger than a double bond, NO^+ has a stronger bond.

For statement b, SF_4 has five electron pairs around the sulfur in the best Lewis structure; it is an exception to the octet rule. Because OF_4 has the same number of valence electrons as SF_4, OF_4 would also have to be an exception to the octet rule. However, Row 2 elements such as O never have more than 8 electrons around them, so OF_4 does not exist. For statement d, two resonance structures can be drawn for ozone:

$$O=O-O \longleftrightarrow O-O=O$$

When resonance structures can be drawn, the actual bond lengths and strengths are all equal to each other. Even though each Lewis structure implies the two O–O bonds are different, this is not the case in real life. In real life, both of the O–O bonds are equivalent. When resonance structures can be drawn, you can think of the bonding as an average of all of the resonance structures.

65. PF_5, 5 +5(7) = 40 valence electrons SF_4, 6 + 4(7) = 34 e^-

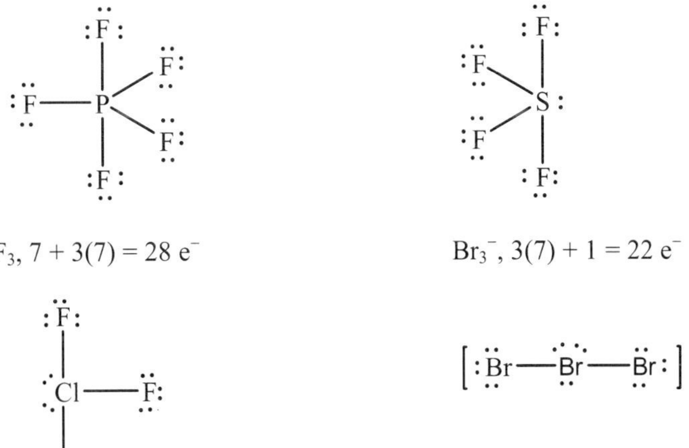

ClF_3, 7 + 3(7) = 28 e^- Br_3^-, 3(7) + 1 = 22 e^-

Row 3 and heavier nonmetals can have more than 8 electrons around them when they have to. Row 3 and heavier elements have empty d orbitals that are close in energy to valence s and p orbitals. These empty d orbitals can accept extra electrons.

For example, P in PF_5 has its five valence electrons in the 3s and 3p orbitals. These s and p orbitals have room for three more electrons, and if it has to, P can use the empty 3d orbitals for any electrons above 8.

67. CO_3^{2-} has $4 + 3(6) + 2 = 24$ valence electrons.

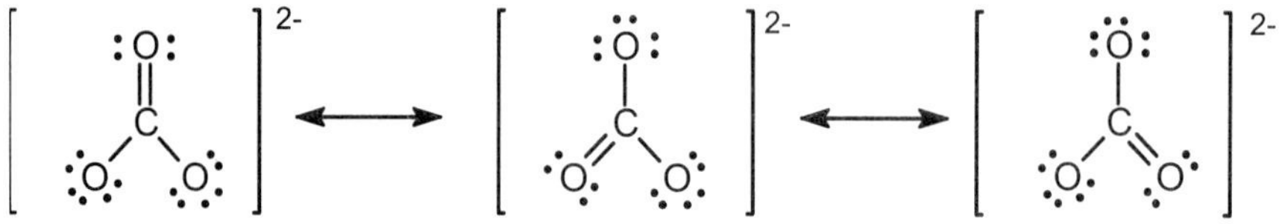

Three resonance structures can be drawn for CO_3^{2-}. The actual structure for CO_3^{2-} is an average of these three resonance structures. That is, the three C–O bond lengths are all equivalent, with a length somewhere between a single and a double bond. The actual bond length of 136 pm is consistent with this resonance view of CO_3^{2-}.

69.

N_2 (10 e⁻): $:N{\equiv}N:$ Triple bond between N and N.

N_2F_4 (38 e⁻): F–N–N–F (each N also bonded to one F) Single bond between N and N.

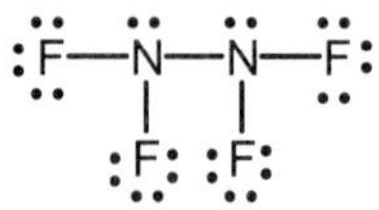

N_2F_2 (24 e⁻): F–N=N–F Double bond between N and N.

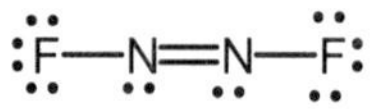

As the number of bonds increase between two atoms, bond strength increases and bond length decreases. From the Lewis structure, the shortest to longest N-N bonds is $N_2 < N_2F_2 < N_2F_4$.

Formal Charge

71. See Exercise 13.52a for the Lewis structures of $POCl_3$, SO_4^{2-}, ClO_4^- and PO_4^{3-}. All of these compounds/ions have similar Lewis structures to those of SO_2Cl_2 and XeO_4 shown below.

a. $POCl_3$: P, FC = 5 – 1/2(8) = +1

b. SO_4^{2-}: S, FC = 6 – 1/2(8) = +2

c. ClO_4^-: Cl, FC = 7 – 1/2(8) = +3

d. PO_4^{3-}: P, FC = 5 – 1/2(8) = +1

e. SO_2Cl_2, $6 + 2(6) + 2(7) = 32\ e^-$

f. XeO_4, $8 + 4(6) = 32\ e^-$

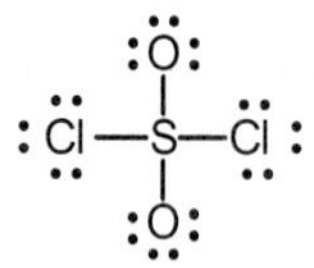

S, FC = 6 – 1/2(8) = +2

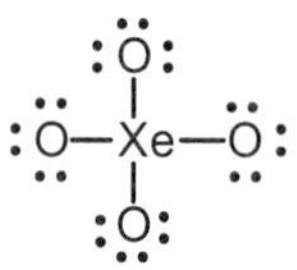

Xe, FC = 8 – 1/2(8) = +4

g. ClO_3^-, $7 + 3(6) + 1 = 26$ e^-

```
[ :Ö—C̈l—Ö: ]⁻
      |
     :Ö:
```

Cl, FC = 7 – 2 – 1/2(6) = +2

h. NO_4^{3-}, $5 + 4(6) + 3 = 32$ e^-

```
      :Ö:
       |
[ :Ö—N—Ö: ]³⁻
       |
      :Ö:
```

N, FC = 5 – 1/2(8) = +1

73. SCl, 6 + 7 = 13; the formula could be SCl (13 valence electrons), S_2Cl_2 (26 valence electrons), S_3Cl_3 (39 valence electrons), etc. For a formal charge of zero on S, we will need each sulfur in the Lewis structure to have two bonds to it and two lone pairs [FC = 6 – 4 – 1/2(4) = 0]. Cl will need one bond and three lone pairs for a formal charge of zero [FC = 7 – 6 – 1/2(2) = 0]. Since chlorine wants only one bond to it, it will not be a central atom here. With this in mind, only S_2Cl_2 can have a Lewis structure with a formal charge of zero on all atoms. The structure is:

```
:C̤̈l——S̤̈——S̤̈——C̤̈l:
```

Molecular Structure and Polarity

75. The first step always is to draw a valid Lewis structure when predicting molecular structure. When resonance is possible, only one of the possible resonance structures is necessary to predict the correct structure because all resonance structures give the same structure. The Lewis structures are in Exercises 13.51, 13.52 and 13.54. The structures and bond angles for each follow.

13.51 a. HCN: linear, 180°

b. PH_3: trigonal pyramid, <109.5°

c. $CHCl_3$: tetrahedral, 109.5°

d. NH_4^+: tetrahedral, 109.5°

e. H_2CO: trigonal planar, 120°

f. SeF_2: V-shaped or bent, <109.5°

g. CO_2: linear, 180°

h and i. O_2 and HBr are both linear, but there is no bond angle in either.

Note: PH_3 and SeF_2 both have lone pairs of electrons on the central atom, which result in bond angles that are something less than predicted from a tetrahedral arrangement (109.5°). However, we cannot predict the exact number. For these cases we will just insert a less than sign to indicate this phenomenon.

13.52 a. All are tetrahedral; 109.5°

b. All are trigonal pyramid; <109.5°

c. All are V-shaped; <109.5°

13.54 a. NO_2^-: V-shaped, ≈ 120°; NO_3^-: trigonal planar, 120°

N_2O_4: trigonal planar, 120° about both N atoms

b. OCN^-, SCN^-, and N_3^- are all linear with 180° bond angles.

77. From the Lewis structures (see Exercise 13.65), Br_3^- would have a linear molecular structure, ClF_3 would have a T-shaped molecular structure, and SF_4 would have a see-saw molecular structure. For example, consider ClF_3 (28 valence electrons):

The central Cl atom is surrounded by five electron pairs, which requires a trigonal bipyramid geometry. Since there are three bonded atoms and two lone pairs of electrons about Cl, we describe the molecular structure of ClF_3 as T-shaped with predicted bond angles of about 90°. The actual bond angles will be slightly less than 90° due to the stronger repulsive effect of the lone pair electrons as compared to the bonding electrons.

79. a. $XeCl_2$ has 8 + 2(7) = 22 valence electrons.

180°

There are five pairs of electrons about the central Xe atom. The structure will be based on a trigonal bipyramid geometry. The most stable arrangement of the atoms in $XeCl_2$ is a linear molecular structure with a 180° bond angle.

b. ICl_3 has 7 + 3(7) = 28 valence electrons.

≈ 90°

≈ 90°

T-shaped; The ClICl angles are ≈ 90°. Since the lone pairs will take up more space, the ClICl bond angles will probably be slightly less than 90°.

c. TeF_4 has 6 + 4(7) = 34 valence electrons.

d. PCl_5 has 5 + 5(7) = 40 valence electrons.

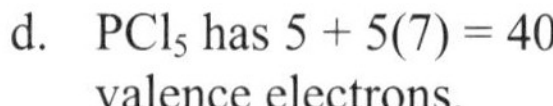

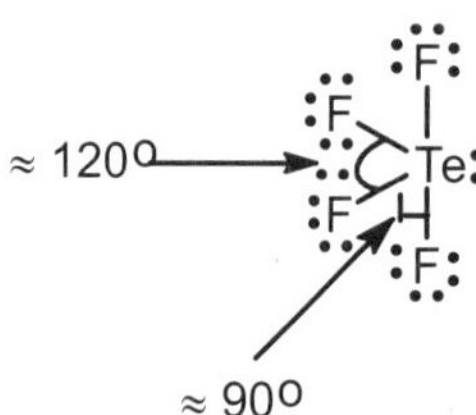

See-saw or teeter-totter or distorted tetrahedron

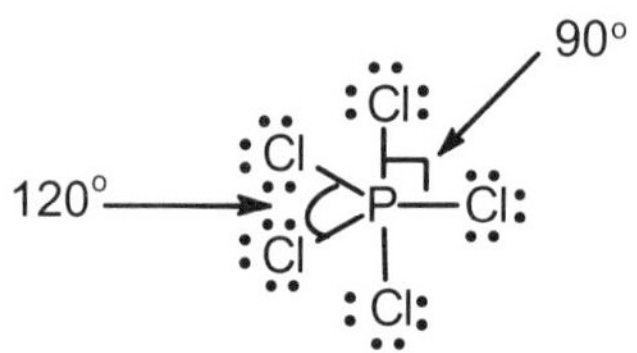

Trigonal bipyramid

All the species in this exercise have five pairs of electrons around the central atom. All the structures are based on a trigonal bipyramid geometry, but only in PCl_5 are all the pairs bonding pairs. Thus PCl_5 is the only one we describe the molecular structure as trigonal bipyramid. Still, we had to begin with the trigonal bipyramid geometry to get to the structures (and bond angles) of the others.

81. Let us consider the molecules with three pairs of electrons around the central atom first; these molecules are SeO_3 and SeO_2, and both have a trigonal planar arrangement of electron pairs. Both these molecules have polar bonds, but only SeO_2 has an overall net dipole moment. The net effect of the three bond dipoles from the three polar Se–O bonds in SeO_3 will be to cancel each other out when summed together. Hence SeO_3 is nonpolar since the overall molecule has no resulting dipole moment. In SeO_2, the two Se–O bond dipoles do not cancel when summed together; hence SeO_2 has a net dipole moment (is polar). Since O is more electronegative than Se, the negative end of the dipole moment is between the two O atoms, and the positive end is around the Se atom. The arrow in the following illustration represents the overall dipole moment in SeO_2. Note that to predict polarity for SeO_2, either of the two resonance structures can be used.

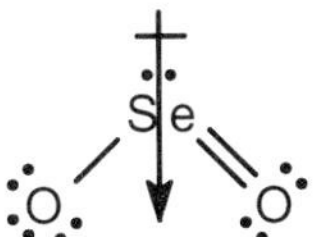

The other molecules in Exercise 13.76 (PCl_3, SCl_2, and SiF_4) have a tetrahedral arrangement of electron pairs. All have polar bonds; in SiF_4 the individual bond dipoles cancel when summed together, and in PCl_3 and SCl_2 the individual bond dipoles do not cancel. Therefore, SiF_4 has no net dipole moment (is nonpolar), and PCl_3 and SCl_2 have net dipole moments (are polar). For PCl_3, the negative end of the dipole moment is between the more electronegative chlorine atoms, and the positive end is around P. For SCl_2, the negative end is between the more electronegative Cl atoms, and the positive end of the dipole moment is around S.

83. The two general requirements for a polar molecule are:

1. Polar bonds

2. A structure such that the bond dipoles of the polar bonds do not cancel

CF_4, 4 + 4(7) = 32 valence electrons XeF_4, 8 + 4(7) = 36 e^-

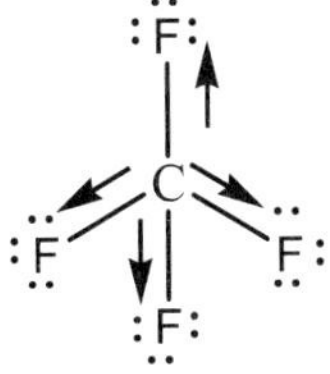

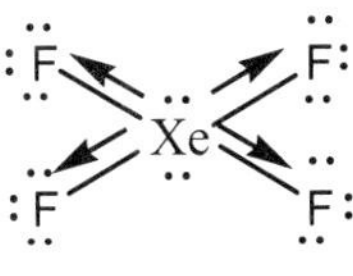

Tetrahedral, 109.5° Square planar, 90°

SF_4, $6 + 4(7) = 34\ e^-$

See-saw, ≈90°, ≈120°

The arrows indicate the individual bond dipoles in the three molecules (the arrows point to the more electronegative atom in the bond, which will be the partial negative end of the bond dipole). All three of these molecules have polar bonds. To determine the polarity of the overall molecule, we sum the effect of all of the individual bond dipoles. In CF_4, the fluorines are symmetrically arranged about the central carbon atom. The net result is for all the individual C–F bond dipoles to cancel each other out, giving a nonpolar molecule. In XeF_4, the 4 Xe–F bond dipoles are also symmetrically arranged, and XeF_4 is also nonpolar. The individual bond dipoles cancel out when summed together. In SF_4, we also have four polar bonds. But in SF_4 the bond dipoles are not symmetrically arranged, and they do not cancel each other out. SF_4 is polar. It is the positioning of the lone pair that disrupts the symmetry in SF_4.

CO_2, $4 + 2(6) = 16\ e^-$ COS, $4 + 6 + 6 = 16\ e^-$

CO_2 and COS both have linear molecular structures with a 180° bond angle. CO_2 is nonpolar because the individual bond dipoles cancel each other out, but COS is polar. By replacing an O with a less electronegative S atom, the molecule is not symmetric any more. The individual bond dipoles do not cancel because the C–S bond dipole is smaller than the C–O bond dipole resulting in a polar molecule.

85. Only statement c is true. The bond dipoles in CF_4 and KrF_4 are arranged in a manner that they all cancel each other out, making them nonpolar molecules (CF_4 has a tetrahedral molecular structure, whereas KrF_4 has a square planar molecular structure). In SeF_4 the bond dipoles in this see-saw molecule do not cancel each other out, so SeF_4 is polar. For statement a, all the molecules have either a trigonal planar geometry or a trigonal bipyramid geometry, both of which have 120° bond angles. However, $XeCl_2$ has three lone pairs and two bonded chlorine atoms around it. $XeCl_2$ has a linear molecular structure with a 180° bond angle. With three lone pairs, we no longer have a 120° bond angle in $XeCl_2$. For statement b, SO_2 has a V-shaped molecular structure with a bond angle of about 120°. CS_2 is linear with a 180° bond angle and SCl_2 is V-shaped but with an approximate 109.5° bond angle. The three compounds do not have the same bond angle. For statement d, central atoms adopt a geometry to minimize electron repulsions, not maximize them.

87. The formula is EF_2O^{2-}, and the Lewis structure has 28 valence electrons.

$$28 = x + 2(7) + 6 + 2, \; x = 6 \text{ valence electrons for element E}$$

Element E must belong to the Group 6A elements since E has six valence electrons. E must also be a Row 3 or heavier element since this ion has more than eight electrons around the central E atom (Row 2 elements never have more than eight electrons around them). Some possible identities for E are S, Se and Te. The ion has a T-shaped molecular structure (see Exercise 13.77) with bond angles of $\approx 90°$.

89. Molecules that have an overall dipole moment are called polar molecules, and molecules that do not have an overall dipole moment are called nonpolar molecules.

a. OCl_2, $6 + 2(7) = 20\ e^-$

V-shaped, polar; OCl_2 is polar because the two O–Cl bond dipoles don't cancel each other. The resulting dipole moment is shown in the drawing.

KrF_2, $8 + 2(7) = 22\ e^-$

Linear, nonpolar; The molecule is nonpolar because the two Kr–F bond dipoles cancel each other.

BeH_2, $2 + 2(1) = 4\ e^-$

H—Be—H

Linear, nonpolar; Be–H bond dipoles are equal and point in opposite directions. They cancel each other. BeH_2 is nonpolar.

SO_2, $6 + 2(6) = 18\ e^-$

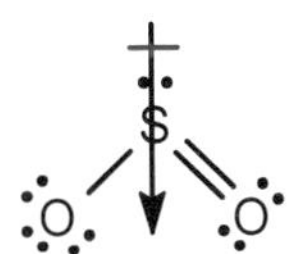

V-shaped, polar; The S–O bond dipoles do not cancel, so SO_2 is polar (has a net dipole moment). Only one resonance structure is shown.

Note: All four species contain three atoms. They have different structures because the number of lone pairs of electrons around the central atom are different in each case.

b. SO_3, $6 + 3(6) = 24\ e^-$

Trigonal planar, nonpolar; bond dipoles cancel. Only one resonance structure is shown.

NF_3, $5 + 3(7) = 26\ e^-$

Trigonal pyramid, polar; bond dipoles do not cancel.

IF_3 has $7 + 3(7) = 28$ valence electrons.

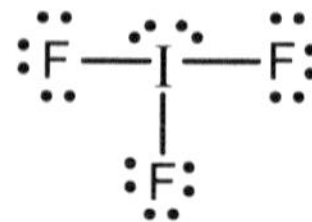

T-shaped, polar; bond dipoles do not cancel.

Note: Each molecule has the same number of atoms but different structures because of differing numbers of lone pairs around each central atom.

c. CF_4, $4 + 4(7) = 32\ e^-$

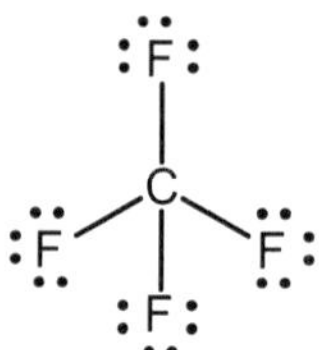

Tetrahedral, nonpolar; bond dipoles cancel.

SeF_4, $6 + 4(7) = 34\ e^-$

See-saw, polar; bond dipoles do not cancel.

KrF_4, $8 + 4(7) = 36$ valence electrons

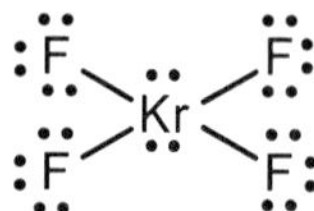

Square planar, nonpolar; bond dipoles cancel.

Note: Again, each molecule has the same number of atoms but different structures because of differing numbers of lone pairs around the central atom.

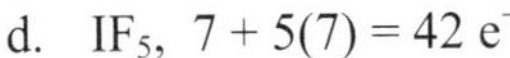

d. IF_5, $7 + 5(7) = 42\ e^-$

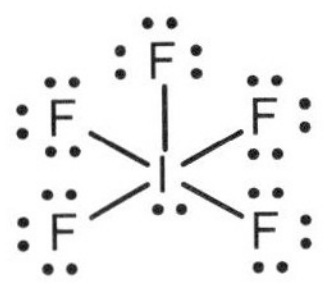

Square pyramid, polar; bond dipoles do not cancel.

AsF_5, $5 + 5(7) = 40\ e^-$

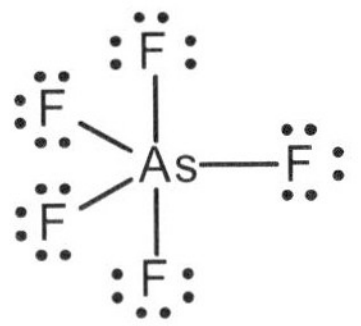

Trigonal bipyramid, nonpolar; bond dipoles cancel.

Note: Yet again, the molecules have the same number of atoms but different structures because of the presence of differing numbers of lone pairs.

91. All these molecules have polar bonds that are symmetrically arranged about the central atoms. In each molecule the individual bond dipoles cancel to give no net overall dipole moment. All these molecules are nonpolar even though they all contain polar bonds.

Additional Exercises

93. CO_3^{2-} has $4 + 3(6) + 2 = 24$ valence electrons.

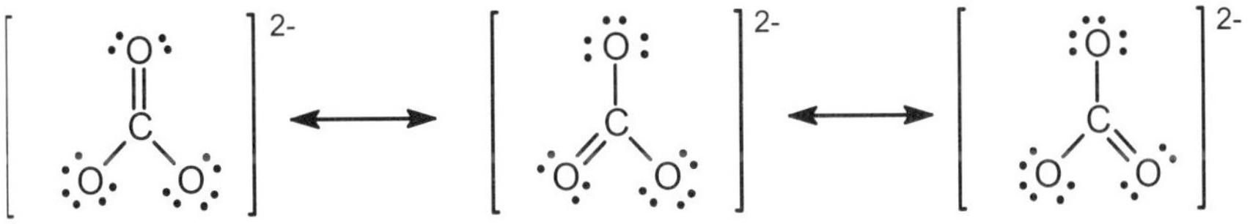

HCO_3^- has $1 + 4 + 3(6) + 1 = 24$ valence electrons.

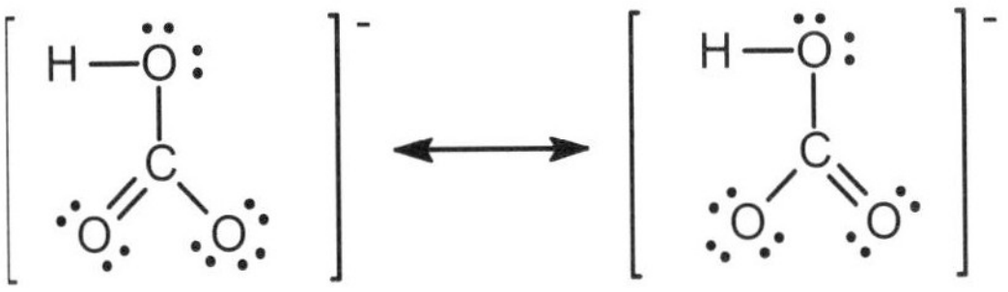

H_2CO_3 has $2(1) + 4 + 3(6) = 24$ valence electrons.

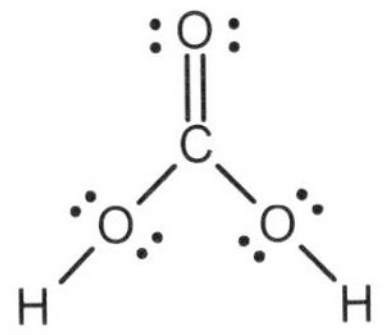

The Lewis structures for the reactants and products are:

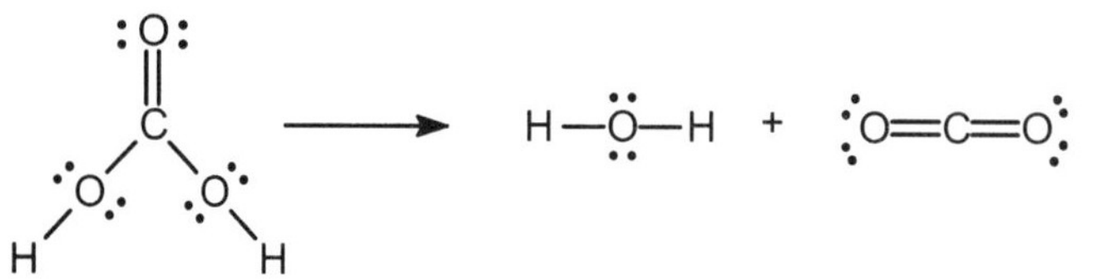

Bonds broken:

2 C–O (358 kJ/mol)
1 O–H (467 kJ/mol)

Bonds formed:

1 C=O (799 kJ/mol)
1 O–H (467 kJ/mol)

$\Delta H = 2(358) + 467 - (799 + 467) = -83$ kJ; the carbon-oxygen double bond is stronger than two carbon-oxygen single bonds; hence CO_2 and H_2O are more stable than H_2CO_3.

95. As the halogen atoms get larger, it becomes more difficult to fit three halogen atoms around the small nitrogen atom, and the NX_3 molecule becomes less stable.

97. The stable species are:

a. NaBr: In $NaBr_2$, the sodium ion would have a 2+ charge, assuming that each bromine has a 1− charge. Sodium doesn't form stable Na^{2+} compounds.

b. ClO_4^-: ClO_4 has 31 valence electrons, so it is impossible to satisfy the octet rule for all atoms in ClO_4. The extra electron from the 1− charge in ClO_4^- allows for complete octets for all atoms.

c. XeO_4: We can't draw a Lewis structure that obeys the octet rule for SO_4 (30 electrons), unlike with XeO_4 (32 electrons).

d. SeF_4: Both compounds require the central atom to expand its octet. O is too small and doesn't have low-energy d orbitals to expand its octet (which is true for all Row 2 elements).

99. a. Radius: $N^+ < N < N^-$; IE: $N^- < N < N^+$

N^+ has the fewest electrons held by the seven protons in the nucleus whereas N^- has the most electrons held by the seven protons. The seven protons in the nucleus will hold the electrons most tightly in N^+ and least tightly in N^-. Therefore, N^+ has the smallest radius with the largest ionization energy (IE), and N^- is the largest species with the smallest IE.

b. Radius: $Cl^+ < Cl < Se < Se^-$; IE: $Se^- < Se < Cl < Cl^+$

The general trends tell us that Cl has a smaller radius than Se and a larger IE than Se. Cl^+, with fewer electron-electron repulsions than Cl, will be smaller than Cl and have a larger IE. Se^-, with more electron-electron repulsions than Se, will be larger than Se and have a smaller IE.

c. Radius: $Sr^{2+} < Rb^{+} < Br^{-}$; IE: $Br^{-} < Rb^{+} < Sr^{2+}$

These ions are isoelectronic. The species with the most protons (Sr^{2+}) will hold the electrons most tightly and will have the smallest radius and largest IE. The ion with the fewest protons (Br^{-}) will hold the electrons least tightly and will have the largest radius and smallest IE.

101. Nonmetals, which form covalent bonds to each other, have valence electrons in the s and p orbitals. Since there are four total s and p orbitals, there is room for only eight valence electrons (the octet rule). The valence shell for hydrogen is just the 1s orbital. This orbital can hold two electrons, so hydrogen follows the duet rule.

103. Assuming 100.00 g of compound: $42.81 \text{ g F} \times \frac{1 \text{ mol F}}{19.00 \text{ g F}} = 2.253 \text{ mol F}$

The number of moles of X in XF_5 is: $2.253 \text{ mol F} \times \frac{1 \text{ mol X}}{5 \text{ mol F}} = 0.4506 \text{ mol X}$

This number of moles of X has a mass of 57.19 g (= 100.00 g – 42.81 g). The molar mass of X is:

$$\frac{57.19 \text{ g X}}{0.4506 \text{ mol X}} = 126.9 \text{ g/mol}$$; This is element I.

IF_5, $7 + 5(7) = 42\ e^-$

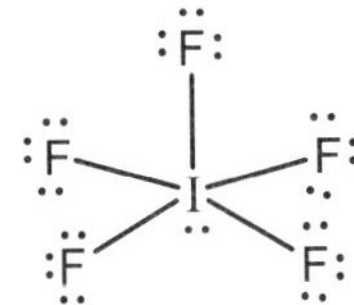

The molecular structure is square pyramid.

105. Yes, each structure has the same number of effective pairs around the central atom, giving the same predicted molecular structure for each compound/ion. (A multiple bond is counted as a single group of electrons.)

Challenge Problems

107. KrF_2, 8 + 2(7) = 22 e^-; from the Lewis structure, we have a trigonal bipyramid arrangement of electron pairs with a linear molecular structure.

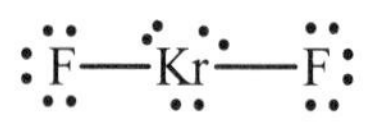

Hyperconjugation assumes that the overall bonding in KrF_2 is a combination of covalent and ionic contributions (see Section 13.12 of the text for discussion of hyperconjugation). Using hyperconjugation, two resonance structures are possible that keep the linear structure.

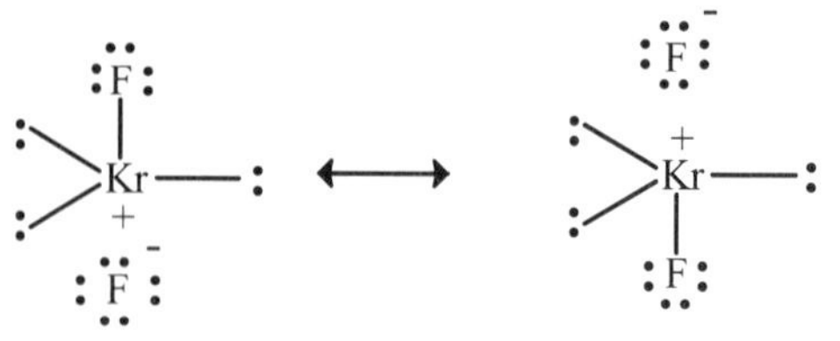

109.

Reaction	ΔH
$2\,Li^+(g) + 2\,Cl^-(g) \rightarrow 2\,LiCl(s)$	$\Delta H = 2(-829\text{ kJ})$
$2\,Li(g) \rightarrow 2\,Li^+(g) + 2\,e^-$	$\Delta H = 2(520.\text{ kJ})$
$2\,Li(s) \rightarrow 2\,Li(g)$	$\Delta H = 2(166\text{ kJ})$
$2\,HCl(g) \rightarrow 2\,H(g) + 2\,Cl(g)$	$\Delta H = 2(427\text{ kJ})$
$2\,Cl(g) + 2\,e^- \rightarrow 2\,Cl^-(g)$	$\Delta H = 2(-349\text{ kJ})$
$2\,H(g) \rightarrow H_2(g)$	$\Delta H = -(432\text{ kJ})$
$2\,Li(s) + 2\,HCl(g) \rightarrow 2\,LiCl(s) + H_2(g)$	$\Delta H = -562\text{ kJ}$

111. a. $N(NO_2)_2^-$ contains 5 + 2(5) + 4(6) + 1 = 40 valence electrons.

The most likely structures are:

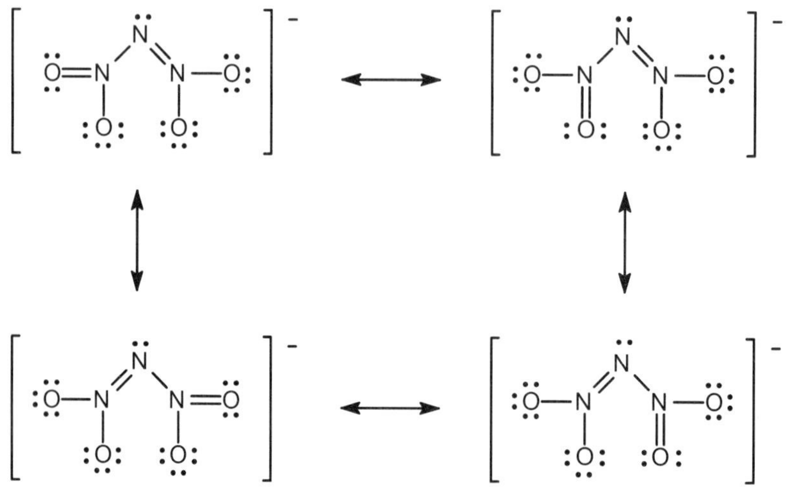

There are other possible resonance structures, but these are most likely.

b. The NNN and all ONN and ONO bond angles should be about 120°.

c. $NH_4N(NO_2)_2 \rightarrow 2\ N_2 + 2\ H_2O + O_2$; break and form all bonds.

Bonds broken:

4 N–H (391 kJ/mol)
1 N–N (160. kJ/mol)
1 N=N (418 kJ/mol)
3 N–O (201 kJ/mol)
1 N=O (607 kJ/mol)

ΣD_{broken} = 3352 kJ

Bonds formed:

2 N≡N (941 kJ/mol)
4 H–O (467 kJ/mol)
1 O=O (495 kJ/mol)

ΣD_{formed} = 4245 kJ

$\Delta H = \Sigma D_{broken} - \Sigma D_{formed}$ = 3352 kJ – 4245 kJ = –893 kJ

d. To estimate ΔH, we completely ignored the ionic interactions between NH_4^+ and $N(NO_2)_2^-$. In addition, we assumed the bond energies in Table 13.6 applied to the $N(NO_2)^-$ bonds in any one of the resonance structures above. This is a bad assumption since molecules that exhibit resonance generally have stronger overall bonds than predicted. All these assumptions give an estimated ΔH value which is too negative.

113. a. i. $C_6H_6N_{12}O_{12} \rightarrow 6\ CO + 6\ N_2 + 3\ H_2O + 3/2\ O_2$

The NO_2 groups have one N–O single bond and one N=O double bond, and each carbon atom has one C–H single bond. We must break and form all bonds.

Bonds broken:

3 C–C (347 kJ/mol)
6 C–H (413 kJ/mol)
12 C–N (305 kJ/mol)
6 N–N (160. kJ/mol)
6 N–O (201 kJ/mol)
6 N=O (607 kJ/mol)

ΣD_{broken} = 12,987 kJ

Bonds formed:

6 C≡O (1072 kJ/mol)
6 N≡N (941 kJ/mol)
6 H–O (467 kJ/mol)
3/2 O=O (495 kJ/mol)

ΣD_{formed} = 15,623 kJ

$\Delta H = \Sigma D_{broken} - \Sigma D_{formed}$ = 12,987 kJ – 15,623 kJ = –2636 kJ

ii. $C_6H_6N_{12}O_{12} \rightarrow 3\ CO + 3\ CO_2 + 6\ N_2 + 3\ H_2O$

Note: The bonds broken will be the same for all three reactions.

Bonds formed:

3 C≡O (1072 kJ/mol)
6 C=O (799 kJ/mol)
6 N≡N (941 kJ/mol)
6 H–O (467 kJ/mol)

ΣD_{formed} = 16,458 kJ

ΔH = 12,987 kJ – 16,458 kJ = –3471 kJ

iii. $C_6H_6N_{12}O_{12} \rightarrow 6\ CO_2 + 6\ N_2 + 3\ H_2$

Bonds formed:

12 C=O (799 kJ/mol)
6 N≡N (941 kJ/mol)
3 H–H (432 kJ/mol)

ΣD_{formed} = 16,530. kJ

ΔH = 12,987 kJ – 16,530. kJ = –3543 kJ

b. Reaction iii yields the most energy per mole of CL-20, so it will yield the most energy per kilogram.

$$\frac{-3543\text{ kJ}}{\text{mol}} \times \frac{1\text{ mol}}{438.23\text{ g}} \times \frac{1000\text{ g}}{\text{kg}} = -8085\text{ kJ/kg}$$

115. The reaction is:

$$1/2\ I_2(s) + 1/2\ Cl_2(g) \rightarrow ICl(g) \qquad \Delta H_f^\circ = ?$$

Using Hess's law:

$1/2\ I_2(s) \rightarrow 1/2\ I_2(g)$	ΔH = 1/2 (62 kJ)	(Appendix 4)
$1/2\ I_2(g) \rightarrow I(g)$	ΔH = 1/2 (149 kJ)	(Table 13.6)
$1/2\ Cl_2(g) \rightarrow Cl(g)$	ΔH = 1/2 (239 kJ)	(Table 13.6)
$I(g) + Cl(g) \rightarrow ICl(g)$	ΔH = –208 kJ	(Table 13.6)
$1/2\ I_2(s) + 1/2\ Cl_2(g) \rightarrow ICl(g)$	ΔH = 17 kJ so ΔH_f° = 17 kJ/mol	

Table of Atomic Masses*

Element	Symbol	Atomic Number	Atomic Mass	Element	Symbol	Atomic Number	Atomic Mass	Element	Symbol	Atomic Number	Atomic Mass
Actinium	Ac	89	(227)†	Gold	Au	79	197.0	Praseodymium	Pr	59	140.9
Aluminum	Al	13	26.98	Hafnium	Hf	72	178.5	Promethium	Pm	61	(145)
Americium	Am	95	(243)	Hassium	Hs	108	(265)	Protactinium	Pa	91	(231)
Antimony	Sb	51	121.8	Helium	He	2	4.003	Radium	Ra	88	226
Argon	Ar	18	39.95	Holmium	Ho	67	164.9	Radon	Rn	86	(222)
Arsenic	As	33	74.92	Hydrogen	H	1	1.008	Rhenium	Re	75	186.2
Astatine	At	85	(210)	Indium	In	49	114.8	Rhodium	Rh	45	102.9
Barium	Ba	56	137.3	Iodine	I	53	126.9	Rubidium	Rb	37	85.47
Berkelium	Bk	97	(247)	Iridium	Ir	77	192.2	Ruthenium	Ru	44	101.1
Beryllium	Be	4	9.012	Iron	Fe	26	55.85	Rutherfordium	Rf	104	(261)
Bismuth	Bi	83	209.0	Krypton	Kr	36	83.80	Samarium	Sm	62	150.4
Bohrium	Bh	107	(264)	Lanthanum	La	57	138.9	Scandium	Sc	21	44.96
Boron	B	5	10.81	Lawrencium	Lr	103	(260)	Seaborgium	Sg	106	(263)
Bromine	Br	35	79.90	Lead	Pb	82	207.2	Selenium	Se	34	78.96
Cadmium	Cd	48	112.4	Lithium	Li	3	6.941	Silicon	Si	14	28.09
Calcium	Ca	20	40.08	Lutetium	Lu	71	175.0	Silver	Ag	47	107.9
Californium	Cf	98	(251)	Magnesium	Mg	12	24.31	Sodium	Na	11	22.99
Carbon	C	6	12.01	Manganese	Mn	25	54.94	Strontium	Sr	38	87.62
Cerium	Ce	58	140.1	Meitnerium	Mt	109	(268)	Sulfur	S	16	32.07
Cesium	Cs	55	132.9	Mendelevium	Md	101	(258)	Tantalum	Ta	73	180.9
Chlorine	Cl	17	35.45	Mercury	Hg	80	200.6	Technetium	Tc	43	(98)
Chromium	Cr	24	52.00	Molybdenum	Mo	42	95.94	Tellurium	Te	52	127.6
Cobalt	Co	27	58.93	Neodymium	Nd	60	144.2	Terbium	Tb	65	158.9
Copper	Cu	29	63.55	Neon	Ne	10	20.18	Thallium	Tl	81	204.4
Curium	Cm	96	(247)	Neptunium	Np	93	(237)	Thorium	Th	90	232.0
Darmstadtium	Ds	110	(281)	Nickel	Ni	28	58.69	Thulium	Tm	69	168.9
Dubnium	Db	105	(262)	Niobium	Nb	41	92.91	Tin	Sn	50	118.7
Dysprosium	Dy	66	162.5	Nitrogen	N	7	14.01	Titanium	Ti	22	47.88
Einsteinium	Es	99	(252)	Nobelium	No	102	(259)	Tungsten	W	74	183.9
Erbium	Er	68	167.3	Osmium	Os	76	190.2	Uranium	U	92	238.0
Europium	Eu	63	152.0	Oxygen	O	8	16.00	Vanadium	V	23	50.94
Fermium	Fm	100	(257)	Palladium	Pd	46	106.4	Xenon	Xe	54	131.3
Fluorine	F	9	19.00	Phosphorus	P	15	30.97	Ytterbium	Yb	70	173.0
Francium	Fr	87	(223)	Platinum	Pt	78	195.1	Yttrium	Y	39	88.91
Gadolinium	Gd	64	157.3	Plutonium	Pu	94	(244)	Zinc	Zn	30	65.38
Gallium	Ga	31	69.72	Polonium	Po	84	(209)	Zirconium	Zr	40	91.22
Germanium	Ge	32	72.59	Potassium	K	19	39.10				

*The values given here are to four significant figures where possible.

†A value given in parentheses denotes the mass of the longest-lived isotope.

Page Numbers of Some Important Tables

Physical Constants

Constant	Symbol	Value
Atomic mass unit	amu	1.66054×10^{-27} kg
Avogadro's number	N	6.02214×10^{23} mol^{-1}
Bohr radius	a_0	5.292×10^{-11} m
Boltzmann's constant	k	1.38066×10^{-23} J K^{-1}
Charge of an electron	e	1.60218×10^{-19} C
Faraday's constant	F	96,485 C mol^{-1}
Gas constant	R	8.31451 J K^{-1} mol^{-1}
		0.08206 L atm K^{-1} mol^{-1}
Mass of an electron	m_e	9.10939×10^{-31} kg
		5.48580×10^{-4} amu
Mass of a neutron	m_n	1.67493×10^{-27} kg
		1.00866 amu
Mass of a proton	m_p	1.67262×10^{-27} kg
		1.00728 amu
Planck's constant	h	6.62608×10^{-34} J s
Speed of light	c	2.99792458×10^{8} m s^{-1}